Feature Papers of Drones-Volume I

Feature Papers of Drones-Volume I

Editors

Diego González-Aguilera
Pablo Rodríguez-Gonzálvez

MDPI • Basel • Beijing • Wuhan • Barcelona • Belgrade • Manchester • Tokyo • Cluj • Tianjin

Editors
Diego González-Aguilera
University of Salamanca
Spain

Pablo Rodríguez-Gonzálvez
University of León
Spain

Editorial Office
MDPI
St. Alban-Anlage 66
4052 Basel, Switzerland

This is a reprint of articles from the Topical Collection published online in the open access journal *Drones* (ISSN 2504-446X) (available at: https://www.mdpi.com/journal/drones/topical_collections/FP_Drones).

For citation purposes, cite each article independently as indicated on the article page online and as indicated below:

LastName, A.A.; LastName, B.B.; LastName, C.C. Article Title. *Journal Name* **Year**, *Volume Number*, Page Range.

Volume 1
ISBN 978-3-0365-6187-5 (Hbk)
ISBN 978-3-0365-6188-2 (PDF)

Volume 1-2
ISBN 978-3-0365-6157-8 (Hbk)
ISBN 978-3-0365-6158-5 (PDF)

Contents

Preface to "Feature Papers of Drones-Volume I"

The present book is divided into two volumes (Volume I: articles 1–23, and Volume II: articles 24–54) which compile the articles and communications submitted to the Topical Collection "Feature Papers of Drones" during the years 2020 to 2022 describing novel or new cutting-edge designs, developments, and/or applications of unmanned vehicles (drones).

Articles 1–8 are devoted to the developments of drone design, where new concepts and modeling strategies as well as effective designs that improve drone stability and autonomy are introduced.

Articles 9–16 focus on the communication aspects of drones as effective strategies for smooth deployment and efficient functioning are required. Therefore, several developments that aim to optimize performance and security are presented. In this regard, one of the most directly related topics is drone swarms, not only in terms of communication but also human-swarm interaction and their applications for science missions, surveillance, and disaster rescue operations.

To conclude with the volume I related to drone improvements, articles 17–23 discusses the advancements associated with autonomous navigation, obstacle avoidance, and enhanced flight planning.

Articles 24–41 are focused on drone applications, but emphasize two types: firstly, those related to agriculture and forestry (articles 24–35) where the number of applications of drones dominates all other possible applications. These articles review the latest research and future directions for precision agriculture, vegetation monitoring, change monitoring, forestry management, and forest fires. Secondly, articles 36–41 addresses the water and marine application of drones for ecological and conservation-related applications with emphasis on the monitoring of water resources and habitat monitoring.

Finally, articles 42–54 looks at just a few of the huge variety of potential applications of civil drones from different points of view, including the following: the social acceptance of drone operations in urban areas or their influential factors; 3D reconstruction applications; sensor technologies to either improve the performance of existing applications or to open up new working areas; and machine and deep learning developments.

Diego González-Aguilera and Pablo Rodríguez-Gonzálvez
Editors

drones

MDPI

Article

Propeller Position Effects over the Pressure and Friction Coefficients over the Wing of an UAV with Distributed Electric Propulsion: A Proper Orthogonal Decomposition Analysis

José Ramón Serrano *, Luis Miguel García-Cuevas, Pau Bares and Pau Varela

CMT—Motores Térmicos, Universitat Politècnica de València, Camino de Vera s/n, 46022 Valencia, Spain; luiga12@mot.upv.es (L.M.G.-C.); pabamo@mot.upv.es (P.B.); pavamar@mot.upv.es (P.V.)
* Correspondence: jrserran@mot.upv.es; Tel.: +34-963877650

Abstract: New propulsive architectures, with high interactions with the aerodynamic performance of the platform, are an attractive option for reducing the power consumption, increasing the resilience, reducing the noise and improving the handling of fixed-wing unmanned air vehicles. Distributed electric propulsion with boundary layer ingestion over the wing introduces extra complexity to the design of these systems, and extensive simulation and experimental campaigns are needed to fully understand the flow behaviour around the aircraft. This work studies the effect of different combinations of propeller positions and angles of attack over the pressure coefficient and skin friction coefficient distributions over the wing of a 25 kg fixed-wing remotely piloted aircraft. To get more information about the main trends, a proper orthogonal decomposition of the coefficient distributions is performed, which may be even used to interpolate the results to non-simulated combinations, giving more information than an interpolation of the main aerodynamic coefficients such as the lift, drag or pitching moment coefficients.

Keywords: distributed electric propulsion; boundary layer ingestion; propeller; fixed wing; proper orthogonal decomposition

Citation: Serrano, J.R.; García-Cuevas, L.M.; Bares, P.; Varela, P. Propeller Position Effects over the Pressure and Friction Coefficients over the Wing of an UAV with Distributed Electric Propulsion: A Proper Orthogonal Decomposition Analysis. *Drones* **2022**, *6*, 38. https://doi.org/10.3390/drones6020038

Academic Editors: Diego González-Aguilera and Pablo Rodríguez-Gonzálvez

Received: 15 December 2021
Accepted: 26 January 2022
Published: 29 January 2022

1. Introduction

The requirement of more efficient and environmentally-friendly unmanned air vehicles (UAVs) is a necessity, as expressed by the National Aeronautics and Space Administration (NASA) in its Environmentally Responsible Aviation (ERA) project that is included in the information Technology Development Solutions (ITDS) [1]. The growth prevision of the global UAV fleet is alarmingly high in the coming years, as shown by the study by the Boston Consulting Group (BCG) [2], where it is estimated that in the year 2050 the fleet of industrial drones in Europe and the United States will exceed one million units.

In recent years, the application of novel technologies has been studied to achieve a higher aerodynamically and propulsively efficient aircraft, in a quest to develop aircraft with less fuel and energy consumption and less pollutant and greenhouse gas emissions.

One of these technologies is distributed electrical propulsion (DEP), which consists of the allocation of the total power required by the aircraft in different propulsive systems throughout the wingspan. This distribution has reported important advantages compared to a classic small aircraft configuration, normally relegated to the use of a single propeller. These benefits include resilience against foreign object damage [3], propulsive efficiency improvements (as the total area swept by the propellers can be increased over what can be done for a few-propellers configurations) [4], noise reduction and spectrum alteration [5,6], improved aerodynamic efficiency by vorticity control and vectored thrust [7,8] or more wing structural stability [9].

The use of DEP in high-weight UAVs is possible thanks to the use of hybrid electric propulsive plants, being this technology widely studied, since by itself it provides advantages from the point of view of fuel economy and related polluting emissions [10,11]

The proper use of DEP is strongly dependent on a correct optimisation of the location of the engines as studied in [5]. If the propulsion plant is placed in front of the wing, the increased lift generated by the propellers at the washed area of the wing allows gaining control authority. However, If the propulsion plant is located behind the wing i.e., near the trailing edge, it is possible to take advantage of the boundary layer formed on the wing to increase both the propulsive efficiency of the engine and the aerodynamic efficiency of the wing. This phenomenon is known as boundary layer ingestion (BLI). BLI's application is based, on the one hand, on reducing the intake speed of the air that the engine ingests, so that it needs less power to produce the same thrust, as described by Budziszewski in [12]. On the other hand, as pointed out in the work of Teperin et al. and Hall et al. [13–15], the ingestion could reduce the lift-induced drag enough to increase aerodynamic efficiency even though the skin friction drag increases due to the re-acceleration.

In this work, an analysis of the pressure and friction coefficients of an aircraft with DEP and BLI will be carried out using computational fluid dynamics (CFD) and proper orthogonal decomposition (POD), a tool that seeks to simplify a complex problem into a series of simpler deterministic functions, allowing the creation of models that facilitate the design.

POD is historically used in fluid mechanics problems, both experimental and numerical, although it has also been used in many other engineering processes in order to always extract dominant characteristics and trends. Some recent studies where the application of POD in fluid fields can be seen are the investigation of Broatch et al. [16] where this tool is used to analyse acoustic fields in radial compressors, or the work of Torregrosa et al. [17] where the modal decomposition helps to characterise the unsteady flow field of a combustion chamber. Like the last study, it is easy to find the application of POD in the field of thermal engines [18–20], but this method is also used to analyse different aerodynamic problems. These include the interpolation of transonic flows [21], the creation of reduced-order models to characterise aircraft flying at high angles of attack [22], or to generate three-dimensional flow models for supersonic aircraft that significantly reduces the need of computationally-expensive high fidelity simulations [23].

The final goal of POD is the decomposition of a space vector into a series of deterministic spatial functions that are modulated through a series of configuration coefficients. Each one of the spatial functions is orthonormal with respect to the others, so they will be independent since their vector product is null. It is possible to order these functions according to what percentage of the problem they explain, which is known as total fluctuating kinetic energy (*TKE*).

The main contribution of the current research paper is the analysis of the effects that the propeller position of an fixed-wing UAV with DEP and BLI induce into the pressure and friction coefficients over its wings. This coefficients were obtained by means of computational fluid dynamics (CFD) simulations, and a modal decomposition by means of POD was performed on that data. The main trends are observed, including the movement of laminar transitional bubbles, changes in the suction peak near the leading edge or reductions and increases in the skin friction coefficient near the trailing edge, all as a function of the propeller positions at different angles of attack. By using the modal decomposition results, it is also possible to interpolate the pressure coefficient and the friction coefficient to non-simulated conditions, thus obtaining more detailed information than just interpolating global aerodynamic coefficients such as the lift or the drag coefficients.

The document is organised as follows. First, in Section 2, the aircraft analysed in this work is described. Then, in Section 3, the main methods and models are presented, explaining both the CFD and the POD approaches. In Section 4, the modal decomposition results for different propeller positions and angles of attack are shown and discussed. Finally, all main results and discussions are summarised in the conclusions in Section 5.

2. Aircraft Description

In this section, the aircraft studied in this work is described. The selection of its different components is the same presented in [24] by Tiseira et al. and [25] by Serrano et al. The aircraft data is shown in Table 1, and Figure 1 shows a schematic view of it.

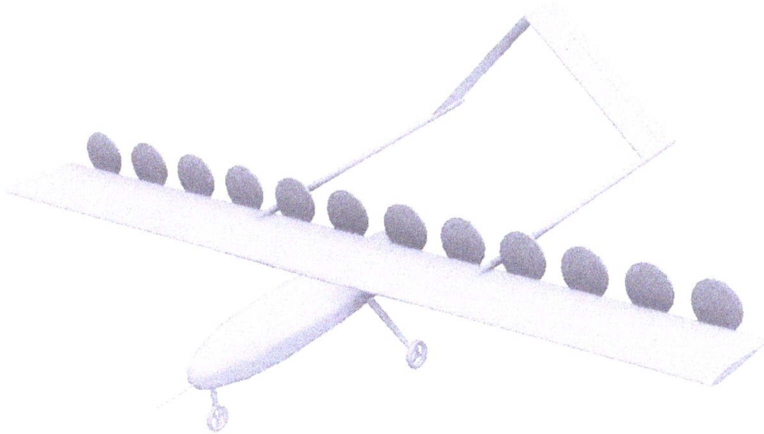

Figure 1. Simulated aircraft, as shown in [24]. This sketch is for a configuration of 12 propellers.

The study uses data from computational fluid dynamics (CFD) simulations. In order to reduce the computational complexity while still achieving accurate and realistic results, all simulations consist of a 2.5D domain.

The domain consists of a single propeller and a portion of the wing affected by that propeller. In this way, the induced drag of the wing caused by three-dimensional effects is not directly simulated. However, the value of the induced drag is calculated based on the lift of the wing portion and an estimated Oswald factor. The details of the domain regarding measures and boundaries will be explained in the next section.

The choice of these two components was carried out taking into account the mission of the aircraft, which is a fixed-wing, long endurance UAV with an *MTOM* of 25 kg, driven by several propellers. The aircraft is similar to the Penguin C from UAV Factory [26] and the TARSIS 25 from AERTEC Solutions [27], and has a wing span of 2 m.

For the wing, a single SD7003 airfoil has been used, designed to operate with low parasitic drag at low Reynolds number, which makes it particularly attractive in the application of civil UAVs. This airfoil has been extensively studied in the literature, and high quality experimental and computational data are readily available in multiple references [28–31].

The propeller chosen for this application was designed by the University of Illinois in Urbana-Champaign. It is the DA4052 model, for which complete geometric data as well as experimental characteristic curves produced in a wind tunnel are found [32]. These data were used to validate the numerical simulations, which were carried out by means of the Blade Element Method [33]. A total of 13 of 80 mm propellers are used for the whole aircraft. The propeller size and the number of propellers were chosen to maximise the specific range of the aircraft, which is the distance flown for each unit of fuel consumption, in a hybrid-electric configuration powered by a four-stroke engine, as described in [24].

In all the simulations, the total thrust generated by the propellers was equal to the total drag of the whole aircraft. For that, the drag was estimated as in Equation (1):

$$D = \frac{1}{2} \cdot \rho_\infty \cdot U_\infty^2 \cdot S \cdot \left(C_{D,0,\text{wing}} + C_{D,0,\text{extra}} + \frac{C_L^2}{\pi \cdot AR \cdot e} \right), \quad (1)$$

where ρ_∞ is the far-field air density, U_∞ is the upstream wind speed, S is the wing surface, $C_{D,0,\text{wing}}$ is the parasitic drag coefficient of the wing and C_L is the lift coefficient. $C_{D,0,\text{extra}}$, on the other hand, is the parasitic drag of the rest of the aircraft: this includes the fuselage, the empennage, and the landing gear. The values of C_L and $C_{D,0,\text{wing}}$ are directly computed using data from the simulations, whereas $C_{D,0,\text{extra}}$ is computed using geometrical information of aircraft with a similar mission. This includes the mentioned TARSIS 25 and Penguin C, as well as values from Harmon and Hiserote's aircraft [10,34]. Finally, the Oswald efficiency factor e is estimated using standard methods as described in [35,36]. The aspect ratio, AR, is set to 10.

Table 1. Aircraft data.

Design Parameters	
Aspect ratio	10
Wing area	0.4 m^2
Wingspan	2 m
Wing chord	0.2 m
Maximum takeoff mass	25 kg
Wing airfoil	SD7003
Propeller radius	40 mm
Number of propellers	13
Aerodynamic Parameters	
$C_{D,0,\text{extra}}$ (fuselage, empennage, others)	0.011
Oswald efficiency factor (e)	0.8

3. Methods

In this section, the computational methods are presented. This includes the CFD setup, the way a wing section is simulated and the method for performing the modal decomposition of the friction and pressure coefficients.

3.1. Computational Domain

All the case studies follow the same computational setup shown in [24,25]. In both cases, 2.5D simulations of a wing section were performed, and the size of the domain was proven to be big enough so the position of the far field boundary conditions did not affect the final solution. The upstream boundary condition is located 20 chords from the wing and uses a free-stream speed condition. Downstream of the wing, a pressure outlet boundary condition is set. The lateral boundaries are set as symmetry boundary conditions. Finally, the surface of the wing is modelled as a non-slip, smooth wall. A sketch of the computational grid is shown in Figure 2.

In all the simulations, the propeller is located over the trailing edge of the wing section, but different positions are studied. The propeller is, in any case, modelled with a single virtual disc actuator, using the Blade Element Momentum Theory (BEMT). This 80 mm of diameter propeller is separated from the trailing edge around 1 mm in the direction of the chord. The relative angle between the profile and the normal to the propeller, the draft angle, is set at 1.5°. These dimensions are shown in Figure 3, and have been chosen based on previous studies [24,25].

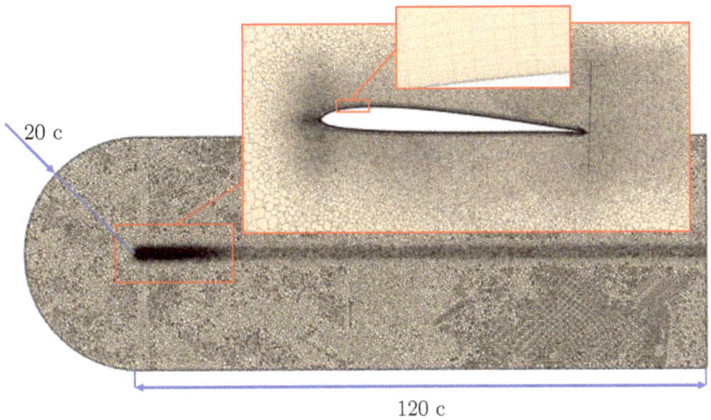

Figure 2. Side view sketch of the computational grid used for the current calculations.

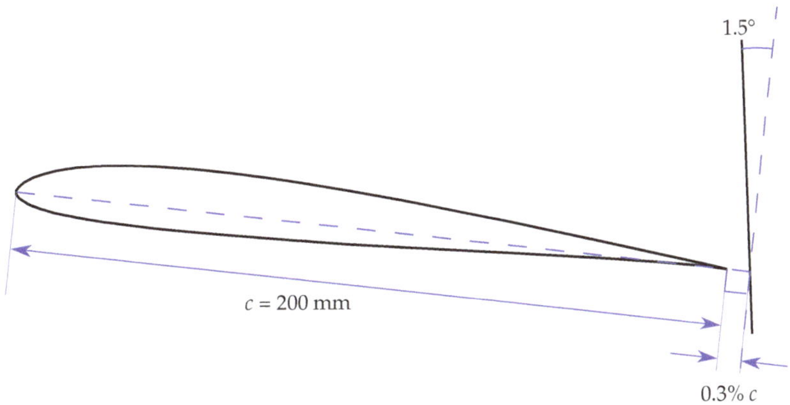

Figure 3. Airfoil side-view with principal heights (not to scale).

The main geometric aspect that varies between cases is the vertical height of the actuator disc. This height, measured as the distance between the propeller shaft and the trailing edge, is expressed as a fraction of the propeller radius. This way, 0% represents that the center of the propeller is aligned with the trailing edge, whereas at the other extreme, at 100%, the entire propeller is above the trailing edge. The latter conditions represent the case with the minimum influence over the pressure side of the wing, whereas the former produce the maximum level of boundary layer ingestion. These two positions are represented in Figure 4, below a sketch of the wing section where the propeller disc is visible.

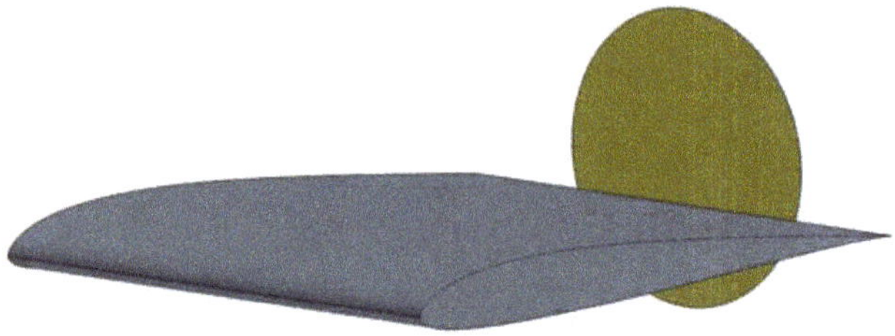

(a)

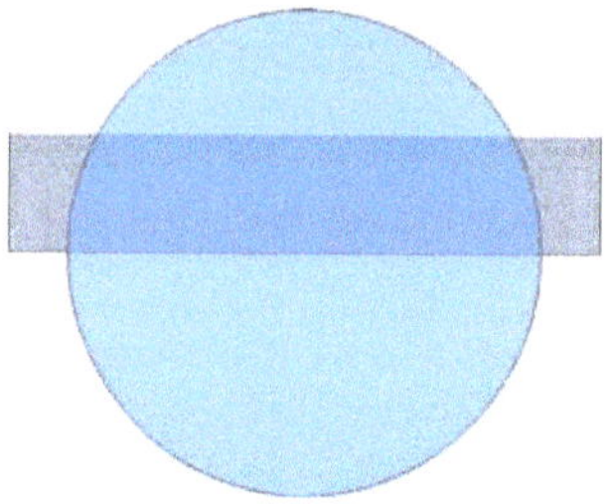

(b)

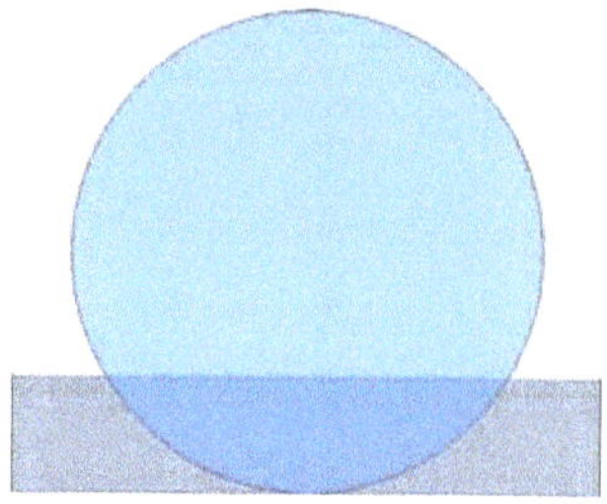

(c)

Figure 4. Maximum and minimum propeller heights above the trailing edge. (**a**) Section of wing simulated with a virtual disc for modelling the propeller. (**b**) 0% position. (**c**) 100% position.

3.2. CFD Methodology

The CFD simulations are carried out in the same way as described in [24,25], using the proprietary software Simcenter STAR-CCM+ with a finite-volume, steady-state, Reynolds-Averaged Navier–Stokes (RANS) equation approach. As the flow speed is relatively small in all the simulated cases, the flow is modelled as incompressible. A Spalart–Allmaras model is chosen to compute the Reynolds stress tensor.

The domain is meshed with a polyhedral mesh except for the boundary layer around the wing: in this zone, a 14-layer prismatic mesh with a geometric grow distribution is applied in a total thickness of 3 mm, which is of the order of the displacement thickness of the boundary layer. This boundary layer mesh ensures a non-dimensional distance from the wall to the first cell centroid y+ less than one in 99% of the wall around the airfoil. Keeping a value of y+ lower than 1 is a requisite of the model to solve the viscous sublayer of the boundary layer without using wall functions. Finally, the mesh size of the polyhedral mesh near the walls has been kept at 1 mm. This mesh size was set after performing a mesh independence study in which the discretisation error was computed by means of a Richardson extrapolation, as seen in [24].

Five different propeller positions were simulated over the trailing edge, between 0% and 100%. For each position, nine angles of attack between 1° and 9° and three different Reynolds: 3×10^5, 5×10^5 and 7×10^5. These Reynolds number correspond to a wind speed at sea level and 15 °C of $22.0\,\mathrm{m\,s^{-1}}$, $36.7\,\mathrm{m\,s^{-1}}$ and $51.4\,\mathrm{m\,s^{-1}}$.

The propeller, modelled with a BEMT actuator disc, uses geometrical data and airfoil drag polar results obtained from a panel method with interactive boundary layer corrections code, XFLR5 [37], which is based on Mark Drela's XFOIL [38]. The rotational speed of the propeller was set so that the total thrust produced by all the propellers was equal to the total drag of the aircraft in each simulation, as shown in Equation (2).

$$T \cdot n_{\text{propellers}} = D = \frac{1}{2} \cdot \rho_\infty \cdot U_\infty^2 \cdot S \cdot \left(C_{D,0,\text{wing}} + C_{D,0,\text{extra}} + \frac{C_L^2}{\pi \cdot Æ \cdot e} \right). \tag{2}$$

The simulations of both the airfoil and the propeller, separately, were validated against experimental and high-fidelity simulations found in the literature.

3.3. POD Application

Once all the simulations were carried out, the pressure coefficient C_p and the friction coefficient C_f were extracted at each simulation. The coefficients were extracted by using a single plane, which divides the actuator disc in half just in the middle of the domain Figure 5. For a fixed Reynolds, each coefficient distribution is a function of the angle of attack α and the relative height of the actuator disc above the wing h.

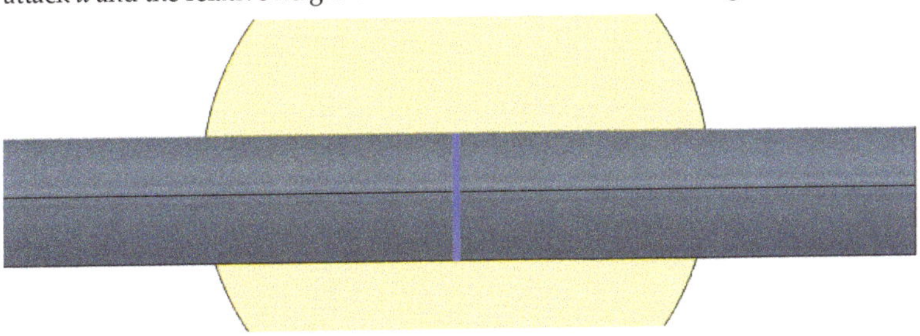

Figure 5. Front section of the analysed airfoil section, marked in blue.

As the process for working with one or the other coefficient is analogous, from now on the equations are written based on the pressure coefficient.

In each simulation, 300 spatial points over the wing surface were taken, and since five positions and nine angles of attack are simulated, 45 different distributions were generated. The combination of these data yields a working matrix of 45 × 300 elements, $\boldsymbol{U}$. Each row corresponds to the coefficient distribution over the airfoil for one combination of angle of attack and propeller position.

Once the matrix was obtained, the coefficient was decomposed into a summation of a series of deterministic spatial functions, also known as spatial modes, (ϕ_k) that depend only on the point of the airfoil where said aerodynamic coefficient is studied. The series is expressed in Equation (3) for the pressure coefficient as a function of the chord position x, the angle of attack α and the relative propeller position h, which is the propeller shaft height over the trailing edge divided by the propeller radius.

$$C_p(x, \alpha, h) = \sum_{k=1}^{\infty} a_k(\alpha, h)\phi_k(x). \tag{3}$$

The modes are in turn modulated by what are known as configuration coefficients (a_k) that depend on the 45 configurations described in the problem.

Once the matrix $\boldsymbol{U}$ was obtained, it was possible to calculate its covariance matrix $\boldsymbol{C}$, an indication of the degree of correlation of the data, as expressed in Equation (4).

$$\boldsymbol{C} = \frac{1}{m-1}\boldsymbol{U}^{\mathrm{T}}\boldsymbol{U}, \tag{4}$$

where m is equal to the total amount of distributions, 45.

After obtaining the covariance matrix, its eigenvalues and eigenvectors were computed, which were then sorted from the largest eigenvalue to the smallest. A set of 300 λ eigenvalues was obtained, as well as 300 eigenvectors. The eigenvalues were arranged in a diagonal matrix $\mathbf{\Lambda}$, whereas the eigenvectors were arranged as columns in a matrix $\mathbf{\Phi}$. The relationship between C, $\mathbf{\Phi}$ and $\mathbf{\Lambda}$ is expressed in Equation (5):

$$C = \mathbf{\Phi}\mathbf{\Lambda}\mathbf{\Phi}^{-1}. \tag{5}$$

By ordering the eigenvalues, it was possible to determine the percentage of the energy content, *TKE*, which explains each mode, as shown in Equation (6). In this way it was possible to assess how many modes explain most of the aerodynamic behaviour of the coefficients.

$$TKE_j = \frac{\lambda_j}{\sum_{j=1}^{300} \lambda_j}. \tag{6}$$

Once the modes were obtained, the configuration coefficients that model the modes were computed through the matrix A expressed in Equation (7), where the original dataset $\boldsymbol{U}$ was projected onto each n mode.

$$A = \boldsymbol{U}\Phi = \begin{pmatrix} a_{11} & \dots & a_{1n} \\ a_{21} & \dots & a_{2n} \\ \vdots & & \vdots \\ a_{m1} & \dots & a_{mn}. \end{pmatrix} \tag{7}$$

In A, each coefficient a_{ij} is the result of projecting the data measured on the airfoil in configuration i over mode j. Thus, each column of matrix A had the configuration coefficients for each of the 45 configurations. Each configuration i corresponds to a combination of propeller position and angle of attack.

Since it is possible to reconstruct the original matrix $\boldsymbol{U}$ through the sum of the contributions of its modes multiplied by the configuration coefficients, it is possible to rewrite $\boldsymbol{U}$ as the sum of the effect of each mode as in Equation (8):

$$\boldsymbol{U} = A\mathbf{\Phi}^{-1} = \sum_{k=1}^{n} \tilde{\boldsymbol{u}}^k = \sum_{k=1}^{n} \begin{pmatrix} \tilde{u}_{11}^k & \dots & \tilde{u}_{1n}^k \\ \tilde{u}_{21}^k & \dots & \tilde{u}_{2n}^k \\ \vdots & & \vdots \\ \tilde{u}_{m1}^k & \dots & \tilde{u}_{mn}^k \end{pmatrix} = \sum_{k=1}^{n} \begin{pmatrix} a_{1k} \\ a_{2k} \\ \vdots \\ a_{mk} \end{pmatrix} \begin{pmatrix} \phi_{1k} & \dots & \phi_{nk}. \end{pmatrix} \tag{8}$$

Therefore, if instead of adding all the modes, only those that represent the most *TKE* of the system are chosen, it is possible to compare the original signal with that reconstructed with a few modes.

4. Results and Discussion

This section presents the different results of the study. It is divided into four parts: the first one, in which the results of the modal decomposition of the pressure coefficient is presented; the second one, in which the same is done for the friction coefficient; a third one, where, using the modal decomposition of the pressure and friction coefficients, the lift and drag coefficients are reconstructed; finally, a subsection in which the pressure and friction coefficients of non calculated cases are computed using a surrogate model.

4.1. Pressure Coefficient Analysis Using POD

As it was drawn as a conclusion in [25], the pressure coefficient is influenced by the position of the propeller over the trailing edge, giving rise to higher suction peaks in the case of higher positions. Furthermore, it is possible to observe that the laminar separation bubble (LSB) occurs closer to the leading edge at higher propeller positions since the influence of the propeller height translates into a change of the circulation around the airfoil, similar to flying at a highest apparent angle of attack.

The C_p is also highly modified depending on the angle of attack, as shown in Figure 6. For a fixed propeller position, the suction peak grows on the suction side or extrados as the angle of attack α increases, contributing positively to the lift of the airfoil. From the suction peak, the flow over the extrados encounters an adverse pressure gradient that decelerates the flow. This sudden increase in pressure leads to instabilities in the laminar boundary layer and causes the LSB to occur near the leading edge so that the higher the angle of attack, the further upstream the bubble is. Regarding the pressure side, when increasing the angle of attack, an overpressure is produced that also contributes, although to a lesser extent, to the lift of the wing. As expected, the behaviour corresponds to what can be observed over an airfoil without BLI, although the actual magnitude and distribution of the pressure coefficient is modified.

Figure 6 exemplifies in solid lines for a propeller height position of 75% the change in the C_p of the suction side (mostly negative values of C_p) and pressure side (mostly positive values of C_p) due to the increase in the angle of attack, along the normalised chord of the airfoil. As a reference, and in dashed lines, the results for a case without BLI is also shown.

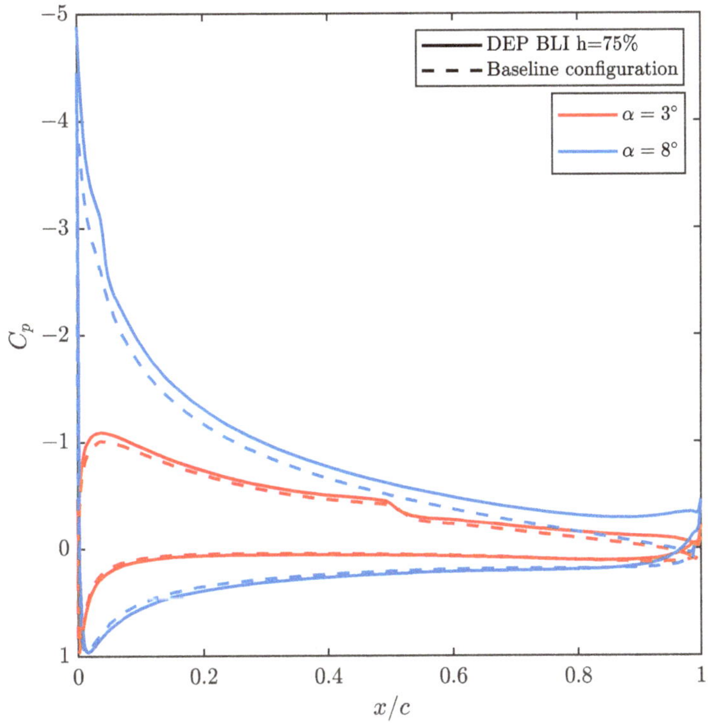

Figure 6. Pressure coefficient distribution at two angles of attack for a Reynolds number of 5×10^5 and a propeller position of 75%.

As mentioned in the previous section, it is possible to know the main effects of the height of the propeller over the trailing edge and the angle of attack affect the C_p through the percentage of *TKE* that their modes explain. To make the analysis more conclusive, the pressure coefficient of both the intrados and extrados are studied separately: in this way, different number of modes can be used to recreate the C_p of each part.

Figure 7 shows the percentage of *TKE* explained by the first 10 *TKE* modes, in red for the pressure side and blue for the suction side.

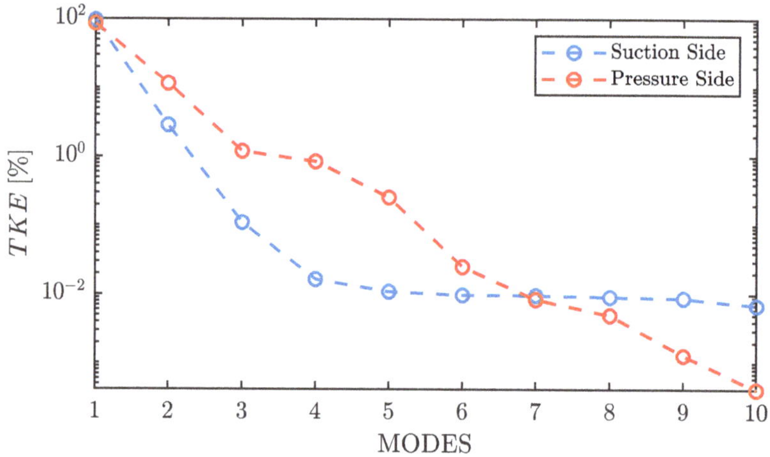

Figure 7. *TKE* of the different C_p modes for both the suction side (in blue) and pressure side (in red).

Since the modes have been ordered as a function of higher to lower eigenvalue, the first modes are those that explain the most energy of the problem. In the case of the suction side, 96% of the energy is explained through the first mode, and 3% is explained by the second, with the contribution of the following modes being marginal. In the pressure side, the first mode explains a lower percentage of energy compared to the suction side, 86%, but it is still much higher than the rest of the modes, whereas the second explains 12% of *TKE*.

In Figures 8 and 9, the first four spatial modes of the C_p of the suction side and the pressure side are represented, respectively. In the extrados, both the first and second modes have a homogeneous shape compared to the lower energy modes. These modes have an important weight in resolving the pressure coefficient near the suction peak, while less energetic modes have a greater influence on the explanation of the pressure behaviour near the trailing edge. In the case of the pressure side, the modes behave in a similar way to the suction side. However, it can be seen that the first three modes provide most information near the leading edge, with the fourth mode being the one with the greatest weight near the trailing edge.

Although the first three modes of the suction side explain 99.5% of the *TKE*, they are not enough to explain the transitional bubble over the airfoil, as can be seen in Figure 10. In this figure, the angle of attack is set to 3° and the relative height of the propeller to 50%. It can be seen that, while the suction peak, which concentrates the largest fraction of *TKE* in the problem, is well defined, from 50% of the chord it is possible to find discrepancies. A greater number of modes are required if the aerodynamic behaviour is to be modelled more realistically. The first three modes are enough to reproduce a shape similar to that of a LSB, albeit with a lower intensity and a different position, moved towards the trailing edge. Regarding the trailing edge behaviour, it is relatively well reproduced in any case, even when only three modes are used to reconstruct the solution. As expected from the results of Figure 7, a reduced number of modes is enough to get most of the behaviour of the pressure coefficient distribution, so the lift coefficient, pitching moment and structural stresses over the skin can be computed with good accuracy from a limited amount of information. Only when detailed information is required, as it is for the position of the LSB, more than three modes are necessary.

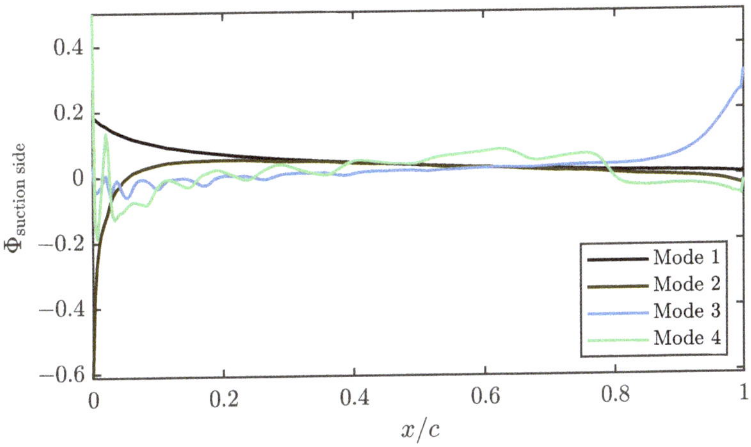

Figure 8. First four modes of the pressure coefficient of the suction side.

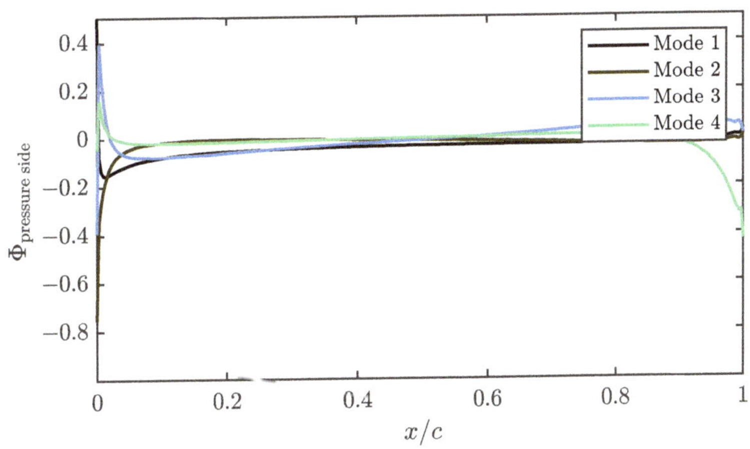

Figure 9. First four modes of the pressure coefficient of the pressure side.

It is possible to obtain more information about the modes by representing the configuration coefficient that modulate them. Figure 11 represents the configuration coefficients of the first 2 modes of the C_p of the extrados as a function of the relative height of the propeller and the angle of attack.

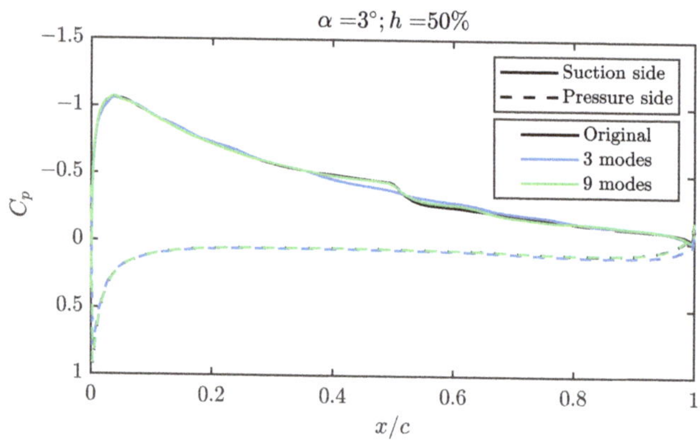

Figure 10. Pressure coefficient over the airfoil for an angle of attack of 3°, a propeller position of 50%. The results for the reconstruction with the first 3 and 9 eigenvectors are also included.

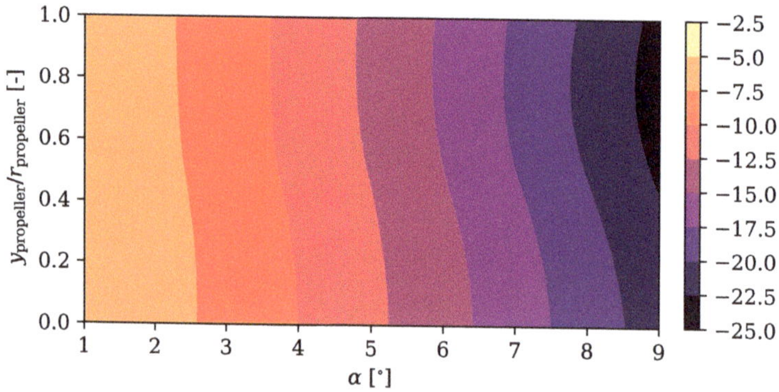

(**a**)

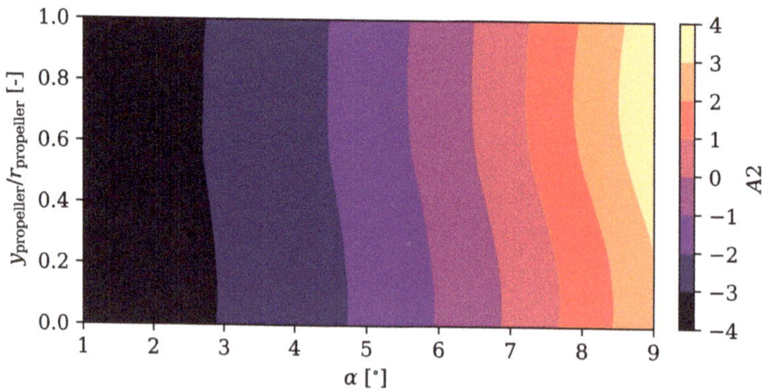

(**b**)

Figure 11. Configuration coefficients for the first two modes of the pressure coefficient over the suction side of the airfoil, as a function of the angle of attack and the relative propeller position. (**a**) Configuration coefficient $A1$ of the first mode of C_p over the suction side. (**b**) Configuration coefficient $A2$ of the second mode of C_p over the suction side.

The first configuration coefficient, $A1$, varies slightly with the height of the propeller for a given angle of attack. From another point of view, for a given relative height, this coefficient is is much more variable with the variation of the angle of attack. The value of the coefficient $A1$ has a negative value for any combination of height and angle of attack, so it influences the mode in all positions, finding the greatest influence at the maximum angle of attack and close to the highest position. At around 75% of relative propeller height, however, the trend is inverted and the value of $A1$ starts to decrease: the propeller is too high and the effects of the boundary layer ingestion are reduced.

This position is represented in Figure 12, where it can be seen that the first mode practically explains all the C_p by itself. Rising the propeller position can be used to produce higher lift: for a given angle of attack, the value of the C_p due to the first mode can be increased 10% to 20% by rising the propeller to around 75% while maintaining the thrust equal to the drag.

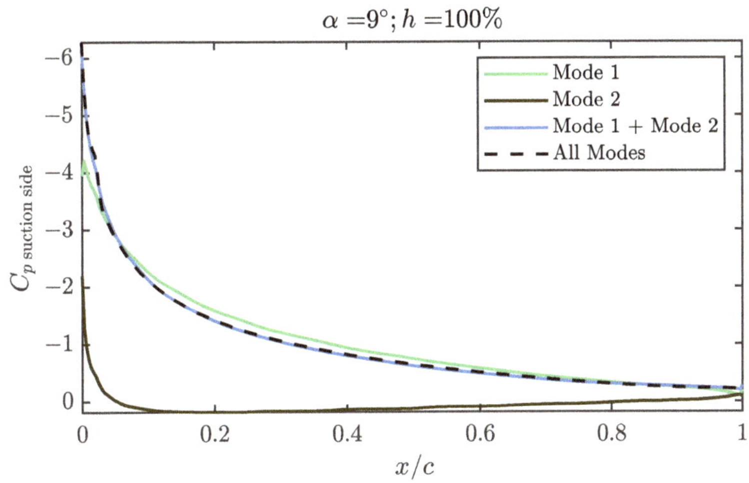

Figure 12. Contribution of the two first modes in the case of maximum relative height of the propeller over the trailing edge and maximum angle of attack simulated.

The configuration coefficient of the second mode, $A2$, have a similar behaviour to the first, that is, it is possible to observe a greater variation with the angle of attack than with the propeller height. However, the coefficient values are far from the first. In this case, the absolute maximum values are lower, which makes sense since this mode explains less energy. Furthermore, the value of the coefficient goes from negative to positive when increasing the angle of attack, reaching a null value for α around 6° to 7°. The variation with the angle of attack seems to be more nonlinear than in the case of the first mode, and makes the suction peak narrower and more intense as α increases. The same can be said for the position of the propeller: as it moves up, the configuration coefficient increases, what moves the suction peak towards the leading edge, makes it narrower and more intense. The shift due to changes in the propeller position is bigger at a height around 50%. This may be explained by the fact that, at lower positions, the propeller affects too much the pressure side of the airfoil, whereas at higher positions the amount of ingested boundary layer is decreased.

The same coefficients are presented in the case of the C_p of the pressure side Figure 13, including in this case the third coefficient.

The first mode of the intrados has a structure similar to that seen in the first mode of the extrados represented in Figure 11. The trend with the angle of attack and the propeller height is similar, although more nonlinear.

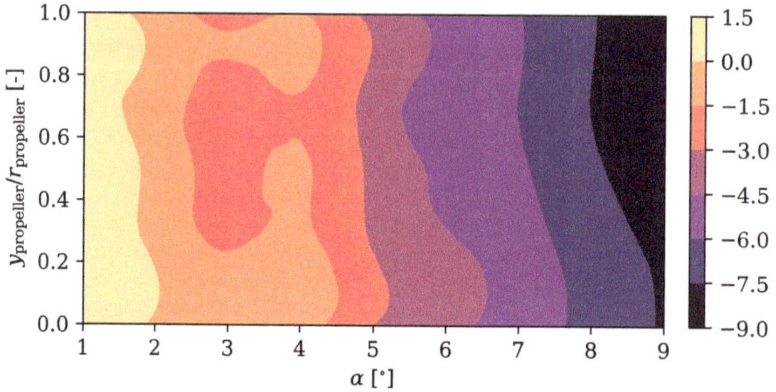

(**a**)

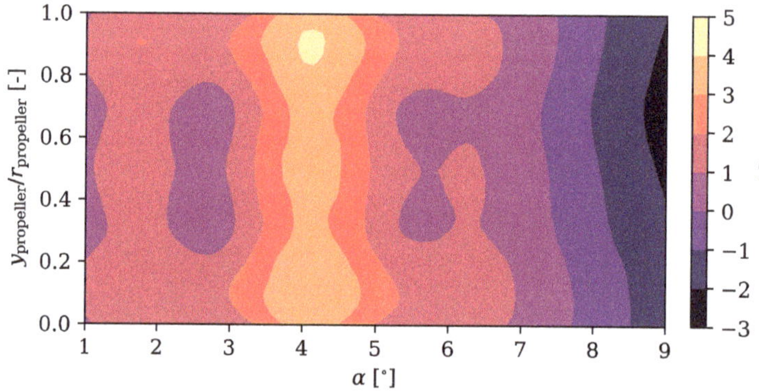

(**b**)

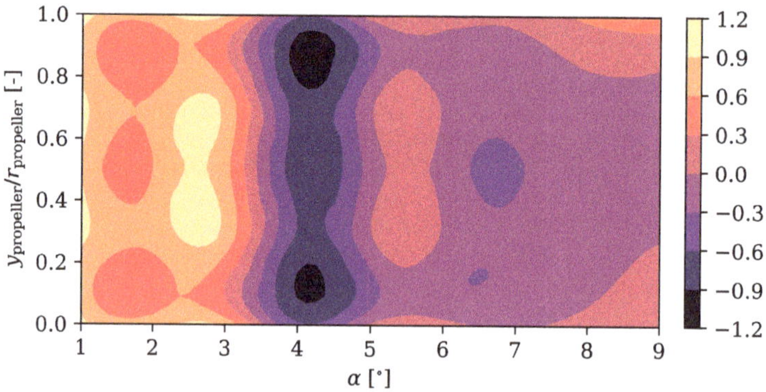

(**c**)

Figure 13. Configuration coefficients for the first three modes of the pressure coefficient over the pressure side of the airfoil, as a function of the angle of attack and the relative propeller position. (**a**) Configuration coefficient $A1$ of the first mode of the pressure coefficient over the pressure side. (**b**) Configuration coefficient $A2$ of the second mode of the pressure coefficient over the pressure side. (**c**) Configuration coefficient $A3$ of the third mode of the pressure coefficient over the pressure side.

The second and third modes (*A*2 and *A*3) have more complex structures and explain more *TKE* than in the case of the suction side, although their effect over the global aerodynamic coefficients is somewhat limited as the absolute value of the C_p over the pressure side is smaller. It is interesting to note the peak in value around an angle of attack of 4°. At that position, they maximise the value of C_p at the leading stagnation point, widening it. At angles of attack smaller or bigger, the bigger value of the configuration coefficient of the third mode makes the stagnation zone narrower. Regarding the propeller position, at low positions the flow accelerates more over the intrados, reducing the pressure coefficient.

4.2. Friction Coefficient Analysis Using POD

The friction coefficient is studied in the same way as C_p. Representing the evolution of this coefficient for a given propeller position 50% on Figure 14, the advancement of the LSB can be observed, marked by the fast growth of C_f followed by a null value of the coefficient, indicating where the bubble is located. In turn, the changes in the intrados are small, and the value of C_f is practically constant throughout the chord.

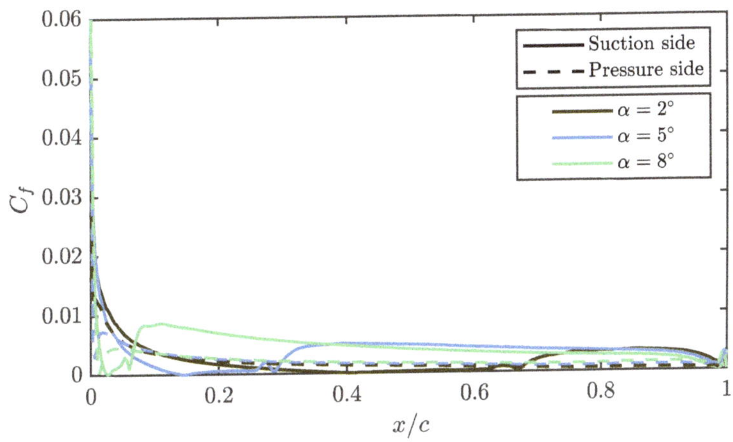

Figure 14. Friction coefficient over the airfoil with the propeller in 50% position.

The first modes represented in Figure 15 are those that have the most energy and explain the problem in a more significant percentage, both in the extrados and the intrados.

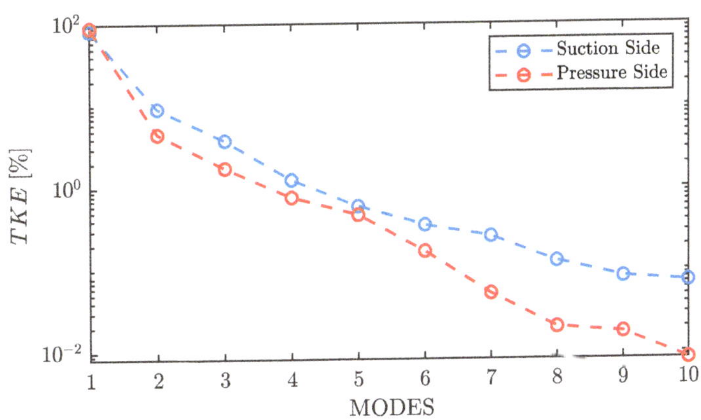

Figure 15. *TKE* of the friction coefficient.

In this case, for the extrados, the energy of the first mode is below 83%, which means that it is necessary to use more modes to accurately represent the problem since they will be more energetic. This can be easily verified in Figure 16a. Even with five modes, there are important differences between the reconstruction and the actual friction coefficient. In order to take into account the behaviour of the transitional LSB, up to ten modes are needed. In the intrados, the C_f has a much more homogeneous shape since the LSB does not appear, therefore fewer modes are required for the reconstruction of this side. Indeed, with only two modes, the error reconstructing the friction coefficient over the pressure side is negligible, as it is shown in Figure 16b.

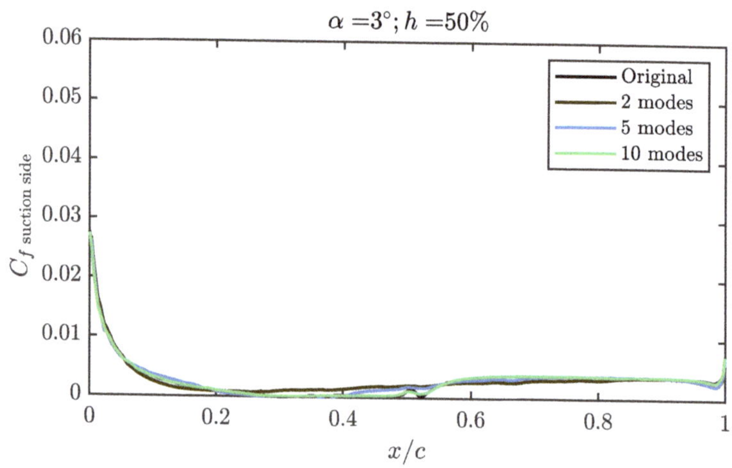

(**a**)

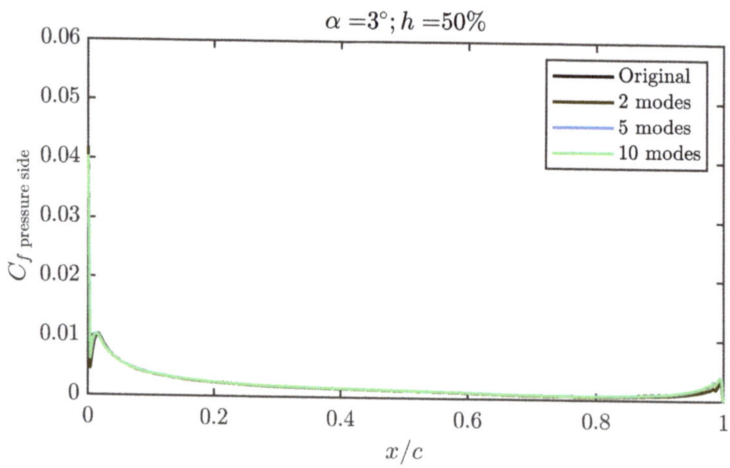

(**b**)

Figure 16. Friction coefficient reconstruction over the airfoil, for an angle of attack of 3° and a propeller relative height of 50%. (**a**) Friction coefficient reconstruction over the suction side of the airfoil. (**b**) Friction coefficient reconstruction over the pressure side of the airfoil.

As was done with the C_p, it is possible to represent the configuration coefficients of the modes in order to obtain more information. In Figure 17 the two first modes of the extrados C_f are represented.

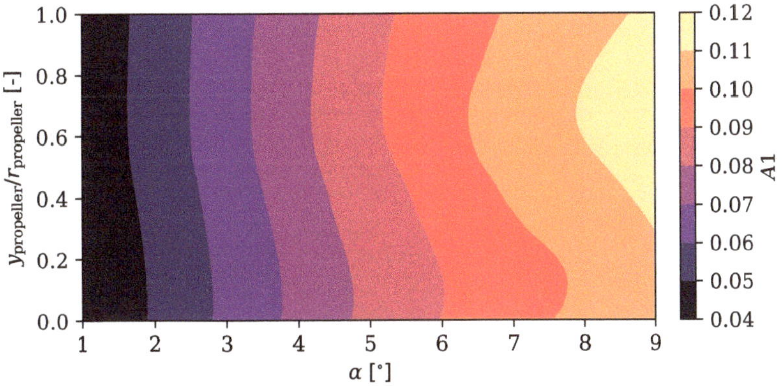

(**a**)

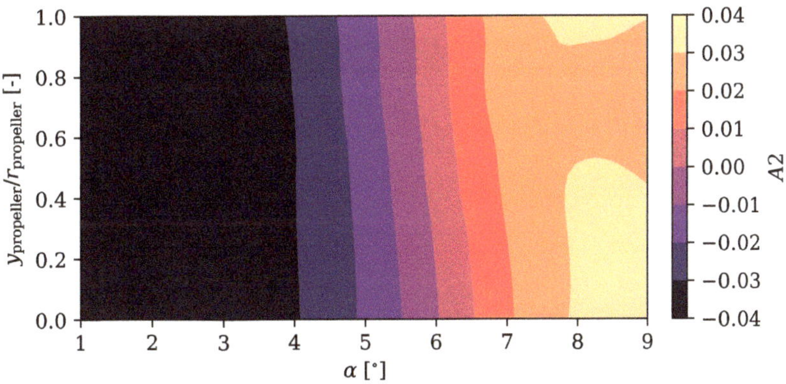

(**b**)

Figure 17. Configuration coefficients for the first two modes of the friction coefficient over the suction side of the airfoil, as a function of the angle of attack and the relative propeller position. (**a**) Configuration coefficient $A1$ of the first mode of the friction coefficient over the suction side. (**b**) Configuration coefficient $A2$ of the second mode of the friction coefficient over the suction side.

The trend of the first coefficient mirrors the one already seen in the case of C_p, getting higher values as the angle of attack increases. Again, a greater variation of the coefficient is observed with the angle of attack than with the height of the propeller. However, in this case, for the highest angles of attack, from 6° onward, the height of the propeller varies to a greater extent the value of the coefficient, obtaining bigger values in the higher positions. This fits with what was stated in [25], where it could be observed that by increasing the relative position of the propeller on the trailing edge, the friction coefficient of the suction side increased, increasing the parasitic drag. Most of the effect can be explained by the first mode alone, which does not carry enough information to resolve the transitional separation bubble: it is expected that the parasitic drag of the wing due to the skin friction over its suction side could be accurately interpolated to other non-simulated conditions even if the LSB is not well resolved. For the second mode, the effect of the propeller position is more limited.

Looking at the configuration coefficients of the pressure side in Figure 18, it is possible to observe that the behaviour of the first coefficient is similar to that of the suction side but with an inverted value. The value of the coefficient decreases as the angle of attack decreases, reaching the minimum at the highest angle of attack and at the highest positions. This behaviour is due to the fact that in the higher positions, a small percentage of the actuator disc influences the pressure side, reaccelerating the flow to a lesser extent and giving rise to a smaller C_f. The parasitic drag due to the friction coefficient over the pressure side of the wing should be more accurately interpolated in this case, as there is no transitional bubble that produces more *TKE* for higher modes.

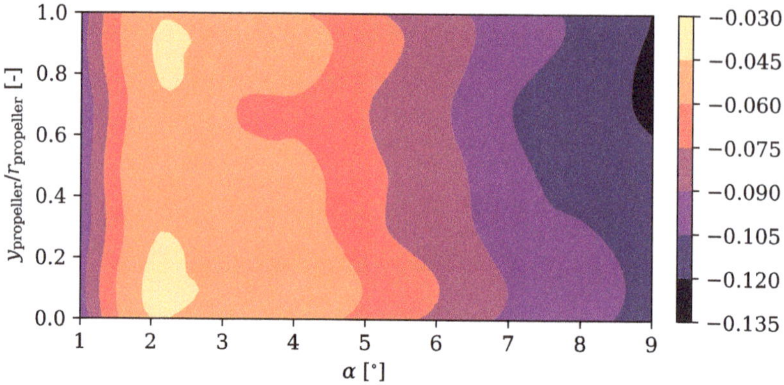

(**a**)

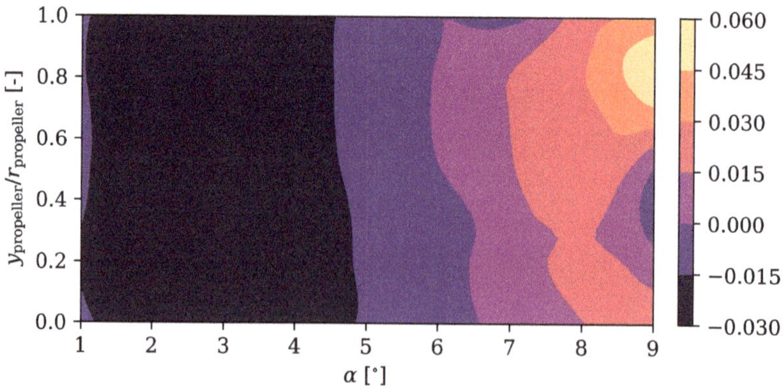

(**b**)

Figure 18. Configuration coefficients for the first two modes of the friction coefficient over the pressure side of the airfoil, as a function of the angle of attack and the relative propeller position. (**a**) Configuration coefficient *A*1 of the first mode of the friction coefficient over the pressure side. (**b**) Configuration coefficient *A*2 of the second mode of the friction coefficient over the pressure side.

4.3. *Lift and Drag Coefficient Analysis and Reconstruction*

It is helpful to know how the coefficients of lift and drag behave as a function of the angle of attack and the position of the propeller. In Figures 19 and 20, these two coefficients are represented as a function of the relative height of the actuator disc and the angle of attack of the aircraft for a Reynolds number of 5×10^5.

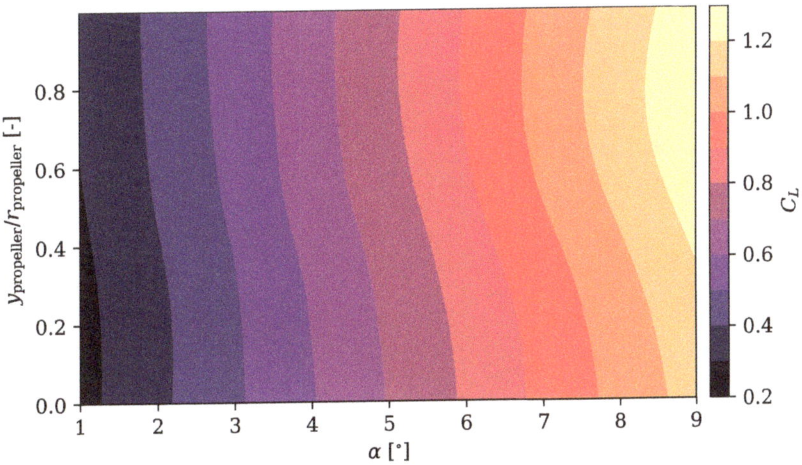

Figure 19. Lift coefficient for each propeller position above the trailing edge and angle of attack.

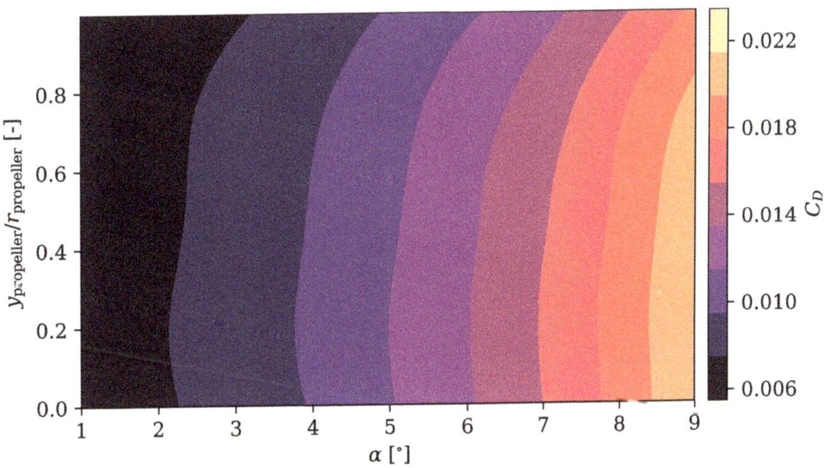

Figure 20. Drag coefficient for each propeller position above the trailing edge and angle of attack.

Both C_L and C_D vary more with the angle of attack than with the propeller height over the trailing edge. Although both increase in magnitude with the angle of attack for a given position, they have opposite trends when the position is varied for a given angle of attack.

For a given angle of attack, as the propeller rises, the lift over the wing is greater while the drag is minimised. That is why the aerodynamic efficiency is maximised in the highest positions for any angle of attack, as shown in [25].

The pressure and friction coefficients have been integrated to obtain the lift and drag coefficients, where this operation has been carried out taking into account a different number of modes. The fractions of the lift coefficient and drag coefficient obtained doing the reconstruction with different number of modes, $C_{L,\text{modes}}/C_L$ and $C_{D,\text{modes}}/C_D$, have been computed and they are shown in Figures 21 and 22. The error results are shown depending on the number of modes used and the position of the propeller for a fixed angle of attack.

It is possible to observe in Figure 21 that the relative error when using two modes is less than 3% in the case of C_L for many configurations. The error in this variable varies little with the position of the propeller, but it does for the angle of attack, where more modes are necessary to reduce the error at a low angle. At high angles of attack, using just 2 modes results in an overestimated C_L, while at low angles of attack, the opposite occurs. As a general rule, at least five modes reduce the error by around 1%.

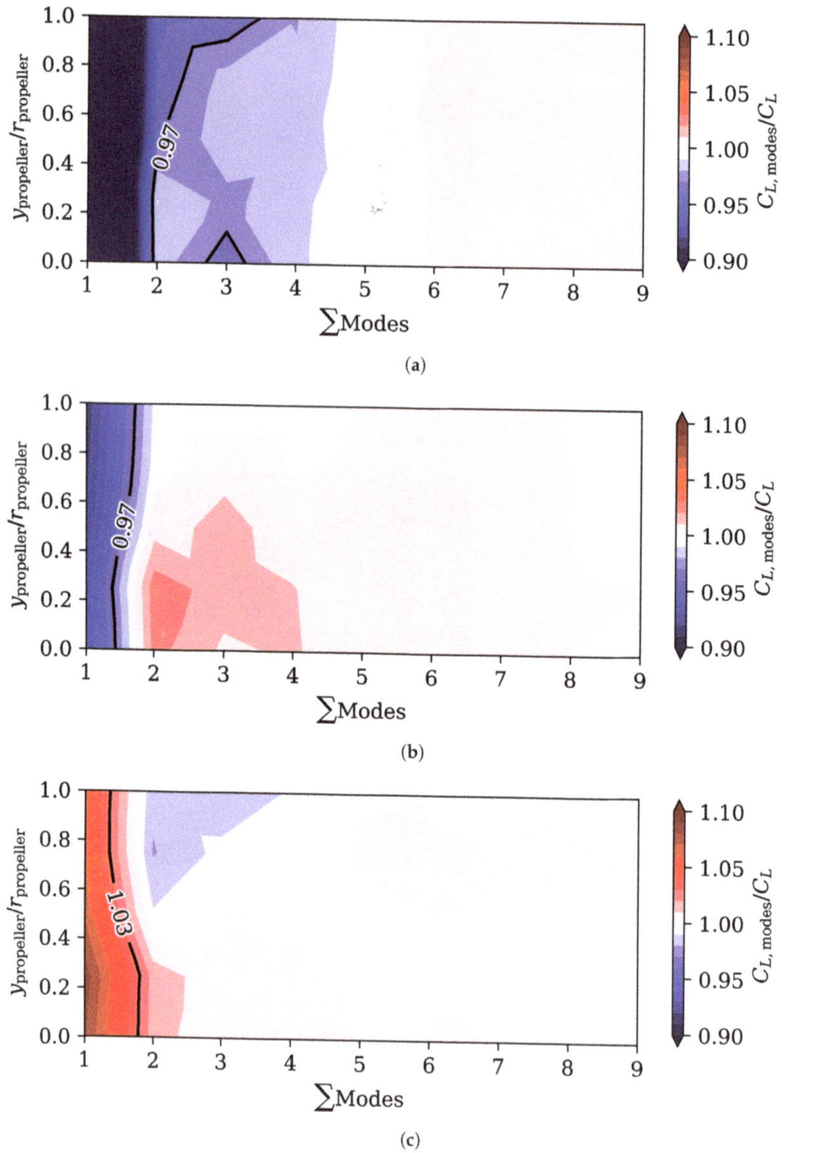

Figure 21. Fraction of lift coefficient computed using from 1 to 9 modes, using both the pressure coefficient and friction coefficient distributions. (**a**) Lift coefficient computed using from 1 to 9 modes for an angle of attack of 1°. (**b**) Fraction of lift coefficient computed using from 1 to 9 modes for an angle of attack of 5°. (**c**) Fraction of lift coefficient computed using from 1 to 9 modes for an angle of attack of 9°.

Figure 22 shows that the drag coefficient needs more modes than the lift coefficient to be reconstructed with low error. If a single mode is used, the error exceeds 50% at any angle of attack. When using 2 or 3 modes, the error remains at 3% and varies from overestimated at low propeller positions to underestimated at higher positions. As in the case of the C_L, more modes are necessary at a lower angle of attack to maintain a C_D relative error under 3%. At least 6 modes are required to represent any angle of attack correctly.

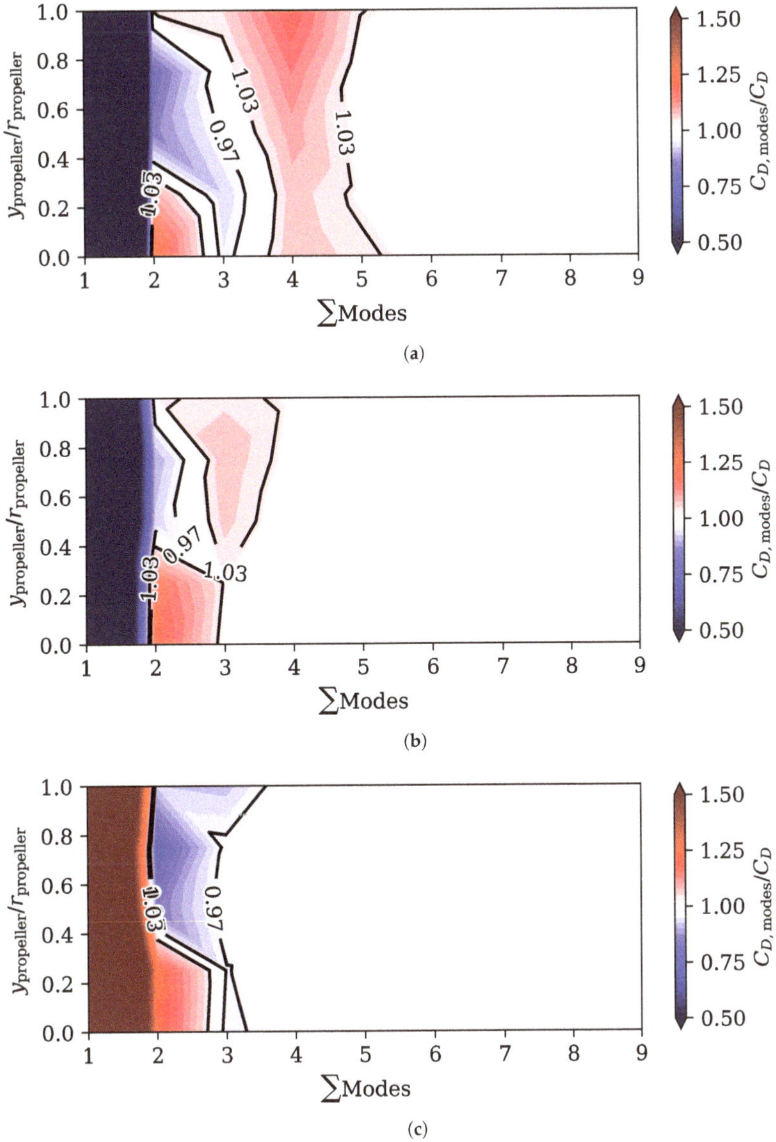

Figure 22. Fraction of drag coefficient computed using from 1 to 9 modes, using both the pressure coefficient and friction coefficient distributions. (**a**) Fraction of drag coefficient computed using from 1 to 9 modes for an angle of attack of 1°. (**b**) Fraction of drag coefficient computed using from 1 to 9 modes for an angle of attack of 5°. (**c**) Fraction of drag coefficient computed using from 1 to 9 modes for an angle of attack of 9°.

4.4. Interpolation of C_p and C_f with a Surrogate Model

The POD results are not only used to analyse the pressure and friction coefficients over the wing. As in other applications, they are also used to reduce the number of high fidelity simulations that it takes to fit a surrogate model. Once the modes for all propeller positions and angle of attack have been calculated, it is possible to interpolate and reconstruct the pressure and friction coefficients for any not-simulated intermediate cases, as shown in Figures 23 and 24.

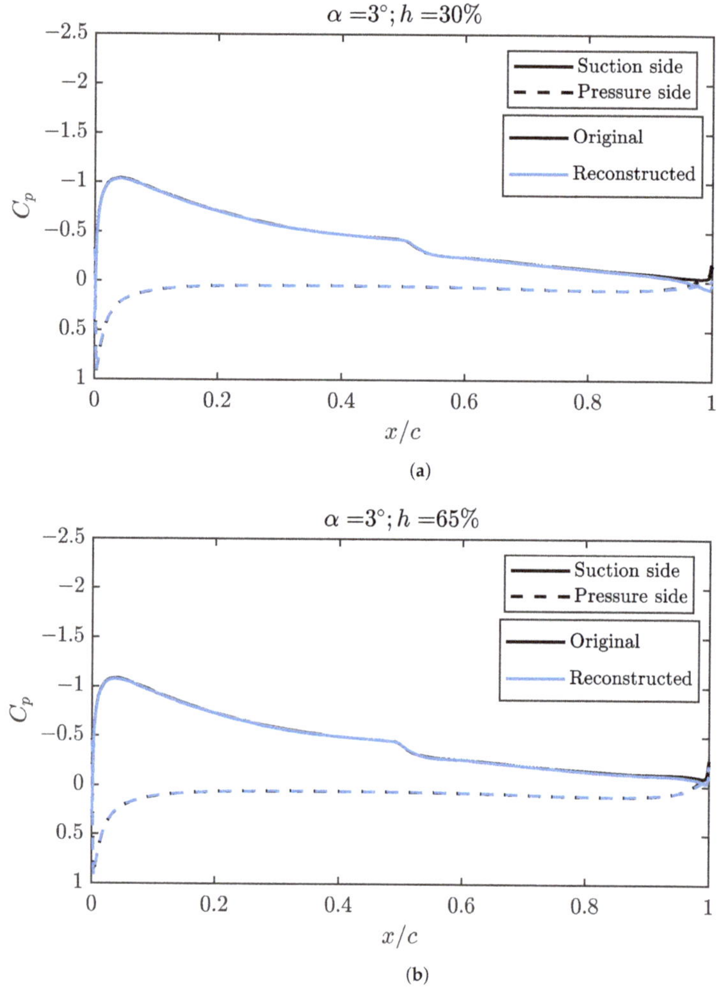

Figure 23. Pressure coefficient reconstructed in a propeller position not used to fit the surrogate model, compared with data from a CFD simulation. (**a**) Pressure coefficient reconstructed an angle of attack of 3° and a propeller position of 30%. (**b**) Pressure coefficient reconstructed for an angle of attack of 3° and a propeller position of 65%.

In Figure 23, the configuration coefficients of the POD are interpolated and the pressure coefficient is computed using the interpolated values and the modes of the decomposition. At 3°, Reynolds of 5×10^5, the positions of 30% and 65% are perfectly reconstructed, except for the C_p closest to the propeller. For practical purposes, the contribution of the pressure distribution over the lift, drag and pitching moment coefficients over the wing are reconstructed with a very small error, but also the stresses over the the skin can be computed with high accuracy.

However, in Figure 24, where an angle of attack not used to fit the interpolator, an adequate reconstruction of the appearance of the recirculation bubble is not obtained, although the rest of the C_p is correctly described. The stresses over the wing skin can still be computed with high accuracy with this interpolation.

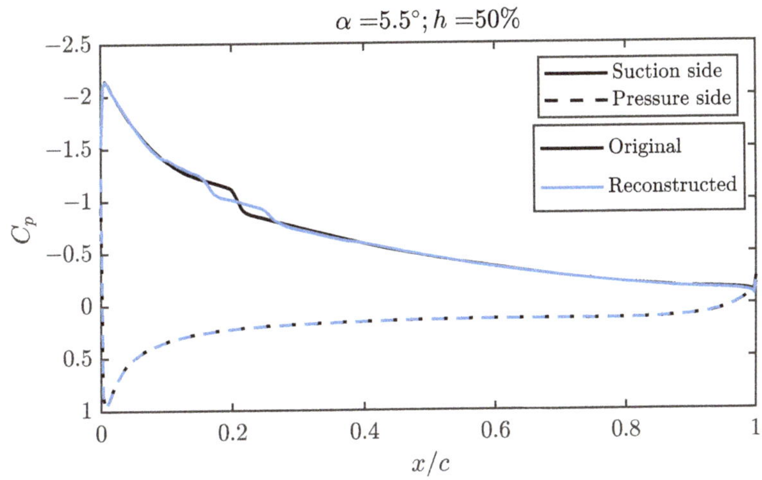

(a)

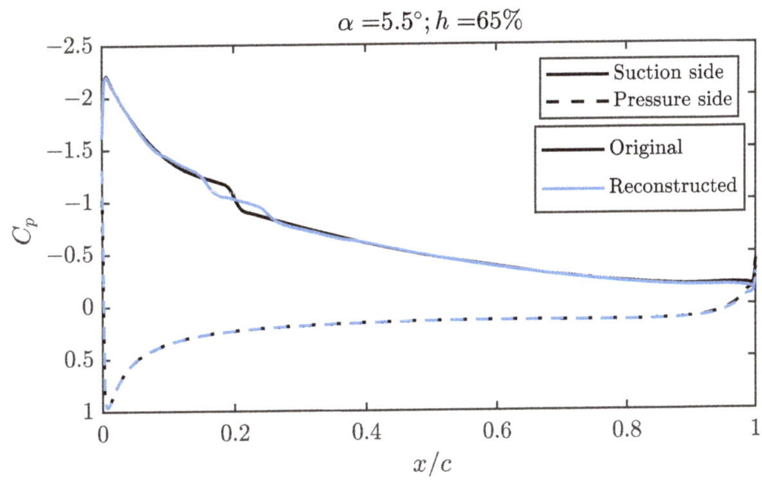

(b)

Figure 24. Pressure coefficient reconstructed in a propeller position and angle of attack not used to fit the surrogate model, compared with data from a CFD simulation. (**a**) Pressure coefficient reconstructed for an angle of attack of 5.5° and a propeller position of 50%. (**b**) Pressure coefficient reconstructed for an angle of attack of 5.5° and a propeller position of 65%.

In the same way, it is possible to represent the C_f in Figures 25 and 26.

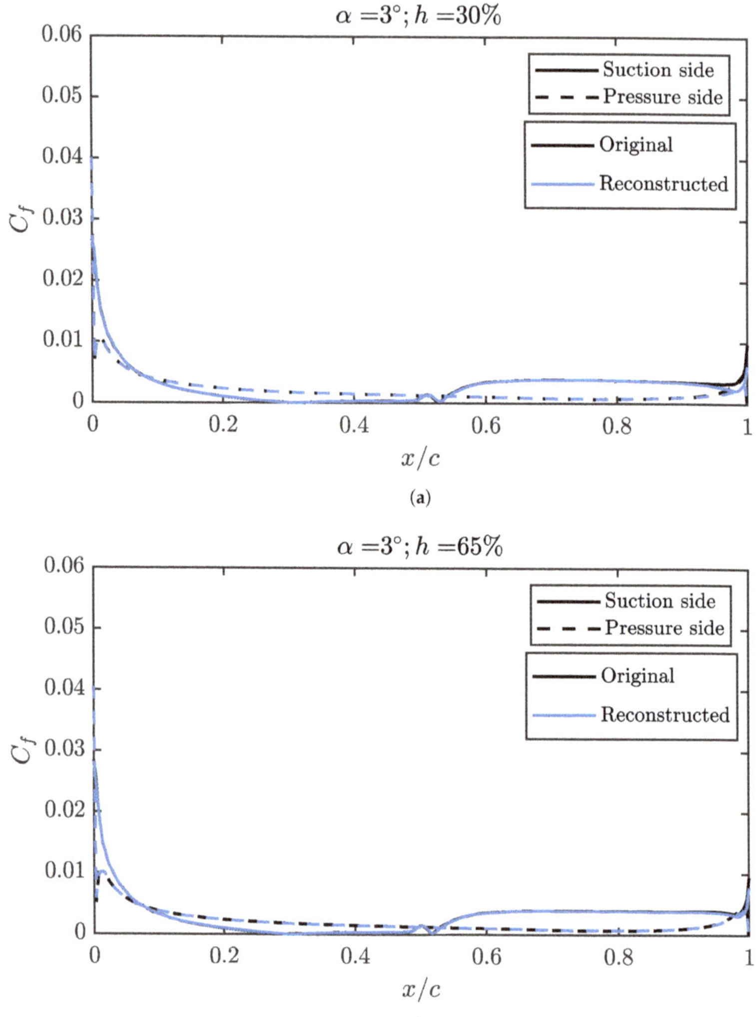

(**a**)

(**b**)

Figure 25. Friction coefficient reconstructed in a propeller position not used to produce the surrogate model, compared with data from a CFD simulation. (**a**) Friction coefficient reconstructed an angle of attack of 3° and a propeller position of 30%. (**b**) Friction coefficient reconstructed for an angle of attack of 3° and a propeller position of 65%.

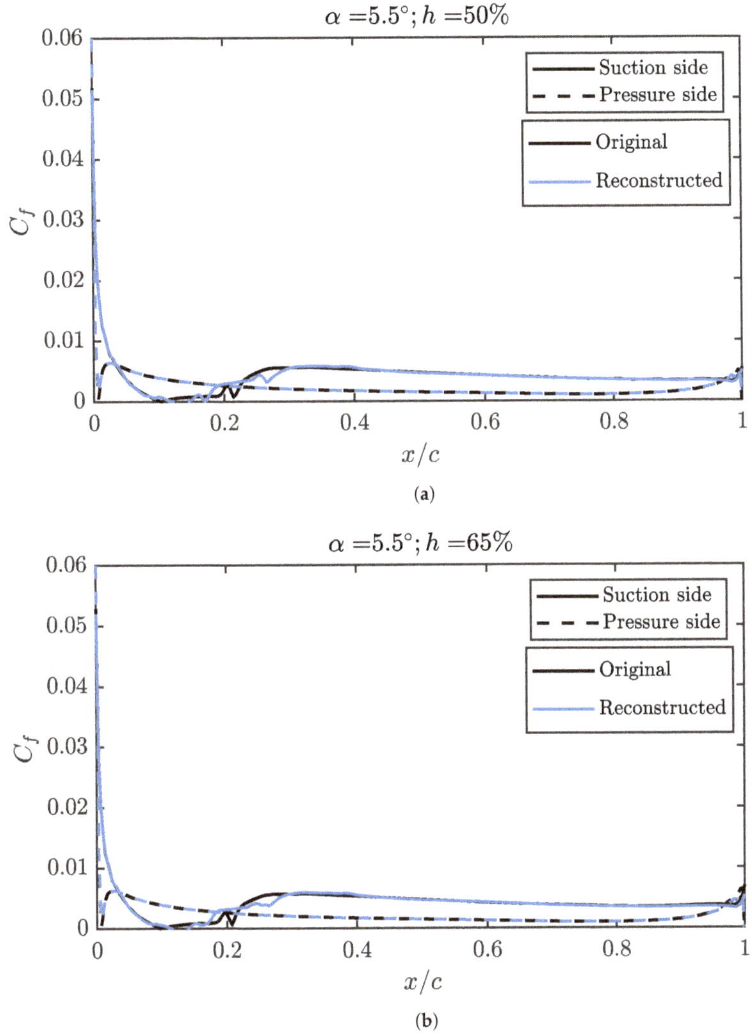

Figure 26. Friction coefficient reconstructed in a propeller position and angle of attack not used to fit the surrogate model, compared with data from a CFD simulation. (**a**) Friction coefficient reconstructed for an angle of attack of 5.5° and a propeller position of 50%. (**b**) Friction coefficient reconstructed for an angle of attack of 5.5° and a propeller position of 65%.

In Figure 25, it can be seen that, for 3°, as it happened with the C_p, the reconstruction of C_f is perfect. In comparison, for 5.5° and in Figure 26, the reconstructed recirculation bubble is delayed with respect to that of the CFD simulations. The rest of the C_f distribution is represented correctly.

The use of POD to reconstruct aerodynamic coefficients with low error interpolating a surrogate model is useful both in the pre-design and while optimising the wings of light UAVs. Light UAVs may be built using materials with very low stiffness, so a precise value of the pressure distribution over the wing is needed to optimise its structure without incurring in non-acceptable deformations of its shape.

5. Conclusions

After computing the pressure and friction coefficients over a wing section for different propeller positions and angles of attack, a modal decomposition using POD was performed. Looking at the data, more than 90% of the *TKE* is explained by the first mode of both the pressure and skin friction coefficients, for both the suction side and the pressure side of the airfoil. This means that, although some details are only described by taking into account higher-order modes, the main behaviour can be analysed by just looking at the first mode. From that, the main parameter affecting the distributions is the angle of attack, although the effect of the propeller position is not negligible.

For the pressure coefficient, rising the position of the propeller increases both the suction over the extrados and the overpressure over the intrados. The effect is similar to increasing the angle of attack, and explains an increment of lift between 10% to 20%. When the position is too high, with the shaft separated from the trailing edge a distance equal to the propeller radius, the trend is inverted and the extra lift starts to decrease. This is consistent with a decrease of the boundary layer ingestion effects.

But not only extra lift is produced: the suction peak is also moved towards the leading edge, decreasing the form drag due to the pressure distribution and creating some leading edge thrust. At the pressure side and higher modes, the trend due to the propeller position is not so obvious.

The nonlinear effects due to the propeller position are more obvious when looking at the skin friction coefficient. The maximum effects are seen for a position of around 70% for the first mode, which carries most of the information to reconstruct the coefficient distribution. For higher modes, no clear trend is found.

In order to reconstruct the LSB in the pressure coefficient over the airfoil, several higher modes are needed. For the friction coefficient, the case is more extreme, and up to ten modes are needed to fully catch the boundary layer transition due to the separation bubble.

By integrating the coefficients of pressure and friction, it is possible to obtain the coefficient of lift and drag for any combination of angle of attack and propeller position. Using at least two modes, it is possible to represent these coefficients below 3% of relative error in many cases, being necessary at least 6 modes for the error to fall below 1%.

The results of the modal decomposition can be used to fit surrogate models for optimisations for which the pressure or friction coefficient distribution is critical, such as when designing low-stiffness wing structures in small UAVs. In those cases, the pressure distribution has to be taken into account to get the exact load distribution over the skin, which may produce prohibitive deformations in some cases. Although some details are not accurately reconstructed in some configurations using this method, such as the exact shape of the transitional bubble, most of the coefficient distribution is obtained with negligible errors.

Author Contributions: Conceptualization, L.M.G.-C., P.V.; Data curation, P.V.; Formal analysis, P.V.; Funding acquisition, J.R.S., L.M.G.-C.; Investigation, P.V.; Methodology, L.M.G.-C., P.V.; Project administration, L.M.G.-C.; Resources, J.R.S.; Software, L.M.G.-C. and P.V.; Supervision, L.M.G.-C.; Validation, L.M.G.-C. and P.V.; Visualization, L.M.G.-C. and P.V.; Writing—original draft, P.V.; Writing—review and editing, J.R.S., L.M.G.-C., P.B. and P.V. All authors have read and agreed to the published version of the manuscript.

Funding: This work was partially funded by the Conselleria d'Innovació, Universitats, Ciència i Societat Digital of the Generalitat Valenciana through grant with expedient number GV/2021/069 of the program for "Grupos de Investigación Emergentes GV/2021". This work is part of the project PID2020-119468RA-I00 funded by MCIN/AEI/10.13039/501100011033. This work is also part of the project EQC2019-006272-P funded by MCIN/AEI/10.13039/501100011033 and by "ERDF A way of making Europe".

Institutional Review Board Statement: Not applicable.

Informed Consent Statement: Not applicable.

Data Availability Statement: The data presented in this study are available on request from the corresponding author.

Acknowledgments: The authors would like to thank L. Ricarte for his contribution to the work during the completion of his master's thesis.

Conflicts of Interest: The authors declare no conflict of interest. The funders had no role in the design of the study; in the collection, analyses, or interpretation of data; in the writing of the manuscript, or in the decision to publish the results.

Abbreviations

The following abbreviations are used in this manuscript:

Abbreviations	
BCG	Boston Consulting Group
BLI	Boundary layer ingestion
BEMT	Blade Element Model Theory
CFD	Computational fluid dynamics
DEP	Distributed electrical propulsion
ERA	Environmentally Responsible Aviation
HE	Hybrid electric
ITDS	Information Technology Development Solutions
LSB	Laminar separation bubble
NASA	National Aeronautics and Space Administration
POD	Proper Orthogonal Decomposition
RANS	Reynols-averaged Navier-Stokes
UAV	Unmanned aerial vehicle

Roman letters	
$\boldsymbol{A}$	Configuration coefficients matrix
a_i	Configuration coefficient of mode i
$\mathcal{R}$	Aspect ratio
b	Wingspan
c	Chord
$\boldsymbol{C}$	Covariance matrix
C_L	Lift coefficient
C_D	Drag coefficient
$C_{D,0,\text{extra}}$	Parasitic drag coefficient of the aircraft without the wing
$C_{D,0,\text{wing}}$	Parasitic drag coefficient of the wing
C_p	Pressure coefficient
C_f	Friction coefficient
D	Drag
e	Oswald efficiency factor
h	Relative height of the propeller shaft
$r_{\text{propeller}}$	Propeller radius
Re	Reynolds
S	Wing surface
T	Thrust
TKE	Total fluctuating kinetic energy
$\boldsymbol{U}$	Dataset matrix
U_∞	Air speed
x	Position across the chord
$y_{\text{propeller}}$	Position of the propeller shaft above the trailing edge

Greek letters	
α	Angle of attack
$\mathbf{\Lambda}$	Eigenvalues matrix
λ	Eigenvalue
$\mathbf{\Phi}$	Eigenvector matrix
ϕ	Eigenvector
ρ	Density

References

1. Nickol, C.L.; Haller, W.J. Assessment of the performance potential of advanced subsonic transport concepts for NASA's environmentally responsible aviation project. In Proceedings of the 54th AIAA Aerospace Sciences Meeting, San Diego, CA, USA, 4–8 January 2016; pp. 1–21. [CrossRef]
2. Amoukteh, A.; Janda, J.; Vicent, J. Drones Go to work. *Harv. Bus. Rev.* **2017**, 77–94.
3. Kim, H.D.; Perry, A.T.; Ansell, P.J. A Review of Distributed Electric Propulsion Concepts for Air Vehicle Technology. In Proceedings of the AIAA/IEEE Electric Aircraft Technologies Symposium, Cincinnati, OH, USA, 9–11 July 2018; pp. 1–21. [CrossRef]
4. Ko, Y.Y.A. The Multidisciplinary Design Optimization of a Distributed Propulsion Blended-Wing-Body Aircraft. Ph.D. Thesis, Virginia Tech, Blacksburg, Virginia, 2003.
5. Moore, K.R.; Ning, A. Distributed electric propulsion effects on traditional aircraft through multidisciplinary optimization. In Proceedings of the AIAA/ASCE/AHS/ASC Structures, Structural Dynamics, and Materials Conference, Kissimmee, FL, USA, 8–12 January 2018. [CrossRef]
6. Kirner, R.; Raffaelli, L.; Rolt, A.; Laskaridis, P.; Doulgeris, G.; Singh, R. An assessment of distributed propulsion: Part B—Advanced propulsion system architectures for blended wing body aircraft configurations. *Aerosp. Sci. Technol.* **2016**, *50*, 212–219. [CrossRef]
7. Stoll, A.M.; Bevirt, J.; Moore, M.D.; Fredericks, W.J.; Borer, N.K. Drag Reduction Through Distributed Electric Propulsion. In Proceedings of the 14th AIAA Aviation Technology, Integration, and Operations Conference, Atlanta, GA, USA, 16–20 June 2014; pp. 1–10. [CrossRef]
8. Stoll, A.M. Comparison of CFD and experimental results of the leap tech distributed electric propulsion blown wing. In Proceedings of the 15th AIAA Aviation Technology, Integration, and Operations Conference, Dallas, TX, USA, 22–26 June 2015; pp. 22–26. [CrossRef]
9. Amoozgar, M.; Friswell, M.I.; Fazelzadeh, S.A.; Khodaparast, H.H.; Mazidi, A.; Cooper, J.E. Aeroelastic stability analysis of electric aircraft wings with distributed electric propulsors. *Aerospace* **2021**, *8*, 100. [CrossRef]
10. Ausserer, J.K.; Harmon, F.G. Integration, validation, and testing of a hybrid-electric propulsion system for a small remotely-piloted aircraft. In Proceedings of the 10th Annual International Energy Conversion Engineering Conference, Atlanta, GA, USA, 30 July–1 August 2012; pp. 1–11. [CrossRef]
11. Kim, C.; Namgoong, E.; Lee, S.; Kim, T.; Kim, H. *Fuel Economy Optimization for Parallel Hybrid Vehicles with CVT*; SAE Technical Paper 1999-01-1148; SAE International: Warrendale, PA, USA, 1999. [CrossRef]
12. Budziszewski, N.; Friedrichs, J. Modelling of a boundary layer ingesting propulsor. *Energies* **2018**, *11*, 708. [CrossRef]
13. Teperin, L. Investigation on Boundary Layer Ingestion Propulsion for UAVs. In Proceedings of the International Micro Air Vehicle Conference and Flight Competition (IMAV), Toulouse, France, 18–21 September 2017; pp. 293–300.
14. Elsalamony, M.; Teperin, L. 2D Numerical Investigation of Boundary Layer Ingestion Propulsor on Airfoil. In Proceedings of the 7th European Conference for Aeronautics and Space Sciences (EUCASS), Milan, Italy, 3–6 July 2017, pp. 1–11. [CrossRef]
15. Hall, D.K.; Huang, A.C.; Uranga, A.; Greitzer, E.M.; Drela, M.; Sato, S. Boundary layer ingestion propulsion benefit for transport aircraft. *J. Propuls. Power* **2017**, *33*, 1118–1129. [CrossRef]
16. Broatch, A.; García-Tíscar, J.; Roig, F.; Sharma, S. Dynamic mode decomposition of the acoustic field in radial compressors. *Aerosp. Sci. Technol.* **2019**, *90*, 388–400. [CrossRef]
17. Torregrosa, A.J.; Broatch, A.; García-Tíscar, J.; Gomez-Soriano, J. Modal decomposition of the unsteady flow field in compression-ignited combustion chambers. *Combust. Flame* **2018**, *188*, 469–482. [CrossRef]
18. Zhu, Z.; Midlam-Mohler, S.; Canova, M. Development of physics-based three-way catalytic converter model for real-time distributed temperature prediction using proper orthogonal decomposition and collocation. *Int. J. Engine Res.* **2021**, *22*, 873–889. [CrossRef]
19. Rulli, F.; Fontanesi, S.; D'Adamo, A.; Berni, F. A critical review of flow field analysis methods involving proper orthogonal decomposition and quadruple proper orthogonal decomposition for internal combustion engines. *Int. J. Engine Res.* **2021**, *22*, 222–242. [CrossRef]
20. Shen, L.; Teh, K.Y.; Ge, P.; Zhao, F.; Hung, D.L. Temporal evolution analysis of in-cylinder flow by means of proper orthogonal decomposition. *Int. J. Engine Res.* **2021**, *22*, 1714–1730. [CrossRef]
21. Malouin, B.; Trépanier, J.Y.; Gariépy, M. Interpolation of transonic flows using a proper orthogonal decomposition method. *Int. J. Aerosp. Eng.* **2013**, *2013*. [CrossRef]

22. Mifsud, M.; Zimmermann, R.; Görtz, S. Speeding-up the computation of high-lift aerodynamics using a residual-based reduced-order model. *CEAS Aeronaut. J.* **2015**, *6*, 3–16. [CrossRef]
23. Mifsud, M.J.; MacManus, D.G.; Shaw, S.T. A variable-fidelity aerodynamic model using proper orthogonal decomposition. *Int. J. Numer. Methods Fluids* **2016**, *82*, 646–663. [CrossRef]
24. Tiseira, A.O.; García-Cuevas, L.M.; Quintero, P.; Varela, P. Series-hybridisation, distributed electric propulsion and boundary layer ingestion in long-endurance, small remotely piloted aircraft: Fuel consumption improvements. *Aerosp. Sci. Technol.* **2022**, *120*. [CrossRef]
25. Serrano, J.R.; Tiseira, A.O.; García-Cuevas, L.M.; Varela, P. Computational Study of the Propeller Position Effects in Wing-Mounted, Distributed Electric Propulsion with Boundary Layer Ingestion in a 25 kg Remotely Piloted Aircraft. *Drones* **2021**, *5*, 56. [CrossRef]
26. UAV Factory USA LLC. Penguin C UAS. Available online: https://www.uavfactory.com (accessed on 12 January 2012).
27. AERTEC Solutions. RPAS TARSIS 25. Available online: https://aertecsolutions.com/rpas/rpas-sistema-aereos-tripulados-remotamente/rpas-tarsis25/ (accessed on 12 January 2012).
28. Lyon, C.A.; Broeren, A.P.; Giguere, P.; Gopalarathnam, A.; Selig, M.S. *Summary of Low-Speed Airfoil Data—Volume 3*; SoarTech Publications: Virginia Beach, VA, USA, 1997; p. 315.
29. Selig, M.S.; Donovan, J.F.; Fraser, D.B. *Airfoils at Low Speeds*; H.A. Stokey: Virginia Beach, VA, USA, 1989; pp. 1–408.
30. Selig, M.S. Low Reynolds Number Airfoil Design Lecture Notes—Various Approaches to Airfoil Design. Available online: https://m-selig.ae.illinois.edu/pubs/Selig-2003-VKI-LRN-Airfoil-Design-Lecture-Series.pdf (accessed on 12 January 2012).
31. Sutton, D.M. Experimental Characterization of the Effects of Freestream Turbulence Intensity on the SD7003 Airfoil at Low Reynolds Numbers. Master's Thesis, Univeristy of Toronto, Toronto, ON, Canada, 2015.
32. Ananda, G.K.; Sukumar, P.P.; Selig, M.S. Measured aerodynamic characteristics of wings at low Reynolds numbers. *Aerosp. Sci. Technol.* **2015**, *42*, 392–406. [CrossRef]
33. Stajuda, M.; Obidowski, D.; Karczewski, M.; Józwik, K. Modified virtual blade method for propeller modelling. *Mech. Mech. Eng.* **2018**, *22*, 603–617. [CrossRef]
34. Harmon, F.G.; Frank, A.A.; Chattot, J.J. Conceptual design and simulation of a small hybrid-electric unmanned aerial vehicle. *J. Aircr.* **2006**, *43*, 1490–1498. [CrossRef]
35. Niţă, M.; Scholz, D. Estimating the Oswald Factor from Basic Aircraft Geometrical Parameters. In *Deutscher Luft- und Raumfahrtkongress*; DGLR: Berlin, Germany, 2012.
36. Hoerner, S. *Fluid-Dynamic Drag*; Hoerner Fluid Dynamics: Bakersfield, CA, USA, 1965; pp. 7-2–7-9.
37. XFLR5. Available online: https://www.xflr5.tech/xflr5.htm (accessed on 12 January 2012).
38. Drela, M. XFOIL Subsonic Airfoil Development System. Available online: https://web.mit.edu/drela/Public/web/xfoil/ (accessed on 12 January 2012).

drones

Article

Optimum Sizing of Photovoltaic-Battery Power Supply for Drone-Based Cellular Networks

Mahshid Javidsharifi [1,*], Hamoun Pourroshanfekr Arabani [2], Tamas Kerekes [1], Dezso Sera [3] Sergiu Viorel Spataru [4] and Josep M. Guerrero [5]

1 Department of AAU Energy, Aalborg University, 9220 Aalborg, Denmark; tak@energy.aau.dk
2 Division of Industrial Electrical Engineering and Automation, Lund University, SE-22100 Lund, Sweden; hamoun.pourroshanfekr_arabani@iea.lth.se
3 Faculty of Engineering, Queensland University of Technology, Brisbane, QLD 4000, Australia; dezso.sera@qut.edu.au
4 Department of Photonics Engineering, Technical University of Denmark, 4000 Roskilde, Denmark; sersp@fotonik.dtu.dk
5 Center for Research on Microgrids (CROM), Department of Energy Technology, Aalborg University, 9220 Aalborg, Denmark; joz@energy.aau.dk
* Correspondence: mja@energy.aau.dk; Tel.: +45-9940-3375

Abstract: In order to provide Internet access to rural areas and places without a reliable economic electricity grid, self-sustainable drone-based cellular networks have recently been presented. However, the difficulties of power consumption and mission planning lead to the challenge of optimal sizing of the power supply for future cellular telecommunication networks. In order to deal with this challenge, this paper presents an optimal approach for sizing the photovoltaic (PV)-battery power supply for drone-based cellular networks in remote areas. The main objective of the suggested approach is to minimize the total cost, including the capital and operational expenditures. The suggested framework is applied to an off-grid cellular telecommunication network with drone-based base stations that are powered by PV-battery systems-based recharging sites in a rural location. The PV-battery system is optimally designed for three recharging sites with three different power consumption profiles with different peak and cumulative loads. Results show that the optimal design of the PV-battery system is dependent on geographical data, solar irradiation, and ambient temperature, which affect the output power of the PV system, as well as the power consumption profile, which affects the required number of PV panels and battery capacity.

Keywords: drones; base stations; cellular networks; photovoltaic system; battery

Citation: Javidsharifi, M.; Pourroshanfekr Arabani, H.; Kerekes, T.; Sera, D.; Spataru, S.V.; Guerrero, J.M. Optimum Sizing of Photovoltaic-Battery Power Supply for Drone-Based Cellular Networks. *Drones* **2021**, *5*, 138. https://doi.org/10.3390/drones5040138

Academic Editor: Diego González-Aguilera

Received: 29 September 2021
Accepted: 17 November 2021
Published: 22 November 2021

1. Introduction

Although the majority of the world's population has access to the internet through mobile networks, there is still a coverage gap in some rural and remote low-income areas [1–3]. Since the average cost of installing network infrastructure in remote areas is double the cost in urban areas, and due to a lower obtained revenue in rural areas, the deployment of fixed terrestrial mobile networks in remote areas is not cost-effective [1–6].

A new component in cellular networks is drones, or unmanned aerial vehicles (UAV), as small cellular base stations for covering scattered populations, especially in remote areas where the deployment of mobile networks is a burdensome task [2–4]. Compared to terrestrial cellular networks, UAV-aided networks are also more flexible. However, the major concerns in drone-based deployment are the limitations and constraints of the drones' batteries [2]. These batteries need to be charged in recharging sites connected to the electricity grid, which in turn, leads to an excess cost [3]. Nevertheless, in some cases, there is no access to a reliable electricity grid in rural areas [5]. To deal with these challenges, renewable energy sources (RESs), such as solar photovoltaic (PV) systems, can be deployed to supply the charging sites of UAVs in rural areas [3]. Besides being able to meet the needs of energy demand, RESs can also reduce the impact of fossil pollutants on the environment and guarantee socio-economic benefits for sustainable development [7,8]. Solar-powered UAV-based cellular telecommunication networks can be particularly significant for regions that have poor grid connectivity while being rich in solar resources (e.g., Sub-Saharan Africa or South Asia) [7].

When designing a PV-battery supply system for base stations in cellular telecommunication networks, dimensioning of the PV-battery system is a challenging issue since different factors, including capital expenditure (CAPEX), operational expenditure (OPEX), the intermittent nature of solar irradiance, varying traffic demand, power autonomy, and physical location constraints should be considered [7]. Power supplying of base stations in mobile networks through PV-battery systems has been investigated in the literature [9–20]. In [9], a cellular network that was powered via both the main electricity grid and a PV-battery system was studied, while the research in [10] was on mobile networks that solely depend on PV-battery power supply. A traffic-aware RES-powered base station was proposed in [11]. Some papers reviewed the sustainable deployment in mobile networks [12–15]. The focus of [16–18] was on modeling the performance of RES-based base stations to determine the size of the PV system and the storage device capacity based on Markovian models.

In order to size the components for an off-grid PV-battery system, the authors in [19] suggested the multi-objective wind-driven optimization (MO-WDO) algorithm. The objective of this paper was to optimize the number of PV panels as well as the capacity of the battery while "the annual total life cycle cost" was minimized. Reference [20] considered a genetic algorithm to solve the designing problem of a PV-battery system for the base station in an off-grid mobile network in Norcia, Italy, with the objective of simultaneously minimizing the investment and operational costs.

The focus of several papers in the literature was on energy management of future green mobile networks [7,19–28]. A review on opportunities, challenges, and perspectives of renewable energy-powered sustainable 5G network infrastructure was conducted in [6]. In [7], a multi-objective optimization framework was presented to minimize the total cost and maximize the power autonomy factor of a mobile network in which the base station is supplied via a PV-battery system or the electricity grid based on the suggested power dispatch strategy. The authors in [21] suggested a cost and energy-effective approach for a hybrid on-grid and solar-powered mobile network in a rural area.

Reference [22] suggested a "distributed online algorithm" to solve the energy management problem of an on-grid cellular network, and the objective was to optimize the data intake levels, energy sharing among base stations, transmit power, energy purchase from the grid, and the battery charging/discharging rate. In order to decrease on-grid energy consumption, [23] proposed "short and long-term reactive and proactive energy management" methods for heterogonous networks (HetNets). "A low complexity energy management" approach was presented in [24] for an on-grid macrocell base station, and the share of energy was with a centralized renewable farm. The goal of this paper was to decrease the grid energy cost.

An energy management approach for renewable-based base stations was suggested in [25]. In order to determine the optimum sizing of the system, the battery aging method was used, and reinforcement learning was applied to maximize battery life. The authors in [26] studied "the green energy provisioning problem" to decrease the CAPEX of renewable resource installations in HetNets while maintaining the quality of service (QoS) in the system. In [27], authors assessed the "power outage probability" of a renewable-based base station to decide on the PV-battery size, while in [28], the size of the battery was provided based on an analytical model of renewable fluctuations in each time scale.

It is recognized from the scope of these papers that almost all of them were focusing on the optimum sizing and energy management of renewable-based base stations in current mobile networks, i.e., without UAV deployment. However, as the importance of the application of UAVs in rural areas without access to a reliable electricity grid is discussed, the main objective of this paper is to optimally size the PV-battery power supply system for drone-based cellular networks to minimize the total cost. Applying UAVs as small cellular base stations is also beneficial in emergency or disaster situations, where the availability to the terrestrial networks is lost [4,29], or in big events such as Olympic Games or concerts where the number of mobile users increases for a short period and the installation of a fixed mobile network is not cost-effective [4,30,31].

Compared to ground-based base stations, UAV-mounted small cellular base stations are able to cover specific sections of an area, have proper channel conditions towards users, and have the possibility of efficiently planning the UAVs' missions to schedule their coverage properly [6]. Consequently, an appealing choice for providing cellular networks in rural areas is small cellular base stations installed on UAVs [1–6]. Different aspects of UAV applications in cellular networks, including UAV mission planning, were studied in [3,32–36] and reveal the significance of the presence of UAVs in future mobile networks.

In [3], energy-efficient mission scheduling of a UAV-based cellular network was proposed that was powered by a PV-Battery system. The main goal was minimizing the energy consumed by UAVs while cellular coverage was provided. In order to formulate the problem, mixed-integer linear programming was applied while a genetic algorithm was defined to solve the problem. The objective of [32] was minimizing the total costs of a UAV-based 5G architecture while considering the site's installation costs, the costs of solar panels, batteries, UAVs, and the optical ring among the installed sites. Each UAV flight was then scheduled either to cover an area or to recharge the UAV at the ground site.

The focus of [33] was on minimizing the energy consumed for moving the UAVs of a 5G deployment during the coverage of a set of rural areas. The authors claimed that the presented formulation and simulation framework was able to cover a set of areas, while minimizing the energy consumed by the UAVs. The main goal of [34] was the energy consumption management of a UAV-based cellular network that provided coverage to rural and low-income areas. The objective was to maximize the energy stored in the UAVs and in the charging sites, and the coverage of the territory was ensured through the scheduling of the UAV missions over space and time. A decomposition-based approach and a genetic algorithm were proposed to deal with the formulized problem. It was observed in [32–34] that the time slot of UAV mission planning was equal to one hour, which led to covering specific areas by UAVs for four consecutive hours, which is non-realistic and not applicable in real-world applications. However, in [3], the mission planning of UAVs was based on a 10-min time slot, which is more realistic.

The authors in [2] investigated the problem of minimizing the installation costs for a UAV-based cellular architecture adapted to serve rural areas. The UAV covering constraints, recharging action over time, and installation constraints on recharging sites were also considered. However, energy-efficient mission planning was not carried out in [2]. In [35], a trade-off between maximizing the throughput provided by the UAVs over a set of areas and minimizing OPEX was presented through a grid-connected microgeneration concept based on minimizing the energy bought from the electricity grid and maximizing the energy sold to the electricity grid in a UAV-based cellular network.

Due to the power consumption of the drones and the mission planning, the determination of optimal sizing of the power supply system of the drone-based cellular telecommunication networks is thoroughly different from conventional architecture [34].

The main contribution of this paper is to derive the power consumption profile of a UAV-based cellular telecommunication network based on the mission planning of UAVs in [3]. However, in [3], the constraints related to the power sources of recharging sites are not considered. Additionally, in [3], the authors did not design the PV-battery systems of the UAV recharging sites, and the results of [32] (namely the number of solar panels and batteries in each site) were used in their research. Moreover, the PVWatts calculator was used to exploit the historical energy production of solar panels. Consequently, a proper approach should be suggested for designing the PV-battery power supply system for recharging sites of UAVs, and a model for PV power generation should be applied such that the available PV generation can be calculated properly.

To achieve the above-mentioned goals, in this paper, we propose an optimization framework to minimize the total financial cost, including *CAPEX* and *OPEX*, of the PV-battery-powered off-grid cellular telecommunication network with drone-based base stations through sizing the PV-battery system based on the derived power consumption profiles for three recharging sites with different peak and cumulative values. We applied the particle swarm optimization (PSO) algorithm to find the best solution to the problem over a set of different scenarios.

The main contributions of this paper can be summarized as follows:

1. Proposing an optimization framework to minimize the total investment and operational costs of a PV-battery-powered off-grid UAV-based cellular telecommunication network in a rural area;
2. Extracting the power consumption profile for recharging stations in the UAV-based cellular telecommunication network based on the results of energy-efficient UAVs' mission planning in [3];
3. Developing a detailed model for PV power generation estimation that compromises PV panels' installation (azimuth and tilt) angles, ambient temperature, and PV module characteristics;
4. Considering the battery's technical constraints in the problem formulation and the battery's economic specifications in the system's operational cost.

The remainder of the paper is as follows: in Section 2, the UAV-aided cellular network configuration is presented. Additionally, the PV-battery energy supply model, as well as the energy consumption profile, are defined in this section. The problem formulation and the considered objective are discussed in Section 3. The suggested optimization algorithm is explained in Section 4. Simulation results are displayed in Section 5. The conclusions are investigated in Section 6.

2. UAV-Aided Cellular Network

Figure 1 shows the drone-based system structure in which two types of places, "areas" and "recharging sites", are considered. A section of the region that should always be covered by a UAV is called an "area", while a recharging site is a sector where PV-battery systems are established. The black line represents the radio link between the low level (LL) 5G base station on the UAV and the high level (HL) 5G ground-based base station [3].

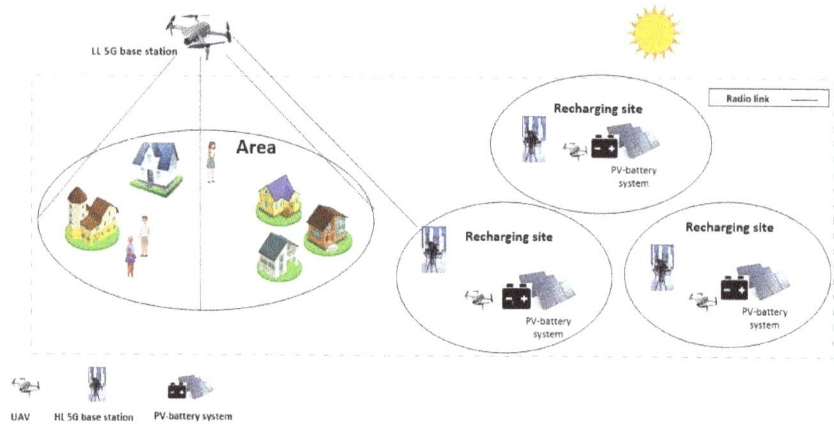

Figure 1. The configuration of an off-grid cellular telecommunication network with drone-based base stations powered by PV-battery systems.

A battery is installed on each UAV, and when it is required, the drone can move to the recharging sites to be charged.

A set of actions are considered for the drones, including (a) STAY, the drone is in idle mode in a recharging site without consuming any energy; (b) Recharging, the drone batteries are recharging at a site; (c) Covering, the drone is covering an area; (d) Moving: the drone is moving from an area to another area or from an area to a recharging site or vice versa.

2.1. PV-Battery System Modelling

The selected mathematical PV model, as well as the considered battery model, are presented in this section.

The following generated PV power model is considered [7]:

$$P_{PV}^{t} = \frac{I^{t}(\theta,\gamma) \times P_{PV}^{Peak} \times (1 + dp(T_{C}^{t} - 25))}{1000} \tag{1}$$

$$I^{t}(\theta,\gamma) = I_{b}^{t}(\theta,\gamma) + I_{d}^{t}(\theta,\gamma) + I_{g}^{t}(\gamma) \tag{2}$$

$$T_{C}^{t} = T_{a}^{t} + \left[\frac{NOCT - 20}{800}\right] \times I^{t}(\theta,\gamma) \tag{3}$$

where P_{PV}^{Peak} is the PV system's rated power, t is the number of the time slot, $I^{t}(\theta,\gamma)$ is the received solar irradiance by the PV system, $I_{b}^{t}(\theta,\gamma)$, $I_{d}^{t}(\theta,\gamma)$ and $I_{g}^{t}(\gamma)$ are respectively the direct-beam, sky-diffuse, and ground-reflected components. dp (%/°C) is the temperature coefficient which is considered −0.4 (for crystalline silicon solar cells technology [7]) in this paper. T_{C}^{t} and T_{a}^{t} are the cell and ambient temperatures, respectively. $NOCT$ is the nominal operating cell temperature, which is considered 45°C [7]. The yearly solar irradiation of the considered case study is shown in Figure 2.

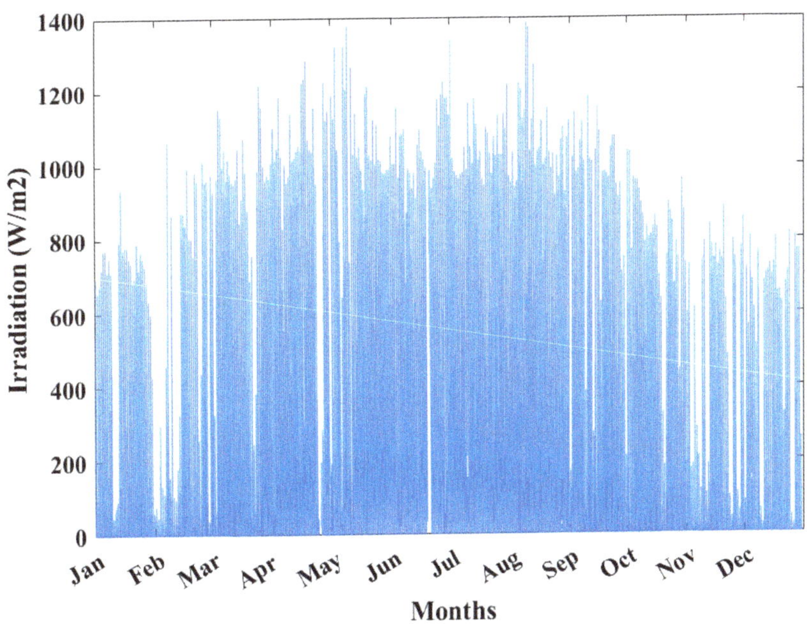

Figure 2. The yearly solar irradiation in the considered case study with 1-min resolution [36].

It is assumed that when a UAV is in a recharging site, the PV panels provide the required energy to charge the UAV's battery. If the generated power of PV panels cannot satisfy the required energy of the UAV's battery, then the batteries of the recharging sites are responsible for fulfilling the remaining power requirement of the UAV's battery. It should be mentioned that when no solar energy is available, for example, during nights or in cloudy weather, the stored energy in the batteries of the PV-battery system is used to satisfy the required energy for recharging the UAV's batteries [7].

The battery charge/discharge rate, as well as the state of charge (SOC) constraints, are considered as follows [7]:

$$SOC_{Batt,\min} \leq SOC^{t}_{Batt} \leq SOC_{Batt,Max} \tag{4}$$

$$SOC^{t}_{Batt} = SOC^{t-1}_{Batt} + \frac{\eta_{Ch} P^{t}_{Ch} \cdot \Delta t}{C_{Batt}} + \frac{1}{\eta_{Dch} \cdot C_{Batt}} P^{t}_{Dch} \cdot \Delta t \tag{5}$$

$$P_{Ch,\min} \leq P^{t}_{Ch} \leq P_{Ch,Max} \tag{6}$$

$$P_{Dch,\min} \leq P^{t}_{Dch} \leq P_{Dch,Max} \tag{7}$$

where $SOC_{Batt,\min}$ and $SOC_{Batt,Max}$ are, respectively, the minimum and maximum SOC of the battery, SOC^{t}_{Batt} is the battery SOC at time t, P^{t}_{Ch} and P^{t}_{Dch} are the battery charge and discharge rates at time t, C_{Batt} is the battery capacity, η_{Ch} and η_{Dch} are, respectively, the battery charge and discharge efficiencies, and Δt is the time slot. $P_{Ch,\min}$ and $P_{Dch,\min}$ are the battery minimum charge and discharge rates, $P_{Ch,Max}$ and $P_{Dch,Max}$ are the maximum battery charge and discharge rates, respectively.

2.2. *Energy Consumption Model*

The considered UAV energy consumption model of [3] is used in this paper to optimally design the PV-battery system of the recharging sites. The UAVs' actions, including recharging, covering, and moving, consume energy. The required energy for recharging the UAVs' battery is supplied by a PV-battery system in the recharging sites.

A rural location with approximately 40,000 inhabitants, containing small houses and without obstacles, was considered in [3]. The studied scenario is mapped in Figure 3. As observed, eight areas and three recharging sites are considered in which three places are simultaneously areas and recharging sites [3]. The power consumption profile of each recharging site was derived based on the mission planning of [3] and presented in Figure 3.

A time slot of 10 min for a one-day (24 h) time span was considered, which is matched with the battery capacity of available UAVs [3]. In Figure 3, the focus is on the first 24 time slots of the day, i.e., from t_1 to t_{24}; however, we assume that the mission planning of the UAVs is carried out in the first 8 h of the day, i.e., from 8 am to 4 pm during which the solar energy production is at its highest values.

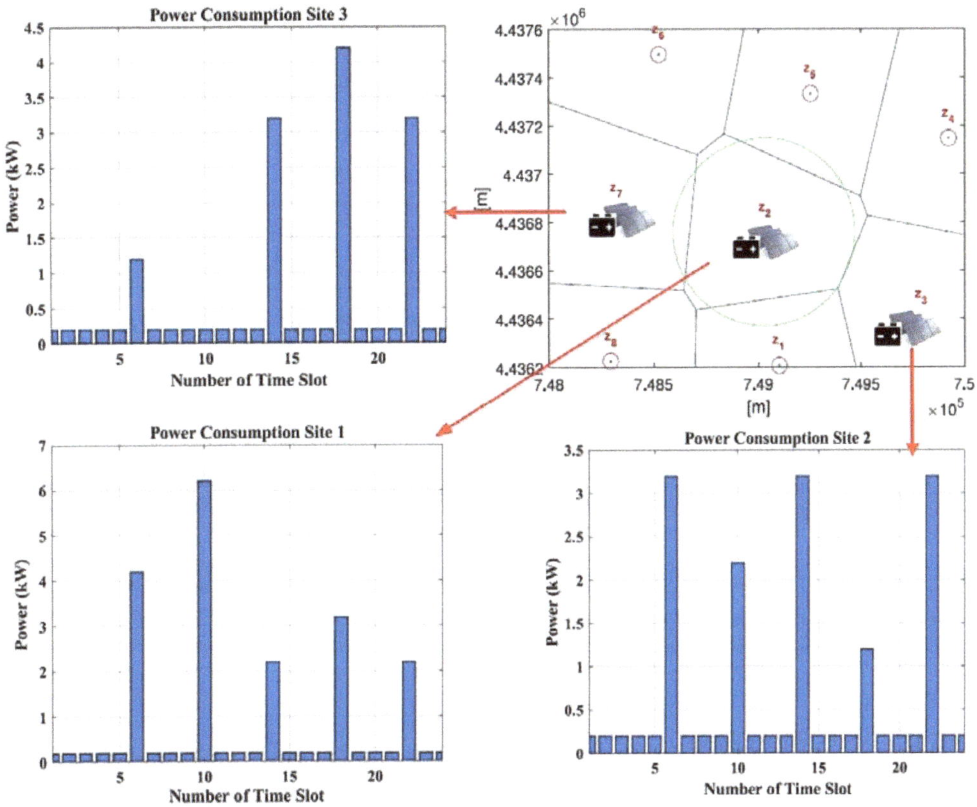

Figure 3. Map of the studied scenario [3] and the derived power consumption profile of each recharging site.

3. Problem Formulation

3.1. Objective Function

In this section, the objective function is presented, which is minimizing the total financial costs of the PV-battery power system, including *CAPEX* and *OPEX*, from the mobile network operators (MNO) point of view:

$$OF = \min(Cost_{Total}) \tag{8}$$

$$Cost_{Total} = CAPEX + OPEX = C_{Batt} \cdot Cost_{Batt} + P_{PV}^{Peak} \cdot Cost_{PV} + \sum_{t=1}^{H} OPEX_{Batt}^{t} \tag{9}$$

where $Cost_{PV}$ is the unit cost in \$/kW, C_{Batt} is the battery capacity, and $Cost_{Batt}$ is the unit cost in \$. H is the horizon of study, $OPEX_{Batt}^{t}$ is the battery operational cost, including the SOC-related ($Cost_{SOC}$) and DoD-related ($Cost_{DoD}$) costs [7]:

$$OPEX_{Batt}^{t} = \sum_{t=1}^{H} Cost_{SOC}^{t} + \sum_{t=1}^{H} Cost_{DoD}^{t} \tag{10}$$

3.2. Technical Constraint

In addition to the battery constraints, the major technical constraint in sizing the PV battery power supply for drone-based cellular networks is the power balance equation as the following [7]:

$$P_{PV}^{t} + P_{Batt}^{t} \geq P_{Load}^{t} \quad (11)$$

where P_{PV}^{t} and P_{Batt}^{t} are the output power of PV panels and the battery charging/discharging power, respectively. P_{Load}^{t} is the power consumption of the recharging site.

4. Optimization Algorithm

Particle swarm optimization (PSO) algorithm was considered in order to choose the optimal number of PV panels and battery capacities to charge the UAVs' batteries in recharging sites. The objective was to minimize the total cost by minimizing the number of PV panels and batteries of the recharging sites.

The advantages of metaheuristic algorithms over conventional mathematical methods are their simple implementation, freeness of gradient calculation, capability of avoiding local optimal points and reaching a global optimal solution, simplicity, and the capability of solving non-convex and nonlinear problems. These benefits led the researchers to apply metaheuristic algorithms in solving optimization problems in a variety of research areas [37–42].

According to the literature, the PSO algorithm is a powerful metaheuristic algorithm, which is efficient for solving optimization problems due to its stability, accuracy, simple formulation and implementation, and fast response [43–45]. The algorithm was first suggested by Kennedy and Eberhart in 1995 [43] and has been extensively applied by researchers in solving a variety of optimization problems [45]. Some current developments of the PSO algorithm were applied to solve problems in a variety of applications [46–48]. The authors in [46] combined an improved PSO algorithm with the continuous high-degree Bezier curve to solve the problem of smooth path planning for mobile robots. In order to tackle the problem of the optimal placement of motion sensors in smart homes and intelligent environments, [47] proposed a combination of the wale optimization algorithm (WOA) and PSO. An inverse kinematics calculation approach based on improved PSO was suggested in [48] for the inverse kinematic operation of general robots.

In the PSO algorithm, the behavior of a group of animals is imitated so that the animals in the group share their individual knowledge to find the best path to follow. The population in PSO is called the swarm, and each member of the swarm is named a particle. The algorithm starts with a random initial population. Each particle moves over the searching space while keeping in mind the best position it has experienced. Particles in the swarm share their best positions with each other and actively update their positions and velocities based on their own best-experienced position and the shared best positions. Accordingly, particles move towards optimal searching space during the searching procedure.

Assuming $f(x)$ as the objective function that is minimized, where $\vec{X}$ is the vector of design variables of x and $d = 1, 2, \ldots, D$ is the counter of design variables where D is the number of design variables. The counter of particles in the population is presented by $i = 1, 2, \ldots, N_{swarm}$, where N_{swarm} is the number of particles or the size of the population. The counter of iterations in the algorithm procedure is defined by $iter = 1, 2, \ldots, iter_{\max}$, where $iter_{\max}$ shows the maximum number of iterations. The velocity $V_i^{(iter)}$ and position $X_i^{(iter)}$ of the ith particle in the $iter$th iteration are updated based on the following:

$$V_i^{(iter+1)} = \omega \cdot V_i^{iter} + c_1 \cdot rand_1(\cdot) \cdot (Pbest_i^{(iter)} - X_i^{(iter)}) + c_2 \cdot rand_2(\cdot) \cdot (Gbest^{(iter)} - X_i^{(iter)}) \quad (12)$$

$$X_i^{(iter+1)} = X_i^{(iter)} + V_i^{(iter+1)} \quad (13)$$

$$X_i^{(iter)} = \left[x_{i,1}^{(iter)}, x_{i,2}^{(iter)}, \ldots, x_{i,d}^{(iter)} \right]_{1 \times D} \quad (14)$$

where ω is the inertia weight, $rand_1(\cdot)$ and $rand_2(\cdot)$ are random numbers in [0,1]. $Pbest_i^{(iter)}$ is the best recorded previous experience of the ith particle in the $iter$th iteration. The best particle among the whole population (the position of the best particle corresponding to the minimum value of the objective function) is presented by $Gbest^{(iter)}$. c_1 and c_2 are constant weights (mostly set equal to two in the literature [49]) of the stochastic acceleration terms that force each particle to move towards $Pbest_i^{(iter)}$ and $Gbest^{(iter)}$.

The steps of the applied PSO algorithm for solving the considered problem are as follows:

Step 1. Specify the geographical location, solar irradiation data, power consumption profile, battery specification, PV panel characteristics, number of design variables, the ranges of design variables, and the termination criterion.

Step 2. Generate initial solutions in the feasible design variable range based on the following:

$$X = X_{\min} + rand(X_{\max} - X_{\min}) \tag{15}$$

where X is an array of design variables considered as a solution including the PV system rated power, battery capacity, and battery charge and discharge rates, $X = \left[P_{PV}^{Peak}, C_{Batt}, P_{Ch}^{t}, P_{Dch}^{t}\right], \quad t = 1, 2, \ldots, T$. $X_{\min}$ and $X_{\max}$ are, respectively, the lower and upper bounds of design variables, and $rand$ is a random value in the range [0,1].

Step 3. Assess the initial particles. For each particle as a solution, the battery constraints are evaluated according to Equations (4)–(7) and the power balance constraint, Equation (10), and calculate the objective function in Equation (9).

Step 4. Sort the initial particles based on the values of the objective function.

Step 5. Set $iter = 1$.

Step 6. Determine $Pbest_i^{(iter)}$ and $Gbest^{(iter)}$ based on the sorted initial particles.

Step 7. Modify the positions of the particles based on Equations (12) and (13).

Step 8. For each modified particle, calculate the objective function after assessing the battery constraints according to Equations (4)–(7) and the power balance constraint in Equation (11).

Step 9. If any of the modified particles result in a lower value of the objective function $Pbest_i^{(iter)}$, replace $Pbest_i^{(iter)}$ by the modified particle. Otherwise, keep $Pbest_i^{(iter)}$.

Step 10. Replace $Gbest^{(iter)}$ by the modified particle if any of the modified particles result in a lower value of the objective function. Otherwise, keep $Gbest^{(iter)}$.

Step 11. If the termination criterion ($iter = iter_{\max}$) is satisfied, terminate the algorithm and go to Step 12; otherwise, set $iter = iter + 1$ and go to Step 6.

Step 12. Print $Gbest^{(iter_{\max})}$, which includes the optimum number of PV panels, number of batteries, and the batteries' charging/discharging values in each time slot and the corresponding objective function value.

The proposed framework is illustrated in Figure 4. The specifications of the PV module and the battery and load profile data are the inputs of the proposed framework. For each combination of PV system rated power and battery capacity, the evaluation of technical constraints and the calculation of total cost are carried out based on the above-mentioned steps. Afterward, if the optimal total cost is achieved, the corresponding combination to the optimal total cost is considered as the optimal solution for the PV-battery system design. Otherwise, the combination of the PV-battery system is changed, and the new combination is evaluated by the PSO algorithm. The proposed framework is general and can be applied for designing PV-battery systems in any application.

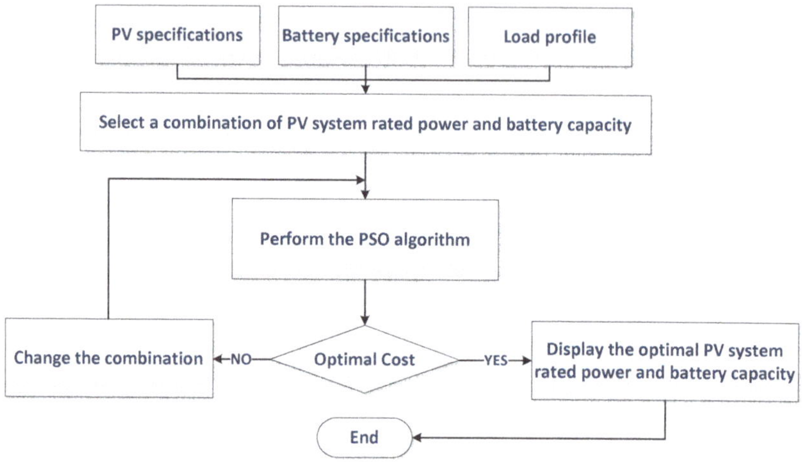

Figure 4. The proposed framework.

5. Simulation Results

In order to justify the effective performance of the suggested framework, a remote area with 40,000 inhabitants was considered (Figure 1), and the 24th of February was considered as the worst-case scenario because the solar irradiation is at its lowest values of the year (see Figure 2). The number of recharging sites is defined based on the "minimum cost algorithm" of [32]. The derived power consumption profile of each recharging site in Figure 3 is assumed as the load that should be satisfied based on (11).

The suggested approach was implemented in a MATLAB environment. The population size (the number of candidate solutions) and number of maximum iterations in the PSO algorithm were 50 and 200, respectively. The system variables and their range are presented in Table 1. The results of the PSO algorithm are compared with the genetic algorithm (GA) in Table 2.

Table 1. The range of system variables.

Variable	Symbol	Unit	Variable Range
Peak PV power	P_{PV}^{Peak}	kW_p	(0–10)
Azimuth angle	θ	degree	(−90–90)
Tilt angle	γ	degree	(0–90)
Battery capacity	C_{Batt}	kWh	(0–35)
Battery state of charge	SOC	%	(10–90)

The optimum number of PV panels, battery capacity, and total cost of the system results for three recharging sites are outlined in Table 2. For the PSO algorithm, the power consumption of recharging Site 1 can be satisfied by 5.6 kW_p PV and 32 kWh battery capacity that led to a total cost of USD 21,264. However, for the GA algorithm in Site 1, the installed PV was 6 kW_p, battery capacity was 32 kWh, and the total cost was USD 21,640. The optimal installed PV system for Sites 2 and 3 were, respectively, 4.8 and 4.4 kW_p when applying the PSO algorithm. Additionally, the required battery capacities for these sites were 29 and 32 kWh, respectively. The value of the resulted total cost of the PV-battery system for site 3 (USD 20,136) was between that of Site 1 (USD 21,264) and Site 2 (USD 19,012).

Table 2. The comparison between the results of PSO and GA algorithms: the optimal system design for each recharging site.

	Recharging Site	Installed PV (kW_p)	Battery Capacity (kWh)	Total Cost (USD)
PSO	Site 1	5.6	32	21,264
	Site 2	4.8	29	19,012
	Site 3	4.4	32	20,136
GA	Site 1	6	32	21,640
	Site 2	6.4	28	20,016
	Site 3	6	30	20,640

When the GA algorithm was applied to solve the problem for Site 2, the required installed PV was 6.4 kW_p, and the battery capacity was 28 kWh. This PV-battery system led to a total cost of USD 20,016. For Site 3, 6 kW_p and 30 kWh battery capacity were required to satisfy the load. The total cost, in this case, was USD 20,640.

To justify the values of Table 2, the feasible solutions of the optimization problem (minimizing the total cost) versus a variety of PV-battery system combinations based on the PV peak power from 1 to 10 kWp with an interval of 0.4 kWp and the battery capacity ranging from 1 to 35 kWh with an interval of 1 kWh are shown in Figures 5–7 for Sites 1, 2, and 3, respectively. According to these figures, it is observed that the PSO algorithm leads to more optimal solutions than GA, as shown in Table 2.

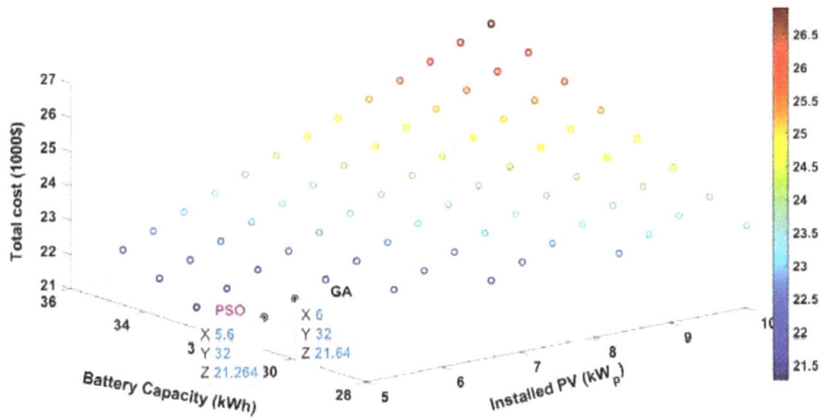

Figure 5. The comparison between PSO and GA algorithms: the feasible solutions of the optimization problem (minimizing the total cost) versus a combination of installed PV and battery capacity for recharging Site 1.

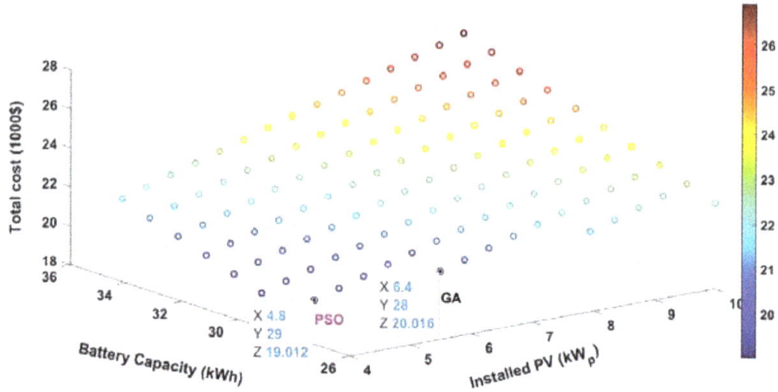

Figure 6. The comparison between PSO and GA algorithms: the feasible solutions of the optimization problem (minimizing the total cost) versus a combination of installed PV and battery capacity for recharging Site 2.

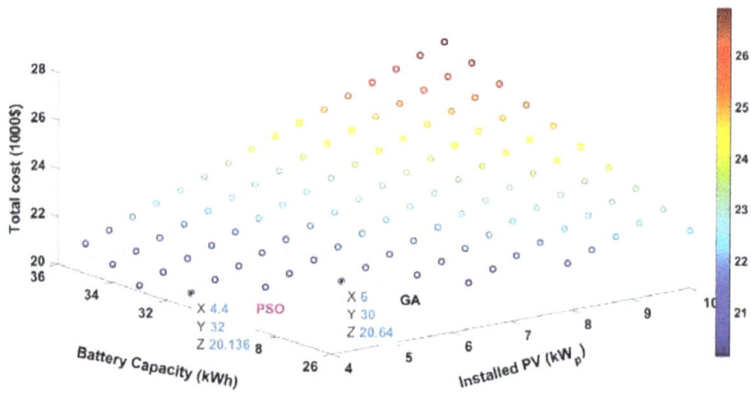

Figure 7. The comparison between PSO and GA algorithms: the feasible solutions of the optimization problem (minimizing the total cost) versus a combination of installed PV and battery capacity for recharging Site 3.

According to Table 2 and Figures 5–7, by comparing the results of PSO and GA algorithms, it can be concluded that the PSO algorithm leads to more cost-efficient solutions for the PV-battery system design of recharging sites.

The obtained results of the PSO algorithm for battery charging/discharging power, the load profile, and the PV system output power in recharging Sites 1, 2, and 3 are demonstrated in Figures 8–10, respectively. Figure 8 shows the battery charging/discharging power, the load profile, and the PV system output power in Site 1. It was observed that in time slots where the power consumption of the site is higher than PV power generation (i.e., time slots 6, 10, 14, 18, 22, 30, 34, 38, 42, and 46 for Site 1 where the UAVs are being recharged in the site) the energy stored in the battery is used to fulfill the load demand. Meanwhile, in the time slots where the available PV power generation was more than the power consumption, the excess power will be stored in the battery.

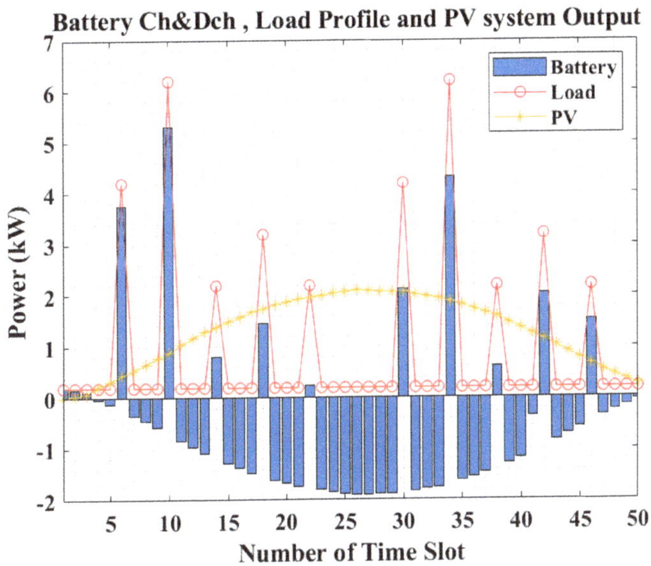

Figure 8. The battery charging-discharging power, the load profile, and the PV system output power in recharging Site 1.

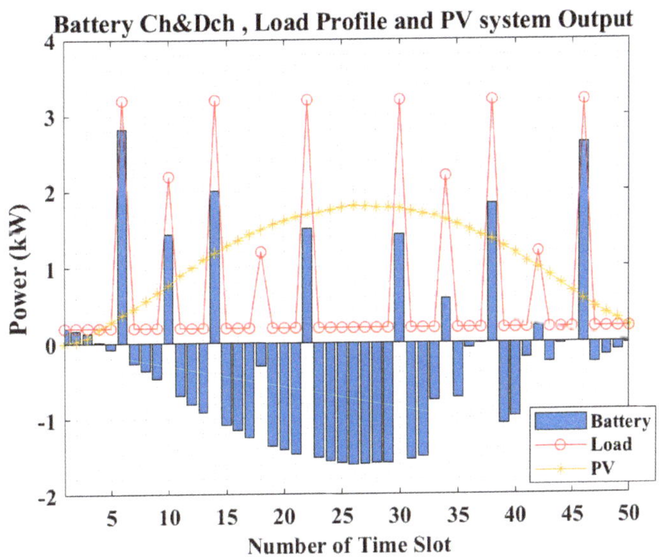

Figure 9. The battery charging-discharging power, load profile, and PV system output power in recharging Site 2.

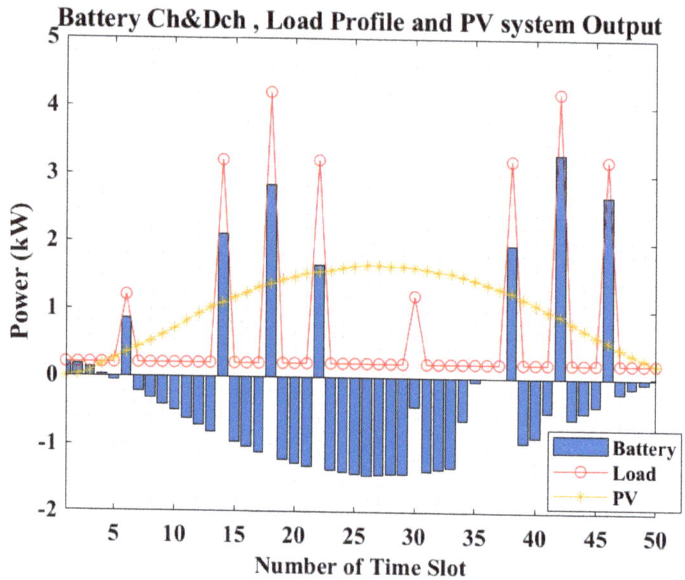

Figure 10. The battery charging-discharging power, load profile, and PV system output power in recharging Site 3.

The battery charging/discharging power, load profile, and PV system output power in recharging Site 2 are shown in Figure 9. By comparing Figures 8 and 9, it is obvious that the peak load of Site 2 (3.2 kW) is lower than that of Site 1 (6.2 kW). Accordingly, in order to satisfy the load, a PV-battery system with lower capacity is required in Site 2, which leads to a lower total cost. However, both cumulative (the cumulative load of Site 1 is 42% more than that of Site 2) and peak loads affect the sizing of the PV-battery system, and the optimum PV-battery dimension is not linearly dependent on only the peak load.

In Figure 10, the battery charging/discharging power, load profile, and PV system output power in recharging Site 3 are illustrated. By comparing Figures 9 and 10, it can be concluded that although the cumulative load of Site 3 is 10% lower than that of Site 2, since the peak power consumption of Site 3 is 33% higher than that of Site 2, designing the PV-battery system leads to a higher cost for Site 3.

6. Discussion and Future Research

In order to design an optimal cost-efficient PV-battery power supply for recharging sites of a UAV-aided cellular telecommunication network in a rural area where there is no access to a reliable electricity grid, we proposed an optimization framework to minimize the total cost of the PV-battery system. The results of optimal mission planning in [3] are considered as the input of our research to achieve the load profile of each recharging site. However, for real-world applications, it is suggested that the mission planning of the UAVs and dimensioning of the PV-battery system be investigated as a united problem and be solved by an appropriate optimization methodology while considering the technical constraints from energy and telecommunication viewpoints simultaneously.

The investigated problem in this paper relies on the irradiation data and the load profile. The duration of the time slot of the extracted load profile in this paper is 10 min. However, the available irradiation data for generation estimation are mostly with a one-hour time resolution. By performing a broad search, irradiation data with a time resolution of 1-min for Milan, Italy, in 2017 [36] were found and used in this paper. However, it is more appropriate that the irradiation data of the studied rural area be used to achieve the most accurate solutions for the considered formulated problem. Another discussable point is that the achieved solutions are based on a daily load profile of 8-h of covering time of UAVs. To reach the solutions for real-world applications, it is suggested to use yearly data for a horizon of study of 10 years. In this case, the replacement and maintenance costs of components can also be considered, which leads us to be able to consider the payback period of the system. However, the required data and computational burden of the problem increase, and this needs more comprehensive considerations. PV-battery-powered UAV-aided telecommunication networks can also be implemented in urban areas where a reliable electricity grid is usually available. In this case, if the system is considered grid-connected, the mobile network operator can profit from selling the excess PV power generation to the main electricity grid instead of dumping surplus PV power.

7. Conclusions

Photovoltaic-battery systems for supplying demand in remote areas have been considered as a promising solution due to low operation costs, being eco-friendly, and easy availability. However, the main obstacle to the high applicability of PV-battery systems in remote areas is the high investment cost. Therefore, many aspects should be investigated before implementing PV-battery systems, such as PV system generation modeling to estimate the output power of PV system, PV panels and battery characteristics, geographical data, technical and economic criteria, and power dispatch strategy. In this paper, an optimization approach is presented to optimally size the PV-battery system for supplying the recharging sites of UAV-based cellular telecommunication networks in a rural off-grid area. The suggested methodology was applied on three UAV recharging sites with different peak and cumulative load profiles. It was observed that the required PV panel number and battery capacity for satisfying each load profile is dependent on both the peak and cumulative loads. The suggested framework was applied for sizing PV-battery systems in ground-based recharging sites for UAVs in cellular networks in off-grid applications. However, a scenario can be defined in which dimensioning of both onboard solar-powered UAVs and ground-based PV-battery systems for recharging sites is considered. In this case, the durability of UAVs' coverage mission increases, and consequently, the quality of service of the telecommunication network will be improved. Additionally, to investigate and compare the economics of different sources of energy, future work can focus on supplying the UAV-aided cellular telecommunication networks with other energy sources such as fuel cells.

Author Contributions: Conceptualization, M.J. and H.P.A.; methodology, M.J., H.P.A., and T.K.; software, M.J. and H.P.A.; validation, M.J., H.P.A. and T.K.; formal analysis, M.J.; investigation, M.J.; writing—original draft preparation, M.J.; writing—review and editing, T.K., D.S., S.V.S. and J.M.G.; supervision, T.K., D.S., S.V.S. and J.M.G. All authors have read and agreed to the published version of the manuscript.

Funding: This research was funded by the European Union's Horizon 2020 research and innovation programme under the Marie Sklodowska-Curie grant agreement No 812991. J.M.G. was supported by VILLUM FONDEN under the VILLUM Investigator Grant (no. 25920): Center for Research on Microgrids (CROM).

Institutional Review Board Statement: Not applicable.

Informed Consent Statement: Not applicable.

Data Availability Statement: Not applicable.

Conflicts of Interest: The authors declare no conflict of interest.

References

1. Bahia, K.; Suardi, S. *The State of Mobile Internet Connectivity 2019*; Tech. Rep.; GSMA Connected Society: London, UK, 2019.
2. Chiaraviglio, L.; Amorosi, L.; Blefari-Melazzi, N.; Dell'Olmo, P.; Mastro, A.L.; Natalino, C.; Monti, P. Minimum Cost Design of Cellular Networks in Rural Areas with UAVs, Optical Rings, Solar Panels, and Batteries. *IEEE Trans. Green Commun. Netw.* **2019**, *3*, 901–918. [CrossRef]
3. Galán-Jiménez, J.; Moguel, E.; García-Alonso, J.; Berrocal, J. Energy-efficient and solar powered mission planning of UAV swarms to reduce the coverage gap in rural areas: The 3D case. *Ad Hoc Netw.* **2021**, *118*, 102517. [CrossRef]
4. Valiulahi, I.; Javidsharifi, M.; Virgili, M.; Masouros, C. D3. 1–Energy Models and Optimisation Framework: Phase 1. Available online: http://painless-itn.com/wp-content/uploads/2020/10/D3.1_final.pdf (accessed on 12 November 2021).
5. Mozaffari, M.; Saad, W.; Bennis, M.; Nam, Y.-H.; Debbah, M. A tutorial on UAVs for wireless networks: Applications, challenges, and open problems. *IEEE Commun. Surv. Tutor.* **2019**, *21*, 2334–2360. [CrossRef]
6. Israr, A.; Yang, Q.; Li, W.; Zomaya, A.Y. Renewable energy powered sustainable 5G network infrastructure: Opportunities, challenges and perspectives. *J. Netw. Comput. Appl.* **2020**, *175*, 102910. [CrossRef]
7. Javidsharifi, M.; Pourroshanfekr, H.; Kerekes, T.; Sera, D.; Spataru, S.; Guerrero, J.M. Optimum Sizing of Photovoltaic and Energy Storage Systems for Powering Green Base Stations in Cellular Networks. *Energies* **2021**, *14*, 1895. [CrossRef]
8. Rokrok, E.; Javidsharifi, M.; Pourroshanfekr, H.; Javidsharifi, B. Adaptive Nonlinear Control Scheme for Three-Phase Grid-Connected PV Central Inverters. In Proceedings of the 29th International Power System Conference (PSC), Terhan, Iran, 27 October 2014; pp. 1–7.
9. Meo, M.; Zhang, Y.; Gerboni, R.; Marsan, M.A. Dimensioning the power supply of a LTE macro BS connected to a PV panel and the power grid. In Proceedings of the 2015 IEEE International Conference on Communications (ICC), London, UK, 8–12 June 2015; pp. 178–184.
10. Marsan, M.G.A.; Bucalo, G.; Di Caro, A.; Meo, M.; Zhang, Y. Towards Zero Grid Electricity Networking: Powering BSs with Renewable Energy Sources. In Proceedings of the 2013 IEEE International Conference on Communications Workshops (ICC), Budapest, Hungary, 9–13 June 2013; pp. 596–601.
11. Ahmed, F.; Naeem, M.; Ejaz, W.; Iqbal, M.; Anpalagan, A.; Kim, H.S. Renewable energy assisted traffic aware cellular base station energy cooperation. *Energies* **2018**, *11*, 99. [CrossRef]
12. Alsharif, M.H.; Kim, J.; Kim, J.H. Green and sustainable cellular base stations: An overview and future research directions. *Energies* **2017**, *10*, 587. [CrossRef]
13. Kusakana, K.; Vermaak, H.J. Hybrid renewable power systems for mobile telephony base stations in developing countries. *Renew. Energy* **2013**, *51*, 419–425. [CrossRef]
14. Aris, A.M.; Shabani, B. Sustainable power supply solutions for off-grid base stations. *Energies* **2015**, *8*, 10904–10941. [CrossRef]
15. Khare, V.; Nema, S.; Baredar, P. Status of solar wind renewable energy in India. *Renew. Sustain. Energy Rev.* **2013**, *27*, 1–10. [CrossRef]
16. Song, J.; Krishnamurthy, V.; Kwasinski, A.; Sharma, R. Development of a Markov-chain-based energy storage model for power supply availability assessment of photovoltaic generation plants. *IEEE Trans. Sustain. Energy* **2012**, *4*, 491–500. [CrossRef]
17. Leonardi, G.; Meo, M.; Marsan, M.A. Markovian Models of Solar Power Supply for a LTE Macro BS. In Proceedings of the 2016 IEEE International Conference on Communications (ICC), Kuala Lumpur, Malaysia, 22–27 May 2016; pp. 1–7.
18. Chamola, V.; Sikdar, B. Outage Estimation for Solar Powered Cellular Base Stations. In Proceedings of the 2015 IEEE International Conference on Communications (ICC), London, UK, 8–12 June 2015; pp. 172–177.
19. Ibrahim, I.A.; Sabah, S.; Abbas, R.; Hossain, M.J.; Fahed, H. A novel sizing method of a standalone photovoltaic system for powering a mobile network base station using a multi-objective wind driven optimization algorithm. *Energy Convers. Manag.* **2021**, *238*, 114179. [CrossRef]
20. Virgili, M.; Forysth, A.J.; James, P. A Multi-Objective Genetic Algorithm Methodology for the Design of Standalone Energy Systems. In Proceedings of the 2021 IEEE Design Methodologies Conference (DMC), Bath, UK, 14–15 July 2021; pp. 1–6.
21. Thakur, R.; Mishra, S.; Murthy, C.S.R. An energy and cost aware framework for cell selection and energy cooperation in rural and remote femtocell networks. *IEEE Trans. Green Commun. Netw.* **2017**, *1*, 423–433. [CrossRef]
22. Du, P.; Ran, L.; Zhai, D.; Ren, R.; Zeng, Q. Admm-based distributed online algorithm for energy management in hybrid energy powered cellular networks. *IEEE Access* **2019**, *7*, 83343–83353. [CrossRef]
23. Fletscher, L.; Suarez, L.; Grace, D.; Peroni, C.; Maestre, J. Energy-aware resource management in heterogeneous cellular networks with hybrid energy sources. *IEEE Trans. Netw. Serv. Manag.* **2019**, *16*, 279–293. [CrossRef]
24. Leithon, J.; Lim, T.J.; Sun, S. Cost-aware renewable energy management with application in cellular networks. *IEEE Trans. Green Commun. Netw.* **2018**, *2*, 316–326. [CrossRef]
25. Mendil, M.; De Domenico, A.; Heiries, V.; Caire, R.; Hadjsaid, N. Battery-aware optimization of green small cells: Sizing and energy management. *IEEE Trans. Green Commun. Netw.* **2018**, *2*, 635–651. [CrossRef]
26. Han, T.; Ansari, N. Provisioning green energy for base stations in heterogeneous networks. *IEEE Trans. Veh. Technol.* **2016**, *65*, 5439–5448. [CrossRef]
27. Chamola, V.; Sikdar, B. Power outage estimation and resource dimensioning for solar powered cellular base stations. *IEEE Trans. Commun.* **2016**, *64*, 5278–5289. [CrossRef]

28. Ghiassi-Farrokhfal, Y.; Keshav, S.; Rosenberg, C.; Ciucu, F. Solar power shaping: An analytical approach. *IEEE Trans. Sustain. Energy* **2015**, *6*, 162–170. [CrossRef]
29. Renga, D.; Meo, M. Dimensioning renewable energy systems to power mobile networks. *IEEE Trans. Green Commun. Netw.* **2019**, *3*, 366–380. [CrossRef]
30. Jing, X.; Sun, J.; Masouros, C. Energy Aware Trajectory Optimization for Aerial Base Stations. *IEEE Trans. Commun.* **2021**, *69*, 3352–3366. [CrossRef]
31. Babu, N.; Papadias, C.B.; Popovski, P. Energy-Efficient 3D Deployment of Aerial Access Points in a UAV Communication System. *IEEE Commun. Lett.* **2020**, *24*, 2883–2887. [CrossRef]
32. Chiaraviglio, L.; Amorosi, L.; Blefari-Melazzi, N.; Dell'Olmo, P.; Natalino, C.; Monti, P. Optimal design of 5G networks in rural zones with UAVs, optical rings, solar panels and batteries. In Proceedings of the 2018 20th International Conference on Transparent Optical Networks (ICTON), Bucharest, Romania, 1–5 July 2018; pp. 1–4.
33. Amorosi, L.; Chiaraviglio, L.; D'Andreagiovanni, F.; Blefari-Melazzi, N. Energy-efficient mission planning of UAVs for 5G coverage in rural zones. In Proceedings of the 2018 IEEE International Conference on Environmental Engineering (EE), Milan, Italy, 12–14 March 2018; pp. 1–9.
34. Amorosi, L.; Chiaraviglio, L.; Galan-Jimenez, J. Optimal energy management of UAV-based cellular networks powered by solar panels and batteries: Formulation and solutions. *IEEE Access* **2019**, *7*, 53698–53717. [CrossRef]
35. Chiaraviglio, L.; D'andreagiovanni, F.; Choo, R.; Cuomo, F.; Colonnese, S. Joint optimization of area throughput and grid-connected microgeneration in uav-based mobile networks. *IEEE Access* **2019**, *7*, 69545–69558. [CrossRef]
36. Available online: http://www.solartech.polimi.it/activities/forecasting/dataset/ (accessed on 23 August 2021).
37. Javidsharifi, M.; Niknam, T.; Aghaei, J.; Mokryani, G.; Papadopoulos, P. Multi-objective day-ahead scheduling of microgrids using modified grey wolf optimizer algorithm. *J. Intell. Fuzzy Syst.* **2019**, *36*, 2857–2870. [CrossRef]
38. Javidsharifi, M.; Niknam, T.; Aghaei, J.; Mokryani, G. Multi-objective short-term scheduling of a renewable-based microgrid in the presence of tidal resources and storage devices. *Appl. Energy* **2018**, *216*, 367–381. [CrossRef]
39. Narimani, M.R.; Vahed, A.A.; Azizipanah-Abarghooee, R.; Javidsharifi, M. Enhanced gravitational search algorithm for multi-objective distribution feeder reconfiguration considering reliability, loss and operational cost. *IET Gener. Transm. Distrib.* **2014**, *8*, 55–69. [CrossRef]
40. Javidsharifi, M.; Niknam, T.; Aghaei, J.; Shafie-khah, M.; Catalão, J.P. Probabilistic Model for Microgrids Optimal Energy Management Considering AC Network Constraints. *IEEE Syst. J.* **2019**, *14*, 2703–2712. [CrossRef]
41. Mahboubi-Moghaddam, E.; Narimani, M.R.; Khooban, M.H.; Azizivahed, A.; Sharifi, M. Multi-objective distribution feeder reconfiguration to improve transient stability, and minimize power loss and operation cost using an enhanced evolutionary algorithm at the presence of distributed generations. *Int. J. Electr. Power Energy Syst.* **2016**, *76*, 35–43. [CrossRef]
42. Pourroshanfekr, H.; Javidsharifi, M.; Javidsharifi, B. Solving Non-convex Economic Dispatch with Valve Point Loading Effect using Imperialist Competitive Algorithm. *Int. J. Manag. Perspect.* **2015**, *4*, 41–45.
43. Eberhart, R.; Kennedy, J. Particle swarm optimization. In Proceedings of the IEEE International Conference on Neural Networks, Perth, WA, Australia, 27 November–1 December 1995; Volume 4, pp. 1942–1948.
44. Ahmadi, S.; Arabani, H.P.; Haghighi, D.A.; Guerrero, J.M.; Ashgevari, Y.; Akbarimajd, A. Optimal use of vehicle-to-grid technology to modify the load profile of the distribution system. *J. Energy Storage* **2020**, *31*, 101627. [CrossRef]
45. Al-Kazemi, B.; Habib, S. Complexity analysis of problem dimension using PSO. In Proceedings of the WSEAS International Conference on Evolutionary Computing, Cavat, Croatia, 12–14 June 2006; pp. 45–52.
46. Song, B.; Wang, Z.; Zou, L. An improved PSO algorithm for smooth path planning of mobile robots using continuous high-degree Bezier curve. *Appl. Soft Comput.* **2021**, *100*, 106960. [CrossRef]
47. Nasrollahzadeh, S.; Maadani, M.; Pourmina, M.A. Optimal motion sensor placement in smart homes and intelligent environments using a hybrid WOA-PSO algorithm. *J. Reliab. Intell. Environ.* **2021**. [CrossRef]
48. Yiyang, L.; Xi, J.; Hongfei, B.; Zhining, W.; Liangliang, S. A General Robot Inverse Kinematics Solution Method Based on Improved PSO Algorithm. *IEEE Access* **2021**, *9*, 32341–32350. [CrossRef]
49. Ridha, H.M.; Gomes, C.; Hazim, H.; Ahmadipour, M. Sizing and implementing off-grid stand-alone photovoltaic/battery systems based on multi-objective optimization and techno-economic (MADE) analysis. *Energy* **2020**, *207*, 118163. [CrossRef]

drones

MDPI

Article

Computational Study of the Propeller Position Effects in Wing-Mounted, Distributed Electric Propulsion with Boundary Layer Ingestion in a 25 kg Remotely Piloted Aircraft

José Ramón Serrano *, Andrés Omar Tiseira, Luis Miguel García-Cuevas and Pau Varela

CMT—Motores Térmicos, Universitat Politècnica de València, 46022 Valencia, Spain; antil@mot.upv.es (A.O.T.); luiga12@mot.upv.es (L.M.G.-C.); pavamar@mot.upv.es (P.V.)
* Correspondence: jrserran@mot.upv.es; Tel.: +34-963877650

Abstract: Distributed electric propulsion and boundary layer ingestion are two attractive technologies to reduce the power consumption of fixed wing aircraft. Through careful distribution of the propulsive system elements, higher aerodynamic and propulsive efficiency can be achieved, as well as a lower risk of total loss of aircraft due to foreign object damage. When used on the wing, further reductions of the bending moment on the wing root can even lead to reductions of its structural weight, thus mitigating the expected increase of operating empty weight due to the extra components needed. While coupling these technologies in fixed-wing aircraft is being actively studied in the big aircraft segment, it is also an interesting approach for increasing the efficiency even for aircraft with maximum take-off masses as low as 25 kg, such as the A3 open subcategory for civil drones from EASA. This paper studies the effect of changing the propellers' position in the aerodynamic performance parameters of a distributed electric propulsion with boundary layer ingestion system in a 25 kg fixed-wing aircraft, as well as in the performance of the propellers. The computational results show the trade-offs between the aerodynamic efficiency and the propeller efficiency when the vertical position is varied.

Keywords: distributed electric propulsion; boundary layer ingestion; propeller; fixed wing

Citation: Serrano, J.R.; Tiseira Izaguirre, A.O.; García-Cuevas, L.M.; Varela Martínez, P. Computational Study of the Propeller Position Effects in Wing-Mounted, Distributed Electric Propulsion with Boundary Layer Ingestion in a 25 kg Remotely Piloted Aircraft. *Drones* **2021**, *5*, 56. https://doi.org/10.3390/drones5030056

Academic Editors: Diego González-Aguilera and Pablo Rodríguez-Gonzálvez

Received: 2 June 2021
Accepted: 25 June 2021
Published: 30 June 2021

1. Introduction

Nowadays, some of the most important challenges related to the development and operation of new small aircraft, whether they are remotely piloted (RPAS) or autonomous, are their safety and environmental impact. Indeed, this is stated by different aviation safety agencies, such as in in the "Study on the societal acceptance of Urban Air Mobility in Europe" by the European Union Aviation Safety Agency (EASA) [1]. This information combined with the available data that predicts a broad growth of this type of aircraft in the coming years, reaching hundreds of thousands of units according to the Single European Sky ATM Research (SESAR) [2], are clear indicators of the need to seek solutions in the design of new aircraft, focused on reducing fuel and energy use and pollutant emissions. One way to achieve the efficiency objectives is by optimising the power train and aerodynamic design, researching more efficient aircraft models, according to NASA in the Information Technology Development Solutions (ITDS) of the Environmentally Responsible Aviation (ERA) project [3].

One of the most studied technologies in recent years to reduce pollutant emissions while maximizing the operational range and endurance of the aircraft is electric hybridisation (HE). It is worth mentioning the research of Auesser et al. [4], where they integrate and validate a parallel HE propulsion system for RPAS; the work of Harmon et al. [5], who proposed an optimisation of both aerodynamics and the HE propulsive system; or Kim et al. [6], who focused in the fuel economy optimisation of parallel HE. Through hybridisation, decoupling between ICE and propeller shaft can be achieved, allowing integration of novel configurations such as distributed electric propulsion (DEP) and boundary layer ingestion (BLI).

DEP consists of separating the total electric power plant and propellers into smaller ones distributed along the wingspan. This distribution presents multiple advantages as increased flow circulation on the wing, increasing this way the aerodynamic efficiency; the use of the power plant as a control platform, reducing the need for aerodynamic actuators; easier maintenance of electric motors and their possible replacement; engine failure is less critical and the noise footprint can be reduced. The advantages of using DEP have been studied in the past years both for conventional and small aircraft [7,8].

The optimisation of a DEP system depends on the location of the electric engines, since its location and distribution have shown to have an important effect on the aircraft aerodynamics. However, there is a position of special interest: By distributing the propellers near the wing trailing edge it is possible to take advantage of the ingestion of the boundary layer produced by the wing. This phenomenon is known as boundary layer ingestion (BLI) and is widely studied in [9–11]. BLI is based on flux re-acceleration around the wing due to the ingestion produced by the propeller. This re-acceleration leads to a wake reduction, with a part of the propeller facing a lower incident air speed compared to what it would have in a classic configuration on the leading edge. In such a way, the engine needs less power for a given thrust as described by Budziszewski in [12], increasing the propulsive efficiency. At the same time, the ingestion could modify the airflow circulation around the airfoil, incrementing the intensity of the suction peak near the leading edge and reducing lift-induced drag, yet increasing the skin friction drag due to the acceleration, which translates into changes in the aerodynamic efficiency. In some studies, such as the work presented by Hall et al. [13], it is pointed out that the aerodynamic efficiency will be higher in the BLI configuration; likewise, Martínez et al. and Teperin et al. [14–16] come to the same conclusion in their BLI research applied to fuselages. By coupling the BLI configuration with DEP, a greater part of the wing and the propulsive plant is affected by the benefits of boundary layer ingestion. This allows, as shown in the work of Goldberg et al. [17], to decrease the fuel consumption compared with a classic distribution.

The main contribution of the current research paper is to analyse the possible positive effects on both aerodynamic and propulsive efficiency thanks to the use of distributed electric propulsion and boundary layer ingestion simultaneously, with a special emphasis on the analysis of the relative position between the propeller and the airfoil trailing edge, in the case of a small fixed-wing aircraft. The research is performed employing computational tools, with experimental validation whenever is possible.

The document is organized as follows. First, in Section 2 the main aircraft configuration is selected. Then, in Section 3, the main methods and models are presented, explaining in detail the modeling of the actuator disc. In Section 4 the different results are computed for different propeller positions, and the best case configuration is compared with a classical configuration without DEP or BLI. Finally, all main results and discussions are summarized in the conclusions in Section 5.

2. Design and Component Selection

In this section, the different design parameters and components are selected in order to prepare the simulations.

One of the main characteristics that limit the design of the aircraft, both aerodynamically and propulsively, is its size. In this case, the size of the aircraft is chosen to take into account the Spanish regulation for RPAS civil use [18] that fixes the maximum takeoff mass (MTOM) of civil RPAS (without special permits) in 25 kg at most, similarly to other RPAS regulations across Europe. Looking for commercial aircraft that meet this restriction, Penguin C from UAV Factory [19] and TARSIS 25 from AERTEC Solutions [20] were selected to set the initial geometric dimensions. However, the wing geometry was simplified, setting a constant chord length airfoil in all the 2 m wingspan with an aspect ratio ($\AE$) equal to 10. A 200 mm SD7003 airfoil was chosen as the main airfoil due to its low parasitic drag at low and medium Reynolds numbers (*Re*). This airfoil is widely studied in aerodynamic research, and it is easy to find experimental and computational data in the literature to validate the simulations performed using it, for example, in [21–24].

Table 1 includes a summary of all the relevant aerodynamic parameters for this study. This includes the parasitic drag coefficient due to the fuselage, empennage and the other aircraft elements, not including the wing, $C_{D,0,\text{extra}}$, and also the Oswald efficiency factor, e. With these parameters, the total drag D can be computed as in Equation (1):

$$D = \frac{1}{2} \cdot \rho_\infty \cdot U_\infty^2 \cdot S \cdot \left(C_{D,0,\text{wing}} + C_{D,0,\text{extra}} + \frac{C_L^2}{\pi \cdot AR \cdot e} \right) \tag{1}$$

where ρ_∞ is the far-field air density, U_∞ is the upstream wind speed, S is the wing surface, $C_{D,0,\text{wing}}$ is the parasitic drag coefficient of the wing and C_L is the lift coefficient. The values of $C_{D,0,\text{wing}}$ and $C_{D,0,\text{extra}}$ are computed using geometrical information of aircraft with a similar mission, including the mentioned Penguin C and TARSIS 25, as well as the Harmon and Hiserote's aircraft [4,5]. The Oswald efficiency factor e is estimated using the methods described in [25,26].

Table 1. Aerodynamic, design parameters and engine data.

Design parameters	
Aspect ratio	10
Wing area	0.4 m^2
Wingspan	2 m
Wing chord	0.2 m
Maximum takeoff mass	25 kg
Aerodynamic parameters	
$C_{D,0,\text{extra}}$ (fuselage, empennage, others)	0.011
Oswald efficiency factor (e)	0.8

Accordingly, the DA4052 double blade propeller designed by UIUC [27] is selected. Complete blade geometry and wind tunnel testing data of this propeller is available, so the simulation results can be validated.

3. Methods

In this section, he computational domain is presented besides the design variables involved in the optimisation and selection of the best DEP and BLI case. Then the different computational models used in order to calculate the performance of different RPAS configurations are presented. First, the computational fluid dynamics (CFD) method used to compute the series hybridisation with DEP and BLI is explained. CFD simulations are performed with different configurations: A section of the wing, a single propeller and a section of the wing with a propeller in the trailing edge. These simulations serve as input to a model to compute the range of the aircraft in different conditions.

3.1. Computational Setup

All the simulations are composed of a large horseshoe domain, a wing section and an actuator disc that simulates the propeller. The actuator disc will be discussed in Section 3.3. The described domain can be observed in Figure 1, where the main dimensions and mesh are displayed.

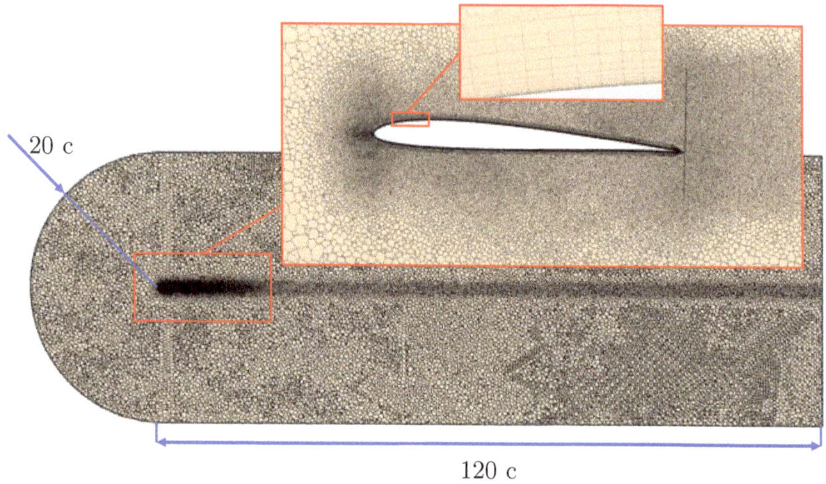

Figure 1. Sketch of the computational grid used for the current calculations.

Using this domain, the upwind boundary condition is set 20 chords away from the wing with a free-stream speed imposed on it. This is also valid for the turbulence intensity and length scale of the flow. Downstream of the wing, the boundary is set at 120 chords as a static pressure outlet. The boundary in the wall of the wing is set as smooth roughness wall with non-slip conditions. The remaining boundaries are modeled assuming that normal derivatives of the variables are zero at these locations, assuming they are far from the body. Boundaries above and below the wing are separated 80 chords in order to ensure that their location do not disturb the results.

To verify that the dimensions of the domain are enough and do not interfere in the solution, a domain independence analysis was performed. Extra cases were carried out, doubling and halving the vertical distance between exterior boundaries. The variation in lift coefficient, C_L, and parasitic drag coefficient, $C_{D,0}$ between the chosen case and the largest domain was less than 1%. Hence, it is considered that the solution obtained is independent of the size of the domain used.

All CFD simulations consist of a three-dimensional wing section with an actuator disc near the airfoil trailing edge. This actuator disc simulates the propeller by means of Blade Element Momentum Theory (BEMT). The described airfoil and propeller are sketched in Figure 2 with their main dimensions.

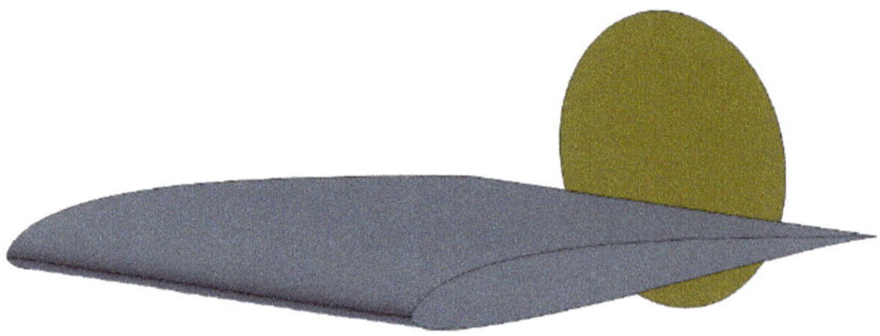

Figure 2. Section of wing simulated with virtual disc.

For each radius, different vertical positions are used in the simulation, which are given by the relative distance between the trailing edge an the center of the actuator disc. The highest position corresponds to 100%, where the propeller is all above the trailing edge, being 0% the lowest position where the shaft of the propeller is aligned with the trailing edge of the airfoil, as can be seen in Figure 3. In this figure, the described positions are shown looking at the wing with a viewing axis parallel to the chord.

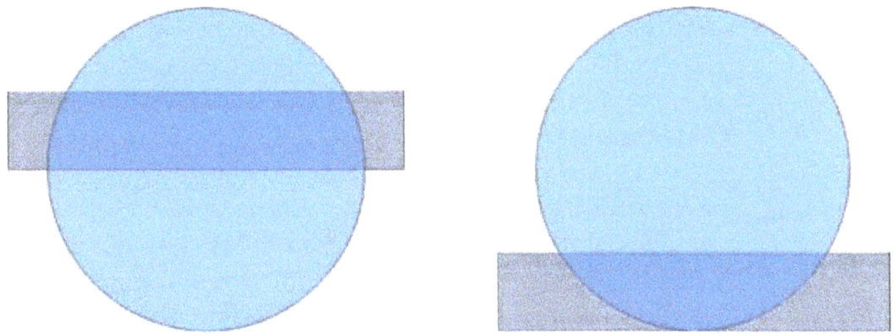

(a) 0% position (b) 100% position

Figure 3. Maximum and minimum propeller position above the trailing edge.

There is a constant length gap between the trailing edge and the actuator disc equal to 1% of the chord length. This gap should be kept as small as possible because, as the airfoil wake moves downstream, it dissipates due to viscous shear stresses, therefore more power will be needed to move the propeller that ingest the wake. As explained in [28], to increase the propulsive efficiency using BLI, the ingestion must take place before the wake dissipates, what is achieved using an small gap.

Next section presents the methods and models used to compute the performance characteristics of the RPAS with different configurations.

3.2. CFD Methodology

All the CFD simulation are numerically solved by mean of a Finite Volume Method using the commercial software Simcenter STAR-CCM+ with steady-state Reynolds-Average Navier-Stokes (RANS) equations approach. Second order methods were used for solving the advection and diffusive terms.

The Spalar–Allmaras turbulence model is chosen to solve the Reynolds stress tensor. This is a one-equation model designed for external aerodynamic applications such as the present case and used extensively in DEP and BLI research [8,29].

As the Mach number is kept below 0.2 in all the simulated cases, the flow can be modelled as incompressible. All domain is meshed with a polyhedral mesh except for the boundary layer around the wing, where a prismatic mesh with a geometric grow distribution is used. The total thickness of the prismatic mesh is 3 mm divided in 14 layers, ensuring an y^+ minor to one in 99% of the wall, able to resolve the viscous subrange of the boundary layer.

The complete mesh is sketched at Figure 1, where different refinement zones and a detail near the airfoil boundary layer can be observed.

A mesh independence analysis has been carried out with an arbitrary angle of attack of 4.8° and a Reynolds number of 3×10^5. For estimating the value of the drag coefficient as function of the base size used in the independence, a generalized Richardson extrapolation was performed, as described by [30]. Final base size was set at 1 mm, guaranteeing an error of less than 2% in drag coefficient. The complete final mesh is composed of more than 3.3×10^6 elements.

3.3. Actuator Disc Setup

As already mentioned in Section 3.1, the propeller is modelled using an actuator disc (also known as virtual disc) approach coupled with a blade-element theory submodel. This approach is much less expensive in simulation time compared to simulating a real propeller and is widely used in DEP and BLI research, as can be seen in [8,14].

Using the blade-element theory (BEMT), the propeller is divided into different sections along the entire blade radius, resulting in two dimensional wings which must be aerodynamically characterised for the model. A tip loss correction factor is fixed as constant and equal to 0 at a relative span of 0.97 due to the three dimensional aerodynamic behaviour of the last fraction of the length of the blade, as advised in [31]. Simplifying the propeller geometry to an actuator disk and solving it by means of the BEMT has some potential issues that have to be taken into account in order to assess the quality of the results. The accuracy of this approach is reduced as sections of the propeller work in stall or transonic conditions. Moreover, the detailed flow behaviour around sections of the propeller blades can not be simulated with this method, so only the results upstream and downstream of the actuator disk are valid: In the actuator disk itself, all the effects are collapsed into a zero-thickness surface. As the present work does not look at the performance parameters in stall or transonic propeller conditions and the detailed flow around the blade airfoils is not studied, the limitations of the actuator disk approach do not affect the main results and conclusions.

The geometry of the blade used is specified in [32] in detail and the different aerodynamic coefficients needed for each section are calculated using a potential flow method with interactive boundary layer correction under XFLR5 [33], which is based on XFOIL [34].

Using the mesh specified in the previous section, the resolution of the actuator disc is fixed to 8 elements in both azimuth and radial directions using an uniform distribution. To carry out the validation of the use of an actuator disc as model for the propeller, the same propeller is simulated with the same domain without any wing thus avoiding the effects of DEP and BLI. The results are compared against experimental data of [32] and shown in Figure 4.

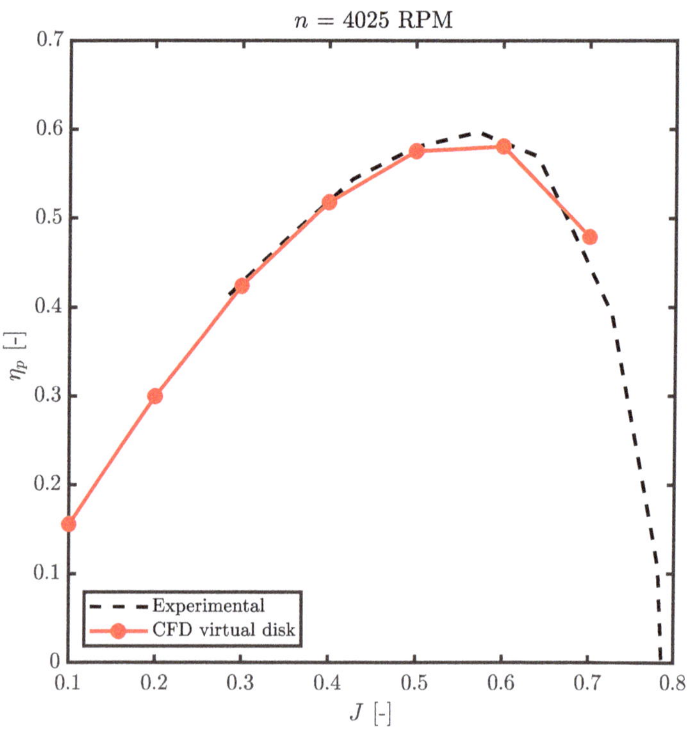

Figure 4. CFD and experimental propulsive efficiency comparison at 4025 rpm.

In Figure 4, the propulsive efficiency eta_p for both the simulated and the experimental propeller are compared in a common range of advance ratio J, where these terms are defined in Equations (2) and (3).

$$\eta_p = \frac{T \cdot U_\infty}{P} \tag{2}$$

where T is the propeller thrust, P is the propeller power and U_∞ the air velocity.

$$J = \frac{U_\infty}{n \cdot 2 \cdot r_{\text{propeller}}} \tag{3}$$

where n is the propeller rotational speed and $r_{\text{propeller}}$ its radius. When expressing the air speed in units of distance divided by seconds, it is customary to express n in Hz.

Fair agreement was found between BEMT + CFD and the experimental results, as both result in similar propulsive efficiency, so all the DEP and BLI cases will be modeled using this approach.

All DEP and BLI cases consist of a single actuator disc, therefore the rotational speed must be set to take into account the thrust produced by all the propellers distributed along the wing.

For simplicity, individual thrust will be considered the same on each propeller, since the flight is considered stationary and leveled. This way, the rotational speed of the propeller can be set in a way that the total thrust produced by all of them would be equal to the total drag of the aircraft in each simulation.

The total drag produced by the entire aircraft is estimated from the simulation force coefficients, taking into account extra parasitic drag coefficient, $C_{D,0,\text{extra}}$, due the non wing produced drag, as shown in Section 2.

In all the simulations, the rotational speed of the propeller is controlled so the total thrust of the aircraft, $T \cdot n_{\text{propellers}}$, is set equal to the total drag, D, as in Equation (4):

$$T \cdot n_{\text{propellers}} = D = \frac{1}{2} \cdot \rho_\infty \cdot U_\infty^2 \cdot S \cdot \left(C_{D,0,\text{wing}} + C_{D,0,\text{extra}} + \frac{C_L^2}{\pi \cdot A\!\!\!R \cdot e} \right) \tag{4}$$

where the propeller thrust, T, is multiplied by the total number of propellers, $n_{\text{propellers}}$. As only one propeller is simulated, the total number of propellers results in a function of the domain width. Since the wingspan is known and all engines are assumed to be equally spaced, the distribution of propellers is uniquely defined. In all the simulations in this document, a total of 13 propellers is considered. Given that the thrust produced by the propellers is equal to the aircraft drag in all the cases, the simulation data can be used and interpolated to predict the performance of the aircraft when flying with different weights and load factors, assuming straight-and-level flight or level turns.

4. Results and Discussion

This section presents the different results of the study. It is divided into two parts: The first one, in which the results of the simulations for different actuator disc positions, at different angles of attack and maintaining the Reynolds number are analysed, so that an optimal design position can be discerned from both the propulsive and aerodynamic point of view; then, the computational fluid dynamics (CFD) analysis of the best case is computed and compared with the base case, showing the improvement of the different efficiencies from an aerodynamic point of view. Each case used 128 cpu-hours to converge.

4.1. Propeller Position Cfd Analysis

CFD simulations for different propeller positions and angles of attack are studied, keeping the same blade radius and distribution. This way, all cases have a propeller radius of 40 mm, a distribution of 13 engines on the wing and a draft angle of 1.5°, only changing the position of the actuator disc over the trailing edge as a geometric parameter.

In this way, it is possible to quantify the effect of the propeller position for different angles of attack on the different aerodynamic and propulsive coefficients. Five different propeller heights relative to the trailing edge, as explained in Section 3.3, are studied for a constant Reynolds equal to 3×10^5. To decide which is the optimal propeller position, the product of propulsive efficiency and aerodynamic efficiency is represented in Figure 5. Bigger distances measured in the direction of the chord were also tested: Increasing them 4% of the chord decreased the aerodynamic and propulsive efficiency product by 4%, whereas incresing this distance a further 4% of the chord decreased the efficiencies product an extra 0.5%.

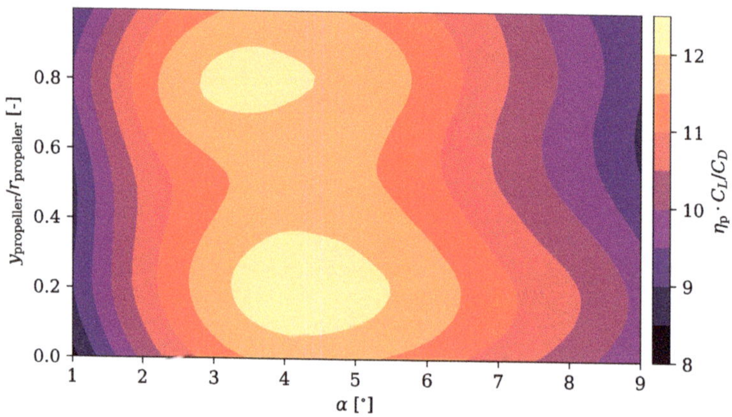

Figure 5. Aerodynamic and propulsive efficiency for each propeller position above the trailing edge and angle of attack.

Figure 5 shows two different positions that optimise the product of efficiencies for a wide range of angles of attack. This is due to the fact that two positions maximise this product but optimising the two efficiencies differently. The improvement in this overall performance in the higher position is due to the improvement in aerodynamic efficiency, while the improvement found in the lowest positions is produced due to the improvement in propulsive efficiency.

To understand how efficiencies are maximised, they are represented separately in Figures 6 and 7.

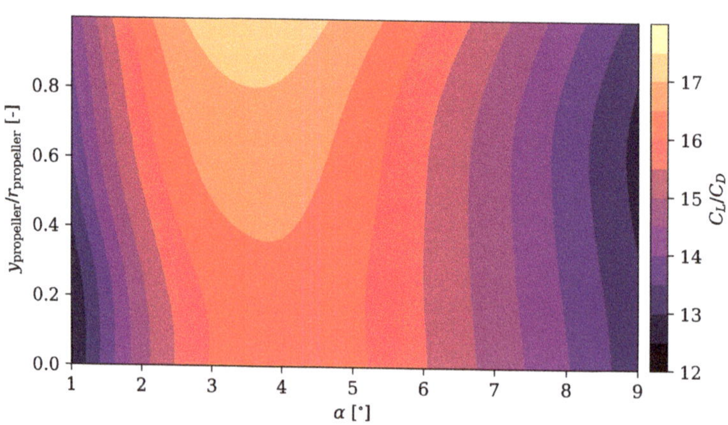

Figure 6. Aerodynamic for each propeller position above the trailing edge and angle of attack.

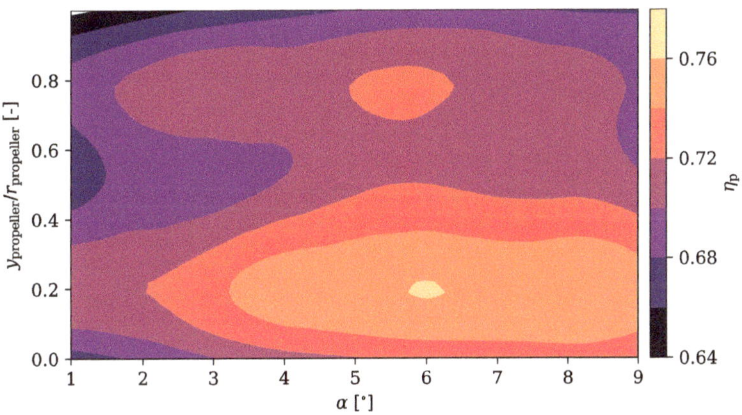

Figure 7. Propulsive efficiency for each propeller position above the trailing edge and angle of attack.

Starting with the aerodynamic efficiency, it increases as the propeller position raises in a wide range of angles of attack. The position of the propeller changes the circulation around the wing section. As this position is higher, the lift increases at the same time as the parasitic drag decreases. This is reflected in the pressure coefficient where if this coefficient is represented in Figure 8 for all positions and a fixed angle, it can be seen that the suction peak is greater as the position moves upwards.

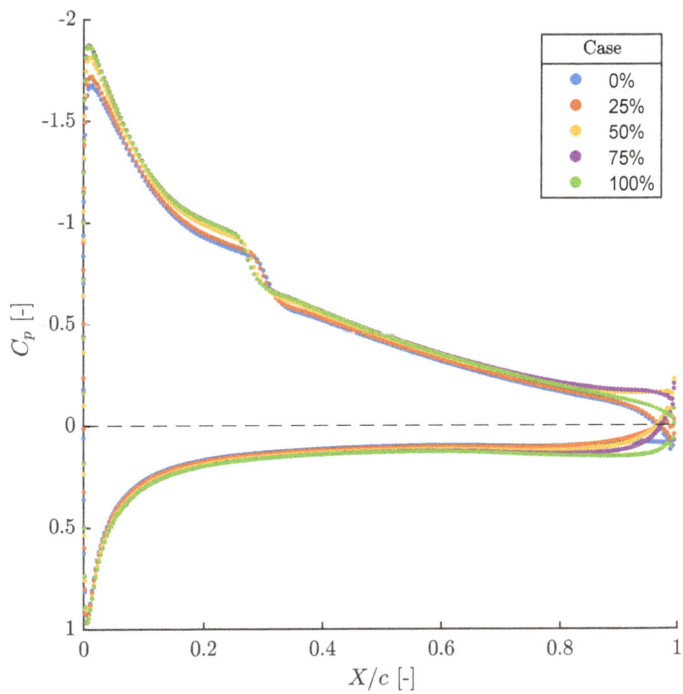

Figure 8. Pressure coefficient around the airfoil at 5° of angle of attack for each propeller position relative to the trailing edge.

The effect of raising the propeller over the trailing edge translates in the pressure coefficient C_p into an effect similar to increasing in the apparent angle of attack, as when flying inside the upwash of other wing. In addition, the elevation of the propeller produces less reacceleration of the pressure side, since a smaller portion of the propeller remains below the trailing edge.

The friction coefficient is plotted in Figure 9. It can be seen that, as the propeller position is higher, the friction in the last part of the suction side increases, thus increasing the parasitic drag associated with this parameter. It should be noted that, in the highest position, the effect of the propeller decreases, which is reflected in a drop in the friction coefficient C_f.

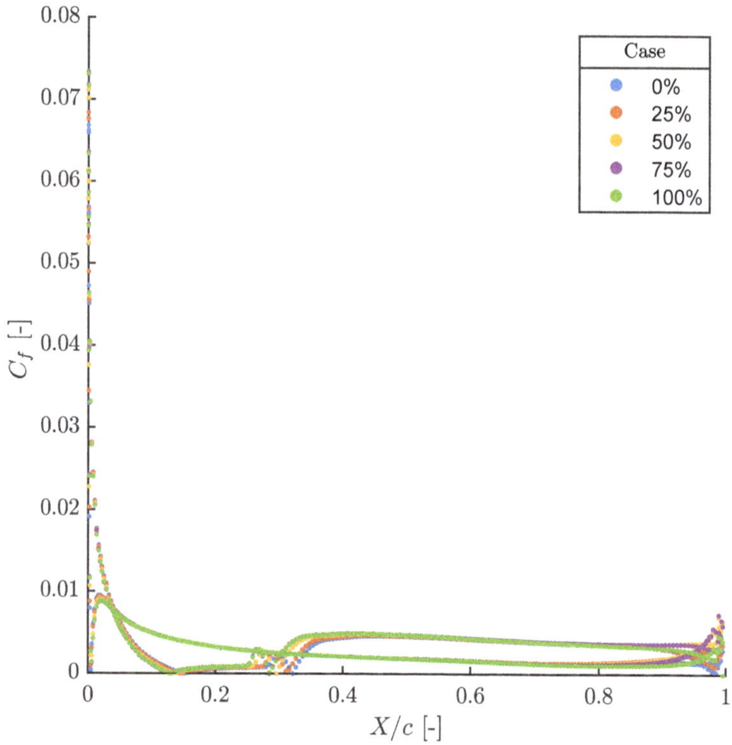

Figure 9. Friction coefficient around the airfoil at 5° of angle of attack for each propeller position relative to the trailing edge.

In the pressure side, a result similar to that of C_p can be obtained, where a higher propeller position implies a lower re-acceleration of the flow and, therefore, a lower C_f.

The effect of the propeller position on the propulsive efficiency results in the combination of several phenomena. The expectation would be to see an increase in this efficiency as the propeller is in the lower positions, that is, more centered on the trailing edge. The propulsive efficiency increase due to boundary layer ingestion can be summed up in that the propeller ingests a flow region whose speed is less than the flight speed, decreasing the power needed to be produced by the propeller for the same required force. As the propeller is positioned lower above the trailing edge, the boundary layer portion that will affect the propeller will be higher. This way, the propulsive efficiency increases. However, since the simulations are not carried out at the same lift coefficient but at the same air speed, the operating point at the same angle of attack varies for the different positions. The change in aerodynamic efficiency translates into different thrust requests for each position, which varies the rotational speed of the propellers. By varying the rotational speed while maintaining the air speed, the advance parameter J changes so that in each position the propeller will work at a different operating point. While the effect of the operation is less than the purely geometric aspect, in this way a high propulsive efficiency is still obtained also with higher propeller positions, but being always greater in the lower positions, with an optimum around the 20 % position.

4.2. Best Case Cfd Analysis

In this study a case with a propeller radius of 40 mm, a distribution of 13 engines on the wing and a draft angle of 1.5° with respect to the airfoil chord is presented. The position above the trailing edge is fixed at 31.5 %. This case maximizes the product of aerodynamic efficiency and propulsive efficiency.

CFD results in near-cruise conditions (i.e., $35\,\mathrm{m\,s^{-1}}$) are analysed, setting for the study an angle of attack of 3° and a Reynolds number equal to 5×10^5. As the angle of attack is low, the boundary layer is fully attached to the airfoil and its growth is moderate as can be seen in Figure 10a, representing the velocity contours in a mid-plane around the wing.

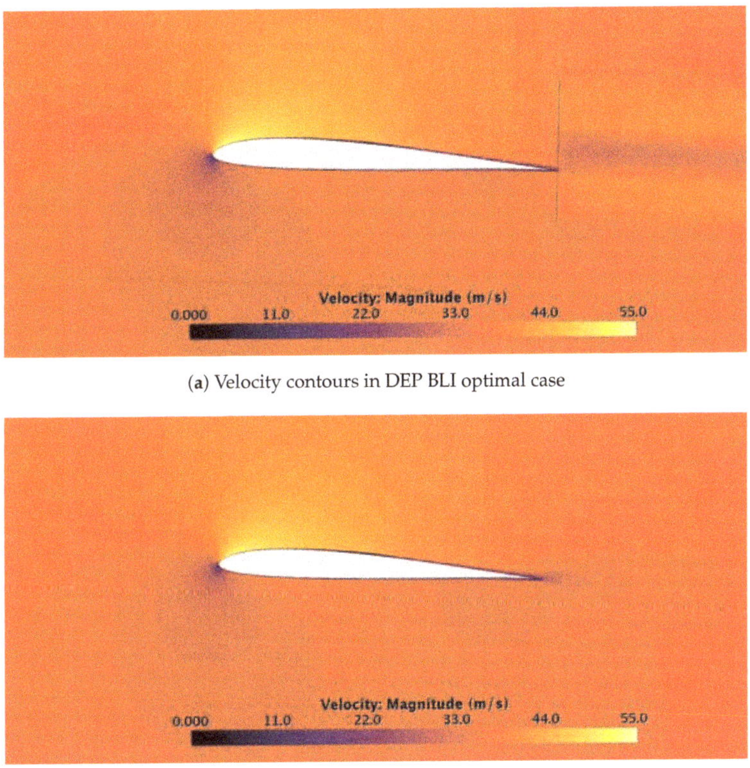

(**a**) Velocity contours in DEP BLI optimal case

(**b**) Velocity contours in base case

Figure 10. *U* contours in mid-plane for best DEP BLI configuration and base case comparison at an angle of attack of3° and a Reynolds number of 5×10^5.

If the BLI case in Figure 10a and the base case in Figure 10b are compared, speed similarities can be observed. Due to the propeller, the case with BLI has a greater reacceleration in the pressure side, at the same time that speeds up the wake behind the propeller. In addition, speed isolines have been added in Figure 11 showing a zoom of the pressure side for a better comparison.

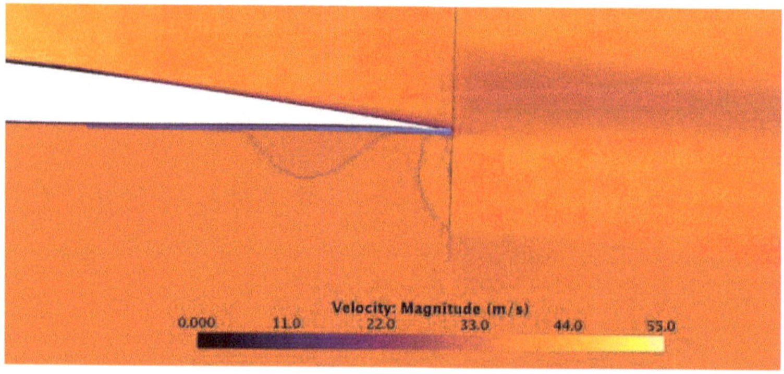

(**a**) Detail of velocity contours in DEP BLI optimal case

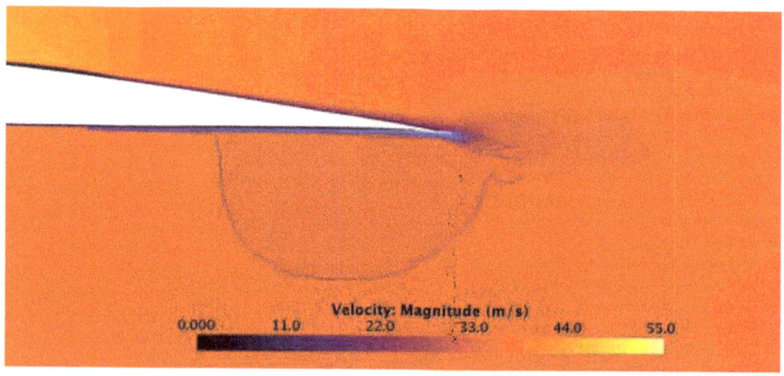

(**b**) Detail of velocity contours in base case

Figure 11. Detail of U in midplane for best DEP BLI configuration and base case with isovelocity lines under the pressure side for better comparison.

In order to compare the product between aerodynamic and propulsive efficiency in both cases, an additional simulation is carried out including only the propeller. This analysis is carried out in this way since in the base case the propulsive system is considered decoupled from the aerodynamic performance of the wing. In the simulation, the same flow velocity and advance parameter J from the DEP-BLI case is imposed.

The aerodynamic and propulsive data for both cases are collected in Table 2. The case with BLI has a higher C_L than the base case, however this increase is accompanied by an increase in $C_{D,0}$ that impairs aerodynamic efficiency by 1%. Nevertheless, the boundary layer bathed area of the actuator disc increases the propulsive efficiency of the BLI case by 8%, improving the product of efficiencies.

Table 2. Coefficient comparison between optimal DEP and BLI case and base case at an angle of attack of 3° and a Reynolds number of 5×10^5.

	C_L	$C_{D,0,\text{wing}}$	C_L/C_D	η_p	$C_L/C_D \cdot \eta_p$
Base case	0.484	0.00769	17.280	0.692	11.963
DEP BLI case	0.505	0.00841	17.080	0.748	12.773

To understand the differences in the aerodynamic coefficients, the pressure coefficient C_p of the case without BLI is plotted against the simulation proposed in Figure 12 in different wing spans.

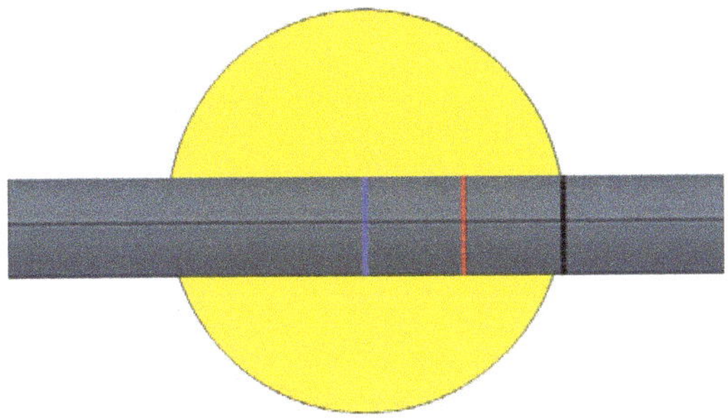

(a) Span positions in frontal view

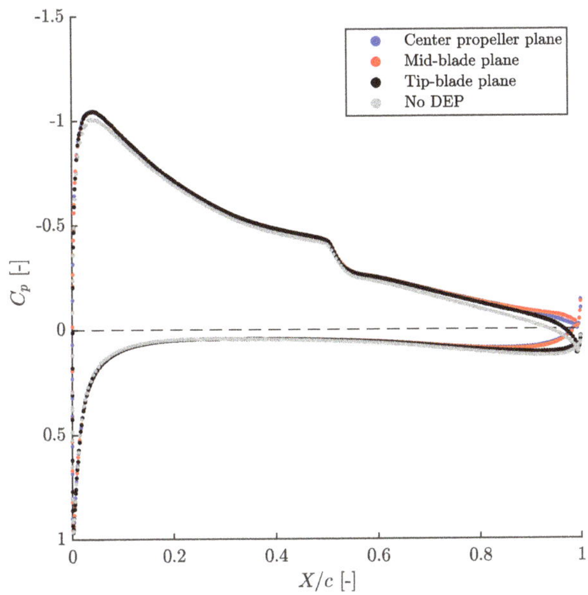

(b) Pressure coefficient distribution at different spans

Figure 12. *Cp* comparison for best DEP BLI configuration in different wing spans and base case middle plane.

Figure 12 shows that the suction peak magnitude in the BLI case is 4% higher than the base case, with increased suction on the entire domain. Simultaneously, in the first half of the pressure side, the case with BLI has a bigger C_p value than the wing alone, which translates to an increase in the lift coefficient.

At 50% of the chord, a transitional laminar separation bubble (LSB) can be observed in the BLI configuration, upstream compared with the base case. The effect of placing the propeller on the trailing edge is similar to increasing the angle of attack of the airfoil since the LSB moves upstream as the angle of attack increases, a common trend that can be observed experimentally, as shown in [27]. The effect is, thus, similar to that of flying with upwash: The resultant of the pressure forces is tilted towards the leading edge, lift is increased and the boundary layer transition appears closer to the leading edge.

In order to confirm the appearance of LSB, the friction coefficient C_f is plotted in Figure 13 for both cases in the mid span.

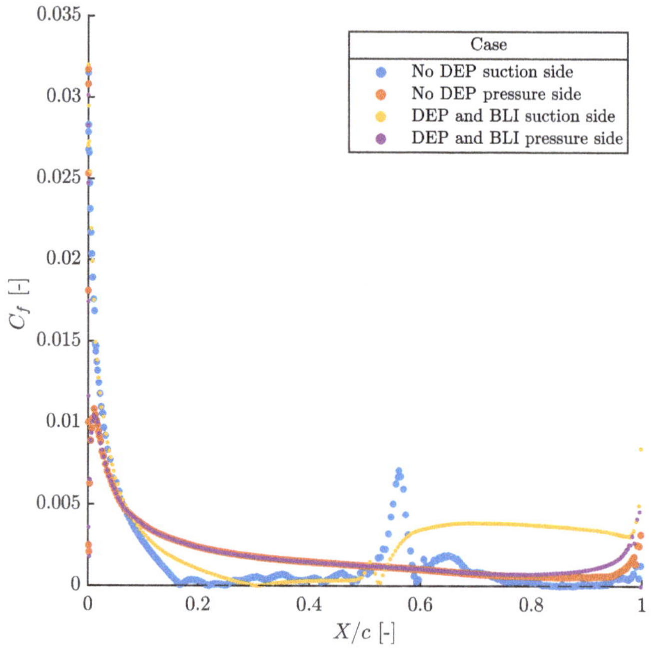

Figure 13. C_f comparison for DEP BLI case and base case without DEP and BLI distinguishing between suction side and pressure side.

Comparing the C_f in the suction side of both cases, it can be verified that the LSB appears earlier in the DEP BLI case resulting in a smaller bubble. As Sutton explains in [24], a smaller laminar separation bubble changes the effective inviscid shape of the flow around the wing, decreasing the pressure drag and allowing higher suction peaks to be achieved. However, as Tepperin has found in [15], the decrease in pressure drag can be outweighed by the increase in friction drag. The air suction near the trailing edge creates a lower pressure zone before the propeller accelerates the flow. Flow acceleration will increase the shear force and, in the same way, the friction drag increases, what can be observed after the appearance of LSB in the case of DEP BLI. Identically, since a part of the propeller disc re-accelerates the flow in the pressure side due to its position, C_f is also higher here.

Near the trailing edge, a discrepancy in C_p is appreciated due to how the surface is influenced by the actuator disc in the BLI case. To understand this effect, the pressure coefficient in Figure 12 is represented in three spans that cross the surface of the wing in the direction of flow: The blue one passes through the propeller separating it in half, the red through half its radius and the black plane corresponding to the blade edge.

The discrepancy of pressure coefficients between the spans can only be seen from 60% of the chord: Upstream, the presence of the propeller has the same effect on the entire surface of the airfoil.

In the outer spans away from the mid-plane, the decrease in pressure is less important and begins to approach the base case. As the effects of the propeller in the behaviour of the flow near the trailing edge are maximised close to its center, its influence in the boundary layer detachment at high angles of attack is expected to present non-uniformities across the wing span.

5. Conclusions

An RPAS with distributed electric propulsion and boundary layer ingestion configuration is feasible employing an hybrid electric power plant, a fuel cell or batteries, so it is a technological configuration that is applicable in an almost powerplant-agnostic manner. This novel configuration leads to improvements in both aerodynamic and propulsive efficiency compared to a classical distribution with one propeller without boundary layer ingestion. These efficiency improvements are achieved as long as the different geometrical parameters of the propeller distribution are set appropriately.

In this work, a methodology to analyse a small, 25 kg aircraft with a DEP BLI configuration has been discussed, whereby using a few simulations of a portion of the wing with a single rotor, the best distribution and overall performance of the aircraft has been estimated. The numerical study has been performed by a three-dimensional finite-volume simulation of the domain around a wing section coupled with a Blade Element Model Theory actuator disc to include the propeller. The rotational speed of the propellers has been set so their thrust equals the total drag of the aircraft, and the simulations have been performed for different angles of attack and vertical separations between the propeller shaft and the wing trailing edge. The simulations have been also carried out for the propeller and the wing separately, validating them against data found in the literature and comparing them with the DEP BLI cases. Results such as the pressure coefficient, skin friction coefficient, aerodynamic efficiency and propulsive efficiency have been obtained for all the simulations, so the overall performance parameters and the main effects of the DEP BLI configuration have been studied. The study of these parameters and effects is the main contribution of this manuscript, as the literature discussing them is still limited for small aircraft.

A distribution of 13 propellers along the wing, with 40 mm blade radius, a draft angle equal to 1.5° and a position of the center of the propeller relative to the trailing edge set at 31.5% has been found to maximise the product of both efficiencies at near-cruise conditions, achieving an improvement of 7%.

The vertical position of the propeller affect in different ways two important performance parameters: The propulsive efficiency and the aerodynamic efficiency. It has been shown that, when the propeller shaft is closer to the trailing edge, the propulsive efficiency is maximised, but there appears some penalties in the aerodynamic efficiency. When the shaft is moved upwards, the propeller section effectively ingesting the boundary layer of the wing is reduced, producing lower values of propulsive efficiency. In that case, however, the form drag of the wing is reduced and an increase in the lift coefficient appears, attaining higher values of aerodynamic efficiency. These opposing trends lead to two different optima: one at a vertical position around 20% of the propeller radius and another one at around 80%.

According to the simulations, moving the propeller upwards reduces its effect in the pressure side of the wing, moves the laminar separation bubble towards the leading edge and increases the suction peak. While the friction coefficient is also increased, the net effect is an increase of the circulation around the wing section and a decrease of the form drag, similar to what may be expected due to upwash effects.

The close proximity of the propellers to the trailing edge of the wing will probably generate interesting challenges due to vibrations, both in the propeller and in the airframe, which is an interesting aspect to study and discuss in future works. Regarding the position of the propellers, it also affects the pitching moment of the whole wing plus propellers group. The final geometry of an aircraft designed with this configuration should be adapted accordingly in order to not affect in a negative way its stability properties.

Author Contributions: Conceptualization, L.M.G.-C.; Data curation, P.V.; Formal analysis, P.V.; Funding acquisition, J.R.S.; Investigation, P.V.; Methodology, L.M.G.-C.; Project administration, L.M.G.-C.; Resources, J.R.S.; Software, L.M.G.-C. and P.V.; Supervision, L.M.G.-C.; Validation, L.M.G.-C. and P.V.; Visualization, L.M.G.-C. and P.V.; Writing—original draft, P.V.; Writing—review and editing, J.R.S., A.O.T., L.M.G.-C. and P.V. All authors have read and agreed to the published version of the manuscript.

Funding: This research was funded by the Agencia Estatal de Investigación of Spain through grant number PID2020-119468RA-I00/AEI/10.13039/501100011033.

Institutional Review Board Statement: Not applicable.

Informed Consent Statement: Not applicable.

Data Availability Statement: The data presented in this study are available on request from the corresponding author.

Acknowledgments: The authors wish to thank P. Raga and P. Quintero for their help with the setup of the computational simulations.

Conflicts of Interest: The authors declare no conflict of interest. The funders had no role in the design of the study; in the collection, analyses, or interpretation of data; in the writing of the manuscript, or in the decision to publish the results.

Abbreviations

The following abbreviations are used in this manuscript:

Abbreviations	
BLI	Boundary layer ingestion
BEMT	Blade Element Model Theory
BSFC	Brake-specific fuel consumption
CFD	Computational fluid dynamics
DEP	Distributed electrical propulsion
EASA	European Union Aviation Safety Agency
ERA	Environmentally Responsible Aviation
HE	Hybrid electric
ICE	Internal combustion engine
ITDS	Information Technology Development Solutions
LSB	Laminar separation bubble
NASA	National Aeronautics and Space Administration
RANS	Reynols-averaged Navier-Stokes
RPAS	Remotely piloted aircraft system
SESAR	Single European Sky ATM Research
TE	Trailing edge
UAV	Unmanned aerial vehicle
Roman letters	
Æ	Aspect ratio
b	Wingspan
c	Chord
C_L	Lift coefficient
C_D	Drag coefficient
$C_{D,0,\text{extra}}$	Parasitic drag coefficient of the aircraft without the wing
$C_{D,0,\text{wing}}$	Parasitic drag coefficient of the wing
C_p	Pressure coefficient
C_f	Friction coefficient
D	Drag
e	Oswald efficiency factor
J	Advance ratio
n	Rotational speed
P	Power
$r_{\text{propeller}}$	Propeller radius
Re	Reynolds
S	Wing surface
T	Thrust
U_∞	Air speed
X	Position across the chord

$y_{\text{propeller}}$	Position of the propeller shaft above the trailing edge
Greek letters	
α	Angle of attack
η_p	Propulsion efficiency
ρ	Density

References

1. EASA Agency. Study on the Societal Acceptance of Urban Air Mobility in Europe. 2021. Available online: https://www.easa.europa.eu/sites/default/files/dfu/uam-full-report.pdf (accessed on 18 June 2021).
2. Single European Sky ATM Research. European Drones Outlook Study. 2016. Available online: http://www.sesarju.eu/sites/default/files/documents/reports/European_Drones_Outlook_Study_2016.pdf (accessed on 10 March 2021).
3. Nickol, C.L.; Haller, W.J. Assessment of the performance potential of advanced subsonic transport concepts for NASA's environmentally responsible aviation project. In Proceedings of the 54th AIAA Aerospace Sciences Meeting, San Diego, CA, USA, 4–8 January 2016; pp. 1–21. [CrossRef]
4. Ausserer, J.K.; Harmon, F.G. Integration, validation, and testing of a hybrid-electric propulsion system for a small remotely-piloted aircraft. In Proceedings of the 10th Annual International Energy Conversion Engineering Conference, IECEC 2012, Atlanta, GA, USA, 29 July–1 August 2012; pp. 1–11. [CrossRef]
5. Harmon, F.G.; Frank, A.A.; Chattot, J.J. Conceptual design and simulation of a small hybrid-electric unmanned aerial vehicle. *J. Aircr.* **2006**, *43*, 1490–1498. [CrossRef]
6. Kim, C.; Namgoong, E.; Lee, S.; Kim, T.; Kim, H. Fuel economy optimization for parallel hybrid vehicles with CVT. In *SAE Technical Papers*; SAE: Warrendale, PA, USA, 1999. [CrossRef]
7. Stoll, A.M.; Bevirt, J.; Moore, M.D.; Fredericks, W.J.; Borer, N.K. Drag Reduction Through Distributed Electric Propulsion. In Proceedings of the 14th AIAA Aviation Technology, Integration, and Operations Conference, Atlanta, GA, USA, 16–20 June 2014; pp. 22–26. [CrossRef]
8. Stoll, A.M. Comparison of CFD and experimental results of the leap tech distributed electric propulsion blown wing. In Proceedings of the 15th AIAA Aviation Technology, Integration, and Operations Conference, Dallas, TX, USA, 22–26 June 2015; pp. 22–26. [CrossRef]
9. Lv, P.; Ragni, D.; Hartuc, T.; Veldhuis, L.; Rao, A.G. Experimental investigation of the flow mechanisms associated with a wake-ingesting propulsor. *AIAA J.* **2017**, *55*, 1332–1342. [CrossRef]
10. Leifsson, L.; Ko, A.; Mason, W.H.; Schetz, J.A.; Grossman, B.; Haftka, R.T. Multidisciplinary design optimization of blended-wing-body transport aircraft with distributed propulsion. *Aerosp. Sci. Technol.* **2013**, *25*, 16–28. [CrossRef]
11. Felder, J.L.; Kim, H.D.; Brown, G.V. Turboelectric distributed propulsion engine cycle analysis for hybrid-wing-body aircraft. In Proceedings of the 47th AIAA Aerospace Sciences Meeting including the New Horizons Forum and Aerospace Exposition, Orlando, FL, USA, 5–8 January 2009; pp. 1–25. [CrossRef]
12. Budziszewski, N.; Friedrichs, J. Modelling of a boundary layer ingesting propulsor. *Energies* **2018**, *11*, 708. [CrossRef]
13. Hall, D.K.; Huang, A.C.; Uranga, A.; Greitzer, E.M.; Drela, M.; Sato, S. Boundary layer ingestion propulsion benefit for transport aircraft. *J. Propuls. Power* **2017**, *33*, 1118–1129. [CrossRef]
14. Teperin, L. Investigation on Boundary Layer Ingestion Propulsion for UAVs. In Proceedings of the International Micro Air Vehicle Conference and Flight Competition (IMAV), Toulouse, France, 18–21 September 2017; pp. 293–300.
15. Elsalamony, M.; Teperin, L. 2D Numerical Investigation of Boundary Layer Ingestion Propulsor on Airfoil. In Proceedings of the 7th European Conference for Aeronautics and Space Sciences (EUCASS), Milan, Italy, 3–6 July 2017; pp. 1–11. [CrossRef]
16. Martínez Fernández, A.; Smith, H. Effect of a fuselage boundary layer ingesting propulsor on airframe forces and moments. *Aerosp. Sci. Technol.* **2020**, *100*, 105808. [CrossRef]
17. Goldberg, C.; Nalianda, D.; MacManus, D.; Pilidis, P.; Felder, J. Installed performance assessment of a boundary layer ingesting distributed propulsion system at design point. In Proceedings of the 52nd AIAA/SAE/ASEE Joint Propulsion Conference, Salt Lake City, UT, USA, 25–27 July 2016; Volume 52, pp. 1–22. [CrossRef]
18. Government of Spain. Real Decreto 1036/2017. In *Boletín Oficial del Estado, BOE-A-2017-15721*; Agencia Estatal Boletín del Estado: Madrid, Spain, 2017.
19. UAV Factory USA LLC. Penguin C UAS. Available online: https://www.uavfactory.com (accessed on 10 March 2021).
20. AERTEC Solutions. RPAS TARSIS 25. Available online: https://aertecsolutions.com/rpas/rpas-sistema-aereos-tripulados-remotamente/rpas-tarsis25/ (accessed on 27 May 2021).
21. Lyon, C.A.; Broeren, A.P.; Giguere, P.; Gopalarathnam, A.; Selig, M.S. *Summary of Low-Speed Airfoil Data*; SOARTECH Publications: Ann Arbor, MI, USA, 1997; Volume 3, p. 315.
22. Selig, M.S.; Donovan, J.F.; Fraser, D.B. *Airfoils at Low Speeds*; H.A. Stokely: Virginia Beach, VA, USA, 1989.
23. Selig, M.S. Low Reynolds Number Airfoil Design Lecture Notes—Various Approaches to Airfoil Design. In *VKI Lecture Series*; The von Karman Institute for Fluid Dynamics and NATO-RTO/AVT: Urbana, IL, USA, 2003; pp. 24–28.
24. Sutton, D.M. Experimental Characterization of the Effects of Freestream Turbulence Intensity on the SD7003 Airfoil at Low Reynolds Numbers. Ph.D. Thesis, University of Toronto, Toronto, ON, Canada, 2015; pp. 1–87.

25. Hoerner, S. *Fluid-Dynamic Drag;* Hoerner Fluid Dynamics: Bakersfield, CA, USA, 1965.
26. Niţă, M.; Scholz, D. Estimating the Oswald Factor from Basic Aircraft Geometrical Parameters. In *Deutsche Gesellschaft für Luft-und Raumfahrt-Lilienthal-Oberth eV*; Hamburg University Of Applied Sciences: Hamburg, Germany, 2012.
27. Ananda, G.K.; Sukumar, P.P.; Selig, M.S. Measured aerodynamic characteristics of wings at low Reynolds numbers. *Aerosp. Sci. Technol.* **2015**, *42*, 392–406. [CrossRef]
28. Smith, L.H. Wake ingestion propulsion benefit. *J. Propuls. Power* **1993**, *9*, 74–82. [CrossRef]
29. Blumenthal, B.T.; Elmiligui, A.A.; Geiselhart, K.A.; Campbell, R.L.; Maughmer, M.D.; Schmitz, S. Computational investigation of a boundary-layer-ingestion propulsion system. *J. Aircr.* **2018**, *55*, 1141–1153. [CrossRef] [PubMed]
30. Roache, P.J. Perspective: A Method for Uniform Reporting of Grid Refinement Studies. *J. Fluids Eng.* **1994**, *116*, 405–413. [CrossRef]
31. Gur, O.; Rosen, A. Comparison between blade-element models of propellers. *Aeronaut. J.* **2008**, *112*, 689–704. [CrossRef]
32. Deters, R.W.; Ananda, G.K.; Selig, M.S. Reynolds number effects on the performance of small-scale propellers. In Proceedings of the 32nd AIAA Applied Aerodynamics Conference, Atlanta, GA, USA, 16–20 June 2014; pp. 1–43. [CrossRef]
33. XFLR5. Available online: http://www.xflr5.tech/xflr5.htm (accessed on 10 March 2021).
34. Drela, M. XFOIL Subsonic Airfoil Development System. Available online: https://web.mit.edu/drela/Public/web/xfoil/ (accessed on 10 March 2021).

drones

MDPI

Article

Biomimetic Drones Inspired by Dragonflies Will Require a Systems Based Approach and Insights from Biology

Javaan Chahl [1,2,*], Nasim Chitsaz [1], Blake McIvor [1], Titilayo Ogunwa [1], Jia-Ming Kok [3], Timothy McIntyre [1] and Ermira Abdullah [4]

1 UniSA STEM, University of South Australia, Mawson Lakes, SA 5095, Australia; nasim.chitsaz@mymail.unisa.edu.au (N.C.); blake.mcivor@mymail.unisa.edu.au (B.M.); titilayo.ogunwa@mymail.unisa.edu.au (T.O.); timothy.mcintyre@unisa.edu.au (T.M.)

2 Joint and Operations Analysis Division, Defence Science and Technology Group, Melbourne, VIC 3207, Australia

3 Aerospace Division, Defence Science and Technology Group, Melbourne,VIC 3207, Australia; jia.kok@dst.defence.gov.au

4 Department of Aerospace Engineering, Faculty of Engineering, Universiti Putra Malaysia, 43400 Seri Kembangan, Malaysia; ermira@upm.edu.my

* Correspondence: javaan.chahl@unisa.edu.au

Citation: Chahl, J.; Chitsaz, N.; McIvor, B.; Ogunwa, T.; Kok, J.-M.; McIntyre, T.; Abdullah, E. Biomimetic Drones Inspired by Dragonflies will Require a Systems Based Approach and Insights from Biology. *Drones* **2021**, *5*, 24. https://doi.org/10.3390/drones5020024

Academic Editor: David R. Green

Received: 1 March 2021
Accepted: 21 March 2021
Published: 27 March 2021

Abstract: Many drone platforms have matured to become nearly optimal flying machines with only modest improvements in efficiency possible. "Chimera" craft combine fixed wing and rotary wing characteristics while being substantially less efficient than both. The increasing presence of chimeras suggests that their mix of vertical takeoff, hover, and more efficient cruise is invaluable to many end users. We discuss the opportunity for flapping wing drones inspired by large insects to perform these mixed missions. Dragonflies particularly are capable of efficiency in all modes of flight. We will explore the fundamental principles of dragonfly flight to allow for a comparison between proposed flapping wing technological solutions and a flapping wing organism. We chart one approach to achieving the next step in drone technology through systems theory and an appreciation of how biomimetics can be applied. New findings in dynamics of flapping, practical actuation technology, wing design, and flight control are presented and connected. We show that a theoretical understanding of flight systems and an appreciation of the detail of biological implementations may be key to achieving an outcome that matches the performance of natural systems. We assert that an optimal flapping wing drone, capable of efficiency in all modes of flight with high performance upon demand, might look somewhat like an abstract dragonfly.

Keywords: drone; dragonfly; insect; biomimetic; biological inspiration; aerodynamics; design

1. Introduction

From emergence until death, the flight performance of the dragonfly is tested. Typically, males establish and fight to maintain a territory with favourable oviposition sites after a short period of orientation following emergence. This involves perpetual, dangerous, aerial combat against male rivals, with only the best aviators achieving the territory needed to breed. Mating requires an aerial pursuit of females that innately stresses the flight performance of males. Upon successfully overcoming the female's defences against weak fliers, successful mating often requires the males to carry the female's inert mass. Throughout all of this aerial combat, aerial predators are a constant threat, requiring defensive air combat maneuvers at regular intervals. Most combat is over a water surface, under which predators lurk and from which a dragonfly is unlikely to extract itself. Finally, feeding is an effort in pursuit and counter-evasion against prey animals, the evolution of which is also locked in a deadly loop with the dragonfly's evolving flight performance. It is a life of aerial combat, pursued for 300 million years, by the oldest extant form of flying insect, the archetypal member of Palaeoptera [1,2].

It can be argued that dragonflies have made the least accommodation for other practicalities of existence of any flying insect. Long, unencapsulated, permanently horizontal wings sprout from a heavily muscled thorax, while their large, high-resolution eyes float on a delicate retractable gimbal neck to stabilise sensory input, all optimised for flight and combat. Dragonflies are so highly adapted to flight that their limbs are more suited to aerial grappling than walking; a grounded dragonfly serves no evolutionary purpose and the bloodless wings of adult insects do not heal. Evolution has committed the dragonfly to its aerial life. The single-role lifestyle of adult Odonata is the motivation for our attempt here to understand their critical system level characteristics, so that we can chart a course to a high performance flapping wing drone.

Dragonflies have been shown in quantitative studies to demonstrate superior flight performance compared to most other insects. Scale must be considered in this assessment, as the Odonata order contains relatively large species, particularly in terms of wingspan. The nature of aerodynamics ensures that larger species with longer wings have a physics advantage in some aspects of flight. Dragonfly mastery of the air has been shown in a number of studies based on high speed video analysis of normal flight, aerial combat, and predation. High-performance turning flight has been shown, with loads in turns exceeding 40 m/s^2 [3] and an ability to takeoff while carrying more than three times their own body weight [4]. Dragonflies have been shown to pursue prey animals using the chasing strategies found in missiles [5] and to have exceptionally high success rates in capturing aerial prey [6,7]. They have been found to exhibit advanced guidance laws that camouflage their apparent motion with respect to their opponent [8,9] using means that have not been convincingly identified. It has been demonstrated that they have a surprisingly efficient glide performance comparable to that of model aircraft with substantially higher Reynolds numbers (*Re*). Observed lift-to-drag ratios range from 3.5 to over 10 [10–12]. Some species of dragonfly have been observed to migrate across hundreds of kilometres [13], a difficult, energetic achievement for such a small organism. Energy scavenging by dragonflies has also been observed: they soar on thermal updrafts and on rising air currents caused by slopes [14]. The evolutionary argument and the measured realities of dragonfly mastery of flight align, indicating that they may be the apex insect flyer. The challenge is to understand the ingredients of this design so that technological systems might be able to achieve some of the same characteristics.

1.1. Biomimetics and Bioinspiration

Biomimetics is a broad approach, without a standard methodology. The range of techniques is a spectrum from "inspiration" to "mimicry". The two extremes on this spectrum have limitations. *Inspiration* might entail simply looking at the outside of a biological system such as a dragonfly, which might help the engineer to decide what application to create: for example, an insect-styled drone. This allows the imagination to be channeled and certainly is a valid engineering approach, complete with expectations and requirements. However, the solution is the result of engineering and does not create a particular head start for the engineer. Inspiration might even create undue constraints, since the operating principles applied might be quite different than in the organism. *Mimicry* creates problems of technological incompatibility between the solutions grown by biology and the highly refined mechanisms in common engineering use. Essentially the components required to make a copy often do not exist. For example, conventional technologies for motion are rotational, yet the known rotating motor in nature is a flagellum with parts on a molecular scale [15], while large motive systems are based on muscle. Artificial muscles, on the other hand, remain elusive. The problems of copying solutions becomes even deeper when organs and cellular mechanisms start to be considered, showing clear limitations to the mimicry approach.

Srinivasan has advocated an approach that accepts the differences in component technology but ensures that the principles of operation are preserved [16]. In many cases, it is inviolable system characteristics that allow the organism to operate, not details of implementation. We will assert that this is the case with dragonflies, despite all appearances. This approach requires an understanding of both the class of technological system being created and of the biological system being mimicked. This argument is supported by the failures of early ornithopters. Despite detailed bird anatomy and observations, there was, at first, no understanding of aircraft. Subsequent work identified the role of mass, balance, and wings, allowing a progression from Cayley [17], to Lilienthal [18], and culminating with the Wright Brothers [19]. In the case of flight, it seems that the path to success required a systematic approach to research into underlying principles and that biomimicry did not provide a substantial shortcut, other than for inspiration.

Dragonfly ocelli are an accessible anecdote of biomimicry used to create a unique technological sensory capability. Dragonflies have evolved a highly specialised second set of eyes and reflexes for stabilisation of flight using the horizon, even in dim light. Evolution has chosen specific spectral bands of ultraviolet and green to maximise discrimination between sky and ground [20]. Construction of functional synthetic ocelli for aircraft required photodiode arrays with sensitivity to appropriate spectral bands; it did not require the building of an eye [21]. The reasons such a device had not been conceived of before were that (i) Stange in the 1970s [22] was not aware that unmanned aircraft was an issue, hence the 20 year gap before a follow-up study [23] and implementations [24], (ii) the world that is visible in the ultraviolet region is outside the human experience, (iii) inertial sensors have been adequate for drones with steady flight patterns and heavily aided by global positioning systems, and (iv) the need for extreme agility and rapid inertial realignment in very small aircraft that exceed current inertial sensor capabilities had not yet appeared. This also shows that knowledge of both the system being mimicked and the class of technological system being created is almost a necessity for the successful adoption of biomimetic concepts.

Aeronautical systems are highly optimised in ways that are not generally appreciated when the rule-based design of conventional aircraft is undertaken. In this case, convention conceals complexity. For the sake of comparison, we present 3D models of a wing, a rotor, and a dragonfly wing in Figure 1. It is likely that all are close to their maximum possible performance, by whatever means that might be measured, but one appears to have more degrees of freedom and is used in more complex ways.

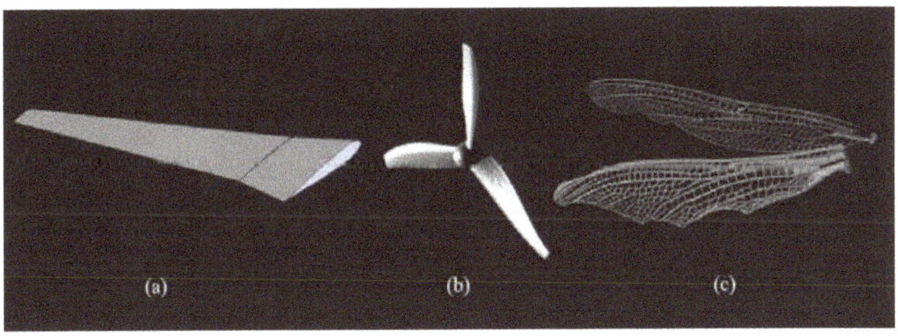

Figure 1. An aircraft wing (**a**), a rotor (**b**), and a dragonfly wing (**c**), each highly optimised for its function.

The remainder of the paper will explore key interrelated systems that are substantially different to fixed wing, multirotor, or wing/rotor chimera craft, but are critical to the flight of a conceptual biomimetic high-performance drone. We will also present experimental evidence showing how critical these functions are. We will propose a technology path to achieving this ambitious goal.

2. Wing Performance and Structure

Micro air vehicles (MAVs) designed to mimic the flight of dragonflies must have well understood aerodynamics. There have been numerous studies on the aerodynamic design and control of dragonfly biomimetic MAVs [25]. Some of the most basic factors include *Re*, wing aspect ratio (*AR*), and wing structural properties.

The flow on and around an aircraft is affected by aircraft size and flight speed. *Re*, the ratio of inertial forces to viscous forces, is a non-dimensional parameter used to scale and compare flow regimes of air vehicles:

$$Re = \frac{U_{ref} L_{ref}}{\nu}, \tag{1}$$

where U_{ref} is the reference velocity, L_{ref} is the reference length, and ν is the kinematic viscosity of the fluid. The smaller the *Re*, the larger the viscous effects. *Re* is proportional to reference length and reference velocity. *Re* decreases with decreasing wing size. Insects operate at low *Re* (10^4 to less than 10^2), where both inertia and viscous effects are important. This regime is characterised by an unsteady laminar flow, involving strong vortices [26]. At low *Re*, the viscous effects are so large that the leading-edge vortex (LEV) is separated, resulting in little lift generation and high drag [27]. Insects take advantage of unsteady aerodynamics associated with low *Re* using a variety of documented mechanisms: a rapid pitch-up, a delayed stall of the LEV, wake capture, clap-and-fling mechanisms, and interactions between the LEV, the trailing-edge vortex (TEV), and the tip vortex [28], to generate the necessary lift. We will continue our analysis without further consideration of these mechanisms because they are each *Re*-specific techniques and do not appear to be necessary to explain the dragonfly design.

Compared to fixed wing flight, the aerodynamics of dragonfly flight is more complex. Researchers have proposed a variety of different methodologies to produce aerodynamic analysis, either using experimental techniques or simulation. One study reconstructed the glide paths of dragonfly flight by filming the flight in a large enclosure [10]. The aerodynamic forces were measured from isolated wings and bodies in a steady air flow at an *Re* from 700–2400 for the wings and 2500–15,000 for the bodies. Using these methods, it was found that dragonfly wings demonstrate superior steady-state aerodynamic properties compared to wings of other insects. In another study, the aerodynamic characteristics of dragonfly wing and conventional airfoils were compared by measuring the lift and drag coefficients at an *Re* of 7880 and 10,000 using a force balance system [29].

Hu et al. [30] investigated the aerodynamic effect of wing–wing interaction on the dragonfly flight using a flapping model dragonfly. The model was submerged in a tank of mineral oil and mounted on a six-component force sensor that measured the forces and torques, comprising wing aerodynamic force, wing mass, and wing inertial force. Results from the experiment showed that wing–wing interaction increases fore wing (FW) lift in forward flight while reducing hind wing (HW) lift at all wing beat phase differences. Using computational fluid dynamics (CFD), it was found that aerodynamic performance improved due to FW–HW interaction during gliding motion for angles of attack ranging from 0 to 25° [31]. CFD has been used to estimate the aerodynamic performance of free flying dragonflies [32]. The spanwise features of vortex interaction between the FW and HW were investigated, and it was found that the flow structure around both wings resulted in increased horizontal force (thrust) generation.

A key difference between dragonflies and flies are their long slender wings. Figure 2 shows a robber fly wing on the left and a dragonfly wing on the right. *AR* is generally a strong indicator of how a wing performs. The *AR* of a wing is defined by the ratio of the wingspan to the mean chord as follows:

$$AR = \frac{b_{ref}^2}{S_{ref}}, \tag{2}$$

where b_{ref} is the reference wingspan, and S_{ref} is the reference wing area.

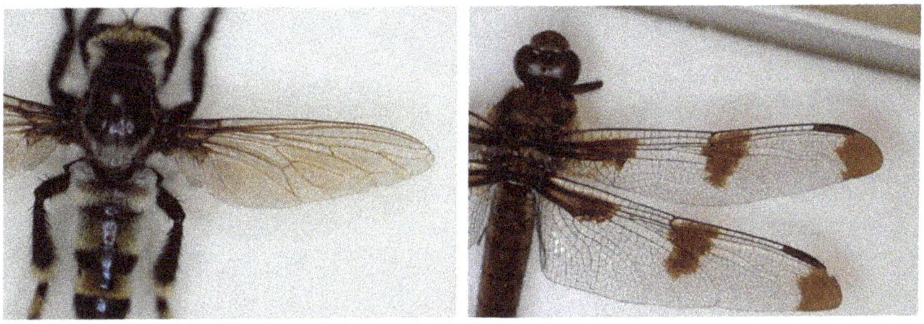

Figure 2. The difference between the wing shapes of two large flying insects—a robber fly, *Phellus piliferus* (**left**), and a dragonfly, *Libellula pulchella* (**right**)—can be described, to a first order approximation, by aspect ratio.

Wingtip vortices have a deleterious effect on wing performance; hence, the amount of energy lost due to induced drag can be reduced by reducing the area of the wing tip, by making it pointed or slender, as in [33]. AR and wing area quantify the shape and size of the wing and consequently are an indication of aerodynamic efficiency. Higher AR is indicative of a wing that is relatively narrower. While flow is attached, reduction in AR causes the lift coefficient (C_L) to reduce for a given angle of attack. Increasing the AR by making a longer and thinner wing improves aerodynamic performance, a tactic commonly employed in modern sailplanes. A longer wing results in a smaller wingtip vortex, which results in a less induced drag, which reduces forward flight energy costs. Reduced induced drag is particularly beneficial for natural flyers that glide at low speed, because at low speed, induced drag dominates the total drag. Long wings are aerodynamically beneficial, but carry obvious consequences for practicality on the ground. It is likely that the evolutionary feature of Odonata having a limited ability to walk is related to their long, fragile, permanently deployed wings.

The wings of dragonflies are a frequently depicted subject of art and jewellery. Amazingly, however, the precise details of a reasonable number of dragonfly's wings have only recently been recorded and analysed. Capturing even one live dragonfly intact is often a significant undertaking for the uninitiated. Capturing a large number of flying insect species in a systematic fashion is a difficult and expensive project, often a lifelong pursuit. Our project of capturing the wing geometry of 75 Odonata species required the use of scientifically and historically significant museum collections. Examples of these fragile specimens are shown on the left of Figure 3, including some broken specimens with limbs, wings, and abdomens detached. A passive optical technique for measurement was developed that could function through the glass display case [34]. Photos were taken from a linear sequence of positions using a robotic slide, shown on the right of Figure 3. Approximately 100 images of 4016 × 6016 pixels were captured and processed to construct a 3D model of the wings. All specimens remained intact and undisturbed [35].

To obtain a 3D image, the structure from motion technique was used by the 3DF Zephyr software [36] to stitch the captured photographs together. The technique captures colour, visible textures, and structures of the wing. The first 3D reconstructions [34] and quantitative comparisons [37] of a substantial number of Odonata species were published in the year 2020, despite scientific interest in the Odonata for millennia. Typical outcomes are shown in Figure 4.

Figure 3. Left: Museum specimens are fragile and cannot be removed from their cases. Intact and damaged specimens are shown. **Right**: A dragonfly wing being scanned by a robotic photogrammetry assistant at the South Australian Museum [34].

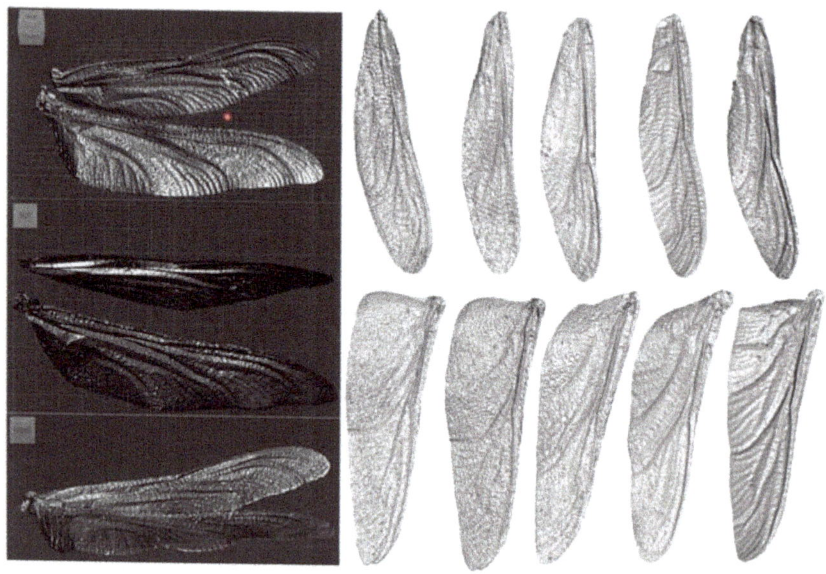

Figure 4. Models of wings reconstructed using an automated photogrammetry technique for measuring insect wing geometry [34].

Dragonfly wings have a comparatively high AR. They are corrugated [38–40], with complex vein patterns along chord and span. The veins contain blood on emergence, which enables the wings to unfurl. Soon after emergence, the blood supply is cut off, so the wings and veins are not living tissue [2]. Spanwise, the corrugations vary and flatten as they reach the wing tip. In [29], Kesel compared the aerodynamic efficiency at low Re of an extruded 2D corrugated aerofoil section and a flat plate section. For an ultra-low Re, between 2000–3000, aerodynamic performance of corrugated aerofoils at several angles of attack from 0° to 15° have been studied in addition to the lift-to-drag ratio [41]. Subsequent work has visualised airflow over the wings using particle image velocimetry (PIV) compared against computational fluid dynamics [42–45]. These flow studies have shown the effect of different levels of structural detail on flow separation of both real wings and manufactured wings, at various angles of attack. The general principle of corrugated wings has been shown to be the formation of profile valleys in which rotating vortices develop. These vortices reduce friction and maintain attached flow at higher angles of attack than on a flat plate or a conventional aerofoil [29]. Aerofoils formed by transects of a dragonfly wing are shown in Figure 5.

Figure 5. Reconstructed aerofoils of *Aeshna cyanea* (Müller, 1764) FW [29].

Typically, corrugation enhances aerodynamic performance, while stiffening the wing. However, in the range of very low *Re*, it has been found that corrugations may produce undesirable aerodynamic effects [46]. Fixed wing aircraft interpretations should be used with caution, however, since drag might be desirable in some modes of flapping flight. CFD was employed to analyse the flow fields and surface pressure distributions of corrugated and flat-plate wings at low *Re* ranging from 200–2400. It was found that the lift coefficient was lower for the corrugated wing, particularly at lower angles of attack. Strongly corrugated wings are definitely stiff, yet aerodynamically they are a solution to a specific range of *Re*. Research continues on the systematic measurement of smaller insect wings. Our comprehensive data set of dragonfly wings allowed for detailed computational flow visualisation and the manufacture of reconstructed 3D test wings for physical flow field visualisation.

Experimental analysis of flow over the full 3D wing, complete with measured micron-scale features, was performed to explore the effect of reconstructed natural surface texture on flight performance at low *Re*. The flow patterns were analysed at an angle of attack of 10° and $Re = 5 \times 10^3$, $Re = 8 \times 10^3$, and $Re = 12 \times 10^3$, showing better performance than that found in previous studies based on extruded 2D corrugated aerofoils [47].

The influence of corrugated FW/HW at low *Re* was investigated by the authors. The results were compared with the same size and planform flat wing. The PIV technique was applied on the manufactured wings. The flow characteristics and boundary layer of the bioinspired wings were measured. Figure 6 shows the instantaneous spanwise vorticity around corrugated profiles and flat profiles at $Re = 3 \times 10^3$ with $\alpha = 0^\circ$ and $\alpha = 10^\circ$. The fabricated corrugated wing provides substantially better aerodynamic performance than a flat wing in terms of delaying flow separation.

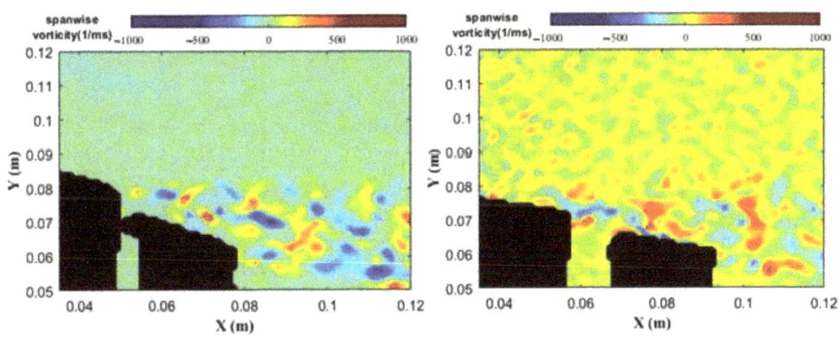

Figure 6. Comparison of span-wise vorticity (color bar) and instantaneous velocity field around the reconstructed FW and HW (**left**) and flat FW and HW (**right**), at $Re = 3 \times 10^3$ and $\alpha = 10^\circ$. The laser light sheet incident from the top side created a shadow underneath the wing.

Analysis of a fully replicated HW, complete with 10 micron scale texture, in gliding flight, at low *Re*, was performed to study its aerodynamic characteristics. Computational results were verified against experimental results with good agreement. Figure 7 compared lift and drag measurements of corrugated and flat wings at $\alpha = 10^\circ$ and $Re = 22 \times 10^3$ (left) and pressure distribution from a corrugated wing at $\alpha = 10^\circ$ and $Re = 6 \times 10^3$ (right).

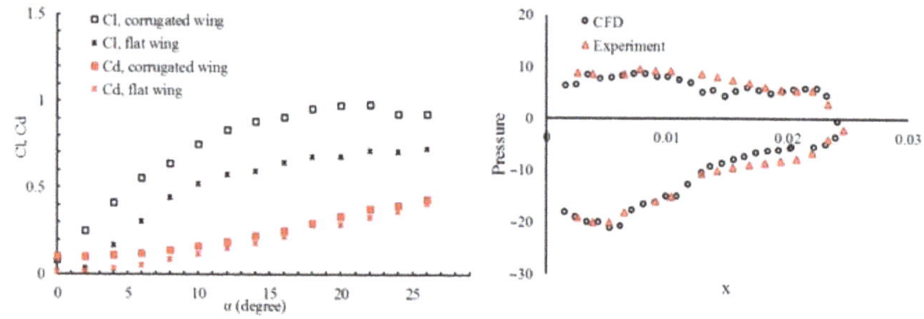

Figure 7. Comparison of aerodynamic lift and drag measurements of corrugated and flat wings at $\alpha = 10°$ and $Re = 22 \times 10^3$ (**left**) and pressure distribution from numerical and experimental results of corrugated wing at $\alpha = 10°$ and $Re = 6 \times 10^3$ (**right**).

Lessons from the Dragonfly Wing

A stiff wing, weighing less than 2% of the body of the animal, achieves a better lift-to-drag ratio than a flat structure and performs at low *Re* where conventional aerofoils are poor. Dragonfly wings appear to be more efficient than expected, particularly at low speed, while still retaining attached flow at high angles of attack, yielding both high lift and high drag on demand. Drag and lift are both associated with the damping property identified in Section 4, which allows nearly all energy put into the wing by the flight motor to be delivered in a controlled fashion to the air. The electron microscope images in Figure 8 show the detail of the dragonfly wing. The precise and self-similar features suggest a highly evolved weight and stiffness optimisation that will be quite hard to replicate and yet may be necessary for an efficient implementation.

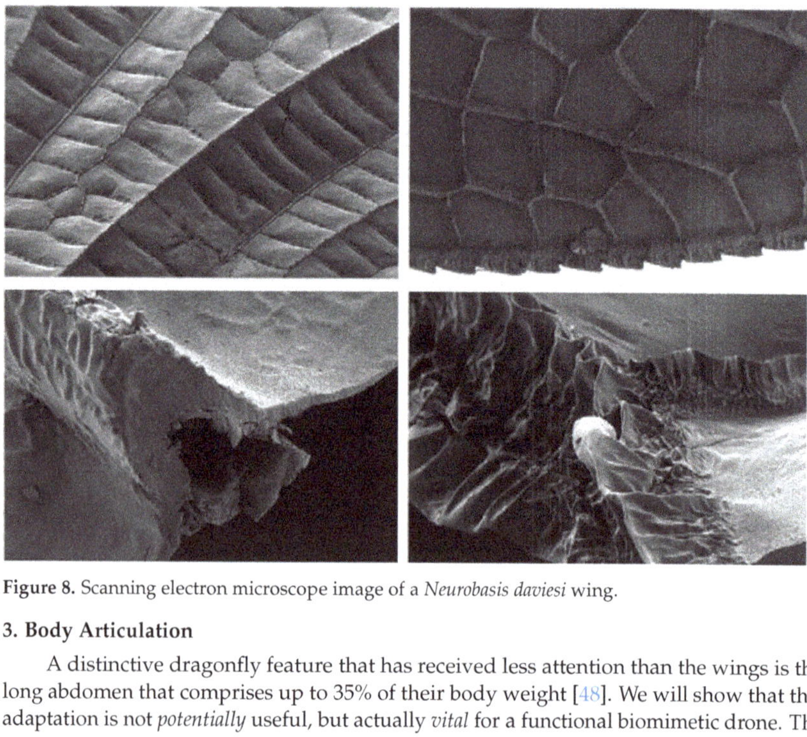

Figure 8. Scanning electron microscope image of a *Neurobasis daviesi* wing.

3. Body Articulation

A distinctive dragonfly feature that has received less attention than the wings is the long abdomen that comprises up to 35% of their body weight [48]. We will show that this adaptation is not *potentially* useful, but actually *vital* for a functional biomimetic drone. The abdomen has evolved to serve many purposes. It houses the digestive tract, has anatomical features related to reproduction, and must bend for mating and oviposition [2]. Less obvious roles are in maintaining balance, enhancing stability, and achieving maneuverability in certain situations. The ability to change posture can modify the position of the centre of mass with respect to the origin of aerodynamic forces on the body [49], allowing body posture to be used for flight control. The use of posture changes for control has been observed in land animals, including geckos [50] and cheetahs [51,52]. In insects, posture changes, in the form of strong abdominal steering reflexes, have been observed in response to mechano-sensory or visual stimuli. For instance, a study presented in [53,54] showed that desert locusts (*Schistocerca gregaria*) respond to an angled wind stimulus during tethered flight with large leg and abdominal movements. Fruit flies (*Drosophila melanogaster*) demonstrate similar responses to visual rotation [55–57]. In addition, modulation of the vertical abdominal angle has been observed in honeybees (*Apis mellifera*) responding to the speed of a translating visual pattern [58]. Hawkmoths (*Manduca sexta*) also show strong abdominal responses to mechanical and visual stimuli [59,60]. Dragonflies have been observed to use head movements to target their prey [5], and several videos by Rüppell show dragonfly abdominal movements during flight [61].

As mentioned earlier, the flight dynamics and aerodynamics of flapping wing systems are more complex (and scientifically contested) than that of fixed and rotary wing aircraft. Mathematical modelling helps gain insight into the complexities associated with flapping flight. Several flight dynamics models for insects have been developed in the literature [60,62–68]; however, most of the models are data-driven for specific insects and, as such, are specific and restrictive in application. Flapping flight dynamics models in the literature, based on laws of classical physics, have neglected effects due to morphological changes. Most studies have focused on hovering or tethered flight. Although hover is important for a dragonfly, it is not their primary mode of flight, as they spend most of their flight time in forward motion [69]. There has been little effort to quantify these effects [70] and little attempt, until recently, to consider the effects of these posture changes on flight performance [71]. The following analysis considers the potential roles of abdominal articulation in an abstracted fixed wing glider at constant speed.

3.1. Equations of Motion

The non-linear equations of motion for a fixed wing aircraft are adapted here to include an abdomen/tail modelled as an added mass, m^T, concentrated at the tip of a massless rod [71]. For the purposes of isolating inertial forces from aerodynamic forces on the tail, the aerodynamic effects of the tail were ignored, as these effects are expected to have little influence on the aerodynamics of the aircraft, compared to the wings, legs, and thorax. The central body (head to thorax region) has six degrees of freedom (DoF) in flight. The abdomen has three additional rotational degrees of freedom, constrained to move with the central body about a joint at *point j* (see Figure 9). Note that, when the reference frame of a vector or tensor is not explicit, the default is the body-fixed reference frame B.

The translational and rotational dynamic equations are represented by Equations (3) and (4), respectively [71,72]:

$$\begin{aligned}\dot{V}_B^I = \frac{1}{m}\Big(F\Big) - \frac{1}{m}\Big(m^T\Big[[R]^{BT}\Big([\dot{\omega}^{TB}]^T \times [\rho_{tj}]^T + [\widetilde{\omega}^{TB}]^T([\widetilde{\omega}^{TB}]^T[\rho_{tj}]^T)\Big),\\ + \Big(\dot{\omega}^{BI} \times ([R]^{BT}[\rho_{tj}]^T + \rho_{jb})\Big) + \Big(\widetilde{\omega}^{BI}\,\widetilde{\omega}^{BI} \times ([R]^{BT}[\rho_{tj}]^T + \rho_{jb})\Big)\\ + \Big(2\,\widetilde{\omega}^{BI}[R]^{BT}([\widetilde{\omega}^{TB}]^T[\rho_{tj}]^T)\Big)\Big]\Big) - \Big(\widetilde{\omega}^{BI}\,V_B^I\Big),\end{aligned} \tag{3}$$

$$\begin{aligned}\dot{\omega}^{BI} = \frac{1}{J_b^{\mathcal{C}}}\Bigg(M - \Bigg[m^T\Big([R]^{BT}[\rho_{tj}]^T + \rho_{jb}\Big)\Big[[R]^{BT}\Big([\dot{\omega}^{TB}]^T \times [\rho_{tj}]^T + [\widetilde{\omega}^{TB}]^T([\widetilde{\omega}^{TB}]^T[\rho_{tj}]^T)\Big) \\ + \Big(\widetilde{\omega}^{BI}\,\widetilde{\omega}^{BI} \times ([R]^{BT}[\rho_{tj}]^T + \rho_{jb})\Big) \\ + \Big(2\,\widetilde{\omega}^{BI}[R]^{BT}([\widetilde{\omega}^{TB}]^T[\rho_{tj}]^T)\Big) + \dot{V}_{\mathcal{B}}^{I} + \widetilde{\omega}^{BI}\,V_{\mathcal{B}}^{I}\Big]\Bigg]\Bigg),\end{aligned} \tag{4}$$

where ρ_{tb} is the position of the tail mass with respect to *point b* in the body frame. ω^{BI} and $\dot{\omega}^{BI}$ are, respectively, the angular velocity and acceleration of the body frame with respect to the inertial frame. $[\omega^{TB}]^T$ and $[\dot{\omega}^{TB}]^T$ are the angular velocity and acceleration of the tail mass with respect to the body frame, respectively. Here, m^T is the mass of the abdomen/tail, and m is the total mass of the aircraft, while F and M represent the total force and moment acting on the aircraft respectively.

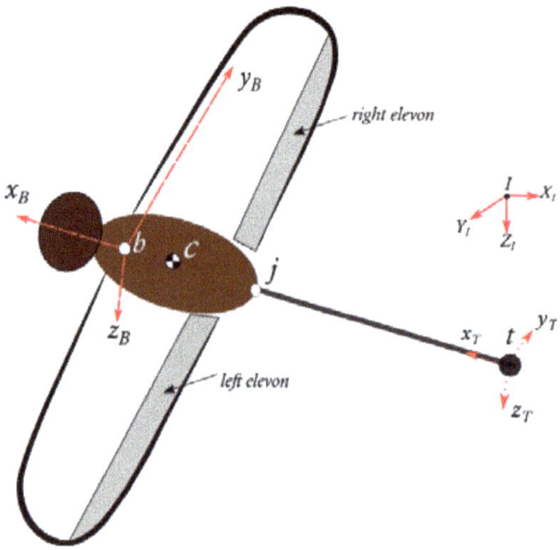

Figure 9. Definition of coordinate systems: Inertial (I), Body (B), and Tail (T).

The change in centre of gravity (cg) position with respect to *point b*, written in the body frame, is expressed as

$$\triangle\rho_{cb} = \frac{m^T\rho_{tb}}{m}. \tag{5}$$

where

$$\rho_{tb} = [R]^{BT}[\rho_{tj}]^T + \rho_{jb}\,. \tag{6}$$

The inertia matrix of the whole system about *point b* is given by

$$J_b^{\mathcal{C}} = J_b^{\mathcal{B}} + J_b^{\mathcal{T}}, \tag{7}$$

where $J_b^{\mathcal{T}}$ is expressed as

$$J_b^{\mathcal{T}} = m^T\Big[(\overline{\rho}_{tb}\,\rho_{tb})\mathbb{I} - \rho_{tb}\,\overline{\rho}_{tb}\Big], \tag{8}$$

and $\mathbb{I}$ is a 3×3 identity matrix.

The elements of the inertia tensor J are

$$[J] = \begin{bmatrix} J_{xx} & -J_{xy} & -J_{xy} \\ -J_{yx} & J_{yy} & -J_{yz} \\ -J_{zx} & -J_{zy} & J_{zz} \end{bmatrix}. \tag{9}$$

3.2. *Flight Simulation for Maneuvering*

A dynamics model can be expressed as a function of its states X and inputs U,

$$\dot{X} = f(X, U) \tag{10}$$

$$y = g(X, U) \tag{11}$$

Depending on the variables of interest, the output vector y elements can be arbitrary combinations of states and inputs. Matlab's trim algorithm, which uses sequential quadratic programming [73], was used to evaluate the desired trimmed flight condition values. The states for the aircraft considered were

$$X = (u, v, w, p, q, r, \phi, \theta, \psi, x, y, H)^T, \tag{12}$$

where (u, v, w) are the body velocity components, (p, q, r) are the body frame angular velocities, and (ϕ, θ, ψ) are the Euler angles that describe the roll, pitch, and yaw of the body, respectively. (x, y, H) represents the position of the body relative to the inertial frame. The control inputs are

$$U = (\delta_e, \delta_a, T_n, \phi_T, \theta_T, \psi_T)^T, \tag{13}$$

where δ_e is the elevator deflection, δ_a is the aileron deflection, and T_n is the thrust, assumed to align with the longitudinal axis, hence producing no moments. $(\phi_T, \theta_T, \psi_T)$, respectively, are the roll, pitch, and yaw of the abdomen/tail, relative to the body frame.

The flight dynamics model of the conceptual aircraft was simulated in the MATLAB/Simulink environment [74] and the physical properties of the conceptual aircraft with articulated tail/abdomen are shown in Table 1. The fore and hind wing areas were combined to form a single wing with elevons. Depending on the desired flight control response, elevons operate in concert to function as elevators and ailerons. The aircraft was initially trimmed for steady level flight using the elevons as an elevator before the maneuver was initiated with tail/abdomen actuation.

Table 1. Aircraft model physical properties [71].

Parameter	Value	Parameter	Value
m^B (kg)	0.325	m^T (kg)	0.06
Body length, l_B (m)	0.3	Tail length, $l_{T_{max}}$ (m)	0.4
Max. body diameter, $d_{B_{max}}$ (m)	0.14	Tail diameter, d_T (m)	0.05
$J_{xx_b}^B$ (kg m^2)	0.00187	b_{ref} (m)	1.4
$J_{yy_b}^B$ (kg m^2)	0.01117	c_{ref} (m)	0.19434
$J_{zz_B}^B$ (kg m^2)	0.00934	S (m^2)	0.26865
cg_b^B (m)	[−0.064; 0; 0.003]	ARP (m)	[0.025; 0; 0]

Simulation Results and Analysis

In steady state flight, translational and rotational velocity components in the body frame are constant, so $\dot{u} = \dot{v} = \dot{w} = \dot{p} = \dot{q} = \dot{r} = 0$. All analyses we present assume no sideslip. Trim analysis was carried out for level cruise at speed $V_0 = 10$ m/s and a height of $H_0 = 100$ m. The tail position was fixed at ($\rho_{tj_0} = 0.264$ m), ($\rho_{jb_0} = 0.3$ m), and the abdomen/tail was undeflected ($\theta_{T_0} = 0$). The model was trimmed without actuator dynamics using the Matlab/Simulink linear analysis tool [74]. The trim results for level cruise are presented in Table 2.

Table 2. Results for steady level flight trim conditions.

V_0 (m/s)	H_0 (m)	$\rho_{jb_{x0}}$ (m)	$\rho_{tj_{x0}}$ (m)	θ_{T_0} (°)	θ_0 (°)	δ_{e_0} (°)	T_{n_0} (N)
10	100	0.264	0.3	0	−0.252	−1.79	0.825

A quasi-steady analysis for a pull down maneuver was performed to confirm the response of the abdominal pitch angle (θ_T) to the control signal input. At altitude 100 m and speed 10 m/s, the duration of the total analysis was 10 s. To examine the longitudinal flight of the aircraft, the abdominal pitch angle control signal was applied as a step response by adding $-5\,^{\circ}$ to the trim value of level cruise. Trimmed level cruise thrust was maintained. Figure 10 shows the time course of flight path, altitude, airspeed, and pitch angle in the simulation. The flight path indicates a decreasing altitude from level cruise.

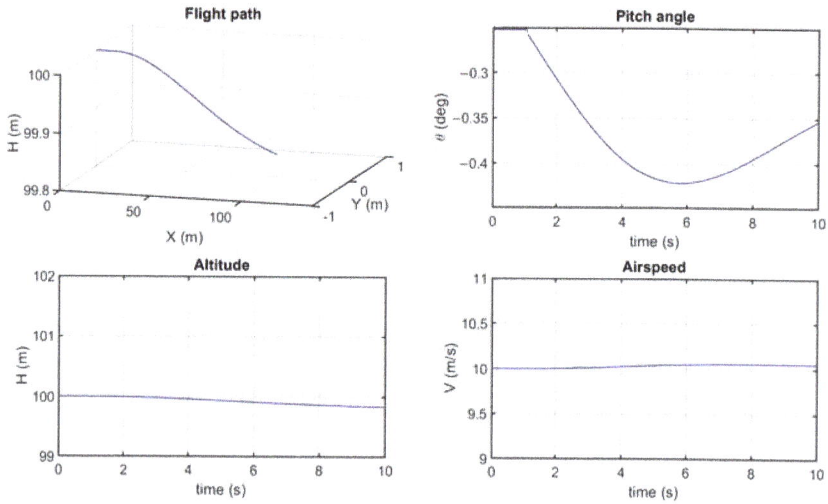

Figure 10. Pull down flight simulation results.

Lateral-directional flight performance was investigated for an abdominal yaw angle (ψ_T) control input. While maintaining all other controls at their level cruise trim values, an abdominal yaw angle doublet input (symmetrical pulse of equal magnitude and opposite in sign) control signal of 0.5°, of duration 0.5 s, was applied 1 s into the simulation. The duration of the analysis was 10 s. Longitudinal and lateral coupling flight performance of the aircraft was examined. The flight path, altitude, airspeed, and Euler angles of the simulation are presented in Figure 11. It is clear that the aircraft responds by starting to turn with decreasing altitude.

3.3. *Integrated Function of Morphological Modifications in Flight*

The wings are not the sole contributors to the maneuverability exhibited by insects. Morphological modifications in flight, such as articulation of the abdomen, are used to improve maneuverability or as an alternative source of control in certain scenarios. For instance, despite not having aerosurfaces for roll, pitch, and yaw control such as conventional aircraft, two modes of turning flight have been observed in dragonflies [75]. In the conventional mode of turn, the dragonfly produces more lift and thrust on one side than the other. It then rolls into a bank such that the lift vector has a sideward component, resulting in a turn equivalent to that of a fixed wing aircraft. The "yaw" turn, which is the second turning mode, is achieved without banking; rather, the longitudinal axis of the dragonfly turns, which has now been modelled and simulated, as shown in Figure 11.

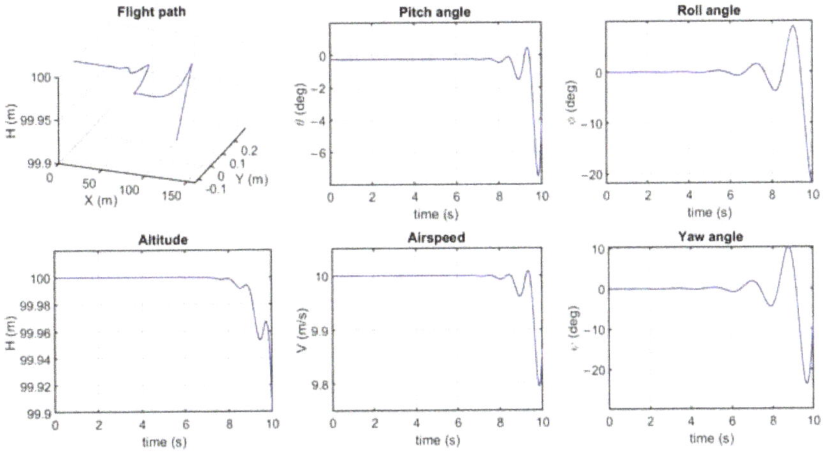

Figure 11. Turning flight simulation results.

Energy savings have been recorded to achieve longitudinal balance and pull up maneuvers by manipulating moments using abdominal articulation as alternative control effectors to conventional aerosurfaces [71]. Overall, a 4–5% average in propulsive power savings was recorded using abdominal movements for the correction of comparatively mild imbalances in steady cruise. Generally, in manned fighter aerial combat, the aircraft with the ability to maintain a higher specific energy and excess power has a greater maneuver advantage [76]. We have reported [71] an average of 1% more specific energy and 1.83% more excess power using abdominal deflection to initiate a steady pull up maneuver, in comparison with using the elevator. In most small electrically powered aircraft, at least 50% of the power available is used for propulsion [77]. Using morphological modifications during flight can save energy and substantially reduce peak power requirements.

3.4. *Lessons from Abdominal Control*

Dragonflies can grapple and carry prey and females. They can also catch prey with their mouth-parts well ahead of their centre of gravity. With no abdomen, they would have no means to immediately balance the load, losing valuable lift simply to induce an aerodynamic torque to offset mass. Dragonfly flight, like all flying wings, is on the edge of being unstable [78]. A forward shift of the centre of gravity with respect to the aerodynamic centre would certainly allow them to recover stability. Analogous situations for a drone can also be conceived of, particularly if it can drop and collect loads, or if it consumes fuel. Careful consideration shows that the dragonfly and insects generally have their shape and articulation for multiple reasons and that dynamic control of balance is particularly useful. It is challenging to conceive of an alternate solution that is more efficient.

4. Wing Beat System Dynamics

Appreciation of the criticality of each part of a dragonfly requires an understanding of power in the system. An understanding of the power and control of dynamics unravels the reason for the evolution of rigid wings [79] and explains why the dragonfly has retained individual control of each wing beat [75,80–84] and why a technological dragonfly should probably use controlled reciprocating actuators rather than rotating actuators.

In previous studies [79], we have shown that the stiffness of the wing–thorax–actuator structure affects the system level performance of the aerial platform. We compared a pliable mechanism operating in a resonant mode with a more rigid mechanism, in the three dominant modes of flight: hover, glide, and manoeuvre. In hover, wings are driven against inertial, aerodynamic, and elastic forces. The power required to sustain hover can be expressed as the sum of power required to overcome aerodynamic forces (P_{Aero}) and the power required to sustain the continuous acceleration and deceleration of the wing, ($P_{Inertia}$), i.e.,

$$P_{Hover} = P_{Aero} + P_{Inertia} \tag{14}$$

Some studies have observed that the use of elastic mechanisms can recover power required to overcome $P_{inertia}$ [85–87]. We have demonstrated a negative impact on performance in gliding and agile flight for systems employing low rigidity and resonant system dynamics to sustain hover. In our study, the inertial and elastic forces were modelled as linear functions, proportional to angular acceleration and position, respectively, i.e.,

$$ElasticForce = K\phi \tag{15}$$

$$InertialForce = I\ddot{\phi} \tag{16}$$

where K and I are the spring constant and inertia, respectively, whilst ϕ, the flapping angle, is the relative angular difference between the minimum and maximum position of the wing during flapping. The aerodynamics was modelled using a quasi-steady blade element model, the details of which are found in [79]. Gains of as low as 14% in hover efficiency were observed through the employment of resonance, which were reduced even further when considering a real implementation. A 50–70% reduction in manoeuvring limits was observed when employing resonance compared to a system without an elastic recovery system (i.e., $K = 0$). Figure 12 shows the instantaneous response of the normalised damping force to a unit force input. From this, we observe that by implementing elastic storage, the ability of the system to modulate aerodynamic forces was reduced, even though these forces are critical to manoeuvrability. Additionally, critical glide speed was reduced. Critical glide speed is the speed at which aeroelastic instabilities begin to occur as wings twist and flutter. Fung [88] presented a set of criteria relating critical speed to the stiffness of the system based on Kussner's formula. According to Fung, critical speed is proportional to the square root of stiffness, $\sqrt{K}$.

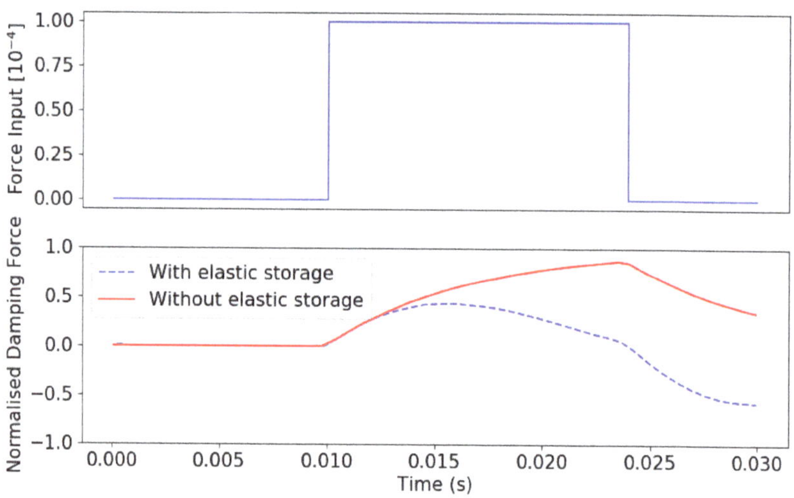

Figure 12. Response of two damped systems with and without elastic storage mechanisms to an instantaneous step input (**Top**). The aerodynamic forces are represented by a quadratic function. The dashed and solid lines represent systems with and without elastic storage, respectively. The damping force (**Bottom**) is the performance metric for determining the manoeuvrability of the system.

These findings support the argument that dragonflies have evolved an apparatus that allows them to efficiently expend energy towards modulating aerodynamic forces as opposed to recovering inertial loss incurred by having heavy wings. The ability to modulate aerodynamic forces is what makes the dragonfly an outstanding natural flyer. This is supported by studies that show that the dragonfly has an aerodynamics-dominated design as opposed to an inertia-dominated design [11,89–91].

In addition to having a relatively rigid wing–actuator system, another factor that allows dragonflies to efficiently modulate aerodynamic forces is the control of wing kinematics. Studies have shown that dragonflies have an exceptional amount of control over the dynamics of their wing to the extent that they are able to actively modulate wing stroke, flapping frequency, stroke plane angle, the upstroke-to-downstroke ratio, the phase between fore and hind wings, and the angle of attack of the wing [75,80–84,92,93]. Observations of dragonflies in nature show that they employ various wing kinematics to effect different manoeuvres. However, most robotic flapping platforms rely on rotary mechanisms with transmissions that highly constrain wing kinematics [94–99], whilst other designs rely on resonant mechanisms that constrain them to efficient flight only when hovering [85,95,100,101].

Therefore, we make the case that the design of a wing–thorax–actuator system that is able to mimic the capabilities of the dragonfly will require relatively rigid structures that do not rely on resonant modes for achieving flight. In addition, it requires a reciprocating actuator, or set of actuators, that have a high level of control over wing kinematics.

Lessons from System Dynamics

We have shown that the balance of aerodynamic and inertial power use in the system is the key to efficiency. In the absence of resonance, efficiency requires that any kinetic energy generated by the wing actuation system should be used to create aerodynamic force. Resonance in the form of elasticity in the wing or flapping mechanism can recover some kinetic energy at the expense of control, as the mechanism then takes on the characteristics of a flywheel that must be modulated to control the system. Low inertia and stiffness are very important, as it supports the per-wingbeat control enjoyed by dragonflies. We noted that flexibility can limit high speed flight by allowing aerostructural divergence.

5. Actuation

Examples of actuation for technological flapping wing systems have commonly centred around rotary electromagnetic actuators (motors), piezoelectric actuators, and dielectric elastomer actuators. While attempting to mimic biology with available technical solutions, these forms of actuation impose constraints that distance the engineered solution from the biological system. Rotary actuators require a transmission to convert rotation to reciprocation. This typically results in mechanisms with a fixed flapping amplitude, where the motor rotation rate controls flapping frequency. Each element in the drive train has rotational inertia, limiting the rate of change of the fixed flapping amplitude. A limited stroke angle and a limited ability to change rate quickly indicate limited control options. Although some success has been achieved with rotational systems, they are so far removed from the way a dragonfly is actuated that further analysis is not useful when considering a biomimetic design [102–105].

The use of servo actuation, which couples the wing rotation to a motor through a gearbox, has become popular and has demonstrated liftoff capability [106,107]. Due to the large torques required to generate flapping motion, the small motors require a reduction gearbox to amplify the torque at the cost of increasing the rotor speed. This increases the inertial load of each wing beat, resulting in a return spring to improve efficiency and allow the system to operate around resonance. This system's limitation is incorporating a large enough return spring without exceeding the gearbox maximum strength. We have already shown that high inertia is a very limiting situation.

Piezoelectric actuators have restricted operational bandwidth, requiring them to be used at their resonant frequency for high efficiency and power density. This also means that the change of operating point to produce more or less power comes at a high price in efficiency [108]. The necessity to operate piezo and electroactive polymer actuators at high voltages comes with additional complexity of control and power electronics. Often, systems under test are tethered, allowing a craft to achieve liftoff without fully addressing the question of drive electronics. The need to generate a high voltage source from a low voltage source costs substantial mass and adds inefficiency. There is a substantial gap between the state of the art of these actuators and a viable flying system.

Our analysis will focus on linear electric motors, the simplest case of which is a solenoid. The electrical requirements of a coil-based actuator are vastly different to that of a piezoelectric actuator. Linear electric motors can be designed to operate at the voltages of common batteries. The power electronics design is also simplified with the use of a simple solid state H-bridge. This converts the DC supply to an AC source, which can be used to control the reciprocation of the actuator. The losses inside the H-bridge transistors are proportional to the resistance of the gates used, which are typically low. However, due to the dominance of ohmic losses in the actuators, and their low impedance, the losses across the gates will be significant.

Insects actuate their wings using muscles that act either directly on the wing base or indirectly by deforming the thoracic exoskeleton [109]. Indirect actuation is found in insects that use asymmetrical flight muscles, allowing them to flap at higher frequencies but with elasticity tuned to improve efficiency [110]. Odonata have a direct connection to symmetrical flight muscles, causing them to flap more slowly but with more control over the entire flapping cycle.

A direct actuation electromechanical system weighing 4.0 g was developed using magnets as a "virtual spring" to allow for high frequency flapping at resonance [111]. This system was capable of lift-off with a lift-to-weight ratio of 1.25; however, this system was uncontrolled [112]. Attempts to use reciprocating electromagnetic actuators to achieve liftoff have been successful while requiring significant electrical power [113]. Although the electrical performance of this actuator was low, it provides a compelling analog to biological muscles and can be designed, tested, and iterated using existing industrial equipment and controlled with simple electronics.

Actuator configurations have typically consisted of a single magnet and coil as shown in Figure 13, left. Our recent work has looked at the performance of actuators with multiple magnets and coils shown in Figure 13, right [114]. These actuators utilise two magnets with axially opposing magnetic fields. The additional coil provides further displacement while reducing peak currents that cause high ohmic losses. The additional magnet provides a higher peak magnetic field, further reducing input current for a fixed force.

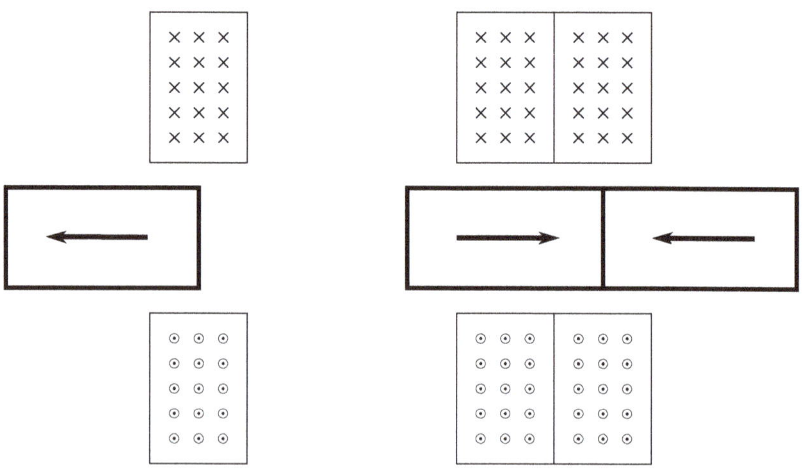

Figure 13. The force profile of a single magnet within a coil (**left**) and the force profile of a dual magnet within a coil (**right**).

5.1. Actuator Modelling

A linear electromagnetic actuator consists of only two major components, a coil of conductive wire and a permanent magnet. When the field of the magnet interacts with the magnetic field of the coil, the actuator generates a force. The force of the actuators can be estimated with an integral model of the actuator adapted from Furlani and Song [115,116] as demonstrated by us in [117]. Their force profile is characterised by

$$K_f(z) = \frac{AB_rB_z(z)}{\mu_0} \tag{17}$$

where B_r is the remnant field of the magnet, $B_z(z)$ is the magnetic field produced by the coil at displacement z, $A = \pi M_d^2/4$, and μ_0 is the permeability of free space. The magnetic field of the coil can be determined by taking the integral of the current loops of the coil along its axial length [115,116], (18)

$$B_z(z) = \int_0^L \int_{\frac{D_1}{2}}^{\frac{D_2}{2}} \frac{\mu_0 N i r^2}{2\pi(r^2+z^2)^{3/2}}\, dr\, dz \tag{18}$$

$$B_z(z) = \frac{\mu_0 N i}{2L(D_1-D_2)}(L+2z)\ln\alpha_1 + \frac{\mu_0 N i}{2L(D_1-D_2)}(L-2z)\ln\alpha_2 \tag{19}$$

where

$$\alpha_1 = \frac{D_2+\sqrt{D_2^2+(L+2z)^2}}{D_1+\sqrt{D_1^2+(L+2z)^2}}$$

$$\alpha_2 = \frac{D_2+\sqrt{D_2^2+(L-2z)^2}}{D_1+\sqrt{D_1^2+(L-2z)^2}}$$

These equations can be used to calculate the force profile of the actuators presented in Figure 13. Figure 14, left, shows the force profile of a 58 mg actuator when the coil current is set to one amp, allowing the force profile to be demonstrated in newtons per amp over the displacement of the magnet relative to the coil. To increase the force output, the actuator would require an increased input current. The losses in the coils, however, are proportional to the square of the current, causing significant losses when high forces are required. A limitation of using a single coil stems from the force profiles crossing zero newtons per amp. Thus, a narrow operating range is required to maintain control, leaving some of the magnetic field unused. Changing to a dual magnet actuator increases the peak force that can be generated per amp. Shown on the right in Figure 14 is the force profile of a two-magnet and two-coil actuator of the same total mass, as the actuator is shown on the left in Figure 14. The resulting force peak is reduced, yet displacement is larger.

We demonstrated that the performance of linear electromagnetic actuators was reduced due to operating significantly above the natural frequency of the mechanism [117]. Tuning this system reduced the power consumption by a factor of 7, from 15 kW/kg [113] (relative to the mass of the craft) to 2 kW/kg. Further modifications of the actuator shown in [114] have demonstrated that, for a fixed mass, an improved design in conjunction with operating at the natural frequency of the system can bring the power-to-weight ratio to between 910 and 260 W/kg. Resonant linear electromagnetic actuators can achieve an efficiency of over 15% at the milligram scale, equivalent to *Drosophila melanogaster* flight muscles, where power output is approximately 60 W/kg for sustained flight with muscle efficiency of around 10% [118].

The argument is illustrated in the approach taken in [119], where the fundamental assumption on wing design is that the operating frequency is the natural frequency of the wing actuation system. This inevitably results in examples where wings with higher inertia can perform better than lower inertia wings [120], which is evidence of conflicting requirements. Electromagnetic actuators allow for the choice to pursue the same avenue of high stiffness mechanisms tuned for resonant operations. However, taking inspiration from nature drives the development of stiff, light wings that reduce the reliance on resonance to broaden the efficient operating range and allow for precise control.

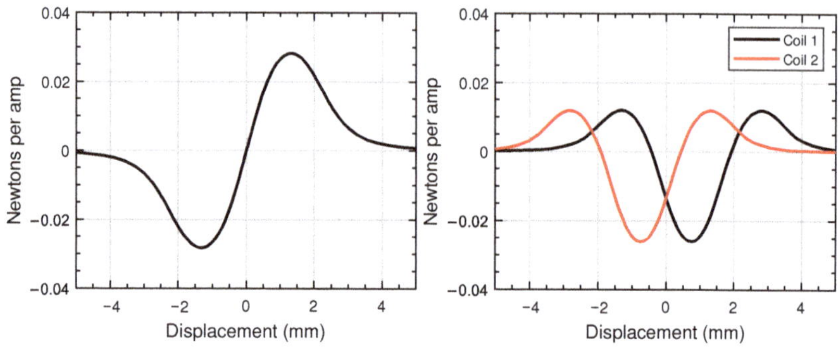

Figure 14. The force profile of a single magnet within a coil (**left**) and the force of a two-magnet and two-coil actuator (**right**).

5.2. Lessons from Actuators

A MAV designed to use the same path to efficient and powerful flight as the dragonfly might be able to use electromechanical actuators that have a large degree of design freedom, minimal transmission losses, low inertia, and very modest electronics requirements. We have established both theoretically and experimentally that performance comparable to insect muscle is attainable, both for efficiency and force development.

No studies have been done with wings of the same weight and stiffness as dragonfly wings. Most artificial wings are around an order of magnitude heavier. The weight difference in these artificial wings dramatically changes the balance of power consumption in the system, tilting them heavily towards needing resonance for efficiency, thus diverging from the dragonfly biological exemplar in a fundamental way.

6. Discussion

The dragonfly is one part of the solution space for making a flapping wing drone. It happens to be apparently aerodynamically efficient and maneuverable, but other factors such as cost and complexity may make it less desirable than other solutions. We will reduce what we know about the solution, drawing from the lessons of each system analysed.

The wings are to be light and stiff, to the limit of available technology, ideally lighter than 2% of the weight of the craft. This maximises the amount of instantaneous control that the actuators have over the wings and increases the speed at which aerostructural divergence limits flight. The wings should have high aerodynamic performance through a combination of a two-dimensional planform, for example a high aspect ratio, and three-dimensional surface features that produce lift and retain an attached flow to high angles of attack. This combination of features leads to a heavily damped, low-inertia load on the actuators.

The actuators managing this wing should operate across a broad range of frequencies without a loss of efficiency. The low residual kinetic energy in the system means that the modulation of individual wing beats becomes achievable, including intermittent flapping, variations in amplitude and frequency, and variations in force profiles within a wing beat. These features indicate an actuator with a substantial scope for control, a direct drive, low inertia, and minimal elastic behaviour at the frequencies in question.

The drone could serve multiple functions, including carrying and dropping payload. It might consume fuel from a fuel cell (the dragonfly analogue is fat deposits). In an outdoor environment, it might become wet. These events will cause shifts in the centre of gravity that need to be compensated with energetically expensive aerodynamic moments from the wings, or an internal shift of mass to re-balance. Flying systems should not carry dead weight for balance (although many do, including many fixed wing drones); yet, perhaps if some vital functions and fuel were formed into a high moment beam with articulation from near the centre of mass, rebalance could be rapidly achieved with little additional weight. Having made such a mechanism, it might as well be used for manoeuvre and for reasserting stability on demand by moving the centre of gravity ahead of the aerodynamic centre. This structure might look like an abdomen. Other concepts might be devised, but it seems very attractive and simple compared to other options.

There is clearly another end of the flapping wing drone solution space (and no doubt a multidimensional space in between). Possibly a smaller "fly-like" solution has wings that are heavier and less efficient, either by design or as a function of scale. The mechanisms might rely on resonance to recover kinetic energy from the wings and might exploit flexible parts of the fuselage to store energy in an elastic form. An increased use of low *Re* aerodynamic energy recapture by wings might be necessary for flight to be possible. Higher wing beat frequencies would compensate for the lack of individual wing control. Wing actuation might be indirect and in a narrow frequency band. The actuated abdomen might remain because of the need for weight shift for steering because of the reduced control over the resonant wing mechanism. This solution would have lower performance and a narrower repertoire of flight modes but would have fewer parts.

Flapping wings for lift and propulsion appear to lead naturally to biomimetic outcomes. Some features that might appear whimsical, such as venation patterns on wings and an articulated aft section, might be a practical necessity in future designs. This might reflect an ingrained influence from the biological systems that surround us. Certainly flapping wing drone designers could consider what sort of solution they are creating and look carefully at the features of existing biological examples. Figure 15 is intended to provoke thought, considering all issues and all roles; if not this, then what?

Figure 15. A whimsical 200 mm wingspan biomimetic dragonfly drone in flight. Large, directly driven, corrugated wings are as efficient as fixed wings on drones with a much larger wingspan. Direct actuation allows for exquisite control of hover, while being able to soar effortlessly on rising currents of air. The abdomen containing the fuel cell doubles as a rudder and ballast, allowing the vehicle to collect and deliver awkward unbalanced loads while flying on the efficient edge of instability.

7. Conclusions

Studies of insects have revealed the principles of most of the systems they use for flight. From this knowledge, it appears that a drone capable of dragonfly-like performance using flapping wings and capable of flexible missions is very likely to have most of the external features of the dragonfly. The wings will be highly optimised with stiff structures and advanced aerodynamic optimisations. The beating of the wings is likely to be directly and individually driven by tuned actuators, also with very little flexibility or inertia. To manage the variations in balance that occur from changes over time or between configurations, a means is required to dynamically alter balance, with the equivalent of an abdomen serving this purpose.

We have shown through analysis of the principles of insect flight systems that future drones that attempt to harness some of the coveted characteristics of dragonflies are likely to be constrained to use many of the same solutions and even their anatomical features.

Author Contributions: Conceptualisation, J.C.; Data curation, N.C.; Formal analysis, Nasim Chitsaz, B.M., T.O., J.-M.K., and E.A.; Funding acquisition, J.C.; Investigation, N.C., B.M., T.O., J.-M.K., and T.M.; Methodology, J.C.; Project administration, T.M.; Resources, J.C. and J.-M.K.; Supervision, J.C. and E.A.; Writing—original draft, J.C.; Writing—review & editing, J.C., N.C., B.M., T.O., J.-M.K., T.M., and E.A. All authors have read and agreed to the published version of the manuscript.

Funding: N.C. and B.M. were supported by the Australian Government Research Training Program (RTP) Scholarship. T.O. was supported by the Australian Government Research Training Program international (RTPi) scholarship. N.C. was supported by the William T. Southcott Scholarship from the University of South Australia.

Institutional Review Board Statement: Not applicable.

Informed Consent Statement: Not applicable.

Data Availability Statement: The 3D geometry of dragonfly wings in zipped files are available at Dryad, https://doi.org/10.5061/dryad.6t1g1jwtt (accessed on 27 March 2021). Each zipped file includes the wing models of one species of Odonata.

Acknowledgments: We would like to thank the South Australian Museum, Adelaide, and the South Australian Research and Development Institute (SARDI) Entomology Science Program for granting access to their collections of dragonflies used in this study.

Conflicts of Interest: The authors declare no conflict of interest.

References

1. Corbet, P.S. *Dragonflies: Behaviour and Ecology of Odonata*; Harley Books: Manunda, Australia, 1999.
2. Silsby, J. *Dragonflies of the World*; CSIRO Publishing: Melbourne, Australia, 2001.
3. Chahl, J.; Dorrington, G.; Premachandran, S.; Mizutani, A. The dragonfly flight envelope and its application to micro uav research and development. *IFAC Proc. Vol.* **2013**, *46*, 231–234. [CrossRef]
4. Marden, J.H. Maximum lift production during takeoff in flying animals. *J. Exp. Biol.* **1987**, *130*, 235–258.
5. Olberg, R.M. Visual control of prey-capture flight in dragonflies. *Curr. Opin. Neurobiol.* **2012**, *22*, 267–271. [CrossRef]
6. Olberg, R.; Worthington, A.; Venator, K. Prey pursuit and interception in dragonflies. *J. Comp. Physiol. A* **2000**, *186*, 155–162. [CrossRef]
7. Combes, S.; Rundle, D.; Iwasaki, J.; Crall, J.D. Linking biomechanics and ecology through predator-prey interactions: Flight performance of dragonflies and their prey. *J. Exp. Biol.* **2012**, *215*, 903–913. [CrossRef]
8. Mizutani, A.; Chahl, J.S.; Srinivasan, M.V. Active motion camouflage by dragonflies. *Nature* **2003**, *423*, 604. [CrossRef]
9. Carey, N.E.; Ford, J.J.; Chahl, J.S. Biologically inspired guidance for motion camouflage. In Proceedings of the 2004 5th Asian Control Conference (IEEE Cat. No. 04EX904), Melbourne, VIC, Australia, 20–23 July 2004; Volume 3, pp. 1793–1799.
10. Wakeling, J.; Ellington, C. Dragonfly flight. I. Gliding flight and steady-state aerodynamic forces. *J. Exp. Biol.* **1997**, *200*, 543–556.
11. Azuma, A.; Watanabe, T. Flight performance of a dragonfly. *J. Exp. Biol.* **1988**, *137*, 221–252.
12. Newman, B. Model test on a wing section of a dragonfly. In *Scale Effects in Animal Locomotion*; Academic Press: London, UK, 1977; pp. 445–477.
13. May, M.L. A critical overview of progress in studies of migration of dragonflies (Odonata: Anisoptera), with emphasis on North America. *J. Insect Conserv.* **2013**, *17*, 1–15. [CrossRef]
14. Scorer, R. The nature of convection as revealed by soaring birds and dragonflies. *Q. J. R. Meteorol. Soc.* **1954**, *80*, 68–77. [CrossRef]

15. Purcell, E.M. The efficiency of propulsion by a rotating flagellum. *Proc. Natl. Acad. Sci. USA* **1997**, *94*, 11307–11311. [CrossRef]
16. Srinivasan, M.V.; Chahl, J.S.; Weber, K.; Venkatesh, S.; Nagle, M.G.; Zhang, S.W. Robot navigation inspired by principles of insect vision. *Robot. Auton. Syst.* **1999**, *26*, 203–216. [CrossRef]
17. Cayley, G. On Aërial Navigation. *Annu. Rep. Aeronaut. Soc. Great Br.* **1876**, *11*, 60–94. [CrossRef]
18. Lilienthal, O. Practical experiments for the development of human flight. In *The Aeronautical Annual*; W.B. Clarke Company: San Jose, CA, USA, 1896; pp. 7–20.
19. McFarland, M.W.; Renstrom, A.G. The Papers of Wilbur and Orville Wright. *Q. J. Curr. Acquis.* **1950**, *7*, 22–34.
20. Stange, G. The Ocellar Component of Flight Equilibrium Control in Dragonflies. *J. Comp. Physiol.* **1981**, *141*, 335–347. [CrossRef]
21. Chahl, J.; Mizutani, A. Biomimetic attitude and orientation sensors. *Sens. J. IEEE* **2012**, *12*, 289–297. [CrossRef]
22. Stange, G.; Howard, J. An Ocellar Dorsal Light Response in a Dragonfly. *J. Exp. Biol.* **1979**, *83*, 351–355.
23. Stange, G.F.; Stowe, S.; Chahl, J.S.; Massaro, A. Anisotropic Imaging in the Dragonfly Median Ocellus: A Matched Filter for Horizon Detection. *J. Comp. Physiol.* **2002**, *188*, 455–467.
24. Thakoor, S.; Morookian, J.M.; Chahl, J.; Hine, B.; Zornetzer, S. BEES: Exploring mars with bioinspired technologies. *Computer* **2004**, *37*, 38–47. [CrossRef]
25. Salami, E.; Ward, T.A.; Montazer, E.; Ghazali, N.N.N. A review of aerodynamic studies on dragonfly flight. *Proc. Inst. Mech. Eng. Part J. Mech. Eng. Sci.* **2019**, *233*, 6519–6537. [CrossRef]
26. Sane, S.P. The aerodynamics of insect flight. *J. Exp. Biol.* **2003**, *206*, 4191–4208. [CrossRef]
27. Cheng, X.; Sun, M. Very small insects use novel wing flapping and drag principle to generate the weight-supporting vertical force. *arXiv* **2018**, arXiv:1807.05629.
28. Shyy, W.; Aono, H.; Kang, C.K.; Liu, H. *An Introduction to Flapping Wing Aerodynamics*; Cambridge University Press: Cambridge, UK, 2013; Volume 37.
29. Kesel, A.B. Aerodynamic characteristics of dragonfly wing sections compared with technical aerofoils. *J. Exp. Biol.* **2000**, *203*, 3125–3135.
30. Hu, Z.; McCauley, R.; Schaeffer, S.; Deng, X. Aerodynamics of dragonfly flight and robotic design. In Proceedings of the 2009 IEEE International Conference on Robotics and Automation, Kobe, Japan, 12–17 May 2009; pp. 3061–3066.
31. Luo, Y.; He, G.; Liu, H.; Wang, Q.; Song, H. *Aerodynamic Performance of Dragonfly Forewing–Hindwing Interaction in Gliding Flight*; IOP Conference Series: Materials Science and Engineering; IOP Publishing: Bristol, UK, 2019; Volume 538, p. 012048.
32. Hefler, C.; Noda, R.; Qiu, H.; Shyy, W. Aerodynamic performance of a free-flying dragonfly—A span-resolved investigation. *Phys. Fluids* **2020**, *32*, 041903. [CrossRef]
33. Bunget, G.; Seelecke, S. BATMAV: A biologically inspired micro-air vehicle for flapping flight: kinematic modeling. In *Active and Passive Smart Structures and Integrated Systems 2008*; International Society for Optics and Photonics: Bellingham, WA, USA, 2008; Volume 6928, p. 69282F.
34. Chitsaz, N.; Marian, R.; Chahl, J. Experimental method for 3D reconstruction of Odonata wings (methodology and dataset). *PLoS ONE* **2020**, *15*, e0232193. [CrossRef]
35. Chitsaz, N.; Chahl, J. Current knowledge of corrugated dragonfly wing structures and future measurement methodology. In Proceedings of the AIAC18: 18th Australian International Aerospace Congress (2019): HUMS—11th Defence Science and Technology (DST) International Conference on Health and Usage Monitoring (HUMS 2019): ISSFD—27th International Symposium on Space Flight Dynamics (ISSFD), Melbourne, Australia, 24–28 February 2019; Engineers Australia, Royal Aeronautical Society: Melbourne, Australia, 2019; pp. 412–417.
36. Oniga, E.; Chirilă, C.; Stătescu, F. Accuracy assessment of a complex building 3d model reconstructed from images acquired with a low-cost Uas. *Int. Arch. Photogramm. Remote. Sens. Spat. Inf. Sci.* **2017**, *42*, 551. [CrossRef]
37. Chitsaz, N.; Marian, R.; Chitsaz, A.; Chahl, J.S. Parametric and Statistical Study of the Wing Geometry of 75 Species of Odonata. *Appl. Sci.* **2020**, *10*, 5389. [CrossRef]
38. Rees, C.J. Form and function in corrugated insect wings. *Nature* **1975**, *256*, 200–203. [CrossRef]
39. Rudolph, R. Aerodynamic properties of Libellula quadrimaculata L.(Anisoptera: Libellulidae), and the flow around smooth and corrugated wing section models during gliding flight. *Odonatologica* **1978**, *7*, 49–58.
40. Okamoto, M.; Yasuda, K.; Azuma, A. Aerodynamic characteristics of the wings and body of a dragonfly. *J. Exp. Biol.* **1996**, *199*, 281–294.
41. Gaurav, S.; Jain, K.K. Numerical investigation of fluid flow and aerodynamic performance of a dragonfly wing section for micro air vehicles (MAVs) application. *Int. J. Innovat. Scient. Res.* **2014**, *92*, 285–292.
42. Flint, T.; Jermy, M.; New, T.; Ho, W. Computational study of a pitching bio-inspired corrugated airfoil. *Int. J. Heat Fluid Flow* **2017**, *65*, 328–341. [CrossRef]
43. Xiang, J.; Du, J.; Li, D.; Liu, K. Aerodynamic performance of the locust wing in gliding mode at low Reynolds number. *J. Bionic Eng.* **2016**, *13*, 249–260. [CrossRef]
44. Khurana, M.and Chahl, J. A computational fluid dynamic study of a bioinspired corrugated aerofoil for micro air vehicles. In Proceedings of the AIAC15: 15th Australian International Aerospace Congress, Melbourne, Australia, 25-28 February 2013; p. 126.
45. Kok, J.; Chahl, J. Optimization of the leading edge segment of a corrugated wing In *Bioinspiration, Biomimetics, and Bioreplication 2014*; International Society for Optics and Photonics: San Diego, CA, USA, 2014; p. 905517.

46. Meng, X.G.; Sun, M. Aerodynamic effects of wing corrugation at gliding flight at low Reynolds numbers. *Phys. Fluids* **2013**, *25*, 071905. [CrossRef]
47. Chitsaz, N.; Siddiqui, K.; Marian, R.; Chahl, J. An experimental study of the aerodynamics of micro corrugated wings at low Reynolds number. *Exp. Therm. Fluid Sci.* **2021**, *121*, 110286. [CrossRef]
48. Heinrich, B.; Casey, T.M. Heat transfer in dragonflies:'fliers' and 'perchers'. *J. Exp. Biol.* **1978**, *74*, 17–36.
49. Dyhr, J.P.; Cowan, N.J.; Colmenares, D.J.; Morgansen, K.A.; Daniel, T.L. Autostabilizing airframe articulation: Animal inspired air vehicle control. In Proceedings of the 2012 IEEE 51st IEEE Conference on Decision and Control (CDC), Maui, HI, USA, 10–13 December 2012; pp. 3715–3720.
50. Jagnandan, K.; Higham, T.E. Lateral movements of a massive tail influence gecko locomotion: An integrative study comparing tail restriction and autotomy. *Sci. Rep.* **2017**, *7*, 1–8. [CrossRef] [PubMed]
51. Kohut, N.J. *Inertial and Aerodynamic Tail Steering of a Meso-Scale Legged Robot*; University of California: Berkeley, CA, USA, 2013.
52. Patel, A. Understanding the Motions of the Cheetah Tail Using Robotics. Ph.D. thesis, University of Cape Town, Cape Town, South Africa, 2015.
53. Camhi, J.M. Yaw-correcting postural changes in locusts. *J. Exp. Biol.* **1970**, *52*, 519–531.
54. Camhi, J.M. Sensory control of abdomen posture in flying locusts. *J. Exp. Biol.* **1970**, *52*, 533–537.
55. Götz, K.G.; Hengstenberg, B.; Biesinger, R. Optomotor control of wing beat and body posture in Drosophila. *Biological Cybern.* **1979**, *35*, 101–112. [CrossRef]
56. Zanker, J.M. How does lateral abdomen deflection contribute to flight control ofDrosophila melanogaster? *J. Comp. Physiol. A* **1988**, *162*, 581–588. [CrossRef]
57. Zanker, J. On the mechanism of speed and altitude control in Drosophila melanogaster. *Physiol. Entomol.* **1988**, *13*, 351–361. [CrossRef]
58. Luu, T.; Cheung, A.; Ball, D.; Srinivasan, M.V. Honeybee flight: A novel 'streamlining'response. *J. Exp. Biol.* **2011**, *214*, 2215–2225. [CrossRef]
59. Hinterwirth, A.J.; Daniel, T.L. Antennae in the hawkmoth Manduca sexta (Lepidoptera, Sphingidae) mediate abdominal flexion in response to mechanical stimuli. *J. Comp. Physiol. A* **2010**, *196*, 947–956. [CrossRef] [PubMed]
60. Dyhr, J.P.; Morgansen, K.A.; Daniel, T.L.; Cowan, N.J. Flexible strategies for flight control: An active role for the abdomen. *J. Exp. Biol.* **2013**, *216*, 1523–1536. [CrossRef] [PubMed]
61. Rüppell, G.; Hilfert-Rüppell, D.; Schneider, B.; Dedenbach, H. On the firing line—Interactions between hunting frogs and Odonata. *Int. J. Odonatol.* **2020**, *23*, 1–19. [CrossRef]
62. Sun, M.; Xiong, Y. Dynamic flight stability of a hovering bumblebee. *J. Exp. Biol.* **2005**, *208*, 447–459. [CrossRef] [PubMed]
63. Mou, X.; Sun, M. Dynamic flight stability of a model hoverfly in inclined-stroke-plane hovering. *J. Bionic Eng.* **2012**, *9*, 294–303. [CrossRef]
64. Liang, B.; Sun, M. Dynamic flight stability of a hovering model dragonfly. *J. Theor. Biol.* **2014**, *348*, 100–112. [CrossRef]
65. Cheng, C.; Wu, J.; Zhang, Y.; Li, H.; Zhou, C. Aerodynamics and dynamic stability of micro-air-vehicle with four flapping wings in hovering flight. *Adv. Aerodyn.* **2020**, *2*, 1–19. [CrossRef]
66. Jang, J.; Tomlin, C. Longitudinal stability augmentation system design for the DragonFly UAV using a single GPS receiver. In Proceedings of the AIAA Guidance, Navigation, and Control Conference and Exhibit, Austin, TX, USA, 11–14 August 2003; p. 5592.
67. Couceiro, M.S.; Ferreira, N.; Tenreiro Machado, J. Modeling and control of a dragonfly-like robot. *J. Control. Sci. Eng.* **2010**, doi:10.1155/2010/643045.
68. Nguyen, Q.V.; Chan, W.L.; Debiasi, M. Design, fabrication, and performance test of a hovering-based flapping-wing micro air vehicle capable of sustained and controlled flight. In Proceedings of the IMAV 2014: International Micro Air Vehicle Conference and Competition 2014, Delft, The Netherlands, 12–15 August 2014; Delft University of Technology: Delft, The Netherlands, 2014.
69. Kok, J.M. Active wing control in a Dragonfly-inspired Micro Air Vehicle. 2016. Available online: https://find.library.unisa.edu.au/primo-explore/fulldisplay?vid=UNISA&search_scope=All_Resources&docid=UNISA_ALMA11153485500001831 (accessed on 16 September 2020).
70. Du, C.; Xu, J.; Zheng, Y. Modeling and control of a dragonfly-like micro aerial vehicle. *Adv. Robot. Autom.* **2015**, *2*, 2.
71. Ogunwa, T.; McIvor, B.; Jumat, N.A.; Abdullah, E.; Chahl, J. Longitudinal Actuated Abdomen Control for Energy Efficient Flight of Insects. *Energies* **2020**, *13*, 5480. [CrossRef]
72. Zipfel, P.H. *Modeling and Simulation of Aerospace Vehicle Dynamics*; American Institute of Aeronautics and Astronautics: Washington, DC, USA, 2007.
73. Constrained Nonlinear Optimization Algorithms—MATLAB & Simulink—MathWorks. Available online: https://au.mathworks.com/help/optim/ug/constrained-nonlinear-optimization-algorithms.html (accessed on 16 September 2020).
74. MATLAB. *9.7.0.1190202 (R2019b)*; The MathWorks Inc.: Natick, MA, USA, 2019.
75. Alexander, D.E. Wind tunnel studies of turns by flying dragonflies. *J. Exp. Biol.* **1986**, *122*, 81–98. [PubMed]
76. Shaw, R.L. Fighter Combat. In *Tactics Maneuvering*; Naval Institute Press: Annapolis, MD, USA, 1985.
77. Dantsker, O.D.; Theile, M.; Caccamo, M. A High-Fidelity, Low-Order Propulsion Power Model for Fixed-Wing Electric Unmanned Aircraft. In Proceedings of the 2018 AIAA/IEEE Electric Aircraft Technologies Symposium (EATS), Cincinnati, OH, USA, 9–11 July 2018; pp. 1–16.

78. Nickel, K.; Wohlfahrt, M. *Tailless Aircraft in Theory and Practice*; Hodder Education: London, UK, 1994.
79. Kok, J.; Chahl, J. Systems-level analysis of resonant mechanisms for flapping-wing flyers. *J. Aircr.* **2014**, *51*, 1833–1841. [CrossRef]
80. Alexander, D.E. Unusual phase relationships between the forewings and hindwings in flying dragonflies. *J. Exp. Biol.* **1984**, *109*, 379–383.
81. Sviderskiĭ, V.; Plotnikova, S.; Gorelkin, V. Structural-functional peculiarities of wing appatatus of insects that have and do not have maneuver flight. *Zhurnal Evoliutsionnoĭ Biokhimii Fiziologii* **2008**, *44*, 545.
82. Ruppell, G. Kinematic analysis of symmetrical flight manoeuvres of Odonata. *J. Exp. Biol.* **1989**, *144*, 13–42.
83. Wang, Z.J. The role of drag in insect hovering. *J. Exp. Biol.* **2004**, *207*, 4147–4155. [CrossRef]
84. Wang, Z.J. Dissecting insect flight. *Annu. Rev. Fluid Mech.* **2005**, *37*, 183–210. [CrossRef]
85. Wood, R.J. The first takeoff of a biologically inspired at-scale robotic insect. *Robot. IEEE Trans.* **2008**, *24*, 341–347. [CrossRef]
86. Sahai, R.; Galloway, K.C.; Wood, R.J. Elastic element integration for improved flapping-wing micro air vehicle performance. *IEEE Trans. Robot.* **2012**, *29*, 32–41. [CrossRef]
87. Baek, S.S.; Ma, K.Y.; Fearing, R.S. Efficient resonant drive of flapping-wing robots. In Proceedings of the IROS 2009. IEEE/RSJ International Conference on Intelligent Robots and Systems, St. Louis, MO, USA, 11–15 October 2009; pp. 2854–2860.
88. Fung, Y.C. *An Introduction to the Theory of Aeroelasticity*; Includes Index; Wiley: New York, NY, USA, 1955; 490p.
89. Sun, M.; Lan, S.L. A computational study of the aerodynamic forces and power requirements of dragonfly (*Aeschna juncea*) hovering. *J. Exp. Biol.* **2004**, *207*, 1887–1901. [CrossRef]
90. Azuma, A.; Azuma, S.; Watanabe, I.; Furuta, T. Flight mechanics of a dragonfly. *J. Exp. Biol.* **1985**, *116*, 79–107.
91. May, M.L. Dragonfly flight: Power requirements at high speed and acceleration. *J. Exp. Biol.* **1991**, *158*, 325–342.
92. Kok, J.; Chahl, J. A low-cost simulation platform for flapping wing MAVs. In *SPIE Smart Structures and Materials+ Nondestructive Evaluation and Health Monitoring*; International Society for Optics and Photonics: San Diego, CA, USA, 2015; p. 94290L.
93. Kok, J.; Lau, G.; Chahl, J. On the Aerodynamic Efficiency of Insect-Inspired Micro Aircraft Employing Asymmetrical Flapping. *J. Aircr.* **2016**, 1–11. doi:10.2514/1.C033356. [CrossRef]
94. Lentink, D.; Jongerius, S.R.; Bradshaw, N.L. The scalable design of flapping micro-air vehicles inspired by insect flight. In *Flying Insects and Robots*; Springer: Berlin/Heidelberg, Germany, 2009; pp. 185–205.
95. Karpelson, M.; Whitney, J.P.; Wei, G.Y.; Wood, R.J. Energetics of flapping-wing robotic insects: towards autonomous hovering flight. In Proceedings of the 2010 IEEE/RSJ International Conference on Intelligent Robots and Systems (IROS), Taipei, Taiwan, 18–22 October 2010; pp. 1630–1637.
96. Ratti, J.; Vachtsevanos, G. A biologically-inspired micro aerial vehicle. *J. Intell. Robot. Syst.* **2010**, *60*, 153–178. [CrossRef]
97. Ratti, J.; Jones, E.; Vachtsevanos, G. Fixed frequency, variable amplitude (FiFVA) actuation systems for micro aerial vehicles. In Proceedings of the 2011 IEEE International Conference on Robotics and Automation (ICRA), Shanghai, China, 9–13 May 2011; pp. 165–171.
98. Ratti, J.; Vachtsevanos, G. Inventing a biologically inspired, energy efficient micro aerial vehicle. *J. Intell. Robot. Syst.* **2012**, *65*, 437–455. [CrossRef]
99. Roll, J.A. Bio-Inspired Flapper with Electromagnetic Actuation. Ph.D. thesis, Purdue University: West Lafayette, IN, USA, 2012.
100. Finio, B.M.; Wood, R.J. Open-loop roll, pitch and yaw torques for a robotic bee. In Proceedings of the 2012 IEEE/RSJ International Conference on Intelligent Robots and Systems (IROS), Vilamoura, Algarve, 7–12 October 2012; pp. 113–119.
101. Karpelson, M.; Wei, G.Y.; Wood, R.J. Milligram-scale high-voltage power electronics for piezoelectric microrobots. In Proceedings of the ICRA'09. IEEE International Conference on Robotics and Automation, Kobe, Japan, 12–17 May 2009; pp. 2217–2224.
102. De Croon, G.; De Clercq, K.; Ruijsink, R.; Remes, B.; De Wagter, C. Design, aerodynamics, and vision-based control of the DelFly. *Int. J. Micro Air Veh.* **2009**, *1*, 71–97. [CrossRef]
103. Ristroph, L.; Childress, S. Stable hovering of a jellyfish-like flying machine. *J. R. Soc. Interface* **2014**, *11*, 20130992. [CrossRef]
104. Palmer, J.L.; Jones, M.B.; Drobik, J. Design Elements of a Bio-Inspired Micro Air Vehicle. *IFAC Proc. Vol.* **2013**, *46*, 235–241. [CrossRef]
105. Phan, H.V.; Kang, T.; Park, H.C. Design and stable flight of a 21 g insect-like tailless flapping wing micro air vehicle with angular rates feedback control. *Bioinspir. Biomim.* **2017**, *12*, 036006. [CrossRef] [PubMed]
106. Campolo, D.; Azhar, M.; Lau, G.K.; Sitti, M. Can DC motors directly drive flapping wings at high frequency and large wing strokes? *IEEE/ASME Trans. Mechatron.* **2012**, *19*, 109–120. [CrossRef]
107. Hines, L.; Campolo, D.; Sitti, M. Liftoff of a motor-driven, flapping-wing microaerial vehicle capable of resonance. *IEEE Trans. Robot.* **2014**, *30*, 220–232. [CrossRef]
108. Steltz, E.; Avadhanula, S.; Fearing, R.S. High lift force with 275 Hz wing beat in MFI. In Proceedings of the 2007 IEEE/RSJ International Conference on Intelligent Robots and Systems, San Diego, CA, USA, 29 October–2 November 2007; pp. 3987–3992.
109. Josephson, R.K. Comparative physiology of insect flight muscle. In *Nature's Versatile Engine: Insect Flight Muscle Inside and Out*; Springer: Berlin/Heidelberg, Germany, 2006; pp. 34–43.
110. Ellington, C.P. Power and efficiency of insect flight muscle. *J. Exp. Biol.* **1985**, *115*, 293–304.
111. Roll, J.A.; Cheng, B.; Deng, X. An electromagnetic actuator for high-frequency flapping-wing microair vehicles. *IEEE Trans. Robot.* **2015**, *31*, 400–414. [CrossRef]

112. Roll, J.A.; Bardroff, D.T.; Deng, X. Mechanics of a scalable high frequency flapping wing robotic platform capable of lift-off. In Proceedings of the 2016 IEEE International Conference on Robotics and Automation (ICRA), Stockholm, Sweden, 16–21 May 2016; pp. 4664–4671.
113. Zou, Y.; Zhang, W.; Zhang, Z. Liftoff of an electromagnetically driven insect-inspired flapping-wing robot. *IEEE Trans. Robot.* **2016**, *32*, 1285–1289. [CrossRef]
114. McIvor, B.A.; Ogunwa, T.T.; Chahl, J.S. Geometry of efficient Electromagnetic Linear actuators for flapping wing MAVs. *Eng. Res. Express* **2020**, *2*, 045027. [CrossRef]
115. Furlani, E.P. *Permanent Magnet and Electromechanical Devices: Materials, Analysis, and Applications*; Academic Press: Cambridge, MA, USA, 2001.
116. Song, C.W.; Lee, S.Y. Design of a solenoid actuator with a magnetic plunger for miniaturized segment robots. *Appl. Sci.* **2015**, *5*, 595–607. [CrossRef]
117. McIvor, B.; Chahl, J. Energy efficiency of linear electromagnetic actuators for flapping wing micro aerial vehicles. *Energies* **2020**, *13*, 1075. [CrossRef]
118. Lehmann, F.O.; Dickinson, M.H. The changes in power requirements and muscle efficiency during elevated force production in the fruit fly Drosophila melanogaster. *J. Exp. Biol.* **1997**, *200*, 1133–1143. [PubMed]
119. Whitney, J.; Wood, R. Conceptual design of flapping-wing micro air vehicles. *Bioinspir. Biomimetics* **2012**, *7*, 036001. [CrossRef] [PubMed]
120. Chen, Y.; Ma, K.; Wood, R.J. Influence of wing morphological and inertial parameters on flapping flight performance. In Proceedings of the 2016 IEEE/RSJ International Conference on Intelligent Robots and Systems (IROS), Daejeon, Korea, 9–14 October 2016; pp. 2329–2336.

drones

MDPI

Article

Suitability of the Reforming-Controlled Compression Ignition Concept for UAV Applications

Amnon Eyal and Leonid Tartakovsky *

Technion—Israel Institute of Technology, Technion City, Haifa 3200003, Israel; amnone@campus.technion.ac.il
* Correspondence: tartak@me.technion.ac.il

Received: 1 September 2020; Accepted: 18 September 2020; Published: 22 September 2020

Abstract: Reforming-controlled compression ignition (RefCCI) is a novel approach combining two methods to improve the internal combustion engine's efficiency and mitigate emissions: low-temperature combustion (LTC) and thermochemical recuperation (TCR). Frequently, the combustion controllability challenge is resolved by simultaneous injection into the cylinder of two fuel types, each on the other edge of the reactivity scale. By changing the low-to-high-reactivity fuel ratio, ignition timing and combustion phasing control can be achieved. The RefCCI principles, benefits, and possible challenges are described in previous publications. However, the suitability of the RefCCI approach for aerial, mainly unmanned aerial vehicle (UAV) platforms has not been studied yet. The main goal of this paper is to examine whether the RefCCI approach can be beneficial for UAV, especially HALE (high-altitude long-endurance) applications. The thermodynamic first-law and the second-law analysis is numerically performed to investigate the RefCCI approach suitability for UAV applications and to assess possible efficiency gains. A comparison with the conventional diesel engine and the previously developed technology of spark ignition (SI) engine with high-pressure TCR is performed in view of UAV peculiarities. The results indicate that the RefCCI system can be beneficial for UAV applications. The RefCCI higher efficiency compared to existing commercial engines compensates the lower heating value of the primary fuel, so the fuel consumption remains almost the same. By optimizing the compression pressure ratio, the RefCCI system efficiency can be improved.

Keywords: Reforming-Controlled Compression Ignition; UAV applications; exergy analysis; hydrogen; dimethyl ether; fuel reactivity control

1. Introduction

According to leading experts, the internal combustion engine (ICE) will keep being a prime propulsion tool for the next decades [1–3]. The ICE is also a major propulsion tool for unmanned aerial vehicle (UAV) applications including drones; while the main challenge and main global trend of the UAV industry are achieving high flight altitudes together with long endurance—HALE (high-altitude long-endurance). Higher flight altitude allows the covering of larger surfaces by one aircraft, more stealth, whereas long endurance makes longer times above an object possible . To be even more valuable for all platforms, the ICE has to significantly improve its energy efficiency to enable longer endurance. In recent decades, much work has been done to meeting this target. Low-temperature combustion (LTC) engines can increase engine efficiency and reduce pollutant emissions simultaneously [4,5]. This is possible since the combustion is simultaneous all over the cylinder resulting in eliminating the compression ratio (CR) restriction owing to the knocking phenomenon characterized in SI engines. In addition, the low-temperature combustion results in lower heat loss to the cylinder walls and NO_x formation. However, the LTC engine has some challenges to be addressed before this approach can be commercial. The major challenges are ignition and combustion-phasing control, high CO and HC emissions, ringing, and limited operation range [4,5]. Several strategies to overcome these challenges have been proposed in recent years. Reactivity controlled compression ignition (RCCI) [6] and reforming-controlled compression ignition (RefCCI) [7–9] are two methods aiming to give a solution to combustion controllability and some additional LTC problems by adapting the charge reactivity to the engine operation regime. This is achieved by using two different fuel types—reactive and nonreactive (for instance: diesel and gasoline, respectively) [8]. Besides, these approaches offer different injection strategies aimed to moderate the heat release rate (HRR). Too high values of HRR might damage the engine structure.

Further efficiency improvement of LTC engines can be achieved by recycling a part of the available exhaust gas enthalpy. Thermochemical recuperation (TCR) is a promising approach enabling the efficient implementation of the latter approach [10]. TCR is based on utilizing the exhaust waste heat to sustain endothermic reactions of a primary fuel reforming in an onboard catalytic reactor [10]. As a result, the heating value of the produced new hydrogen-rich fuel (reformate) rises, with subsequent benefits in terms of efficiency and emissions mitigation [11,12]. Some recently published works suggest using TCR for reformate production from alcohol fuels, such as methanol and ethanol in SI engines. A novel method of high-pressure thermochemical recuperation (HP-TCR) preventing maximum power loss and abnormal combustion was developed by the Technion researchers [13]. This method is based on the direct injection of the gaseous reformate into the engine cylinders. To mitigate the work required to compress the reformate and allow its direct injection, which would make this method worthless, the authors suggested pressurizing the liquid primary fuel before its reforming [14]. Hence, the latter occurs at elevated pressure. Poran et al. [14] experimentally investigated the first ever prototype of ICE combined with the HP-TCR. The primary fuel was methanol and the burned reforming products comprised mainly from CO_2 (up to 25% mol.) and hydrogen (up to 75% mol.). They reported a 19% to 30% relative gain in the indicated efficiency, and a more than 90% reduction in NO_x, CO, and HC emissions compared to the gasoline-fuel reference case. The above-described HP-TCR method is a part of the RefCCI concept resulting in higher efficiency compared to RCCI because of waste heat recovery and hydrogen combustion.

A primary fuel fitting well for use in ICE with HP-TCR (it) is dimethyl ether (DME) [10]. DME is a nontoxic, low carbon intensity, noncorrosive, and not harmful to the ozone layer alternative fuel [15]. Its combustion normally produces lower particle, hydrocarbon, NOx, and carbon monoxide emissions [16]. DME has also challenging aspects that are related to its physical properties. DME is a gas at ambient conditions, which requires a sealed fuel supply system. DME viscosity is 20 times lower than that of diesel. This may cause leakage and lubrication problems in fuel pumps and injectors [16]. Therefore, the engine fuel system should be adapted to DME. DME has a low reforming temperature of 250–350 °C, which enables efficient recovery of ICE exhaust gas energy. The main physical properties of DME, hydrogen, and some other fuels are listed in Table 1.

Table 1. Physical properties of selected fuels.

	Hydrogen	Dimethyl-ether	Methanol	Gasoline
Chemical formula	H_2	CH_3OCH_3	CH3OH	-
Molecular mass (g/mol)	2.02	46.07	32.4	-
Chemical exergy [MJ/kg]	116.69	30.71	22.48	47.4 (octane)
Lower heating value (MJ/kg)	119.96 [1]	28.9 [2]	19.92	44.0
Stoic. air to fuel ratio (kg_{air}/kg_{fuel})	34.3	9.0	6.45	14.7
Flammability limits by λ [3]	0.14–10.08	0.19–1.99	0.25–1.95	0.26–1.62

[1] [17]; [2] [18]; [3] atmospheric pressure

To overcome the challenges of catalyst coking, to avoid the need for multiple reformers, and enable the use of a single primary fuel, the reforming-controlled compression ignition (RefCCI) approach was suggested and analyzed by the authors in [7] and [9]. The RefCCI method couples the benefits of the HP-TCR and the advanced combustion concept, the reactivity-controlled compression ignition (RCCI) developed by researchers from the University of Wisconsin [6]. According to the RefCCI method, a part of the primary fuel is reformed using the steam reforming of DME (SRD) in a catalytic reactor [7] to create a hydrogen-rich reformate. Both the nonreformed high-reactivity primary fuel (DME) and the low-reactivity gaseous reformate (mainly H_2) are injected directly into the cylinder in a controllably varied ratio. The simulation results [7] showed that, as a result of the heat recovery effect, only efficiency improvement of up to 5.4% is achieved compared to the basic engine without TCR. The predicted efficiency of the RefCCI engine was approximately 50% for a very basic engine configuration.

The second-law analysis is an accepted tool of recognizing ways to improve efficiency [19]. Mapping the exergy destruction causes allows redesign of the tested system to minimize exergy destruction [20] and consequently to improve system efficiency [21]. This approach is very common in many different engineering areas [22] such as gas turbines [23,24], power plants with CO_2 capture [25], and organic Rankine cycle-based waste energy recovery [26]. The second-law analysis concept is also extensively applied in the ICE field. It begins with fundamental research [27–29] and continues to the analysis of engineering applications [30–33].

The second-law analysis of the RefCCI engine [9] showed that approximately 33% of the fuel exergy consumed by the system is destroyed owing to the irreversible process in the combustion chamber. An additional 5% is destroyed in the reforming system. Chemical reaction and in-cylinder gas to walls heat interaction are the major contributors to exergy destruction in the engine, accounting for approximately 45% and 38%, respectively. After the optimization of parameters affecting the exergy destruction, the obtained results revealed indicated efficiency improvement above 7%, and even higher if the compression ratio is increased.

The main goal of the reported study is to examine the suitability of the RefCCI concept to UAV applications with an emphasis on HALE (high-altitude long-endurance) platforms. Ambient conditions at high altitude require some adjustments to the RefCCI system. First, the known problem of limited power owing to reduced air density should be considered. Second, since the autoignition principle is applied in the RefCCI method, the pressure after compression has a very important role in enabling the charge compression ignition. Too low pressure may result in misfiring since the autoignition delay becomes longer with pressure reduction [5]. Turbocharging enables both challenges to be resolved. An intercooler usually is applied as well to reduce the compressed air temperature and increase further its density.

Considering the mentioned above, the issue that should be examined whether there is sufficient available exhaust gas energy for simultaneous powering the turbocharger turbine and sustaining the efficient reforming process. The enthalpy required by the latter increases so much as the required hydrogen flow rate is higher. This happens as the average pressure and temperature in the cylinder rise, for example with load increase. Boosting the air in the compressor on one hand increases the intake pressure and consequently, the average pressure and, on the other hand, decreases the average temperature during combustion owing to the air dilution effect. Therefore, the combined effect of over-boosting should be examined. If the engine demand for hydrogen is higher with boosting, the lack of available energy to operate both systems may be a realistic scenario. However, in this case, the produced turbine power is reduced resulting in lower boosting until the system arrives at a certain point enabling operation of both systems. Nevertheless, as mentioned above, if the intake pressure is too low combustion may not occur. This consequence can restrict the altitude range. Thus, this work investigates the ability of the boosted RefCCI system to operate at the HALE UAV conditions. The focus is on altitudes from 20,000 to 40,000 ft (approximately 6.1 to 12.2 km).

2. Methods

2.1. RefCCI Concept Description

The RefCCI concept uses a combination of RCCI [6] and HP-TCR [13] advantages with DME as a primary fuel and an onboard single-stage reformer filled with a bifunctional catalyst. In the RefCCI system, the engine exhaust gases are exploited to sustain endothermic reactions of fuel reforming in a single catalytic reformer. Three main reactions occur simultaneously in this reformer throughout the steam reforming of the DME (SRD) process. Each of them is supported by one of the catalyst functions (acid or metal): DME-hydrolysis (Equation (1), the opposite reaction to methanol dehydration), methanol steam reforming (SRM; Equation (2)), and reverse water gas shift (r-WGS; Equation (3)). These reactions and their enthalpy values are listed below:

$$CH_3CH_{3_{(g)}} + H_2O_{(g)} \rightleftarrows 2CH_3OH_{(g)} \quad \Delta h = 24[kJ/kg], \tag{1}$$

$$2CH_3OH_{(g)} + 2H_2O_{(g)} \rightleftarrows 2CO_2 + 6H_2 \quad \Delta h = 50[kJ/kg], \tag{2}$$

$$CO_2 + H_2 \rightleftarrows H_2O_{(g)} + CO \quad \Delta h = 41[kJ/kg], \tag{3}$$

The sum of the first two reactions gives:

$$CH_3OH_{(g)} + 3H_2O_{(g)} \rightleftarrows 2CO_2 + 6H_2 \quad \Delta h = 139[kJ/kg], \tag{4}$$

The produced hydrogen-rich reformate is injected into the engine cylinder directly. DME is injected into the engine cylinder separately from the reformate. Combustion characteristics of DME and hydrogen differ substantially (the first is a high-reactive fuel and the second is a low-reactive fuel). The low-reactivity fuel has a shorter ignition delay than the high-reactivity fuel. Therefore, the increase of ratio between the injected low-reactivity primary fuel (DME) and the hydrogen-rich reformate results in advanced igniting timing and vice versa. Thus, varying this ratio can solve the controllability issue of compression autoignition (CAI) engines.

2.2. Model Description

The model used in this work is based on the previously developed two models described in [7,9]. The model is 1-D, homogenous combustion calculated using a reduced mechanism based on the work of Kaiser et al. [34]. The vaporizer, reformer, and intercooler are considered as adiabatic. The saturated water and DME thermodynamic properties are based on data from [18,35], respectively. The superheated water and DME vapors thermodynamic properties (considered as an ideal gas) were taken from NASA's database [36]. The dependency of the ambient conditions on the flight altitude is shown in Figure 1.

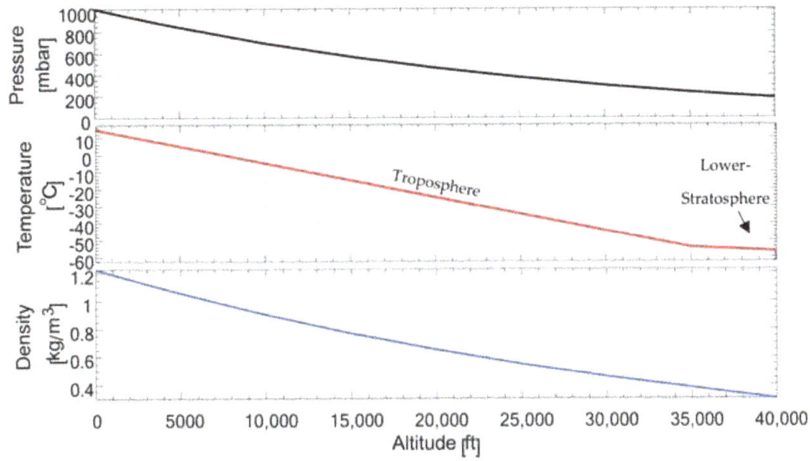

Figure 1. Pressure, temperature, and air density as a function of altitude over sea level. Based on [37].

The model numerically analyzes the RefCCI system (Figure 2) from both the first and the second law perspectives. As seen, the primary fuel (DME) and water are compressed and enter into the intercooler from the cool side. Then they enter into the vaporizer (pt. 1) to complete the evaporation process. The gas mixture flows into the reformer (pt. 2) and a low-reactivity hydrogen-rich reformate is obtained. The reformate is cooled down in the intercooler. In parallel, DME is injected directly into the engine cylinder. The reformer and vaporizer receive heat from the exhaust. The rest of the exhaust enthalpy is utilized to operate the turbine supplying power to the turbocharger compressor.

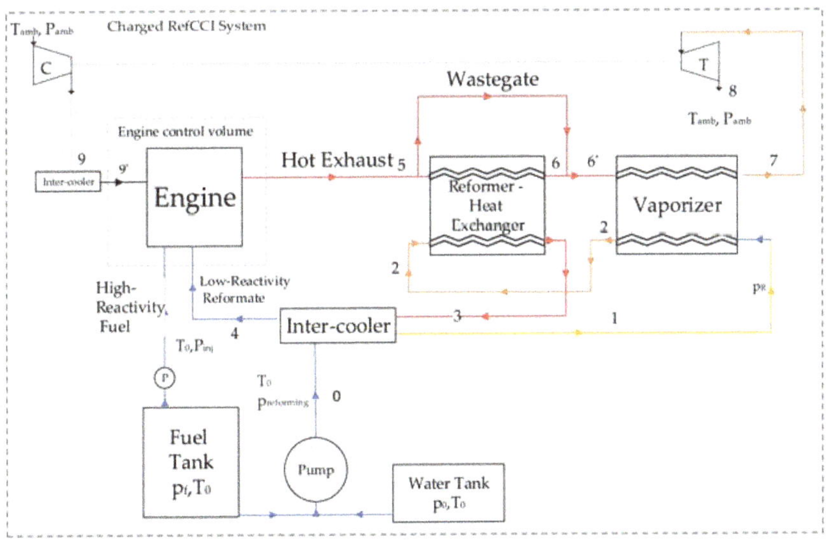

Figure 2. Schematic layout of the boosted reforming-controlled compression ignition (RefCCI) engine (T—turbine; C—compressor; P—dimethyl ether (DME) injection pump).

To simplify the model calculation, it is assumed that the DME and water stream in different pipes and mixing before entering the reformer (pt.2). Based on the first-law calculation, the model finds all states in the systems. Since DME has a lower saturation temperature than water, it evaporates first, probably almost entirely in the intercooler. The maximal temperature that the model allows in pt. 2 is 600 K even if the vaporizer can warm it to greater temperature. However, pt. 2 temperature may be lower than 600 K when no energy is available to warm up the flow to this temperature or $T_{6'}$ is lower than 600 K. The chemical reaction rate is calculated according to the model proposed by Oar-Artera et al. [38]. The bifunctional catalyst in this chemical model is $CuFe_2O_4/\gamma\text{-}Al_2O_3$. This chemical model, in particular its mathematical description part, is explained in more detail in [7].

A full explanation about assumptions made, mathematical descriptions, and other useful details about the model can be found in [7,9]. The modeled engine specification is listed below in Table 2.

Table 2. The modeled engine specification.

Type	Four-Stroke/Single-Cylinder
Displaced volume	367 cm ^{3}
Stroke	73 mm
Bore	80 mm
Connecting rod length	126 mm
Compression ratio	16:1
Number of valves	2
Exhaust valve open	38° BBDC @ 0.15 mm lift
Exhaust valve close	4° ATDC @ 0.15 mm lift
Intake valve open	28° BTDC @ 0.15 mm lift
Intake valve close	4° BBDC @ 0.15 mm lift

The previously developed model was modified for UAV propulsion conditions. The turbine and compressor sub-models are universally applicable and are not based on a specific commercial device, but the efficiency of 0.72 was selected for both and the turbine pressure ratio was obtained from Bernoulli's principle where a wastegate is utilized to control this pressure ratio. The turbine power is then calculated using Equation (5).

$$\dot{W}_{turbine} = \eta_T c_p T_{in,T} \dot{m}_{exhaust} \left[1 - \left(\frac{p_{out,T}}{p_{in,T}} \right)^{\left(\frac{\gamma-1}{\gamma}\right)} \right], \tag{5}$$

Here, $p_{in,T}$ is the turbine inlet or engine exhaust pressure (pt. 7- Figure 2), $p_{out,T}$ is the turbine exit pressure assumed to be the ambient pressure, c_p is the average specific heat of gas streaming through the turbine, $T_{in,T}$ is the gas temperature at the turbine inlet, $\dot{m}_{exhaust}$ is the exhaust mass flow rate, η_T is the turbine efficiency, and γ is the exhaust gas heat capacity ratio.

The power produced by the turbine is supplied to the compressor and, as a result, a compression pressure ratio is calculated using Equation (6).

$$\dot{W}_{compresor} = \frac{c_p T_{in,C}}{\eta_c} \dot{m}_{intake} \left[1 - \left(\frac{p_{out,C}}{p_{in,C}} \right)^{\left(\frac{\gamma-1}{\gamma}\right)} \right], \tag{6}$$

Here, $p_{in,C}$ is the compressor inlet pressure assumed to be the ambient pressure (pt. 7- Figure 2), $p_{out,C}$ is the compressor exit pressure, c_p is the average specific heat of air streaming through the compressor, $T_{in,C}$ is the air temperature at the compressor inlet, $\dot{m}_{intake}$ is the intake air mass flow rate, η_C is the compression efficiency, and γ is the intake air heat capacity ratio. Note that in this work a simple turbocharging system is thermodynamically modeled. In reality, a two-stage turbocharging could be required, as discussed in [39], to enable efficient system operation in real high-altitude flight conditions.

To reduce the reforming average temperature, the reformer wastegate opening degree (Figure 2) is controlled by a PID controller. The selected set-point is 600 K at the reformate exit (pt. 6, Figure 2), which is the temperature resulting in high fuel conversion according to chemical equilibrium considerations.

Too high exhaust-side reformer temperature may result in extensive exergy destruction owing to a large temperature difference.

The surrounding reference state chosen for this analysis is the sea-level typical state meaning 25 °C and 1013 mbar. The environment is composed of N_2, O_2, $H_2O_{(g)}$, CO_2, and other components in the molar fractions 75.67%, 20.35%, 3.12%, 0.03%, and 0.83%, respectively [35].

Notably, the general engine model uses existing and well-verified sub-models of the commercial GT-Suite SW. The latter is an extensively validated commercial SW widely used worldwide in engine research and development. Therefore, it was not essential to validate the engine model. As regards the reformer model, its validation for the methanol steam reforming reaction (SRM) was performed and described in detail in the previous publication of the Technion research group [14].

2.3. Second-Law Analysis

In each RefCCI component (Figure 2), the exergy destruction is calculated according to its thermodynamic classification and operating conditions. The reformer, vaporizer, turbine, compressor, and both intercoolers are assumed to be adiabatic and in steady-state control volumes. Among these parts, chemical reactions take place only in the reformer. In the model, the exergy destruction is calculated either by Equation (7) or by exergy balance—Equation (8).

$$\dot{B}_d = T_0\left(\frac{dS}{dt} - \sum_j \frac{\dot{Q}_j}{T_j} - \sum_{in} \dot{m}_{in}s_{in} + \sum_{out} \dot{m}_{out}s_{out}\right), \quad (7)$$

Here, T_0 is the reference temperature, $\dot{Q}_j$ is the heat interaction rate of the system with the walls, and T_j is the wall j temperature.

$$\dot{B}_d = \sum_j\left(1 - \frac{T_0}{T_j}\dot{Q}_j\right) - \left(\dot{W} - p_0\frac{dV}{dt}\right) + \sum_{in} \dot{m}_{in}b_{flow,in} - \sum_{out} \dot{m}_{out}b_{flow,out} - \frac{dB}{dt} \quad (8)$$

where, b_{flow} is the specific flow exergy including both thermomechanical and chemical exergy, b_{TM} and b_{ch}, respectively. To calculate exergy destruction in the reformer, intercooler, and evaporator the exergy balance was applied. Since these systems are assumed to be adiabatic in steady-state operation mode, only the exergy flows are considered.

For the combustion chamber, the integration of Equation (7) over an entire cycle results in the total exergy destruction in the combustion chamber per engine cycle. Thus, the engine exergy destruction rate in steady-state ($\dot{B}_{d,engine}$) is obtained by multiplying this term by the engine speed (RPM) and dividing by 120. The exergy destruction in the combustion chamber is a result of the three main irreversible processes: chemical reaction, heat interaction, and injection. When engine valves are closed, Equation (7) degenerates so only the entropy change in time and heat interaction take place. It is reasonable to relate the chemical reaction exergy destruction ($\dot{B}_{d,chem}$) to the change in entropy between EOI and EVO and exergy destruction due to heat interaction ($\dot{B}_{d,heat\ interaction}$) to the sum of heat interactions between the combustion chamber and its walls during the same interval. The exergy destruction produced during injection is the total engine exergy destruction subtracted by the two mentioned above parameters.

For the turbine and compressor, the exergy destruction is obtained from the exergy balance derivative of Equation (8) as described in Equation (9).

$$\dot{B}_d = \dot{m}(b_{in} - b_{out}) - \dot{W}, \quad (9)$$

Here, $\dot{m}$, b_{in}, b_{out}, and $\dot{W}$ are the mass flow rate, inlet specific exergy, outlet specific exergy, and mechanical power of the turbine/compressor, respectively.

The engine second-law efficiency is defined as follows:

$$\eta_{engine}^{2-law} = \frac{\dot{W}}{\dot{m}_{DME,DI} b_{DME}^{ch} + \dot{m}_{reformate} b_{reformate}}, \tag{10}$$

Here, $\dot{m}_{DME,DI}$ is the DME consumption of the DME directly injected into the cylinder, $\dot{m}_{reformate}$ is the reformate consumption, and $b_{reformate}$ is the reformate exergy containing the chemical and thermomechanical exergy. Evidently, the chemical constituent is dominant.

The RefCCI system second-law efficiency is defined as:

$$\eta_{RefCCI,sys}^{2-law} = \frac{\dot{W}}{\dot{m}_{DME} b_{DME}^{ch}}, \tag{11}$$

Here, $\dot{m}_{DME}$ is the entire DME consumption and b_{DME}^{ch} is the DME chemical exergy (Table 1). Similarly, the system efficiency is defined as:

$$\eta_{RefCCI,sys} = \frac{\dot{W}}{\dot{m}_{DME} LHV_{DME}}, \tag{12}$$

Where, LHV_{DME} is the DME lower heating value (Table 1).

3. Results and Discussion

3.1. Boosted RefCCI Analysis

3.1.1. The Effect of Boosting on RefCCI Engine Behavior

Since the RefCCI system works according to the autoignition principle, the conditions in the vicinity of TDC must be sufficient to enable the autoignition of the fuel–air mixture. In particular, the initial pressure at the beginning of the compression stroke has a crucial role since it determines the maximal pressure before TDC. If this pressure is too low, autoignition will not occur. At high altitude, the surrounding pressure is very low (Figure 1), and therefore boosting is required. Besides, to gain the advantages of low-temperature combustion (LTC), such as RefCCI, it would be beneficial to have the air–fuel mixture as lean as possible. This means that extra air should enter the cylinder which requires extra boosting. However, extra boosting might cause negative phenomena such as elevated backpressure in the exhaust. Hence, the compression pressure ratio has a range in which the engine can work and this range varies depending on engine operating mode and flight altitude. Note that, in this work, the boosting system is not studied, and the compressor and turbine efficiency are fixed at 0.72. However, the effect of changing the compression pressure ratio on the RefCCI system performance is examined.

Figure 3 shows this trade-off. When the compression pressure ratio increases, the air–fuel equivalence ratio (λ) expectedly increases as well. The extra air entering into the cylinder decreases the maximal temperature and consequently heat losses to walls. As a result, the indicated gross efficiency consistently increases when the pressure becomes higher. However, the higher pressure-ratio is, the higher the pumping losses. This results in a trend change and an optimum appearance when the indicated net efficiency is considered. Eventually, for the case illustrated in Figure 3 (30,000 ft), the optimal pressure ratio is 3.25. The system efficiency is higher than the engine efficiency by approximately 1.75–2 percentage points (absolute) owing to waste heat recovery. Another interesting trend that is unique to a boosted RefCCI compared with a naturally aspirated RefCCI [7,9] is that the required H_2/DME ratio does not extendedly modify. This is probably because of the opposite trends that boosting causes. As much as more air enters into the cylinder, the average in-cylinder pressure during the compression stroke is higher resulting in lower required fuel reactivity (higher

H_2/DME ratio). However, since the air is also a diluter, higher dilution results in the lower average temperature, which demands higher fuel reactivity. The simulation results show that at a flight altitude of 40,000 ft, the minimal compression pressure ratio that achieves the required in-cylinder pressure before ignition (to allow autoignition) is 3.25 compared to 1.5 at 20,000 ft. In fact, the intake pressure enabling autoignition is around 0.7 bar, reflecting approximately a minimum of 35 bar maximal pressure before combustion.

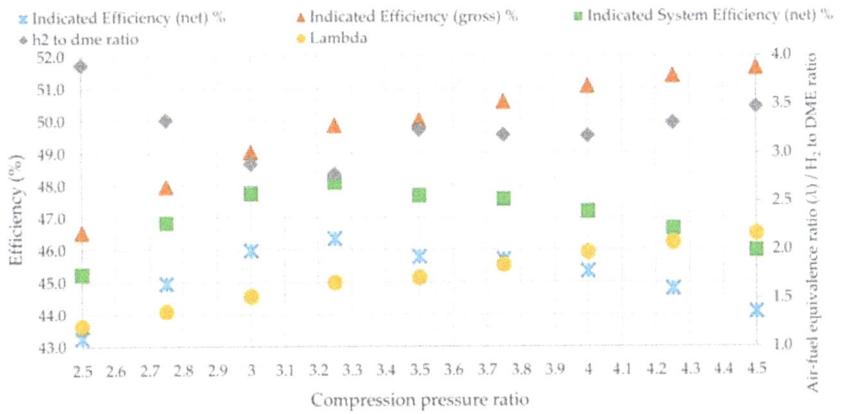

Figure 3. Engine performance, the air–fuel equivalence ratio, and H_2/DME ratio vs. compression pressure ratio. Engine regime: is 12.75 kW for a single-cylinder, 5000 RPM, and 30,000 ft altitude.

3.1.2. Exergy and Energy Analysis

To understand better the reasons for process irreversibility in the RefCCI system, it is useful to map the exergy and energy flow inside the system. After that, by utilizing tools that can reduce the process irreversibility, the system efficiency can be enhanced. Figure 4 shows an example of an exergy and energy flow map for a single case of engine operation at 40,000 ft. altitude, while the compression pressure ratio is 3.25 and the engine power is 12.75 kW for a single cylinder (defined as 50% of the maximal power) and engine speed 5000 RPM.

The main exergy destruction source is the engine itself. The exergy destruction in the reformer and vaporizer is significant as well. Therefore, to improve the system efficiency this destruction should be reduced as much as possible. In this case (Figure 4), the relatively low compression ratio does not allow the supply of extra air, so the air–fuel equivalence ratio (λ) is very close to 1. This causes some bad consequences affecting energy utilization in the system. On the one hand, low λ causes high in-cylinder temperature increasing heat losses. Only part of this heat can be utilized to produce work, while the exergy transferred by heat interaction ($\sum_j \left(T_0/T_j - 1\right)Q_j$) is the maximum work value that can be produced (according to Carnot). Evidently, additional tools are necessary to harvest this work. Consequently, the high in-cylinder temperature results in high exhaust gas temperature and therefore higher temperature difference between the exhaust side and the fuel/reformate side both in the reformer and the vaporizer. On the second hand, increasing λ requires higher compression pressure ratio which inevitably leads to higher exhaust backpressure. The latter may result in lower cycle work.

When analyzing the sources of exergy destruction in the cylinder, chemical and heat exergy destruction are the major contributors. Two main factors affect the chemical exergy destruction during combustion. First, the fuel properties: as much as the fuel contains more hydrogen, the chemical exergy destruction is lower since, for a constant amount of air, the number of product molecules becomes lower compared to the number of the reactants (air/fuel) molecules [40]. Second, the amount of air participating in combustion: the higher mass is, the higher the entropy change rate. Figure 5 illustrates this trend. Note that, in this graph and from now on, the relative difference (Ψ_i) in the parameter i value appearing on the vertical axis is defined as the following:

$$\Psi_i = \frac{\psi_i - \psi_{ref}}{\psi_{ref}} \tag{13}$$

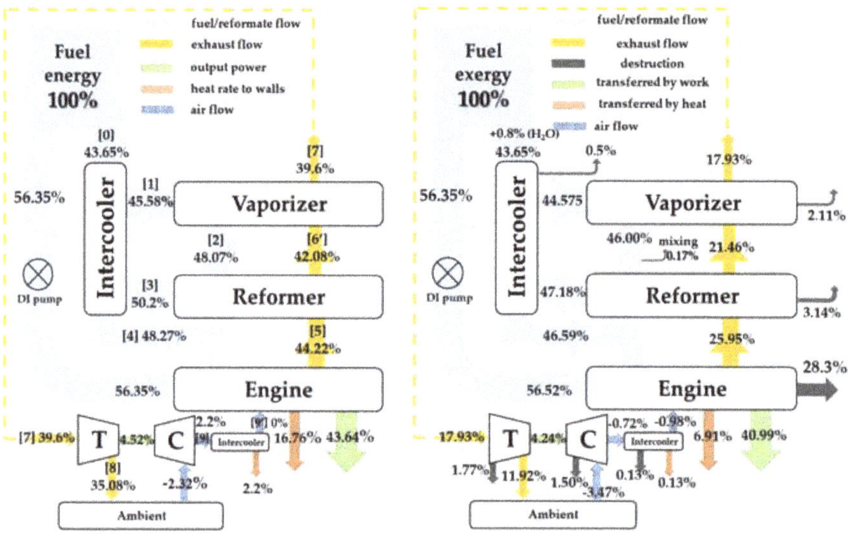

Figure 4. Exergy and exergy flow map in the boosted RefCCI system. The exergy and energy flow rates are normalized by the exergy or energy fuel consumption, respectively. The numbers in the square brackets represent the system points as designated in Figure 2. The flight altitude is 40,000 ft, engine power is 12.75 kW for a single cylinder (defined as 50% of the maximal power), the engine speed is 5000 RPM, and compression pressure ratio is 3.25.

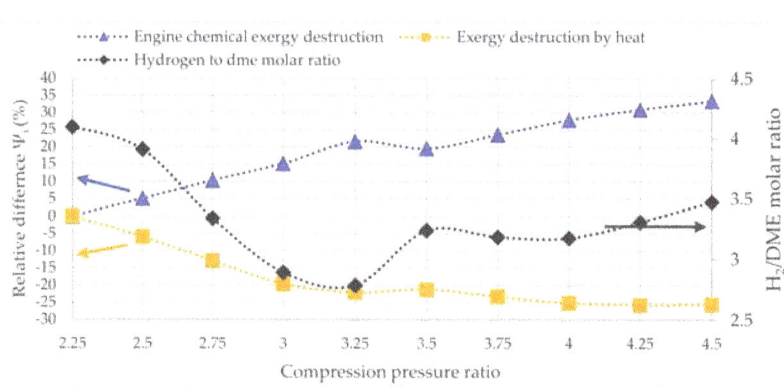

Figure 5. Sources of exergy destruction in the RefCCI engine depending on the compression pressure ratio. Altitude: 30,000 ft; engine speed: 5000 RPM; indicated power: 12.75 kW.

Where Ψ_i is the predicted parameter i value and ψ_{ref} is its value for the reference case. The reference case is the lowest investigated value (unless it is defined differently), meaning the relative difference for the reference case is by definition zero.

The H_2/DME ratio value shown in Figure 5 is the value required to maintain the engine combustion process in such a way that CA50 would be the same and equal 10 degrees ATDC. Generally, the chemical destruction increases with the increase of the compression pressure ratio. However, its slope varies depending on the H_2/DME ratio. In the range of 2.25–3.25, when the H_2/DME ratio decreases, the chemical exergy destruction curve slope is sharper than between 3.5–4.5 when the H_2/DME ratio increases.

Increasing compression pressure ratio has some more effects on the boosted RefCCI system. It reduces exhaust temperature and consequently decreases reformer exergy destruction. It causes more recovery of waste energy by the boosting system and, as a result, less energy/exergy flows through the system exhaust. However, the increase of the compression pressure ratio is not unlimited. It may result in the need for more complex compression procedures such as two-stage compression with its accompanying problems [39]. In Figure 6 the results of the simulation for 40,000 ft, 5000 RPM, and 50% power are presented. The results match the mentioned above qualitative considerations. As explained, the system exhaust exergy flow (pt. 8 in Figure 2) significantly decreases (up to 90%) when the pressure ratio increases. In parallel, the system intake exergy flow, which is negative because the surrounding pressure and temperature are lower than in the reference state, becomes lower meaning more air-flow enters the system. The exergy transferred by heat decreases since the average temperature in the cylinder is lower. Consequently, the second-law efficiency increases with the pressure ratio increase meaning DME exergy consumption is lower. The reformer and vaporizer exergy destruction both decrease since the exhaust temperature is lower resulting in lower heat transfer temperature difference.

In the previous study [9], the authors investigated the influence of various parameters on exergy destruction in the RefCCI system. Some of them could affect the RefCCI system performance somewhat differently at UAV-relevant conditions. In addition to the previously studied parameters, there are some additional ones, which are unique for UAVs. For instance, since the UAV engine typically operates at higher speeds than a typical automotive engine, for the same load, the required H_2/DME ratio and, as a result, hydrogen consumption would be lower. Therefore, either the reformer can be smaller or the temperature difference inside the reformer can be lower, resulting in reduced irreversibility. Another parameter, the range of which changes in UAV applications, is the boosted air temperature after intercooling (hereinafter, intercooling temperature). At high altitudes, the ability to cool the compressed air is enhanced. On the one hand, extra cooling of the air enables charging more air into the cylinder resulting in lower heat losses. On the other hand, it increases the chemical exergy destruction. In addition, the heat rejected from the intercooler is lost. Figure 7 shows the influence of the reduction in the intercooling temperature on exergy destruction and second-law efficiency of the boosted RefCCI system. The reference case is a temperature of 25 °C, while all other parameters remain constant. As can be seen from Figure 7, the exergy destruction in the engine itself grows slightly with the decrease of the intercooling temperature. This is a result of the increase in chemical exergy destruction, which is higher than the reduction of exergy destroyed owing to heat transfer to the cylinder walls. However, the reduction in exergy flowing out of the system leads to improvement of the second-law efficiency by almost 4% at the intercooling temperature of −25 °C compared to the 25 °C case (as a reminder, the ambient air temperature at the flight altitude of 40,000 ft is below −50 °C).

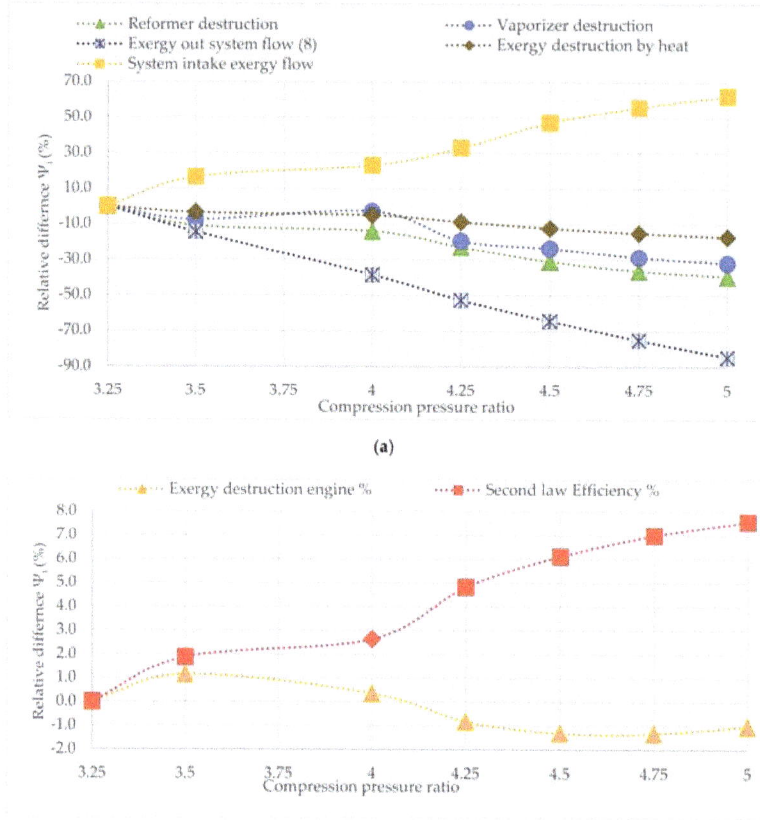

(**a**)

(**b**)

Figure 6. The influence of increasing the compression pressure ratio on exergy flow and destruction in the charged RefCCI system. Altitude: 40,000 ft; rotary speed: 5000 RPM; indicated power: 12.75 kW. (**a**)- large-scale changes), (**b**)- small-scale changes)

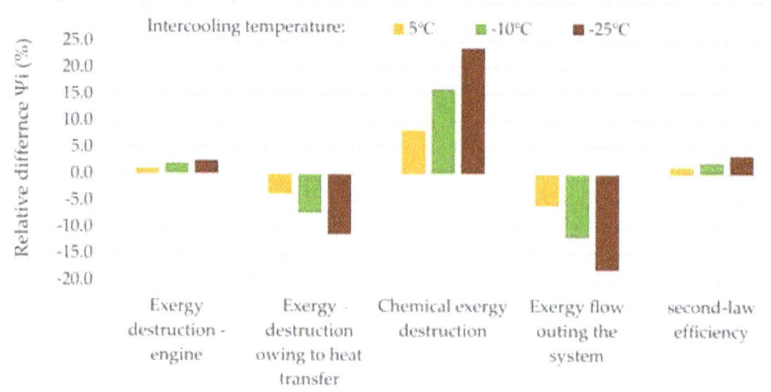

Figure 7. The influence of the intercooling temperature on major exergy destruction sources and exergy flow in the boosted-RefCCI system. Altitude: 40,000 ft; rotary speed: 5000 RPM; indicated power: 12.75 kW. The reference case is 25°C.

When optimizing the boosted RefCCI system according to the findings of this and the previous work [9], it is possible to enhance the system performance significantly. Table 3 shows a comparison between the two cases, the reference case, which is presented in Figure 4, and the optimized case. The parameters changed are also shown in the Table. The results show that although the engine exergy destruction slightly increases, the significant reduction in the reforming exergy destruction and the exergy lost with the exhaust flow result in an increase of almost 4% in the second-law efficiency of the boosted RefCCI system. We believe that extensive optimization including more parameters than those investigated in this work can lead to even higher improvement.

Table 3. Exergy flow and destruction in the boosted-RefCCI system for the nonoptimized and optimized cases. The optimization was done according to the findings in this and previous work.

Relative difference Ψi (%)	Reformer destruction −33.4%	Vaporizer destruction −59.0%	Intercooler destruction −41.3%	Engine exergy destruction +3.7%	Exergy flowing out the system −20.6%	Intake exergy flow 12.0%	Second-law efficiency 3.8%
Parameter	**Injection/ reforming pressure (bar)**	**Injection end of DME (deg. BTDC)**		**Reformer heat transfer area (m^2)**		**Intercooling temperature(°C)**	**Compression pressure ratio**
Reference/ optimized	25/45	60/40		0.63/0.79		25/(−25)	3.25/3.25

3.2. Comparison between the RefCCI Concept and HP-TCR and a Commercial SI Engine

In this section, the investigated boosted RefCCI approach is compared with the HP-TCR SI engine and a commercial Rotax 914 engine. The Rotax 914 is a gasoline SI four-cylinder engine. The prediction of its performance was carried out based on the model described in [41]. The HP-TCR method is explained in detail and its model is described in [13]. All models were adjusted to enable a comparison with the investigated boosted RefCCI engine. Both HP-TCR and RefCCI models simulate the engine modified from the Lester Petter AD1 commercial engine. Both concepts enable the use of a high compression ratio. The RefCCI since it is based on compression autoignition, and HP-TCR since the combusted fuel in this concept is hydrogen, which has strong resistance to the ringing phenomenon. This ability was proven in [42]. To enable a comparison between the engines of different displacement, the indicated mean effective pressure (IMEP) value was kept the same. The compared propulsion technologies are summarized in Table 4.

Table 4. Summary of the compared propulsion technologies.

Modeled Engine	Concept	Compression Ratio	Primary Fuel (LHV– MJ/kg)	Combusted Fuel	Remarks
[1] Rotax 914	SI	9:1	Gasoline (44)	Gasoline	CR is limited by ringing
[2] Lester Petter AD1	HP-TCR	16:1	Methanol (19.9)	Hydrogen	CR is greater owing to hydrogen knock resistance character
[3] Lester Petter AD1	RefCCI	16:1	DME (28.9)	DME+ hydrogen	Compression autoignition

The models above are based on models presented in [1] [41]; [2] [13]; [3] [7,9].

In the UAV field, reducing the aircraft mass has an important significance. Thus, the main parameter that should be examined is ISFC. Fuel consumption depends on both engine efficiency and fuel heating value. The HP-TCR and RefCCI concepts are beneficial from the efficiency point of view than the commercial SI Rotax 914 engine, since both enable the use of higher CR and combust hydrogen, the fuel with much higher burning velocity (to remind: the latter enables getting closer to the ideal Otto cycle). However, the conventional Rotax 914 engine uses gasoline as the primary fuel, and gasoline has a considerably higher lower heating value (LHV) than methanol and DME (44 MJ/kg compared to 19.92 and 28.9, respectively—see Table 1). Thus, even if the efficiency of HP-TCR and RefCCI are higher than the SI Rotax 914, still their ISFCs may be larger. Nevertheless, the HP-TCR and

RefCCI have other advantages that in the future might be crucial such as using non-fossil alternative fuel and cleaner combustion.

For the purpose of comparison, equal conditions in intake were set—1.08 bar and 298 K. Three regimes were tested, all at 5000 RPM, where the indicated power was set to three values: 50%, 39%, and 29.5% of the maximal power defined as 25.5 kW per cylinder for the HP-TCR and RefCCI. The Rotax 914 displacement volume is 82.5% of the HP-TCR and RefCCI engine. Thus, the maximal power was reduced according to this value to keep the IMEP value unchanged.

Figure 8 shows the results for the tested cases. The efficiency of the RefCCI system is relatively higher than that of the HP-TCR by approximately 10% on average and that of the Rotax 914 by approximately 50% on average. In on-land applications, this is a huge advantage. Yet, in the aerial application, the fuel mass carried by the platform should be minimal and thus the fuel consumption is the main consideration. The results show that the ISFC of the RefCCI system and the Rotax 914 engine are very close (approximately 1.2% only higher in the RefCCI vs. the Rotax 914). This means that the efficiency enhancement compensates for the lower heating value of DME compared to gasoline. The HP-TCR system has fairly high efficiency but also high ISFC meaning that it is less applicable for UAV applications.

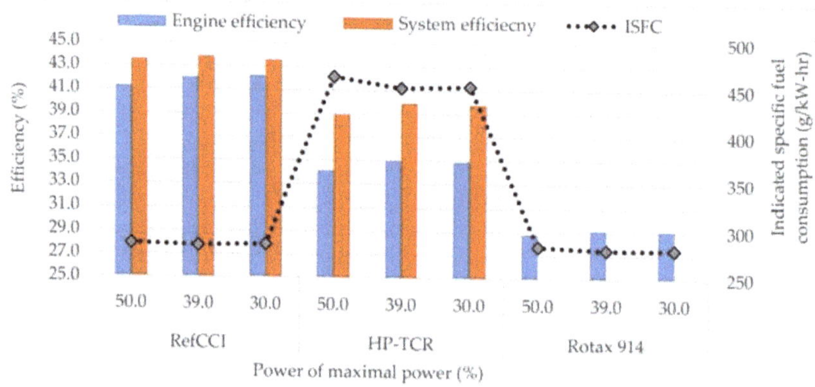

Figure 8. Comparison of three possible propulsion tools for UAV applications.

It should be noted that both HP-TCR and RefCCI require water for the reforming process [7]. This water was not considered as part of the fuel mass consumption since the water can be easily condensed and reused from the exhaust flow. In the RefCCI, to produce two molecules of hydrogen, a single molecule of steam is required ($CH_3OCH_3 + 3H_2O \rightleftarrows 6H_2 + 2CO_2$) and incomplete combustion from a molecule of hydrogen a single molecule of steam is formed ($6H_2 + 3[O_2 + 3.76N_2] + 2CO_2 \rightarrow 6H_2O + 2CO_2 + 11.28N_2$). In total, the ratio between the combustion-formed water molecules to the required for reforming molecules is two. In the HP-TCR, this ratio is 3. To condense this steam, even if the air-fuel equivalence ratio is high and as a result, the partial pressure of the exhaust steam is very low, the surrounding temperature is low enough to enable condensation in any case. Keep in mind that the steam partial pressure decreases with the increase of flight altitude, but the surrounding temperature decreases as well.

Besides DME, other fuels may be considered as a primary fuel for the RefCCI engine. For aerial applications, the primary fuel should have as high LHV as possible and together with the high system efficiency, the fuel mass consumption might be even lower than that achieved in the RefCCI engine with DME as the primary fuel. However, to the best of the authors' knowledge, no relevant works dealing with this issue have been published yet, and further research is required.

Some challenges still delay the practical penetration of the RefCCI method into UAV applications. The major challenge is ringing, which is not unique for UAV only. However, it may be more significant in UAV applications. Ringing owing to rapid heat release rate may be more severe in UAVs because of reduced air mass entering the engine. Intensive research is ongoing nowadays to find solutions to this problem. Among possible solutions, suiting appropriate injection strategy enabling to restrain the heat release rate should be mentioned. Evidently, a practical implementation of the RefCCI engine in the UAV platform would be easier if an alternative non-fossil primary fuel with an LHV similar to that of diesel is applied.

4. Summary and Conclusions

In this work, the suitability of the reforming-controlled compression ignition for UAV application was examined. First, the influence of boosting on the RefCCI engine behavior was discussed, then an exergy analysis to map the sources of exergy destruction in the RefCCI system was executed, and possibilities of performance improvement were analyzed. A comparison of the RefCCI approach with the high-pressure thermochemical recuperation SI engine and the commercial SI Rotax 914 engine were carried out.

The main findings are listed below:

1. Increasing compression pressure ratio consistently increases the gross indicated efficiency of the RefCCI system, but this has also an opposite effect reflected in the pumping work increase. The net indicated efficiency has a maximum value where the first effect is the most dominant.
2. The necessary fuel reactivity (H_2/DME ratio) varies only slightly with the increase of the compression pressure ratio and actually shows some volatility because of its opposite effects on autoignition timing. Increasing the compression pressure ratio, on the one hand, increases average pressure before combustion, but, on the other hand, decreases average temperature owing to the excess air dilution effect.
3. The RefCCI method seems to be a good alternative candidate for UAV application since, despite the relatively low heating value of DME, the high system efficiency leads to fuel consumption very close to that of the conventional gasoline engine.
4. The results of the RefCCI system performance optimization show that a significant reduction in the reforming exergy destruction and the exergy lost with the exhaust flow can be achieved. This results in an improvement of almost 4% in the second-law efficiency of the boosted RefCCI system. We believe that extensive optimization including more parameters than those investigated in this work can lead to even higher improvement.

Besides DME, other fuels may be considered as a primary fuel for the RefCCI engine. For aerial applications, the primary fuel should have as high LHV as possible, and, together with the high RefCCI system efficiency, lower fuel mass consumption compared to other engine technologies may be achieved with cleaner combustion.

Experimental research of the RefCCI engine is necessary to better understand complicated processes taking place in the system in their mutual relationship. Developing injection strategies to restrain ringing in the engine cylinder is vital.

Author Contributions: Conceptualization, L.T. ; methodology, A.E. and L.T.; software, A.E.; validation, A.E.; formal analysis, A.E.; investigation, A.E.; resources, L.T.; data curation, L.T.; writing—original draft preparation, A.E.; writing—review and editing, L.T.; visualization, A.E.; supervision, L.T.; project administration, L.T.; funding acquisition, L.T. All authors have read and agreed to the published version of the manuscript.

Funding: This research was funded by the Israel Science Foundation, grant 2054/17 and Israel Ministry of Energy, grant 218-11-026.

Conflicts of Interest: The authors declare no conflict of interest.

Nomenclature

ATDC	after top dead center
BBDC	before bottom dead center
BDC	bottom dead center
BTDC	before top dead center
CAI	compression auto ignition
CR	compression ratio
DME	dimethyl ether
EOI	end of injection
EVO	exhaust valve open
HALE	high altitude long endurance
HP-TCR	high-pressure thermochemical recuperation
HRR	heat release rate
ICE	internal combustion engine
IMEP	indicated mean effective pressure
ISFC	indicated specific fuel consumption
LHV	lower heating value
LTC	low-temperature combustion
r-WGS	reverse water gas shift
RCCI	reactivity-controlled compression ignition
RefCCI	reforming-controlled compression ignition
SI	spark ignition
SRD	steam reforming of dimethyl ether
SRM	methanol steam reforming
b_{ch}	specific chemical exergy
$\dot{B}_{\mathrm{d}}$	exergy destruction rate
b_{flow}	specific flow exergy
b_{TM}	specific thermo-mechanical exergy
$\dot{m}$	mass flow rate
s	specific entropy
γ	heat capacity ratio
η_{C}	compressor efficiency
$\eta_{\mathrm{engine}}^{2-\mathrm{law}}$	engine second-law efficiency
$\eta_{\mathrm{Ref\,CCI,sys}}$	RefCCI system efficiency
$\eta_{\mathrm{RefCCI,sys}}^{2-\mathrm{law}}$	RefCCI system second-law efficiency
η_{T}	turbine efficiency
ψ_{i}	predicted parameter *i* value
Ψ_{i}	relative difference in parameter *i* value
ψ_{ref}	predicted parameter reference value

References

1. Leadership Summit (panelists: Dante Boutell, Toyota Motor North America Inc.; Dave Filipe, Ford Motor Company; Timothy Frazier, Cummins Inc.; John Heywood, MIT.; John Juriga, Hyundai Motor Group; Jeff Lux, FCA US LLC). Today, Tomorrow, and the Future of Propulsion Systems? In Proceedings of the SAE World Congress, Detroit, MI, USA, 11 April 2019.
2. Senecal, K. End of the ICE Age? The Premature Burial of Internal Combustion. In Proceedings of the SAE High Efficiency IC Engine Symposium, Detroit, MI, USA, 8 April 2019.
3. Reitz, R.D.; Ogawa, H.; Payri, R.; Fansler, T.; Kokjohn, S.; Moriyoshi, Y.; Agarwal, A.K.; Arcoumanis, D.; Assanis, D.; Bae, C.; et al. IJER editorial: The future of the internal combustion engine. *Int. J. Engine Res.* **2020**, *21*, 3–10. [CrossRef]
4. Yao, M.; Zheng, Z.; Liu, H. Progress and recent trends in homogeneous charge compression ignition (HCCI) engines. *Prog. Energy Combust. Sci.* **2009**, *35*, 398–437. [CrossRef]
5. Saxena, S.; Bedoya, I.D. Fundamental phenomena affecting low temperature combustion and HCCI engines, high load limits and strategies for extending these limits. *Prog. Energy Combust. Sci.* **2013**, *39*, 457–488. [CrossRef]
6. Kokjohn, S.L.; Hanson, R.M.; Splitter, D.A.; Reitz, R.D. Fuel reactivity controlled compression ignition (RCCI): A pathway to controlled high-efficiency clean combustion. *Int. J. Engine Res.* **2011**, *12*, 209–226. [CrossRef]
7. Eyal, A.; Tartakovsky, L. *Reforming-Controlled Compression Ignition—A method combining benefits of Reactivity-Controlled Compression Ignition and High-Pressure Thermochemical Recuperation*; SAE International: Warrendale, PA, USA, 2019; SAE Technical Paper 2019-01-0964.
8. Eyal, A.; Tartakovsky, L. *Reforming Controlled Homogeneous Charge Compression Ignition—Simulation Results*; SAE International: Warrendale, PA, USA, 2016; SAE Technical Paper 2016-32-0014.
9. Eyal, A.; Tartakovsky, L. Second-law analysis of the reforming-controlled compression ignition. *Appl. Energy* **2020**, *263*, 114622. [CrossRef]
10. Tartakovsky, L.; Sheintuch, M. Fuel reforming in internal combustion engines. *Prog. Energy Combust. Sci.* **2018**, *67*, 88–114. [CrossRef]
11. Tartakovsky, L.; Baibikov, V.; Gutman, M.; Mosyak, A.; Veinblat, M. *Performance Analysis of SI Engine Fueled by Ethanol Steam Reforming Products*; SAE International: Warrendale, PA, USA, 2011; SAE Technical Paper 2011-01-1992.

12. Tartakovsky, L.; Amiel, R.; Baibikov, V.; Fleischman, R.; Gutman, M.; Poran, A.; Veinblat, M. *SI Engine with Direct Injection of Methanol Reforming Products—First Experimental Results*; SAE International: Warrendale, PA, USA, 2015; SAE Technical Paper 2015-32-0712.
13. Poran, A.; Tartakovsky, L. Energy Efficiency of a Direct-Injection Internal Combustion Engine with High-Pressure Methanol Steam Reforming. *Energy* **2015**, *88*, 506–514. [CrossRef]
14. Poran, A.; Thawko, A.; Eyal, A.; Tartakovsky, L. Direct injection internal combustion engine with high-pressure thermochemical recuperation–Experimental study of the first prototype. *Int. J. Hydrog. Energy* **2018**, *43*, 11969–11980. [CrossRef]
15. Cocco, D.; Tola, V.; Cau, G. Performance evaluation of chemically recuperated gas turbine (CRGT) power plants fuelled by di-methyl-ether (DME). *Energy* **2006**, *31*, 1446–1458. [CrossRef]
16. Semelsberger, T.A.; Borup, R.L.; Greene, H.L. Dimethyl ether (DME) as an alternative fuel. *J. Power Sources* **2006**, *156*, 497–511. [CrossRef]
17. McAllister, S.; Chen, J.Y.; Fernandez-Pello, A.C. *Fundamentals of Combustion Processes*; Springer: New York, NY, USA, 2011.
18. Teng, H.; McCandless, J.; Schneyer, J. *Thermodynamic Properties of Dimethyl Ether—An Alternative Fuel for Compression-Ignition Engines*; SAE International: Warrendale, PA, USA, 2004; SAE Technical Paper 2004-01-0093.
19. Razmara, M.; Bidarvatan, M.; Shahbakhti, M.; Robinett, R.D., III. Optimal exergy-based control of internal combustion engines. *Appl. Energy* **2016**, *183*, 1389–1403. [CrossRef]
20. Bejan, A. *Advanced Engineering Thermodynamics*; John Wiley & Sons: Hoboken, NJ, USA, 2016.
21. Ma, B.; Yao, A.; Yao, C.; Wu, T.; Wang, B.; Gao, J.; Chen, C. Exergy loss analysis on diesel methanol dual fuel engine under different operating parameters. *Appl. Energy* **2020**, *261*, 114483. [CrossRef]
22. Dincer, I.; Cengel, Y.A. Energy, Entropy and Exergy Concepts and Their Roles in Thermal Engineering. *Entropy* **2001**, *3*, 116–149. [CrossRef]
23. Yazdi, M.R.M.; Aliehyaei, M.; Rosen, M.A. Exergy, Economic and Environmental Analyses of Gas Turbine Inlet Air Cooling with a Heat Pump Using a Novel System Configuration. *Sustainability* **2015**, *7*, 14259–14286. [CrossRef]
24. Jin, Y.; Du, J.; Li, Z.; Zhang, H. Second-Law Analysis of Irreversible Losses in Gas Turbines. *Entropy* **2017**, *19*, 470. [CrossRef]
25. Siefert, N.S.; Narburgh, S.; Chen, Y. Comprehensive Exergy Analysis of Three IGCC Power Plant Configurations with CO2 Capture. *Energies* **2016**, *9*, 669. [CrossRef]
26. Tian, Z.; Yue, Y.; Zhang, Y.; Gu, B.; Gao, W. Multi-Objective Thermo-Economic Optimization of a Combined Organic Rankine Cycle (ORC) System Based on Waste Heat of Dual Fuel Marine Engine and LNG Cold Energy Recovery. *Energies* **2020**, *13*, 1397. [CrossRef]
27. Dunbar, W.R.; Lior, N. Sources of combustion irreversibility. *Combust. Sci. Technol.* **1994**, *103*, 41–61. [CrossRef]
28. Khaljani, M.; Saray, R.K.; Bahlouli, K. Evaluation of a combined cycle based on an HCCI (Homogenous Charge Compression Ignition) engine heat recovery employing two organic Rankine cycles. *Energy* **2016**, *107*, 748–760. [CrossRef]
29. Szybist, J.P.; Chakravathy, K.; Daw, C.S. Analysis of the impact of selected fuel thermochemical properties on internal combustion engine efficiency. *Energy Fuels* **2012**, *26*, 2798–2810. [CrossRef]
30. Li, Y.; Jia, M.; Chang, Y.; Kokjohn, S.L.; Reitz, R.D. Thermodynamic energy and exergy analysis of three different engine combustion regimes. *Appl. Energy* **2016**, *180*, 849–858. [CrossRef]
31. Zheng, J.; Caton, J.A. Second law analysis of a low temperature combustion diesel engine: Effect of injection timing and exhaust gas recirculation. *Energy* **2012**, *38*, 78–84. [CrossRef]
32. Tahmasebzadehbaie, M.; Sayyaadi, H.; Sohani, A.; Pedram, M.Z. Heat and mass recirculations strategies for improving the thermal efficiency and environmental emission of a gas-turbine cycle. *Appl. Therm. Eng.* **2017**, *125*, 118–133. [CrossRef]
33. Chuahy, F.D.; Kokjohn, S.L. High efficiency dual-fuel combustion through thermochemical recovery and diesel reforming. *Appl. Energy* **2017**, *195*, 503–522. [CrossRef]
34. Kaiser, E.W.; Wallington, T.J.; Hurley, M.D.; Platz, J.; Curran, H.J.; Pitz, W.J.; Westbrook, C.K. Experimental and modeling study of premixed atmospheric-pressure dimethyl ether–air flames. *J. Phys. Chem A* **2000**, *104*, 8194–8206. [CrossRef]

35. Moran, M.J.; Shapiro, H.N.; Boettner, D.D.; Bailey, M.B. *Fundamentals of Engineering Thermodynamics*, 8th ed.; John Wiley & Sons: Hoboken, NJ, USA, 2012.
36. NASA, Glenn Research Center, Chemical Equilibrium with Applications. Available online: https://www.grc.nasa.gov/www/CEAWeb/ceaThermoBuild.htm (accessed on 21 August 2019).
37. Engineering ToolBox—Resources, Tools and Basic Information for Engineering and Design of Technical Applications! Available online: https://www.engineeringtoolbox.com/standard-atmosphere-d_604.html (accessed on 10 August 2020).
38. Oar-Arteta, L.; Aguayo, A.T.; Remiro, A.; Arandia, A.; Bilbao, J.; Ana, G. Gayubo. Kinetics of the steam reforming of dimethyl ether over CuFe2O4/γ-Al2O3. *Chem. Eng. J.* **2016**, *306*, 401–412. [CrossRef]
39. Fass, Y.; Tartakovsky, L. Limitations of Two-Stage Turbocharging at High Flight Altitudes. *SAE Int. J. Engines* **2018**, *11*, 511–524. [CrossRef]
40. Chakravarthy, V.K.; Daw, C.S.; Pihl, J.A.; Conklin, J.C. Study of the theoretical potential of thermochemical exhaust heat recuperation for internal combustion engines. *Energy Fuels* **2010**, *24*, 1529–1537. [CrossRef]
41. Amiel, R.; Tartakovsky, L. *Effect of Flight Altitude on the Knock Tendency of SI Reciprocating Turbocharged Engines*; SAE International: Warrendale, PA, USA, 2016; SAE Technical Paper 2016-32-0006.
42. Thawko, A.; Persy, S.A.; Eyal, A.; Tartakovsky, L. *Effects of Fuel Injection Method on Energy Efficiency and Emissions of SI Engine Fed with A Hydrogen-Rich Reformate (No. 2020-01-2082)*; SAE International: Warrendale, PA, USA, 2020; SAE Technical Paper 2020-01-2082.

drones

MDPI

Article

Modeling and Investigations on Surface Colors of Wings on the Performance of Albatross-Inspired Mars Drones and Thermoelectric Generation Capabilities

Devyn Rice [1], Samah Ben Ayed [2], Stephen Johnstone [3] and Abdessattar Abdelkefi [1,*]

[1] Department of Mechanical and Aerospace Engineering, New Mexico State University, Las Cruces, NM 88003, USA; drice20@nmsu.edu

[2] Department of Engineering Technology and Surveying Engineering, New Mexico State University, Las Cruces, NM 88003, USA; benayed@nmsu.edu

[3] Los Alamos National Laboratory, Los Alamos, NM 87545, USA; Stephen.e.johnstone@nasa.gov

* Correspondence: abdu@nmsu.edu

Received: 28 July 2020; Accepted: 13 August 2020; Published: 16 August 2020

Abstract: Thermal effects of wing color for Albatross-inspired drones performing in the Martian atmosphere are investigated during the summer and winter seasons. This study focuses on two useful consequences of the thermal effects of wing color: the drag reduction and the thermoelectric generation of power. According to its color, each wing side has a certain temperature affecting the drag. Investigations of various configurations have shown that the thermal effect on the wing boundary layer skin drag is insignificant because of the low atmospheric pressure. However, the total drag varies as much as 12.8% between the highest performing wing color configuration and the lowest performing configuration. Additionally, the large temperature differences between the top and the bottom wing surfaces show great potential for thermoelectric power generation. The maximum temperature differences between the top and bottom surfaces for the summer and winter seasons are, respectively, 65 K and 30 K. The drag reduction and the power generation via thermoelectric generators both contribute to enhancing the endurance of drones. Future drone designs will benefit from increased endurance through optimizing the wing color configuration.

Keywords: drones; Martian atmosphere; thermal analysis; aerodynamics; thermoelectric generators

1. Introduction

Over the past few decades, various government and private programs have been developed for space exploration. Among the high interest exploration targets is the planet Mars. NASA has performed many missions in the past and will continue with new mission in the future including the upcoming Mars 2020 mission [1]. The private company, SpaceX, hopes to establish a supply depot on Mars by the year 2024 as a cornerstone for future manned missions to the planet [2]. As Mars begins to generate greater interest in the near future, planetary exploration will become much more important. The current methods used to explore the Martian surface include rover, lander, and satellite [3]. In the past few years, the concept of using drones for planetary exploration has gained traction and many design concepts have been developed [3]. To design a drone suitable for Mars exploration, it is important to understand the Martian atmosphere and other properties for flight. Compared to Earth, the atmosphere of Mars is much thinner, and the temperature is also colder [4]. Atmospheric properties such as density, pressure, and temperature were obtained at varying altitudes from the high-speed entry of a probe into the atmosphere [5].

Due to the extreme conditions and challenges associated with flying in the Martian atmosphere, maximizing the efficiency of a drone is essential [3,4]. Drones require high endurance to be able to perform many defined missions for planetary exploration. The endurance of a drone is primarily limited by the available energy [6]. There are various propulsion systems and fuel/oxidizers that provide energy for drones, and there are additional methods, such as solar cells, batteries, etc., that provide energy to the drone [7,8]. Solar cells have been gaining popularity for use in space exploration for several systems. Orbitals and drones have had studies performed for Solar Electric Propulsion (SEP) applications [9,10]. When considering enhancing the endurance of a drone, there are various methods to take into consideration. Optimizing the propulsion system to increase the total available energy is one option. In the case of an electric propulsion system, recharging the battery by various means can enhance the endurance of the drone. Solar panel methods have been researched as a possible option to power or partially power drones [11–14]. Additionally, thermoelectric generators are another possible method to harness energy via converting the thermal energy from the temperature difference between two surfaces to electrical power [15]. Another option for enhancing the endurance of a drone is to optimize the structure so that energy can be best conserved which will result in a reduction of necessary power [16–21]. Control algorithms have also a great impact on the energy consumption [22,23].

Researchers have explored various techniques to decrease and manage the required power for drones. An example is to propose new design methods for drones, which improve their aerodynamic and mechanical efficiencies and consequently reduce the power consumption. Therefore, design methods enhance the performance and efficiency of drones which will extend their endurance and improve their utility in complex environments. Reduction of drag can be considered one of the main factors during the design process to conserve energy for the mission. In other words, drag reduction of fixed wing drones is crucial to flight efficiency on Mars [24].

The structure of a fixed wing drone can come in many different shapes and sizes. Often, when trying to optimize a drone, inspiration can be sought from nature. As seen from nature, on Earth there exists different modes of flight depending on the species of bird including flapping, hovering, gliding, and soaring [25]. Specifically, soaring and gliding are flight modes that are used by fixed wing drones. The nature of a bird's mode of flight was used in a study to design a drone capable of wing morphing. Two different flight conditions were achieved through either sweeping or spanning the wing [26]. Another study was previously performed to analyze the thermal effects of wing color on the resulting heated boundary layer flow over the wing of an Albatross [27]. The study analyzed different wing color configurations to identify the most efficient configuration. The results verify the drag efficiency of the natural wing color configuration of the Albatross (which is black on the top surface and white on the bottom surface) compared to other color configurations. To design an efficient drone capable of flying in the Martian atmosphere, a similar study will be performed to analyze the thermal effects of wing color on the heated boundary layer of an Albatross wing shape within the Martian atmosphere. To the authors' knowledge, a thermal boundary layer analysis of wing color configurations for drag efficiency calculations of a drone based on wing color on Mars has not been previously performed. This study will provide information on the surface temperature of a wing during different seasons and for different color configurations. An optimal wing color configuration will be determined to achieve maximum drag efficiency. Additionally, due to a possible temperature difference between the top and bottom wing surfaces, an investigation into the application of thermoelectric generators (TEGs) will be performed. Thermoelectric generators are composed of semiconductor material that produces thermoelectric modules. These modules are between ceramic wafers and then connected to an electrical load resistance. Electrical power is generated from the temperature difference as a result that can be explained through the Seebeck effect. The specific type of TEG considered for this study is the BiTe which is considered suitable for low-temperature heat recovery applications [28–30].

The paper is organized as follows: the thermal model describing the energy balance, wing color configurations, and the surrounding temperatures and heat fluxes is discussed in Section 2. Section 3 considers both the wing surfaces to investigate the skin drag forces and the effects of the angle of attack. A 3D aerodynamic analysis is performed to determine the total drag forces in Section 4. Section 5 studies the application of thermoelectric generators for drones in the Martian atmosphere. A summary and conclusions are drawn in Section 6.

2. Mathematical Model

An analytical model is developed using an energy balance to determine the wing surface temperature and consequently the skin friction drag. Atmospheric properties, such as sky, ground, and ambient temperatures are determined and applied to the energy balance. Each section of the energy balance is discussed in detail for the following sections: solar energy absorption, convective heat transfer, and radiative heat transfer. The conductive heat transfer will be defined in Section 4.

2.1. Energy Balance

To understand the thermal effects of different wing colors, a thermal analysis is performed using an energy balance and including the influences of solar energy absorption, convection heat loss, and radiative heat loss. The energy balance on a drone fixed wing can be expressed as:

$$P_{solar} = P_{convective} + P_{radiative} + P_{conductive} \quad (1)$$

The various sources of energy acting on the wing can be seen in Figure 1. The absorbed solar radiation is balanced by the convective, radiative and conductive transfer. Determining the various fluxes in the energy balance requires information from the Martian atmosphere. In this study, the ambient temperature, pressure, wind speed, and other atmospheric properties are obtained from the dataset collected by Viking Lander 1 and the entry probe (VL1), which is located at 22.37 N latitude and 47.97 W longitude on the surface of Mars [31].

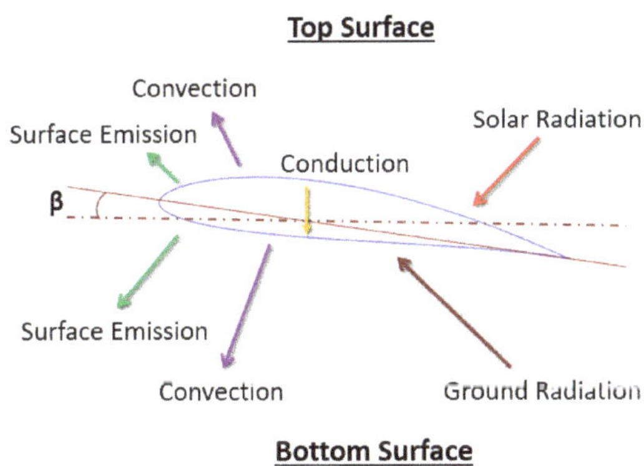

Figure 1. Heat flux on a wing.

The top and the bottom wing surfaces experience different heat fluxes, which can be determined by the view factors, surface temperatures, surface absorptance, and surface emissivity. The total energy balance equations for both the top and the bottom surfaces of the wing are given in Equations (2) and (3), respectively [32].

$$\alpha_t \cdot G_b \cdot \cos\theta + \alpha_t \cdot F_d^t \cdot G_{dh} + \alpha_t \cdot al \cdot F_{al}^t \cdot (G_{bh} + G_{dh}) - h \cdot (T_{s-top} - T_{amb}) - \varepsilon_t \cdot \sigma \cdot F_d^t \cdot (T_{s\text{-}top}^4 - T_{sky}^4) - \varepsilon_t \cdot \sigma \cdot F_{al}^t \cdot (T_{s-top}^4 - T_{grd}^4) - \frac{1}{R_{eff}} \cdot (T_{s\text{-}top} - T_{s\text{-}bottom}) = 0 \quad (2)$$

$$\begin{aligned}\alpha_b\cdot F_d^b\cdot G_{dh} \quad &+\alpha_b\cdot al\cdot F_{al}^b\cdot(G_{bh}+G_{dh})-h\cdot(T_{s\text{-bottom}}-T_{amb})\\ &-\varepsilon_b\cdot\sigma\cdot F_d^b\cdot(T_{s\text{-bottom}}^4-T_{sky}^4)\\ &-\varepsilon_b\cdot\sigma\cdot F_{al}^b\cdot(T_{s\text{-bottom}}^4-T_{grd}^4)\\ &-\tfrac{1}{R_{eff}}\cdot(T_{s\text{-bottom}}-T_{s\text{-top}})=0\end{aligned} \tag{3}$$

In the energy balance equations, the meanings of the different symbols are detailed in the nomenclature, using subscripts t and b to denote top and bottom wing surface, respectively. The convective heat transfer coefficient, denoted h is obtained from calculating the Nusselt Number on a flat plate in laminar regime as detailed in [33] using Mars atmospheric properties and assuming that the dominant gas is CO_2. These properties were obtained from [34]. The atmospheric density was determined by using ideal gas law at 8.5 mbar of atmospheric pressure.

2.2. Sky, Ground, and Ambient Temperatures

As mentioned previously, this study includes the summer and winter seasons. As such, the sky, ground, and ambient temperatures for both seasons are needed. The information for a summer day corresponds to a solar longitude of 140° and for a winter day corresponds to a solar longitude of 280°. Figure 2 shows the ambient temperatures based on the season. The ambient temperature for summer is higher than winter and has a larger variation over the day. The summer daily variation has a range of almost 50 °C, while the winter only has a daily variation of nearly 15 °C [31]. Figure 3 shows the sky and ground temperatures as reported by Matz et al. [32] at the same location on the same day. The ground temperature follows a similar trend where the summer temperatures are higher and vary more throughout the day compared to the winter ground temperatures. The sky temperatures, however, have the opposite trend. The winter sky temperature is higher and varies more compared to the summer sky temperatures.

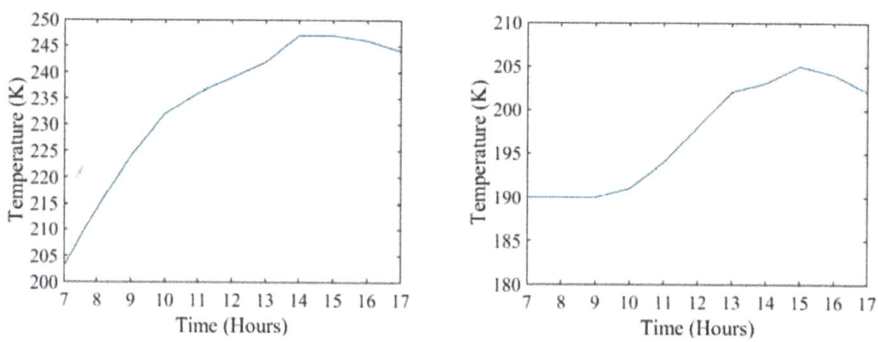

(a) (b)

Figure 2. Ambient temperature for (**a**) summer and (**b**) winter in the Martian atmosphere [31].

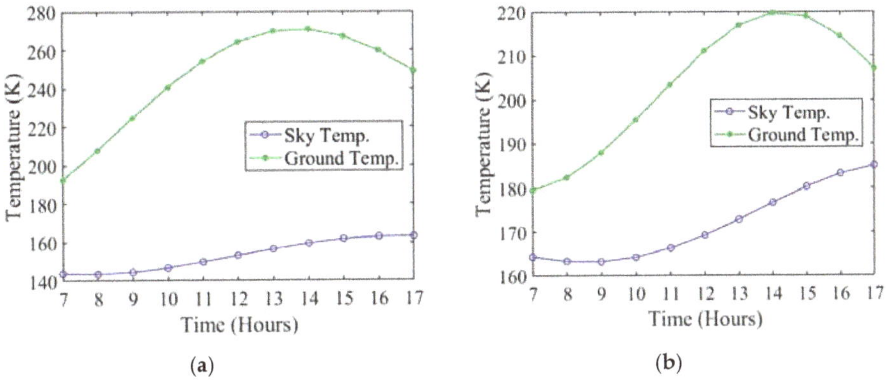

Figure 3. Sky and ground temperatures for (**a**) summer and (**b**) winter in the Martian atmosphere [32].

2.3. Solar Energy Absorption

Being inspired by migrating birds' flight on Earth, different color combinations are investigated. It is assumed that the top and bottom surfaces have different colors and therefore different absorptivity values. The different color configurations considered during this study are shown in Figure 4. The four configurations are on the top and bottom surfaces, respectively, black–white, black–black, white–black, and white–white. The solar irradiance that acts on the top surface of the wings includes direct beam, diffuse, and albedo radiations. The wing surface exposure to different radiation fluxes is governed by the view factors to the sky and the ground. The absorbed energy on the two wing surfaces can be written as [32]:

$$P_{\text{solar-top}} = \alpha_t \cdot G_b \cdot \cos\theta + \alpha_t \cdot F_d^t \cdot G_{dh} + \alpha_t \cdot al \cdot F_{al}^t \cdot (G_{bh} + G_{dh}) \quad (4)$$

$$P_{\text{solar-bottom}} = \alpha_b \cdot F_d^b \cdot G_{dh} + \alpha_b \cdot al \cdot F_{al}^b \cdot (G_{bh} + G_{dh}) \quad (5)$$

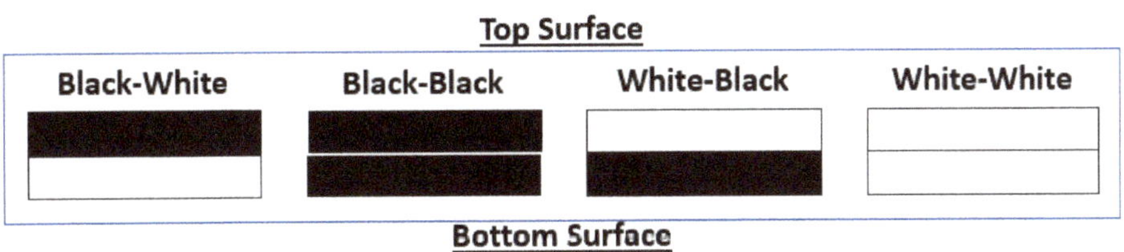

Figure 4. Wing color configurations in this study.

The view factors for the top and bottom surfaces are needed and are expressed by Equations (6)–(9), where β is the angle of attack. The view factor for the top surface to the sky is F_d^t while the view factor for the top surface to the ground is F_{al}^t. For the bottom surface, the view factor for the bottom to the sky F_d^b is identical to F_{al}^t. Likewise, the view factor from the bottom surface to the ground F_{al}^b is identical to F_d^t. In this case, the drone is considered to be flying very close to the ground because some of the atmospheric properties used were measured by VL1 at ground level. It is noted that at a zero-degree angle of attack Equations (2) and (3) are decoupled because the view factor from the top surface to the sky is 1 and the view factor for the top surface to the ground is zero.

$$F_d^t = \frac{1 + \cos\beta}{2} \quad (6)$$

$$F_{al}^t = \frac{1 - \cos\beta}{2} \quad (7)$$

$$F_d^b = F_{al}^t \quad (8)$$

$$F_{al}^b = F_d^t \quad (9)$$

The absorptivity values for α_t and α_b used in this study are 0.88 and 0.23, which correspond to the absorptance of anodize black paint and biphenyl white solid paint, respectively. Both are often used for space applications [35]. Depending on several variables, such as the solar angle of incidence shown in Figure 5, the optical depth of the Martian atmosphere, and the season, the solar irradiance changes. NASA researchers from Lewis Research Center in Ohio have derived a set of equations to determine the solar irradiance on Mars [36]. One of the most crucial elements for determining the solar irradiance is the position of the wing with respect to the sun. The solar incidence angle θ is used to describe this relation and can be determined from Equation (10), as follows [37]:

$$\cos\theta = \sin\varphi\sin\delta\cos\beta - \sin\delta\cos\varphi\cos\lambda\sin\beta + \cos\varphi\cos\delta\cos\omega\cos\beta + \cos\delta\sin\varphi\cos\lambda\cos\omega\sin\beta + \cos\delta\sin\lambda\sin\omega\sin\beta \quad (10)$$

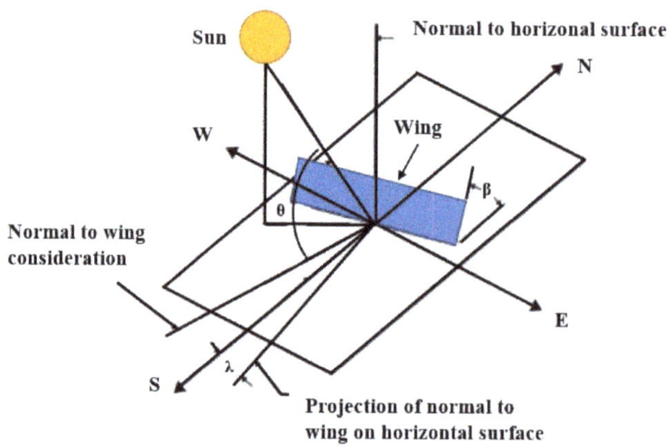

Figure 5. Solar angles.

To determine the solar incidence angle, positional information about the drone must be known, such as the latitude φ, declination angle δ, hour angle ω, angle of attack β, and the wing azimuth angle λ. The wing azimuth angle can be considered as the direction of flight. In this study, the drone is assumed to be flying directly southward. The hour angle only considers daylight hours from 7:00 to 17:00. Additionally, the hour angle refers to solar Martian time (0:00–24:00). The declination angle depends on the season which is described by the areocentric longitude L_s. The latitude selected for the study is the location of the rover which is where the atmospheric properties are obtained. The values for latitude, declination angle, tilt angle, and wing azimuth angle that are used in this study are shown in Table 1 for both summer and winter seasons.

Table 1. Location parameters [36].

	Summer (L_s = 140°)	**Winter (L_s = 280°)**
Latitude φ	22°	22°
Declination angle δ	17°	−24°
Tilt angle β	0°	0°
Wing Azimuth angle λ	0°	0°

After determining the solar incidence angle, the solar irradiance can then be determined. The direct beam irradiance G_b, diffuse irradiance on a horizontal surface G_{dh}, and the global irradiance on a horizontal surface G_h which are used to determine the beam, diffuse, and albedo irradiance, respectively, can be determined from Equations (11)–(13) [36]. The optical depth τ varies depending on the date which is represented by the areocentric longitude L_s. The optical depth is a result of the atmospheric dust particles in the atmosphere [38]. The values for the solar irradiance variables that are used for this study are shown in Table 2. The normalized net flux function $f(\theta, \tau)$ values [36] used during this study that correspond to the solar incidence angle and optical depth are presented in Table 3.

$$G_b = G_{ob} \cdot e^{\left(\frac{-\tau}{\cos\theta}\right)} \tag{11}$$

$$G_{dh} = G_h - G_{bh} \tag{12}$$

$$G_h = \frac{G_{ob} \cdot \cos\theta \cdot f(\theta, \tau)}{1 - al} \tag{13}$$

Table 2. Solar irradiance variables [36].

	Summer	Winter
Areocentric longitude	140°	280°
Optical depth	0.4	2.7
Mars eccentricity	0.093377	0.093377
Albedo	0.1	0.1

Table 3. Normalized net flux function [36].

Hour	f(θ,τ) (Summer)	f(θ,τ) (Winter)
7	0.667	0.136
8	0.75	0.175
9	0.802	0.28
10	0.815	0.325
11	0.824	0.36
12	0.827	0.37
13	0.824	0.36
14	0.815	0.325
15	0.802	0.28
16	0.75	0.175
17	0.667	0.136

The solar irradiance at the top of the atmosphere G_{ob}, which can be seen in Equation (14) [36], is needed along with the optical depth τ, solar incidence angle θ, and the normalized net flux function $f(\theta, \tau)$ to determine the direct beam irradiance and the global irradiance. Mars eccentricity b is around 0.093377, as can be seen in Table 2. The diffuse irradiance can be determined by subtracting the direct beam irradiance on a horizontal surface G_{bh} from the global irradiance on a horizontal surface G_h. The equation for direct beam irradiance on a horizontal surface can be expressed as shown in Equation (15) [36].

$$G_{ob} = \frac{590 \cdot [1 + b \cdot \cos(L_s - 248°)]^2}{(1 - b^2)^2} \tag{14}$$

$$G_{bh} = G_{ob} \cdot \cos\theta \cdot e^{\left(\frac{-\tau}{\cos\theta}\right)} \tag{15}$$

2.4. Radiative Heat Loss

Equations (16) and (17) represent the radiative heat loss for the top and bottom wing surfaces, respectively. The top and bottom wing surfaces will differ depending on the emissivity ε and the shape factors. For a zero-degree angle of attack, the top wing surface only loses energy to the sky while the bottom wing surface only loses energy to the ground. The values used for emissivity are based on anodize black paint and biphenyl white solid paint, respectively. For both white and black colors, the emissivity is 0.88 [24]. The sky and ground temperatures can be found in Figure 3a,b, respectively.

$$P_{\text{radiative-top}} = \varepsilon_t \cdot \sigma \cdot F_{dt} \cdot (T^4_{\text{s-top}} - T^4_{\text{sky}}) + \varepsilon_t \cdot \sigma \cdot F_{alt} \cdot (T^4_{\text{s-top}} - T^4_{\text{grd}}) \tag{16}$$

$$P_{\text{radiative-bottom}} = \varepsilon_b \cdot \sigma \cdot F_{db} \cdot (T^4_{\text{s-bottom}} - T^4_{\text{sky}}) + \varepsilon_b \cdot \sigma \cdot F_{alb} \cdot (T^4_{\text{s-bottom}} - T^4_{\text{grd}}) \tag{17}$$

2.5. Thermal Boundary Layer Analysis

For simplicity, the wing curvature is ignored, and a flat plate is considered for the boundary layer analysis. Figure 6 represents a schematic view of the boundary layer over a flat plat [25].

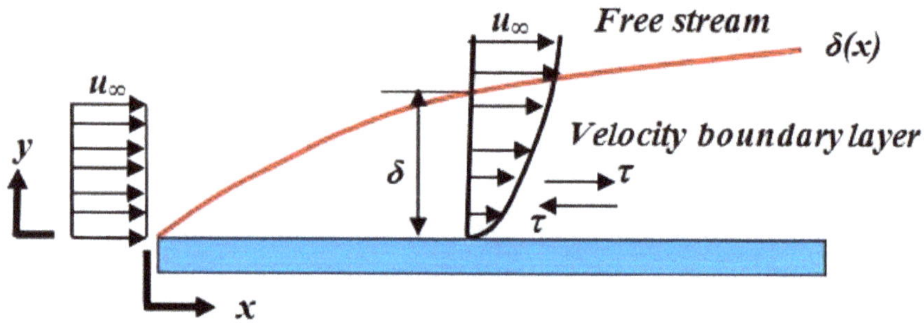

Figure 6. Considered boundary layer over a flat plate [25].

The thermal boundary layer effects on the skin friction drag are determined using Equation (18) [39], where the wingspan ws, is 3.5 m and the chord length, L, is 0.22 m [7]. To determine the viscosity μ, Sutherland's formula, which can be found in Equation (19), is used [40]. T_0 and μ_0 are the reference temperature and viscosity, respectively, which are 293.14 K and 1.48×10^{-5} $\frac{kg}{m \cdot s}$ for the Martian atmosphere [41]. The Sutherland constant C, for CO_2 is 240 K [41] and T_s is the wing surface temperature. The density ρ is determined using ideal gas law considering CO_2 in Equation (20). The ideal gas constant for the Martian atmosphere R, is 192.1 $\frac{J}{kg \cdot K}$ [30]. The pressure P is found from the information collected by VL1.

$$D = 0.664 \cdot ws \cdot \sqrt{\rho \cdot \mu \cdot u^3 \cdot L} \tag{18}$$

$$\mu = \mu_0 \cdot \left(\frac{T_0 + C}{T_s + C}\right) \cdot \left(\frac{T_s}{T_0}\right)^{3/2} \tag{19}$$

$$\rho = \frac{P}{192.1R \cdot T_s} \tag{20}$$

3. Wing Surface Analysis

In this section, the whole wing (top and bottom surfaces) is considered for the study. The wing structure will be considered as a hollow wing composed of a hollow center with a top layer and a bottom layer of material. The model of the overall wing can be seen in Figure 7a,b. From Figure 7b, it can be observed that the conductive heat transfer will cover three layers of material; the first layer of material, the interior which is air (CO_2 for the Martian atmosphere), and the third layer of material.

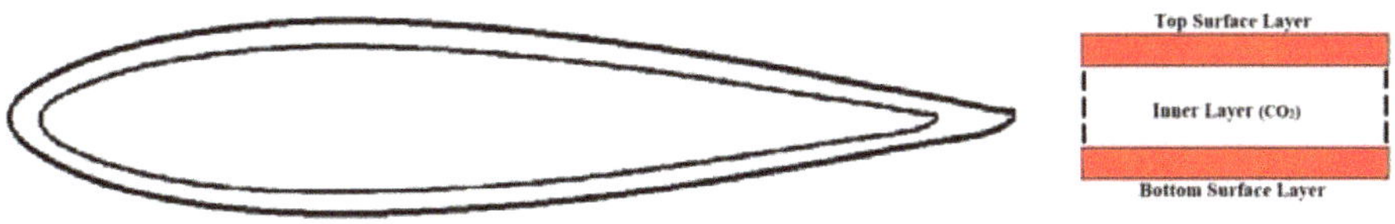

(a) (b)

Figure 7. (**a**) Hollow airfoil and (**b**) conduction model.

The expression for heat conduction is given in Equation (21), where the conduction is determined from the temperature difference between the top surface $T_{s\text{-top}}$, and the bottom surface $T_{s\text{-bottom}}$ divided by the effective resistance of all materials. The effective resistance is presented in Equation (22) where $L_{surface}$ is the depth of the wing material, L_{air} is the average depth of the hollow part of the wing, $K_{surface}$ denotes the thermal conductivity of the material, K_{air} represents the thermal conductivity of the Martian atmosphere, $A_{surface}$ is the area of the material, and A_{air} denotes the area of the hollow part.

$$P_{conductive} = \frac{T_{s\text{-top}} - T_{s\text{-bottom}}}{R_{eff}} \tag{21}$$

$$R_{eff} = \frac{2L_{surface}}{K_{surface}A_{surface}} + \frac{L_{air}}{K_{air}A_{air}} \tag{22}$$

Air (CO_2) acts as an insulator in this case which prevents conductive heat transfer from occurring between the top surface and the bottom surface. The surface temperature for the top and bottom surfaces with and without considering conductive heat fluxes are shown in Figure 8 for summer.

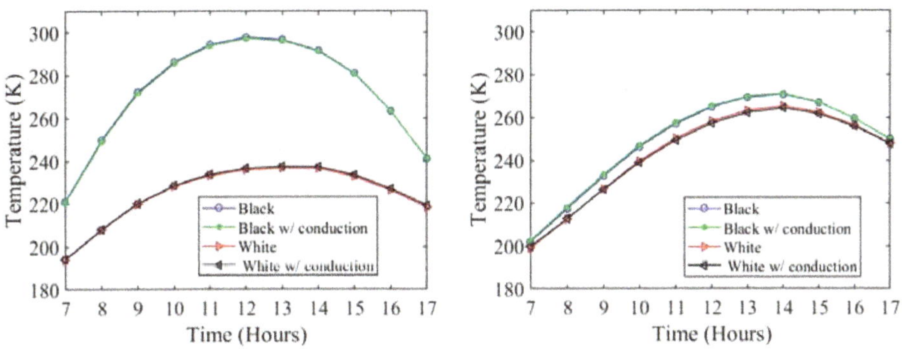

(a) (b)

Figure 8. Surface temperature with and without conduction for (**a**) top surface and (**b**) bottom surface.

Air has a very low thermal conductivity, which will result in a very high thermal resistance. The heat conducted through the wing can be considered negligible compared to the total energy balance. This allows Equations (2) and (3) to be simplified into Equations (23) and (24). The two new energy balance equations become uncoupled without the conductive heat transfer and are used for the rest of this study.

$$\alpha_t \cdot G_b \cdot \cos z + \alpha_t \cdot F_d^t \cdot G_{dh} + \alpha_t \cdot al \cdot F_{al}^t \cdot (G_{bh} + G_{dh}) - h \cdot (T_{s\text{-top}} - T_{amb}) - \varepsilon_t \cdot \sigma \cdot F_d^t \\ (T_{s\text{-top}}^4 - T_{sky}^4) - \varepsilon_t \cdot \sigma \cdot F_{al}^t \cdot (T_{s\text{-top}}^4 - T_{grd}^4) = 0 \tag{23}$$

$$\alpha_b \cdot F_d^b \cdot G_{dh} + \alpha_b \cdot al \cdot F_{al}^b \cdot (G_{bh} + G_{dh}) - h \cdot (T_{s\text{-bottom}} - T_{amb}) - \varepsilon_b \cdot \sigma \cdot F_d^b \cdot (T_{s\text{-bottom}}^4 - T_{sky}^4) - \varepsilon_b \cdot \sigma \cdot F_{al}^b \cdot (T_{s\text{-bottom}}^4 - T_{grd}^4) = 0 \quad (24)$$

When considering an angle of attack of zero degrees, the equations are completely decoupled and thus can be solved individually. Therefore, the effects of the conduction are not considered in further analysis. As can be seen in Figure 9a,b, the solar absorption for the top surface (both white and black colors) are higher than the bottom surface. This is due to the direct beam irradiance that is only experienced by the top surface. Additionally, the solar absorption during summer is higher than during the winter due to the high solar intensity that occurs during the summer period of the Martian solar cycle.

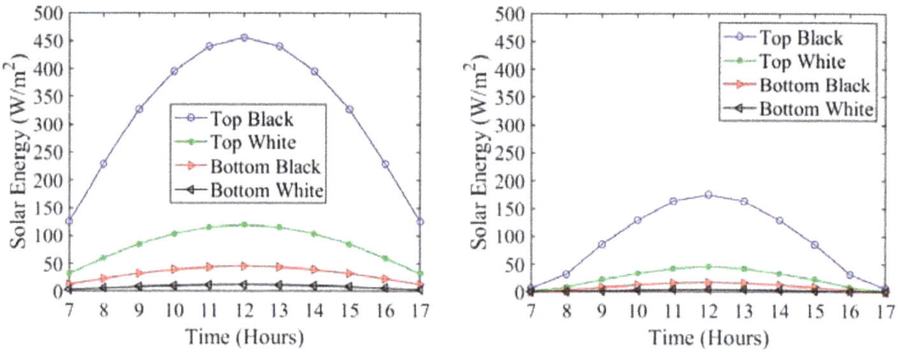

(a) (b)

Figure 9. Surface solar absorption for (**a**) summer and (**b**) winter.

The two new energy balance Equations (23) and (24) can then be used to determine the surface temperature for both the bottom and top wing surfaces for black and white wing colors as presented in Figure 10. During the summer and winter seasons, the wing surface experiences the highest temperatures when a black color is applied to the top surface. The lowest wing temperatures occur when white paint is used on the top surface. As expected, due to the high solar absorption during summer, the overall temperatures on the wing surfaces are higher during summer compared to winter.

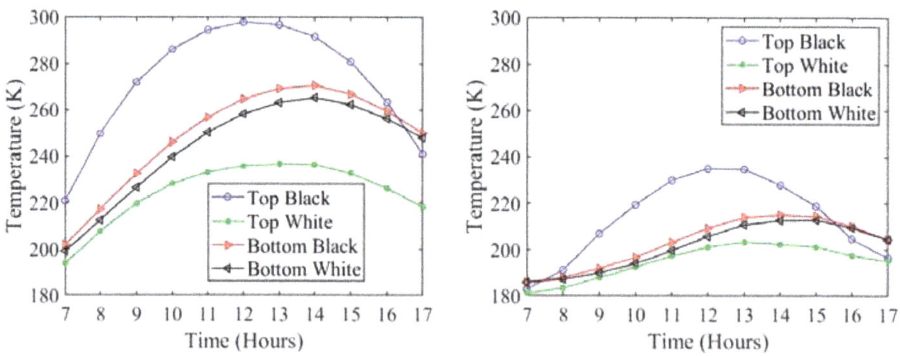

(a) (b)

Figure 10. Surface temperature for (**a**) summer and (**b**) winter.

Using the surface temperature, Equations (18)–(22) can be solved to find the viscosity, density, and drag. The viscosity for both wing surfaces and for black and white colors are plotted in Figure 11a,b, respectively. The viscosity follows a similar trend to the surface temperature. The higher surface temperature wing color configuration results in a higher viscosity along the boundary. The density, which can be seen in Figure 12a, is lower along the boundary layer when the surface temperature is higher. The total drag of the combined wing color configurations is plotted in Figure 13a,b for summer and winter, respectively. The four different wing color combinations are as follows: black on top and black on bottom (black–black), black on top and white on bottom (black–white), white on top and black on bottom (white–black), and white on top and white on bottom (white–white). It follows from Figure 13a, during the summer, that the highest drag color configuration is the white–white, and the lowest drag is the black–black. During the winter, it is shown in Figure 13b that the highest drag is from the color configuration black–black and the lowest drag is from the color configuration white–white.

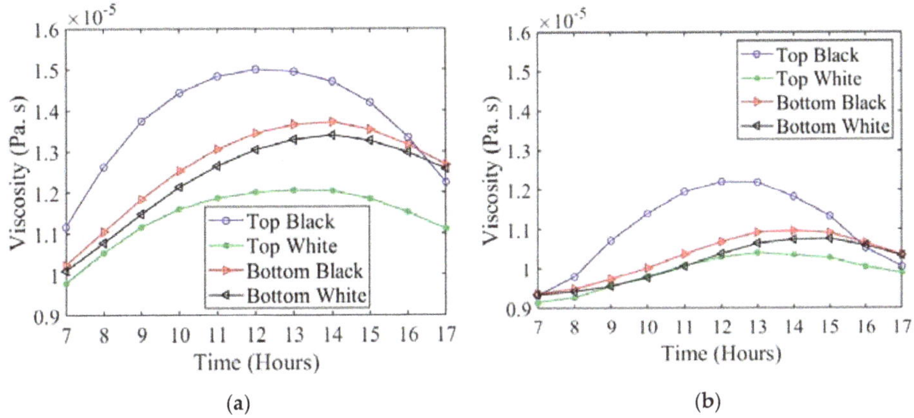

Figure 11. Surface viscosity for (**a**) summer and (**b**) winter.

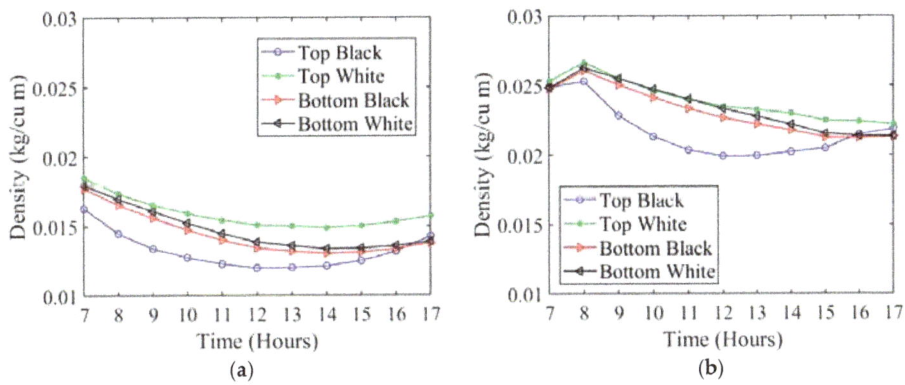

Figure 12. Surface density for (**a**) summer and (**b**) winter.

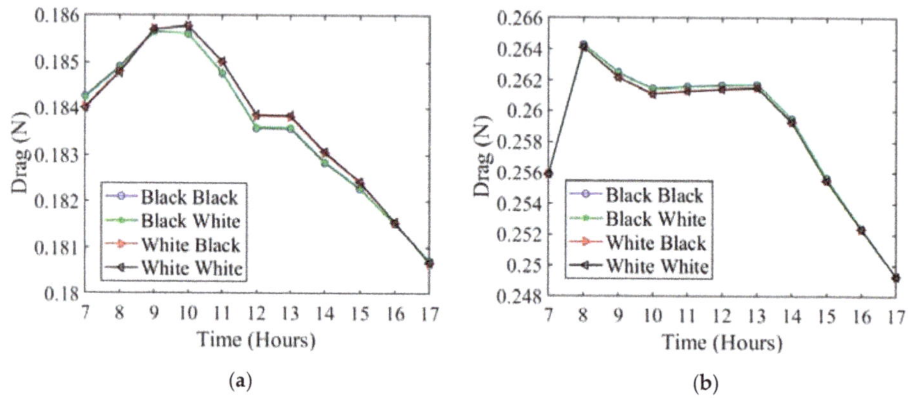

Figure 13. Surface drag for (**a**) summer and (**b**) winter in the Martian atmosphere.

Comparing the relative change between the highest drag color configuration and the lowest drag color configuration results in Figure 13a,b. During the summer the highest drag color configuration is white–white and the lowest is the black–black color configuration. During the winter, the trend is reversed with the highest drag color configuration being black–black and the lowest color configuration white–white.

During the summer season, the optimal wing color configuration is black–black which results in a decrease of drag by a maximum of 0.125% compared to the white–white wing color configuration. During the winter season, the optimal wing color configuration is white–white which has a maximum drag reduction of 0.15% compared to the black–black configuration, as shown in Figure 14.

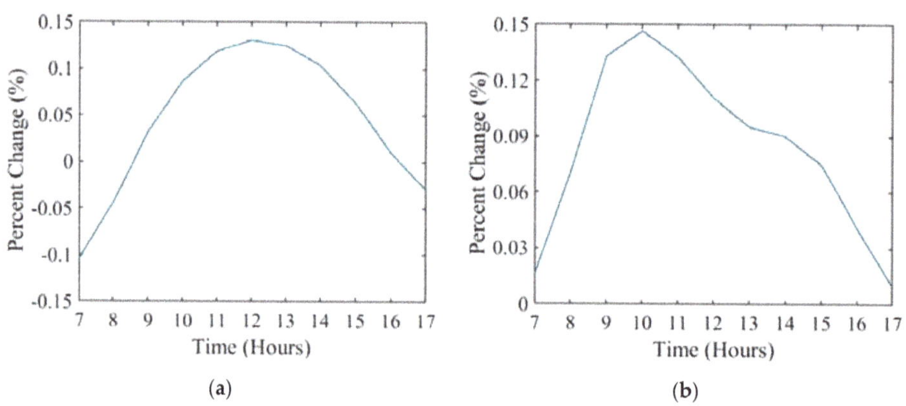

Figure 14. Relative change for (**a**) summer and (**b**) winter in the Martian atmosphere.

Until now, the results have been generated using a zero-degree angle of attack (AoA). To understand the effects of the angle of attack, the drag as a function of angle of attack at solar noon is plotted in Figure 15. The change in drag during the summer period is near constant over a range of 10 degrees AoA. During the winter, the maximum change in drag occurs for the white–black color configuration with a change percentage of 0.001%. It can be concluded that the change in angle of attack is negligible to determining the drag force. This led to using a constant zero-degree angle of attack for the study.

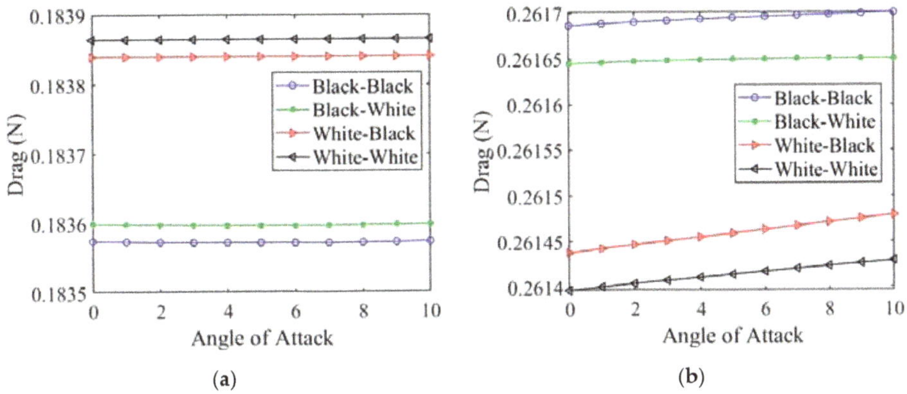

Figure 15. Drag as a function of angle of attack for (**a**) summer and (**b**) winter.

4. 3D Aerodynamic Analysis

The skin friction drag has already been determined based on the boundary layer analysis in Section 3. In this section, the aerodynamic performance for the whole wing will be analyzed by determining the total drag. The drag will be determined using an aerodynamic analysis software named XFLR5. The total drag includes the form drag, the induced drag, and the parasitic drag. The software used to analyze the 3D wing geometry can determine the drag coefficients. The study considers a constant zero-degree angle of attack under similar conditions as the boundary layer analysis from Section 3 and the two results are compared. Four analysis methods, namely, Lifting Line Theory (LLT), Horseshoe Vortex (VLM1), Ring Vortex (VLM2), and 3D panel method are used in order to estimate the total drag of an Albatross-inspired drone in the Martian atmosphere. All of these methods consider the assumption of inviscid flow.

The LLT method uses the assumption that the lift coefficient as a function of the angle of attack is linear. One major assumption for the LLT method is that the surfaces are in the X-Y plane and that the dihedral angles and the sweep are not needed for the lift distribution calculations. The lift of the wing for this theory is determined from the incremental vortices shed along the trail span of the wing along the freestream direction [42].

The VLM1 and VLM2 methods allow for more freedom in the choice of wing geometry. This includes winglets, high dihedral angles, low aspect ratio, and sweep. For the VLM methods, the general idea is to model the perturbation of the wing planform using a vortices' distribution sum. The lift coefficient is computed by integrated surface forces, such as the moment coefficients. The VLM methods calculate the lift distribution, induced angles, and induced drag. Because of this, the methods are independent of the wing speed. The VLM methods assume small angles of attack, so stall angles should be avoided when computing results. The main difference between the VLM1 and VLM2 methods is that the VLM1 method does not consider side slip [42].

The 3D panel method is able to account for the thickness of the wing unlike the VLM methods that only consider the mean camber line. Because of this, the 3D panel method can allow for an understanding of the center of pressure distribution over the top and bottom surfaces of the wing. The 3D panel method uses an approach that sum the doublets and sources distributed of the surface of the wing from the perturbations generated. The 3D panel method can be used to polish the results of the other three methods, so this is often considered to be the most accurate method. The 3D panel method will be considered as the standard for this study that the other three methods will be compared to [42].

As mentioned in the previous section, the 3D aerodynamic analysis methods all have inviscid assumptions. To account for the viscosity in the 3D analysis, a 2D viscous analysis is performed for the selected airfoil to obtain 2D viscous information. The 2D information is then interpolated and then incorporated into the 3D analysis and this is how the viscous effects are included in the analysis. The 2D boundary layer problem is solved using a software called XFoil by using an iterative method along with the inviscid velocity field as an input. The method provides a viscous velocity from the boundary layer solution which is used as an input for the potential flow solver. This iterative method is called the "Interactive Boundary Layer" [42]. The wing shape considered for the 3D analysis is that of an Albatross. Figure 16 shows the analyzed wing shape and Table 4 lists the geometric properties.

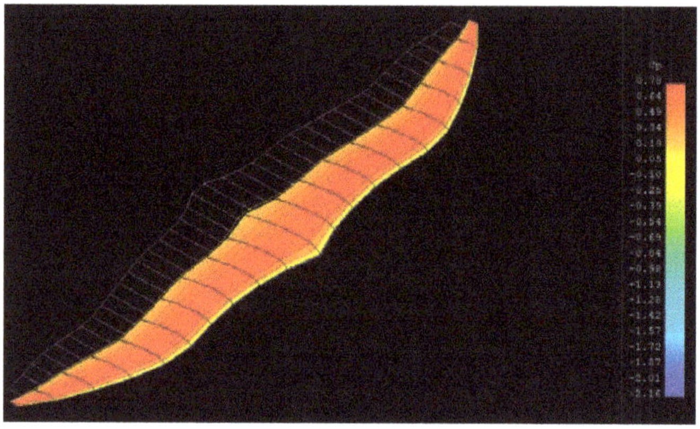

Figure 16. 3D wing shape.

Table 4. Wing geometry of the drone.

Wing Span (m)	Wing Area (m^2)	Mean Aerodynamic Chord (m)	Aspect Ratio
3.5	0.69	0.22	17.6

To solve the 2D and 3D analyses, certain properties are needed, namely, the Reynolds number and wing velocity. The wing velocity and the Reynolds number for the 3D analysis are the same as for the boundary layer analysis. The Reynolds number is calculated based on the viscosity and density found in Section 3. The total drag varies depending on the analysis method. As mentioned previously, the 3D panel method is considered to be the most accurate method and the other three methods are compared to the 3D panel method to determine their accuracy. The number of panels on the wing is determined in order for the different methods of analysis to converge. Figure 17 compares the total drag for the different analysis methods for both summer and winter. Inspecting the plotted curves in these figures, it is clear that VLM1 and VLM2 methods have comparable results to the 3D panel method. The LLT method varies greatly compared to the other three methods. In Tables 5 and 6, the percent difference between the different methods compared to 3D panel method is shown. The percent difference between VLM1 and VLM2 compared to the 3D panel method is 3~4%, which shows good agreement. However, the LLT method compared to the 3D panel method has a greater than 50% error, which shows that the LLT method is not capable of handling the complex geometry of the Albatross-like wing shape.

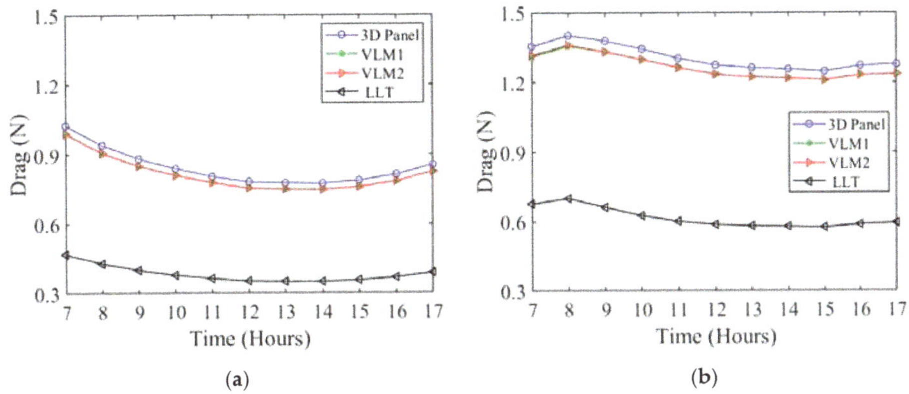

Figure 17. Total drag comparison of the different analysis method for black–black color configuration for (**a**) summer and (**b**) winter.

Table 5. Drag percent difference compared to 3D panel method during the summer.

	Black–Black (%)	Black–White (%)	White–Black (%)	White–White (%)
VLM1	3.49	3.51	3.50	3.50
VLM2	3.50	3.53	3.52	3.51
LLT	54.44	54.44	54.41	54.41

Table 6. Drag percent difference compared to 3D panel method during the winter.

	Black–Black (%)	Black–White (%)	White–Black (%)	White–White (%)
VLM1	3.22	3.22	3.39	3.38
VLM2	3.16	3.16	3.25	3.21
LLT	52.96	52.85	52.11	51.90

The total drag forces for the wing at a constant angle of attack and at different times throughout the day is calculated using the 3D panel method. The total drag includes the form drag, the induced drag, and the parasitic drag. Figure 18 shows the total drag dependent on the hour for the 3D panel method. Compared to the drag determined using the boundary layer analysis, the drag calculated using the 3D panel method is much larger, by a factor of 5 on average. During the winter, the drag is much higher than during the summer. This remains consistent with the trend found for the skin friction drag. Additionally, the black–black wing color configuration is the most optimal case that results in the least amount of drag for both summer and winter. The percent change between the most optimal wing color configuration (black–black) and the least optimal color configuration (white–white) is much more significant when considering the total drag instead of just the skin drag. For summer the percent change is 12.6% and for the winter the change is 6.8%. This demonstrates that the effects of the skin drag are significantly less compared to the total drag.

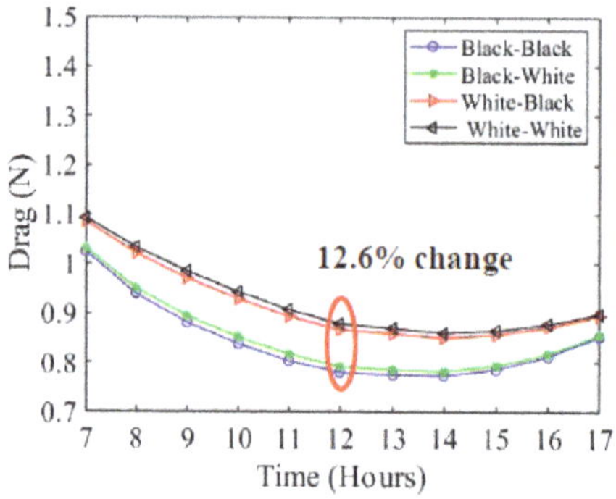

(**a**)

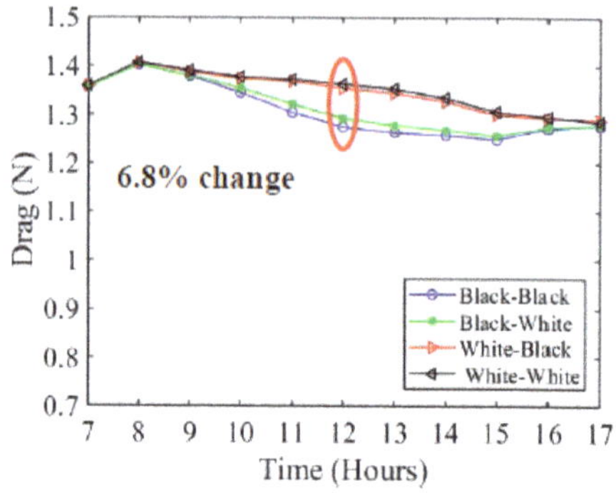

(**b**)

Figure 18. Total drag as a function of time using 3D panel method for (**a**) summer and (**b**) winter.

5. Thermoelectric Energy Generators in Martian Atmosphere

The basis of a thermoelectric generator is the ability to convert heat energy into electrical power when there exists a temperature difference between two surfaces. Most TEGs are created using pairs of thermocouples that are connected in series (electrical) and parallel (thermal) in between two plates. A representation of a typical TEG is presented in Figure 19. The power produced from the TEG is the difference between the heat supply P_h and heat removal P_c which is also equal to the product of the voltage across the external load resistance V and the current I. The energy balance can be expressed as:

$$P_{TEG} = P_h - P_c = V \cdot I \tag{25}$$

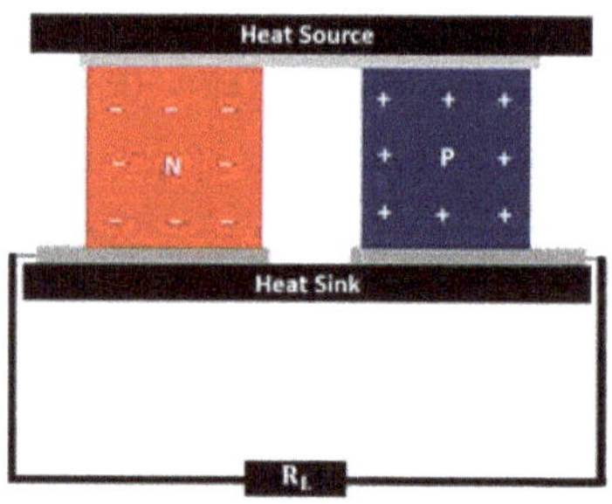

Figure 19. Schematic of a thermoelectric generator.

The derived equation for the electrical harvested power is written as [15]:

$$P_{TEG} = R_L \cdot \frac{\alpha^2 \cdot (T_h - T_c)^2}{(R_i - R_L)^2} \tag{26}$$

It follows from Equation (26) that the electrical power is dependent on the Seebeck coefficient α, the internal resistance R_i, the external resistance R_L, and the temperature difference between the hot surface T_h and the cold surface T_c. Depending on the specific TEG and the manufacturer, the coefficients will differ. It is often assumed that the internal properties of a TEG device are constant and independent of temperature. However, it has been demonstrated through experimental data by Hsu et al. [28] that the internal properties vary as a dependent on the hot side temperature. For the experiment, Hsu et al. considered a TEG module (TMH400302055, Wise Life Technology, Taiwan) and varied the hot side temperature to determine the effects on the Seebeck coefficient and the internal resistance. Expressions for both the internal resistance and the Seebeck coefficient are determined by [15]. The expressions are validated to the experimental data by Hsu et al. [28]. The expressions for the Seebeck coefficient and the internal resistance are given by:

$$\alpha(T) = a_1 T^2 + a_2 T + a_3 \quad (27)$$

$$R_i(T) = b_1 T + b_2 \quad (28)$$

The temperature range that is experienced by the wing surfaces on Mars are 180–300 K. In Figure 20, the internal properties for the TEG under consideration are determined by using the expressions above for the Seebeck coefficient and the internal resistance.

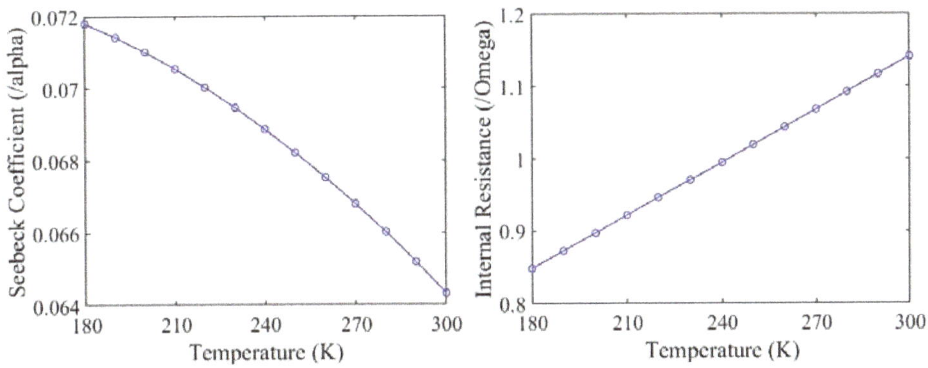

(a) (b)

Figure 20. Internal properties of a TEG device over a temperature range for (**a**) Seebeck coefficient and (**b**) internal resistance.

An analysis is performed to determine the harvested power over a range of values for the external resistance, Seebeck coefficient, and internal resistance. These coefficients were determined over a range of temperatures from 180–300 K. According to Figure 20, the Seebeck coefficient ranges from 0.064–0.072 (V/K) and the internal resistance ranges from 0.85–1.15 (Ω). However, the set of temperatures were obtained using data from one specific day each during the summer and winter. Consequently, to account for a certain amount of variation, a Seebeck coefficient ranging from 0.05–0.08 (V/K) and an internal resistance ranging from 0.8–1.3 (Ω) will be used for the parametric study and performance analysis. Figure 21 shows a 3D plot representing the harvested power as a function of the external resistance and the Seebeck coefficient with a fixed value for the internal resistance during summer for each color configuration. Figure 21a–d are for the color configurations black–white, black–black, white–black, and white–white, respectively. Each color configuration has a different time of day for which the maximum power occurs. This is because the maximum power is occurring when the temperature difference between the top and bottom surfaces is greatest. For the maximum harvested power, the time of day for each color configuration (a–d) is 10:00, 10:00, 15:00, and 15:00 solar time.

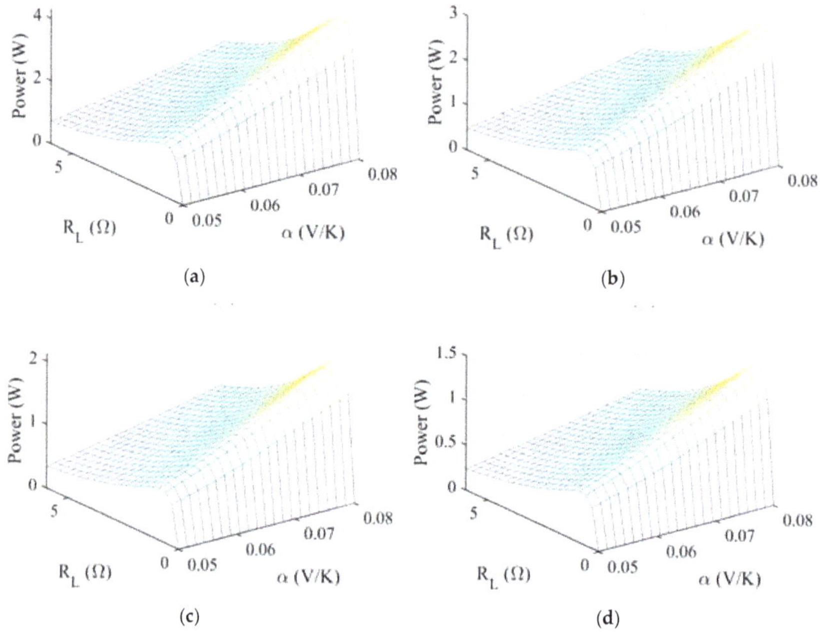

Figure 21. Harvested power as a function of the external resistance and Seebeck coefficient when Ri = 0.8 Ω for (**a**) Black–white (10:00), (**b**) Black–black (10:00), (**c**) White–black (15:00), and (**d**) White–white (15:00).

The optimal power generation for winter and summer is shown in Figure 22. The corresponding time of day for each color configuration is also shown. The black–white color configuration has the largest power generation while the white–white color configuration has the lowest. During the earlier part of the day, when the top surface of the wing is black, the TEG experiences higher power generation. We can derive the optimal power generation $Q_{P\text{-}Optimal}$ from Equation (26) using the expression:

$$\frac{\partial P_{TEG}}{\partial R_L} = 0 \tag{29}$$

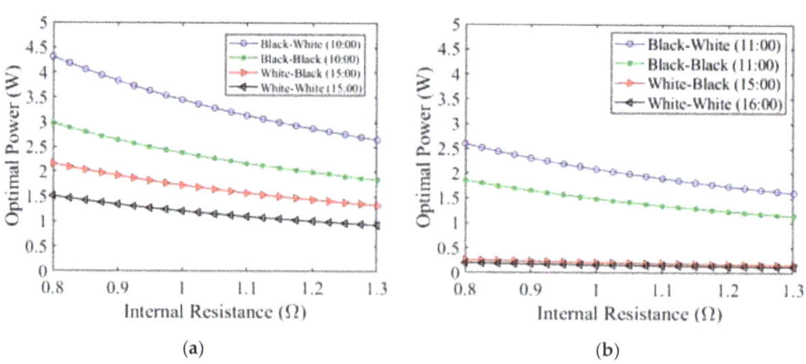

Figure 22. Optimal harvested power as a function of internal resistance for (**a**) summer and (**b**) winter when the Seebeck coefficient is equal to 0.08.

The resulting optimal power generation equation is:

$$P_{TEG\text{-}Optimal} = \frac{\alpha^2 \cdot (\Delta T)^2}{4 \cdot R_i} \tag{30}$$

In Table 7, the optimal power generation along with the maximum temperature difference and the time of day for each color configuration for both summer and winter is given. The optimal wing color configuration that generates the most harvested power is the black–white configuration for both summer and winter. This is due to having the highest temperature difference. The white–white configuration has the lowest power generation and the lowest temperature difference. Overall, the TEG performs better during the summer than during the winter.

Table 7. Optimal power comparison (Time, ΔT_{max}, Power).

	Black–White	Black–Black	White–Black	White–White
Summer	(10:00, 46.3 °C, 4.3 W)	(10:00, 38.4 °C, 2.9 W)	(15:00, 32.6 °C, 2.1 W)	(15:00, 27.2 °C, 1.5 W)
Winter	(11:00, 36.1 °C, 2.6 W)	(11:00, 30.5 °C, 1.8 W)	(15:00, 11.6 °C, 0.3 W)	(16:00, 10.0 °C, 0.2 W)

6. Conclusions

To enhance the design of an efficient drone capable of flying in the Martian atmosphere, a parametric study was performed to analyze the thermal effects of wing color on the heated boundary layer of a wing. This study provided information on the optimal wing color configuration during different seasons on Mars considering the color configurations of black–black, black–white, white–black, and white–white. A model based on an energy balance, and including wing color configurations, sky, ground, and ambient temperatures, solar energy absorption, convective heat transfer, and radiative heat loss, was developed and discussed. The adiabatic case, which only includes the top wing surface and drag effects, was analyzed. The effects of conduction and angle of attack on the total energy balance were studied. They both had a negligible effect on the thermal energy balance. The drag effects considering the color configurations on both sides of the wing were investigated. The first method analyzed the boundary layer and the skin friction drag, while the second method included a 3D analysis that investigated the total drag. The skin drag did not significantly change as a result of the color configuration. However, the total drag experienced a change percent of 12.8% during summer and 6.8% during winter. This shows that the skin drag does not play a significant role in the total drag on Mars due to the low atmospheric pressure. The study performed on the thermoelectric generator yielded interesting results. Due to the high temperature differences on the top and bottom wing surfaces that occur on Mars, TEGs are capable of high power generation. The black–white configuration had the largest temperature difference and resulted in the greatest power generation for both respective seasons, summer and winter.

Author Contributions: Conceptualization, D.R., S.B.A., S.J., and A.A.; methodology, D.R., S.B.A., and A.A.; software, D.R.; formal analysis, D.R. and S.B.A.; investigation, D.R.; resources, A.A.; writing—original draft preparation, D.R.; writing—review and editing, S.B.A., S.J., and A.A.; supervision, S.B.A., S.J., and A.A.; project administration, A.A.; funding acquisition, S.J. and A.A. All authors have read and agreed to the published version of the manuscript.

Funding: This research was funded by New Mexico Consortium and Los Alamos National Laboratory.

Acknowledgments: The authors D. Rice and A. Abdelkefi would like to acknowledge the financial support from New Mexico Space Grant Consortium. Additionally, the authors would like to thank Mostafa Hassanalian for his fruitful discussions in this subject.

Conflicts of Interest: The authors declare no conflict of interest.

Nomenclature

A = surface area
al = Albedo
b = Mars eccentricity
C = Sutherland's constant
D = Drag force
$f(z,\tau)$= Normalized net flux
F_{dt}= View factor top-side to the sky
F_{alt}= View factor top-side to the ground
F_{db}= View factor bottom-side to the sky
F_{alb}= View factor bottom side to the ground
G_b= Direct beam irradiance on surface
G_{bh}= Direct beam irradiance horizontal surface
G_{dh}= Diffuse irradiance on horizontal surface
G_h= Global irradiance on horizontal surface
G_{ob}= Beam irradiance at top of the Atmosphere
h = heat transfer coefficient
I = Current
K = Atmospheric thermal conductivity
L = Chord length
L_s= Areocentric Longitude
Nu = Nusselt number
P = Atmospheric Pressure
Pr = Prandtl number
P_h and P_c = Hot and cold sources
$P_{Conductive}$ = Conductive heat transfer
$P_{convective}$= Convective heat loss
$P_{radiative}$= Radiative heat loss
P_{solar}= Solar energy absorbed
P_{TEG}= Electrical Power
R = Ideal gas constant
R_i = Internal resistance
Re = Reynolds number
R_{eff} = Effective conductive resistance of the wing
R_L = External resistance
T_0= Reference temperature for Sutherland's
T_{amb}= Ambient temperature
T_{grd}= Ground temperature
T_s= Wing surface temperature
T_{sky}= Sky temperature
u = velocity
V = Voltage across external load resistance
ws = Wing span
α = Seebeck coefficient
α_t= Top surface absorptance
α_b= Bottom surface absorptance
β = Angle of attack
δ = Declination angle
ε_t= Top surface emissivity
ε_b= Bottom surface emissivity
θ = Solar incidence angle
λ = Wing azimuth angle (direction)

μ_0= Reference viscosity for Sutherland′s Formula
μ = Boundary layer viscosity
μ_a= Atmospheric viscosity
ρ = Boundary layer density
ρ_a= Atmospheric Density
σ = Stefan-Boltzmann Constant
τ = Optical depth
φ = Latitude
ω = Hour angle

References

1. Mars 2020 Perseverance Rover. Available online: https://mars.nasa.gov/programmissions/missions/future/mars2020/ (accessed on 2 February 2020).
2. SpaceX—Missions: Mars. Available online: https://www.spacex.com/mars (accessed on 2 February 2020).
3. Hassanalian, M.; Rice, D.; Abdelkefi, A. Evolution of space drones for planetary exploration: A review. *Prog. Aerosp. Sci.* **2018**, *97*, 61–105. [CrossRef]
4. Hassanalian, M.; Rice, D.; Johnstone, S.; Abdelkefi, A. Performance analysis of fixed wing space drones in different solar system bodies. *Acta Astronaut.* **2018**, *152*, 27–48. [CrossRef]
5. Intrieri, P.F.; De Rose, C.E.; Kirk, D.B. Flight characteristics of probes in the atmospheres of Mars, Venus and the outer planets. *Acta Astronaut.* **1977**, *4*, 789–799. [CrossRef]
6. Abdelkefi, A.; Ghommem, M. Piezoelectric energy harvesting from morphing wing motions for micro air vehicles. *Theor. Appl Mechs Lett.* **2013**, *3*, 052004. [CrossRef]
7. Nam, T. A Generalized Sizing Method for Revolutionary Concepts under Probabilistic Design Constraints. Ph.D. Thesis, Georgia Institute of Technology, Atlanta, GA, USA, May 2007.
8. Kuhl, C. Design of a Mars Airplane Propulsion System for the Aerial Regional-Scale Environmental Survey (ARES) Mission Concept. In Proceedings of the 44th AIAA/ASME/SAE/ASEE Joint Propulsion Conference Exhibit, Hartford, CT, USA, 21–23 July 2008; AIAA: Reston, VA, USA, 2008.
9. Díaz, F.C.; Carr, J.; Johnson, L.; Johnson, W.; Genta, G.; Maffione, P.F. Solar electric propulsion for human mars missions. *Acta Astronaut.* **2019**, *160*, 183–194. [CrossRef]
10. Woolley, R.C.; Baker, J.D.; Landau, D.F.; Nicholas, A.K. Cargo logistics for a Notional Mars Base using Solar Electric Propulsion. *Acta Astronaut.* **2019**, *156*, 51–57. [CrossRef]
11. Klesh, A.T.; Kabamba, P.T. Solar-powered aircraft: Energy-optimal path planning and perpetual endurance. *J. Guid. Control. Dyn.* **2009**, *32*, 1320–1329. [CrossRef]
12. Huang, Y.; Chen, J.; Wang, H.; Su, G. A method of 3D path planning for solar-powered UAV with fixed target and solar tracking. *Aerosp. Sci. Technol.* **2019**, *92*, 831–838. [CrossRef]
13. Wu, M.; Shi, Z.; Xiao, T.; Ang, H. Energy optimization and investigation for Z-shaped sun-tracking morphing-wing solar powered UAV. *Aerosp. Sci. Technol.* **2019**, *91*, 1–11. [CrossRef]
14. Wu, M.; Shi, Z.; Ang, H.; Xiao, T. Theoretical study on energy performance of a stratospheric solar aircraft with optimum Λ-shaped rotatable wing. *Aerosp. Sci. Technol.* **2020**, *98*, 105670. [CrossRef]
15. Abdelkefi, A.; Alothman, A.; Hajj, M.R. Performance analysis and validation of thermoelectric energy harvesters. *Smart Mater. Struct.* **2013**, *22*, 095014. [CrossRef]
16. Edgerton, K.; Thronberry, G.; Takeshita, A.; Hocut, C.M.; Shu, F.; Abdelkefi, A. Numerical and experimental comparative performance analysis of emerging spherical-caged drones. *Aerosp. Sci. Technol.* **2019**, *96*, 105512. [CrossRef]
17. Zhao, A.; Zou, H.; Jin, H.; Wen, D. Structural design and experimental verification of an innovative whole adaptive variable camber wing. *Aerosp. Sci. Technol.* **2019**, *89*, 11–18. [CrossRef]
18. Zhao, A.; He, D.; Wen, D. Structural design and experimental verification of a novel split aileron wing. *Aerosp. Sci. Technol.* **2020**, *98*, 105635. [CrossRef]
19. Panagiotou, P.; Yakinthos, K. Aerodynamic efficiency and performance enhancement of fixed-wing UAVs. *Aerosp. Sci. Technol.* **2020**, *99*, 105575. [CrossRef]
20. Hassanalian, M.; Quintana, A.; Abdelkefi, A. Morphing and growing micro unmanned air vehicle: Sizing process and stability. *Aerosp. Sci. Technol.* **2018**, *78*, 130–146. [CrossRef]

21. Hassanalian, M.; Thronberry, G.; Abdelkefi, A. Wing shape and dynamic twist design of bioinspired nano air vehicles for forward flight purposes. *Aerosp. Sci. Technol.* **2017**, *68*, 518–529. [CrossRef]
22. Wang, Y.; Zhou, Y.; Lin, C. Modeling and control for the mode transition of a novel tilt-wing UAV. *Aerosp. Sci. Technol.* **2019**, *91*, 593–606. [CrossRef]
23. Zhen, Z.; Tao, G.; Xu, Y.; Song, G. Multivariate adaptive control based consensus flight control system for UAVs formation. *Aerosp. Sci. Technol.* **2019**, *93*, 105336. [CrossRef]
24. Hassanalian, M.; Rice, D.; Abdelkefi, A. Aerodynamic Performance Analysis of Fixed Wing Space Drones in Different Solar System Bodies. In Proceedings of the AIAA Aerospace Sciences Meeting, Kissimmee, FL, USA, 8–12 January 2018; AIAA: Reston, VA, USA, 2018.
25. Hassanalian, M.; Abdelmoula, H.; Ayed, S.B.; Abdelkefi, A. Thermal impact of migrating birds' wing color on their flight performance: Possibility of new generation of biologically inspired drones. *J. Therm. Biol.* **2017**, *66*, 27–32. [CrossRef]
26. Quintana, A.G.; Hassanalian, M.; Abdelkefi, A. Aerodynamic Analysis of A Morphing Drone with Spanning and Sweeping in Transition Modes. In Proceedings of the 2018 AIAA/AHS Adaptive Structures Conference, Kissimmee, FL, USA, 8–12 January 2018; AIAA: Reston, VA, USA, 2018.
27. Hassanalian, M.; Throneberry, G.; Ali, M.; Ayed, S.B.; Abdelkefi, A. Role of wing color and seasonal changes in ambient temperature and solar irradiation on predicted flight efficiency of the Albatross. *J. Therm. Biol.* **2018**, *71*, 112–122. [CrossRef] [PubMed]
28. Hsu, C.T.; Huang, G.Y.; Chu, H.S.; Yu, B.; Yao, D.J. Experiments and simulations on low-temperature waste heat harvesting system by thermoelectric power generators. *Appl. Energy* **2011**, *88*, 1291–1297. [CrossRef]
29. Champier, D.; Bedecarrats, J.P.; Rivaletto, M.; Strub, F. Thermoelectric power generation from biomass cook stoves. *Energy* **2010**, *35*, 935–942. [CrossRef]
30. Gou, X.; Xiao, H.; Yang, S. Modeling, experimental study and optimization on low-temperature waste heat thermoelectric generator system. *Appl. Energy* **2010**, *87*, 3131–3136. [CrossRef]
31. "Hourly" Averaged Surface Meteorology Data-CERTIFIED. Available online: https://atmos.nmsu.edu/data_and_services/atmospheres_data/MARS/viking/surface_met.html (accessed on 2 February 2020).
32. Matz, E.; Appelbaum, J.; Taitel, Y.; Flood, D.J. Solar cell temperature on Mars. *J. Propuls. Power* **1998**, *14*, 119–125. [CrossRef]
33. Bergman, T.L.; Incropera, F.P.; DeWitt, D.P.; Lavine, A.S. *Fundamentals of Heat and Mass Transfer*; John Wiley Sons: Hoboken, NJ, USA, 2011; pp. 434–485.
34. Chapman, A.J. *Heat Transfer*, 4th ed.; Prentice Hall: Macmillan, NY, USA, 1984.
35. Solar Absorptance and Thermal Emittance of Some Common Spacecraft Thermal-Control Coatings. Available online: https://ntrs.nasa.gov/citations/19840015630 (accessed on 2 February 2020).
36. Appelbaum, J.; Flood, D.J. Solar radiation on Mars. *Sol. Energy* **1990**, *45*, 353–363. [CrossRef]
37. Mitchell, J.W.; Braun, J.E. *Heating, Ventilation, and Air Conditioning in Buildings*; Wiley: Hoboken, NJ, USA, 2013; pp. 185–220.
38. Haranas, I.; Pagiatakis, S. Satellite orbit perturbations in a dusty Martian atmosphere. *Acta Astronaut.* **2012**, *72*, 27–37. [CrossRef]
39. Bird, R.B.; Warren, E.S.; Edwin, N.L. *Transport Phenomena*; John Wiley Sons: Hoboken, NJ, USA, 2006.
40. Smits, A.J.; Dussauge, J.P. *Turbulent Shear Layers in Supersonic Flow*; Springer Science Business Media: Berlin, Germany, 1996.
41. Wei, T.; Xiaofeng, Y.; Yewei, G.; Yanxia, D. Aerodynamic Prediction and Performance Analysis for Mars Science Laboratory Entry Vehicle. *World Acad. Sci. Eng. Technol. Int. J. Mech. Aerosp. Ind. Mech. Manuf. Eng.* **2015**, *9*, 1040–1045.
42. Deperrois, A. Software Package, XFLR5 Analysis of Foils and Wings Operating at Low Reynolds Numbers. Guidelines for XFLR5. Available online: https://engineering.purdue.edu/~aerodyn/AAE333/FALL10/HOMEWORKS/HW13/XFLR5_v6.01_Beta_Win32%282%29/Release/Guidelines.pdf (accessed on 2 February 2020).

drones

MDPI

Article

Unmanned Autogyro for Mars Exploration: A Preliminary Study

Enrico Petritoli * and Fabio Leccese

Dipartimento di Scienze, Università degli Studi "Roma Tre", Via della Vasca Navale n. 84, 00146 Rome, Italy; fabio.leccese@uniroma3.it
* Correspondence: enrico.petritoli@uniroma3.it; Tel.: +39-06-5733-7347

Abstract: Starting from the Martian environment, we examine all the necessary requirements for a UAV and outline the architecture of a gyroplane optimized for scientific research and support for (future) Mars explorers, highlighting its advantages and criticalities. After a careful trade-off between different vehicles suitable for a typical mission, some parameters are established to optimize the size and performance. In the second part, the project of the *Spider* gyroplane and the methodology used to balance the longitudinal masses are presented; in the third part, the parameters of the aerodynamic forces acting on the aircraft are highlighted to be able to focus them during the fluid dynamics simulations.

Keywords: Mars; UAV; autogyro; gyroplane

Citation: Petritoli, E.; Leccese, F. Unmanned Autogyro for Mars Exploration: A Preliminary Study. *Drones* **2021**, *5*, 53. https://doi.org/10.3390/drones5020053

Academic Editors: Diego González-Aguilera and Pablo Rodriguez-Gonzalvez

Received: 28 April 2021
Accepted: 17 June 2021
Published: 18 June 2021

1. Introduction

"A journey of a thousand [miles] starts with a single step." (Lao-Tsu)

Man has always turned his gaze to the sky: while the stars took the shape of the constellations and gave rise to myths, the planets soon became the abode of the gods. Their movement independent of the Earth's rotation suggested a completely different nature than the stars: they were soon considered places such as our planet and, therefore, either inhabited or habitable.

The first successful landing on another planet was made by the Soviet *Venera 7* probe on Venus on 15 December 1970, while Mars, after a partial landing of *Mars 3*, was conquered only in 1976 by two NASA *Viking* landers. Only recently, thanks to advances in technology, has it been possible to send the *Ingenuity* helicopter to wander around the *Perseverance* rover. Now the road to the Martian atmosphere is opened to UAVs: these vehicles allow a greater panoramic view than a rover while maintaining the possibility of examining, in detail and closely, details of Mars that could be of extreme scientific interest.

1.1. State of Exploration of Mars

The exploration of planets is currently entrusted to probes and automatic machines that have the task of leading the way to human colonization. Automatic exploration systems have reached all the inner planets and several outer planets: in the last two decades, the attention has been focused on Mars. The continuous progress of aerospace and electronic technology has put the colonization of the "red planet" among the works of human ingenuity that can be completed in a few years and no longer a science fiction dream. Precursors of this last step are the probes and the rovers: multi-wheel robots that can make simple decisions and perform a series of tests: due to the harsh Martian soil, their movements are very cautious and, for some activities, dependent on the day-night cycle. In Figure 1, we can see the tortuous path taken by the Curiosity Mars rover: a journey of a few kilometers.

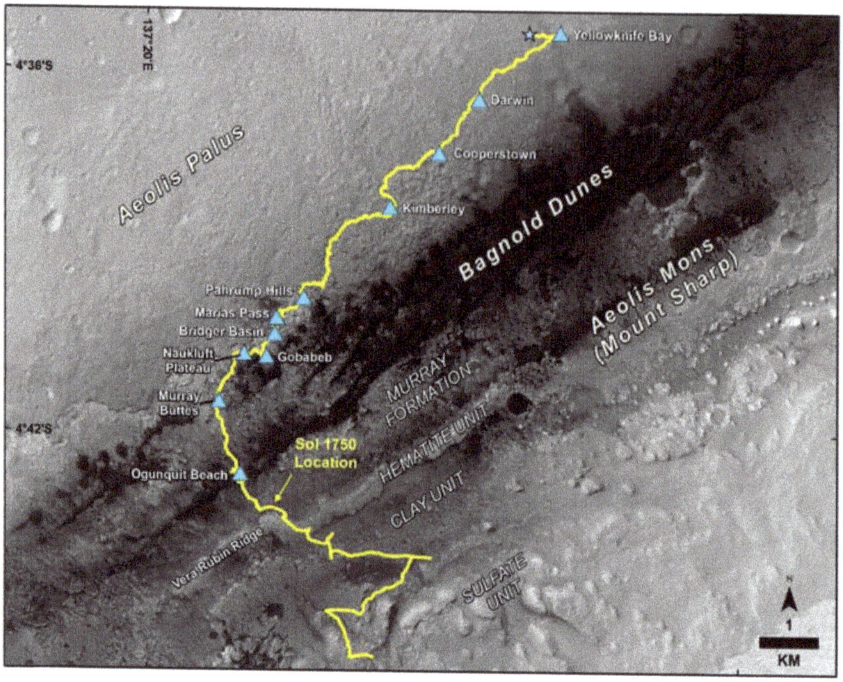

Figure 1. This map shows the route driven by NASA's Curiosity Mars rover, from the location where it landed in August 2012 to its location in July 2017 (Sol 1750), and its planned path to additional geological layers of lower Mount Sharp (image and caption: © NASA).

This type of vehicle is, therefore, very suitable for close-up exploration of the land, as it can capture every detail of the terrain but is unable to have a panoramic view of the landscape. On the other side, there are the observation satellites in Martian orbit: the Mars Reconnaissance Orbiter (with the HiRISE—High-Resolution Camera—onboard) is in orbit 450 km from the surface; although it is able to map the ground with great precision (HiRISE can produce images from which topography can be calculated to an accuracy of 0.25 m). It is absolutely evident there is a lack of a vehicle that is able to place itself reasonably far from the ground to be able to see a wider horizon but, at the same time, capable of grasping details and, if necessary, overlooking points that could be of extreme interest. Finally, the speed factor is important to quickly reach the points of greatest interest with respect to the point of arrival on the Planet.

1.2. Mission Concept

The operating conditions of a possible automatic flying vehicle (UAV: Unmanned Aerial Vehicle) on Mars are quite complex and, in some cases, they collide with each other. Mainly they are:

- The UAV must be rather simple and robust as it will first have to withstand the stresses due to launch, space flight, entry into the Martian atmosphere, and then, it will have to be deployed as more parts will surely have traveled folded.
- The vehicle must be extremely reliable as there is no maintenance required: many electronic systems and all critical mechanical systems must be redundant.
- The vehicle has a limited life span thus, it will be necessary to optimize his work to have specific and scheduled tasks.

- Its work is mainly carried out during the day, as it is equipped with solar cells to recharge the batteries. It is also equipped with multispectral sensors optimized for Martian light.
- The vehicle is meant to work with a rover, which, of course, will arrive at the same time. However, the two vehicles are not in symbiosis: the UAV will be able to fly away from the rover even for a considerable distance and for several days.
- Drones must, therefore, be able to store the large mass of scientific data it is intended to collect. Later, it will reach the rover and will download the data: it will then be the task of the land vehicle to act as a relay.

From these first considerations, it can be deduced that the vehicle must be a fair compromise between gross weight and payload.

1.3. Why an Autogyro?

An *autogyro* (from Greek αὐτός and γύρος, "self-turning"), also known as a *gyroplane* or *gyrocopter*, is a type of rotorcraft that uses an unpowered rotor in free autorotation to develop lift [1]. Forward thrust is provided independently, usually by an engine-driven propeller. While like a helicopter rotor in appearance, the autogyro's rotor must have air flowing across the rotor disc to generate rotation and the air flows upwards through the rotor disc rather than down [2]. These types of flying machines were successful in the 1920s and 1930s when the performance of fixed-wing aircraft was far from flattering. Following the war events, it gave a strong impulse to the development of the helicopter, a vehicle that monopolized the technological development of the rotary wing [3].

Table 1 shows the comparison between the main flight characteristics of airplanes, helicopters, and autogyros: obviously, it should not be understood as a ranking in which we try to establish which is the "best flying machine," but to stigmatize the different characteristics in order to find the most suitable for the needs of the mission [4].

Table 1. Comparison of the main flight characteristics of airplanes, helicopters, and autogyros.

Characteristic	Type of Vehicle		
	Airplane	*Helicopter*	*Autogyro*
Stability	Good	Poor	Particularly good, even at low speeds
VTOL	Possible but limited	Engine-powered main rotor	Yes: Pre-rotators
Hovering	Not Possible	Possible	Extremely low flight speed, almost hovering [a]
Stall	Stalls: recovery maneuver required	Stalls: autorotation maneuver required	Impossible: full controllable even at even in the absence of thrust
Maneuvering	Easy	Complex	Easy

[a] Horizontal speed not negligible but extremely low.

The airplane is an excellent platform for instruments as it exhibits excellent stability (even in gusts of wind) and extreme maneuverability: unfortunately, it has a high stall speed due to its architecture. It would be possible to equip an aircraft with STOL (Short Take-Off and Landing) or VTOL (Vertical Take-Off and Landing) features at the price of a very high weighting of the structure, as they require the addition of "ad hoc" engines or aerodynamic surfaces [5].

The performance of the helicopter is very close to our desired in terms of stall speed: unfortunately, it turns out to be a very unstable platform and subject, as everyone knows, due to high vibrations. Furthermore, it is a mechanically very complex machine like its maneuvering [6].

In Figure 2, we can see graphs showing horsepower per tons of gross weight versus speed for different families of transport or utility vehicles.

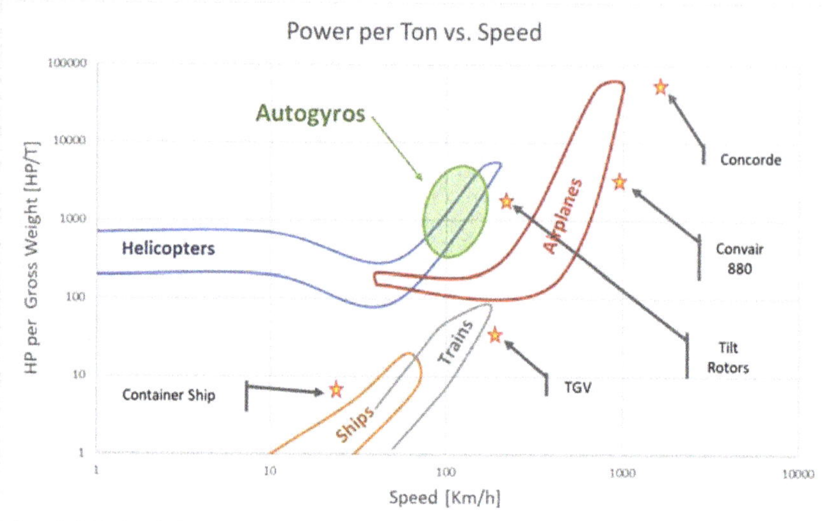

Figure 2. Most common vehicles: power per ton vs. speed.

The behavior of ships and trains is immediately evident, which in the face of rather low power transport goods at a relatively low speed: even if it is not visible in the graph, all this is accomplished with a general economy of the service.

Helicopters, on the other hand, have a "useful speed" even zero as they are often called upon to perform missions such as "sky cranes", that is to say, for lifting antennas or power lines above rivers, valleys, etc.

The gyroplanes (green zone) have been highlighted in the green zone.

2. Materials and Methods

2.1. The Environment: Mars

Mars is the planet that has always attracted the main scientific interest from the international community and the various space agencies, as demonstrated by the great commitment made in its study since the beginning of the exploration of the solar system. The planet has been studied for centuries, but only the beginning of the space age has allowed us to understand in detail many of its characteristics, until then only hypothesized. The space probes in orbit and the vehicles on the ground made it possible to collect a great variety of information, from the composition and internal structure to the interaction of the upper atmosphere with the solar wind (see Table 2).

Table 2. Mars/Earth Comparison [1].

Parameter	Mars	Earth	Ratio Mars/Earth
Mass (10^{24} kg)	0.64171	59.724	0.107
Equatorial radius (km)	3396.2	6378.1	0.532
Polar radius (km)	3376.2	6356.8	0.531
Surface gravity (m/s^2)	3.71	9.80	0.379
Surface Atmosphere density (kg/m^3)	0.020	1.2210	0.01638
Surface Speed of Sound (m/s)	240	340	0.705
Solar irradiance (W/m^2)	586.2	1361.0	0.431
Number of natural satellites	2	1	

[1] Source National Space Science Data Center (NSSDC).

The Atmosphere

The atmosphere has been the object of particular attention on our part because it is the gaseous medium through which our vehicle moves: since it is not equipped (because of the absence of oxygen in the Martian atmosphere) with an internal combustion engine but an electric one, we will only consider the aerodynamic and physics-chemical interactions with the *gyro*.

The Martian atmosphere (see Table 3) has a mass much less than that of the Earth: a reference value for the surface pressure can be considered 6.1 mbar, 3 orders of magnitude lower than the corresponding Earth value. This value is highly variable due to the great topographical differences of the surface. The temperature during the polar night can drop below the CO_2 condensation point, resulting in the deposition of a relevant fraction of the atmospheric mass and the annual bimodal pressure cycle observed by landers and orbiters. The study on the isotopic and elementary ratios of atmospheric gaseous species [2] has shown that the primitive atmosphere of Mars, formed after the conclusion of the T-Tauri phase of the solar life cycle, was affected by a considerable process of gas removal due to different causes. The proposed values for the surface pressure primitive of Mars are all around 1 bar. The reduction of the atmospheric mass has mainly affected the currently predominant component that is carbon dioxide, primarily because of meteorite bombardment and atmospheric escape. Several authors have also highlighted the possible role played by the inclusion of carbon in carbonate minerals in a water-rich environment. Thus far, carbonates have not been detected on Mars in appreciable quantities, and, therefore, this effect, even if present, should be of secondary importance. In the atmosphere, both CO and O_2 are observed, both derived from the dissociation of CO_2.

Table 3. Martian Atmosphere [1].

Parameter	Value	Notes
Surface pressure:	6.36 mb	@ mean radius: variable from 4.0 to 8.7 mb depending on the season
Surface density	0.020 kg/m^3	
Scale height	11.1 km	
Total mass of the atmosphere	2.5 × 1016 kg	
Average temperature	210 K (−63 °C)	
Diurnal temperature range	184 K to 242 K (−89 to −31 °C)	
Wind speeds	2–7 m/s (summer) 5–10 m/s (fall) 17–30 m/s (dust storm)	
Atmospheric composition (by volume)	Carbon Dioxide (CO_2)—95.1% Nitrogen (N_2)—2.59% Argon (Ar)—1.94% Oxygen (O_2)—0.16% Carbon Monoxide (CO)—0.06%	Major (%)
	Water (H_2O)—210 Nitrogen Oxide (NO)—100 Neon (Ne)—2.5 Hydrogen-Deuterium-Oxygen (HDO)—0.85 Krypton (Kr)—0.3 Xenon (Xe)—0.08	Minor (ppm)

[1] Source: National Space Science Data Center (NSSDC).

The role of water on Mars has always been the subject of a large literature: in particular, H_2O vapor exhibits highly variable atmospheric concentrations in space and time. The seasonal variations were observed in detail by the Mars Atmospheric Water Detector (MAWD) aboard the Viking 1 Orbiter and were confirmed by observations of the Thermal Emission Spectrometer (TES) instrument aboard the NASA MGS (Mars Global Surveyor) probe. Water vapor showed a maximum concentration during the waning phase of the northern hemisphere polar cap, when a fraction of the H_2O, trapped as ice during the winter, sublimated into the atmosphere while a secondary maximum was present during the hemisphere spring. The surface plays a role both as a source and as a deposit for the water present in the atmosphere. The surface layers play an important role due to their hygroscopic character deriving both from the morphology (aggregates of mineral dust of various sizes, from centimeter to sub-millimeter) and from the mineralogical composition of the soils, on both seasonal and daily time scales. Several meters below the surface, some observational evidence of the MARSIS radar instrument (Mars Advanced Radar for Subsurface and Ionosphere Sounding) placed on board of the Mars Express probe, and observations of the Shallow Subsurface Radar (SHARAD), instrument onboard the Mars Reconnaissance Orbiter probe (MRO), suggested that there may be a large water reservoir, in the form of a permanent ice layer, probably a region rich in crystalline water dispersed in a matrix of mineral grains. To date, the presence of the underground water source has not yet been confirmed. Possible evidence in this sense was represented by the presence of "rampart" craters, peculiar Martian impact craters characterized by evidence of erosion by fluid agents, and by observations of the hydrogen concentration on Mars of the GRS experiment (Gamma Ray Spectrometer) on board of the Mars Odyssey probe.

In addition to gases and water ice clouds, the Martian atmosphere contains a considerable amount of mineral dust, the presence of which produces various observable phenomena. Among these, there were the dust devils, columns of dust with a diameter of the order of 10s of meters and over 6 kilometers in height that were raised from the surface by winds localized near the surface and dust storms, which can hide large areas surface to observations, for periods ranging from several days to covering the entire planet for several months, as observed during the summer of 2007 by instruments aboard the Mars Express and Mars Reconnaissance Orbiter missions.

2.2. *Mars Environment: Pro and Cons*

Although the Martian atmosphere, for density, gaseous composition, and solar radiation, does not allow human life if not adequately protected by pressure suits, the Navier-Stokes equations remain valid thus an aircraft can fly. Obviously, the diversity of the environment requires a profound redesign of the vehicle: this is to define unequivocally that, for example, a helicopter designed for the Earth will never be able to fly over Mars or vice versa [7–11]. The positive and negative factors of operating in the Martian atmosphere with a gyroplane will now be examined.

2.2.1. Pro and Cons

In this section, by examining the differences in the Martian environment, we see what the advantages and disadvantages of an air vehicle can be.

Cons

- The main problems for a flying vehicle in the Martian atmosphere derive from its very low density (about 1.7%), as the expression of lift is:

$$L = \frac{1}{2}\rho v^2 C_L S \tag{1}$$

$$\begin{cases} \rho_E = 1.2210 \quad \mathrm{Kg/m^2} \\ \rho_M = 0.0020 \quad \mathrm{Kg/m^2} \end{cases} \tag{2}$$

$$\frac{L_M}{L_E} = \frac{\rho_M}{\rho_E} = 0.001638 \tag{3}$$

- Because the speed of sound is much lower (about 70% of that of the Earth), the tips of the blades approach the transonic regime much faster with the same rotor revolutions.

$$\frac{v_M^{sound}}{v_E^{sound}} = \frac{240 \text{ m/s}}{340 \text{ m/s}} = 0.705 \tag{4}$$

- Blades and wings operate in low Reynolds number flows: thus, they will need to be shorter and wider (thicker chord), but the profile drag is much higher.

$$R_e = \frac{\rho v l_{ch}}{\mu} \tag{5}$$

In Equations (1)–(3) we see the expression of lift, in which:
L, L_E, L_M is the lift of the vehicle, on Earth, on Mars.
ρ, ρ_E, ρ_M is density of atmosphere, on Earth, on Mars.
v is the speed of the vehicle.
In Equation (4), we have the comparison of the speed of sounds:
v_E^{sound}, v_M^{sound} is the speed of sound on Earth, on Mars.
In Equation (5), we see the expression of the R_e (Reynolds number) in which:
μ is the dynamic viscosity.
l_{ch} is the length of the chord of the wing profile.
In the face of the aforementioned strong difference in density, we do not have an equal lowering of the dynamic viscosity, thus that the operating conditions on Mars are found at low Reynold numbers.

Pros

- Gravity on Mars is just under a third of that on Earth.

$$\begin{cases} g_E = 9.80 \quad \text{m/s}^2 \\ g_M = 3.71 \quad \text{m/s}^2 \end{cases} \tag{6}$$

- It is possible to build vehicles with a much lighter structure.
- The aerodynamic coefficient of drag is significantly lower.

$$\frac{D_M}{D_E} = \frac{\rho_M}{\rho_E} = 0.001638 \tag{7}$$

where D_E, D_M is the drag of the vehicle, on Earth, on Mars.

2.2.2. The Tale of Two Planets

We can operate the first approach in general sizing by the similarity between the two planets since, regardless of the environment in which they operate, the balance laws of forces (and energies) must be the same.

The balance of forces for the vehicle on Earth (moving at a constant speed and at a constant altitude) must be (see Figure 3a/left):

$$\begin{cases} T_E = D_E \\ L_E = W_E \end{cases} \tag{8}$$

where:

T_E is the necessary thrust provided by the engine on Earth.
D_E is the drag of the vehicle in the Earth's atmosphere.
L_E is the total lift of the vehicle in the Earth's atmosphere.

W_E is the total weight of the vehicle on Earth.

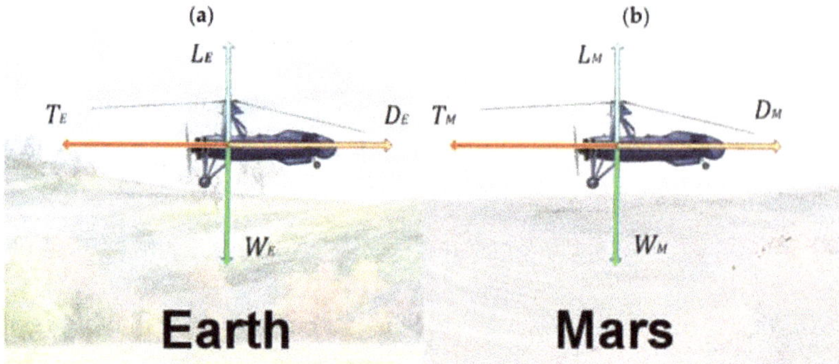

Figure 3. The balance of forces for the vehicle on Earth (**a**/**left**) and on Mars (**b**/**right**).

Starting from the principle that thrust is only dependent on engine technology, we expand the other members, the drag is:

$$D_E = \frac{1}{2}\rho_E {V_E}^2 C_{D_E} S_E \tag{9}$$

where:

ρ_E is the atmosphere density of Earth.
V_E is the velocity of the vehicle on Earth.
C_{D_E} is the coefficient of drag of the vehicle on Earth.
S_E is the lift surface of the vehicle.

For the lift, we have:

$$L_E = \frac{1}{2}\rho_E {V_E}^2 C_{L_E} S_E \tag{10}$$

where:

C_{L_E} is the coefficient of lift of the vehicle on Earth.

For the weight, we have:

$$W_E = M \cdot g_E \tag{11}$$

where:

M is the mass of the vehicle.
g_E is the surface acceleration of gravity on Earth.

The balance of forces for the vehicle on mars (moving at a constant speed and at a constant altitude) must be (see Figure 3b)

$$\begin{cases} T_M = D_M \\ L_M = W_M \end{cases} \tag{12}$$

where:

T_M is the necessary thrust provided by the engine on Mars.
D_M is the drag of the vehicle in the Mars' atmosphere.
L_M is the total lift of the vehicle in the Mars' atmosphere.
W_M is the total weight of the vehicle on Mars.

Starting from the principle that thrust is only dependent on engine technology, we expand the other members, and the drag is:

$$D_M = \frac{1}{2}\rho_M {V_M}^2 C_{D_M} S_M \tag{13}$$

where:

ρ_M is the density of Mars' atmosphere.
V_M is the velocity of the vehicle on Mars.
C_{D_M} is the coefficient of drag of the vehicle on Mars.
S_M is the lift surface of the vehicle.

For the expression of the lift, we have:

$$L_M = \frac{1}{2}\rho_M V_M{}^2 C_{L_M} S_M \tag{14}$$

where:

C_{L_M} is coefficient of lift of the vehicle on Mars.

For the weight, we have:

$$W_M = M \cdot g_M \tag{15}$$

where:

g_M is the surface acceleration of gravity on Mars

Let us start working on the weight of the vehicle: with the same mass, we have that:

$$\begin{cases} M = \frac{W_E}{g_E} \\ M = \frac{W_M}{g_M} \end{cases} \tag{16}$$

Then we have:

$$\frac{W_E}{g_E} = \frac{W_M}{g_M} \tag{17}$$

and

$$\frac{W_M}{W_E} = \frac{g_E}{g_M} = 2.641 \tag{18}$$

This means that all things being equal, the lower Martian gravity allows us to fly vehicles just under 3 times heavier (large): let us see how this affects the other parameters. For the necessary lift, we have:

$$\frac{L_M}{L_E} = 2.641 \tag{19}$$

hence:

$$\frac{\rho_M V_M{}^2 C_{L_M} S_M}{\rho_E V_E{}^2 C_{L_E} S_E} = 2.641 \tag{20}$$

Considering the density, we have:

$$\frac{V_M{}^2 C_{L_M}\, S_M}{V_E{}^2 C_{L_E}\, S_E} = 2.641 \cdot \frac{\rho_E}{\rho_M} = 161.233 \tag{21}$$

Explicating with respect to the velocities:

$$\frac{V_M}{V_E} = \sqrt{161.233 \cdot \frac{C_{L_E}}{C_{L_M}} \cdot \frac{S_E}{S_M}} \tag{22}$$

It is, therefore, possible to establish an optimal surface to evaluate the main parameters of a vehicle operating on Mars in an analog way with one optimized to operate on Earth.

The surface described by expression (22) is represented in Figure 4 below. The physical meaning was twofold: first of all, to design a vehicle that is optimized for the Martian environment, one must remain inside and on the borders of the surface described. Any departure from this will, therefore, lead to a penalty either in performance or otherwise such as for example in the ability to carry a heavy payload [12–15]. In other words, if you try to magnify a drone's ability by moving away from the surface, this requires compensation thus, we can rightly consider it an excellent surface. The second meaning is the ability to be able to design a vehicle for Mars in analogy with a similar vehicle for Earth, i.e., the expression (22) is a sort of "transfer function" for the design between the two planets [16–18].

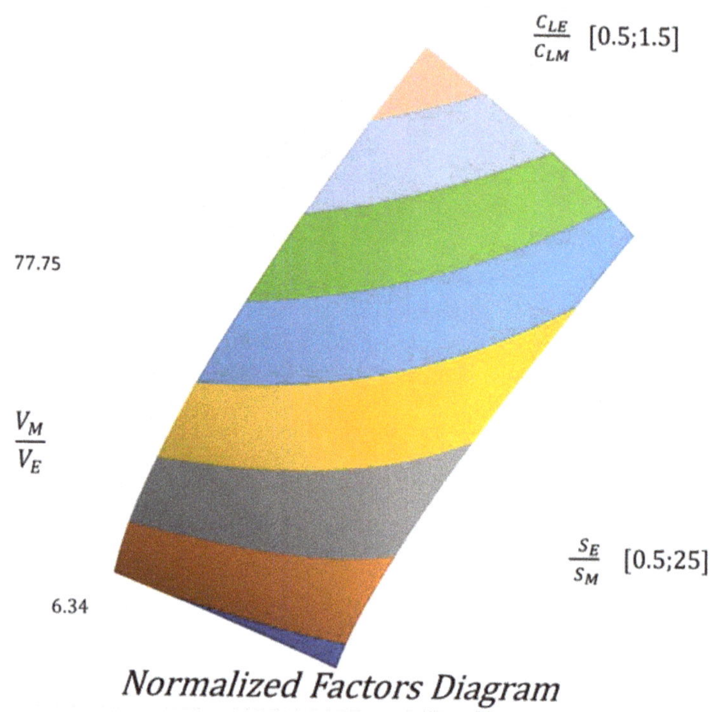

Figure 4. Normalized surface of the factors in expression (22).

2.3. *The Gyroplane*

Based on all the considerations of the previous paragraphs, our working group has decided to develop a drone project for Martian exploration based on the architecture of the gyroplane since, according to the technologies present at state of the art, it is the one that satisfies all mission requirements.

2.3.1. The Mission: "Fly Like a Butterfly and Sting Like a Bee"

A gyroplane is not a hybrid; it is not a cross between various architectures: it is a vehicle with peculiar characteristics and capabilities that, in a Martian environment, are particularly highlighted. First of all, it is a much faster vehicle than a helicopter, which allows it, for the same amount of time, to explore a considerably greater area. Secondly, it can have a significantly heavier scientific payload for the same MTOW (Maximum Take-Off Weight). It is also less prone to gusts of wind, thus becoming a rather independent aerial platform with respect to the environment. It certainly does not have the vertical take-off and landing capacity of the helicopter, but this is compensated by extraordinarily strong STOVL (Short Take-Off and Vertical Landing) characteristics, being able to land and take off in just over 20 m. Certainly, the necessary spaces are widely available on Mars.

Since, as it is known, it was not possible to study a vehicle that can do everything at the same time, it was, therefore, necessary to define one or more missions in order to "tailor" the architecture and payload, and, therefore, we have predicted the following scenarios:

Pathfinder

In this first scenario. The vehicle proceeds as a forerunner of a possible land rover. Since the speed of the latter is very low and its proceeding on the Martian soil very cautious due to the impossibility of having a wide "observable horizon", the aerial drone would be its "eye in the sky". Therefore, its task should be mainly to find a safe and obstacle-free path for the Martian rover and, in the eventuality, to provide surveys and photographic evidence of interesting details of areas or rock formations that may be the subject of further study.

Precision Aerial Photogrammetry

The photographic mapping of Mars is mainly entrusted to satellite vehicles, which have already provided a good mapping of the planet but suffer from two major penalties: the lack of flexibility as the area framed in the FOV (field of view) is obviously linked to the trajectory orbital and the inability to further detail a particular area worthy of interest. All this, combined with the fact of not being able to have a resolution higher than a quarter of a meter, makes it the ideal means to create general maps but not detailed ones. Furthermore, it would be impossible to observe the same area through different hours of the day or the seventh, for example, the melting of an area of the ice cap: the two successive observations would be temporally spaced by two orbital passages. In this case, the gyroplane could overcome all these limitations.

Cargo Shuttle

Considering the ambitious human colonization project, a gyroplane is an ideal vehicle to support a human base on Mars. It can carry out SAR (Search And Rescue) missions of isolated astronauts, moving essential supplies (for example, batteries, oxygen, etc.) in a short time, or acting as an audio/video relay for telecommunications by starting to orbit in a circle at high altitude.

These are just a few scenarios in which a gyroplane can make a strong contribution.

2.3.2. Spider

Our group has, therefore, given the name of *Spider* (see Figure 5) to the gyroplane for the evident similarities with the insect. The estimated characteristics of the vehicle are shown in Table 4 (see).

Figure 5. Spider Gyroplane: a prospective view.

Table 4. Spider: characteristics and performances (estimated).

Max Takeoff Weight	Main Rotor Diameter	Length [1]	Wingspan	Cruise Speed	Range [2]	Endurance [2]
10 kg	2.25 m	1.75 m	1.85 m	45 km/h	~10 km	1.5 h

[1] Without rotor. [2] At cruise speed.

The gyroplane is composed of a few simple elements that almost all perform aerodynamic and structural functions (see Figure 6). The central body is composed of a carbon fiber cylinder, which contains the guidance and control systems, the OBC (On-Board Computer), which oversees all the autonomous piloting functions. The 2 wings are connected to the cylinder (cantilever); on their upper surface, there is a strip of solar cells, which has the task of recharging the batteries. At the end of each wing, there is a tip-tank that supports the main landing gear wheel: this has two positions retracted and deployed (see Figure 7). At about one-third of the wingspan, there are the 2 tail booms, which in turn support the V-shape tail wings; each of them has a strip of solar cells on the surface, which, like those of the wings, help recharge the batteries. Near the tail, there are also the 2 wheels of the rear landing gear. In the upper part, there is the "saddle" that supports the rotor, made up of 5 blades, each with double hinges (swing and flapping—not visible in the image). In the front, protected by a radome, there is an icosahedron with pentagonal faces: on each, there are a series of optical and electromagnetic sensors for exploration and navigation.

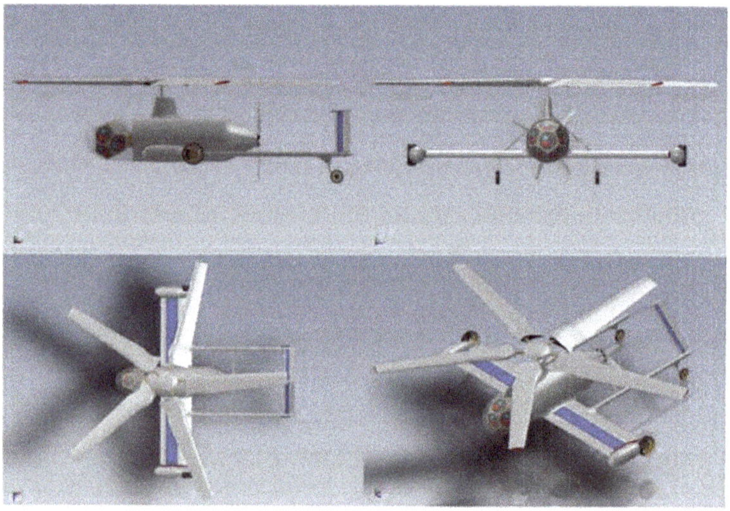

Figure 6. Four views of the drone.

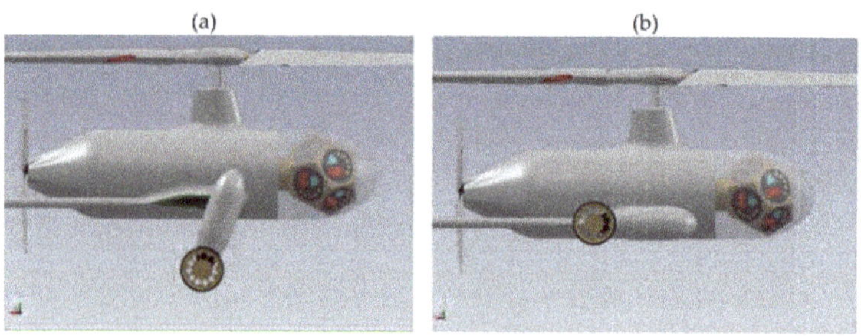

Figure 7. Main landing gear and tip tanks: deployed (**a**/**left**) and retracted (**b**/**right**).

In the rear, conical part, there is the electric motor for the propeller (composed by seven blades) in a pushing configuration.

The choice of the particular type of fletching is due to the fact that the architecture allows them to be immersed in the direct flow of the propeller, thus that they are effective even at very low speed.

2.3.3. Longitudinal Balance

To correctly manage the positioning of vehicle systems and subsystems, it was necessary to accurately define the balance of forces and momentum in the vertical plane. Figure 8 (below) shows the forces acting on the vehicle in stationary conditions.

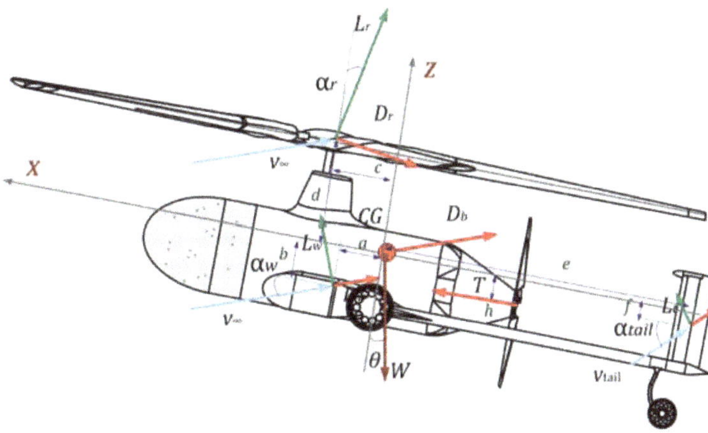

Figure 8. The balance of forces and momentum.

In this section, we considered the dynamic balance of forces on the vertical plane (X, Z) and momentum with respect to the Y-axis. In these conditions, the gyroplane proceeds at a constant speed. In the following discussion, the variation of density of the air with the variation of the altitude will not be considered, nor of the relative variation in propeller and rotor efficiency. We will consider these constant elements with reasonable approximation in a non-negligible altitude interval.

By definition:

$$\begin{cases} \sum F_x = 0 \\ \sum F_z = 0 \\ \sum M_y = 0 \end{cases} \quad (23)$$

Considering the X-axis:

$$\sum F_x = L_r \sin\alpha_r + D_r \cos\alpha_r + L_w \sin\alpha_w - D_w \cos\alpha_w + L_t \sin\alpha_t - D_t \cos\alpha_t + T - W\sin\theta - D_b \cos\alpha_w \quad (24)$$

where:

L_r is the lift force of the rotor.
D_r is the drag force of the rotor.
L_w is the lift of the wing.
D_w is the drag of the wing.
L_t is the lift of the tail.
D_t is the drag of the tail.
α_r is the angle between v_∞ and the rotor plane.
α_w is the angle between v_∞ and the wing.
α_t is the angle between v_{tail} (due to downwash) and the tail.

θ is the angle of body axes and ground.
W is the weight of the vehicle.
T is the thrust.

For the balance of the forces on the Z-axis w have:

$$\sum F_z = L_r \cos\alpha_r + D_r \sin\alpha_r + L_w \cos\alpha_w + D_r \sin\alpha_w + L_t \cos\alpha_t + D_t \sin\alpha_t - W\ \cos\theta + D_b \sin\alpha_w \quad (25)$$

The balance around the Z-axis is:

$$\begin{aligned} \sum M_y = d{\cdot}L_r \sin\ \ & \alpha_r + c{\cdot}L_r \cos\alpha_r + d{\cdot}D_r \cos\alpha_r - c{\cdot}D_r \sin\alpha_r + b \\ & L_w \sin\alpha_w - b{\cdot}D_w \cos\alpha_w + a{\cdot}L_w \cos\alpha_w + b{\cdot}D_w \sin\alpha_w - e \\ & L_t \sin\alpha_t + f{\cdot}L_t \cos\alpha_t - f{\cdot}D_t \cos\alpha_t - e{\cdot}D_t \sin\alpha_t + T{\cdot}h \end{aligned} \quad (26)$$

Now considering a little α_w, α_t and α_r, considering θ, h, f negligible for the expression (24)–(26) we have:

$$0 = L_r\alpha_r + D_r + L_w\alpha_w - D_w + L_t\alpha_t - D_t + T - D_b \quad (27)$$

$$0 = L_r + D_r\alpha_r + L_w + D_r\alpha_w + L_t + D_t\alpha_t - W + D_b\alpha_w \quad (28)$$

$$0 = d{\cdot}L_r\alpha_r + c{\cdot}L_r + d{\cdot}D_r - c{\cdot}D_r\alpha_r + b{\cdot}L_w\alpha_w - b{\cdot}D_w + a{\cdot}L_w + b{\cdot}D_w\alpha_w - e{\cdot}L_t\alpha_t - e{\cdot}D_t \quad (29)$$

Thus the (23) becomes:

$$\begin{cases} -T = L_r\alpha_r + D_r + L_w\alpha_w - D_w + L_t\alpha_t - D_t - D_b \\ W = L_r + D_r\alpha_r + L_w + D_r\alpha_w + L_t + D_t\alpha_t - W + D_b\alpha_w \\ 0 = d{\cdot}L_r\alpha_r + c{\cdot}L_r + d{\cdot}D_r - c{\cdot}D_r\alpha_r + b{\cdot}L_w\alpha_w - b{\cdot}D_w + a{\cdot}L_w + b{\cdot}D_w\alpha_w - e{\cdot}L_t\alpha_t - e{\cdot}D_t \end{cases} \quad (30)$$

By grouping all the friction factors not dependent on the angles:

$$D_x^{tot} = -D_r + D_w + D_t + D_b \quad (31)$$

For the first member of (30), we have:

$$T = L_r\alpha_r + L_w\alpha_w + L_t\alpha_t + D_x^{tot} \quad (32)$$

Posing:

$$L_y^{tot} = L_r + L_w + +L_t \quad (33)$$

For the second member of (30), we have:

$$W = +D_r\alpha_r + D_r\alpha_w + D_t\alpha_t + D_b\alpha_w + L_y^{tot} \quad (34)$$

Posing $\alpha_r \approx \alpha_w \approx \alpha_t = \delta\alpha$, for the third member of the (30), we have:

$$0 = L_r(d{\cdot}\delta\alpha + c) + D_r(d - c{\cdot}\delta\alpha) + L_w(b{\cdot}\delta\alpha + a) + b{\cdot}D_w(\delta\alpha - 1) - e{\cdot}L_t\delta\alpha - e{\cdot}D_t \quad (35)$$

Explicating L_r:

$$L_r = -\frac{+D_r(d - c{\cdot}\delta\alpha) + L_w(b{\cdot}\delta\alpha + a) + b{\cdot}D_w(\delta\alpha - 1) - e{\cdot}L_t\delta\alpha - e{\cdot}D_t}{d{\cdot}\delta\alpha + c} \quad (36)$$

Taking into account dynamically the forces that contribute to the balance of forces, always remembering to check the positioning on the surface depicted in Figure 4, with the expressions (32), (34), and (35), we can move on to the physical sizing of the drone. Remember that a move away from the aforesaid surface entails a deviation from the optimal design conditions and that this is paid in terms of lowering the yield of some parameters.

2.3.4. Simulation

Now, it is necessary to analytically define the parameters to be searched for in the fluid dynamics simulation; for the lift of the wing, we have:

$$L_w = \frac{1}{2}\rho_m v^2 S_w C_L^w \tag{37}$$

where:

ρ_m: mars atmosphere density
S_w: wing surface
v: relative speed (refer to air)
C_L^w: coefficient of lift of the wing

According to Taylor's method, the last member can be separated in:

$$C_L^w = C_{L_0}^w + C_{L_{\alpha w}}^w \alpha_w \tag{38}$$

where:

$C_{L_0}^w$: coefficient of lift at $\alpha_w = 0$
$C_{L_{\alpha w}}^w$: coefficient of lift at $\alpha_w \neq 0$

Then, the expression (36) becomes:

$$L_w = \frac{1}{2}\rho_m v^2 S_w \left(C_{L_0}^w + C_{L_{\alpha w}}^w \alpha_w \right) \tag{39}$$

Similarly, for the lift of the rotor we have:

$$L_r = \frac{1}{2}\rho_m v^2 S_r \left(C_{L_0}^r + C_{L_{\alpha r}}^r \alpha_r \right) \tag{40}$$

where:

S_r: rotor disc surface
$C_{L_0}^r$: coefficient of lift at $\alpha_r = 0$
$C_{L_{\alpha r}}^r$: coefficient of lift at $\alpha_r \neq 0$

For the lift of the tail, we have

$$L_t = \frac{1}{2}\rho_m v^2 S_t C_{L_{\alpha_t}}^t \alpha_t \tag{41}$$

where:

S_t: tailplane surface
$C_{L_0}^t = 0$: the wing profile is symmetrical.
$C_{L_{\alpha_t}}^t$: coefficient of lift at $\alpha_t \neq 0$

The lift of the fuselage can be considered negligible ($L_w \approx 0$) compared to the other forces involved.

For the drag of the wing, we have:

$$D_w = \frac{1}{2}\rho_m v^2 S_w \left(C_{D_0}^w + C_{D_{\alpha w}}^w \alpha_w \right) \tag{42}$$

where:

$C_{D_0}^w$: coefficient of drag at $\alpha_w = 0$
$C_{D_{\alpha w}}^w$: coefficient of drag at $\alpha_w \neq 0$

Similarly, for the drag of the rotor we have:

$$D_r = \frac{1}{2}\rho_m v^2 S_r \left(C^r_{D_0} + C^r_{D_{\alpha_r}} \alpha_r \right) \tag{43}$$

where:

$C^r_{D_0}$: coefficient of drag at $\alpha_r = 0$
$C^r_{D_{\alpha_r}}$: coefficient of drag at $\alpha_r \neq 0$

For the drag of the tail, we have:

$$D_t = \frac{1}{2}\rho_m v^2 S_t \left(C^t_{D_0} + C^t_{D_{\alpha_t}} \alpha_t \right) \tag{44}$$

where:

$C^t_{D_0}$: coefficient of drag at $\alpha_t = 0$
$C^t_{D_{\alpha_t}}$: coefficient of drag at $\alpha_t \neq 0$

For the drag of the body, we have:

$$D_b = \frac{1}{2}\rho_m v^2 S_b \left(C^b_{D_0} + C^b_{D_{\alpha w}} \alpha_w \right) \tag{45}$$

where:

S_b: exposed surface of the body
$C^b_{D_0}$: coefficient of drag at $\alpha_w = 0$
$C^b_{D_{\alpha w}}$: coefficient of drag at $\alpha_w \neq 0$

By grouping the expressions (38), (39), (40), (41), (42), (43) and (44) in matrix form, we can write:

$$\begin{bmatrix} L_w \\ L_r \\ L_t \\ D_w \\ D_r \\ D_t \\ D_b \end{bmatrix} = \frac{1}{2}\rho_m v^2 \begin{bmatrix} S_w \\ S_r \\ S_t \\ S_w \\ S_r \\ S_t \\ S_b \end{bmatrix} \mathbb{C} \tag{46}$$

where for $\mathbb{C}$ we have the following diagonal matrix:

$$\mathbb{C} = \begin{bmatrix} C^w_{L_0} + C^w_{L_{\alpha w}} \alpha_w & & & & & & \\ & C^r_{L_0} + C^r_{L_{\alpha_r}} \alpha_r & & & & & \\ & & C^t_{L_{\alpha_t}} \alpha_t & & & & \\ & & & C^w_{D_0} + C^w_{D_{\alpha w}} \alpha_w & & & \\ & & & & C^r_{D_0} + C^r_{D_{\alpha_r}} \alpha_r & & \\ & & & & & C^t_{D_0} + C^t_{D_{\alpha_t}} \alpha_t & \\ & & & & & & C^b_{D_0} + C^b_{D_{\alpha w}} \alpha_w \end{bmatrix} \tag{47}$$

We evaluated the vehicle's coefficients of $\mathbb{C}$ by inserting the CAD model (developed with SolidWorks®) into its fluid dynamics application (see Figure 9), testing several angles of attack in order to define all the parameters of the matrix and thus be able to obtain a dynamic model of our vehicle.

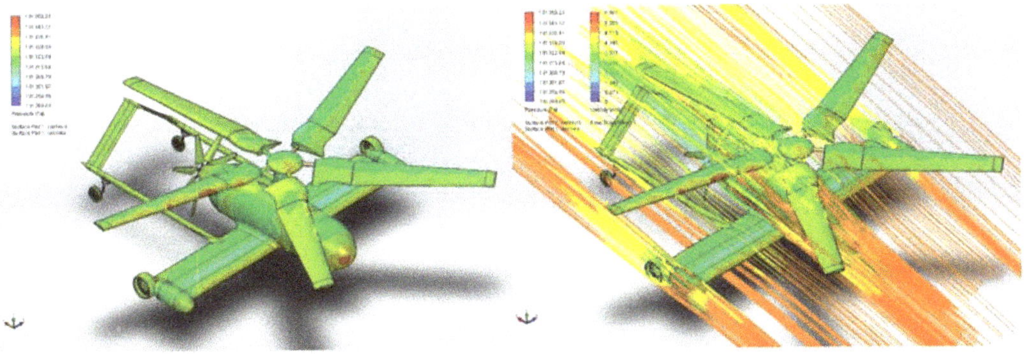

Figure 9. Two images taken from a more extensive simulation of the behavior of the Spider drone in the Martian atmosphere (v_∞ = 10 m/s). On the right, the pressure gradient on the vehicle surface, while on the left the flow lines in the same conditions.

3. Conclusions

The exploration of Mars by means of aerial drones is an opportunity that is opening with the *Ingenuity* helicopter: everything suggests that it will be followed by many others. In our work, we first wanted to reiterate the state of the art of planetary exploration, which, at the moment, is focused on rovers, highlighting their criticalities and limitations by briefly examining the *Curiosity Mars Rover* case study. Next, we examine all the constraints that a flying vehicle mission imposes in the Martian atmosphere, finding all the functional requirements that cannot be absent.

From a trade-off with other flying vehicles, we arrive at the result that an autogyro proves to be the most suitable for that role, both for the payload and for the reasonable shortness of take-off and landing spaces.

From the examination of the Martian environment, several problems have been highlighted, such as low density, the need to operate at low Reynolds numbers, and the fact that in those conditions, the speed of sound is rather low enough to operate even at low rpm the extremity of the blades are nearly in the transonic zone. On the other hand, gravity is much lower, and this allows for much heavier payloads with the same performance.

From the operational similarity between two flying vehicles on Mars and on Earth, we come to define an "optimal surface" that identifies the limitations and characteristics of the operating conditions of an aerial drone.

In the second part, we illustrated our study for a gyroplane: we defined a series of missions such as *pathfinder*, *precision aerial photogrammetry*, and *cargo shuttle*, which, now, seem to be the most useful ones to be entrusted to an autonomous vehicle. We presented our "spider" vehicle and defined the longitudinal balance, as this type of aircraft is particularly sensitive to the position of the center of gravity; therefore, consequently, its positioning affects the internal arrangement of the avionics, services and all the necessary equipment, as well as the payload. In the last part, we defined, which are the aerodynamic parameters necessary to fully define a simulation. This work is only a preliminary part of a larger study for the design of a Martian flying vehicle.

Author Contributions: Conceptualization, E.P. and F.L.; methodology, E.P.; validation, F.L.; formal analysis, E.P.; investigation, F.L.; resources, F.L.; data curation, E.P.; writing—original draft preparation, E.P.; writing—review and editing, F.L.; supervision, F.L.; funding acquisition, F.L. All authors have read and agreed to the published version of the manuscript.

Funding: This research received no external funding.

Institutional Review Board Statement: Not applicable.

Informed Consent Statement: Not applicable.

Data Availability Statement: Not applicable.

Conflicts of Interest: The authors declare no conflict of interest.

Abbreviations

Symbols and Coefficients.

L	Lift
ρ_M	Air Density Mars
ρ_E	Air Density Earth
C_L	Coefficient of Lift
D	Drag
C_D	Coefficient of Drag
T	Thrust
S	Surface (aerodynamic)
α	Angle of Attack (local)
g_M	acceleration of gravity on Mars
g_E	acceleration of gravity on Earth
M	Mass of the vehicle
θ	angle between ground
W	Weight
W_M	Weight on Mars
W_M	Weight on Earth

References

1. Capon, P.T. Cierva's First Autogiros. *Aeropl. Mon.* **1979**, *7*, 234–240.
2. NASA. Planetary Fact Sheet—Metric. Available online: https://nssdc.gsfc.nasa.gov/planetary/factsheet/ (accessed on 1 March 2021).
3. Thomson, D.G.; Houston, S. *The Aerodynamics of Gyroplanes*; University of Glasgow: Glasgow, UK, 2010. [CrossRef]
4. Glauert, H. A General Theory of the Autogyro. In *Aeronautical Research Committee Reports and Memoranda No. 1111*; Cranfield University: Cranfield, UK, 1926.
5. Wheatley, J.B. *An Aerodynamic Analysis of the Autogyro Rotor with a Comparison between Calculated and Experimental Results*; NACA: Moffett Field, CA, USA, 1934; Volume 487.
6. Wheatley, J.B. *An Analytical and Experimental Study of the Effect of Periodic Blade Twist on the Thrust, Torque and Flapping Motion of an Autogyro Rotor*; US Government Printing Office: Washington, DC, USA, 1937; p. 591.
7. Houston, S.S. Validation of a rotorcraft mathematical model for autogyro simulation. *J. Aircr.* **2000**, *37*, 403–409. [CrossRef]
8. Singh, J.; Dadhich, A. Optimization of Autogyro for Preliminary Development of Personal Flying Vehicle. *Int. J. Innov. Res. Sci. Eng. Technol.* **2017**, *6*.
9. Wu, W. A Controller for Tracking Steep Glide Slopes for an Unmanned Gyroplane. *Br. J. Appl. Sci. Technol.* **2015**, *10*, 1–11. [CrossRef]
10. Leishman, J.G. Development of The Autogyro: A Technical Perspective. *J. Aircr.* **2004**, *41*, 765–781. [CrossRef]
11. Petritoli, E.; Leccese, F. Unmanned Autogyro for Advanced SAR Tasks: A Preliminary Assessment. In Proceedings of the 2020 IEEE 7th International Workshop on Metrology for AeroSpace (MetroAeroSpace), Pisa, Italy, 22–24 June 2020; pp. 615–619. [CrossRef]
12. Vainilavičius, D.; Augutis, V.; Malcius, M.; Bezaras, A. Analysis of autogyro rotor balancing and vibration. In Proceedings of the 2015 IEEE Metrology for Aerospace (MetroAeroSpace), Benevento, Italy, 4–5 June 2015; pp. 416–420. [CrossRef]
13. Wang, R.; Wu, L.; Bing, F. Automatic Landing Control Design of Gyroplane. In Proceedings of the 2019 IEEE 3rd Advanced Information Management, Communicates, Electronic and Automation Control Conference (IMCEC), Chongqing, China, 11–13 October 2019; pp. 1480–1484. [CrossRef]
14. Young, L.A.; Aiken, E.; Lee, P.; Briggs, G. Mars rotorcraft: Possibilities, limitations, and implications for human/robotic exploration. In Proceedings of the 2005 IEEE Aerospace Conference, Big Sky, MT, USA, 5–12 March 2005; pp. 300–318. [CrossRef]
15. Young, L.A.; Aiken, E.W.; Gulick, V.; Mancinelli, R.; Briggs, G.A. Rotorcraft as Mars Scouts. In Proceedings of the IEEE Aerospace Conference, Big Sky, MT, USA, 9–16 March 2002; Volume 1, pp. 1–378. [CrossRef]
16. Pergola, P.; Cipolla, V. Mission architecture for Mars exploration based on small satellites and planetary drones. *Int. J. Intell. Unmanned Syst.* **2016**. [CrossRef]
17. Braun, R.D.; Wright, H.S.; Croom, M.A.; Levine, J.S.; Spencer, D.A. Design of the ARES Mars airplane and mission architecture. *J. Spacecr. Rocket.* **2006**, *43*, 1026–1034. [CrossRef]
18. Serna, J.G.; Vanegas, F.; Gonzalez, F.; Flannery, D. A review of current approaches for UAV autonomous mission planning for Mars biosignatures detection. In Proceedings of the 2020 IEEE Aerospace Conference, Big Sky, MT, USA, 7–14 March 2020; pp. 1–15.

drones

MDPI

Article

Safety Enhancement of UAVs from the Signal Processing's Perspectives: A Bird's Eye View

Chiman Kwan

Signal Processing, Inc., Rockville, MD 20850, USA; chiman.kwan@signalpro.net

Abstract: Unmanned air vehicles (UAVs) or drones have gained popularity in recent years. However, the US Federal Aviation Administration (FAA) is still hesitant to open up the national air space (NAS) to UAVs due to safety concerns because UAVs have several orders of magnitude of more accidents than manned aircraft. To limit the scope in this paper, we focus on large, heavy, and expensive UAVs that can be used for cargo transfer and search and rescue operations, not small radio-controlled toy drones. We first present a general architecture for enhancing the safety of UAVs. We then illustrate how signal processing technologies can help enhance the safety of UAVs. In particular, we provide a bird's eye view of the application of signal processing algorithms on condition-based maintenance, structural health monitoring, fault diagnostics, and fault mitigation, which all play critical roles in UAV safety. Some practical applications are used to illustrate the importance of the various algorithms.

Keywords: UAVs; safety; drones; signal processing; condition-based maintenance; structural health monitoring; fault diagnostics; fault mitigation; contingency planning

Citation: Kwan, C. Safety Enhancement of UAVs from the Signal Processing's Perspectives: A Bird's Eye View. *Drones* **2021**, *5*, 16. https://doi.org/10.3390/drones5010016

Received: 20 January 2021
Accepted: 22 February 2021
Published: 26 February 2021

1. Introduction

The use of unmanned aerial vehicles (UAV) in the military and industry today is becoming more and more widespread. However, perhaps due to lower manufacturing standards and budget limitations, the mishap rates in unmanned aerial vehicles (UAVs) are several orders of magnitude greater than manned aviation [1]. Considering these high mishap rates, the US Department of Transportation's Federal Aviation Administration (FAA) has initiated several programs and partnerships to enhance the safety and reliable operation of UAVs [2].

In this paper, we focus on large, heavy, and expensive (millions of dollars) UAVs such as cargo transport and search and rescue aircraft, not radio-controlled amateur drones. In general, UAV safety can be improved from the following perspectives. First, the UAV manufacturers need to use durable engines and communication equipment and strong structural materials. Reliable communication equipment will ensure drone safety, as a lost link between UAV and ground station is very dangerous [3]. Durable engines and strong materials will ensure reliable flight in rough weather conditions. Second, advanced conditioned-based maintenance (CBM) practice should be deployed [4–6]. Compared with traditional periodic preventive maintenance, CBM can be cost-effective but may require additional sensors to monitor some critical components such as the engine. Third, structural health monitoring (SHM) [7,8], especially the non-destructive type of SHM, should be frequently used to monitor cracks, loosened fasteners, etc.

The aforementioned ideas mostly require both hardware and software, and have been routinely used in UAVs to maintain reliability and safety. In addition to the above safety enhancement practices, we would like to emphasize that signal processing algorithms also play important roles in other critical areas. Figure 1 shows the general architecture of an integrated system that can enhance the safety of UAVs. Some monitoring operations such as SHM, CBM, are done off-line on the ground. The off-line process involves data collection and processing. Others such as sensor and actuator fault diagnostic, fault-tolerant control are done online. The sensor's measurements (angles, speed, etc.) are all collected in real-time and processed in real-time. Contingency planning is done off-line. Moreover, as shown in Figure 1, we can clearly see the use of signal processing algorithms in fault detection, fault isolation, fault magnitude reconstruction, and fault mitigation. First, robust fault diagnosis algorithms [9–18] perform accurate fault detection and isolation. Failures in sensors and actuators can cause system instability. Diagnosing those faults will improve the overall safety of UAVs. Second, in fault mitigations, robust [19–23] and fault-tolerant controllers [24–31] are also critical for UAV safety. Robust controllers can tolerate some small perturbations in the UAVs due to aging or external disturbances. Fault-tolerant controllers can perform control reconfiguration to directly address sensor and actuator malfunctions during flight. Third, one critical aspect of fault mitigation is contingency planning for engine failures, which is the last line of defense for UAV safety. That is, all the aforementioned practices have been tried, but nothing works and the UAV is eventually on its course to crash due to engine loss. Under such an emergency, the goal of contingency planning is either to help UAV operators to glide the UAV or will make the UAV autonomously land itself to a crashing/ditching site or local airport runway if there is no reliable communication link available. In the case of an emergency due to full loss of thrust, wind plays a critical role with respect to reachability to the emergency landing site [32]. Upon full loss of thrust due to engine failure, because of the wind impact, the UAV may not reach the designated landing site and crash into populated areas causing loss of lives. Thus, the wind impact on reachability needs to be addressed in path planning for engine loss contingencies. It is also important that in the event of an emergency that might happen at high altitudes, the UAV should choose a forced landing path that does not violate no-fly zones or stormy weather air zones not to further complicate the situation.

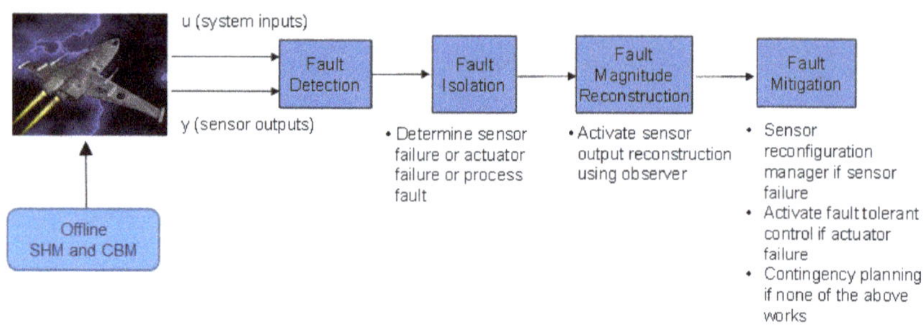

Figure 1. Enhancing UAV safety from the perspectives of signal processing algorithms.

The contributions of our paper is as follows:

- We provide a bird's eye view of the importance of signal processing algorithms in enhancing the safety of UAVs.
- In each area, we highlight some recent advances in the literature.

Our paper is organized as follows. In Sections 2 and 3, we briefly mention CBM and SHM practices. In Sections 4 and 5, we focus on fault diagnostics and fault magnitude estimation. In Section 6, we address fault mitigation using robust and fault-tolerant controllers. Section 7 discusses the last line of defense: contingency planning for an emergency landing. Finally, some concluding remarks, limitations, and future directions are given in Section 8.

2. Condition Based Maintenance (CBM)

Traditionally, mechanical components such as engines, bearings, and gearboxes are maintained by using preventive maintenance, which performs periodic checks on the components. For example, engine oil for cars is replaced with new oil every three thousand miles or every six months. Preventive maintenance is certainly effective and has been widely used. However, one drawback is that it may not be cost-effective. When a car is not being used frequently, periodic maintenance may be wasteful because the oil quality may still be good.

In recent years, people have been advocating condition-based maintenance [33,34], which means some maintenance decisions should be made based on the condition of the component rather than time. Two survey papers [35,36] provided a good review of the relevant works in this area. Some papers have proposed the use of wireless sensor networks (WSN) for quantifying system conditions [37,38]. Figure 2 illustrates how WSN can be used for induction motor monitoring.

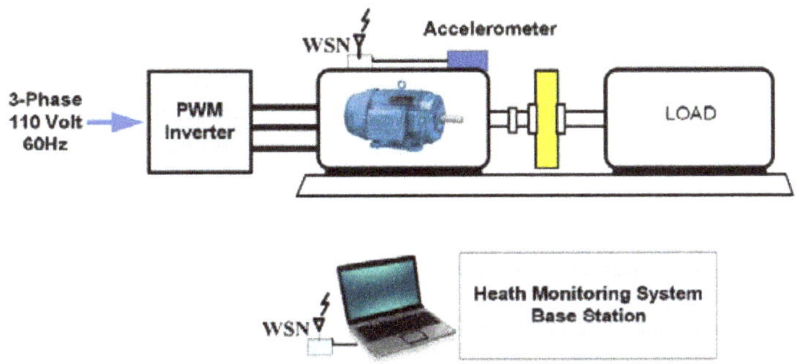

Figure 2. The use WSN for induction motor monitoring.

It should be noted the Internet of Things (IOT) is also a hot area that can be beneficial to CBM. In [39], some discussions talked about the various applications of IOT in CBM. IOT can also help online fault diagnosis as well.

3. Structural Health Monitoring (SHM)

As shown in Figure 3, Structural health monitoring (SHM) [7,8] is one way of nondestructive inspection (NDI). It can be done passively or actively. The purpose is to detect structural defects such as cracks on the wing, loosened bolt, and fasteners, etc.

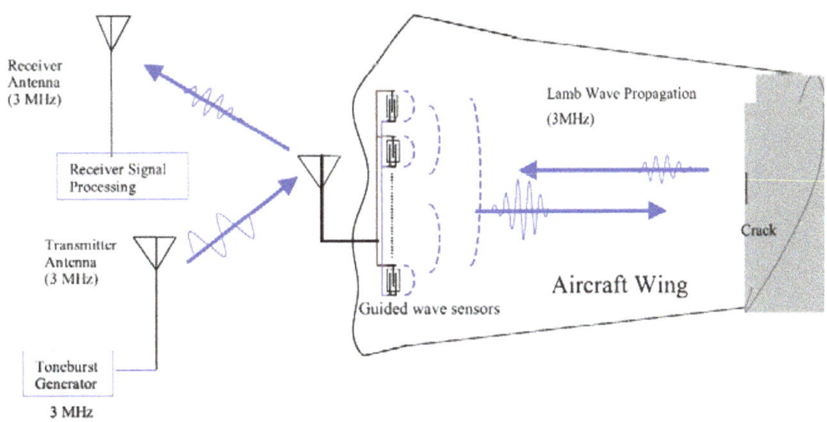

Figure 3. Schematic view of the direct analog RF coupling approach for wireless aircraft wing inspection.

In our opinion, SHM is also one form of CBM. To illustrate the basic idea of SHM, one can refer to Figure 4, which shows an array of sensors/actuators to monitor an aircraft wing panel. Ultrasonic pulses are generated from each sensor, and the rest of the array is receiving the transmitted signals. Once all the sensors are scanned, the collected signals from all the elements will be fed into an algorithm for fault localization. In [8], a RAPID algorithm was developed, which has been widely used in many papers.

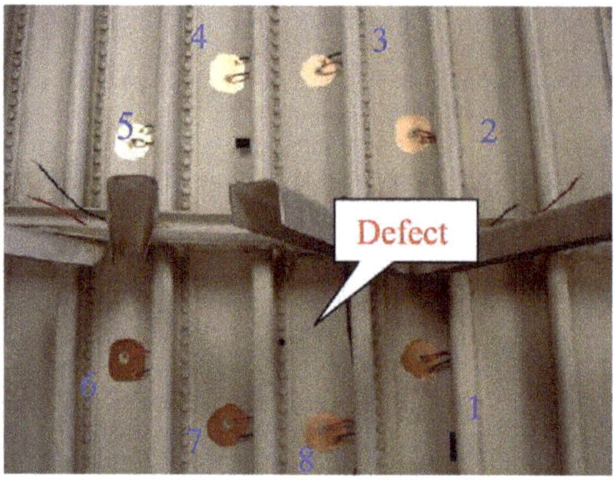

Figure 4. Application of ultrasonic sensor/actuator array to monitor aircraft wing panel.

The interrogation methods in [7,8] are bulky. Recently, there are new and recent advances in wireless sensing for SHM. As shown in Figure 5, passive low-power wireless node can be installed onto structures. Details can be found in [40].

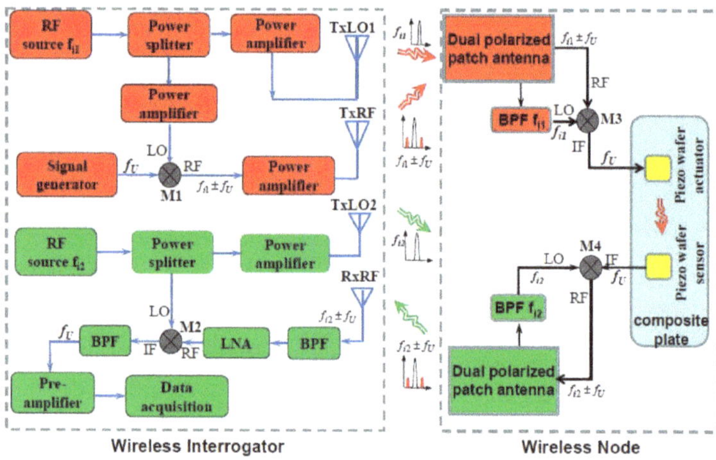

Figure 5. Recent SHM using wireless sensors for active crack detection.

4. Sensor and Actuator Fault Diagnostic Algorithms

Sensors and actuators are critical components in complex systems. For instance, in an airplane, effective flight control is impossible if sensors and/or actuators are malfunctioning [41,42]. Sensors and actuators can fail, and their failures have a significant impact on the performance of a system. In the worst case, the failure even can affect the safe operation of the system, leading to a catastrophic event. Sensor failures may include precision degradation, drift, frozen reading, and complete failure [43]. Similarly, actuator failures may include limited range of motion, e.g., valve stiction and complete failure [44]. It is a challenging task to detect sensor and actuator failures [41,42] because sensor outputs contain information from a multitude of sources: normal system outputs, faulty sensor signals, and signals due to noise and external disturbances. Isolating different signals requires the utilization of the input-output relationship of the system. Conventional approaches to increasing the reliability of aircraft systems include installing redundant sensors, which will add more weights, costs, complexity, and most importantly, additional reliability problems.

Given a linear multi-input multi-output (MIMO) system, the inputs, internal states, and outputs can be described by

$$\begin{aligned} \mathbf{x}_{k+1} &= \mathbf{A}\mathbf{x}_k + \mathbf{B}\mathbf{u}_k + \mathbf{B}\mathbf{M}_u\Delta\mathbf{u}_k + \omega_k \\ \mathbf{y}_k &= \mathbf{C}\mathbf{x}_k + \mathbf{D}\mathbf{u}_k + \mathbf{D}\mathbf{M}_u\Delta\mathbf{u}_k + \mathbf{M}_y\Delta\mathbf{y}_k + \varepsilon_k \end{aligned} \tag{1}$$

where $\mathbf{u}_k$, $\mathbf{y}_k$, and $\mathbf{x}_k$, are system inputs, outputs, and internal states, respectively. $\Delta\mathbf{u}_k$, $\Delta\mathbf{y}_k$ denote actuator and sensor faults. We would like to detect if the magnitude of $\Delta\mathbf{u}_k$ or $\Delta\mathbf{y}_k$ is non-zero. If the answer is positive, we have to estimate the magnitude and identify the failure direction matrix $\mathbf{M}_u$ or $\mathbf{M}_y$. Finally, the magnitudes of failed sensors or actuators will be reconstructed.

Sensor fault detection and diagnosis have been a research topic for decades, and many articles have been published. Interested readers are referred to the survey paper by Frank [45]. While fault detection is relatively easy, isolation of multiple faults is still a challenge to many existing schemes. Recent advances in fault diagnosis use kernel partial least square [46] and nonlinear techniques [47].

5. Sensor Magnitude Reconstruction

After the identification of faulty sensors/actuators, one has to estimate the fault magnitudes. Detailed procedures can be found in [48]. With estimated fault magnitudes, one can correct measurements of the faulty sensors. The measurements in faulty sensors are affected both by the actuator fault $\Delta\mathbf{u}_k$ and sensor faults $\Delta\mathbf{y}_k$. In order to get the corrected measurements in faulty sensors, one can design a Kalman filter,

$$\begin{aligned} \hat{\mathbf{x}}_{k+1} &= \mathbf{A}\hat{\mathbf{x}}_k + \mathbf{B}\mathbf{u}_k + \mathbf{K}(\mathbf{y}_k - \mathbf{M}_y\Delta\hat{\mathbf{y}}_k - \mathbf{D}\mathbf{M}_u\Delta\hat{\mathbf{u}}_k - \mathbf{C}\hat{\mathbf{x}}_k - \mathbf{D}\mathbf{u}_k) \\ \hat{\mathbf{y}}_k &= \mathbf{C}\hat{\mathbf{x}}_k + \mathbf{D}\mathbf{u}_k \end{aligned} \tag{2}$$

where $\hat{\mathbf{y}}_k$ is the corrected measurement of faulty sensors. $\mathbf{A}$, $\mathbf{B}$, $\mathbf{C}$, $\mathbf{D}$ are the system matrices in the state-space model. $\mathbf{K}$ is the Kalman gain, $\mathbf{M}_u$ and $\mathbf{M}_y$ are the fault direction matrices [48].

Nowadays, UAV formation (swarm) control practice uses the Real-time Kinematic Global Navigation Satellite System-Inertial Navigation System (RTK-GNSS-INS) [49] for routing and navigation. Moreover, inter-UAV sensors such as Automatic Dependent Surveillance-Broadcast (ADS-B) [50], which provides altitude, aircraft flight ID, and vertical airspeed, have been used in advanced systems. ADS-B reduces the risk of runway collisions, even at night or during heavy rainfall. ADS-B applications being developed now will give pilots indications or alerts of potential collisions. Such systems may be used to dissolve the problem related to faulty sensors.

Here we describe some results of detecting GNSS sensor failure and how to use reconstructed sensor information to control the formation of UAVs. We will illustrate the effectiveness of a fault-tolerant formation control algorithm [30] using the rotary UAV developed by UC Berkeley. Two types of formations are considered: mesh and triangle. For each formation, we performed two types of flight: straight flight and 90-degree turning flight. In each case, we assume a fault occurs to the fifth UAV at 10 s. A GNSS/INS (Inertial Navigation System) combination is usually used in UAV navigation systems. In the presence of GNSS failure, the position information provided by the INS will diverge due to error accumulation. Motivated by this observation, the fault model under consideration simulates slowly divergent position measurements. A fault isolation scheme was designed to capture this GNSS failure. Then an observer was designed and used to estimate the positions of the UAV. Figure 6 shows the results, which clearly demonstrated that the reconstruction can still maintain the UAV formation.

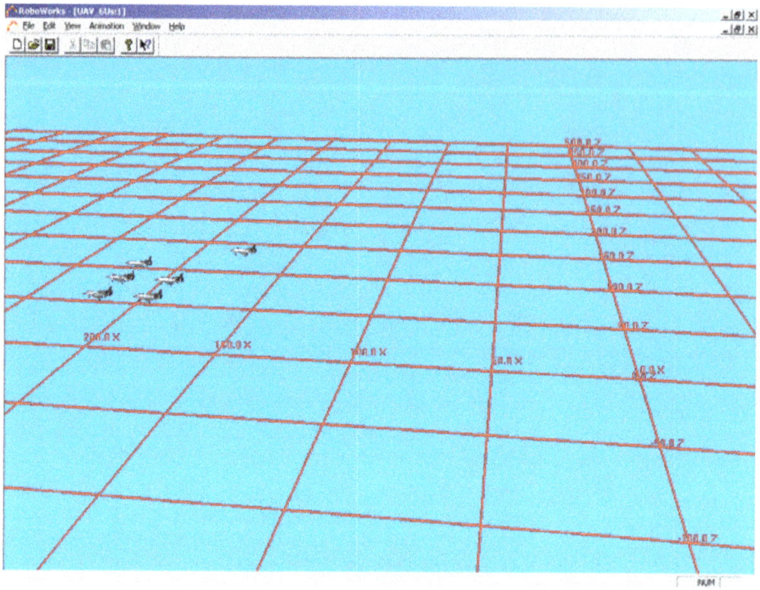

(a)

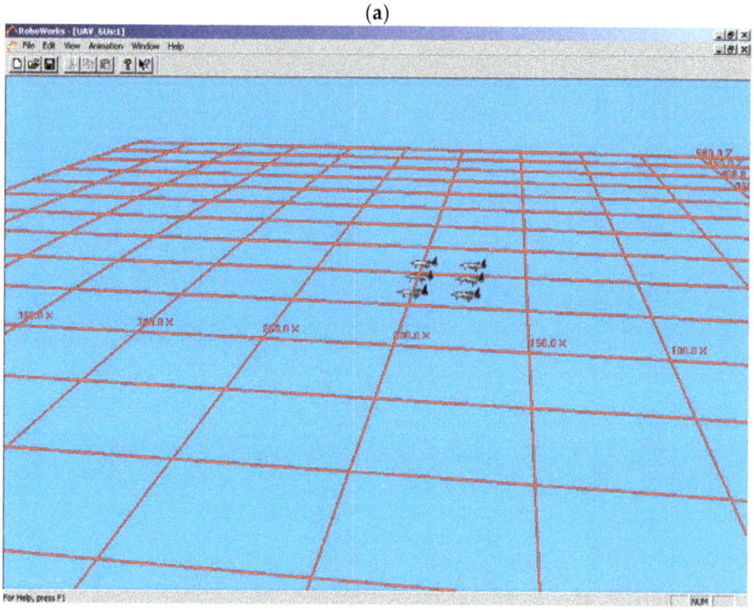

(b)

Figure 6. Comparison of results with fault tolerance and without fault tolerance. (**a**) Without fault tolerance (one UAV is drifting away); (**b**) with fault tolerance.

6. Robust and Fault-Tolerant Control

As shown in Figure 7, there can be quite a few actuators for controlling the aircraft. For example, the rudder controls the yaw channel, elevators control the pitch channel, and flaps control the roll channel. From Table 1, we can see that some actuators can be used as secondary actuators for some channels. This shows that different combinations of actuators can serve as backup actuators for different roll, pitch, and yaw channels. In general, fault tolerant control refers to the selection of substitute actuators when primary actuators fail to function.

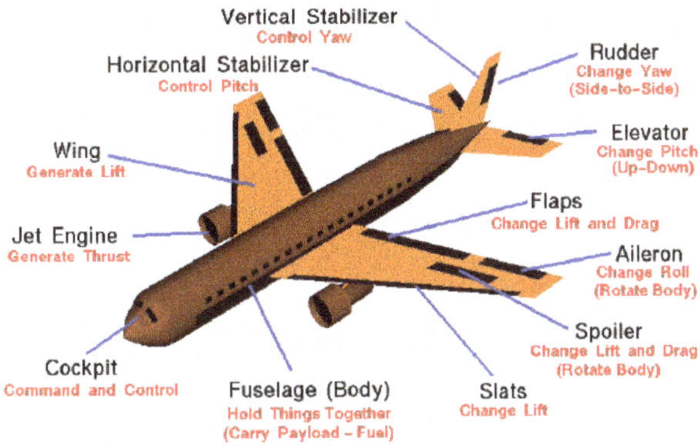

Figure 7. Various actuators in a hypothetical aircraft.

Table 1. Primary, secondary and tertiary actuators for the roll, pitch, and yaw channels.

	Primary Actuator	Secondary Actuator	Tertiary Actuator
Roll Channel	Aileron	Rudder & Asymmetric Engine Thrust	N/A
Pitch Channel	Tailplane	Symmetric Aileron	Symmetric Engine Thrust
Yaw Channel	Rudder	Asymmetric Engine	

A robust controller [19–23] can deal with parametric uncertainties to some extent. However, when some faults such as sensor and actuator faults occur, robust controllers are not strong enough to handle such situations. A fault-tolerant controller [24–31] is extremely useful for guaranteeing closed-loop control performance.

Figure 8 shows a schematic diagram of how fault-tolerant control works. First, a fault diagnosis system that performs on-line fault detection and isolation. Second, a controller suite consists of a primary nominal control system used under normal operating conditions (without faults) and a secondary adaptive fault-tolerant control system engaged only after fault detection. Third, a reconfiguration supervisor makes decisions regarding control system reconfiguration and control reallocation using the fault information provided by the diagnostic module.

Here, we briefly summarize one application of our neural net (NN) controller. Our controller suite includes the nominal controller and the NN-based adaptive fault-tolerant controller. The nominal controller is used when the system is in a fault-free condition. The NN adaptive controller is activated after a fault is detected to compensate for the effect of the fault and to maintain acceptable control performance even in the presence of a fault. The fault-tolerant controller performance was demonstrated by using a well-known RCAM (Research Civil Aircraft Model) developed by the Group for Aeronautical Research and technology in Europe (GARTEUR). The design of the nominal controller is based on standard approximate dynamic inversion. Figure 9 illustrates the proposed nonlinear adaptive control architecture: the aircraft (f), the command filter to provide desired handling qualities, approximate dynamic inverse ($\hat{f}^{-1}$), a conventional linear tracking controller as described above, and an online learning neural network to correct for errors and uncertainty in association with the inversion model.

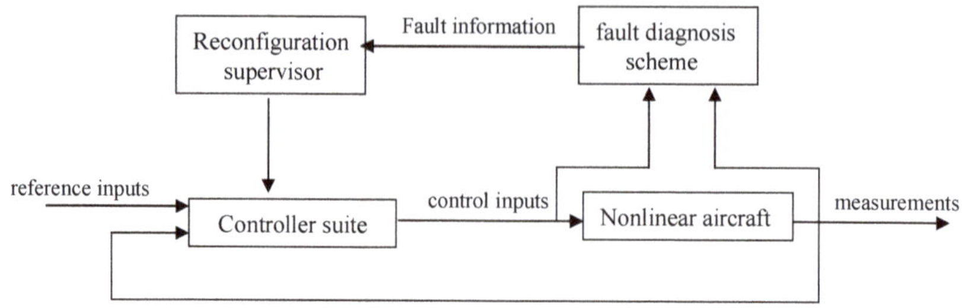

Figure 8. Illustration of fault-tolerant control strategy.

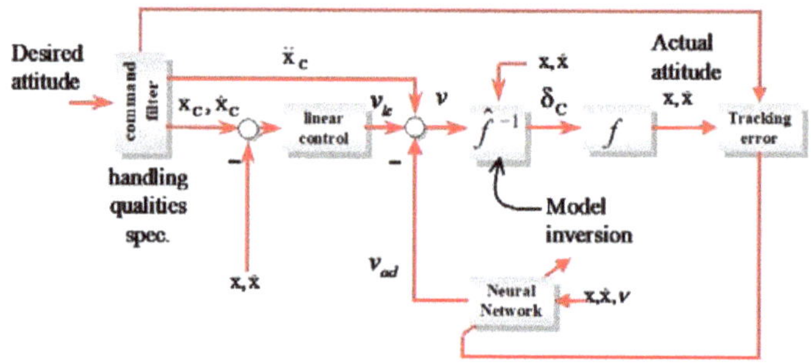

Figure 9. Neural Network Augmented Model Inversion Control Architecture.

One of the key advantages of the proposed fault-tolerant control scheme is its capability to handle any occurrences of new or unanticipated faults. The neural network-based adaptive controller activated after fault detection is still capable of compensating for the effect of the fault on-line and to maintain acceptable control performance before further pilot intervention.

Figures 10 and 11 give the control performances of the nominal controller and nonlinear adaptive controller engaged after a fault detection, respectively. We can clearly see the benefit of controller reconfiguration using on-line diagnostic information.

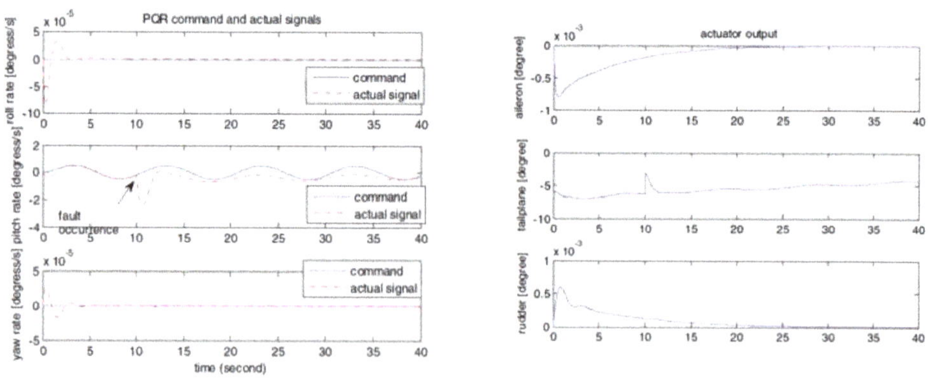

Figure 10. Tracking performance of the nominal controller when a aileron failure occurs at t = 10 s.

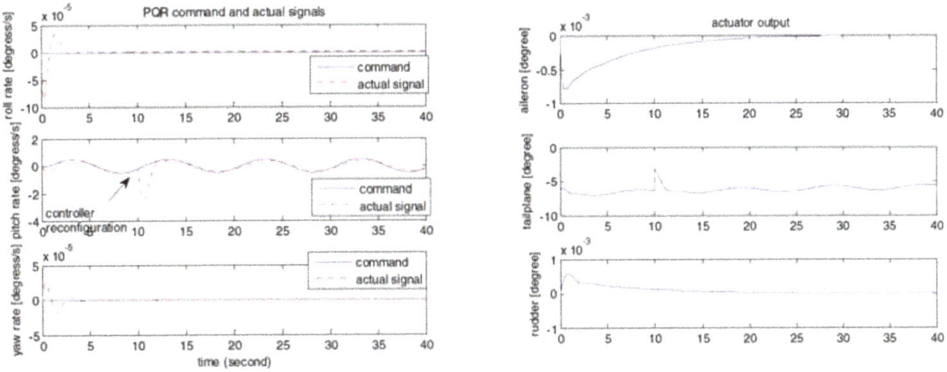

Figure 11. Tracking performance with controller reconfiguration after detection of a tailplane failure.

In recent years, there has been some new progress in fault-tolerant control. In [51], a hybrid fault-tolerant controller was proposed to handle control surface damages. Compared to the nonlinear control approach in [18], the controller in [51] was based on linearized models. In [52], a nonlinear sliding mode controller was proposed to deal with actuator failure in quadrotor UAVs. The inner loop was the attitude control and the outer loop was the position control. Sliding mode control was applied to both inner and outer loops. The control is challenging because if one or two control actuators fail, the system is under-actuated. A special form of sliding mode control was proposed based on back-stepping. In [53], an observed-based fault-tolerant controller was proposed for carrier-based UAVs. Some states in the UAVs are assumed to be unmeasurable and hence the control problem is challenging. The controller was nonlinear and closed-loop stability was given.

7. Contingency Planning

This is the last defense for fault mitigation in UAVs. The UAV has lost its engine and it is on its way to crash. Emergency landing via the parachute systems is part of the contingency planning for some small-to-medium sized UAVs [54]. However, for big and heavy drones like Global Hawk (12 tons) [55], parachute is not feasible. Can we still do something to minimize the damage? In other words, for fixed wing UAVs, the hanging time of some UAVs such as Global Hawk can still be 30 min or more due to its high flying altitude and large wing span. Moreover, if one plans ahead, the UAVs can still glide to some safe landing places such as airports, non-populated places such as beaches, waterways, grassy areas, etc. A well-known example is the US Airways Flight 1549, which avoided a crash landing by gliding onto the Hudson River.

In our recent papers [56–60], we have provided detailed procedures for contingency planning for engine failures.

Figure 12 shows the workflow of contingency plan generation. It is an off-line process. Given a UAV and its associated flying capabilities (wingspan, gliding speed, descending rate, etc.) and also the theater of operations, we need the following two major steps: preprocessing and contingency plan generation. We will summarize those two steps in the next two sub-sections.

7.1. Preprocessing

There are multiple modules in the preprocessing step. First, landing place selection is needed. Based on the UAV's size and gliding speed, an appropriate landing site needs to have enough length. Potential landing sites include airport runways, beaches, waterways, etc. In [58], we developed a landing site selection algorithms based on Google maps. Figure 13 shows a landing site.

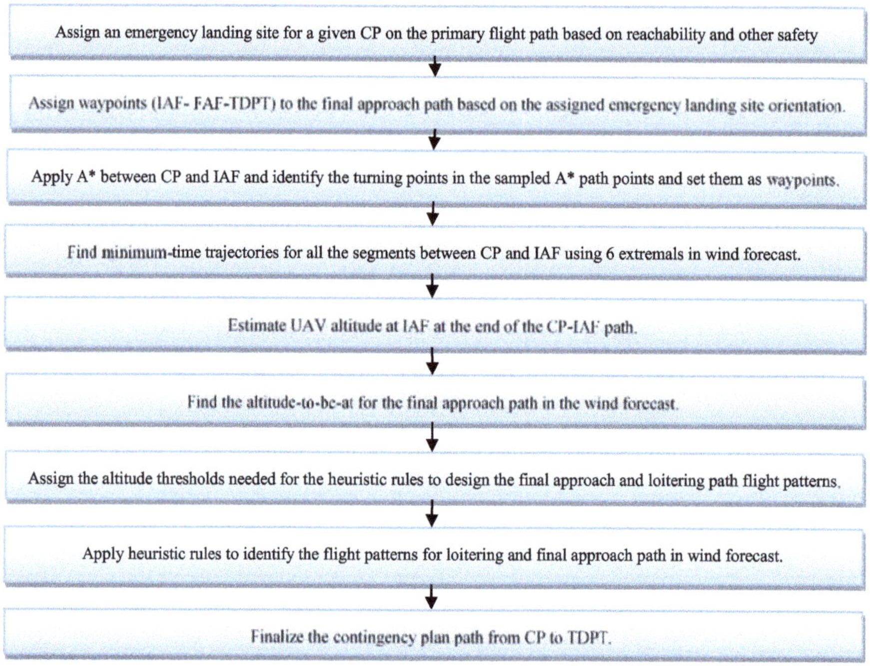

Figure 12. Contingency plan generation workflow [56].

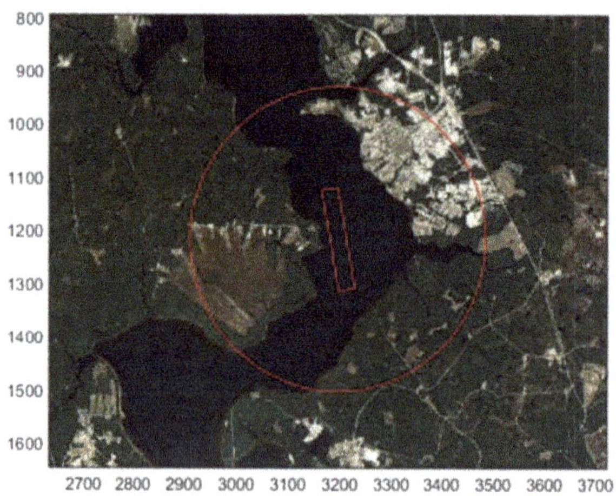

Figure 13. Landing site candidate-1 (green dotted line passes through the center of the landing site along its length).

Second, for each landing site, we need to assign some waypoints such as Touch-down point (TDPT), Initial Approach Fix (IAF), and Final Approach Fix (FAF). Detailed procedures are provided in [56].

7.2. Contingency Plan Generation

Given a primary flight plan containing hundreds or even thousands of waypoints, we need to generate a contingency plan for each waypoint. An example of a contingency plan is shown in Figure 14.

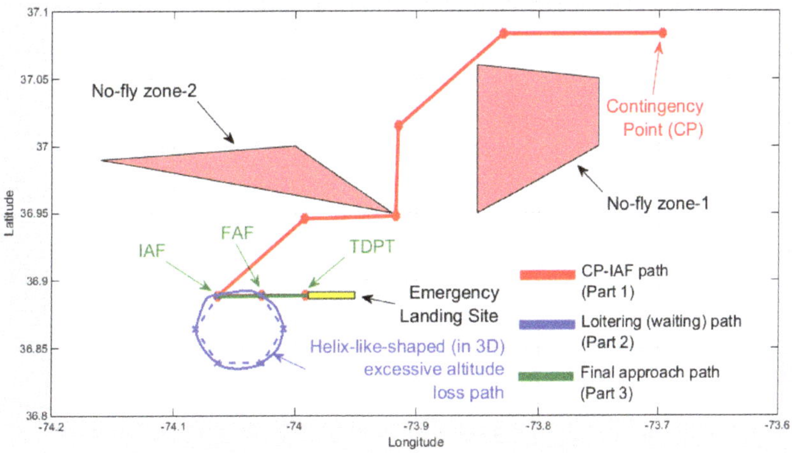

Figure 14. Contingency plan example.

Suppose an engine failure occurs near a waypoint in the primary flight path, we will apply A* path planning algorithm [56] to generate a contingency path between the contingency point (CP) and the IAF. Some no-fly zones need to be bypassed. From Figure 14, the red line section shows the A* generated plan. In the contingency plan, when the UAV reaches the IAF, there may still be excessive altitude to lose. We have developed time-constrained path generation algorithm to lose excessive altitude [57]. Wind speed needs to be taken into account. Some extremal paths can be seen in Figure 15. RSR means right-straight-right; RSL means right-straight-left.

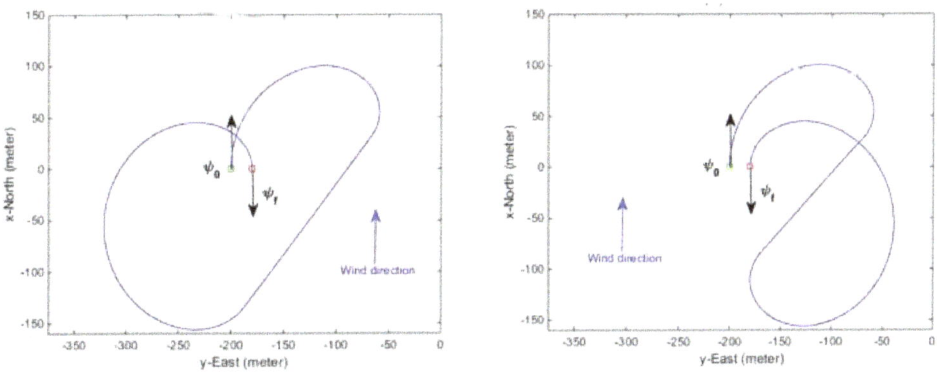

(a) RSR (b) RSL

Figure 15. Two extremal paths. Other paths such as LSR, LSL, RSL, and RSR can be found in [57].

There are also some additional processing steps to deal with different excessive altitudes. Details can be found in [56].

One limitation of our contingency planning approach is that if the wind conditions deviate too much from the forecast conditions, which can happen in practice, then the pre-planned contingency paths may need to change on the fly. More research is needed in this direction.

7.3. Emergency Landing in Hudson River

Here, we demonstrate how we generate a contingency plan for US Airways Flight 1549, which lost both engines due to a bird strike. The plane was on route from New York City's LaGuardia Airport to Seattle, Washington and ditched into Hudson River on 15 January, 2009. In the climb phase right after its takeoff, it struck a flock of Canada geese and lost its engine power. The pilots Mr. Sullenberger and Mr. Skiles glided the plane to a ditching in the Hudson River. At the time of the bird strike, Flight 1549's airspeed was about 200 knots. The highest altitude right before the plane started sinking was 3034 feet (925 m). At this altitude, the plane was located at coordinates: Latitude: 40.861666 degrees, Longitude: −73.879722 degrees. Time was 3:27:29 p.m. The wind amplitude was around 13.4 knots. The wind direction was 320 degrees. Assuming the coordinates at the highest altitude corresponds to the coordinates of CP, we found the heading angle at CP using the waypoint coordinates at the highest altitude, the waypoint right before that, and the wind information.

We applied our contingency plan generation tool to this incident. Figures 16 and 17 show the top view and 3D view of the generated plan.

In the movie "Sully", various flight simulator runs also showed that it was possible to return to LaGuardia and Teterboro airports. Why did the pilots choose to land on Hudson river? The reason is that the pilots need to follow some procedures to make sure both engines were lost. During this period of checking, the plane has lost quite some altitude. Consequently, there was not enough altitude to glide it back to the airports. Based on his experience, the pilot, Sullenberger, made some rough calculations and decided to land on Hudson Bay; the decision saved over 150 people onboard.

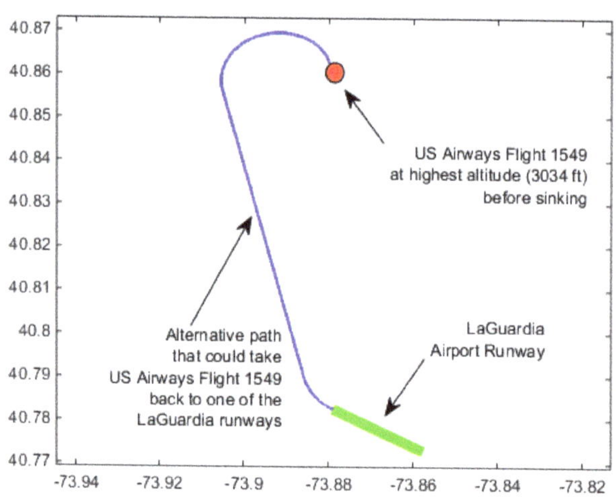

Figure 16. Contingency plan for US Airways Flight 1549 to return to LaGuardia airport: Top down view.

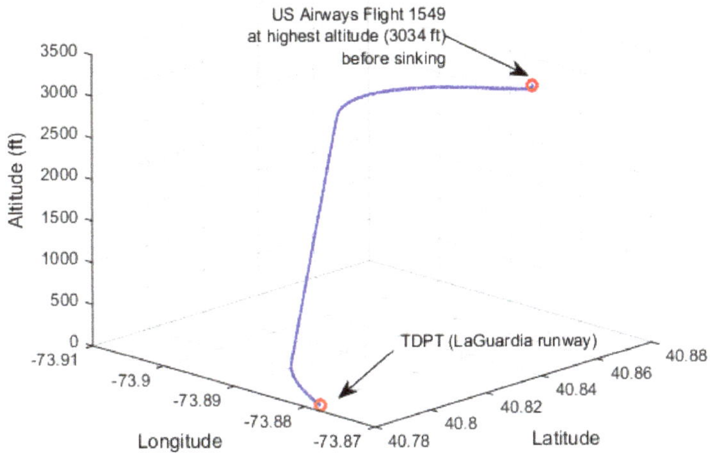

Figure 17. Contingency plan for US Airways Flight 1549 to return to LaGuardia airport: 3D view.

8. Conclusions

UAVs are gaining popularity. However, the safety of UAVs is not on par with manned aircraft. In this paper, we present safety enhancements of UAVs using signal processing algorithms, which can help condition-based maintenance, structural health monitoring, sensor and actuator fault diagnostics, fault magnitude reconstruction, fault-tolerant control, and contingency plan generation. Some recent advances in the aforementioned areas are also highlighted.

It is important to emphasize that UAV safety requires an integrated approach that contains all of the above. More research and development effort is needed to produce an integrated safety system for UAVs. For instance, the contingency plans are generated off-line, and if wind conditions deviate a lot from the forecast data during actual flights, our current system may not be able to handle that scenario. One potential future direction in contingency planning is to deal with highly dynamic windy conditions. Another direction is to devise some online contingency planning strategies to complement the off-line generated contingency plans.

Funding: This project was supported in part by US government under the PPP program. The views, opinions and/or findings expressed are those of the author(s) and should not be interpreted as representing the official views or the US Government.

Institutional Review Board Statement: Not applicable.

Informed Consent Statement: Not applicable.

Conflicts of Interest: The author declares no conflict of interest.

References

1. Tvaryanas, A.P.; Thompson, B.T.; Constable, S.H. US military unmanned aerial vehicle mishaps: Assessment of the role of human factors using HFACS. In *USAF 311th Performance Enhancement Directorate*; Technical Report; HSW-PE-BR-TR-2005-0001; Brooks City-Base: San Antonio, TX, USA, 2005.
2. Programs, Partnerships and Opportunities. US Department of Transportation, Federal Aviation Administration. Available online: https://www.faa.gov/uas/programs_partnerships (accessed on 23 February 2018).
3. Kwan, C.; Ayhan, B. Enhancing Safety of UAVs in National Airspace. In Proceedings of the IEEE 9th Annual Computing and Communication Workshop and Conference, Las Vegas, NV, USA, 7–9 January 2019.
4. Choi, C.; Lee, W. Design and evaluation of voltage measurement-based sectoral diagnosis method for inverter open switch faults of permanent magnet synchronous motor drives. *IET Electr. Power Appl.* **2012**, *6*, 526–532. [CrossRef]

5. Kiani, M.; Lee, W.; Kenarangui, R.; Fahimi, B. Detection of Rotor Faults in Synchronous Generators. In Proceedings of the IEEE International Symposium on Diagnostics for Electric Machines, Power Electronics and Drives, Cracow, Poland, 6–8 September 2007; pp. 266–271.
6. Zhang, G.; Kwan, C.; Xu, R.; Vichare, N.; Pecht, M. An Enhanced Prognostic Model for Intermittent Failures in Digital Electronics. In Proceedings of the IEEE Aerospace Conference, Big Sky, MT, USA, 3–10 March 2007; pp. 1–8.
7. Zhao, X.; Qian, T.; Mei, G.; Kwan, C.; Zane, R.; Walsh, C.; Paing, T.; Popovic, Z. Active health monitoring of an aircraft wing with an embedded piezoelectric sensor/actuator network: II. Wireless approaches. *J. Smart Mater. Struct.* **2007**, *16*, 1218–1225. [CrossRef]
8. Zhao, X.; Gao, H.; Zhang, G.; Ayhan, B.; Yan, F.; Kwan, C.; Rose, J.L. Active health monitoring of an aircraft wing with embedded piezoelectric sensor/actuator network: I. Defect detection, localization and growth monitoring. *J. Smart Mater. Struct.* **2007**, *16*, 1208–1217. [CrossRef]
9. Kwan, C.; Xu, R. A note on simultaneous isolation of sensor and actuator faults. *IEEE Trans. Control Syst. Technol.* **2004**, *12*, 183–192. [CrossRef]
10. Xu, R.; Kwan, C. Robust Isolation of Sensor Failures. *Asian J. Control* **2008**, *5*, 12–23. [CrossRef]
11. Qin, S.J.; Li, W. Detection, identification, and reconstruction of faulty sensors with maximized sensitivity. *AIChE J.* **1999**, *45*, 1963–1976. [CrossRef]
12. Dunia, R.; Qin, S.J. Joint diagnosis of process and sensor faults using principal component analysis. *Control Eng. Pract.* **1998**, *6*, 457–469. [CrossRef]
13. Li, W.; Yue, H.; Valle-Cervantes, S.; Qin, S. Recursive PCA for adaptive process monitoring. *J. Process Control* **2000**, *10*, 471–486. [CrossRef]
14. Lee, J.-M.; Qin, S.J.; Lee, I.-B. Fault detection and diagnosis based on modified independent component analysis. *AIChE J.* **2006**, *52*, 3501–3514. [CrossRef]
15. Qin, S.J. Survey on data-driven industrial process monitoring and diagnosis. *Annu. Rev. Control* **2012**, *36*, 220–234. [CrossRef]
16. Gertler, J.; Singer, D. Augmented Models for Statistical Fault Isolation in Complex Dynamic Systems. In Proceedings of the American Control Conference, Boston, MA, USA, 19–21 June 1985; pp. 317–322.
17. Gertler, J.J. A numerical-structural approach to failure detection and isolation in complex plants. In Proceedings of the 1986 25th IEEE Conference on Decision and Control, Athens, Greece, 2 April 2007; pp. 1576–1580.
18. Zhang, X.; Liu, Y.; Rysdyk, R.; Kwan, C.; Xu, R. An Intelligent Hierarchical Approach to Actuator Fault Diagnosis and Accommodation. In Proceedings of the 2006 IEEE Aerospace Conference, Big Sky, MT, USA, 4–11 March 2006.
19. Kuljaca, O.; Swamy, N.; Lewis, F.; Kwan, C. Design and implementation of industrial neural network controller using backstepping. *IEEE Trans. Ind. Electron.* **2003**, *50*, 193–201. [CrossRef]
20. Tao, G.; Kokotovic, P. *Adaptive Control of Systems with Actuator and Sensor Nonlinearities*; John Wiley & Sons: Hoboken, NJ, USA, 1996.
21. Yeung, K.S.; Cheng, C.C.; Kwan, C. A Unifying Design of Classical and Sliding Controllers. *IEEE Trans. Autom. Control* **1993**, *38*, 1422–1426. [CrossRef]
22. Ioannou, P.; Sun, J. *Robust Adaptive Control*; Dover Publications: Mineola, NY, USA, 2012.
23. Kwan, C.; Xu, H.; Lewis, F.L. Robust Spacecraft Attitude Control Using Adaptive Fuzzy Logic. *Int. J. Syst. Sci.* **2000**, *31*, 1217–1225. [CrossRef]
24. Ciuryla, M.; Liu, Y.; Farnsworth, J.; Kwan, C.; Amitay, M. Flight Control Using Synthetic Jets on a Cessna 182 Model. *J. Aircr.* **2007**, *44*, 642–653. [CrossRef]
25. Zhao, D.; Polycarpou, M.M. Distributed Fault Accommodation for a Class of Interconnected Nonlinear Systems with Event-Triggered Inter-Communications. In Proceedings of the 2020 International Joint Conference on Neural Networks (IJCNN), Glasgow, UK, 19–24 July 2020.
26. Zhang, K.; Jiang, B.; Yan, X.; Mao, Z.; Polycarpou, M.M. Fault-Tolerant Control for Systems with Unmatched Actuator Faults and Disturbances. In *IEEE Transactions on Automatic Control*; Early Access; IEEE: Piscataway, NJ, USA, 2020.
27. Khalili, M.; Zhang, X.; Cao, Y.; Polycarpou, M.M.; Parisini, T. Distributed Fault-Tolerant Control of Multiagent Systems: An Adaptive Learning Approach. *IEEE Trans. Neural Netw. Learn. Syst.* **2020**, *31*, 420–432. [CrossRef]
28. Keliris, C.; Polycarpou, M.M.; Parisini, T. An Adaptive Approach to Sensor Bias Fault Diagnosis and Accommodation for a Class of Input-Output Nonlinear Systems. In Proceedings of the IEEE Conference on Decision and Control (CDC), Miami Beach, FL, USA, 17–19 December 2018.
29. Khalili, M.; Zhang, X.; Cao, Y.; Polycarpou, M.M.; Parisini, T. Distributed adaptive fault-tolerant leader-following formation control of nonlinear uncertain second-order multi-agent systems. *Int. J. Robust Nonlinear Control* **2018**, *28*, 4287–4308. [CrossRef]
30. Zhang, X.; Xu, R.; Kwan, C.; Haynes, L.; Yang, Y.; Polycarpou, M. Fault tolerant formation flight control of UAVs. *Int. J. Veh. Auton. Syst.* **2004**, *2*, 217. [CrossRef]
31. Polycarpou, M.; Zhang, X.; Xu, R.; Yang, Y.; Kwan, C. A Neural Network Based Approach to Adaptive Fault Tolerant Flight Control. In Proceedings of the 2004 IEEE International Symposium on Intelligent Control, Taipei, Taiwan, 4 September 2004; pp. 61–66.
32. Coombes, M.; Chen, W.H.; Render, P. Reachability Analysis of Landing Sites for Forced Landing of a UAS in Wind Using Trochoidal Turn Paths. In Proceedings of the International Conference on Unmanned Aircraft Systems, Piscataway, NJ, USA, 9–12 June 2015; pp. 62–71.

33. Wongsaichua, W.; Lee, W.; Oraintara, S.; Kwan, C.; Zhang, F. Integrated high speed intelligent utility tie unit for disbursed/renewable generation facilities. *IEEE Trans. Ind. Appl.* **2005**, *41*, 507–513. [CrossRef]
34. Jain, H.; Korkua, S.; Lee, W.J.; Kwan, C. Detection and Severity Classification of Rotor Imbalance Faults in Induction Machines. In Proceedings of the IEEE-IAS Annual Conference, Houston, TX, USA, 3–7 October 2010.
35. Higgs, P.A.; Parkin, R.; Jackson, M.; Al-Habaibeh, A.; Zorriassatine, F.; Coy, J. A Survey on Condition Monitoring Systems in Industry. In Proceedings of the ASME 7th Biennial Conference on Engineering Systems Design and Analysis, Manchester, UK, 19–22 July 2004; Volume 3, pp. 163–178.
36. Mohammed, A.; Norman, E.S.; Abouel, N.; Adel, A.S.; Husam, K. Overview of predictive condition based maintenance research using bibliometric indicators. *J. King Saud Univ. Eng. Sci.* **2019**, *31*, 355–367.
37. Korkua, S.K.; Lee, W.-J.; Kwan, C. Design and Implementation of Zigbee based Vibration Monitoring and Analysis for Electrical Machines. In Proceedings of the International Conference on Wireless Networks—ICWN'11, Las Vegas, NV, USA, 18–21 July 2011.
38. Korkua, S.; Jain, H.; Lee, W.-J.; Kwan, C. Wireless Health Monitoring System for Vibration Detection of Induction Motors. In Proceedings of the IEEE-IAS, ICPS Annual Conference, Tallahassee, FL, USA, 9–13 May 2010.
39. Lai, C.T.A.; Jiang, W.; Jackson, P.R. Internet of Things enabling condition-based maintenance in elevators service. *J. Qual. Maint. Eng.* **2019**, *25*, 563–588. [CrossRef]
40. Kwan, C.; Huang, H.; Islam, M.M.; Ayhan, B. Unpowered Wireless Ultrasound Generation and Sensing for Structural Health Monitoring of Composites. In Proceedings of the IEEE International Conference on Prognostics and Health Management, Seattle, WA, USA, 11–13 June 2018.
41. Xu, R.; Zhang, G.; Zhang, X.; Kwan, C.; Semega, K. Sensor Validation Using Nonlinear Minor Component Analysis. In Proceedings of the Third International Symposium on Neural Networks, Lecture Notes in Computer Science, Chengdu, China, 28 May–1 June 2006.
42. Zhang, X.; Polycarpou, M.M.; Xu, R.; Kwan, C. Actuator Fault Diagnosis and Accommodation for Improved Flight Safety. In Proceedings of the IEEE International Symposium on Intelligent Control and Mediterranean Conference on Control and Automation Conference, Limassol, Cyprus, 27–29 June 2005; pp. 640–645.
43. Dunia, R.; Qin, J.; Edgar, T.F.; McAvoy, T.J. Identification of faulty sensors using principal component analysis. *AIChE J.* **1996**, *42*, 2797–2812. [CrossRef]
44. Deluca, A.; Mattone, R. Actuator Failure Detection and Isolation Using Generalized Momenta. In Proceedings of the IEEE Conference on Robotics and Automation, Taipei, Taiwan, 14–19 September 2003; pp. 634–639.
45. Frank, P. Fault diagnosis in dynamic systems using analytical and knowledge-based redundancy: A survey and some new results. *Automatica* **1990**, *26*, 459–474. [CrossRef]
46. Yi, J.; Huang, D.; He, H.; Zhou, W.; Han, Q.; Li, T. A Novel Framework for Fault Diagnosis Using Kernel Partial Least Squares Based on an Optimal Preference Matrix. *IEEE Trans. Ind. Electron.* **2017**, *64*, 4315–4324. [CrossRef]
47. Yi, J.; Wu, L.; Zhou, W.; He, H.; Yao, L. A Sparse Dimensionality Reduction Approach Based on False Nearest Neighbors for Nonlinear Fault Detection. *IEEE Trans. Syst. Man Cybern. Syst.* **2019**, 1–13. [CrossRef]
48. Li, W.; Kwan, C. A Novel Approach to Sensor and Actuator Integrity Monitoring. In Proceedings of the IEEE Conference on Decision and Control, Las Vegas, NV, USA, 12–14 December 2016; pp. 2140–2145.
49. Ando, T.; Kugimiya, W.; Hashimoto, T.; Momiyama, F.; Aoki, K.; Nakano, K. Lateral Control in Precision Docking Using RTK-GNSS/INS and LiDAR for Localization. *IEEE Trans. Intell. Veh.* **2020**. [CrossRef]
50. ADS-B. Available online: https://www.faa.gov/nextgen/programs/adsb/faq/ (accessed on 13 February 2021).
51. Ergöçmen, B.; Yavrucuk, İ. Active Hybrid Fault Tolerant Flight Control of an UAV under Control Surface Damage. In Proceedings of the American Control Conference (ACC), Denver, CO, USA, 1–3 July 2020; pp. 4169–4174.
52. Zhu, Z.; Cao, S. Back-stepping sliding mode control method for quadrotor UAV with actuator failure. *J. Eng.* **2019**, *2019*, 8374–8377. [CrossRef]
53. Zheng, F.; Zhen, Z.; Gong, H. Observer-based backstepping longitudinal control for carrier-based UAV with actuator faults. *J. Syst. Eng. Electron.* **2017**, *28*, 322–337.
54. Al-Madani, B.; Svirskis, M.; Narvydas, G.; Maskeliūnas, R.; Damaševičius, R. Design of Fully Automatic Drone Parachute System with Temperature Compensation Mechanism for Civilian and Military Applications. *J. Adv. Transp.* **2018**, *2018*, 1–11. [CrossRef]
55. Global Hawk. Available online: https://www.reuters.com/article/us-mideast-iran-usa-factbox/factbox-the-global-hawk-drone-shot-down-by-iran-idUSKCN1TL29K (accessed on 13 February 2021).
56. Ayhan, B.; Kwan, C.; Budavari, B.; Larkin, J.; Gribben, D. A Preflight Contingency Planning Approach for Fixed Wing UAVs in Full Loss of Thrust Using Wind Forecast. *Sensors* **2018**, *19*, 227. [CrossRef]
57. Ayhan, B.; Kwan, C. Time-Constrained Extremal Trajectory Design for Fixed-Wing Unmanned Aerial Vehicles in Steady Wind. *J. Guid. Control Dyn.* **2018**, *41*, 1569–1576. [CrossRef]
58. Ayhan, B.; Kwan, C.; Um, Y.-B.; Budavari, B.; Larkin, J. Semi-Automated Emergency Landing Site Selection Approach for UAVs. *IEEE Trans. Aerosp. Electron. Syst.* **2019**, *55*, 1892–1906. [CrossRef]

59. Ayhan, B.; Kwan, C.; Budavari, B.; Larkin, J.; Gribben, D. Path Planning for UAVs with Engine Failure in the Presence of Winds. In Proceedings of the Industrial Electronics Society, IECON 2018-44th Annual Conference of the IEEE, Washington, DC, USA, 21–23 October 2018.
60. Ayhan, B.; Kwan, C. A Comparative Study of Two Approaches for UAV Emergency Landing Site Surface Type Estimation. In Proceedings of the Industrial Electronics Society, IECON 2018-44th Annual Conference of the IEEE, Washington, DC, USA, 21–23 October 2018.

drones

MDPI

Article

Drones in B5G/6G Networks as Flying Base Stations

Georgios Amponis [1,2], Thomas Lagkas [1,*], Maria Zevgara [2], Georgios Katsikas [3], Thanos Xirofotos [3], Ioannis Moscholios [4] and Panagiotis Sarigiannidis [5]

1 Department of Computer Science, International Hellenic University, 65404 Kavala, Greece; geaboni@cs.ihu.gr or gamponis@k3y.bg
2 K3Y Ltd., 1612 Sofia, Bulgaria; mzevgara@k3y.bg
3 UBITECH Ltd., 15231 Athens, Greece; gkatsikas@ubitech.eu (G.K.); txirofotos@ubitech.eu (T.X.)
4 Department Informatics & Telecommunications, University Peloponnese, 22100 Tripolis, Greece; idm@uop.gr
5 Department of Electrical and Computer Engineering, University of Western Macedonia, 50100 Kozani, Greece; psarigiannidis@uowm.gr
* Correspondence: tlagkas@cs.ihu.gr

Abstract: Advances in the fields of networking, broadband communications and demand for high-fidelity low-latency last-mile communications have rendered as-efficient-as-possible relaying methods more necessary than ever. This paper investigates the possibility of the utilization of cellular-enabled drones as aerial base stations in next-generation cellular networks. Flying ad hoc networks (FANETs) acting as clusters of deployable relays for the on-demand extension of broadband connectivity constitute a promising scenario in the domain of next-generation high-availability communications. Matters of mobility, handover efficiency, energy availability, optimal positioning and node localization as well as respective multi-objective optimizations are discussed in detail, with their core ideas defining the structure of the work at hand. This paper examines improvements to the existing cellular network core to support novel use-cases and lower the operation costs of diverse ad hoc deployments.

Keywords: cellular-enabled drones; aerial base stations; 5G/B5G/6G cellular networks

Citation: Amponis, G.; Lagkas, T.; Zevgara, M.; Katsikas, G.; Xirofotos, T.; Moscholios, I.; Sarigiannidis, P. Drones in B5G/6G Networks as Flying Base Stations. *Drones* **2022**, *6*, 39. https://doi.org/10.3390/drones6020039

Academic Editors: Diego González-Aguilera and Pablo Rodríguez-Gonzálvez

Received: 15 December 2021
Accepted: 30 January 2022
Published: 5 February 2022

1. Introduction

Next-generation cellular communications constitute a key enabler of the greater adoption of next-generation Internet of Things (NG-IoT)-based technologies, by allowing an increase in the number of interconnected orders of magnitude, offering high data rates and near real-time responsiveness as well as addressing various requirements of NG-IoT [1]. Elevated security, better quality of service (QoS), reduced end-to-end delay and higher data-rates are directly correlated to the utilization of higher frequencies, which in turn demand more power and introduce additional dependencies and overhead at the hardware and software levels. The aforementioned parameters, requirements and considerations significantly limit the available spectrum of competent low-power devices and introduce power, effort and networking overhead, especially in ad hoc and remote sensing applications. Millimeter wave (mmWave) communications, multiple-input–multiple-output (MIMO) and non-orthogonal multiple access (NOMA) are some examples of the technological novelties introduced by 5G and highlighted in the 6G standard currently under development. Novel orchestration mechanisms specific to next-generation cellular networks allow for and push towards a more intelligent edge, with an increasing number of functionalities being implemented in an ad hoc, distributed manner.

It is important to note that 5G and 6G introduce the requirement for a previously unseen densification of networks. This is particularly challenging and constitutes a challenge directly addressable via aerial and ad hoc communications. FANETs have the potential of bringing about this technological revolution by means of the intelligent relaying and provision of broadband in otherwise isolated areas and cut-off hubs. Individual UAVs can constitute aerial base stations as a means of serving local wireless networks, e.g., wireless sensor networks (WSNs). Correspondingly, networks of flying base stations constitute great candidates for the units responsible for the opportunistic interconnection of isolated nodes and hubs by employing context-aware routing and swarm topology formation in a 3D aerial grid. The necessary performance optimizations of such networks are in need of scalable and decentralized architectural approaches.

The ongoing development and wide adoption of 5G have driven the design of wireless systems beyond 5G (B5G), including the sixth generation (6G). This new generation of cellular communications should be able to unlock the full potential of the numerous autonomous services that encompass both past and emerging trends. More specifically, 6G should bring novel, breakthrough wireless technologies and innovative network architectures into focus.

The sixth generation of cellular communications has the potential to offer extreme data rates to address the massive connectivity aspect and enable an extremely high throughput, even under extreme conditions or in emergency scenarios in which node density, spectrum and infrastructure availability, and traffic patterns may vary. Additionally, B5G/6G networks will be pivotal in achieving a high degree of immersion and capacity whilst also offering a uniform and highly deterministic quality of experience, required by novel applications. A key application area of next generation networks is delivering real-time feedback-based services enabled by near-zero latency to fulfill the requirements of said novel applications.

As highlighted in [2], airborne communication base stations are envisaged to constitute a pivotal component of the B5G/6G cellular architecture, as mentioned due to their flexible and highly mobile nature. As networks and UEs become increasingly mobile, it is absurd to keep gNBs and relaying equipment solely statically, especially for providing coverage, e.g., in hotspots and in areas with sub-optimal infrastructure (in environments recovering from disaster, rural/suburban areas that suffer due to lack of financial incentives for network operators, etc.).

The first large-scale attempt by the industry was made by Google, aiming to address lack of infrastructure to provide Internet access for the currently non-covered population. Project Loon aspired to provide connectivity in remote or rural areas using stratosphere balloons. These balloons would hover at a height of approximately 20 km. The core idea revolved around using wind waves blowing in the right direction to steer the balloon to the area in need of coverage, and establishing a stratosphere-layer mobile ad hoc network [3]. Nevertheless, high operational costs rendered this idea non-viable in the commercial landscape, and the project was shut down. Similarly, in recent years, there has been a substantial amount of research in the domain of satellite-enabled broadband provision. The authors in [4] reviewed the potential of the usage of satellites for the provision of 5G NR channels. The authors investigated issues arising from a severe Doppler shift and impact on reception and demodulation on a user-equipment level. Several severe limitations render satellite-enabled NR channels unpractical for mobile or real-time applications. Dedicated receivers and demodulation hardware are required, along with software and hardware methods to mitigate issues related to the random-access process.

Increased ease in establishing a direct LOS link with ground users and other cellular infrastructures can support B5G/6G networks in formulating more reliable links and offering a wider coverage area, immune to reflection-induced or environmental losses. These advantages of airborne BSs combined with the requirements set by the oncoming 6G networks have led to the investigation of a spectrum of drone-enabled and cellular-specific communication networks, namely air-to-ground channel characteristics, the optimal positioning of drones (either as sole relays or parts of a swarm), as well as flight trajectory optimizations.

An important and rather novel improvement of 6G is the direct incorporation of artificial intelligence (AI) in the network core as a means of supporting seamless data-centric and context-aware services capable of great degrees of self-optimization [5]. Additionally, 6G will enable the modern cellular communications landscape to meet the desired high levels of end-to-end reliability and correspondingly low end-to-end latency to support ultra-high mobility scenarios, such as flying vehicles. B5G and 6G communications are envisaged to support wireless federated learning (WFL) [6] through the joint optimization of resources targeting a reduction in the delay for novel federated applications [7]. Inherent support for AI within the network core will be pivotal in supporting novel applications in a spectrum of domains, with a great example being drone-enabled intelligent surveillance and machine vision-supported remote sensing [8].

B5G/6G, a major driving force behind the vision of 6G, involves the deployment of connected and autonomous vehicle systems (CAVs) and drone communications. Research efforts in the field of CAV and drone-based communication systems have been steadily increasing in both academia and industry, targeting strict requirements, especially ultra-low latency and unprecedented communication reliability. As the industry is shifting towards wireless, real-time and high-throughput networking, drone base stations are envisaged to constitute pivotal assets.

Table 1 showcases the main differences between 5G and 6G networks and the main improvements with regard to their core attributes [9].

Table 1. Comparison of 5G and 6G attributes.

Attribute	5G	6G
Peak Frequency	110 GHz (W-band)	10 THz
Peak Spectral Efficiency	30 bps/Hz	100 bps/Hz
Peak Data-rate	20 Gbps	1000 Gbps
End-to-End Latency	10 ms	1 ms
Connection Density	1 million per sq. kilometer	10 million per sq. kilometer
Supported Node Mobility	500 km/h	1000 km/h

Table 2 compares the present work to already existing surveys' drone-BS related papers. The present work is focused not only on surveying and reviewing the current state of the art, but also documenting what is missing from the current research landscape. This paper also contributes to identifying the challenges directly associated with the NR landscape, examining the usability of aerially supported communication frameworks, offered benefits, the implications and challenges of such technological leaps, mainly revolving around resource allocation and power consumption, node mobility and path formation, positioning, security and offered QoS. Our approach proves to be the most complete in terms of the variables considered for the survey comparison.

By reading the presented work, the reader will have gained applicable knowledge in the domain of next-generation ad hoc communications, as well as the capability to critically compare and review related literature, thus supporting further research in this rapidly evolving field. Furthermore, the aerially supported applications of 5G communications are disseminated, and implications of secure, resource-aware and intelligent orchestration are examined. As the presented work constitutes an output of the 5G-INDUCE H2020 project, parallels are drawn—where applicable—between the examined use cases and challenges.

The layout of the rest of this paper is as follows: Section 2 follows this introduction, which documents the evolution of cellular communications and the potential offered by 5G/B5G/6G networks; Section 3 introduces the reader to the concept of using unmanned aerial vehicles to extend cellular connectivity and is divided into two main subsections: Section 3.1, which describes the various potential use cases of this technology and Section 3.2, which analyzes the main challenges currently faced. Section 4 discusses the overall developments enabling the acceleration of aerially supported next-generation cellular communications. Section 5 concludes this paper with comments regarding further work in this field and possible extensions of existing research. Figure 1 provides a high-level view of the paper's structure, discussed topics and overall flow.

Table 2. Related research.

	Consideration and/or Analysis of:					
Related Work	**Drone-BSs**	**5G/B5G/6G**	**Energy Availability**	**Path Planning**	**BS Positioning**	**Drone-BS Use Cases**
Nikooroo et. al.	✓	✓	✓	-	✓	-
Mach et. al.	✓	-	✓	-	✓	-
Plachy et. al.	✓	-	-	-	✓	-
Zhao et. al.	✓	-	-	-	✓	-
Fotouhi et. al.	✓	-	-	-	-	✓
Becvar et. al.	✓	-	✓	-	-	✓
Bayerlein et. al.	✓	-	-	✓	-	-
Zhang et. al.	✓	-	-	-	✓	-
Alzenad et. al.	✓	-	✓	-	✓	-
Bushnaq et. al.	✓	-	-	-	✓	✓
Mozaffari et. al.	✓	✓	-	-	-	-
Our work	✓	✓	✓	✓	✓	✓

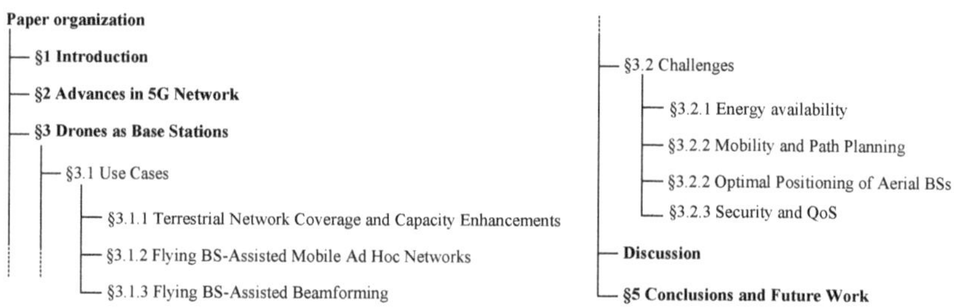

Figure 1. High-level structure of the presented work.

2. Advances in 5G Networks

The core elements of the 3GPP 5G architecture are defined in ETSI TS 123 501 V15.2.0 (2018-06). As demonstrated in Figure 2, the core 5G services (implemented in the form of network functions (NFs)), are the network slice selection function (NSSF), the network exposure function (NEF), the network repository function (NRF), the policy control function (PCF), the unified data management (UDM), the application function (AF), the authentication server function (AUSF), the access and mobility management function (AMF), the session management function (SMF), the user plane function (UPF), the data network (DN), the radio access network (RAN), and the user equipment (UE).

In the context of the presented work, the most important components can be narrowed down to the AMF, the RAN and the UE. The AMF is responsible for registration and connection management, as well as ensuring reachability and managing UE mobility. As demonstrated in Figure 1, the supported mobility for 5G networks reaches up to 500 km/h, and up to 100 km/h for the upcoming 6G networks. Handling node mobility is enabled by this network function. The RAN utilizes radio transceivers (gNodeB/gNB instances) to facilitate cellular connectivity; gNBs provide the New Radio (NR) user plane and control plane protocol interfaces with the UE. According to 3GPP, a device utilized by an end-user to facilitate communication with another user or service is a UE, which is in turn connected to the gNB.

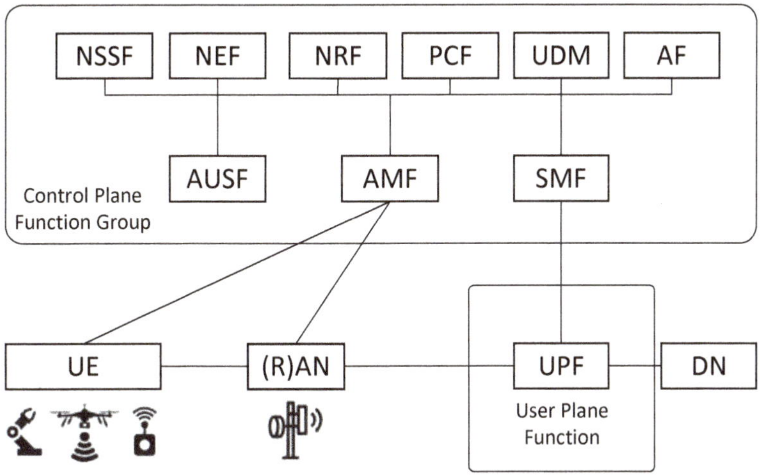

Figure 2. 3GPP-compliant 5G architecture.

It can be assumed that drones have a dual function in this architecture. On the one hand, drones constitute consumers of the services offered by the 5G core (5GC) and can be considered as the devices with the end-user is in direct interface with, while on the other hand, drones utilized as flying base stations are implementing the services offered by the gNBs as they offer the end-users a connection to the 5GC and the respective NF services. Since drones are typically not in a direct interface with the rest of the 5G core network architecture, they can be better described as UEs implementing (part of) the gNB services to serve other nodes in terms of communication enhancements and range extension for both terrestrial and mobile ad hoc networks. The authors in [10] considered the possibility of using low-cost solutions to realize a flying 5G UPF, to assist the attachment of mobile devices to the network core, whose functionality is also implemented within the drone itself. This method promises to enable easier offloading. However, this method poses several security risks, as it exposes core network services to end-users and potential attackers. As mentioned, due to the presented architecture being highly compartmentalized, a high degree of distribution can be achieved. This directly supports the dockerization and offloading of NFs and tasks via resource-aware orchestrators, thus enabling traditionally non-cellular infrastructure (namely drones or other edge devices) to implement functionalities of the core cellular network.

Existing technologies fall short in terms of fully leveraging smart 5G infrastructure capabilities. Even the most mature orchestration technologies to date are typically restricted to serve rather simple 5G slicing requests, resulting in:

1. Partial or no support for advanced services, namely the use-cases mentioned in the presented paper and an advanced slicing mechanism associated with application offloading;
2. The inability to discover and directly expose the entire range of smart 5G infrastructure capabilities, which may be advertised by an underlying network function virtualization orchestrator (NFVO) platform.

The intelligent OSS of the 5G-INDUCE project, is of particular interest to the currently examined set of applications, as it will be capable of exposing the core network capabilities to the end-users at a high application level without revealing any potentially confidential low-level (infrastructure-related) information. This is particularly valuable for usage in the networks established in an ad hoc or on-demand manner. By offering novel intelligent orchestration services, 5G-INDUCE aims to support drone-assisted network performance and coverage monitoring for critical scenarios. The main goal of the 5G-INDUCE project is thus to enable an entirely new spectrum of ad hoc and high-fidelity services.

3. Drones as Base Stations

As highlighted in [11], drones are relatively low-cost solutions that can efficiently extend next-generation connectivity in areas that are normally either unreachable via traditional infrastructure or momentarily unreachable due to, e.g., network disruptions. Similarly, drone base stations (BSs) can increase the quality of a connection between an ad hoc/remote node and a given cellular base station. Drone BS are also attractive solutions for providing reliable, broadband and wide-area temporary wireless connectivity during special events or harsh scenarios, namely natural disasters [12], smart farming [13], sporting events, and many more scenarios where the permanent installment of cellular infrastructure is unnecessary and comes with considerable monetary overhead.

The authors in [14] stated that, in addition to the above, drone BSs at high altitudes are expected to provide long-term and cost-effective connectivity for rural areas. The integration of drone BSs with other physical layer techniques such as mmW and massive MIMO as well as cognitive radios is a promising solution for providing data-intensive services and is expected to create new challenges for next-generation flying. The optimal positioning of drone BSs is one of the critical challenges to be overcome in dense deployment scenarios. For that purpose, the optimal positioning of the drone BSs is an issue in need of tackling. Optimal positioning is one of the most critical challenges and must be addressed in dense B5G/6G deployment scenarios.

The utilization of drones as airborne base stations enables a service provider to offer enhancements of connectivity and capacity of already existing terrestrial wireless networks, with cellular ones being the main areas of interest. Compared to conventional terrestrial base stations, the advantage of using UAVs as flying base stations is their ability to dynamically readjust their altitude, and their comparative ease in terms of establishing direct LOS links to terrestrial gNBs or UE instances. Due to their inherent characteristics in terms of mobility and flexible altitude adaptation, cellular-enabled drone base stations can effectively support existing cellular systems by providing additional communication capacity to areas and ensuring network coverage in difficult-to-reach rural areas, as such deployments are naturally three-dimensional and offer unrestricted mobility. There exist numerous types of UAVs to facilitate connectivity and undertake the role of a base station, with each having its advantages and disadvantages in terms of mobility, autonomy, maneuverability and maximum payload. Table 3 compares the main types of available UAVs.

Table 3. Comparison of UAV types.

UAV Type	Stationary Flight (Yes/No)	Typical Battery Life (mins)	Typical Velocity (m/s)	Typical Payload (kg)
Multi-rotor	yes	=<15	=<11	=<2.5
Fixed-wing	no	=<60	=<22	=<14
Baloon	yes	=<60	=<2.5	=<4.5

3.1. Use Cases

The researchers in [15] noted that mobile 5G is a key driver of network services in different industries, and therefore accelerates the digital transformation of the respective services. 5G provides more advanced and enhanced capabilities compared to 4G, and following the same pattern, 6G will introduce the entire industry to a new era. New industry-specific standards for next-generation cellular networks are already being incorporated in 3GPP Releases, with 3GPP Release 15 defining new radio (NR) and Packet Core evolution as a means of establishing fully interoperable deployments. The next 3GPP Release (Release 16) aspires to support gigahertz cellular communications as prescribed by IMT-2020, whilst also incorporating new communication standards for smart cars and Industry 4.0 factories. Considering the aforementioned remarks, airborne BSs will mostly be used for relaying information and facilitating range extension where cellular infrastructure fails to deliver. The most prominent use-cases for drones functioning as base stations are:

- The extension of terrestrial network coverage and capacity;
- The assistance of mobile ad hoc networks (MANETs);
- Beamforming applications.

3.1.1. Terrestrial Network Coverage and Capacity Enhancements

The first drone-enabled base station scenario revolves around terrestrial cellular network enhancements. In this scenario, it is assumed that cellular-enabled drones function as service providers; aerial nodes functioning as cellular base stations can provide pivotal improvements to ultra-dense small cell networks, which are highlighted in B5G/6G next-generation communications. Regarding next-generation mmW communications, drone-enabled flying base stations find great applicability in establishing short-term line-of-sight (LOS) links among gNBs and UEs. Thus, the coverage and capacity of wireless networks can be effectively enhanced, whilst next-generation communications can be supported in a more effective manner, especially in dense cells. Additionally, as mentioned in Section 3, MIMO-based techniques have the potential to formulate an entirely new and dynamically reconfigurable enhanced cellular network, capable of providing never-seen-before high capacity services. A great example of on-demand terrestrial network coverage and capacity enhancements is the establishment of high-throughput links in first-response and emergency scenarios, in which the existing infrastructure is either damaged or inadequate; in this case, aerial base stations can be used to alleviate the load on the terrestrial cellular grid, or provide broadband connectivity where no infrastructure was available in the first place. Figure 3 showcases a possible application of on-demand terrestrial network coverage enhancement in the case of a wildfire. In this scenario, the line-of-sight between the radio antenna (gNB) and the corresponding UEs associated with the end user in need (firefighter) is blocked. Temporarily deploying a drone BS helps alleviate this issue by introducing a new path, thereby establishing a temporary yet direct and reliable link between the end user and cellular infrastructure.

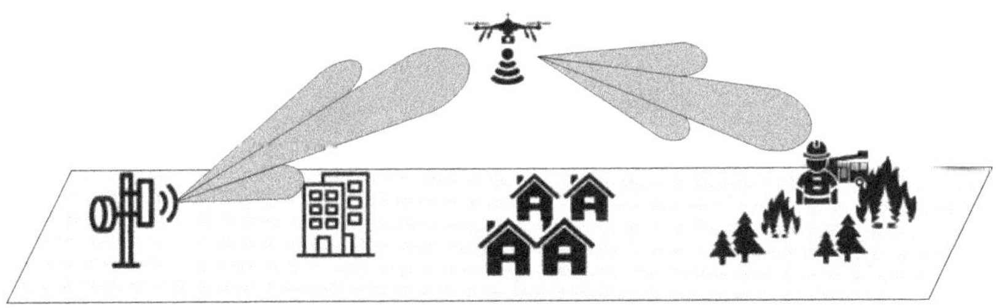

Figure 3. Terrestrial network coverage enhancement: a drone BS-supported firefighting scenario.

Moreover, network enhancements find substantial applicability in the formation of reliable communication links in rural and/or remote areas, with little-to-no existing infrastructure; this scenario supports the rejuvenation of rural areas as well as smart farming, and can assist in endeavors to close the digital gap. The use of aerial base stations is a promising solution to a number of challenges associated with terrestrial IoT networks as well, namely wireless sensor networks (e.g., ad hoc WSNs). Drone BSs can be deployed to provide reliable and energy-efficient uplink and downlink for device-to-device IoT communications, due to drones being effectively deployed in a manner aimed at reducing the shadowing and blockage effects, which constitute major causes of signal attenuation and losses in wireless links. Connectivity enhancements and terrestrial network capacity increases cannot be achieved without sufficiently optimizing the positioning of all communicating nodes in the

3D space. The efficient positioning of drones is thus of the utmost significance in terms of connecting immobile UEs throughout the course of the connectivity-supporting process. This is particularly true in the case of low-power IoT devices, which require their respective gateways to be placed on locations such that the aforementioned devices can successfully establish a connection to the cellular network using the minimum required transmission power [16]. In this manner, drone BSs can support massive B5G/6G IoT diverse networks by constantly updating their relative positions to ensure optimal connectivity and reduce the need for permanent cellular infrastructure installations.

3.1.2. Flying BS-Assisted Mobile Ad Hoc Networks

Given their mobility and LOS-establishing capabilities, UAVs can support mobile ad hoc networks (MANETs) on the terrestrial plane, and more specifically vehicular ad hoc networks (VANETs). With the advent of smart and self-driving cars, the requirement for constant, uninterrupted and real-time communication is more pressing than ever. FANETs and singular UAVs have repeatedly proven their usability in terms of supporting device-to-device communications, and given the increased support for high mobility in B5G/6G networks (see Table 1), they are excellent candidates for facilitating real-time information exchange and message broadcasting among mobile networked peers. A good example of this type of communications is UAV-enabled safety-related information broadcasting across numerous vehicles without a direct LOS or sufficient network coverage [17]. Aerial BSs can also enhance the reliability of device-to-device and inter-vehicular links by mitigating issues caused by interferences, which are a result of an increased number of re-transmissions. Figure 4 demonstrates an inter-vehicular communication scenario supported by drone-BSs.

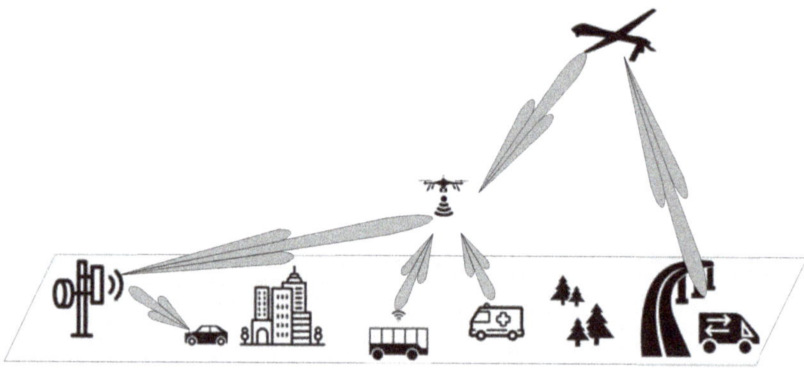

Figure 4. Drone-BS-assisted vehicular communications scenario.

Furthermore, airborne BSs offer a new non-terrestrially confined spectrum of diverse networking opportunities to boost reliability and last-mile connectivity in the networks at hand. An intelligent approach to accommodating mobile ad hoc networking is the clustering of ground UE instances into mobile swarms. This can be implemented to enable drone BSs to potentially communicate with a singular networked UE instance functioning as a "representative" (cluster head) of the mobile terrestrial cluster; the cluster head is responsible for distributing messages via broadcasting, and respectively, providing gateway services to the rest of the cluster nodes. Following this approach, the connectivity of mobile terrestrial ad hoc networks can be significantly improved by adopting cluster-based approaches and leveraging the unique mobility characteristics of drones. To that end, as is the case with the terrestrial network enhancement use-case scenario, drones will significantly increase the quality and reliability of their offered services by considering efficient placement and swarm nodes' relative positioning, where applicable. The matter of optimal positioning and path planning are discussed in detail in Section 3.2.3.

3.1.3. Flying BS-Assisted Beamforming

In this scenario, drones can be viewed as airborne antenna elements, that can be conjointly utilized for performing massive MIMO, 3D MIMO and mmW communications. Beamforming using aerial elements can be pivotal in reducing intercell interference by forming distinct beams to simultaneously cover desired grid elements.

Research in recent years has been focused on full-dimension MIMO communications, where the horizontal and vertical dimensions of a cellular network alike, are utilized for beamforming. This supports the formulation of distinct beams in the cellular network, and can be resorted to as a means of minimizing intercell interference, which is projected to be a major issue in 6G communications. Three-dimensional beamforming solutions show great potential in offering significantly higher system throughput, all whilst supporting a greater number of UE instances, effectively tackling issues associated with high network density in, e.g., urban or industrial environments, which are key targeted use-cases for next-generation cellular networks. The aforementioned improvements render the evolution of MIMO communications from 2D to 3D not only beneficial but potentially mandatory for the wide applicability of B5G/6G communications. As highlighted in [18], cellular-enabled aerial BSs can be key enablers of high-density cellular networks by supporting intercell interference mitigation and management by beamforming the horizontal and vertical channel planes. Figure 5 showcases the potential for intercell interference mitigation for leveraging drone BSs, thus paving the path towards high-density next-generation cellular networks.

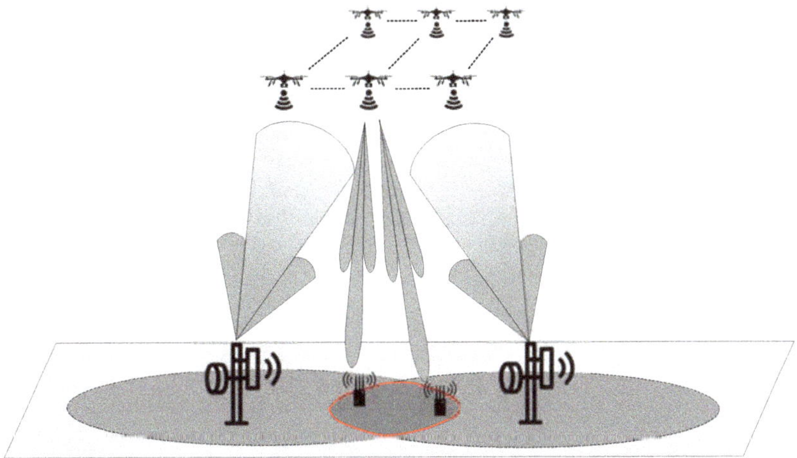

Figure 5. Beamforming: a drone BS-supported intercell interference mitigation scenario.

Drone-supported 3D MIMO is more suitable for high-density scenarios, and even more so in cases in which UE instances are distributed across a three-dimensional grid with different elevation angles (assuming a direct LOS with the respective cellular base station). As aerial base stations are typically significantly elevated in comparison to terrestrial UE, their respective altitude and elevation angles' differences can be relatively easily distinguished and considered as routing, relaying and dynamic antenna re-positioning metrics. Additionally, as elevation differences easily support direct LOS conditions, aerial base stations enable efficient and effective beamforming in a three-dimensional grid. When compared to conventional (terrestrially deployed and static) antenna array systems, a drone-based antenna array has the following advantages:

- The number of antenna elements is not limited by spatial constraints;
- Beamforming gain can be increased on-demand by adjusting array element (drone) spacing;
- Drones' mobility allows for effective beam-steering in virtually any 3D direction;

- Utilization of drone swarms within an array formation can provide the capability to form a massive antenna array which can virtually accommodate any arbitrary shape and perform beamforming;
- Energy optimization or tethering (discussed in Section 3.2.1) can increase reliability to near terrestrial-node levels.

The authors in [19] engaged in a thorough analysis of the potential usage of drone swarm-based scattering for adaptive beamforming to efficiently relay data streams to locations normally not inside the transmitting node's range of communication or LOS. The proposed swarm-based beamforming methodology revolves around a ground-based transmitter radiating towards a swarm of drones, where each "antenna element" (individual drone) carries a half-wavelength resonant wire functioning as a scattering object, which in turn works as a reflector antenna radiating the BS's transmitted signal. The researchers investigated potential methods for optimizing the process of adjusting the altitude of drones, and as such, the phase of the scattered field from each array element to enable support for the formulation of desired radiation patterns. A strong argument for the usage of quadcopters as beamforming elements is the fact that because all signal phase shifting is implemented by adjusting the drone locations, no excessive, computationally intense or additional phase-shifting circuitry is required on-board. The researchers' optimization algorithm was able to form a beam composed of two main lobes, with a low level of side-lobes, which could be steered in the desired direction for different patterns.

Similarly, the researchers in [20] attempted to produce a directional beam to increase a network's quality of service, more specifically concerning cellular downlink. The novelty of their approach stems from the fact that the authors aimed to maximize coverage whilst considering human body-induced losses and model respective optimization procedures. The researchers used a uniform linear array of antennas at the transmitter and designed the optimized beam direction to maximize the number of covered users while considering the QoS constraint in the network. The significance of the authors' work stems from the fact that their developments are centered around the usage of mmW communication frameworks while considering the channel blockage effects of medium-to-large bodies; this enables the facilitation of optimizations aware of the environmental parameters and characterizing the respective link.

3.2. Challenges

This subsection is dedicated to the detailed analysis of the main identified challenges hindering the wider adoption of aerially enabled cellular base stations. The identified challenges can be narrowed down two five main issues, namely energy availability, mobility and path planning, positioning of nodes, security and privacy issues, and the offered quality of service. As the nature of the interfaces among the relaying equipment and the next-generation cellular network core is highly compartmentalized (assuming an ETSI TS 123 501 V15.2.0-compliant 5G architecture), drone BSs belong to the RAN layer as UE instances, as highlighted in Section 3. All challenges, especially security and quality of service, are affected by this compartmentalized, "blackbox" approach, as the 5G core has no authoritative access to the radio-layer.

3.2.1. Energy Availability

An important issue with drone-based BSs is energy availability. As relays in B5G and 6G cellular networks are expected to relay greater volumes of information and provide reliable and sufficient QoS, throughput and minimal latency, the energy expenditure of relays rises exponentially. The energy expenditure of UAVs functioning as relays is mainly associated with:

- Energy consumed for the purpose of flying and hovering above a desired location;
- Energy consumed for communication and on-board processing;

A substantial amount of research has gone into designing energy-efficient routing and communication schemes to prolong battery lifespan via the usage of, e.g., received signal strength indication (RSSI) and the drone position data [21]. However, since the propulsion energy is significantly more than the payload energy, energy-efficient communication will not highly affect the flight time, which is of utmost importance for establishing a network of aerial cellular base stations. It can be safely assumed that for the purpose of enabling long flight times, battery technology has to keep up with the newly introduced requirements set by B5G/6G communications.

In all cases, the goal of energy-aware optimizations in regard to BS communication and mobility metrics is to minimize transmission power, whilst considering a predefined constraint for the minimum data rate which UEs require in all cases [22]. This type of optimization balances energy availability and the offered QoS. Energy consumption is in turn constrained by either maintaining at the same coverage area, or maximizing; in turn, QoS is constrained by either maintaining its current value or maximizing it.

In that spirit, the authors in [2] proposed a tether-based UAV setup for the deployment of drones as aerial base stations in 6G networks and engaged in an analysis of the mobility-endurance trade-off. The methodology proposed in the context of tethering 5G/6G enabled quadcopters solves the inherent limitations of the short flight time and reliable backhaul links. The ground station supplies the airborne BS with energy whilst also providing the data link via a psychical tether. It is safely concluded that the most important defining factor for the offered QoS and overall functionality is the placement of individual drone relays. This is especially true for tethered UAVs, as they have substantial limitations in terms of horizontal mobility, thus reducing reachable positions in the 3D space. The authors in [23] analyzed hybrid drone-specific power supply systems that combine batteries with other types of energy sources, namely fuel cells, solar cells, supercapacitors, tethering and laser-enabled in-flight recharging. The aforementioned technologies are envisaged to help alleviate performance and autonomy issues, thus bringing about a drone-supported next-generation cellular landscape.

In the context of eliminating unnecessary energy expenditure, the authors in [22] proposed a solution for the elimination of the redundant movement of aerial BSs. The energy-saving framework they proposed relies on deriving new positions for the base stations considering the overall mobility and movement patterns of UEs. The authors considered that there exist UE mobility scenarios, in which it is not necessary to reposition the base station, mainly due to slight QoS variations falling within an acceptable spectrum. This multi-objective optimization problem can be summarized as "reducing globally-spent BS hovering energy, whilst maintaining networking capacity approximating the one associated with optimal node positions". It is suggested, that for maximizing energy availability of non-tethered (free-flying drone BSs), it is suggested that the aerial nodes do not precisely follow the best QoS-enabling position, but rather stay in a pre-computed distance from the optimum. Following this non-QoS-centric approach, the energy consumed for the purpose of flying can be kept to a minimum; this is implemented without sacrificing mobility, but establishing a threshold of acceptability for the loss of QoS.

As mentioned earlier, tethering techniques can be resorted to in order to remove battery-life constraints. The researchers in [24] proposed a new drone-based mobile relaying system, in which a laser beacon is employed to wirelessly charge the energy-constrained UAV relay. This approach of preserving energy and increasing flight time is of great interest, as it can potentially help solve the dilemma of choosing between mobility and energy availability. The authors aimed to conjointly maintain a global optimum, considering both required transmission power and mobility-specific parameters. The authors proposed two algorithms to solve this optimization problem, and showed that the laser beacon wavelength and environmental parameters such as weather conditions greatly impact both data and power transmission efficiency.

3.2.2. Mobility and Path Planning

In light of the aforementioned remarks, it can be assumed that node mobility capabilities and swarm-wide path planning is of utmost importance for the effective relaying of cellular communications for all three examined scenarios (terrestrial and mobile network coverage and capacity enhancements and beamforming applications) in Section 3.1. The mobility-awareness and optimal positioning of aerial base stations is an intricate and rather convoluted problem [25], with its complexity being a direct function of an entire spectrum of highly volatile parameters, namely:

- The number of aerial BSs participating in the relaying;
- The type of interfaces among these participating nodes (inter-drone relaying, conjoint formation of array antennas);
- The elevation, angle, position and velocity of each node relative to the respective gNB;
- Energy availability, expected energy expenditure and estimated uptime for new links;
- The topology of the terrain and potential blockages in LOS.

The researchers in [26] developed "SEDMAG", a 6G-specific path-planning algorithm for usage with aerial IoT nodes in both single-node and swarm deployments, mainly revolving around monitoring and extensible to cellular-relaying and WSN data harvesting scenarios. The authors' work was envisioned to reduce localization latency as well as energy overhead, by trading off precision in localization; this renders the *SEDMAG* algorithm useful for non-beamforming-related scenarios, where accuracy and localization precision are of utmost importance for phase shifting. Using a smart search algorithm and graph reduction, the researchers managed to provide better positioning efficiency when compared to non-dynamic path planning algorithms, namely *Zcurve*, but also dynamic path planning algorithms such as *DREAMS* in terms of errors and energy consumption. In line with resource-aware orchestration in next-generation cellular networks, the authors proposed a smart load-balancing approach capable of balancing drones load in swarm deployment scenarios and resulting in the reduction in localization delay. The method of function for SEDMAG and its respective derivatives is dividing the area of interest into equal smaller areas and assigning an aerial node to each said smaller area. The SEDMAG algorithm is tasked with positioning the drone(s) so that all areas of interest are adequately covered; it then computes the shortest path for the drone to visit all areas required to be covered by the network and decides on the order in which they are to be visited. The *SSEDMAG* algorithm (a derivative of SEDMAG) utilizes an intelligent searching approach to shorten the drone's flight path. Similarly, *SSEDMAG-reduced* aims to further reduce the drone's trajectory by re-applying SSEDMAG on the reduced flight path. Lastly, the *SSEDMAG-reduced-balanced* algorithm is the one finding the greatest applicability in swarm deployment scenarios, as it is tasked with balancing the tasks and overall load among drones, aiming to achieve a global localization optimum for sufficient coverage given a minimum number of trajectory shifting and path alterations.

3.2.3. Optimal Positioning

Researchers in [27] analyzed the optimal positioning of UAV BSs in conjunction with transmission power allocation, user clustering and next-generation NOMA networks. The authors attempted to elicit a means of optimally clustering UEs and positioning drone BSs so that only a minimum required transmission power is utilized for successful link establishment, whilst ensuring that QoS is above a given minimum threshold at all times. The proposed multi-objective optimization method allows for a significant increase in the duration of aerial coverage of a given terrestrial area. Overall, the proposed solution can be divided into three distinct sub-tasks:

- Task 1: Select the optimal clusters of a given number of UEs to be simultaneously served by a NOMA network;
- Task 2: Allocating the optimal transmission power to each node;
- Task 3: Determining the position of the flying BS in the 3D space.

Correspondingly, the identified constraints are:

- The available propulsion energy;
- The guaranteed minimum capacity for each mobile user.

Finally, the direct optimization goal is to increase the duration of communication coverage in NOMA. The authors measured an increase in the order of 67–270% when compared to existing solutions, assuming a constant propulsion power.

The authors in [28] considered the scenario of connectivity enhancement in the case of on-demand ad hoc deployments of cellular networks. To that end, the researchers considered the optimal positioning of drones in swarm deployments so that they can mitigate interferences and offer on-demand communication-extending services to a substantial amount of UEs. It is important to note the highly dynamic and volatile nature of the networking requirements of end-users in such ad hoc deployments. The authors of [28] focused on the issue of maximizing the offered QoS (and consequently user satisfaction), by proposing an algorithm capable of:

- Associating UEs with the best-suited aerial BS;
- Finding optimal positions of all aerial BSs.

Additionally, the authors considered the performance of the genetic and particle swarm optimization algorithms, which are evaluated in terms of performance, accuracy and offered QoS (calculated as available data rates). The authors showed that the particle swarm optimization algorithm is substantially less complex compared to the genetic algorithm, while the latter one is more efficient in its utilization of nodes. The tradeoff seems to be: time complexity vs. efficacy.

The researchers in [29] considered the usage of multiple drone-mounted radio heads as a means of providing on-demand connectivity and dynamic cloud radio access networking capabilities. To facilitate this, the authors realized an optimal drone positioning mechanism to address the requirement for transmission power minimization. The researchers divided the multi-objective optimization problem into two distinct ones, one per each axis of movement:

- Horizontal positioning of drones (minimization of distance sum);
- Vertical positioning of drones (maximization of coverage).

For the first problem (horizontal positioning optimization), the authors resorted to the Weiszfeld algorithm to compute and output the point that minimizes the total distance to be covered by the sum of (re)transmissions [30]. Similarly, the second problem (vertical positioning optimization) was approached with the goal of calculating the optimal elevation angle and the terrestrially projected radius of the coverage area per drone. The resulting algorithm proved to be capable of offering the minimum required transmission power while maintaining a pre-defined acceptable performance threshold in terms of end-user connectivity.

3.2.4. Security and QoS

NG communications inevitably give rise to various security concerns. As a substantially greater and ever-increasing amount of data is being relayed per unit of time, the infiltration of a NG network is automatically rendered more rewarding. Furthermore, the advent of novel time-sensitive use cases are exponentially increasing reliance on low-latency and near real-time communications; this implies that network stability becomes mandatory, and QoS must be kept above a certain threshold at all times. Should a network service be interrupted, various negative cascading events would take place, especially considering smart vehicles and generally VANET-supporting NG services, as shown in Section 3.1.2.

Security and QoS in next-generation cellular networks are closely correlated, as they are associated with the interfaces established between relaying nodes and the cellular core (namely the 5G core). The usage of drones as means of connectivity extension poses a considerable security threat. As such devices constitute 5G/B5G user equipment instances, they belong to the RAN layer of the 5G architecture. This means that security features implemented on a 5G-core level have little to no authority with regard to handling UE authorization, security and QoS provision. Drones such as UEs are directly exposed (both physically and from a networking point of view) to end users. A potential security breach in the RAN-exposed NFs is a serious security rise, as it can escalate rather quickly. Authentication among the networked drone BSs is even more challenging to implement in a reliable manner [31]. More specifically, the following 5G components and network functions are exposed to the UE:

- gNB, the terrestrial cellular base station (RAN);
- AMF, the mobility management function of (core network).

It becomes evident that the creation of an abstraction layer to secure QoS and security is mandatory for a sustainable and reliable transition to aerially supported high-density cellular networks. This abstraction layer comprises additional security functions and orchestration mechanisms that improve the architecture of cellular communications [32]. In this domain, the 5G-INDUCE project aims to facilitate the incorporation of additional features for security, and cognitive capabilities that cannot be supported by existing environments to ensure QoS above an end-user-defined threshold. Special focus is given to QoS assurance in conjunction with security, by leveraging the smart exposure of application-specific interfaces to end-users. With throughput reaching the scales of dozens of Gbps, end-to-end latency being reduced to the scale of a few milliseconds, and network availability being stretched to its limit, monitoring link- and network-layer metrics to accommodate a targeted QoS is extremely challenging to do while preserving privacy, security and keeping computational overhead to a minimum.

4. Discussion

In the presented work, we discussed matters of 5G and B5G/6G communications in a spectrum of scenarios, resource utilization, various types of optimizations and the elicitation of efficient, energy-aware and QoS-preserving methods of providing high-quality services to a set of UEs. We identified the three main scenarios which are expected to significantly benefit from drones in next-generation cellular networks functioning as flying base stations.

To this end, the 5G-INDUCE project is envisaged to offer a variety of services and components as add-ons to the 5G core architecture presented in Section 2 and described on a high-level (considering interfaces among the management and network orchestration (MANO) and the virtual/physical infrastructure and the NFVO) in Figure 6. 5G-INDUCE offers a full-stack NetApp management platform to orchestrate services and functionalities, mainly in the industrial domain. Orchestration enhancement can support data confidentiality, securely encrypt critical infrastructure management and monitoring, and reliable operator–drone communication interfaces. The scenarios described in Sections 3.1.1 and 3.1.2 and (to a lesser degree) Section 3.1.3, strongly relate to the targeted NetApps of the 5G-INDUCE project and are aligned with its goal of establishing easily extensible yet secure and QoS-aware next-generation cellular connectivity in critical scenarios. All aforementioned use-cases rely on novel orchestration algorithms for the deployment of services over containerized realms.

The challenges currently faced by the entire research and industrial landscape range from security and privacy, to licensing and AI-related issues, namely explainability and legislative/ethical concerns in regard to automated piloting, no-fly-zones and potential collisions; all these areas require more research in the near future, as little to no work has targeted the aforementioned parameters in a networking context. It can be easily deduced that the complexity of said issues will increase, even more so with the increase in network heterogeneity and the additional requirements entailed. Low-earth orbit satellites (and constellations thereof) also seem to be a rather promising technology in terms of supporting ubiquitous connectivity for NR networks. With this in mind, future protocols, frameworks, and even hardware modulators and demodulators shall be designed to support satellite-to-drone connectivity where applicable. Consequently, standardization for all developments targeting the aforementioned challenges is going to be a direct focus of all relative standardization groups and institutes in the coming years.

Said challenges are to be addressed through the upcoming cellular networks' services' ability to compartmentalize all their components and to only expose the desired entry-points to the end-users. This can be managed by combining application orchestration with network services orchestration mechanisms, so as to allow the higher layer to manage the deployment and lifecycle of the services at hand, and the lower orchestration layer to optimally interact with the network and computational resources of the nodes. The upcoming compartmentalized orchestration mechanisms will thus have to build on existing knowledge and go beyond the ETSI NFV MANO, following the respective 3GPP specifications, while being aligned to the expected 3GPP Release 16 standards, as mentioned in Section 3.1. Overall, the extension of connectivity and the provision of services on-demand seems to be pivotal in the formulation of novel core orchestrator architectures. All the scenarios described in this paper strongly rely on the establishment of good-quality and secure links between gNBs and UE instances for effectively relaying and seamless coordination among them, especially in the case of beamforming. Thus, future research must be as focused on service-provision optimizations as it must be focused on improving the existing 5G core service-based architecture.

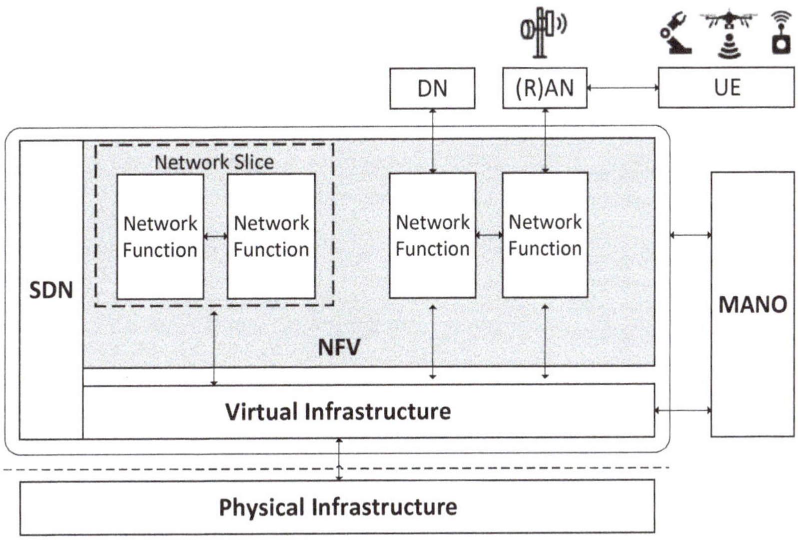

Figure 6. High-level overview of the virtual-physical infrastructure compartmentalization and key component interfaces.

5. Conclusions and Future Work

This paper discussed a number of potential use-cases for drone base stations in B5G/6G networks and the implications of their usage in various environments. Matters of resource allocation, optimal positioning and channel provision are discussed in detail, in order to establish a common discussion ground for future developments in regard to cellular connectivity enhancements. Moreover, this paper has discussed improvements in 5G orchestration mechanisms as a means of achieving substantial improvements in terms of throughput, task allocation optimization mechanisms and swarm positioning optimizations. Given ongoing developments in cross-layer metric utilization for task and resource orchestration, future developments can potentially revolve around the environment-aware predictive deployment of drone base stations using machine learning, as has been already proposed in [33]. On a technical level, the formulation of multi-objective optimization problems in the near future will be capable of accommodating and considering various weights and newly introduced realistic constraints to deduce the optimal solution in a machine learning-enabled manner; this will significantly boost developments in the domain of path planning and optimal positioning, as discussed in Section 3.2.3. Advances in orchestration and network function compartmentalization will support efforts in complementing the terrestrial cellular system in the face of adverse events to offload excess traffic or increase cellular range.

Most of the developments in the context of the usage of B5G/6G-enabled drones as base-stations can be expressed as direct functions of localization and optimal relative positioning efficacy. As routing algorithms increase in efficiency, aerial ad hoc deployments will be capable of formulating three-dimensional grids capable of offering more efficient relaying services. This use case scenario is discussed in detail by the authors in [34]. More specifically, the authors discussed how an efficient drone base-station 3D placement algorithm will support efforts to maximize the total number of UE instances whilst utilizing the minimum required power. Resource-aware routing is an ongoing development in the research community, with a relative study of resource-aware cross-layer routing for FANETs performed by the authors of the present paper in [35]. Cross-layering (the utilization of, e.g., physical/MAC-layer parameters in the context of network-layer routing) has the potential to increase performance for the physical and link-layer processing, which will, in turn, enable lower costs for operators in terms of extremely dense deployments namely in Industry 4.0 use-cases [36] or emergency applications [37]. Finally, the seamless mobility and integration of heterogeneous links for collaboration in the same ad hoc cellular network, will be supported by multi-connectivity and a cell-less architecture envisaged in [38] and enabled the utilization of novel scheduling algorithms and a new core network design. Regarding the optimization of the resource allocation and positioning of aerial nodes, in future work, the problem of optimal bandwidth allocation should be considered, along with the issue of accommodating highly mobile UEs, as well as the effective mitigation of interference among BS nodes and securing the underlying cellular infrastructure by means of high degrees of compartmentalization and the strategic exposure of network resources to UEs.

Author Contributions: Conceptualization, T.L., P.S., G.K. and G.A.; methodology, G.A., T.L. and I.M.; software, G.A. and T.L.; validation, T.L., G.A. and P.S.; formal analysis, G.A., T.L. and P.S.; investigation, G.A., T.L., M.Z., G.K., T.X., I.M., P.S.; resources, T.L., P.S. and T.X.; data curation, T.L., I.M. and G.K.; writing—original draft preparation, G.A. and T.L.; writing—review and editing, T.L., G.A. and M.Z.; visualization, G.A.; supervision, T.L. and P.S.; project administration, T.L. and P.S. All authors have read and agreed to the published version of the manuscript.

Funding: The project leading to this application has received funding from the European Union's Horizon 2020 research and innovation programme under grant agreement No 101016941.

Institutional Review Board Statement: Not applicable.

Informed Consent Statement: Not applicable.

Data Availability Statement: No new data were created or analyzed in this study. Data sharing is not applicable to this article.

Conflicts of Interest: The authors declare no conflict of interest.

References

1. Pliatsios, D.; Goudos, S.K.; Lagkas, T.; Argyriou, V.; Boulogeorgos, A.A.A.; Sarigiannidis, P. Drone-Base-Station for Next-Generation Internet-of-Things: A Comparison of Swarm Intelligence Approaches. *IEEE Open J. Antennas Propag.* **2022**, *3*, 32–47. [CrossRef]
2. Kishk, M.; Bader, A.; Alouini, M.S. Aerial Base Station Deployment in 6G Cellular Networks Using Tethered Drones: The Mobility and Endurance Tradeoff. *IEEE Veh. Technol. Mag.* **2020**, *15*, 103–111. [CrossRef]
3. Kaur, S.; Randhawa, S. Google LOON: Balloon-powered internet for everyone. *AIP Conf. Proc.* **2018**, *2034*, 020006. [CrossRef]
4. Saarnisaari, H.; de Lima, C.M. 5G NR over Satellite Links: Evaluation of Synchronization and Random Access Processes. In Proceedings of the 2019 21st International Conference on Transparent Optical Networks (ICTON), Angerrs, France, 9–13 July 2019; pp. 1–4. [CrossRef]

5. Rekkas, V.P.; Sotiroudis, S.; Sarigiannidis, P.; Karagiannidis, G.K.; Goudos, S.K. Unsupervised Machine Learning in 6G Networks -State-of-the-art and Future Trends. In Proceedings of the 2021 10th International Conference on Modern Circuits and Systems Technologies (MOCAST), Thessaloniki, Greece, 5–7 July 2021; pp. 1–4. [CrossRef]
6. Bouzinis, P.S.; Diamantoulakis, P.D.; Karagiannidis, G.K. Wireless Federated Learning (WFL) for 6G Networks[4]Part I: Research Challenges and Future Trends. *IEEE Commun. Lett.* **2022**, *26*, 3–7. [CrossRef]
7. Bouzinis, P.S.; Diamantoulakis, P.D.; Karagiannidis, G.K. Wireless Federated Learning (WFL) for 6G Networks—Part II: The Compute-Then-Transmit NOMA Paradigm. *IEEE Commun. Lett.* **2022**, *26*, 8–12. [CrossRef]
8. Chaschatzis, C.; Karaiskou, C.; Mouratidis, E.G.; Karagiannis, E.; Sarigiannidis, P.G. Detection and Characterization of Stressed Sweet Cherry Tissues Using Machine Learning. *Drones* **2022**, *6*, 3. [CrossRef]
9. Rekkas, V.P.; Sotiroudis, S.; Sarigiannidis, P.; Wan, S.; Karagiannidis, G.K.; Goudos, S.K. Machine Learning in Beyond 5G/6G Networks—State-of-the-Art and Future Trends. *Electronics* **2021**, *10*, 2786. [CrossRef]
10. Fattore, U.; Liebsch, M.; Bernardos, C.J. UPFlight: An enabler for Avionic MEC in a drone-extended 5G mobile network. In Proceedings of the 2020 IEEE 91st Vehicular Technology Conference (VTC2020-Spring), Antwerp, Belgium, 25–28 May 2020; pp. 1–7. [CrossRef]
11. Bariah, L.; Mohjazi, L.; Muhaidat, S.; Sofotasios, P.C.; Kurt, G.K.; Yanikomeroglu, H.; Dobre, O.A. A Prospective Look: Key Enabling Technologies, Applications and Open Research Topics in 6G Networks. *IEEE Access* **2020**, *8*, 174792–174820. [CrossRef]
12. Selim, M.Y.; Kamal, A.E. Post-Disaster 4G/5G Network Rehabilitation Using Drones: Solving Battery and Backhaul Issues. In Proceedings of the 2018 IEEE Globecom Workshops (GC Wkshps), Abu Dhabi, United Arab Emirates, 9–13 December 2018; pp. 1–6. [CrossRef]
13. Livanos, G.; Ramnalis, D.; Polychronos, V.; Balomenou, P.; Sarigiannidis, P.; Kakamoukas, G.; Karamitsou, T.; Angelidis, P.; Zervakis, M. Extraction of Reflectance Maps for Smart Farming Applications Using Unmanned Aerial Vehicles. In Proceedings of the 2020 12th International Symposium on Communication Systems, Networks and Digital Signal Processing (CSNDSP), Porto, Portugal, 20–22 July 2020; pp. 1–6. [CrossRef]
14. Mozaffari, M.; Saad, W.; Bennis, M.; Nam, Y.H.; Debbah, M. A Tutorial on UAVs for Wireless Networks: Applications, Challenges, and Open Problems. *IEEE Commun. Surv. Tutorials* **2019**, *21*, 2334–2360. [CrossRef]
15. Slalmi, A.; Chaibi, H.; Chehri, A.; Saadane, R.; Jeon, G. Toward 6G: Understanding network requirements and key performance indicators. *Trans. Emerg. Telecommun. Technol.* **2021**, *32*, e4201. [CrossRef]
16. Saym, M.M.; Mahbub, M.; Ahmed, F. Coverage Maximization by Optimal Positioning and Transmission Planning for UAV-Assisted Wireless Communications. In Proceedings of the 2021 International Conference on Science Contemporary Technologies (ICSCT), Dhaka, Bangladesh, 5–7 August 2021; pp. 1–4. [CrossRef]
17. Su, Y.; LiWang, M.; Hosseinalipour, S.; Huang, L.; Dai, H. Optimal Position Planning of UAV Relays in UAV-assisted Vehicular Networks. In Proceedings of the ICC 2021—IEEE International Conference on Communications, Virtual, 14–23 June 2021; pp. 1–6. [CrossRef]
18. Seifi, N.; Zhang, J.; Heath, R.; Svensson, T.; Coldrey, M. Coordinated 3D Beamforming for Interference Management in Cellular Networks. *Wirel. Commun. IEEE Trans.* **2014**, *13*, 5396–5410. [CrossRef]
19. Egarguin, N.J.A.; Jackson, D.R.; Onofrei, D.; Leclerc, J.; Becker, A. Adaptive Beamforming Using Scattering From a Drone Swarm. In Proceedings of the 2020 IEEE Texas Symposium on Wireless and Microwave Circuits and Systems (WMCS), Waco, TX, USA, 26–28 May 2020; pp. 1–6. [CrossRef]
20. Vaezy, H.; Salehi Heydar Abad, M.; Ercetin, O.; Yanikomeroglu, H.; Omidi, M.J.; Naghsh, M.M. Beamforming for Maximal Coverage in mmWave Drones: A Reinforcement Learning Approach. *IEEE Commun. Lett.* **2020**, *24*, 1033–1037. [CrossRef]
21. Spyridis, Y.; Lagkas, T.; Sarigiannidis, P.; Argyriou, V.; Sarigiannidis, A.; Eleftherakis, G.; Zhang, J. Towards 6G IoT: Tracing Mobile Sensor Nodes with Deep Learning Clustering in UAV Networks. *Sensors* **2021**, *21*, 3936. [CrossRef] [PubMed]
22. Becvar, Z.; Mach, P.; Nikooroo, M. Reducing Energy Consumed by Repositioning of Flying Base Stations Serving Mobile Users. In Proceedings of the 2020 IEEE Wireless Communications and Networking Conference (WCNC), Seoul, Korea, 25–28 May 2020; pp. 1–7. [CrossRef]
23. Boukoberine, M.N.; Zhou, Z.; Benbouzid, M. Power Supply Architectures for Drones—A Review. In Proceedings of the IECON 2019—45th Annual Conference of the IEEE Industrial Electronics Society, Lisbon, Portugal, 14–17 October 2019; Volume 1, pp. 5826–5831. [CrossRef]
24. Zhao, M.M.; Shi, Q.; Zhao, M.J. Efficiency Maximization for UAV-Enabled Mobile Relaying Systems With Laser Charging. *IEEE Trans. Wirel. Commun.* **2020**, *19*, 3257–3272. [CrossRef]
25. Pijnappel, T.R.; van den Berg, J.L.; Borst, S.C.; Litjens, R. Drone-Assisted Cellular Networks: Optimal Positioning and Load Management. In Proceedings of the 2021 IEEE 93rd Vehicular Technology Conference (VTC2021-Spring), Helsinki, Finland, 25–28 April 2021; pp. 1–6. [CrossRef]
26. Kouroshnezhad, S.; Peiravi, A.; Haghighi, M.S.; Jolfaei, A. Energy-Efficient Drone Trajectory Planning for the Localization of 6G-Enabled IoT Devices. *IEEE Internet Things J.* **2021**, *8*, 5202–5210. [CrossRef]
27. Nikooroo, M.; Becvar, Z. Optimal Positioning of Flying Base Stations and Transmission Power Allocation in NOMA Networks. *IEEE Trans. Wirel. Commun.* **2021**, 1. [CrossRef]
28. Plachy, J.; Becvar, Z.; Mach, P.; Marik, R.; Vondra, M. Joint Positioning of Flying Base Stations and Association of Users: Evolutionary-Based Approach. *IEEE Access* **2019**, *7*, 11454–11463. [CrossRef]

29. Pliatsios, D.; Sarigiannidis, P.; Goudos, S.K.; Psannis, K. 3D Placement of Drone-Mounted Remote Radio Head for Minimum Transmission Power Under Connectivity Constraints. *IEEE Access* **2020**, *8*, 200338–200350. [CrossRef]
30. Pliatsios, D.; Sarigiannidis, P.; Moscholios, I.D.; Tsiakalos, A. Cost-efficient Remote Radio Head Deployment in 5G Networks Under Minimum Capacity Requirements. In Proceedings of the 2019 Panhellenic Conference on Electronics Telecommunications (PACET), Volos, Greece, 8–9 November 2019; pp. 1–4. [CrossRef]
31. Feng, C.; Liu, B.; Guo, Z.; Yu, K.; Qin, Z.; Choo, K.K.R. Blockchain-based Cross-domain Authentication for Intelligent 5G-enabled Internet of Drones. *IEEE Internet Things J.* **2021**, 1. [CrossRef]
32. Odarchenko, R.; Usik, P.; Volkov, O.; Simakhin, V.; Gospodarchuk, O.; Burmak, Y. 5G Networks Cyberincidents Monitoring System for Drone Communications. In Proceedings of the 2019 IEEE 5th International Conference Actual Problems of Unmanned Aerial Vehicles Developments (APUAVD), Kiev, Ukraine, 22–24 October 2019; pp. 165–169. [CrossRef]
33. Zhang, Q.; Saad, W.; Bennis, M.; Lu, X.; Debbah, M.; Zuo, W. Predictive Deployment of UAV Base Stations in Wireless Networks: Machine Learning Meets Contract Theory. *IEEE Trans. Wirel. Commun.* **2021**, *20*, 637–652. [CrossRef]
34. Alzenad, M.; El-Keyi, A.; Lagum, F.; Yanikomeroglu, H. 3-D Placement of an Unmanned Aerial Vehicle Base Station (UAV-BS) for Energy-Efficient Maximal Coverage. *IEEE Wirel. Commun. Lett.* **2017**, *6*, 434–437. [CrossRef]
35. Amponis, G.; Lagkas, T.; Sarigiannidis, P.; Vitsas, V.; Fouliras, P.; Wan, S. A survey on FANET routing from a cross-layer design perspective. *J. Syst. Archit.* **2021**, *120*, 102281. [CrossRef]
36. Nikam, V.B.; Arora, A.; Lambture, D.; Zaveri, J.; Shinde, P.; More, M. Optimal Positioning of Small Cells for Coverage and Cost Efficient 5G Network Deployment: A Smart Simulated Annealing Approach. In Proceedings of the 2020 IEEE 3rd 5G World Forum (5GWF), Bangalore, India, 10–12 September 2020; pp. 454–459. [CrossRef]
37. Hunukumbure, M.; Tsoukaneri, G. Cost Analysis for Drone Based 5G eMBB Provision to Emergency Services. In Proceedings of the 2019 IEEE Globecom Workshops (GC Wkshps), Waikoloa, HI, USA, 9–13 December 2019; pp. 1–5. [CrossRef]
38. Giordani, M.; Polese, M.; Mezzavilla, M.; Rangan, S.; Zorzi, M. Toward 6G Networks: Use Cases and Technologies. *IEEE Commun. Mag.* **2020**, *58*, 55–61. [CrossRef]

drones

Article

On the Performance of a UAV-Aided Wireless Network Based on NB-IoT

Silvia Mignardi *, Riccardo Marini, Roberto Verdone and Chiara Buratti

WiLab, CNIT/DEI, University of Bologna, 40136 Bologna, Italy; r.marini@unibo.it (R.M.); roberto.verdone@unibo.it (R.V.); c.buratti@unibo.it (C.B.)
* Correspondence: silvia.mignardi@unibo.it

Abstract: In recent years, interest in Unmanned Aerial Vehicles (UAVs) as a means to provide wireless connectivity has substantially increased thanks to their easy, fast and flexible deployment. Among the several possible applications of UAV networks explored by the current literature, they can be efficiently employed to collect Internet-of-Things (IoT) data because the non-stringent latency and small-size traffic type is particularly suited for UAVs' inherent characteristics. However, the implications coming from the implementation of existing technology in such kinds of nodes are not straightforward. In this article, we consider a Narrow Band IoT (NB-IoT) network served by a UAV base station. Because of the many configurations possible within the NB-IoT standard, such as the access structure and numerology, we thoroughly review the technical aspects that have to be implemented and may be affected by the proposed UAV-aided IoT network. For proper remarks, we investigate the network performance jointly in terms of the number of successful transmissions, access rate, latency, throughput and energy consumption. Then, we compare the obtained results on different and known trajectories in the research community and study the impact of varying UAV parameters such as speed and height. Moreover, the numerical assessment allows us to extend the discussion to the potential implications of this model in different scenarios. Thus, this article summarizes all the main aspects that must be considered in planning NB-IoT networks with UAVs.

Keywords: UAVs; NB-IoT; network analysis; simulations

Citation: Mignardi, S.; Marini, R.; Verdone, R.; Buratti, C. On the Performance of a UAV-Aided Wireless Network Based on NB-IoT. *Drones* **2021**, *5*, 94. https://doi.org/10.3390/drones5030094

Academic Editor: Vishal Sharma

Received: 30 July 2021
Accepted: 4 September 2021
Published: 9 September 2021

1. Introduction

Machine-type communications have witnessed a renewed interest from the scientific community thanks to advancements in technology and the plethora of new applications involving them. Statistical reports already proved a steady increase in the number of machine-type or Internet of Things (IoT) links (see, e.g., https://www.statista.com/statistics/802690/worldwide-connected-devices-by-access-technology/ or https://www.ericsson.com/en/mobility-report/reports/november-2019/iot-connections-outlook, accessed on 7 September 2021.), but the massive presence of IoT devices might not be the only major challenge to be addressed for the future. Other key challenges lie on the more differentiated and stringent requirements on communication performance—the demand—imposed by the several applications and use cases possible. These may include autonomous vehicles, wearables, industrial IoT for Industry 4.0, data monitoring, alarm detection, municipality services and many others, in which one can observe that commonalities are few. These aspects call for new paradigms to network design. To avoid the densification and deployment of new terrestrial bases needing huge investments in capital and operational expenditures, a viable and largely foreseen solution can be found in mobile base stations (BSs).

Unmmaned aerial vehicles (UAVs) (a.k.a. drones), where mobile BSs can be mounted, represent very interesting means to add the required flexibility and scalability for the future networks. UAVs have, in fact, the potential to fly on-demand and exactly where is needed. Moreover, they are not tied to roads, not affected by traffic congestion and can feature good connectivity with both on-ground users and terrestrial BSs (i.e., backhaul), thanks to the large probability of being in the line of sight (LoS). The usage of UAVs as mobile BSs is especially suited for massive machine-type communication (mMTC) and IoT links, since IoT nodes are mostly static (do not change their position over time) and their traffic demand is usually predictable [1]. The knowledge of these two basic inputs allows to make decisions in advance on the trajectory to follow to maximize IoT service, plan periodical flights when the demand arises, modify the UAV behaviour if needed, and so on. This may be further relevant if these mobile BSs have to direct themselves in remote and different areas where the nodes are placed. To consider at our best a scenario with the plurality of IoT applications mentioned before, as pictured in Figure 1, we consider a massive number of IoT devices scattered in different zones of a service area, requesting to transmit a periodical packet. This is a realistic scenario, and perfect for UAV services during a single flight.

Figure 1. Illustrative example of a scenario with multiple IoT applications and nodes served by UAV BSs (UABs).

Thanks to their versatility, flying network nodes like UAVs have gained an ever-increasing interest from researchers concerning standardization bodies. The 3rd Generation Partnership Project (3GPP), after considering the feasibility of UAVs being user equipment (UE) (i.e., end users of the cellular network), started approaching on-board radio access for UAVs (denoted with the acronym UxNB) at the beginning of Rel. 17 [2,3]. Because of the several activities considering the aid of UAVs to cellular users, we focus on the cellular radio access technology, as stated in the 3GPP documents [4]. In fact, if a UAV could have installed the same radio-frequency equipment with a similar protocol stack to target both IoT nodes and broadband users, it would be convenient for mobile operators. To this purpose, there exist a number of technologies targeting IoT applications which follow the fourth-generation (4G) numerology [5]. In particular, we will focus on NarrowBand-IoT (NB-IoT) for the design intended to target low-end IoT applications with low data rates, delay tolerance, massive connections, and extremely wide coverage [6]. To this end, this work can be considered an extension of [1], where an initial and simple approach was proposed to UAV-aided NB-IoT networks.

In this article, we throroughly study different approaches to aerial support for NB-IoT networks, in order to provide a general overview of the challenges and potentials of these systems. To properly assess the network performance of UAVs serving NB-IoT nodes, we jointly consider as performance metrics the percentage of completely served nodes, the throughput provided and the latency which has to be expected and the IoT nodes' energy consumption. Our key contributions can be summarized as follows:

- We propose a UAV-aided NB-IoT scenario with several hundreds of nodes, located in different parts of the area as to simulate diverse applications. Differently from other papers in literature, the NB-IoT technology is considered in details as specified by the 3GPP documents and studied in all of its features, including the signaling procedures and the parameters' implications of the three NB-IoT coverage classes;
- we investigate the performance in terms of number of completely served nodes, achieved network throughput, perceived latency and energy consumption of nodes;
- we analyze the impact of using different UAV trajectories and the effect of varying UAV parameters, such as speed and height.

The article is organized as follows. In Section 2, we give an overview of the literature about UAV-aided networks in IoT applications. NB-IoT and its features are presented in Section 3. Section 4 describes the scenario and the network model. Final simulation results are reported and discussed in Section 5. Finally, Section 6 concludes the article.

2. Literature Overview

Initial studies on UAV-aided networks focused on the key link-level considerations, and specifically on the characterization of path loss and its impact on the so-called air-to-ground (ATG) channel [7,8]. These activities on UABs were followed by the aim of finding an acceptable trade-off between coverage, capacity and connectivity, as in [9]. To be more specific, in [8] the effect of the user-UAV incident angle w.r.t. the ground plane as a function of drone height is studied. It is defined as elevation angle , and its aperture may determine a link being or not being free from obstructions and in LoS conditions. In the remainder of this article, we refer to the authors' ATG model. In recent years, the interest in UAV-aided networking gradually increased, introducing studies ranging from UAV deployment issues to grant adequate coverage [10,11] to more complex problems of UAV trajectory design for fair and satisfactory quality of service to ground users [12–14]. Differently from our objective, the majority of papers addresses a general user or device, forgetting the implications and limitations of the specific protocol procedures. For example. dynamic trajectories for the 3D space are studied in [12] with the purpose to connect IoT nodes at their activation time. The authors jointly optimize the transmission power of ground nodes, the overall energy spent in movement and the choice of the next stop of each UAV. Some activities deal with the definition of an optimal trajectory for UABs. In [13,14] the trajectory is optimized with the aim of maximizing the minimum user rate. Optimization algorithms have also been studied for UAV placement to achieve optimal energy efficiency [15,16] by minimizing power consumption and transmission delays, which are interesting requirements for IoT applications. However, they do not consider specific protocol constraints or overhead as it is in 3GPP standards.

Similarly to this study, some other works related to UAVs are targeting the IoT field. As for recent examples, references [17,18] employ UAVs to gather IoT data from remote areas in which devices experience low connectivity or capacity issues. In [17], the authors address the trajectory and resource allocation optimization for time-constrained IoT nodes, while the authors in [18] optimize the UAB 3D placement and resource allocation minimizing the total transmission power of IoT devices and balancing the multi-UABs' tasks. Reference [19] focuses on the modelling and optimization of the carrier sensing-based medium access control (MAC) layer protocol of UAV-aided IoT networks; this paper analyses the achievable throughput as the performance metric and CSMA/CA as the MAC protocol. Moreover, reference [20] studies different mechanism for the energy and delay aware task assignment of UAVs with onboard IoT nodes. More specific to the technology investigated hereby, reference [21] studies trajectories for energy minimization in a NB-IoT context, reference [22] investigates connectivity strategies for a specific NB-IoT application, and [23] introduces a coverage analysis for UAV-aided NB-IoT networks. The NB-IoT protocol is considered in these articles, too, since it belongs to the machine-type emerging technologies [6] and targets low data rates, delay tolerance, massive connections, and extremely wide coverage [24]. However, these works usually lack either a fine-grained protocol study or focus on the optimization of one metric above the others. On the contrary, we aim at providing a more general overview of potentials and challenges of UAV-aided NB-IoT networks, with a deeper focus on protocol details rather than the optimization of a single performance metric. Moreover, optimization frameworks struggle to handle large input instances (e.g., a massive number of nodes in the scenario) because of the excessive computation times, in terms of days or weeks. For example, references [13,14] consider less than 10 users in the service area. Furthermore, in [25], it is optimized the 3D locations of UAVs for wireless powered NB-IoT. Another work worth to mention is [26]. The authors propose a NB-IoT model to collect underground soil parameters in potato crops using a UAV-aided network. The analysis in this case is mostly application-dependent, and therefore differs from our general evaluation with different metrics.

This activity can be considered an extension of [1]. To the best of the authors' knowledge, the literature still lacks the detailed model and protocol analysis of similar scenarios and setups. Therefore, the focus of this work is to extend and further discuss the system dynamics of NB-IoT networks served by an UAV, rather than compare our approach with other research activities. This study helps us to extract the major impacts of the overall protocol stack of the NB-IoT technology on UAV-aided networks.

3. The Narrowband-IoT Technology

The NB-IoT technology is intended to address the needs of mMTC, and has been standardized by 3GPP initially on Rel. 13 in 2016, with new functionalities introduced in the subsequent releases. It will be briefly described in this section, taking as references [5,27]. Particular emphasis is given to the uplink, which is the scenario of interest for this paper. The NB-IoT technical solution originates from the Long Term Evolution (LTE) technology, which has been substantially simplified to reduce the overheads, minimizing complexity, cost and consumption, and, at the same time, it keeps its usual mechanisms, such as synchronization, radio access and resource assignment. In addition, the NB-IoT technology features substantial flexibility allowing to deploy NB-IoTs cells by running a software update on already existing LTE cells. Indeed, three deployment options are available: Standalone, to reuse 200 kHz GSM carriers; guard-band, to exploit the guard band of two adjacent LTE carriers; and in-band, where one LTE Physical Resource Block (PRB) is reserved for NB-IoT within an LTE carrier bigger than 1.4 MHz. NB-IoT implements several mMTC-oriented enhancements, such as narrowband transmission and enhanced power saving techniques to increase battery life of UEs. In addition, up to 2048 and 128 repetitions in DL and UL, respectively, can be used to exploit the time variation of the radio channel, so that each replica can be decoded separately, or multiple replicas can be combined to further increase the reception probability. Furthermore, NB-IoT's coverage enhancement is provided by defining three coverage classes: Normal, robust, and extreme. Classes are differentiated through thresholds based on Reference Signals Received Power (RSRP), defined to introduce three levels of coverage extension. Such thresholds depend on the cell deployment, the propagation environment (i.e., outdoor, indoor, deep-indoor, underground), and the spatial distribution of devices. Network parameters as the number of repetitions can be tuned separately for each class.

Before Rel. 15 introducing time division duplex (TDD) operation mode, the frequency division duplex (FDD) mode was the only option for NB-IoT. In this paper we consider the latter, since it is the primary mode used in most commercial networks and enables the maximum performance. The FDD mode implies different frequency bands to be used for UL and DL transmissions. The channel bandwidth of NB-IoT is 180 kHz, that corresponds to one LTE PRB. In UL, subcarrier spacing of either 15 or 3.75 kHz are possible, thus providing either 12 or 48 possible subcarriers within a 180 kHz resource block. The 15 kHz spacing allows transmission of either single or multitone (over up to 12 carriers) signals, while only single-carrier transmissions are possible for 3.75 kHz grid. On the contrary, in DL, only the 15 kHz resource grid is used.

3.1. Random Access Procedure

The random access (RA) procedure of a NB-IoT UE, which is necessary to allow an uplink packet transmission, is composed of multiple steps. To start with, the UE scans the channels for the synchronization signals. When it gets correctly synchronized to the base station, the UE obtains first the Master and then a number of Secondary Information Blocks (MIB and SIBs, respectively), containing all the relevant information about the network, the cell and its resource allocations. Then, to connect to the cell, the RA procedure starts by sending a preamble (or Msg1) during one of the periodic random access windows (NPRACH). After sending Msg1, the UE waits for Msg2 from the eNB. Note that, at this stage, the network is not aware of possible overlapping among UEs' transmissions, which may happen if more than one UE chooses the same random preamble. With Msg2, the eNB notifies that it has received the specific preamble sequence and allocates NPUSCH resources for Msg3 for UEs which transmitted that specific sequence. In Msg3, each of the UEs sends different ID data and, in this case, collisions may actually happen, since UEs which used the same preamble sequence will use the same resources to send their Msg3. If the eNB can correctly receive at least one of the Msg3 packets, it will respond with Msg4 to complete the procedure. If the access procedure is carried out correctly by an UE, the resources for the UL and downlink (DL) transmissions are scheduled by the cell eNB with data integrity insured through the Hybrid Automatic Repeat Request (HARQ) processes.

3.2. Energy Consumption

Clearly, the exchange of these signalling messages will affect the energy consumed by the nodes. Therefore, NB-IoT has to introduce appropriate techniques to power savings while keeping the synchronization with the cellular network. To this purpose, NB-IoT introduces predefined intervals for discontinuous reception (DRX), enhanced DRX (eDRX) and power saving mode (PSM) to reduce at minimum the energy consumption when its demand is fulfilled. In our case, considering all the different power saving techniques is not desirable. In fact, the UAB is a mobile BS, and therefore spends a limited time interval in which grants connectivity to a group of NB-IoT nodes. Therefore, if a node has a packet (or a queue of them) ready to be transmitted, we assume attempts the RA procedure as soon as it gets under the UAB coverage, to go in PSM immediately after.

3.3. Uplink Channels, Parameters, and Implications

Only two channels are defined in the UL, the narrowband physical random access channel (NPRACH) and the narrowband physical uplink shared channel (NPUSCH). The NPRACH is used to trigger the RA procedure. It is composed of a contiguous set of either 12, 24, 36, or 48 subcarriers with 3.75 kHz spacing, which are repeated with a predefined periodicity, that may take several discrete values between 40 ms and 2560 ms. The RA procedure starts with the transmission of a preamble, with a duration of either 5.6 ms or 6.4 ms (Format 0 and 1, respectively) depending on the size of the cell, and can be repeated up to 128 times to improve coverage. A preamble is composed of four symbol groups, each transmitted on a different subcarrier. The initial subcarrier is chosen randomly by the UE, while the following ones are determined according to a specific sequence depending on the first one. Two UEs selecting the same initial subcarrier will thus collide for the entire sequence, as described in Section 3. A special mechanism can help resolving the collisions and thus the access probability of a node u can be approximated as Equation (1):

$$P_{\text{acc},u} = e^{-\frac{n_{\text{RA}}}{N_{\text{SC}}}} \tag{1}$$

where n_{RA} is the number of nodes entering RA and N_{RU} are the total available subcarriers.

In case of standalone deployment, the NPUSCH occupies all the UL resources left available after the allocation of the NPRACH. NPUSCH is used for UL data and UL control information. The eNB decides how many resources to allocate to the UEs depending on the amount of data to be sent, the modulation-coding scheme (MCS) used and the number of repetitions needed to correctly receive the data. The minimum resource block which can be allocated, referred to as a resource unit (RU), depends on the UE capabilities and the configured numerology. Specifically, in the case of 3.75 kHz subcarrier spacing and single-tone operation, which is the configuration assumed in this paper, the RU is 32 ms long. The number of the RUs (ranging from one to ten) to be allocated depends on the size of the transport block size (up to 2536 bits in Rel. 15) and the MCS chosen to meet the required success probability. In addition, the eNB specifies the desired number of repetitions. Thus, the total number of resources allocated to a UE is equal to:

$$N_{\mathrm{RU_{TOT}}} = N_{\mathrm{RU}} \cdot N_{\mathrm{REP}} \tag{2}$$

where N_{RU} is the number of resources needed to send a packet of size B depending on the chosen MCS and N_{REP} is the number of transmission repetitions the UE is configured to send.

Without the loss of generality, in what follows we imply 3.75 kHz subcarrier spacing with 48 carriers allocated for RACH. Furthermore, we need to set a number of other uplink parameters to define a resource unit and resource availability on the NPUSCH. For example, we have to define the setting for the MCS used to determine the Transport Block Size (TBS), that is the number of bits which can be transmitted given a certain number of RUs assigned in the NPUSCH. In our model, it holds $I_{TBS} = I_{MCS} = 6$, and, for clarity, we report in Table 1 only the possible TBSs for the selected scheme.

Table 1. NB-IoT UL TBS in bits.

RUs Number	2	4	5	6	8	10
Max. Packet Size for $I_{MCS} = 6$ [bits]	176	392	504	600	808	1000

Moreover, we need to set the uplink parameters for the three different coverage classes. For simplicity, we refer to them as coverage extension (CE), so that the Normal becomes CE0, the Robust CE1 and the Extreme CE2. These are characterized for a different number of repetitions, subcarrier NPRACH assignment, periodicity and receiver sensitivity. Please note that the optimal choice of these parameters' values is out of the scope of this paper, and most of them are taken from [5]. Table 2 summarizes our settings, that should be handled easy by a UAB.

Table 2. Coverage Parameters.

	CE0	CE1	CE2
$P_{ce,\min}$ [dBm]	−101	−111	−121
NPRACH Periodicity [ms]	320	320	320
NPRACH Subcarriers	24	12	12
NPRACH Format	0	0	1
NPRACH Repetitions	2	8	32
NPUSCH Repetitions	2	8	32

4. System Model

The scenario taken into consideration has the scope to recreate a realistic deployment of IoT nodes, denser in service areas and absent in other locations. For example, IoT devices can be positioned at smart traffic junctions, in city parks, at waste collection points, in the parking lots, or into buildings, to name just a few. Thus, practically, we can state we have clusters of IoT nodes, characterized by close vicinity when they implement the same application requirements.

The spots scattered with NB-IoT nodes are considered not to be served adequately by the terrestrial infrastructure, and, for this reason, an unmanned aerial BS (UAB) equipped with NB-IoT radio access is sent to supply the service instead.

4.1. Network Scenario

To be specific, we model the scenario using a Poisson Cluster Process, namely the Thomas cluster process (TCP) [28], as proposed in [1] and conventionally done in the literature (see, e.g., [29]). The TCP is a stationary and isotropic Poisson cluster process generated by a set of offspring points independently and identically distributed (i.i.d.) around each point of a parent Poisson Point Process (PPP) [28]. In particular, the locations of parent points are modeled as a homogenous PPP, with intensity λ_{p}, around which offspring points are distributed according to a symmetric normal distribution with variance σ^2 and mean value m. As a conquence, the intensity of the offspring points can be written as $\lambda = \lambda_{\text{p}} \cdot m$. In our scenario, offspring points represent the IoT nodes asking for service, while parent points are only reference coordinates for cluster centers.

We simulate a square area of size $L \times L$ m^2, where offspring points are located according to the description above. We picture a sample scenario together with possible UAB trajectories in the following (see Figures 2–4 below). We consider a single UAB to decrease capital and operational expenditures, and to simplify the final numerical evaluation. Please note that, in this model, the extension to the case of multiple UABs does not necessitate additional complex settings, and therefore it is not a major focus of this work. However, for the sake of completeness, it will be discussed in the following.

We assume the UAV starts its flight from a fixed position, which can be considered as a recharge station, where it has to come back at the end of the trajectory. In this way, it can recharge or change its battery for the next flight. In this scenario, we assume that the capacity of the UAV battery is sufficient to enable a full round trip over any trajectory. Provided that the UAB has no heavy payload other than the RF equipment and the flight time is no longer than half an hour (that is always our case), this is reasonable [30,31]. The UAB is assumed to fly at a constant altitude from the ground between 200 m and 300 m (not violating the regulations in EU—[32]).

4.2. Channel Model

Motivated by the short-sized traffic demand, we assume the backhaul link UAB-terrestrial BS undergoes free-space propagation, and the capacity achieved is sufficient for both the UAB control links (for manouvrability and command and control signals) and data forward. In this article, the propagation model affects the UAB-ground node link, and is therefore known as ATG channel. We compute the received power, P_{rx} as a function of the transmit power, P_{tx}, as: $P_{rx}[\text{dBm}] = P_{tx}[\text{dBm}] - A_L[\text{dBi}] - L_{\text{ATG}}[\text{dB}]$. The ATG propagation can statistically model the loss value, L_{ATG}, as the reference considered here for drones in urban environment [7,33]. According to this model, connections between drone and nodes can either be LoS or Non-LoS (NLoS). For NLoS links, the signals travel in LoS before interacting with objects located close to the ground which result in shadowing effects. We denote as p_{LoS} the probability of connection being LoS. The probability p_{LoS} at a given elevation angle, θ, is computed according to the following equation

$$p_{\text{LoS}} = \frac{1}{1 + \alpha \exp(-\beta[\frac{180}{\pi}\theta - \alpha])} \quad (3)$$

with α and β being environment-dependent constants, i.e., rural, urban, etc, and adopted as given in [7,8]. Equation (3) determines for every link if it is in LoS or NLoS condition, impacting then the value of ξ_{LoS} in Equation (4). The LoS path loss model is given by:

$$L_{\text{ATG}}(d)[\text{dB}] = 20\log\left(\frac{4\pi f_c d}{c}\right) + x_{\text{LoS}}\xi_{\text{LoS}} + (1 - x_{\text{LoS}})\xi_{\text{NLoS}} + \eta \quad (4)$$

where x_{LoS} equals 1 in the case of realization of LoS links and 0 otherwise. ξ represents the shadowing coefficient which depends on LoS or NLoS conditions as well and is set as described in [7,8]. Then, c is the speed of light, f_c is the center frequency, and d is the transmitter-receiver distance in meters. An additional penetration loss, η, as for in indoor monitoring or basement applications is considered.

If the received power is above the receiver sensitivity, we consider the node to be in the connectivity range of the UAB. Because of the fact that NB-IoT may have three coverage classes, we have three sensitivity thresholds, one for each class ce, denoted as $P_{ce,\text{min}}$. Once the device is connected and synchronized to its coverage class signalling, it can attempt to access the channel through the NB-IoT NPRACH (see Section 3.3), so that, if succeeded, it may be given resources to transmit its data. The number of resources assigned determines the packet size that the node can transmit in the given time window (see Section 3.3 and Table 1 for scheduling details). Note that, since the IoT nodes are the devices more limited in their characteristics, for example considering the maximum transmit power, we assume:

- the connectivity range is defined by the uplink,
- the downlink control communication is error-less.

4.3. Traffic Model and Metrics

Each node will then request to the UAB to transmit one uplink packet of size B. We assume NB-IoT nodes are already synchronized to the cellular network; they start their operations from the RA, exchanging the required NB-IoT signalling messages, and, if it is completed correctly, the required uplink resources are scheduled in NPUSCH. As one can imagine, the procedure is robust but its completion is not guaranteed. In fact, the main obstacles may be found in the UAB movement, channel fluctuations and collisions. The last two are application and environment dependent, while the first may be subject to a proper trade-off of the different latency, energy consumption, throughput and success rate metrics.

To later assess this trade-off, let us formulate here the performance metrics which are node-dependent; network-dependent metrics are formalized in the following. In the following, pedex u will indicate a generic node in the network.

Latency, that is the interval elapsing between the instant when the node has a packet to transmit to the service completion, is computed as:

$$\Delta\tau_u = \tau_{u,\text{tx}} - \tau_{u,\text{start}} \tag{5}$$

where $\tau_{u,\text{tx}}$ is the time instant in which the node transmits its packet and $\tau_{u,\text{start}}$ when this data is first available. For ease of evaluation, we assume $\tau_{u,\text{start}}$ equals the time in which the UAB starts its trajectory for all nodes u. Then, the throughput, t_u, achieved by a single node u whose transmission is considered successful depends on $\Delta\tau_u$ as:

$$t_u = \frac{B}{\Delta\tau_u} \tag{6}$$

If a node u_n is not able to transmit its packet demand, it would be $t_{u_n} = 0$ b/s. Finally, the energy consumption has to account for the signalling related to the RA procedure. It holds:

$$\begin{aligned} E_u =& E_{\text{tx},u} + E_{\text{rx},u} + E_{\text{idle},u} + E_{\text{sleep},u} = \\ =& V\ I_{\text{tx}}\ T_{\text{tx},u} + V\ I_{\text{rx}}\ T_{\text{rx},u} + V\ I_{\text{idle}}\ T_{\text{idle},u} + V\ I_{\text{sleep}}\ T_{\text{sleep},u} \end{aligned} \tag{7}$$

where V indicates the voltage with which the IoT node is powered, I_{tx} and I_{rx} the current needed in transmission and reception mode, respectively, I_{idle} the current present when the node stays in idle, and I_{sleep} the current during PSM. Similarly, we indicate with $T_{\text{tx},u}$, $T_{\text{rx},u}$, $T_{\text{idle},u}$, $T_{\text{speed},u}$ the times spent by each node u for being in the corresponding operation mode. Of course, this depends on the message exchange described in Section 3 (that includes an alternating of transmission and reception modes). Some of these parameters are fixed and shown in Table 3 [34].

Table 3. Energy Consumption Parameters.

Parameter	I_{tx} [mA]	I_{rx} [mA]	I_{idle} [mA]	I_{sleep} [mA]	V [V]
Value	220	46	6	0.003	3.6

4.4. Different Trajectories for UABs

As mentioned, we analyze multiple possible trajectories for one UAB flying over clusters of IoT nodes. These trajectories follow a predefined path, since IoT nodes are placed in fixed positions and usually have a traffic demand that is easily predicted or periodical. In this way, we can avoid static positioning of multiple drones, that would require increasing capital expenses. Moreover, a static deployment of hovering UABs has further energy consumption issues from the UAB side, that may be less under control. We consider the following possible trajectory design:

- Circular path;
- Paparazzi-like trajectory;
- Flight following the solution of a Traveling Salesman Problem (TSP) over clusters' parent points.

Each of these trajectories has its pros and cons. Thanks to the wide coverage which can be achieved by the three coverage classes of NB-IoT, the circular path might be an option for its short path length. On the other side, if IoT nodes are not adequately covered, the paparazzi-like trajectory is able to scan the entire area. However, since the UAB has to serve clusters of fixed nodes, there is a third option. We can consider the locations of the parent points as reference coordinates to model the trajectory as a TSP [35], as also proposed in [1]. In this way, we can observe the effectiveness of these choice compared to other known alternatives. The TSP determines, for a finite set of points whose pairwise distances are known, the shortest route connecting all points. The circular path has a radius length equal to half the length of the circumscribed circle of the service area. To better adapt the circular trajectory to the nodes deployment, we consider as the perimeter of the service area the maximum extension of nodes' location in all directions, centering it and the circular path consequently. A similar implementation is repeated for the paparazzi trajectory, which considers also a sensing radius for the UAV to define the width of the serpentine, fixed to 500 m. Examples of trajectories and cluster positions for a scenario snapshot are represented in Figures 2–4.

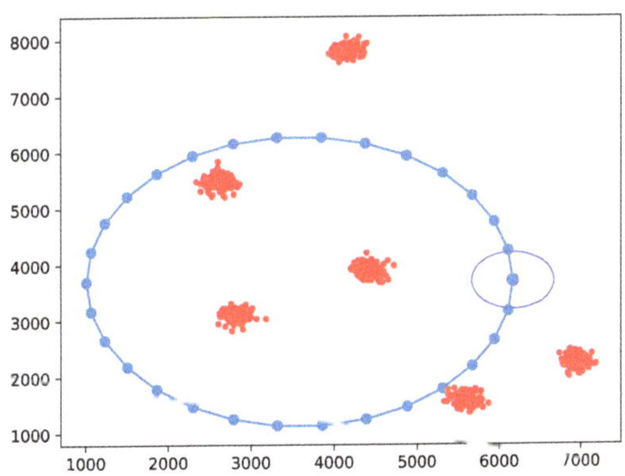

Figure 2. Example of circular path (blue line) and clusters with nodes (red dots) location; the dark blue circle indicates the start of the trajectory.

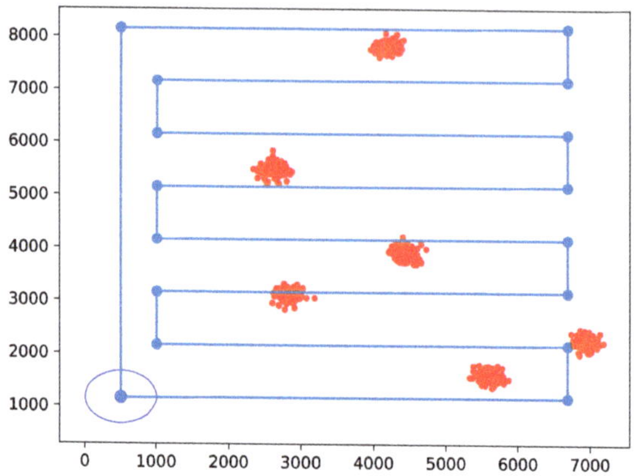

Figure 3. Example of Paparazzi path (blue line) and clusters with nodes (red dots) location; the dark blue circle indicates the start of the trajectory.

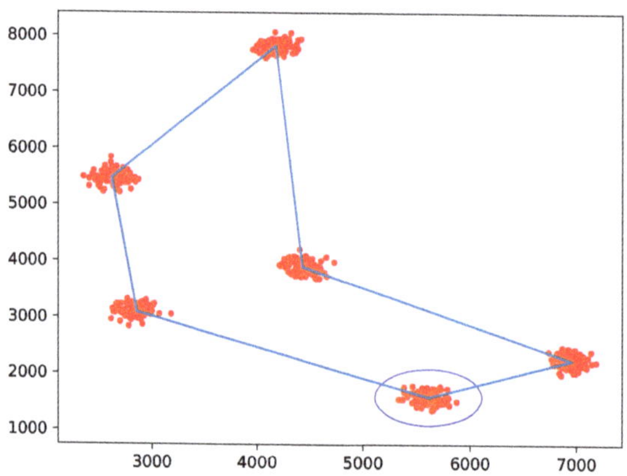

Figure 4. Example of TSP-driven path (blue line) and clusters with nodes (red dots) location; the dark blue circle indicates the start of the trajectory.

As one can easily see, the performance of the considered network depends both on the UAB mobility pattern and the UAB NB-IoT cell configuration. In the following section we consider how the respective parameters (e.g., UAB height and speed, NB-IoT coverage class configuration and cluster dimension) affect the performance.

5. Results and Discussion

In this section, we will analyze our NB-IoT network performance and discuss the achieved results. Simulations are carried out with the parameters listed in Table 4. As previously mentioned, we are jointly studying a number of metrics, being:

- the access rate, R_{acc};
- the percentage of nodes completely served, S_{suc};
- the mean latency perceived by nodes, $\Delta\tau_{avg}$;
- the network throughput, T_{net};
- the mean energy consumption of NB-IoT nodes, E.

These metrics are computed via the following formulas:

$$R_{acc} = \frac{n_{suc}}{n_{att}} \tag{8}$$

$$S_{suc} = \frac{n_{suc}}{n_{tot}} \cdot 100 \tag{9}$$

$$\Delta\tau_{avg} = \frac{\sum_{u=1}^{n_{tot}} (\tau_{u,tx} - \tau_{u,start})}{n_{tot}} \tag{10}$$

$$T_{net} = \sum_{u=1}^{n_{tot}} t_u \tag{11}$$

$$E = \frac{\sum_{u=1}^{n_{tot}} E_u}{n_{tot}} \tag{12}$$

We denote as access rate, R_{acc} in Equation (8), the fraction of the number of succeeded transmissions, n_{suc}, over the overall attempted by anyone of the IoT nodes, n_{att}. In this way, we analyze how frequently a node u may get access to the channel. When the R_{acc} value gets close to zero, the number of access attempts has to increase before success. Equation (9) let us identify how effective is the UAB service. In fact, S_{suc} counts the number of nodes successfully transmitting their traffic demand over the total number of present IoT nodes, n_{tot}, in percentage. Then, Equation (10) computes the latency or delay for the node u spanning from the time in which the request is started, $\tau_{u,start}$, until when the transmission succeeded, $\tau_{u,tx}$. If a node is not able to transmit its packet, we do not consider it in the average latency computation. Equation (11) computes the overall network throughput by summing the one of each node u, t_u. Finally, Equation (12) averages the energy consumed by all nodes present in the service area, n_{tot}.

Table 4. Radio and Network Parameters.

Parameter	Value
Area side, L	10 km
Mean number of nodes per cluster, m	100
Intensity of parent points, λ_p	5
Locations' standard deviation, σ	100
Packet size, B	500 bits
UL transmit power, P_{tx}	14 dBm
Antennas loss, A_L	2.5 dBi
Penetration loss, η	40 dB
Noise power, P_N	30×10^{-17} W
α	9.6117
β	0.1581
Channel bandwidth, B_c	180 KHz
Subcarrier spacing	3.75 kHz
Available subcarriers, N_{SC}	48
Carrier frequency, f_c	800 MHz
RU duration	32 ms
MCS index, I_{MCS}	6

To start with, Figure 5 shows the percentage of IoT nodes served by the UAB, that means their traffic demand is completely fulfilled. Its value is represented while varying the UAB speed, v, and for the different trajectories. This picture has some relevant outcomes. First, the performance is not the same when varying speed; for each trajectory, it drops fairly below the 50% of served nodes. This is clearly related to the time interval in which the UAB is able to maintain a robust radio connection with each node. In fact, if the received power falls below the receiver sensitivity before the signalling is completed and the node is scheduled an uplink resource on the NPUSCH (for example, if the UAB has flown away too fast), it will not be able to successfully transmit its packet. This effect becomes relevant when discussing latency, because the value of v creates a trade-off between low application delays and successful transmissions. At first glance, it is also evident that the TSP trajectory is able to serve a higher number of nodes for every UAB speed, making more robust what first proposed in [1]. In fact, this trajectory ensures the UAB gets in close vicinity with each cluster and each node, while minimizing the distance travelled. On the opposite, the circular trajectory decreases the percentage of successful transmissions of 50% with respect to TSP and Paparazzi ones, on average. The coverage extension and the time spent over each node, does not allow in this case a sufficient service. The analysis on the other metrics would allow us to finally assess and discuss the final trajectory-dependent performance.

Figure 6 represents the access rate, R_{acc}, again while varying the UAB speed for different trajectories. By looking at increasing speeds and for the same trajectory, we have the value of R_{acc} dropping for each trajectory at 25 m/s, as it was for the service in Figure 5. With the UAB driving faster, the occasions to attempt channel access decrease. From Equation (8), we observe this metric evaluates together the number of overall attempts tried by nodes and, from these, the ones which are successful. This might explain why the Paparazzi and circular trajectories have higher access rates; these can be achieved if the total number of attempts is small (the denominator) with respect to other trajectories. If a lower number of nodes try to access at the same time to the channel (maybe for connectivity issues) the probability of successful transmission increases (see Equation (1)).

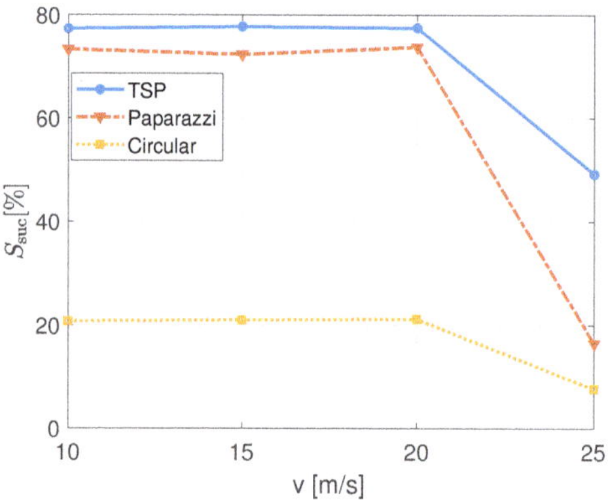

Figure 5. Percentage of served IoT nodes for different UAB speeds and trajectories with h = 200 m.

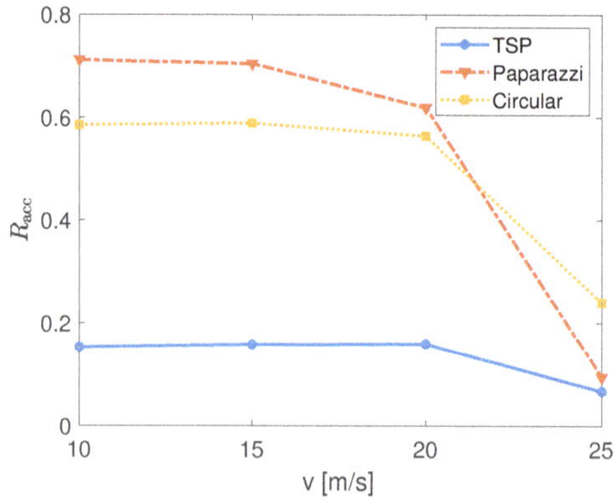

Figure 6. Access rate of IoT nodes for different UAB speeds and trajectories with h = 200 m.

Then, another figure of merit is the latency with which packets are transmitted (on average), $\Delta\tau_{avg}$, in Figure 7. As expected, the latency decreases with increasing speed for each trajectory; the NB-IoT nodes which can successfully transmit their packet are served faster because the UAB will reach them with a smaller delay. Furthermore, if we focus on the latency of the different trajectories, we notice the TSP path performs better, just followed by the circular one. In fact, the distances covered by these two trajectories are fairly smaller with respect to the other. On the contrary, for its characteristic of scanning the entire service area, the Paparazzi trajectory employs much larger delays than the other two.

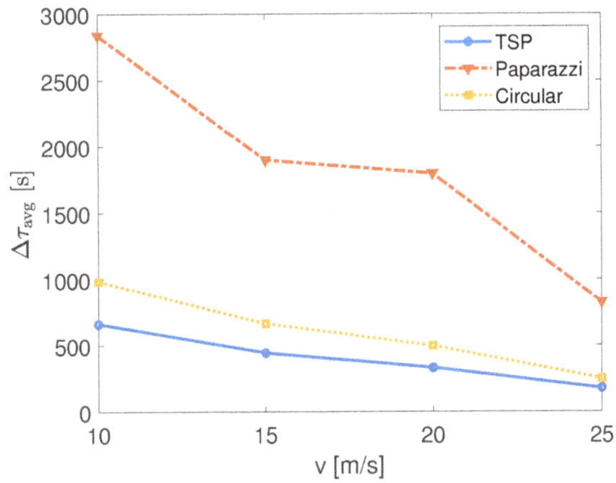

Figure 7. Perceived latency of IoT nodes for different UAB speeds and trajectories with $h = 200$ m.

Furthermore, we are interested in the network throughput, T_{net}, that can be achieved. Numerical results are presented in Figure 8. As formulated in Equation (11), these results will also depend on the ones of Figures 5 and 7 because of the dependence on service completion and latency in Equation (6). Interestingly, we can see three different trends with varying UAB speeds, v, for the three different trajectories: (i) The TSP has a maximum, (ii) the circular does not change significantly, while (iii) the Paparazzi shows an improvement with increasing velocities. The maximum shown by the network throughput in the TSP trajectory corresponds exactly to the service-latency trade-off with UAB speed mentioned before. However, because of the low total service offered, the circular trajectory is hardly showing any maxima. Similarly, the maximum cannot be appreciated for the Paparazzi path, but the network throughput appears to be increasing with increasing speed. In fact, we observed that, at lower UAB speeds as 10 to 15 m/s, the average latency is so large (almost one or a half hour) that the maximum cannot occur for the chosen UAB speed range. As expected, the TSP trajectory shows a relevantly higher throughput with respect to the other two, up to 16 kb/s. Moreover, because of a much larger average latency, the Paparazzi trajectory achieves an even lower throughput than the circular one (that had a much worse percentage of served nodes).

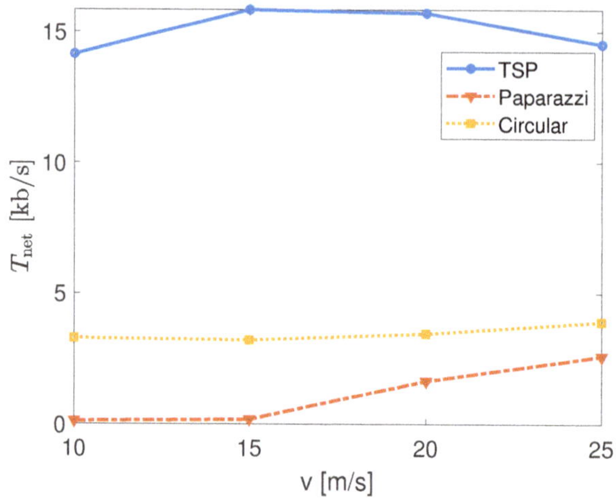

Figure 8. Network throughput for different UAB speeds and trajectories with $h = 200$ m.

As last performance metric, we study the impact of trajectories on the average energy consumption, E, in Figure 9. For decreasing values of the UAB speed, v, the energy consumed by NB-IoT nodes tend to decrease until few hundreds of mJ. Even if this appears to be a positive effect, we should consider the motivations behind this behaviour.

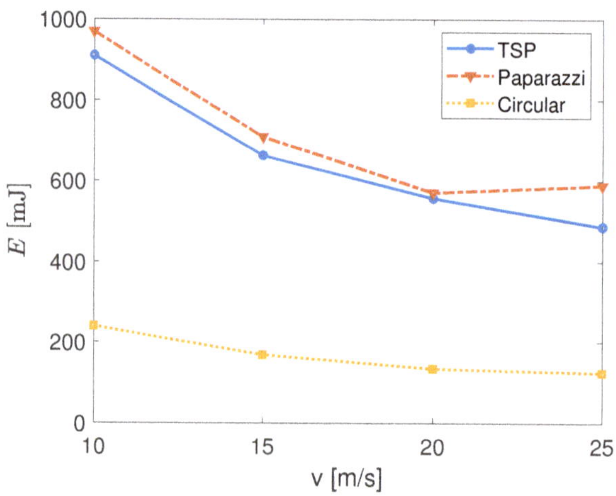

Figure 9. Energy consumption of IoT nodes for different UAB speeds and trajectories with $h = 200$ m.

To avoid cluttering, we now show the impact of two different UAB heights, h, for the same trajectory. The TSP path is the trajectory chosen for its better results in terms of almost all metrics.

In this case, the percentage of served NB-IoT nodes with varying UAB speed, v, is plotted in Figure 10. The behaviour of the two curves with respect to speed is the same as already discussed in Figure 5. However, we can see a slight difference from before, that is a lesser sharp decrease of successful transmissions when the speed is 25 m/s. We observe that for lower speeds the system performs better at lower UAB heights, h, while for higher speeds increasing heights improve the service. One might expect that, since for higher altitudes the transmitter-receiver distance increases on average, the curve with the larger UAB height would perform always worse than the other (as it happens for values of v lower than or equal to 20 m/s). However, this does not consider the NB-IoT signalling messages procedure and timing. In fact, for increasing values of h, not only the transmitter-receiver distance gets higher, but also the coverage range of the UAB increases (given by the angle of incidence of the UAB with the ground, or elevation angle in [7,8]). Consequently, it increases the time interval during which, on average, the NB-IoT node remains in the coverage range of the UAB. As shown in Figure 10, this effect can be more appreciated with increasing values of speed.

What has been discussed above is confirmed in Figure 11, showing the average energy consumption, E, with varying UAB speeds and heights. Being an average, its value depends on the number of nodes able to reach the UAB connectivity and enter the RA (i.e., signalling) procedure. The curves' decreasing trend with speed is the same for the two h values, and was already discussed for the TSP in Figure 9. A higher altitude lets a larger number of nodes being in connectivity with the UAB, and try and start the RA procedure. This fact increases the overall energy consumption, regardless of successful transmission or not. With lower speeds, more nodes would be able to complete the signalling procedure, whether with higher speeds the energy consumed could be only for the transmission of NB-IoT Msg1 and/or Msg2.

Our analysis on the energy consumption can be validated by Figure 12. It represents the access rate, R_{acc}, for different UAB heights and speeds in the TSP trajectory. Here, the curves trend with respect the to speed, v, is the same, as for Figure 11, and the curve with the higher height, h, has a lower access rate. This means that we have a larger number of attempts that does not correspond to the same number of successful transmissions. In fact, also due to the average larger transmitter-receiver distances with a higher-height UAB, it gets more difficult for nodes to complete the RA procedure.

Related to average latency, $\Delta\tau_{avg}$, we observe a quite less significant impact. In Figure 13, the value of $\Delta\tau_{avg}$ is affected more by increasing UAB speeds rather than increasing UAB heights.

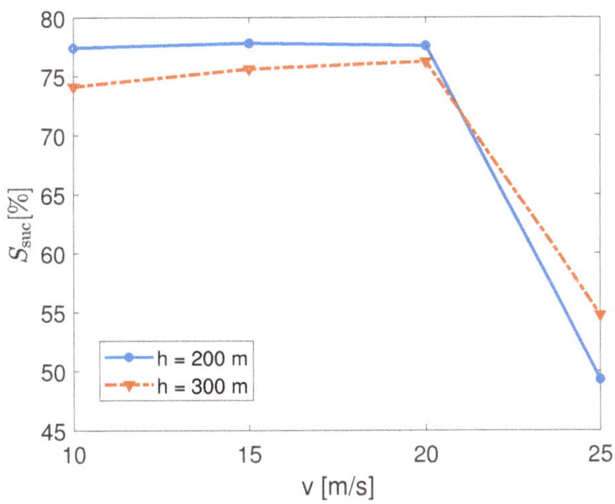

Figure 10. Percentage of served IoT nodes for different UAB speeds and heights in the TSP path.

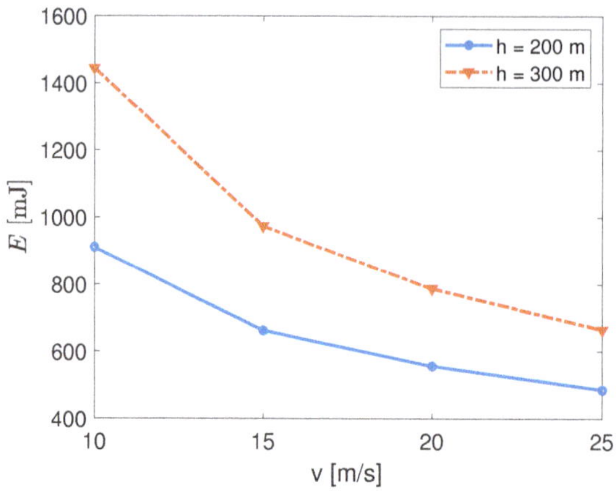

Figure 11. Energy consumption of IoT nodes for different UAB speeds and heights in the TSP path.

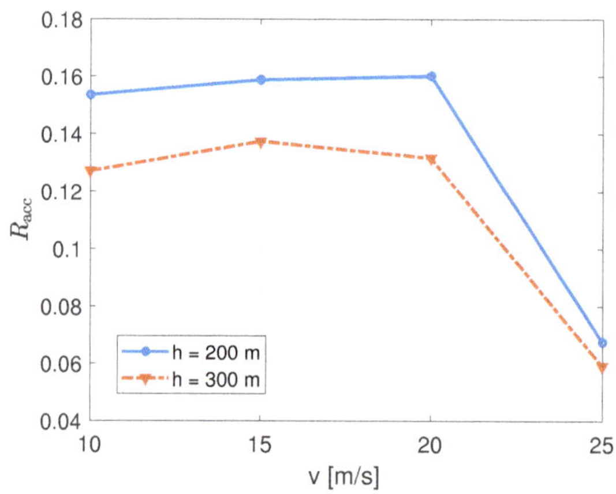

Figure 12. Access rate for different UAB speeds and heights in the TSP path.

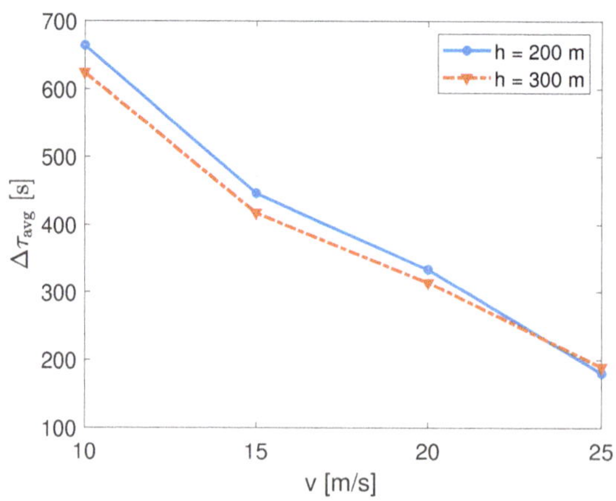

Figure 13. Perceived latency of IoT nodes for different UAB speeds and heights in the TSP path.

To summarize our findings into three main points, which reflect the previously mentioned key contributions, we can state that:

- Simulations performed have accounted for all the described NB-IoT standard parameters, from the random access procedure to the precise scheduling in uplink transmission with the three NB-IoT coverage classes;
- we jointly investigated a number of performance metrics related to the IoT field, namely (i) the number of completely served nodes, (ii) the achieved network throughput, (iii) the perceived latency and (iv) energy consumption of nodes. Numerical values achieved are related to the scenario under consideration and may scale accordingly. Different IoT applications may prefer one metric over the others as its main requirements, but our results show that pursuing one usually improves also the other metrics;
- as evinced from the analyzed results, the TSP trajectory performs better than the others in terms of successful transmissions, latency and throughput, making it the most promising among the three for cellular IoT (e.g., NB-IoT) applications. As expected, the prevailing trajectory in terms of performance is scenario-dependent. Moreover, in our results, one can extract the behaviour of UAV networks with respect to speed and height; in particular, the UAV speed should be taken under control, since it might abruptly decrease the performance metric if kept too high.

Thanks to our number of outcomes and simulations (and the variety of real scenarios that could apply to our system model), we can discuss results further and broaden our findings to different cases.

In the numerical evaluation, we first identify which are the relevant factors an operator must take into account when deploying this kind of system. First, the speed undergoes a trade-off between average perceived latency and throughput from one side to total number of successfully served nodes on the other. In this sense, the NB-IoT protocol plays a relevant role, since too higher speeds do not allow the completion of the RA procedure, and therefore the scheduling of uplink resources in the NPUSCH. Moreover, though a larger altitude would grant an increased NB-IoT nodes connectivity, increases the transmitter-receiver distance, which has again a negative impact on the successful completion of the RA procedure.

From our results, we can also infer conclusions on the trajectory selection. We can state that the circular trajectory, as would be for any path that travels the perimeter of a convex figure, is neither an effective trajectory in terms of latency nor a robust choice for the service of a massive number of IoT nodes. In fact, it is not able to follow the particular deployment of nodes for a given service area. On the other hand, a Paparazzi trajectory seems a good alternative, since it ensures to scan the overall service area. This might be a fair solution if a mobile operator does not know a priori the location of nodes. However, because of the fact that this information is usually easy to retrieve and that the Paparazzi trajectory becomes expensive in terms of energy consumption and latency, is probably not desirable. This confirms the expectations on the robustness of the TSP path.

If we want to achieve the 100% successfully served nodes, the use of multiple UABs might be necessary. However, this can apply only for those trajectories reaching full nodes connectivity, as for the case of TSP and Paparazzi. In fact, an n-th UAB, following the path of its predecessors, will address only those nodes not previously served. This fact would lighten the access procedure, since there will be less nodes contending and more radio resources available in a smaller time interval. In fact, a similar behaviour with respect to speed in Figures 5 and 6 suggests that the number of access attempts may be a tighter bottleneck than the time needed for signalling. Multiple UABs may lower the traffic load allowing a larger number of nodes to get access to the channel.

6. Conclusions

In summary, we proposed a thorough network performance evaluation of an UAB-aided NB-IoT network through detailed simulations. We considered the different aspects of the NB-IoT protocol, including the signalling granting resources for uplink transmission and the NB-IoT coverage classes. Then, we jointly evaluated the system performance on the service offered, access rate, average latency, network throughput and energy consumption metrics. UAB speed and height reveal a noticeable impact on the final performance, requiring a performance trade-off on the different metrics. Finally, we also observed the implications of different trajectory selection. A trajectory given by the TSP solution is the most suitable for clustered environments. The presented approach applies to IoT applications not constrained in time and does not consider the UAV battery expiration before the end of the trajectory. Indeed, this would modify all the trajectories to include the time dimension meeting the application requirements and an energy threshold before which the UAV needs to recharge its batteries. This study is left for further works, together with the introduction of multiple UAB service.

Author Contributions: Conceptualization, C.B. and S.M.; methodology, C.B. and R.M.; software, R.M. and S.M.; validation, S.M., R.M. and C.B.; investigation, S.M.; data curation, S.M.; writing—original draft preparation, S.M. and R.M.; visualization and supervision, C.B. and R.V.; funding acquisition, R.V. All authors have read and agreed to the published version of the manuscript.

Funding: This research received no external funding.

Institutional Review Board Statement: Not applicable.

Informed Consent Statement: Not applicable.

Conflicts of Interest: The authors declare no conflict of interest.

References

1. Mignardi, S.; Mikhaylov, K.; Cacchiani, V.; Verdone, R.; Buratti, C. Unmanned Aerial Base Stations for NB-IoT: Trajectory Design and Performance Analysis. In Proceedings of the 2020 IEEE 31st Annual International Symposium on Personal, Indoor and Mobile Radio Communications, London, UK, 31 August–3 September 2020; pp. 1–6.
2. 3GPP TS 22.289 V17.1.0. Enhancement for Unmanned Aerial Vehicles. September 2019.
3. 3GPP TS 22.125 V17.1.0. Unmanned Aerial System (UAS) Support in 3GPP. December 2019.
4. 3GPP TS 36.300 V15.4.0. Evolved Universal Terrestrial Radio Access (E-UTRA) and Evolved Universal Terrestrial Radio Access Network (E-UTRAN); Overall Description. December 2018.
5. Liberg, O.; Sundberg, M.; Wang, E.; Bergman, J.; Sachs, J. *Cellular Internet of Things: Technologies, Standards and Performance*; Academic Press: Cambridge, MA, USA, 2017.
6. Xia, N.; Chen, H.H.; Yang, C.S. Emerging technologies for machine-type communication networks. *IEEE Netw.* **2019**, *34*, 214–222. [CrossRef]
7. Al-Hourani, A.; Kandeepan, S.; Jamalipour, A. Modeling air-to-ground path loss for low altitude platforms in urban environments. In Proceedings of the 2014 IEEE Global Communications Conference, Austin, TX, USA, 8–12 December 2014.
8. Al-Hourani, A.; Kandeepan, S.; Lardner, S. Optimal LAP Altitude for Maximum Coverage. *IEEE Wirel. Commun. Lett.* **2014**, *3*, 569–572. [CrossRef]
9. Yanmaz, E. Connectivity versus area coverage in unmanned aerial vehicle networks. In Proceedings of the 2012 IEEE International Conference on Communications (ICC), Ottawa, ON, Canada, 10–15 June 2012; pp. 719–723. [CrossRef]
10. Sharma, V.; Bennis, M.; Kumar, R. UAV-Assisted Heterogeneous Networks for Capacity Enhancement. *IEEE Commun. Lett.* **2016**, *20*, 1207–1210. [CrossRef]
11. Lyu, J.; Zeng, Y.; Zhang, R.; Lim, T.J. Placement optimization of UAV-mounted mobile base stations. *IEEE Commun. Lett.* **2016**, *21*, 604–607. [CrossRef]
12. Mozaffari, M.; Saad, W.; Bennis, M.; Debbah, M. Mobile Unmanned Aerial Vehicles (UAVs) for Energy-Efficient Internet of Things Communications. *IEEE Trans. Wirel. Commun.* **2017**, *16*, 7574–7589. [CrossRef]
13. Wu, Q.; Zeng, Y.; Zhang, R. Joint trajectory and communication design for multi-UAV enabled wireless networks. *IEEE Trans. Wirel. Commun.* **2018**, *17*, 2109–2121. [CrossRef]
14. Wu, Q.; Zhang, R. Common Throughput Maximization in UAV-Enabled OFDMA Systems with Heterogeneous Delay Requirements. *IEEE Trans. Commun.* **2018**, *66*, 6614–6627. [CrossRef]
15. Zeng, Y.; Zhang, R. Energy-efficient UAV communication with trajectory optimization. *IEEE Trans. Wirel. Commun.* **2017**, *16*, 3747–3760. [CrossRef]
16. Lu, J.; Wan, S.; Chen, X.; Chen, Z.; Fan, P.; Letaief, K.B. Beyond Empirical Models: Pattern Formation Driven Placement of UAV Base Stations. *IEEE Trans. Wirel. Commun.* **2018**, *17*, 3641–3655. [CrossRef]
17. Samir, M.; Sharafeddine, S.; Assi, C.M.; Nguyen, T.M.; Ghrayeb, A. UAV Trajectory Planning for Data Collection from Time-Constrained IoT Devices. *IEEE Trans. Wirel. Commun.* **2020**, *19*, 34–46. [CrossRef]
18. Liu, Y.; Liu, K.; Han, J.; Zhu, L.; Xiao, Z.; Xia, X.G. Resource Allocation and 3-D Placement for UAV-Enabled Energy-Efficient IoT Communications. *IEEE Internet Things J.* **2021**, *8*, 1322–1333. [CrossRef]

19. Li, B.; Guo, X.; Zhang, R.; Du, X.; Guizani, M. Performance Analysis and Optimization for the MAC Protocol in UAV-Based IoT Network. *IEEE Trans. Veh. Technol.* **2020**, *69*, 8925–8937. [CrossRef]
20. Motlagh, N.H.; Bagaa, M.; Taleb, T. Energy and Delay Aware Task Assignment Mechanism for UAV-Based IoT Platform. *IEEE Internet Things J.* **2019**, *6*, 6523–6536. [CrossRef]
21. Song, K.; Zhang, J.; Ji, Z.; Jiang, J.; Li, C. Energy-Efficiency for IoT System With Cache-Enabled Fixed-Wing UAV Relay. *IEEE Access* **2020**, *8*, 117503–117512. [CrossRef]
22. Kavuri, S.; Moltchanov, D.; Ometov, A.; Andreev, S.; Koucheryavy, Y. Performance Analysis of Onshore NB-IoT for Container Tracking During Near-the-Shore Vessel Navigation. *IEEE Internet Things J.* **2020**, *7*, 2928–2943. [CrossRef]
23. Büyükyavuz, M.C.; Demirtaş, M.; Kargı, E.; Arslan, C. Narrow-Band IOT NPRACH Channel Coverage Analysis for Unmanned Air Vehicle Base Stations. In Proceedings of the 2020 28th Signal Processing and Communications Applications Conference (SIU), Gaziantep, Turkey, 5–7 October 2020; pp. 1–4. [CrossRef]
24. Zayas, A.D.; Merino, P. The 3GPP NB-IoT system architecture for the Internet of Things. In Proceedings of the 2017 IEEE International Conference on Communications Workshops (ICC Workshops), Paris, France, 21–25 May 2017; pp. 277–282.
25. Ye, H.T.; Kang, X.; Joung, J.; Liang, Y.C. Joint uplink and downlink 3D optimization of an UAV swarm for wireless-powered NB-IoT. In Proceedings of the 2019 IEEE Global Communications Conference (GLOBECOM), Waikoloa, HI, USA, 9–13 December 2019; pp. 1–6.
26. Castellanos, G.; Deruyck, M.; Martens, L.; Joseph, W. System Assessment of WUSN Using NB-IoT UAV-Aided Networks in Potato Crops. *IEEE Access* **2020**, *8*, 56823–56836. [CrossRef]
27. Feltrin, L.; Tsoukaneri, G.; Condoluci, M.; Buratti, C.; Mahmoodi, T.; Dohler, M.; Verdone, R. Narrowband IoT: A Survey on Downlink and Uplink Perspectives. *IEEE Wirel. Commun.* **2019**, *26*, 78–86. [CrossRef]
28. Haenggi, M. *Stochastic Geometry for Wireless Networksk*; Cambridge University Press: Cambridge, UK, 2012.
29. Saha, C.; Afshang, M.; Dhillon, H.S. 3GPP-Inspired HetNet Model Using Poisson Cluster Process: Sum-Product Functionals and Downlink Coverage. *IEEE Trans. Commun.* **2018**, *66*, 2219–2234. [CrossRef]
30. Hajj, J.; Mazur,M.; Wiśniewski, A.; Khoury, R. Telecom Operators in the Age of Drones: Preparing for the New Era. Strategy & Technical Report. 2017. Available online: https://www.strategyand.pwc.com/media/file/Telecom-operators-in-the-age-of-drones.pdf (accessed on 7 September 2021).
31. Dormehl, L. 7 Drones That Can Stay Airborne for Hours—And the Tech That Makes It Possible. Available online: https://www.digitaltrends.com/cool-tech/drones-with-super-long-flight-times/ (accessed on 7 September 2021).
32. Drone Laws: Regulations by State and Country. Available online: https://uavcoach.com/drone-laws/ (accessed on 7 September 2021).
33. Mozaffari, M.; Saad, W.; Bennis, M.; Debbah, M. Drone Small Cells in the Clouds: Design, Deployment and Performance Analysis. In Proceedings of the 2015 IEEE Global Communications Conference (GLOBECOM), San Diego, CA, USA, 6–10 December 2015; pp. 1–6. [CrossRef]
34. u-blox. SARA-N2 Series and Chipsets,Datasheet. Available online: https://www.u-blox.com/sites/default/files/SARA-N2_DataSheet_%28UBX-15025564%29.pdf (accessed on 7 September 2021).
35. Dantzig, G.; Fulkerson, R.; Johnson, S. Solution of a large-scale traveling-salesman problem. *J. Oper. Res. Soc. Am.* **1954**, *2*, 393–410. [CrossRef]

drones

MDPI

Article

Communication Aware UAV Swarm Surveillance Based on Hierarchical Architecture

Chengtao Xu, Kai Zhang, Yushan Jiang, Shuteng Niu, Thomas Yang and Houbing Song *

Department of Electrical Engineering and Computer Science, Embry-Riddle Aeronautical University, Daytona Beach, FL 32114, USA; xuc3@my.erau.edu (C.X.); zhangk3@my.erau.edu (K.Z.); jiangy2@my.erau.edu (Y.J.); shutengn@my.erau.edu (S.N.); yang482@erau.edu (T.Y.)
* Correspondence: songh4@erau.edu

Abstract: Multi-agent unmanned aerial vehicle (UAV) teaming becomes an essential part in science mission, modern warfare surveillance, and disaster rescuing. This paper proposes a decentralized UAV swarm persistent monitoring strategy in realizing continuous sensing coverage and network service. A two-layer (high altitude and low altitude) UAV teaming hierarchical structure is adopted in realizing the accurate object tracking in the area of interest (AOI). By introducing the UAV communication channel model in its path planning, both centralized and decentralized control schemes would be evaluated in the waypoint tracking simulation. The UAV swarm network service and object tracking are measured by metrics of communication link quality and waypoints tracking accuracy. UAV swarm network connectivity are evaluated over different aspects, such as stability and volatility. The comparison of proposed algorithms is presented with simulations. The result shows that the decentralized scheme outperforms the centralized scheme in the mission of persistent surveillance, especially on maintaining the stability of inner UAV swarm network while tracking moving objects.

Keywords: persistent surveillance; hierarchical architecture of UAV teaming; communication aware UAV formation; dynamical object tracking; intercommunication quality

Citation: Xu, C.; Zhang, K.; Jiang, Y.; Niu, S.; Yang, T.; Song, H. Communication Aware UAV Swarm Surveillance Based on Hierarchical Architecture. *Drones* **2021**, *5*, 33. https://doi.org/10.3390/drones5020033

Academic Editors: Diego González-Aguilera and Pablo Rodríguez-Gonzálvez

Received: 22 March 2021
Accepted: 27 April 2021
Published: 30 April 2021

1. Introduction

Emerging new technologies on unmanned aerial vehicles (UAVs), such as 24/7 persistent surveillance [1], high-resolution sensing [2], autonomous UAV powered by machine learning [3,4], and secure UAV communication [5–8], have led to the changes of industries, academia, and government in investigating the strategies of rescuing, monitoring, and emergency communication construction. Persistent surveillance of UAVs gives the possibility of realizing continuous sensing coverage in an area of interest (AOI). High-resolution sensing on a different class of UAVs provides various levels of detail in the surveillance field. However, due to the multiple purposes of continuous monitoring, it is challenging for a single UAV agent to complete sensing tasks in a large-scale AOI. Therefore, the cooperative strategy using multiple UAVs is proposed to enhance the efficiency of continuous AOI surveillance. Meanwhile, in emergency scenarios such as wildfires and earthquakes, a minor delay of information sharing between multiple UAVs indicates more redundancy for agencies to process real-time tasks.

Swarm refers to a concept in which a group of coordinated, intelligent systems or agents are designed to achieve the same goal in a complex environment. It is not a simple accumulation of agents but a more redundant and efficient framework to achieve a high quality of service. UAV swarm has become a prevalent field of research in realizing persistent surveillance. Unlike research in the coordination between swarm agents in group path planning and barrier avoidance, AOI's continuous coverage with the UAV swarm faces various challenges in changing environments. For example, (a) the efficient surveillance task division within sensing group [9];

(b) optimization of the consuming energy of UAV [10–12]; (c) adjustment of surveillance plan based on the occasional natural disaster which has high priority [13].

To realize persistent surveillance with coordinated UAVs, three strategies have been considered in recent years: (i) on-duty UAVs replacement scheme with different recharging station distributions [14]; (ii) energy-efficiency path planning [15–18]; (iii) novel structures of UAV teaming [19]. UAV distribution's hierarchical architecture helps plan the efficient and adaptive surveillance while the mission is affected by the changing monitoring map which is caused by the weather condition or unseen factors. UAVs teaming's hierarchical structure intends to divide the surveillance task into different platforms such as ground station, high altitude UAV, and UAV sensing swarm. Each platform is in charge of different functions such as control, motion coordination, data transferring, and data package routing.

Figure 1 describes the hierarchical structure of the UAV surveillance team. Each platform marks its capabilities based on its difference in computation resources, communication system characteristics, and energy supply. High altitude UAVs equipped with a high-resolution camera are usually applied in the field sensing and the data exchange with ground stations since it has a relatively stable power supply compared to low altitude UAVs. However, due to its high altitude operation zone, the ground device with low transmission power makes it difficult to access its network service. To realize the economical and efficient operation of rescue missions and provide network service, a buffer layer of UAV swarm between high altitude UAV and ground objects extends this system's ability on connection service and real-time tracking. With many agents within the UAV swarm layer, an adaptive formation policy could be developed to fit in various requests. In [20], a distributed UAV surveillance scheme is applied in monitoring an AOI in terms of the event location. The covered area of each agent changes with the location of events. A ground station with a stable power supply is used as the terminal of data exchange. It obtains sensing and network service data from UAVs swarm or high altitude UAV in a specific design. A distance that exceeds the UAV swarm's communication range is considered impractical with a lower transmission power of low altitude UAV.

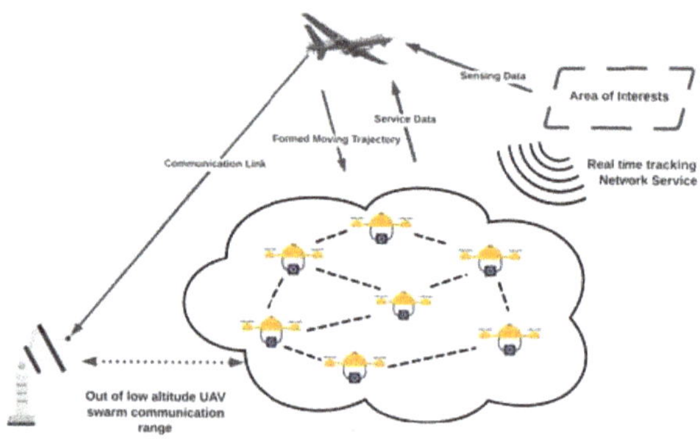

Figure 1. Hierarchical structure of UAV swarm based sensing and monitoring.

In [21], Li discussed a distributed coordination scheme of UAV swarm based on flock forming framework proposed by Reynold [22] in 1987, who described the basic flocking model consisting of three simple steering behaviors, separation, alignment, and cohesion. To realize the flock of UAV swarm, centralized control policies have been proposed [23–25] to realize implicit leader-follower teaming of UAVs. In [26], the authors consider a team distribution of leader-follower formation with a centralized control policy, in which the leader UAV collects the relative position of other following agents. The simulation results show its effectiveness on block avoidance with continuous surveillance function. Feedback formation control scheme provides another avenue in trajectory formation with consensus velocity and thrust [27–29].

In persistent surveillance provided by the proposed UAV teaming structure, long-term monitoring and emergency communication and survivor tracking are the two application scenarios we considered in reality. Long-term monitoring requires the cooperation between low-altitude UAV swarm and high altitude platforms. To maximize the coverage area, the attraction values of the map cells are usually updated randomly [30–32]. The swarm searches on the existing map to find the shortest route to the target points. Meanwhile, the general attraction rate of each map cell changes regarding the swarm flying trajectory. Low altitude swarm extends the data-gathering capability of high-altitude UAVs by following a specific flying trajectory. The flying trajectories usually do not have priorities compared to the surveillance data quality under conditions.

However, the spreading patterns of natural disasters such as wildfires, earthquakes, and mudslides should be considered in the path planning. For example, in a wildfire, the percentage of forest coverage in the AOI influences the fire spreading rate between neighboring trees, which then affects the spreading time and direction [33]. A residential area in the direction of fire spreading needs longer staying time of UAV swarm to provide the network service, especially considering this in the phase of general path planning. The fire transition rate is the parameter dependent on the wind speed and forest coverage rate, assuming the residential area is under the direction of wind. Therefore, the assigned flying trajectory from high altitude UAV should follow the direction of wind. In our communication aware UAV swarm path planning, we would investigate the link quality variance with such a change of the mission's path planning.

We propose a framework (Figure 2) based on the hierarchical UAVs teaming architecture to perform a navigation task via the trajectory design according to changing map and decentralized control of low altitude sensing UAV swarm. It achieves navigation and trajectory functions by changing maps and decentralized control scheme of low altitude UAV swarm. Each UAV agent updates the state of communication link connecting the neighboring nodes. The consensus of agents' kinetic variable is based on its communication link quality, which forms a decentralized control scheme not relying on the leading device's coordination. With the remote sensing data on changing maps, the high altitude UAV generates an optimized path based on the low altitude swarm state. The swarm agents track each assigned waypoint by correcting its geometry center point. Real-time object tracking and network service data would be collected and transmitted to the terminal ground station by platforms with a greater power rate. The distributed control scheme of low altitude UAV swarm suggests that no coordination delay between each waypoint tracking and the idle link between low and high altitude platforms could enhance the package delivery rate or throughput on two layers. The simulation results show the ability of fast convergence of the link quality between agents and disturbance tolerance while following the assigned waypoints.

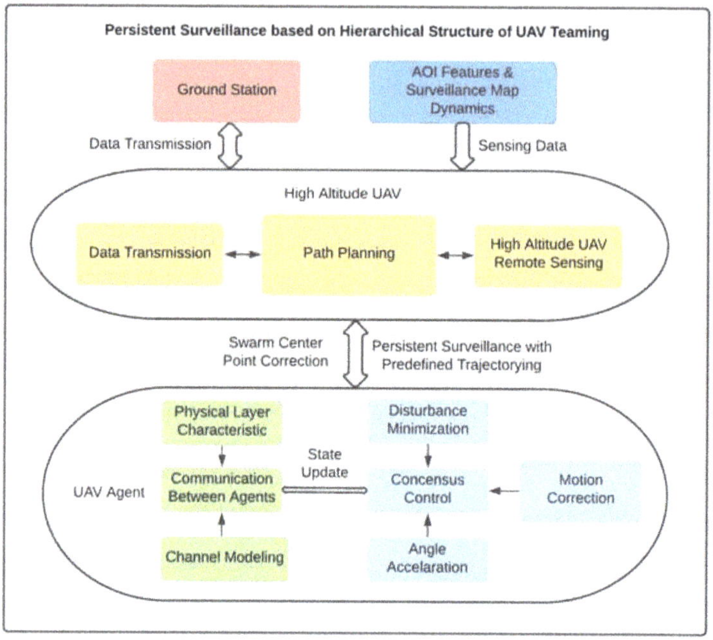

Figure 2. Function modules in each layer of platforms with hierarchical structure.

This paper studies the communication-aware formation of low altitude UAV swarm in applying persistent surveillance within hierarchical structure UAV teaming. Our schemes are proposed based on centralized and decentralized control policies considering the communication link quality between agents, and the waypoint tracking accuracy of the UAV swarm. The predefined trajectory for environment exploration would be discussed in response to different types of events. Exploring the desired area with different surveillance plans would result in various general real-time data transmission rates within the swarm network. The contributions of this work are summarized as follows:

(1) We compared the centralized and decentralized control scheme in studying the UAV swarm's waypoint tracking function while considering its inner communication link quality, which is a practical technical challenge needed to be solved to realize applicable UAV swarm-based persistent surveillance.
(2) We compared the performance of UAV swarm trajectory following algorithm based on the centralized and decentralized position updating strategy to achieve better waypoint tracking accuracy.
(3) We evaluated the robustness of swarm internal communication link, formation stability with the disturbance in chasing the wanted waypoint.

The rest of the paper focuses on the dynamics and communication of low altitude UAV swarm, and the contents are organized as follows. The channel state model of low altitude UAV agent and consensus control of low altitude UAV swarm are presented and discussed in detail in Section 2. Section 3 gives the simulation results of centralized and distributed control on waypoint tracking and disturbance tolerance of low latitude UAV swarm on following the trajectory with a changing sensing environment. Finally, we present the conclusions and present our future work in Section 4.

2. Problem Formulation

2.1. Communication Channel State for UAV Agent Formation Control

We assume that multiple inputs and multiple outputs (MIMO) antennas are adopted on low-altitude UAV agents to access the AOI ground users. As shown in Figure 3, the sensing and network service 'cloud' is formed by the UAV swarm nodes. Each sensing node exchanges the information with its neighboring nodes with one or multiple communication channels. This one or multiple communication channels forms the communication between agents. The quality of such a communication link could reveal the overall performance of swarm network service, which is also called quality of service (QoS). Our model neglects the cochannel interference on the signal receiver side, which is caused by the difference in transmission power.

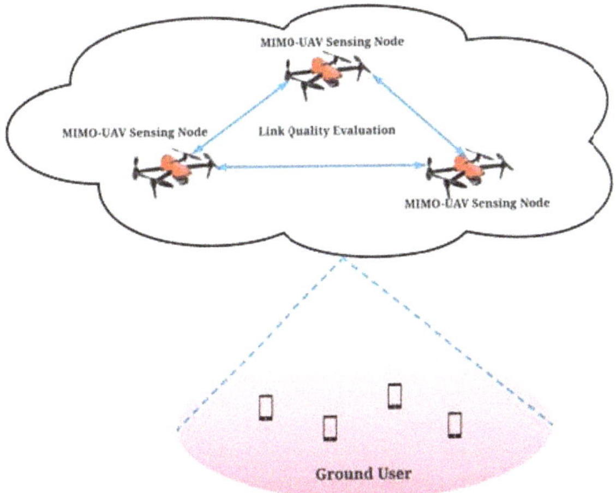

Figure 3. Low altitude UAV swarm communication and sensing network.

Moreover, we assume the adaptive filter on a single receiver's side could suppress the cochannel signal and environment noise. Therefore, to evaluate the communication channel quality between 2 UAV agents, we simplified the problem to purely measure the single input and single output (SISO) transceiver and receiver link. One UAV could maintain multiple SISO links with its neighboring UAV sensing nodes.

We denote the position of each UAV as q_i in an multiagent system with n agents, the dynamics of each UAV is given by $\dot{q}_i = u_i$, where $\dot{q}_i, u_i \in R^3$. It simplifies the single UAV dynamics, which considers the angle accelerations provided on yaw, roll, pitch. u_i denotes the control of the ith UAV based on the current channel state with its neighboring agent. r_{ij} indicates the distance between agent i and j for rigid formation [34], which represents the neighboring agents keep a prescribed desired distance with each other.

$$r_{ij} = \left\| q_i - q_j \right\|_2 \tag{1}$$

The reception probability of data package transmission is defined in Equation (2), a critical wireless channel parameter for a SISO communication link. It is used to evaluate the communication channel state between each neighboring UAV. As the SISO model shows in Figure 4, α denotes the antenna characteristics, which stays the same on the sides of transceiver and receiver. β is the required data rate fed into the digital signal processor (DSP). v is the path loss exponent related to the environment. r_0 is the distance of causing near field effects between neighboring antennas. The data package reception probability indicates the rate of successful communication between transceiver and receiver. Therefore, to evaluate the SISO communication channel quality between ith and jth UAV, firstly we have the reception probability denoted by

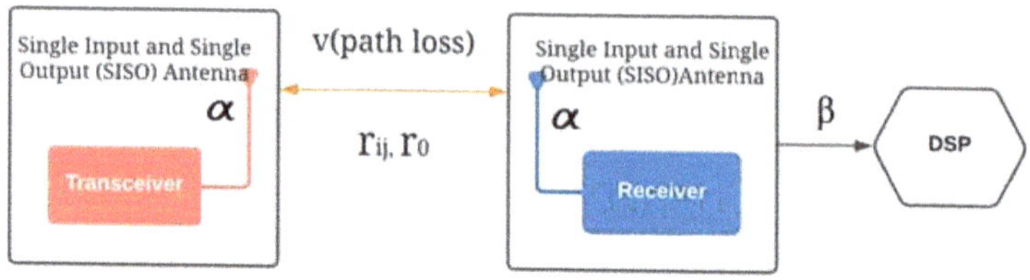

Figure 4. SISO communication channel model between UAV agents.

$$a_{ij} = exp(-\alpha(2^{\beta} - 1)(\frac{r_{ij}}{r_0})^{v}) \tag{2}$$

where a_{ij}, the package reception probability measuring the possibility that the receiver receiving sensing information accurately from transceiver. In the swarm's dynamics, it evaluates the probability of influence from transceiver to the receiver side.

In the domination of modeling SISO antenna, the far field is majorly considered with the path loss effect of the channel, which decreases the packages reception rate with the increase of radio propagation distance. However, in forming a UAV swarm, the near field effects happen more easily between agents. It refers to the severe mutual interference of SISO antenna that significantly degrades the communication performance of swarm communication network. A simplified approximation model used in describing the near fields effects is defined as:

$$g_{ij} = \frac{r_{ij}}{\sqrt{r_{ij}^2 + r_0^2}} \tag{3}$$

here, r_{ij} denotes the distance between ith and jth agent, r_0 refers the distance of causing the near field effects between agents. When $r_{ij} \rightarrow 0$, g_{ij} goes to 0, which characterizes the interference effect in the near field. When $r_{ij} \gg r_0$, g_{ij} is nearly 1, which implies the near field effects can be ignored in the antenna far field. Then, communication performance indicator between neighboring UAVs can be defined as:

$$\phi_{ij} = -\beta\Big(\big(vr_{ij}\big)^{v+2} + \big(\beta v r_{ij}^{v} + r_0^{v}\big) r_0^2\Big) \frac{e^{-\beta}}{r_0^{v}} g_{ij} \tag{4}$$

To define the communication aware formation controller, the potential function is used in evaluating the interaction between neighboring agents. The artificial potential function is defined here for measuring the difference between performance indicator and its reference:

$$\psi(r_{ij}) = \phi^{*} - \phi_{ij} \tag{5}$$

where the function ϕ^{*} is the reference value of the optimized value of the communication performance indicator, then, the gradient of ϕ_{ij} is computed as:

$$\nabla_{q_i}\phi(r_{ij}) = P(a_{ij}) \cdot e_{ij} \tag{6}$$

where $e_{ij} = (q_i - q_j)/r_{ij}$, and P is function of package acceptance rate for determine the neighboring nodes of agent i. Therefore, the gradient of velocity or control of UAV agent i can be written as:

$$u_i = \nabla_{q_i} V_i = \phi_{ij}\psi(r_{ij}) \tag{7}$$

The neighbor of the ith UAV is defined as those channels having a package acceptance rate larger than the threshold th. The receiver throws the transmitted package away when the channel quality lower than th since the potential data lose with bad channel quality. So the function $P(a_{ij})$ can be defined as

$$P(a_{ij}) = \begin{cases} 1, & a_{ij} \geq th \\ 0, & a_{ij} < th \end{cases} \tag{8}$$

In [35], the author designed the formation controller based on the variance of communication performance indicator, in which each agent follows the control input's gradient direction to reach the lower value of ϕ_{ij}.

2.2. Cooperative Control of UAV Swarm in Trajectory Following with Centralized Agent Speed Control

Unlike target point searching of a single UAV agent, the UAV swarm's group behavior is hard to control in adjusting the swarm center to reach the target point. Simultaneously, the UAV agent also has to maintain the communication link performance to satisfy the package reception rate condition on the receiver side. Here, we discuss centralized position updating methods for the control of UAV agent's speed.

2.2.1. Target Point Searching

For low altitude UAV swarm shown in Figure 5, assuming one target point within the AOI is assigned by the high altitude platform as the surveillance task in a given time. Maintaining the distance between each UAV by controlling communication link quality, the swarm center is expected to get closed to the wanted mission point. Therefore, it is necessary to calculate the difference between swarm center and target point position as one input of the swarm control function. To calculate the center of swarm C_t at time t, we have

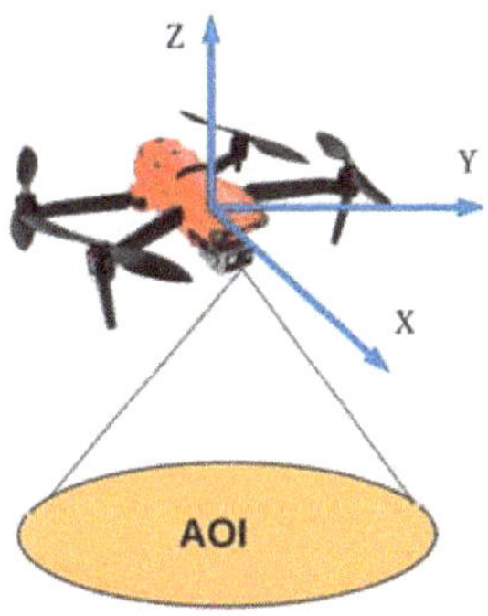

Figure 5. Low altitude UAV agent model.

$$C_t = \{x_t, y_t, z_t\} \tag{9}$$

$$x_t = \frac{\sum_{i=1}^{n} x_i}{n}, y_t = \frac{\sum_{i=1}^{n} y_i}{n}, z_t = \frac{\sum_{i=1}^{n} z_i}{n}, \tag{10}$$

where n is the swarm size, x_i is the position of ith low Altitude UAV on x axis, y_i is the position of ith low altitude UAV on y axis, z_i is the position of ith low altitude UAV on z axis. The target point position is denoted as

$$T = \{x_T, y_T, z_T\} \tag{11}$$

Distance between center of swarm at time t and target point can be denoted as:

$$r_{C_t} = ||C_t - T||_2. \tag{12}$$

In order to optimize the search efficiency, an artificial potential function is designed to evaluate the distance between target point and the center of swarm.

$$\eta(r_{C_t}) = -\delta(r_{C_t}), \forall(i,j) \in n \tag{13}$$

The target point searching controller makes the group move in the direction of minimizing the difference between swarm center and target point.

$$\arg\min_{r_{C_t}} \chi(r_{C_t}) = \left\{ r_{C_t} \mid \chi(r_{C_t}) = \min_{r'_{C_t}} \chi(r'_{C_t}) \right\} \tag{14}$$

The gradient of $\delta(r_{C_t})$ is computed as:

$$\nabla_{C_t}\delta(r_{C_t}) = \frac{C_t - T}{r_{C_t}} \tag{15}$$

Therefore, the target searching control function can be defined as:

$$u_{C_t} = \nabla_{C_t}\delta(r_{C_t}) \tag{16}$$

Therefore, the control function of agent i at moment t could be written as

$$u_t = u_{C_t} + u_i \tag{17}$$

2.2.2. Waypoint Tracking

We assume multiple waypoints are on the assigned surveillance mission by high altitude UAV. The UAV swarm is assigned a time to reach the waypoint in AOI. With disruptions from environment and swarm agent position adjustment caused by maintaining the communication link's stability, the assigned waypoint reaching time could not be enough for the UAV swarm to move toward the new waypoint. Therefore, we propose an alternate centralized waypoint tracking scheme in the path planning of the UAV swarm.

We consider two waypoints, k and l, that are assigned consecutive waypoints. The corresponding positions are T_k and T_l. The distance between center of swarm and node k and l can be denoted as:

$$\begin{aligned} r_{tk} &= \|C_t - T_k\|_2 \\ r_{tl} &= \|C_t - T_l\|_2. \end{aligned} \tag{18}$$

The artificial potential functions are designed to evaluate the distance between center point and two waypoints.

$$\begin{aligned} \kappa(r_{tk}) &= -\delta(r_{tk}) \\ \kappa(r_{tl}) &= -\delta(r_{tl}) \end{aligned} \tag{19}$$

To reach both two waypoints, we need

$$\arg\min_{r_{ik}} \chi(r_{tk}) = \left\{ r_{tk} \mid \chi(r_{tk}) = \min_{r'_{ik}} \chi(r'_{tk}) \right\} \tag{20}$$

$$\arg\min_{r_{il}} \chi(r_{tl}) = \left\{ r_{tl} \mid \chi(r_{tl}) = \min_{r'_{il}} \chi(r'_{tl}) \right\}. \tag{21}$$

Considering the controller functions between center point to both waypoints are nearly same, we have

$$\nabla_{C_t}\delta(r_{tl}) \approx \nabla_{C_t}\delta(r_{tk}) = m_{c_t}, \tag{22}$$

where the m_{Ct} is the average of two control inputs, it could be denoted as

$$m_{c_i} = \frac{2C_t - (T_k + T_l)}{r_{tk} + r_{tl}} u_{kl} = m_{c_i}. \tag{23}$$

The control of reaching the following waypoint is

$$u = u_{kl} + u_i. \tag{24}$$

The agent i's control includes the speed updating component from maintaining the neighboring communication link quality and thrust of reaching waypoints.

2.3. Cooperative Control of UAV Swarm in Trajectory Following with Decentralized Agent Speed Control

Different from target points searching, a trajectory is formed by a set of continuous waypoints, which is hard to find the average value of waypoints to adjust the swarm center to locate on the path. Such a center position updating of the swarm is recognized as a centralized control scheme in realizing the waypoint tracking function, which is less effective for the path planning of a large-scale UAV swarm. Therefore, we let the position updating be made by every agent rather than leveraging an identical control setting over agents based on the difference between the swarm center and the predefined waypoint.

The distance from ith agent to waypoint at time t is denoted as

$$r_i = \|q_i - f(t)\|_2, \tag{25}$$

in which $r_i, f(t) \in R^3$ represents the distance from agent i to the waypoint of trajectory at time t and trajectory formed by high altitude UAV in terms of surveillance needs.

The artificial potential function under this distributed formation control is

$$\varphi(r_i) = -\varsigma(r_i)\varsigma(r_i) = q_i - f(t). \tag{26}$$

where $f(t)$ is used as a reference value for agent ith position.

The control function with agent based position difference could be written as:

$$S_i = \frac{\varphi(r_i)}{r_i}. \tag{27}$$

Therefore, we have the control on the agent i by following the waypoint on a trajectory:

$$u = S_i + u_i \tag{28}$$

3. Simulation Results and Analysis

This section simulates the model based on the proposed target point searching and waypoints tracking algorithms in MATLAB from a centralized and decentralized control scheme. Pseudo codes are presented to explain the simulation results. Evaluation metrics such as communication indicator value, the distance between agents, and distance from swarm center to the wanted position are defined and used on evaluating the performance of UAV swarm in the mission of persistent surveillance. In the simulation, we assume the weather condition is ideal for UAV operation, which means the wind flow does not affect the UAV dynamics. There are no projectiles, such as birds or barriers, in the environment, which could block the flying path of UAV. For UAV swarm's actual application in persistent surveillance, battery capacity is essential to accomplish the mission. Increasing the UAV agent's control gain will increase the device's power consumption, which lowers the efficiency of the UAV operation. Here, in the simulation, each UAV agent's battery capacity is large enough to satisfy the flight operation time.

To realize the low latency communication within the UAV swarm network, especially in providing emergency network service, a stable communication link between agents helps avoid the data package transmission congestion. Under the requirement of high information exchange rate in swarm networks, a sudden lousy communication quality with existing routing algorithms [36,37] could cause the uncontrollable package flooding phenomenon. Therefore, we consider an accumulated index in evaluating the UAV swarm communication quality

$$\zeta_i(t) = \left\{ t \mid \sum_{j=1}^{n} \phi_{ij}, i \in n, j \in n \right\}. \tag{29}$$

in which, ζ is calculated as the sum of communication indicator in terms of all the channel links between ith UAV and its neighboring agents. A sudden bad channel state between ith UAV and one near UAV agent could cause the increment of indicator value. Therefore, the sum of the communication indicator's value could reveal QoS fluctuation with UAV swarm dynamics.

Of the near-field effects introduced in our SISO model for evaluating the communication link quality, a closer distance between UAV agents would give a more significant value of the communication indicator $\phi_{i,j}$, which suggests the significant portion of interference in the data exchanging between UAVs. In our model, the control portion of maintaining a stable communication link suppresses this interference. Therefore, the drone within the swarm could maintain the distance to prevent them colliding with each other. From the experiment results, the swarm can also maintain a stable distribution in the 3D space. Although it does not organize the team with a leader and follower format within the swarm surveillance group, to stable geometry distribution of swarm agent, we choose the distance from agent 1 to other UAVs to measure the movement's dynamical stability of the swarm. It is also observed from the experiment results that the control of maintaining the appropriate distance between UAV agents has higher priorities than reaching the wanted surveillance location.

$$r_{1j} = \|q_1 - q_j\| \quad j \in s. \tag{30}$$

The difference between swarm geometry center and waypoint expresses our method's tracking accuracy, in which this metric has the same expressing format as Formula (12).

With limited assigned waypoint transition time by higher altitude platforms, a trade-off between tracking and maintaining communication link stability is inevitable. However, it could be regulated by the value of the package acceptance rate threshold.

Therefore, three evaluation parameters are proposed to evaluate the performance of UAV swarm waypoint tracking algorithm. r_{C_t} measures the accuracy of the swarm waypoint tracking function. Since agent one usually operates in the geometry center of 7 UAV swarm, r_{1j} is adopted to reveal the swarm's stability while approaching the wanted waypoint. $\zeta_i(t)$, the sum of the communication indicator value through the seven UAV agents, is used on revealing the fluctuation of communication link while the tracking waypoint of the UAV swarm is changing.

The following simulation will use several UAV agents to experiment with proposed algorithms. Seven UAV agents are specially picked as the benchmark of swarm agent quantity setting. Compared to the number of agents lower than seven, it has relatively suitable complexity in describing UAV swarm dynamics. For example, especially on maintaining communication link performance, four UAV agents setting in the experiment cannot present the problem of centralized formation control with the increasing of swarm agent quantity. It is also not suitable for deploying UAV swarm in the ongoing surveillance mission required to provide high-speed network accessing service. In the end, the simulation of a decentralized control scheme would focus on the scalability aspects of UAV swarm in waypoint tracking.

3.1. Target Point Searching

The agents' speeds are considered to be adjusted with two components: maintaining communication link quality and gradually approaching the assigned target point. The process is expressed in Algorithm 1.

Algorithm 1: Target Point Searching

```
Result: Maintaining communication link while approaching the target point
initialization:
T = (10,10,10)% Assigning Target Points
q_i = random value < 100
speed_i = 0
while t < assigned surveillance time do
    Sum of agent position = Σ_{i=1}^{m} q_i
    C_t = sum of agent position / swarm size
    Difference between center and target point D_t = C_t − T
    while i < swarm size do
        while j < swarm size (j! = i) do
            r_ij = norm (q_i − q_j)
            a_ij = exp(−α(2^β − 1)(r_ij / r_0)^v)
            if a_ij >= P then
                φ_ij = −β((v r_ij)^(v+2) + (β v r_ij^v + r_0^v) r_0^2) (e^(−β) / r_0^v) g_ij
            else
                φ_ij = 0
            end
        end
        e_ij = (q_i − q_j) / r_ij
        ∇_{q_i} φ(r_ij) = φ(r_ij) · e_ij
        u_c = (C_t − T) / ||C_t − T||_2
        u_i = u_c + ∇_{q_i} φ(r_ij) % calculate the control of each agent
        speed_i = speed_i + u_i
        q_i = q_i + speed_i * step_size
    end
end
```

In each time step, the UAV's control would be adjusted based on the communication indicator ϕ_{ij}, which depends on the package acceptance rate threshold P. The lower the package acceptance rate P holds, the greater the control value to maintain the team format on symmetrical distribution. An initial randomized swarm agent position did not show the constraints of the proposed algorithm in the following simulation.

In Figure 6, each line of color represents a moving UAV trajectory, which gradually gets close to the target point. It starts with a randomized position in the space. In the early phase, due to the enormous control brought by the difference between the swarm center and the target point, each agent's velocity update concentrates on getting close to the target. While the group center is getting close to the wanted position, the trajectory with curve shows the phase of UAV revising its position to build up a better communication link. When both criteria become stable, the final result shows that all seven agents could be evenly distributed on the space with (10, 10, 10) as their center point. Meanwhile, as shown in Figure 7, agent 7′s (the centering agent) trajectory shows the adjustment of UAV swarm could be brought by the mismatch of swarm center and target point.

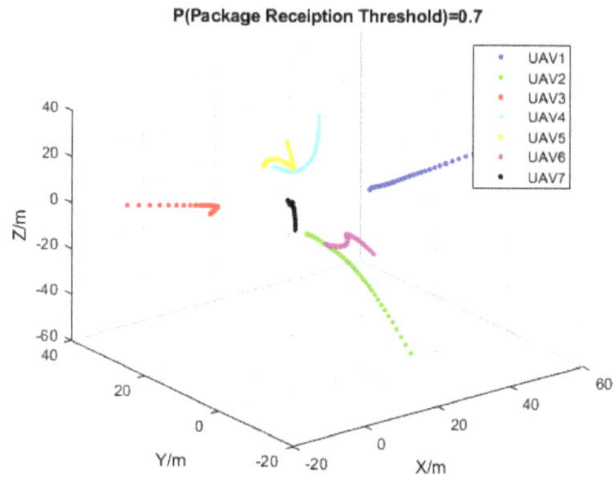

Figure 6. UAV swarm searching path to target point (10, 10, 10).

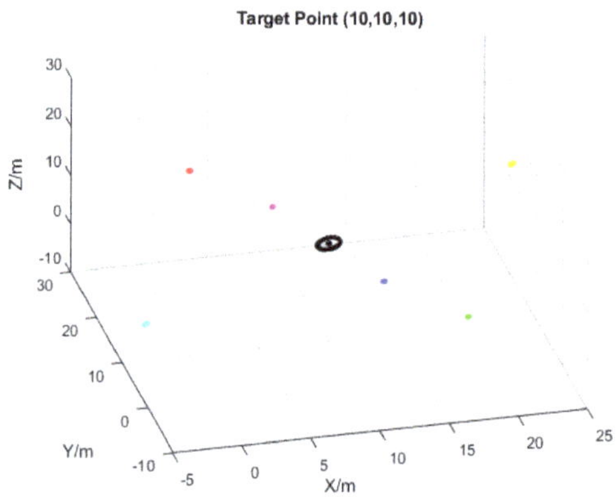

Figure 7. Position maintaining of swarm on the top of surveillance target.

Besides, based on the simulation result, we found out that $P = 0.7$ could be the optimal value of the application rate threshold in this algorithm, which gives the optimal control factor to revise the position of UAV to achieve the convergence of communication indicator value. In Figure 8, the fast convergence of ϕ_{ij} is based on such an threshold value.

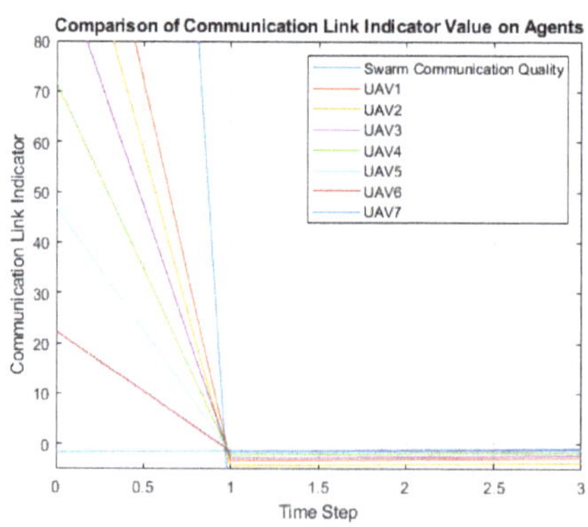

Figure 8. Convergent communication indicator of each UAV agent.

3.2. Waypoint Tracking Based on Centralized Control Scheme

In the early phase of the UAV swarm persistent surveillance, after reaching the initial target point, to monitor the AOI, the low altitude UAV swarm is assigned to go over waypoints predetermined by the high altitude UAV platform. Here, we assume the waypoints are a set of points evenly distributed on the surveillance area, which requires the UAV swarm to reach the point in a given amount of time decided by the path planning platform. High altitude UAV gives those waypoints with its remote sensing data based on the multi types of sensors attached to it in real applications. UAV swarm assigned to specific waypoints is desired to realize the function of real-time tracking of an object, providing network service. Generally, a transition time of moving the swarm from one waypoint to the other depends on the task priorities and distance from the current waypoint [38–40]. Therefore, the transition time between waypoints is an essential aspect of evaluating the waypoint tracking performance. With insufficient transition time, the policy of UAV swarm tracking should be revised as well.

Figure 9 shows the trajectory of UAV swarm agent. Swarm center and waypoint are updating with the proposed target searching function in waypoint tracking. The red line represents the trajectory of the swarm center at each simulation time step. Waypoint distribution is designed as a hexagon shape to verify the swarm's ability to expand its surveillance region in AOI. As expected, the swarm center trajectory would pass every waypoint and form the same shape as a hexagon. However, the UAV agent's trajectory and the swarm center do not present the ability to track those waypoints with high accuracy.

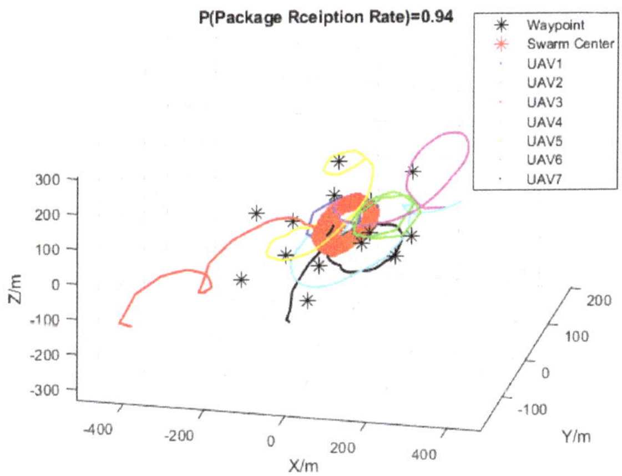

Figure 9. Swarm trajectory, center point, waypoint updating with centralized position updating policy.

In Figure 10, the distance between agent one and others varies in the form of erratic curves. The distance between UAV1 and UAV3, UAV1 and UAV7 increases with adding new waypoints in the surveillance area. To monitor AOI's wider area, the distribution of waypoints could only be more sparse in our simulation. As time goes by, the distance between UAV3, UAV7, and the remaining swarm agent cannot support the swarm in forming an effective surveillance network. Eventually, the lost control UAV3 and UAV7 will lose the connection with the surveillance group. In Figure 11, the swarm center to waypoint, yet, show a decreasing of the distance difference on time step smaller than 200. The outer hexagon layer of the waypoint reduces the tracking accuracy of the UAV swarm after 200-time steps.

The target searching algorithm is less effective in realizing the waypoint tracking of UAV in the early phase of surveillance. Therefore, we proposed the following Algorithm 2 to solve swarm losing control in the transition of waypoints. The alternative waypoint tracking algorithm is based on the centralized control scheme. It compresses the update portion of the communication indicator value because the algorithm block limitation is in one page.

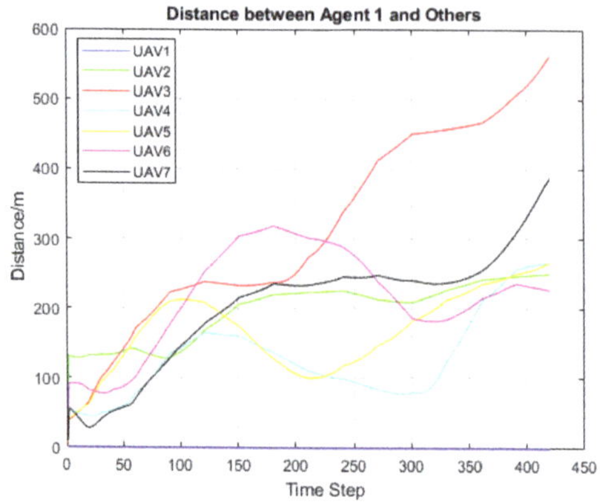

Figure 10. Distance between Agent 1 and others with centralized control policy.

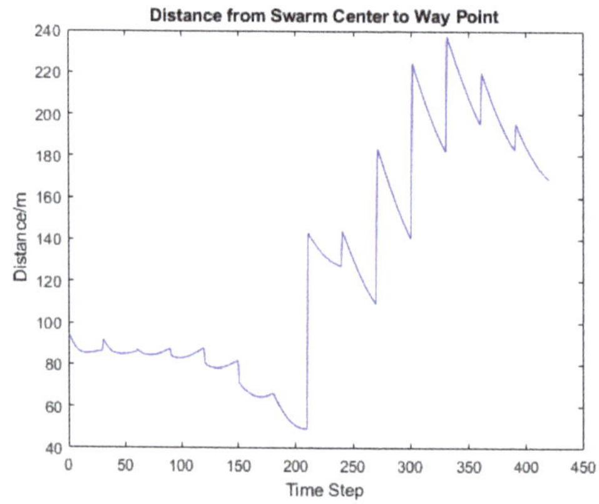

Figure 11. Distance from swarm Center to Waypoint with centralized control policy.

In Algorithm 2, before moving forward from one waypoint to the others, ε is used to evaluate whether the next waypoint is reachable for the swarm agent. Otherwise, a replaced waypoint closer to the swarm's current position would be applied in adjusting the speed of each UAV.

In Figure 12, even the swarm shows similar tracking accuracy, the UAV agent trajectory shows the convergence of distance between UAV agents. A more reliable channel link between UAV agents could be built with a distance from 20 to 50 m, which prevents the collision between agents. The distance varies between agents when tracking the near waypoint. It could be seen from Figure 13. Besides the swarm center adjustment, the communication aware control intends to pull its neighboring agent together in the process of waypoint transition, making this alternated position control scheme reliable. The alternative waypoint makes the control priorities become maintaining the stable distribution of the UAV swarm. In Figure 14, instead of presenting a divergence trend shown in Figure 11, while the waypoint moves from the inner hexagon circle to the outer one, the new control scheme suppresses the divergence trend and keeps their distance difference under 160 m. The communication link quality is still not promised, as shown in Figure 15.

Algorithm 2: Waypoint Tracking

Result: Maintaining communication link while transmitting between waypoints.
initialization:
$q_i = (x_i, y_i, z_i)$% Randomized Position Value
$speed_i = 0$
while *number of investigated waypoint < number of planned waypoints* **do**
  Sum of agent position = $\sum_{i=1}^{m} q_i$
  $C_t = \frac{\text{sum of agent position}}{\text{swarm size}}$
  $T(k+1)$ = Assigned waypoint position at time k + 1
  $T(k)$ = Assigned waypoint position at time k
  Difference between center and current waypoint $T(k)$: $D_t = \|C_t - T(k)\|_2$
  % In tracking next waypoint
  if $D_t < \varepsilon$ **then**
    while *t < assigned surveillance time* **do**
      $e_{ij} = (q_i - q_j)/r_{ij}$
      $\nabla_{q_i}\phi(r_{ij}) = \phi(r_{ij}) \cdot e_{ij}$
      $u_c = \frac{C_t - T(k+1)}{\|C_t - T(k+1)\|_2}$
      $u_i = u_c + \nabla_{q_i}\phi(r_{ij})$ % calculate the control of each agent
      $speed_i = speed_i + u_i$
      $q_i = q_i + speed_i * step\_size$
    end
  else
    % Compute middle point of straight line between current waypoint and next waypoint
    $T_{avg} = \frac{T_k + T_{k+1}}{2}$
    $e_{ij} = (q_i - q_j)/r_{ij}$
    $\nabla_{q_i}\phi(r_{ij}) = \phi(r_{ij}) \cdot e_{ij}$
    $u_c = \frac{C_t - T_{avg}}{\|C_t - T_{avg}\|_2}$
    $u_i = u_c + \nabla_{q_i}\phi(r_{ij})$ % calculate the control of each agent
    $speed_i = speed_i + u_i$
    $q_i = q_i + speed_i * step\_size$
  end
end

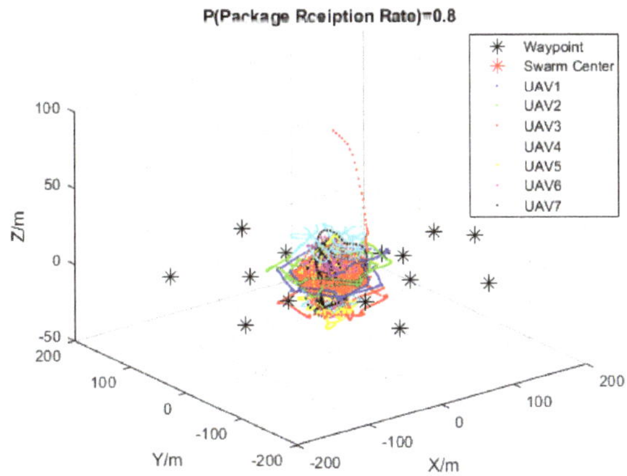

Figure 12. Swarm dynamics with an alternated centralized control algorithm.

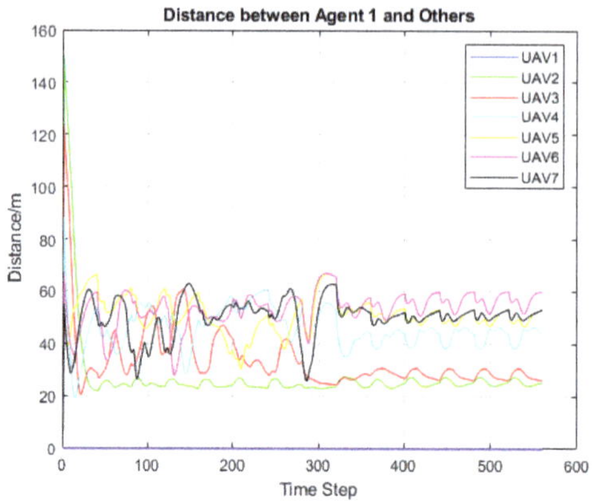

Figure 13. Distance between agent 1 and Others with alternative centralized policy.

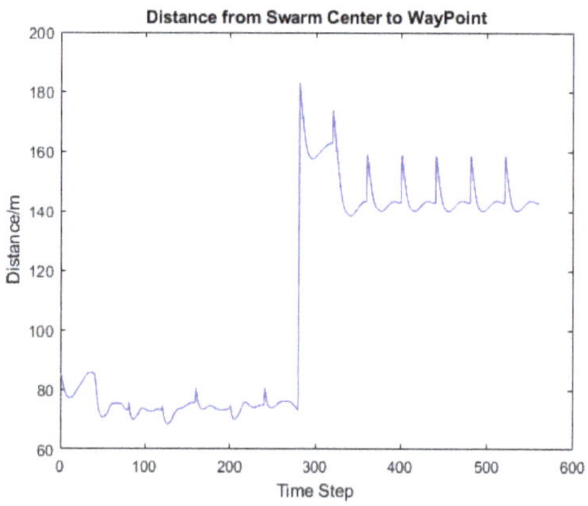

Figure 14. Distance between Swarm Center and Waypoint with alternated centralized policy.

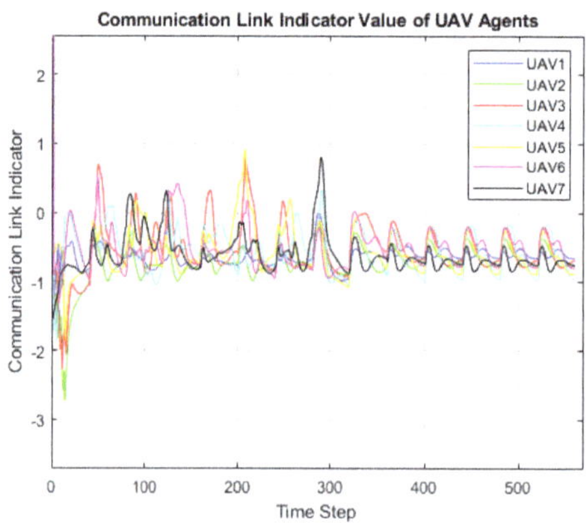

Figure 15. Communication link indicator variance with alternative centralizing waypoint tracking algorithm.

3.3. Decentralized Control Scheme of UAV Swarm in Waypoint Tracking

An assigned trajectory is combined with dense waypoints, requiring UAV agent to have accurate control on adjusting its position to maintain the communication quality and waypoint tracking accuracy in surveillance. The centralized swarm center position comparing the algorithm above is not effective in working in such a task. Therefore, instead of giving each agent the same control inputs based on the swarm center position comparing result, the decentralized control policy compares the UAV agent position with the assigned waypoint in the trajectory at each time step. The algorithm is explained as follows.

Algorithm 3: Decentralized Control Scheme on Waypoint Tracking

Result: Maintaining communication link while following the assigned waypoint
initialization:
$q_i = (x_i, y_i, z_i)$% Randomized Position Value
$speed_i = 0$
while *Past Surveillance Point < Assigned Surveillance Point* **do**
  while *t < assigned surveillance time* **do**
    $T = f(t)$
    Sum of agent position = $\sum_{i=1}^{m} q_i$
    Difference between agent position and target point $D_t = q_i - T$
    while *i < swarm size* **do**
      while *j < swarm size (j! = i)* **do**
        r_{ij} = norm $(q_i - q_j)$
        $a_{ij} = \exp\left(-\alpha(2^\beta - 1)\left(\frac{r_{ij}}{r_0}\right)^v\right)$
        if $a_{ij} >= P$ **then**
          $\phi_{ij} = -\beta\left((vr_{ij})^{v+2} + \left(\beta v r_{ij}^v + r_0^v\right) r_0^2\right) \frac{e^{-\beta}}{r_0^v} g_{ij}$
        else
          $\phi_{ij} = 0$
        end
        $e_{ij} = (q_i - q_j)/r_{ij}$
        $\nabla_{q_i}\phi(r_{ij}) = \phi(r_{ij}) \cdot e_{ij}$
        $speed_i = speed_i + \nabla_{q_i}\phi(r_{ij})$
      end
      $u_c = \frac{D_t}{\|D_t\|_2}$
      $u_i = u_c + \nabla_{q_i}\phi(r_{ij})$ % calculate the control of each agent
      $speed_i = speed_i + u_i$
      $q_i = q_i + speed_i * step\_size$
    end
  end
end

Each UAV agent updates its control portion without relying on consensus control introduced by the difference between UAV swarm center and waypoint position. It neglects to collect each agent's position information in the swarm and calculates the control input needed for each one of them, which helps solve the problem brought by the information exchange delay and instability due to the loss of control command data. Meanwhile, the computation efficiency for UAV forming a surveillance team is improved by assigning the position comparing process on each independent platform within the swarm. The control portion relies on two aspects, the agent's position difference to waypoint and the communication link quality. The designed algorithm maintains each agent's communication link quality.

Meanwhile, link quality stability promises the distance between agents is also stable and will not collide. In other words, the mechanism makes the agent group stay steady on the top of the wanted waypoint. When the waypoint update is assigned, the position comparing mechanism will move the whole group to the new surveillance area again. The following simulation results would explain these effects.

Figures 16–18 show the moving trajectory of seven UAV agents and the swarm center with each time step. Each UAV agent owns the unique dynamical behavior and moves forward with keeping the distance to maintain the communication link's stability. The swarm center could perfectly follow the straight-line trajectory predefined by the high altitude UAV platform. Simulation result from Figure 18 shows swarm adjusts its center after each waypoint updating. With enough transition time given between each waypoint, the distance difference could be lower than 10 m. Most of the time, the center position difference keeps between 20 to 30 m, which is caused by not enough assigned transition and surveillance time. It could be seen from Figure 17 that the UAV agent keeps the distance between each at the range from 20 m to 50 m, which could be an appropriate length for an amateur drone to hold the communication link with its neighboring agent. For example, the furthest range of existing amateur UAV platforms (e.g., RyzeTech Tello Quadcopter, SNAPTAIN A10 MiniFoldable Drone) is 300 m to 400 m. The range of causing near field interference of SISO antenna was assumed to be 2 m based on the relation between radio wavelength and antenna diameter. Therefore, a communication range between UAV agents from 20 m to 50 m gives a promising distance for high data rate communication. Furthermore, the distance variance between the agent one and others stays in the range of 10 m.

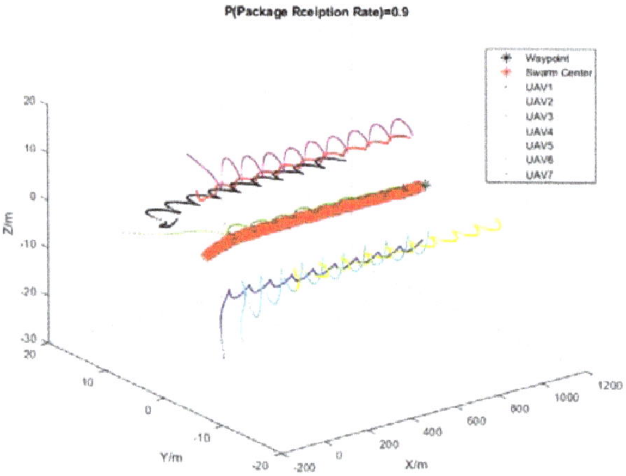

Figure 16. Swarm trajectory, center point updating with straight-line trajectory under decentralized control policy.

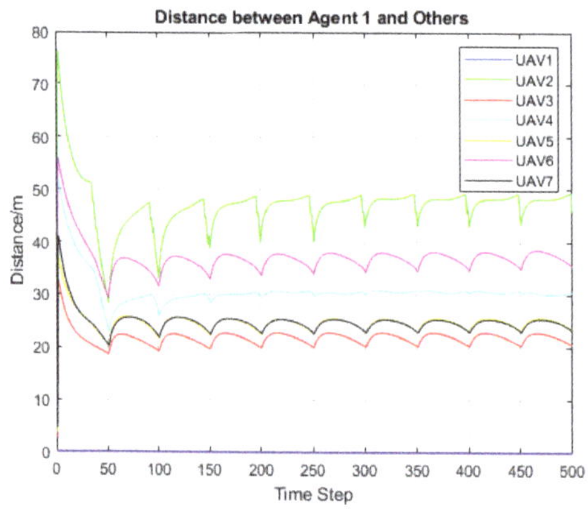

Figure 17. Distance between agent 1 and others with decentralized control policy.

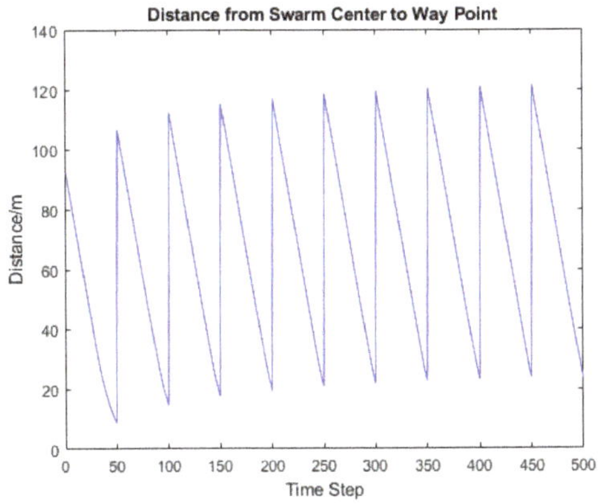

Figure 18. Distance between swarm center and waypoint with decentralized control policy.

Figure 19 represents the consistent result with Figure 17. Ten in total communication link disturbance is caused by the waypoint updating and the swarm center transition. The indicator value happens to be stable after the initial transition of the waypoint finishes. For example, it appears at the time step $t = 50, 100, 150$ that the center of the swarm has the lowest distance to the waypoint in Figure 18. Meanwhile, the communication indicator with high highest value starts dropping and approaches the value with stable communication quality to its neighboring agent. Here, a delay effect of such decentralized control seems to take place. Assuming the difference between agent position and waypoint is still large enough, the agent could get affected from both of the waypoints when the first waypoint transition time is not finished. With the increasing historical waypoint on the surveillance map, we wonder if it could cause the disturbance over communication links within the UAV swarm network. The following results would explain it.

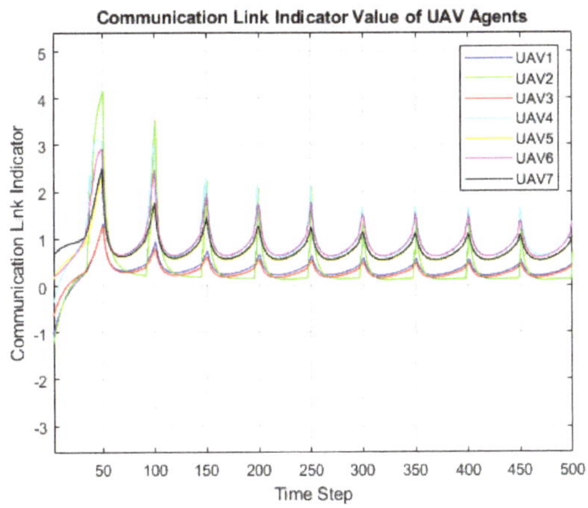

Figure 19. Communication indicator variance with decentralized control policy.

As shown in Figure 20, we simulate the swarm track waypoints in a hexagon distribution with fixed transition time to each waypoint and high swarm position control gain, respectively. Our control scheme shows its constraints on realizing its functions in accurate swarm center point tracking with short transition time. In other words, the gain of position updating control for UAV swarm is not enough. The limited coverage area reduces the performance of the swarm in providing continuous surveillance service. The distance from agent 1 to the others maintains at the range from 20 m to 40 m, which is still an appropriate range for building high throughput rate of data exchange network within the UAV swarm. The swarm center gets further away from the waypoint as time goes on due to not enough gain over control. The swarm center shows the trend of approaching the updated waypoint after time step 400. Compared to the successful waypoint tracking realized with straight-line trajectory yet, the hexagon waypoint tracking with less comprehensive surveillance coverage has a more stable communication link. From the figure, we conclude that the mobility of low altitude UAV swarm is sacrificed by maintaining the stability of swarm communication network.

We did the experiment with increasing the gain for each agent's position control. The performance of UAV reaching designed surveillance waypoint becomes better; however, the high control gain increases the battery consumption in real swarm operation, which is another essential factor affecting successful rate of the UAV swarm's surveillance mission. Here, the power efficiency is sacrificed by the control efficiency. Moreover, the high gain's instability could make the swarm system less controllable on building a stable UAV swarm network. The instability trend of swarm's internal communication link suggests the high QoS variance within the UAV network. As shown in our SISO model and the simulation result above, the higher value of SISO communication indicator suggests the more substantial mutual interference between neighboring antennas. A high package loss rate happens in this situation. However, the oversaturated waypoint tracking gain on control essentially increases the accuracy of waypoint tracking.

Considering each UAV platform's battery consumption, we would also like to avoid the high gain on the agent's position control. Other factors that could also affect the surveillance mission performance include the high altitude UAV's assigned surveillance time and the waypoint distribution. Here, we use the same hexagon distributed waypoints example with optimized transition time and control gain to express this idea. As shown in Figures 21–23, all metrics on the tracking accuracy and agent's communication indicator value are being kept on the relatively low value compared to the results of overfed gain above. The high accurate waypoint tracking could be seen from Figure 22. The distance value is all lower or near 10 m after fast position updating regarding the assigned waypoint. The communication channel quality is controlled by the after adjusting swarm center position. As time goes by, the indicator value stays on the origin after 9th waypoint updates. Figure 21 shows the UAV agent has the distance to the others with appropriate range value.

Figure 24 compares the waypoint tracking accuracy from three algorithms: (1) Centralized Policy (CP); (2) Alternative Tracking with Centralized Policy (ATCP); (3) Decentralized Waypoint Tracking Algorithm. This is based on the different gain value on communication aware control and waypoint tracking control. Decentralized control algorithm could be divided into (a) Decentralized Policy with High Gain on Tracking (DPHGT); (b) Decentralized Policy with Optimal Gains (DPOG); (c) Decentralized Policy without Gain Control (DPWGC). Here, tracking accuracy evaluates the percentage of the reduced distance between swarm center and waypoint position with half of the transition time. Therefore, we have

$$\text{Tracking Accuracy}(\%) = \frac{d_{\text{median}}}{d_{\text{initial}}} \tag{31}$$

where the d_{medium} denotes the median of the distance between swarm center and waypoint within one transition period, d_{initial} denotes the initial value of the distance between swarm center and waypoint in the waypoint transition period. With more accurate waypoint tracking ability, the faster convergence of swarm center and waypoint distance the algorithm would reach. In Figure 24, the DPHGT has the highest tracking accuracy because of the significant gain value on the UAV agent's position control. CP has the lowest tracking accuracy because the UAV agent loses control while following multiple waypoints. DPOG, with the optimally designed gain control on both aspects of communication and waypoint tracking, has a similar tracking accuracy compared to DPHGT. ATCP shows better tracking accuracy after 8th round of waypoint updating. DPWGC shows less accurate waypoint tracking performance without the adjustment of control gain in following the hexagon distributed waypoints.

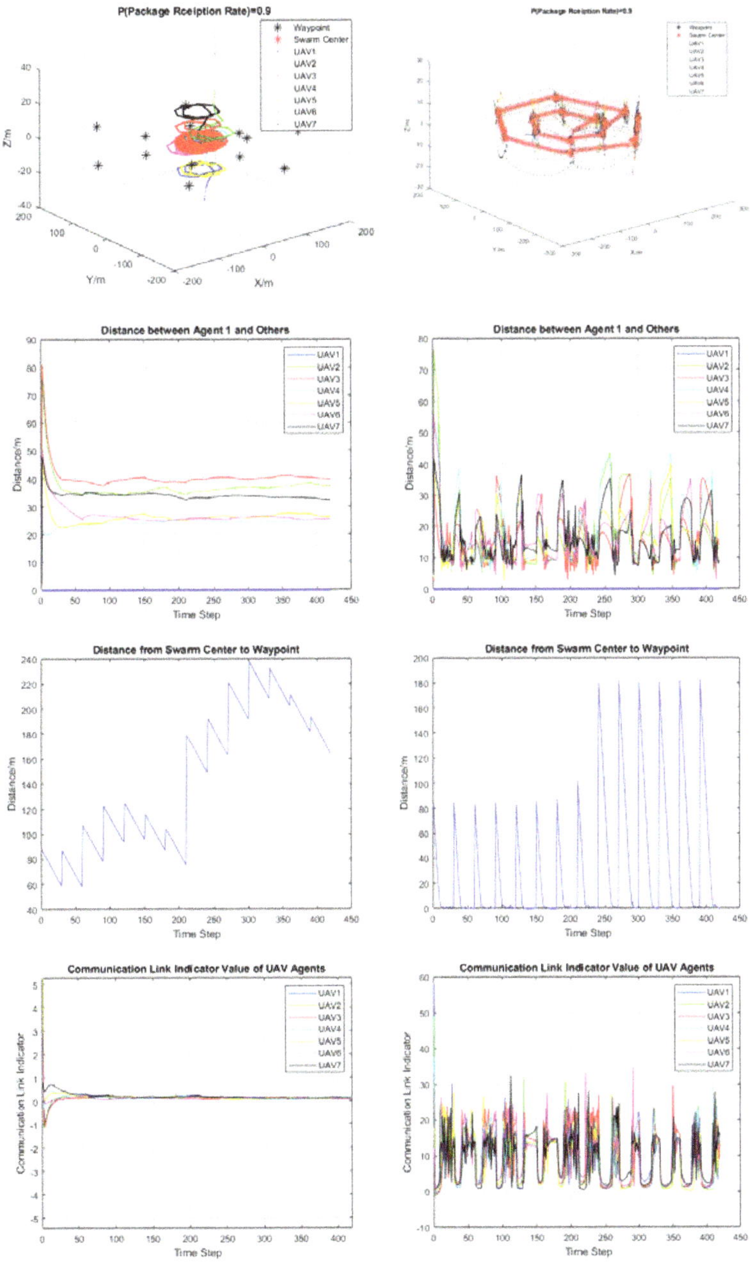

Figure 20. Comparison of UAV Swarm Dynamics, Distance between Agents, Distance between Swarm Center and Waypoints, Communication Indicator on not enough transition time (or low position approaching control gain) (left column), high position approaching control gain (right column), respectively.

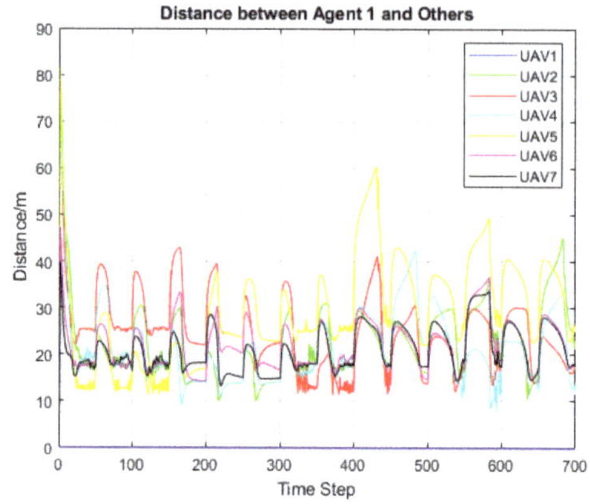

Figure 21. Distance between agent 1 and others based on optimized gain and transition time.

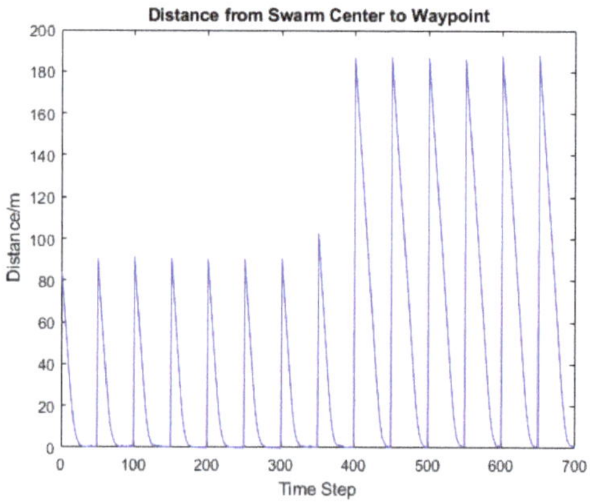

Figure 22. Distance between swarm center and waypoint based on optimized gain and transition time.

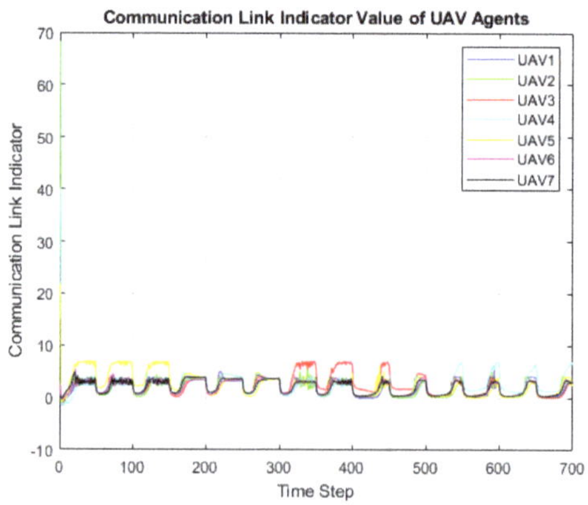

Figure 23. Communication indicator variance based on optimized gain and transition time.

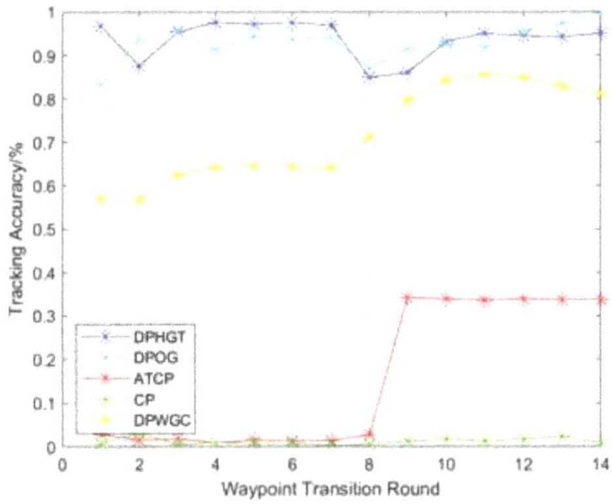

Figure 24. Waypoint Tracking Accuracy with Different Algorithms.

The average distance from agent 1 to the others measures the general performance of formation control on each algorithm. A larger average distance value suggests that the algorithm is less performed in forming a surveillance swarm team. It can be represented as

$$D_{avg} = \frac{\sum_{n=1}^{\text{swarm size}} D_{1i}}{swarm\ size} \tag{32}$$

in which, D_{1i} represents the distance between agent one and agent i. *swarm size* is chosen as 7 to match our simulation setup. As we can see in Figure 25, the divergence of CP suggests the swarm agent lost control in the process of following surveillance waypoints. DPOG reaches the lowest average distance between UAV agents but also prevents the collision between them.

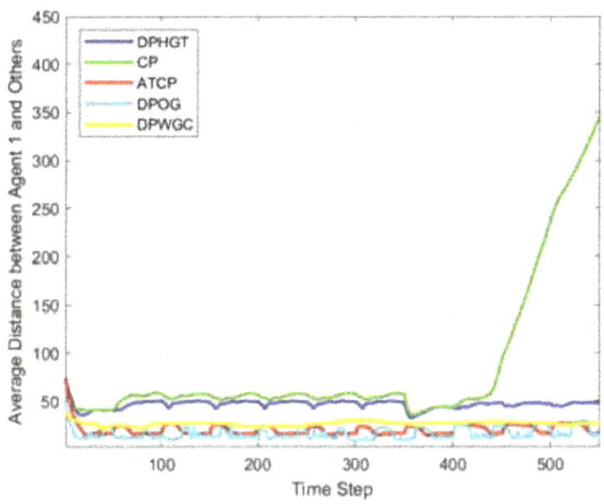

Figure 25. Average Distance to UAV Agent 1 with Different Algorithms.

4. Conclusions

This paper proposes a communication-aware formation control algorithm to realize high accuracy waypoint tracking with the application in persistent surveillance based on UAV teaming's hierarchical architecture. A framework of such UAV and ground station teaming scheme is proposed in the surveillance strategy. We employed three algorithms to achieve high accuracy waypoint tracking while maintaining communication link quality based on the centralized and decentralized control scheme. The simulation shows that the centralized control scheme reached a fast convergence of maintaining the communication link performance and UAV swarm dynamics stability. In the waypoint tracking, we discussed an alternative centralized control scheme to cope with the divergence of swarm formation. Although the results show that the alternative centralized scheme could realize a good formation control scheme and network service, the waypoint tracking accuracy of the swarm is reduced, making this scheme hard to implement in a real persistent surveillance mission. Therefore, the decentralized control scheme divides the position updating process from the center to each agent. The simulation results demonstrate the superiority of such decentralized control policy on maintaining the network service quality in terms of communication link stability. Meanwhile, the agents' distance suggests that the UAV swarm's team formation stays in a stable state.

In the simulation result based on the decentralized swarm formation scheme, we studied the trade-off between choosing the waypoint tracking accuracy and maintaining the communication link quality. We discussed the relationship between these two parameters and the battery capacity, algorithm computation efficiency, and thrust of control. With not enough waypoint set on the map for low altitude UAV swarm, the sensing coverage area could not be expanded, which reduces the success rate in persistent surveillance missions. This problem could be solved by adding gain over the position updating control on each agent. However, higher gain on the position updating control cuts down UAV swarm's stability, maintaining an effective communication link with its neighboring agents. Both the balanced gain over two portions on communication quality and tracking accuracy should be considered. In the end, the simulation results showed the optimized gain value, which helps enhance the performance of UAV swarm in persistent surveillance.

In our future research of communication link performance in UAV swarm dynamics, the UAV MIMO antenna's cochannel noise is an inevitable technical challenge to build an effective and efficient UAV swarm network. Cochannel noise is intended to be suppressed with the variance of signal transmission power and arrival angle, especially concerning the application of a millimeter-wave communication system. A more accurate UAV agent dynamic model is desired to simulate its kinetic move within the swarm. Environment interference would be less considered due to the relatively short communication distance between swarm agents. Therefore, millimeter-wave has fewer path loss effects in the process of signal transmission.

Similarly, multipath effects caused by the reflection of wave propagation should be neglected because the UAV swarm operates mostly in the outdoor environment and the existing signal filtering algorithm could eliminate these noises. Through applying persistent surveillance using hierarchical structure UAV teaming with the high altitude UAV platform, the dynamics behavior and communication process and quality evaluation could be more complicated. The simulation of integrating the low altitude UAV swarm and fixed-wing high altitude UAV platform should be done in the future.

Author Contributions: Conceptualization, C.X.; methodology, C.X. and Y.J.; software, C.X.; validation, C.X.; formal analysis, C.X.; writing—original draft preparation, C.X.; writing—review and editing, C.X., K.Z., and S.N.; supervision, T.Y. and H.S.; project administration, H.S.; funding acquisition, H.S. All authors have read and agreed to the published version of the manuscript.

Funding: This work was supported in part by the National Science Foundation under Grant No. 1956193.

Institutional Review Board Statement: Not applicable.

Informed Consent Statement: Not applicable.

Data Availability Statement: The data and simulation code presented in this study are available on request from the corresponding author. The data are not publicly in this study due to project restrictions.

Acknowledgments: We thank Sergey V Drakunov from Embry-Riddle Aeronautical University Department of Engineering Physics for providing insight and expertise in the EP707 course: nonlinear system control, which greatly assisted this research.

Conflicts of Interest: The authors declare no conflict of interest.

References

1. Valenti, M.; Dale, D.; How, J.; de Farias, D.P.; Vian, J. Mission health management for 24/7 persistent surveillance operations. In Proceedings of the AIAA guidance, Navigation and Control Conference and Exhibit, Hilton Head, SC, USA, 20–23 August 2007; p. 6508.
2. Aasen, H.; Honkavaara, E.; Lucieer, A.; Zarco-Tejada, P.J. Quantitative remote sensing at ultra-high resolution with UAV spectroscopy: A review of sensor technology, measurement procedures, and data correction workflows. *Remote Sens.* **2018**, *10*, 1091. [CrossRef]
3. Yang, Y.; Zhang, K.; Liu, D.; Song, H. Autonomous UAV Navigation in Dynamic Environments with Double Deep Q-Networks. In Proceedings of the 2020 AIAA/IEEE 39th Digital Avionics Systems Conference (DASC), San Antonio, TX, USA, 11–15 October 2020; pp. 1–7.
4. Xu, C.; Zhang, K.; Song, H. UAV Swarm Communication Aware Formation Control via Deep Q Network. In Proceedings of the 2020 IEEE 39th International Performance Computing and Communications Conference (IPCCC), Austin, TX, USA, 6–8 November 2020; pp. 1–2.
5. Wang, X.; Feng, W.; Chen, Y.; Ge, N. UAV swarm-enabled aerial CoMP: A physical layer security perspective. *IEEE Access* **2019**, *7*, 120901–120916. [CrossRef]
6. Li, B.; Fei, Z.; Zhang, Y.; Guizani, M. Secure UAV communication networks over 5G. *IEEE Wirel. Commun.* **2019**, *26*, 114–120. [CrossRef]
7. Wang, J.; Liu, Y.; Niu, S.; Song, H. Extensive Throughput Enhancement For 5G Enabled UAV Swarm Networking. *IEEE J. Miniat. Air Space Syst.* **2021**. [CrossRef]
8. Wang, J.; Liu, Y.; Niu, S.; Song, H. Beamforming-Constrained Swarm UAS Networking Routing. *IEEE Trans. Netw. Sci. Eng.* **2020**. [CrossRef]
9. Wu, H.; Li, H.; Xiao, R.; Liu, J. Modeling and simulation of dynamic ant colony's labor division for task allocation of UAV swarm. *Phys. A Stat. Mech. Its Appl.* **2018**, *491*, 127–141. [CrossRef]
10. Ruetten, L.; Regis, P.A.; Feil-Seifer, D.; Sengupta, S. Area-Optimized UAV Swarm Network for Search and Rescue Operations. In Proceedings of the 2020 10th Annual Computing and Communication Workshop and Conference (CCWC), Las Vegas, NV, USA, 6–8 January 2020; pp. 0613–0618.
11. Zhou, X.; Wu, Q.; Yan, S.; Shu, F.; Li, J. UAV-enabled secure communications: Joint trajectory and transmit power optimization. *IEEE Trans. Veh. Technol.* **2019**, *68*, 4069–4073. [CrossRef]
12. Li, M.; Cheng, N.; Gao, J.; Wang, Y.; Zhao, L.; Shen, X. Energy-efficient UAV-assisted mobile edge computing: Resource allocation and trajectory optimization. *IEEE Trans. Veh. Technol.* **2020**, *69*, 3424–3438. [CrossRef]
13. Erdelj, M.; Król, M.; Natalizio, E. Wireless sensor networks and multi-UAV systems for natural disaster management. *Comput. Netw.* **2017**, *124*, 72–86. [CrossRef]
14. Shakhatreh, H.; Khreishah, A.; Chakareski, J.; Salameh, H.B.; Khalil, I. On the continuous coverage problem for a swarm of UAVs. In Proceedings of the 2016 IEEE 37th Sarnoff Symposium, Newark, NJ, USA, 19–21 September 2016; pp. 130–135.
15. Monwar, M.; Semiari, O.; Saad, W. Optimized path planning for inspection by unmanned aerial vehicles swarm with energy constraints. In Proceedings of the 2018 IEEE Global Communications Conference (GLOBECOM), Abu Dhabi, United Arab Emirates, 9–13 December 2018; pp. 1–6.
16. Li, Z.; Chen, M.; Pan, C.; Huang, N.; Yang, Z.; Nallanathan, A. Joint trajectory and communication design for secure UAV networks. *IEEE Commun. Lett.* **2019**, *23*, 636–639. [CrossRef]
17. Changsheng, Y.; Zhang, R. 3D trajectory optimization in Rician fading for UAV-enabled data harvesting. *IEEE Trans. Wirel. Commun.* **2019**, *18*, 3192–3207.
18. Chen, W.; Liu, B.; Huang, H.; Guo, S.; Zheng, Z. When UAV swarm meets edge-cloud computing: The QoS perspective. *IEEE Netw.* **2019**, *33*, 36–43. [CrossRef]
19. Wang, R.; Du, J.; Xiong, Z.; Chen, X.; Liu, J. Hierarchical Collaborative Navigation Method for UAV Swarm. *J. Aerosp. Eng.* **2021**, *34*, 04020097. [CrossRef]
20. Jing, T.; Wang, W.; Wang, T.; Li, X.; Zhou, X. Dynamic Control Scheme of Multiswarm Persistent Surveillance in a Changing Environment. *Comput. Intell. Neurosci.* **2019**. [CrossRef] [PubMed]
21. Wei, L. Persistent surveillance for a swarm of micro aerial vehicles by flocking algorithm. *Proc. Inst. Mech. Eng. Part G J. Aerosp. Eng.* **2015**, *229*, 185–194.
22. Reynolds, C.W. Flocks, herds and schools: A distributed behavioral model. In Proceedings of the 14th Annual Conference on Computer Graphics and Interactive Techniques, New York, NY, USA, August 1987; pp. 25–34.
23. Loayza, K.; Lucas, P.; Peláez, E. A centralized control of movements using a collision avoidance algorithm for a swarm of autonomous agents. In Proceedings of the 2017 IEEE Second Ecuador Technical Chapters Meeting (ETCM), Salinas, Ecuador, 16–20 October 2017; pp. 1–6.
24. McCune, R.; Purta, R.; Dobski, M.; Jaworski, A.; Madey, G.; Madey, A.; Wei, Y.; Blake, M.B. Investigations of dddas for command and control of uav swarms with agent-based modeling. In Proceedings of the 2013 Winter Simulations Conference (WSC), Washington, DC, USA, 8–11 December 2013; pp. 1467–1478.
25. Wei, Y.; Blake, M.B.; Madey, G.R. An operation-time simulation framework for UAV swarm configuration and mission planning. *Procedia Comput. Sci.* **2013**, *18*, 1949–1958. [CrossRef]
26. He, L.; Bai, P.; Liang, X.; Zhang, J.; Wang, W. Feedback formation control of UAV swarm with multiple implicit leaders. *Aerosp. Sci. Technol.* **2018**, *72*, 327–334. [CrossRef]

27. Olfati-Saber, R.; Fax, J.A.; Murray, R.M. Consensus and cooperation in networked multi-agent systems. *Proc. IEEE* **2007**, *95*, 215–233. [CrossRef]
28. D'Amato, E.; Notaro, I.; Mattei, M.; Tartaglione, G. Attitude and position estimation for an UAV swarm using consensus Kalman filtering. In Proceedings of the 2015 IEEE Metrology for Aerospace (MetroAeroSpace), Benevento, Italy, 4–5 June 2015; pp. 519–524.
29. Li, Y.; Li, H.; Ding, X.; Zhao, G. Leader-follower consensus of multiagent systems with time delays over finite fields. *IEEE Trans. Cybern.* **2018**, *49*, 3203–3208. [CrossRef] [PubMed]
30. Pham, H.X.; La, H.M.; Feil-Seifer, D.; Nefian, A. Cooperative and distributed reinforcement learning of drones for field coverage. *arXiv* **2018**, arXiv:1803.07250.
31. San, J.V.; Santos, M.; Andújar, J.M. Intelligent UAV map generation and discrete path planning for search and rescue operations. *Complexity* **2018**. [CrossRef]
32. Albani, D.; Nardi, D.; Trianni, V. Field coverage and weed mapping by UAV swarms. In Proceedings of the 2017 IEEE/RSJ International Conference on Intelligent Robots and Systems (IROS), Vancouver, BC, Canada, 24–28 September 2017; pp. 4319–4325.
33. Ntinas, V.G.; Moutafis, B.E.; Trunfio, G.A.; Sirakoulis, G.C. Parallel fuzzy cellular automata for data-driven simulation of wildfire spreading. *J. Comput. Sci.* **2017**, *21*, 469–485. [CrossRef]
34. Li, H.; Peng, J.; Liu, W.; Gao, K.; Huang, Z. A novel communication-aware formation control strategy for dynamical multi-agent systems. *J. Frankl. Inst.* **2015**, *352*, 3701–3715. [CrossRef]
35. Baillieul, J.; Antsaklis, P.J. Control and communication challenges in networked real-time systems. *Proc. IEEE* **2007**, *95*, 9–28. [CrossRef]
36. Oubbati, O.S.; Chaib, N.; Lakas, A.; Bitam, S.; Lorenz, P. U2RV: UAV-assisted reactive routing protocol for VANETs. *Int. J. Commun. Syst.* **2020**, *33*, e4104. [CrossRef]
37. Kashyap, A.; Ghose, D.; Menon, P.P.; Sujit, P.B.; Das, K. UAV aided dynamic routing of resources in a flood scenario. In Proceedings of the 2019 International Conference on Unmanned Aircraft Systems (ICUAS), Atlanta, GA, USA, 11–14 June 2019; pp. 328–335.
38. Fu, Y.; Ding, M.; Zhou, C.; Hu, H. Route planning for unmanned aerial vehicle (UAV) on the sea using hybrid differential evolution and quantum-behaved particle swarm optimization. *IEEE Trans. Syst. Man Cybern. Syst.* **2013**, *43*, 1451–1465. [CrossRef]
39. Roberge, V.; Tarbouchi, M.; Labonté, G. Comparison of parallel genetic algorithm and particle swarm optimization for real-time UAV path planning. *IEEE Trans. Ind. Inf.* **2012**, *9*, 132–141. [CrossRef]
40. Phung, M.D.; Quach, C.H.; Dinh, T.H.; Ha, Q. Enhanced discrete particle swarm optimization path planning for UAV vision-based surface inspection. *Autom. Constr.* **2017**, *81*, 25–33. [CrossRef]

drones

MDPI

Article

StratoTrans: Unmanned Aerial System (UAS) 4G Communication Framework Applied on the Monitoring of Road Traffic and Linear Infrastructure

Robert Guirado [1,2], Joan-Cristian Padró [1,3,*], Albert Zoroa [1], José Olivert [4], Anica Bukva [5] and Pedro Cavestany [5]

1 Exodronics SL, Tech Department, Mas Vinyoles, 08572 Sant Pere de Torelló, Spain; roberto.guirado.linan@estudiantat.upc.edu (R.G.); tech@exodronics.com (A.Z.)
2 Campus Nord UPC 1-3, Universitat Politècnica de Catalunya, Edifici B3, 08034 Barcelona, Spain
3 Departament de Geografia, Edifici B, Universitat Autònoma de Barcelona, 08193 Bellaterra, Spain
4 EACOM SA, 08035 Barcelona, Spain; eacom@eacomsa.com
5 Eurecat, Centre Tecnològic de Catalunya, Multimedia Technologies, 08005 Barcelona, Spain; anica.bukva@eurecat.org (A.B.); pedro.cavestany@eurecat.org (P.C.)
* Correspondence: JoanCristian.Padro@uab.cat

Abstract: This study provides an operational solution to directly connect drones to internet by means of 4G telecommunications and exploit drone acquired data, including telemetry and imagery but focusing on video transmission. The novelty of this work is the application of 4G connection to link the drone directly to a data server where video (in this case to monitor road traffic) and imagery (in the case of linear infrastructures) are processed. However, this framework is appliable to any other monitoring purpose where the goal is to send real-time video or imagery to the headquarters where the drone data is processed, analyzed, and exploited. We describe a general framework and analyze some key points, such as the hardware to use, the data stream, and the network coverage, but also the complete resulting implementation of the applied unmanned aerial system (UAS) communication system through a Virtual Private Network (VPN) featuring a long-range telemetry high-capacity video link (up to 15 Mbps, 720 p video at 30 fps with 250 ms of latency). The application results in the real-time exploitation of the video, obtaining key information for traffic managers such as vehicle tracking, vehicle classification, speed estimation, and roundabout in-out matrices. The imagery downloads and storage is also performed thorough internet, although the Structure from Motion postprocessing is not real-time due to photogrammetric workflows. In conclusion, we describe a real-case application of drone connection to internet thorough 4G network, but it can be adapted to other applications. Although 5G will -in time- surpass 4G capacities, the described framework can enhance drone performance and facilitate paths for upgrading the connection of on-board devices to the 5G network.

Keywords: UAS; drones; traffic monitoring; 4G/LTE; VPN; real-time video; computer vision

Citation: Guirado, R.; Padró, J.-C.; Zoroa, A.; Olivert, J.; Bukva, A.; Cavestany, P. StratoTrans: Unmanned Aerial System (UAS) 4G Communication Framework Applied on the Monitoring of Road Traffic and Linear Infrastructure. *Drones* **2021**, *5*, 10. https://doi.org/10.3390/drones5010010

Academic Editor: Diego González-Aguilera
Received: 28 December 2020
Accepted: 25 January 2021
Published: 28 January 2021

Publisher's Note: MDPI stays neutral with regard to jurisdictional claims in published maps and institutional affiliations.

1. Introduction

Drones are one of the most fastest growing business of the decade 2010–2020, either for professional or personal uses, and scientific purposes [1–3]. New applications in unmanned aerial systems (UAS) are expected to become increasingly useful to accelerate productivity, save costs and risks, and to extend the current limitations of these systems, adapting them to the next decade. The technological challenge of this study is to develop a flexible solution to remotely and in real time control fixed-wing aircrafts by means of the 3G-4G LTE telecommunication network, that is ready to be adapted to 5G. The innovation aims at providing high control range Beyond the View Line of Sight of the pilot (BVLOS), accordingly with the aircraft flight autonomy. Furthermore, this will allow the integration of deep learning tools, such as Computer Vision (CV) technologies, and big data analysis, adapted to the new autonomous drone fleet paradigm.

In recent years, drone communications have been increasingly addressed by both the academia and the industry [4]. Unmanned vehicles and battery technologies have evolved to provide increasing endurance, currently allowing flight distances that exceed the range of the local radio link with the Ground Control Station (GCS)—where the pilot is usually located—especially when dealing with fixed wing drones. To fix this issue many researchers and companies have proposed alternatives to the local link that involve the connection of the UAS to the internet [5], thus being accessible from almost any connected device. For instance, flying ad hoc networks (FANETs) have become a hot topic since they represent a step further, enabling communications from single drones to multiple drones [6]. However, a challenge that is still pending to be solved is the BVLOS communication, which inherently implies long-range links. Moreover, if the desired solution requires high-capacity links, such as real-time video applications, the use of cellular network (i.e., 4G) is one of the best choices we can aim at. In this direction, [7] analyzed the main advantages and drawbacks of such schemes and provided some recommendations. [8] further discusses technological challenges including interferences and mobility issues and gives potential solutions. This also caused the appearance of dedicated protocols and communication network architecture solutions for a wide range of drone use cases. The use of cellular networks requirements for the connectivity of drones was already analyzed in 2015 [9] but it is fully in development since the implementation of 4G technology in most developed countries (around 2016), which has entailed further network design proposals [10] and even drone-to-drone communications [11]. This technology allows the transmission of relatively large data, exceeding mere telemetry packets, being able to handle video and images especially. Recent implementations like LARUS [12] have used 4G/LTE networks to connect various drones for long-range rescue missions [13]. However, some threats are still concerning the community, such as the on-board energy consumption, the handover rate between antennas, or the bandwidth [14]. Some alternatives to the use of the terrestrial antenna network as the main way to connect the drone to the internet are focusing upwards, to the High-Altitude Pseudo Satellites (HAPS) [15,16] or to satellite constellations [17], but these are in experimental stage and currently there are not available options.

The problem addressed in this work is to supplement the drone-to-local GCS data link with a drone-to-remote server data link. This new framework, based on 4G communications to directly link the drone to internet, allows the real-time processing of the drone acquired data in powerful servers located in the headquarters of the data user (Figure 1).

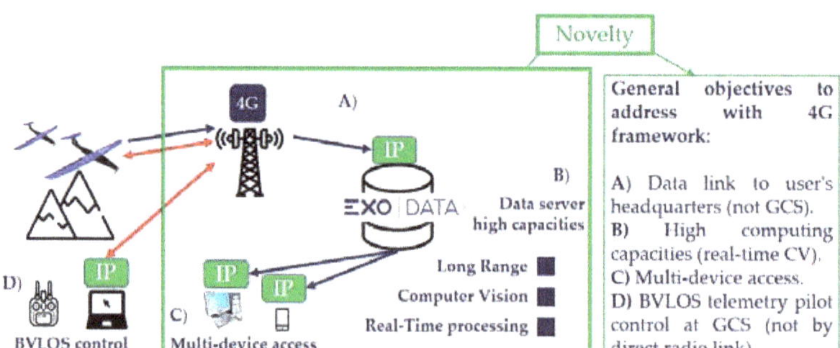

Figure 1. General objectives to address with the 4G communication framework. Note that this framework is extensive to many applications, although in this study we apply it to traffic monitoring.

In this work, we provide application details of our own-operated UAS, while discussing implementation and component details in the system and sharing the resultant complete architectural scheme. The initial and main requirement was the unlimited range of communication between the drone and the ground segment (including the pilot GCS and the data processing center), accounting for telemetry and live video to monitor transportation infrastructures and traffic, namely StratoTrans project [18]. This requirement involves overpassing the local radio-link communications between the GCS and the drone Flight Controller (FC). The objective is tackled by using the 4G network, which allows to link the drone with the GCS and with any device connected to internet wherever it is located, being the latter the main interest of the presented framework. The purpose of the StratoTrans project is to run CV algorithms over the video acquired to extract object identification using convolutional networks [19,20], tracking trajectories [21,22], and velocity estimation [23,24]. Therefore, the drone acquired video is sent to a data server to be processed and finally it is visualized in the operator's headquarters computer or in a user's mobile device.

2. Materials and Methods

2.1. Methodological Framework for UAS—4G Communications

4G communications make sense when the aircraft cannot maintain the radio link with the GCS, typically when a limit distance is exceeded or when there are topographic occlusions between them. Therefore, we consider BVLOS conditions when it is necessary to use such cellular network. The presented solution can transmit imagery (both First Person View (FPV) orientation or zenithal captured) and telemetry data from the aircraft via the 4G network. The imagery, which can be real-time video, is intended to be viewed in the GCS o in any device with internet connection.

The hardware and system architecture inside the drone include a set of navigation sensors (GNSS, IMU, barometer, compass, and pitot tube) that feed data to the FC, and a telemetry module with an antenna to radio-link the FC with the GCS. In order to add 4G capabilities, an extra microcomputer, a modem and a SIM card are needed to connect the drone to Internet. Even though the extra hardware could be powered through the FC, it is worth noting that we power it with an independent power circuit using a power module and a regulator in between. This avoids the supply through the FC and helps to keep the independence between functions, for safety reasons. Regarding the video cameras, these must be compatible with the microcomputer ports. In the case of FPV video cameras, they must comply with specific regulation features such as the Field of View (FOV) and frames per second (fps), while for downwards Earth Observation (EO) cameras the fit for purpose spatial and spectral resolution and geometric quality are the critical issue.

The system network architecture is based on a Virtual Private Network (VPN) composed of three main IP address endpoints: 1/The on-board drone location. 2/The GCS pilot location. 3/The remote server location. Each of these three VPN endpoints have access to internet and are securely linked between them, with an internal fixed identifier, as shown (Figure 2).

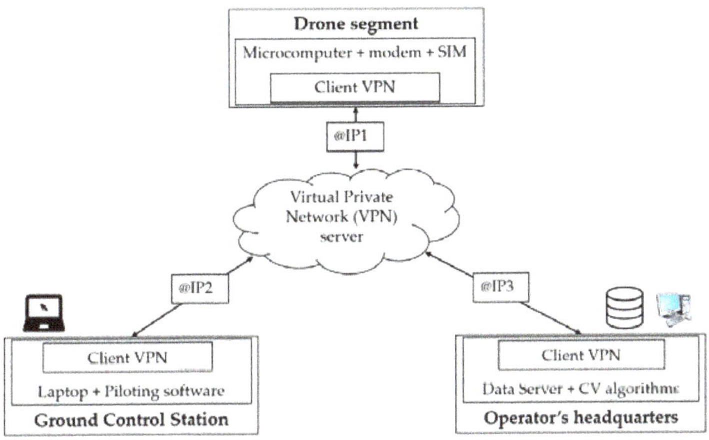

Figure 2. Virtual Private Network (VPN) architecture scheme.

A dedicated software is installed in an on-board microcomputer, which reads the telemetry from the Flight Controller (FC), the video from a camera, and is accessible to the VPN endpoints for its visualization in a web interface. In addition, it can stream both telemetry and video data to the VPN endpoints, which can then locally reproduce the data by reading directly from the network ports. The microcomputer acts both as a hub between the drone input data to be sent (FC and cameras), and as a bridge between the drone and the VPN.

Regarding the streaming protocols, TCP/IP is the recommended protocol to securely transmit the telemetry without losing data packages, while Real Time Streaming Protocol (RTSP) through UDP is the best option to transmit the video and assure a combined low latency. The telemetry data encoding and packaging is done by following the Micro Air Vehicle Link (MAVlink) protocol, which can be decoded and plotted in widely used GCS software.

2.2. *Hardware, Software, and Methodology under Use Case*

Many test flights have been performed at Barcelona Drone Center (BDC) [25] testing site facilities. BDC, located in Moià (Catalonia, Spain), manages a Temporary Segregated Area (TSA) of 2500 ha specifically conceived for BVLOS flights. Their facilities are located in a rural area, without the ubiquitous 4G antennas and signal overlapping of urban areas, making it an excellent site to test coverage and performance indicators. Also, other flights were carried out in different locations of Catalonia to apply the developments with different traffic and infrastructure scenarios, such as roundabouts and roads.

The implemented hardware and system architecture (namely EXO Data) have been integrated and developed by Exodronics, in collaboration with other institutions under the StratoTrans project. EACOM [26] has collaborated with telecommunications issues; UAVMatrix [27] provided a reference commercial product to develop 4G drone-to-ground services; the server to store the mission's telemetry data and the video hosting, has been developed in collaboration with NEXIONA [28]; the CV algorithms specifically designed for tracking elements from the video imagery where developed by Eurecat Multimedia department [29]. The aircraft used to carry out the telecommunication experiments is an EXO C2-L+ fixed wing developed by Exodronics [30]. This lightweight platform (1.3 kg Maximum Take-off Weight (MTOW)) provides around 75 min flight autonomy, which is appropriate to test under BVLOS conditions. It is, then, a specific case example of where it is appropriate to use such cellular network, capable to transmit FPV video, EO imagery and telemetry data from the aircraft via the 4G network. However, for hovering traffic monitoring in roundabouts, we used multicopter platforms (Figure 3).

The specific hardware housed within the drone is a Pixhawk 2.1 Cube Black Flight Controller [31,32], and a telemetry module with an antenna to locally radio-link (Spread Spectrum at 868 Mhz) the FC with the GCS. In the EXO Data implementation, an extra microcomputer (Raspberry Pi Zero W [33]), a 4G/LTE USB-stick a modem with a data rate of 150 Mbps [34], and a multioperator SIM card [35,36] are used to connect the drone to Internet, with an aggregated extra weight of 40 g. The onboard power system is based on a Li-ion 4S 3.4 Ah battery and a power module that feeds independently the FC and the EXO Data hardware. The EXO Data system video cameras, compatible with the microcomputer ports, are a ZeroCam [37] for FPV video, and for downwards Earth Observation video cameras we use a PiCam V2 [38] or a remote sensing camera with video functionalities with a HDMI—CSI converter connected to the Raspberry video input port. The video encoding and packaging can be carried out with H264 compression (1:125 ratio), accounting for a video quality of 1280 $\times$ 720 pixels, 720 p and 30 fps, which can be decoded and reconstructed in widely used video software (e.g., VLC, Gstreamer). The complete hardware implementation in the EXO C2-L+ is shown as follows (Figure 4).

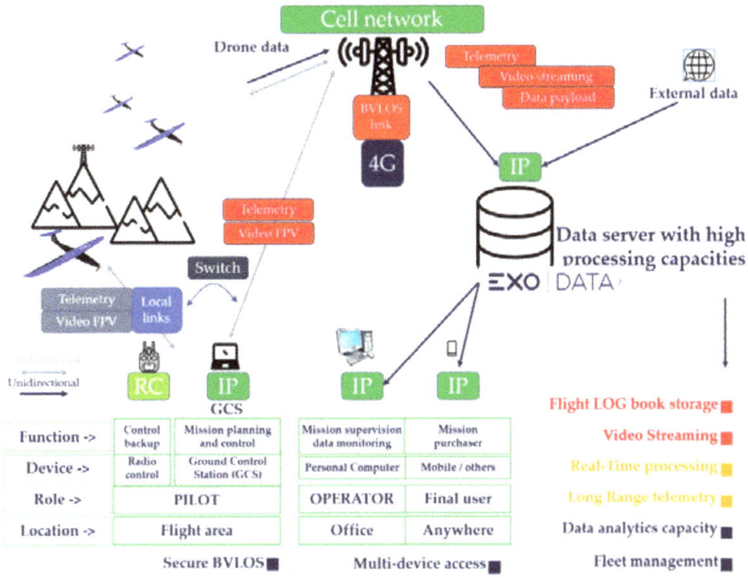

Figure 3. Complete scheme of the 4G cell network functions and connections.

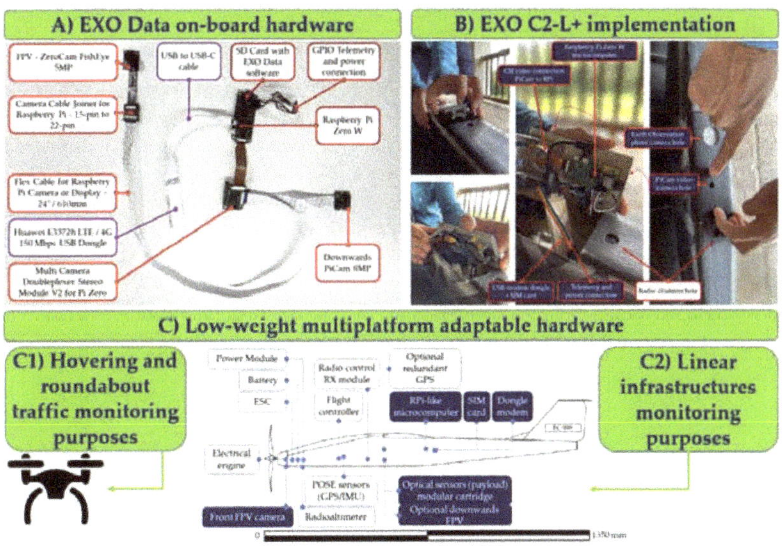

Figure 4. (**A**) On board hardware devices. (**B**) Complete hardware implementation in the EXO C2-L+ fixed wing drone. (**C**) The implementation is adaptable to multicopters and fixed-wing drones, useful for different purposes.

The implementation uses a secured VPN [39] to switch the 3 endpoints: The drone, the GCS and the EXO Data server. The telemetry link is based on a MAVLink decoders, with an open-source software in the GCS (Mission Planner [40]) that can be linked to the onboard FC thorough local radio-link or thorough TCP/IP protocol. Moreover, a specifically developed software [28] in the EXO Data server is used to store the telemetry and imagery data in separated dockers. The complete implementation is detailed as follows (Figure 5).

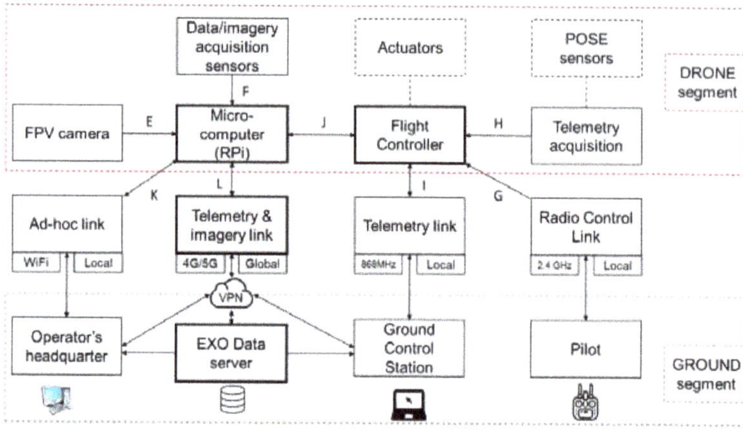

Figure 5. Logical block diagram of drone elements and connections. Links description. E: First Person View (FPV) video (medium bandwidth). F: Downwards HD video (high bandwidth). G: Telemetry control (low bandwidth). H/I: Telemetry data (low bandwidth). J: Control commands (low bandwidth). K: Local bidirectional configuration/testing link. L: Local bidirectional configuration/testing link + FPV video (medium bandwidth) + Downwards HD video (high bandwidth) + Telemetry data (low bandwidth).

The Raspberry Pi, which runs a software based in GStreamer and ZeroTier, captures video from the Pi Camera through a CSI interface, and captures telemetry data from the PixHawk 2 Cube controller, through the GPIO interface. Additionally, a USB interface connects the Raspberry with the 4G modem. Finally, a PC acts as a GCS and receives the data through the VPN tunnel. For instance, it could play in real-time the video using GStreamer, VLC or MissionPlanner. In this practical case, the pilot is located near the ground control station, which is also acting as an EXO Data server, so there are only two VPN endpoints. As an extra interface, we enable a second gate to obtain access to the Raspberry through the ad-hoc WiFi network using the SSH protocol. This way, when it is necessary to setup the software before flight, the 4G connection does not have to be used.

Finally, video frames are processed in the EXO Data server, where Faster-RCNN algorithms are located to detect and classify vehicles [21], to track its trajectories applying the IOU-tracker algorithm and the *intersection over union* concept [22], and extracting the relative movement of the drone respect to the ground objects to determine its speed [23].

3. Results and Discussion

The implementation, namely EXO Data, is based on existing commercial products [27], but with important evolutions. The solution links the on-board drone hardware and the ground segment, composed by the GCS (link A) and the novelty of the data server with extra processing capacities for the traffic monitoring (link B), both with different requirements (Figure 6).

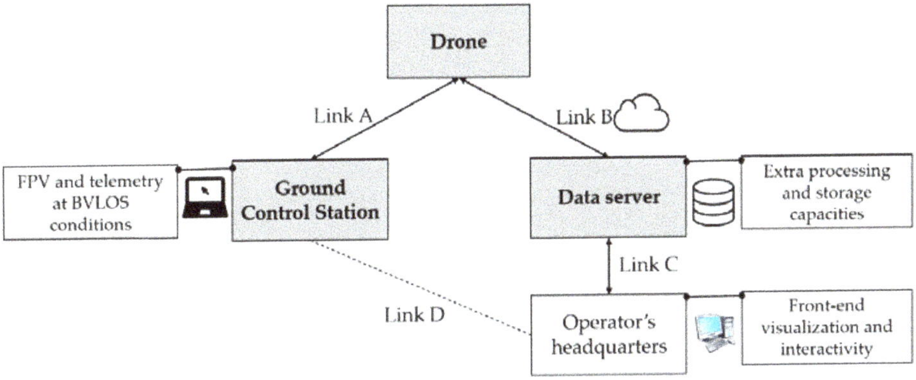

Figure 6. Example of EXO Data logical scheme. Links description. Link A: FPV, telemetry and HD video (optional). Link B: FPV (optional), telemetry and HD video. Link C: CV processed HD video, telemetry insights and Data Base storage. Link D: Telemetry and FPV requests (optional).

3.1. Telecommunications Analysis and Applied Results

The throughput of the 4G network is not symmetrical, having less available throughput in the upstream channel than in the downstream channel. In the case of streaming video transmission, the limiting factor is the throughput in the upstream channel, from the drone to the ground points. The standard specifies a maximum of 100 Mbps in the downstream channel and 50 Mbps in the upstream channel. However, since these transmission systems share the medium with several users, the actual throughput that we are going to obtain will depend on factors such as the load of the network at that time and the interference caused by such. In practice, we found a range from very low rates to peaks of 15 Mbps. Taking as an example a 720 p video at 30 fps, the calculation of the storage needs, in MB, assuming 3 colors per pixel and 8 bits per color and a 1:125 compression rate for one minute of video can be calculated as follows (Equation (1)):

$$C = \Delta c \cdot \Delta r \cdot fps \cdot Nb \cdot \lambda \cdot t \cdot CCR \quad (1)$$

where C is the storage need in MB, Δc is the number of columns, Δr is the number of rows, *fps* is the frames-per-second rate, *Nb* is the number of bits resolution of each color band, λ is the number of bands per pixel, and *CCR* is the Code Compression rate.

One minute of video will require approximately 40 MB of memory. Therefore, a bandwidth of at least 5 Mbps will be necessary to transmit this video in streaming with the appropriate quality. Taking as an example a video in 4 K quality and 30 fps, one minute of video will require approximately 170 MB of memory. We consider that a bandwidth of at least 25 Mbps will be necessary to transmit this video in streaming. This is a high consumption of data, which makes it unaffordable for common users, but with the upcoming 5G technology this scenario can change, as noticed in [41].

Latency can be the most critical point, especially in first person view (FPV) video, which must be below 250 ms following the current regulations in most European countries [42]. Depending on the position of the VPN root servers, the latency can reach 100 ms only due to the system architecture. This, added to the latency of the electronics (between 30 ms and 50 ms) and the one introduced by the 4G network, can in some cases cause problems for the video in real time (UDP protocol). In practice, our tests show that although there are some small micro-cuts, the quality of the FPV is good enough at a resolution of 720 p and 30 fps and the latency is under the 250 ms requirement. However, this value is very volatile and dependent on the signal quality. These results agree with the possibilities described in [9] and further developed in [5,10].

The major concern is the 4G network telecommunication feasibility for such applications, especially the link cuts due to sporadic coverage losses. We carried out theorical and practical experiments to demonstrate its feasibility. Typical antennas deployed for the 4G network have vertical radiation patterns with half-power beam widths that are around 7°. The antennas are subjected to both mechanical and electrical inclinations (tilt) to get them to focus on the coverage area and avoid overreaching with neighboring stations. We made a trigonometric approach, in order to estimate the gain of the antenna as a function of the horizontal distance from the drone to the base station. The vertical angles between drone and antenna, are set between 5° and 20° (taking into consideration the tilt, 5° is considered the maximum of the secondary lobe and 20° is considered out of the secondary lobe). At distances in the range of hundreds of meters in the horizontal plane, the gain of the antenna is expected to be more than 20 dB below its maximum gain. As the distance increases, the vertical angle decreases and the gain of the antenna gets closer to its maximum gain, compensating the loss due to free space propagation. Figure 7 illustrates the above described:

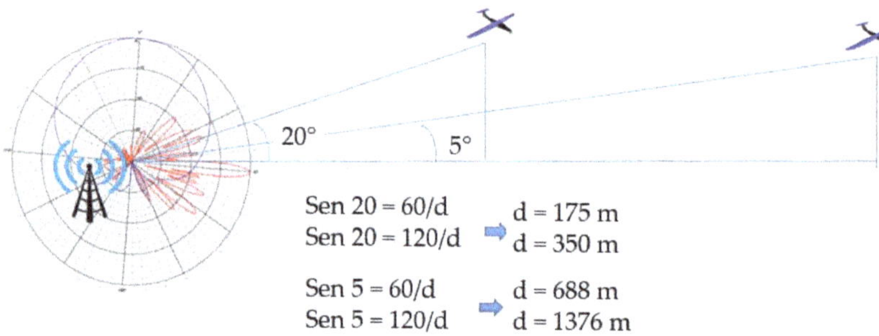

Figure 7. 4G antenna beam lobule distribution pattern: Theorical connectivity range from 4G antennas. (Source: Modified from Kathrein [43]).

Taking a scenario considering the most restrictive frequency (2.6 GHz), a carrier power of 20 Watt, the gain of a typical base station antenna in the direction of maximum radiation (18.5), the free space loss the level of the main lobe to the secondary, and a distance of 1376 m, the resulting expected signal level at the drone antenna is −59.79 dBm. A good signal level in 4G is considered above −75 dBm so there are more than 15 dB of margin (75 dBm–59.79 dBm), that would increase the distance beyond the 2 km. The frequency of 2.6 Ghz is likely to be used in an urban area or an environment with small cells, but for rural environments it is much more likely to use lower frequencies (in the band of 800 Mhz) that would increase the coverage range even further.

Therefore, the drone can receive a good quality 4G signal at distances of around 2 km from the network antenna node, being the lack of coverage more probable in the GCS that in the drone, as explained below. In practice, we found that at flying heights under 120 m the coverage was better than at ground level. The explanation for such behavior is that the topography and the surface elements (vegetation, buildings) are not shadowing the signal when the drone is in the air, while it was compromised in ground before taking off. Moreover, if the 4G network antenna covering the flight area is located on a hill exceeding the height of the drone top altitude (as usual, for instance, in rural areas), the main antenna lobule is covering the drone 4G receiver and the signal is excellent. Nevertheless, if the 4G network antenna is located within a basin position (not common), or the drone if flying at higher altitudes than 120 m above ground level (commonly not allowed), the coverage quality drops. However, the movement of the drone will cause at some point to enter a zero radiation from the network antenna node. This will force the start of a handover protocol to transfer communication to another node on the network, which will interrupt for some seconds the communication. The antenna handovers and the performance indicators depending on the coverage have been analyzed in [7,14], with similar results, but they are dealing with copters and in urban areas with more 4G signal density. In other circumstances, such as in [13], the coverage is worse than in our study, but they are supported with mobile ground station to aid the drone.

There are some telecommunication companies that have agreements with the main telecommunication operators and provide multioperator SIM cards (e.g., Wirelesslogic [36]). Having a multioperator SIM card in the drone on-board modem enables us to connect the operator network with the best receiving signal at the drone antenna. In most developed countries, there are three or more physical network managers (e.g., Vodafone, Orange, Telefonica) and they share the infrastructure. Nevertheless, there will always be a signal better than the others, for instance due to network management parameters, so the multioperator feature adds a significant value. The multioperator SIM analyses the coverage when it becomes active and selects the operator with the best signal to perform the specific flight.

3.2. Traffic Monitoring Applied Results

The video acquired by the drone sensor is transmitted by the 4G network thorough the VPN and received in the EXO Data server endpoint. There, the video is stored in a docker, namely the video log repository. Tacking advantage of the higher computing capacities of the data server (compared with GCS and with the on-boards RPi), the video can be processed with the computer vision algorithms to detect traffic-flow events obtaining near-real time information.

In this work we have used the state-of-the-art algorithm for object detection, Faster R-CNN [21], trained on the Visdrone dataset (http://aiskyeye.com/) for the detection of six previously defined vehicle classes: Pedestrian, bicycle, car, van, bus, and truck. The algorithm processes an input video frame by frame. The output of the algorithm is a bounding box with the coordinates of each detected vehicle and the probability that the detected vehicle belongs to the corresponding vehicle class. The threshold for probability is set to 70%, meaning that only the vehicles that are detected with the probability higher than 70% are considered. After detecting the vehicles, we have applied the IOU-tracker algorithm [22] (https://github.com/bochinski/iou-tracker), assigning a unique ID to each detected vehicle. Finally, the algorithm implemented for speed estimation is based on Li et al. proposal [23]. The imagery analyzed was taken by a static drone where the surface of the road was flat and the camera was mounted with a zenithal orientation, so the distortion given by perspective is minimal.

The three algorithms streamline the process of manually annotating data by a traffic analyst, allowing, among others, the automatic computation of a roundabout matrix. The main applications of these CV algorithm are (Figure 8):

(a) Generation of in-out matrices in roundabouts per vehicle type.
(b) Estimation of vehicle velocities and trajectories.
(c) Heatmap generation per each vehicle class.

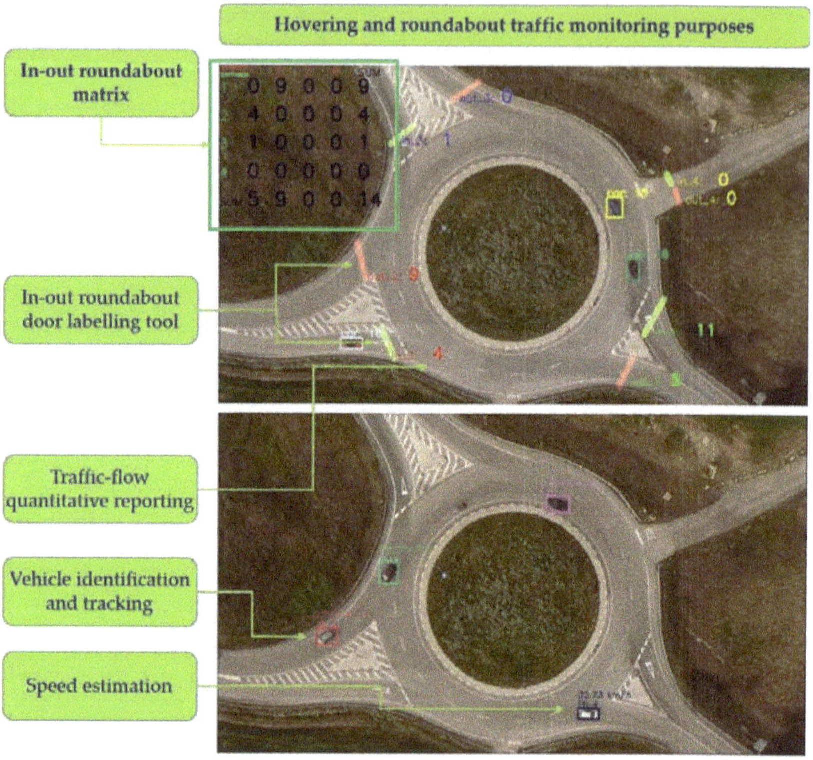

Figure 8. Traffic flow Computer Vision algorithms process the video received in EXO Data server. This allows near real-time information for traffic management, such as in-out roundabout matrix (segmented by object class) and speed estimation.

3.3. Linear Infrastructure Monitoring Applied Results

The photographic imagery acquired by the drone sensor is transmitted by the 4G network thorough the VPN and received in the EXO Data server endpoint. There, the imagery is stored in a docker, namely the photo log repository.

The retrieval of information related to the road infrastructure and its affectation zone is not required in real-time. Moreover, mapping processing to obtain Digital Surface Models (DSM) or orthomosaic products is heavy time consuming and needs all the individual images to mosaic them. Also, it needs geometric [44], radiometric [45] and thematic [46] quality control. Commonly a Structure from Motion (SfM) processing of the images (alignment, dense point cloud generation, mosaicking of images, orthomosaic generation) is used with photogrammetric software, and Remote Sensing and Geographic Information System (RS&GIS) software to extract new information and to combine the drone acquired data with existing cartographic databases.

In our implementation, we used Mapir Survey 3 RGB sensor and Flir Vue Pro R 640 thermal sensor to acquire imagery over linear infrastructures. We used Metashape Photoscan photogrammetric software to perform the SfM processing, and for terrain modelling and land use land cover classification we used QGIS software.

The availability of linear infrastructure networks shapefiles in most of the official mapping agencies avoids the manual digitalization of a given road. Nevertheless, the operator must define the buffer distance from the road axis to delimit the affectation area to be monitored. Once the imagery is processed, georeferenced and clipped with the area of interest, the road manager can exploit and analyze the information (Figure 9).

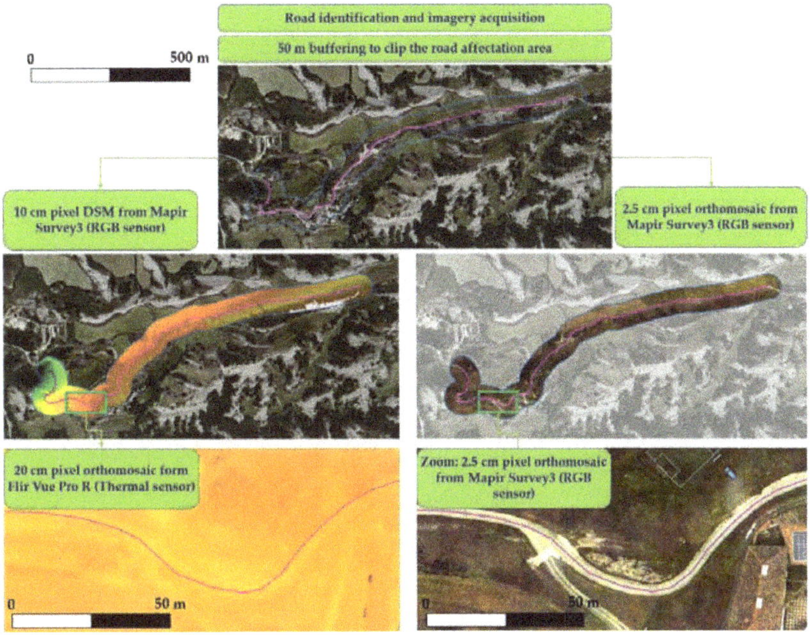

Figure 9. The drone tracks a flight plan following the linear infrastructure and takes images. The imagery is saved in EXO Data dockers and postprocessed using photogrammetric, remote sensing and GIS software. The infrastructure manager can retrieve information for conservation, maintenance or exploitation purposes.

The main applications of these mapping products, are:

(a) Map update of constructed elements surrounding the linear infrastructure.
(b) Detection of horizontal signing and concrete degradation.
(c) Monitoring of conservation/maintenance works.
(d) Locate wildlife paths or other thermal indicators of fauna activity.

4. Conclusions

This work provides a framework and system specification to use the currently deployed 4G telecommunication network to transmit real-time telemetry and video data from a UAS to a control ground station. Results are presented that show how data is being collected and exploited to improve traffic and road infrastructure monitoring. The on-board payload contains, apart from the drone, its flight controller (FC) and its positioning systems (GNSS), an extra microcomputer capable of reading telemetry data from the FC and video from a peripherical camera. A USB-modem and a SIM card connect the drone to the 4G network. The dedicated software allows the tunneling of such data though a Virtual Private Network to the public internet and towards the desired ground station. We have shown the logical schematic of such system and the main parameters needed for it to work, as well as experimental data carried out in a test site to prove the feasibility of the system. The on-board system combined with the 4G terrestrial antennas network have a limited capacity of performance, which was found to limit the upload stream to 9 Mbps and the download stream to 15 Mbps. Also, the 4G coverage is better when the drone is in the air than close to/on the ground, i.e., less than 400 ft (120 m) over the terrain, due to the 4G network antennas lobule shape and the absence of topographic shadows.

The complete 4G framework for drone telemetry and video is a solution for BVLOS operations that overpasses the classical radio-link range. Moreover, we foresee that this study that can be a base for an upcoming upgrade to 5G networks, since the challenges that cannot be reached with 4G technology in terms of data consumption and latency will be reached with the next communication paradigm. Data collection continues and further enhancements of the traffic monitoring tools are anticipated during the final year of the StratoTrans project [18].

Author Contributions: Conceptualization, J.-C.P.; Methodology, J.-C.P. and R.G.; formal analysis, EACOM, J.O.; investigation, R.G., J.-C.P. and A.Z.; UAS data acquisition and flight planning, J.-C.P. and A.Z.; data curation, A.Z.; writing—original draft preparation, R.G.; writing—review and editing, J.-C.P.; visualization, R.G.; supervision, J.-C.P.; project administration, J.-C.P.; Computer Vision formal analysis and supervision, A.B. and P.C.; funding acquisition J.-C.P. All authors have read and agreed to the published version of the manuscript.

Funding: This work was funded by Exodronics SL and was partially supported by the Spanish Government through the Centro para el Desarrollo Tecnológico Industrial (CDTI) under the StratoTrans Project [IDI-20191270] (Proyecto cofinanciado por el Centro para el Desarrollo Tecnológico e Industrial (CDTI) en el marco de la convocatoria de Proyectos de I + D de Transferencia Tecnológica CERVERA).

Institutional Review Board Statement: Not applicable.

Informed Consent Statement: Not applicable.

Acknowledgments: This work was boosted by the R&D department of Exodronics SL, and partially supported by the Spanish Government through the Centro para el Desarrollo Tecnológico Industrial (CDTI) under the StratoTrans Project [IDI-20191270]. We would like to acknowledge the collaborators in the StratoTrans Project, the EURECAT, the Technology center of Catalunya and its Multimedia Technologies Unit with the large experience in Computer Vision research and development, the EACOM Company for formal analysis, the NEXIONA Company for the development of the EXO Data server, and Barcelona Drone Center for the care in the facilities use. Finally, we acknowledge the UAV Matrix Company for the UAV Cast tool and the Wirelesslogic Company for the multioperator SIM card testing.

Conflicts of Interest: The authors declare no conflict of interest.

References

1. Whitehead, K.; Hugenholtz, C.H. Remote sensing of the environment with small unmanned aircraft systems (UASs), part 1: A review of progress and challenges. *J. Unmanned Veh. Syst.* **2014**, *2*, 69–85. [CrossRef]
2. Whitehead, K.; Hugenholtz, C.H.; Myshak, S.; Brown, O.; LeClair, A.; Tamminga, A.; Barchyn, T.; Moorman, B.; Eaton, B. Remote sensing of the environment with small unmanned aircraft systems (UASs), part 2: Scientific and commercial applications. *J. Unmanned Veh. Syst.* **2014**, *2*, 86–102. [CrossRef]
3. Aasen, H.; Honkavaara, E.; Lucieer, A.; Zarco-Tejada, P.J. Quantitative remote sensing at ultra-high resolution with UAV spectroscopy: A review of sensor technology, measurement procedures, and data correction workflows. *Remote Sens.* **2018**, *10*, 1091. [CrossRef]
4. Gharibi, M.; Boutaba, R.; Waslander, S.L. Internet of drones. *IEEE Access* **2016**, *4*, 1148–1162. [CrossRef]
5. Yan, C.; Fu, L.; Zhang, J.; Wang, J. A Comprehensive Survey on UAV Communication Channel Modeling. *IEEE Access* **2019**, *7*, 107769–107792. [CrossRef]
6. Sharma, V. Advances in Drone Communications, State-of-the-Art and Architectures. *Drones* **2019**, *3*, 21. [CrossRef]
7. Ivancic, W.D.; Kerczewski, R.J.; Murawski, R.W.; Matheou, K.; Downey, A.N. Flying Drones Beyond Visual Line of Sight Using 4g LTE: Issues and Concerns. In Proceedings of the 2019 Integrated Communications, Navigation and Surveillance Conference (ICNS), Herndon, VA, USA, 9–11 April 2019; pp. 1–13. [CrossRef]
8. Muruganathan, S.D.; Lin, X.; Maattanen, H.L.; Zou, Z.; Hapsari, W.A.; Yasukawa, S. An Overview of 3GPP Release-15 Study on Enhanced LTE Support for Connected Drones. *arXiv* **2018**, arXiv:1805.00826.
9. Sundqvist, L. Cellular Controlled Drone Experiment: Evaluation of Network Requirements. Master's Thesis, Aalto University, Espoo, Finland, 2015. Available online: https://aaltodoc2.org.aalto.fi/handle/123456789/19152 (accessed on 8 December 2020).
10. Azari, M.M.; Rosas, F.; Pollin, S. Cellular Connectivity for UAVs: Network Modeling, Performance Analysis, and Design Guidelines. *IEEE Trans. Wirel. Commun.* **2019**, *18*, 3366–3381. [CrossRef]
11. Azari, M.M.; Geraci, G.; Garcia-Rodriguez, A.; Pollin, S. UAV-to-UAV Communications in Cellular Networks. *IEEE Trans. Wirel. Commun.* **2020**, *19*, 6130–6144. [CrossRef]
12. LARUS Research Project Website. 2019. Available online: http://larus.kn.e-technik.tu-dortmund.de (accessed on 8 December 2020).

13. Güldenring, J.; Gorczak, P.; Eckermann, F.; Patchou, M.; Tiemann, J.; Kurtz, F.; Wietfeld, C. Reliable Long-Range Multi-Link Communication for Unmanned Search and Rescue Aircraft Systems in Beyond Visual Line of Sight Operation. *Drones* **2020**, *4*, 16. [CrossRef]
14. Azari, M.M.; Arani, A.H.; Rosas, F. Mobile Cellular-Connected UAVs: Reinforcement Learning for Sky Limits. *arXiv* **2020**, arXiv:2009.09815.
15. Chmielewski, P.; Wróblewski, W. Selected issues of designing and testing of a HALE-class unmanned aircraft. *J. Mar. Eng. Technol.* **2017**, *16*, 365–376. [CrossRef]
16. GMV. Pseudo-Satellites, a World of Solutions and Applications. 2019. Available online: https://www.gmv.com/en/Company/Communication/News/2019/10/Hapsview.html (accessed on 8 December 2020).
17. Singh, L.A.; Whittecar, W.R.; DiPrinzio, M.D.; Herman, J.D.; Ferringer, M.P.; Reed, P.M. Low cost satellite constellations for nearly continuous global coverage. *Nat. Commun.* **2020**, *11*, 200. [CrossRef] [PubMed]
18. Exodronics. Cervera-CDTI StratoTrans Project. 2019. Available online: https://exodronics.com/cervera-cdti-programme/ (accessed on 8 December 2020).
19. Shelhamer, E.; Long, J.; Darrell, T. Fully convolutional networks for semantic segmentation. *IEEE Trans. Pattern Anal. Mach. Intell.* **2017**, *39*, 640–651. [CrossRef] [PubMed]
20. Redmon, J.; Divvala, S.; Girshick, R.; Farhadi, A. You Only Look Once: Unified, Real-Time Object Detection. In Proceedings of the IEEE Conference on Computer Vision and Pattern Recognition (CVPR), Las Vegas, NV, USA, 26 June–1 July 2016; pp. 779–788. [CrossRef]
21. Ren, S.; He, K.; Girshick, R.; Sun, J. Faster R-CNN: Towards Real-Time Object Detection with Region Proposal Networks. *IEEE Trans. Pattern Anal. Mach. Intell.* **2017**, *39*, 1137–1149. [CrossRef] [PubMed]
22. Bochinski, E.; Senst, T.; Sikora, T. Extending IOU Based Multi-Object Tracking by Visual Information. In Proceedings of the 15th IEEE International Conference on Advanced Video and Signal Based Surveillance (AVSS), Auckland, New Zealand, 27–30 November 2018; pp. 1–6. [CrossRef]
23. Li, J.; Chen, S.; Zhang, F.; Li, E.; Yang, T.; Lu, Z. An Adaptive Framework for Multi-Vehicle Ground Speed Estimation in Airborne Videos. *Remote Sens.* **2019**, *11*, 1241. [CrossRef]
24. Dahl, M.; Javadi, S. Analytical Modeling for a Video-Based Vehicle Speed Measurement Framework. *Sensors* **2020**, *20*, 160. [CrossRef]
25. Barcelona Drone Center (BDC). Barcelona Drone Center Test Site. 2020. Available online: https://www.barcelonadronecenter.com/uav-test-site/ (accessed on 8 December 2020).
26. Eacomsa. Eacomsa Telecomunicacions. 2020. Available online: https://www.eacomsa.com/ (accessed on 8 December 2020).
27. UAVmatrix. UAV Cast-pro. 2020. Available online: https://uavmatrix.com/ (accessed on 8 December 2020).
28. Nexiona. Miimetiq Composer IoT. 2020. Available online: https://nexiona.com/miimetiq-composer/ (accessed on 8 December 2020).
29. Eurecat. Audiovisual Technologies. 2020. Available online: https://eurecat.org/en/field-of-knowledge/audiovisual-technologies/ (accessed on 8 December 2020).
30. Exodronics. EXO C2-L Fixed Wing Drone. 2019. Available online: https://exodronics.com/exo-c2-l/ (accessed on 8 December 2020).
31. ProfiCNC. Pixhawk 2.1 Cube Black. 2020. Available online: http://www.proficnc.com/ (accessed on 8 December 2020).
32. Ardupilot. The Cube Overview. 2020. Available online: https://ardupilot.org/copter/docs/common-thecube-overview.html (accessed on 8 December 2020).
33. Raspberry Pi. Raspberry Pi Zero W. 2020. Available online: https://www.raspberrypi.org/products/raspberry-pi-zero-w/ (accessed on 8 December 2020).
34. Huawei. HUAWEI 4G Dongle E3372. 2020. Available online: https://consumer.huawei.com/en/routers/e3372/specs/ (accessed on 8 December 2020).
35. Wirelesslogic. SIMpro Management Platform. 2020. Available online: https://www.wirelesslogic.com/simpro/ (accessed on 8 December 2020).
36. Wirelesslogic. Case Study. Exodronics: Optimizing Drones through IoT Connectivity. 2020. Available online: https://www.wirelesslogic.com/case-study/exodronics/ (accessed on 8 December 2020).
37. Raspberry Pi. ZeroCam FishEye. 2020. Available online: https://raspberrypi.dk/en/product/zerocam-fisheye/ (accessed on 8 December 2020).
38. Raspberry Pi. Camera Module V2. 2020. Available online: https://www.raspberrypi.org/products/camera-module-v2/ (accessed on 8 December 2020).
39. ZeroTier. ZeroTier VPN Server. Available online: https://www.zerotier.com/ (accessed on 8 December 2020).
40. Ardupilot. Mission Planner. 2020. Available online: https://ardupilot.org/planner/ (accessed on 8 December 2020).
41. Ferro, E.; Gennaro, C.; Nordio, A.; Paonessa, F.; Vairo, C.; Virone, G.; Argentieri, A.; Berton, A.; Bragagnini, A. 5G-Enabled Security Scenarios for Unmanned Aircraft: Experimentation in Urban Environment. *Drones* **2020**, *4*, 22. [CrossRef]
42. European Union Aviation Safety Agency (EASA). Drones—Regulatory Framework Background. 2019. Available online: https://www.easa.europa.eu/easa-and-you/civil-drones-rpas/drones-regulatory-framework-background (accessed on 8 December 2020).
43. Thiele, L.; Wirth, T.; Börner, K.; Olbrich, M.; Jungnickel, V.; Rumold, J.; Fritze, S. Modeling of 3D field patterns of downtilted antennas and their impact on cellular systems. In Proceedings of the ITG International Workshop on Smart Antennas (WSA), Berlin, Germany, 16–19 February 2009.

44. Padró, J.C.; Muñoz, F.J.; Planas, J.; Pons, X. Comparison of four UAV georeferencing methods for environmental monitoring purposes focusing on the combined use with airborne and satellite remote sensing platforms. *Int. J. Appl. Earth Obs. Geoinf.* **2019**, *79*, 130–140. [CrossRef]
45. Padró, J.C.; Muñoz, F.J.; Avila, L.A.; Pesquer, L.; Pons, X. Radiometric Correction of Landsat-8 and Sentinel-2A Scenes Using Drone Imagery in Synergy with Field Spectroradiometry. *Remote Sens.* **2018**, *10*, 1687. [CrossRef]
46. Padró, J.C.; Carabassa, V.; Balagué, J.; Brotons, L.; Alcañiz, J.M.; Pons, X. Monitoring opencast mine restorations using Unmanned Aerial System (UAS) imagery. *Sci. Total Environ.* **2018**, *657*, 1602–1614. [CrossRef] [PubMed]

drones

Article

SuSy-EnGaD: Surveillance System Enhanced by Games of Drones

Daniel H. Stolfi [1,*], Matthias R. Brust [1], Grégoire Danoy [1,2] and Pascal Bouvry [1,2]

1 SnT, University of Luxembourg, 6, Avenue de la Fonte, L-4364 Luxembourg, Luxembourg; matthias.brust@uni.lu (M.R.B.); gregoire.danoy@uni.lu (G.D.); pascal.bouvry@uni.lu (P.B.)
2 FSTM/DCS, University of Luxembourg, 6, Avenue de la Fonte, L-4364 Luxembourg, Luxembourg
* Correspondence: daniel.stolfi@uni.lu

Abstract: In this article, we propose SuSy-EnGaD, a surveillance system enhanced by games of drones. We propose three different approaches to optimise a swarm of UAVs for improving intruder detection, two of them featuring a multi-objective optimisation approach, while the third approach relates to the evolutionary game theory where three different strategies based on games are proposed. We test our system on four different case studies, analyse the results presented as Pareto fronts in terms of flying time and area coverage, and compare them with the single-objective optimisation results from games. Finally, an analysis of the UAVs trajectories is performed to help understand the results achieved.

Keywords: swarm robotics; unmanned aerial vehicle; evolutionary game theory; evolutionary algorithm; surveillance system; multi-objective optimisation

Citation: Stolfi, D.H.; Brust, M.R.; Danoy, G.; Bouvry, P. SuSy-EnGaD: Surveillance System Enhanced by Games of Drones. *Drones* **2022**, *6*, 13. https://doi.org/10.3390/drones6010013

Academic Editors: Diego González-Aguilera and Pablo Rodríguez-Gonzálvez

Received: 1 December 2021
Accepted: 3 January 2022
Published: 6 January 2022

1. Introduction

Unmanned aerial vehicles (UAVs) have a wide range of applications [1] such as wildfire monitoring [2], smart farming [3], aerial surveillance [4], environmental monitoring [5], goods transportation [6] and structural damage mapping [7]. Particularly, surveillance systems [8] require resilience and flexibility to address issues such as drone failures and adverse communications. They can be achieved using swarm intelligence [9] as a way of modifying the collective behaviour of the swarm by using individual parameters. The members' interactions usually follow local rules based on pheromones and probabilities [10], upper confidence trees [11] or finite games [12], among others [13]. These strategies can be based on competitions [14] or collaborations [15] between members, having each approach its own advantages and disadvantages [16]. The associated high dimensional search space makes these problems good candidates to be solved using an intelligent bio-inspired technique, such as evolutionary algorithms [17].

Game-based approaches are commonly present in UAV swarms as a strategy of cooperation between members. Games have been used for cooperative search and surveillance [18], energy-efficient communications [19], beyond-visual-range air combat missions [20], Vehicular Ad hoc NETworks (VANET) communications under adverse network conditions [21], etc. In Evolutionary Game Theory (EGT) [22], group interactions are modelled with the assumption that the surviving strategy is the one which reports outcomes higher than other possible strategies.

In this research work, we propose a surveillance system based on a swarm of UAVs patrolling an area divided into concentric security rings. UAVs in inner rings move slower than those in outer rings but consume also less energy, which allows them to fly for a longer period of time. This surveillance scheme features a compromise between maximum flying time (most of UAVs in the innermost ring) and maximum area coverage (a smart strategy to fly by different rings according to each UAV's battery state). Consequently, our contributions can be summarised as follows:

1. A new Surveillance System Enhanced by Games of Drones (SuSy-EnGaD), based on cooperative UAVs that explore the area of interest arranged in concentric rings.
2. Three different approaches to obtain the optimal strategy taking into account maximum flying time and area coverage.
3. Two bio-inspired evolutionary algorithms adapted to this problem based on the well-known NSGA-II and a genetic algorithm.

The remainder of this paper is organised as follows. In the next section, we review the state of the art related to our work. In Section 3, the SuSy-EnGaD architecture is presented. Our optimisation algorithms are explained in Section 4 and the case studies are described in Section 5. The experimentation done using simulations is presented in Section 6, as well as the discussion of results. After that, Section 7 brings conclusions and future work.

2. Related Work

In this section, we review research works related to our proposal. First, we go through the multi-objective optimisation of UAV related problems. Second, energy optimisation proposals are analysed. Third, surveillance systems using UAVs are reviewed. And finally, cooperative and competitive strategies for UAV swarms are commented.

Many problems related to UAVs are often modelled and solved as multi-objective ones. In [23], a multi-objective path planning framework is proposed to explore a suitable path for a UAV operating in a dynamic urban environment. The authors use safety index maps to capture static (offline search) and dynamic (online search) obstacles. This path planning problem is addressed taking into account two objectives: shorten the travel time and avoid obstacles. A multi-objective optimisation algorithm to allocate tasks and plan paths for a team of UAVs is presented in [24]. A genetic algorithm is used to minimise the mission completion time and can be tuned to prioritise coverage or connectivity. Results obtained via simulations indicate that by transmitting lower rate notifications in the network, the mission time can be shortened. In [25] a multi-objective path planning approach based on the Crowd Distance-based NSGA-II (CDNSGA-II) method is proposed to find an optimal collision-free path for UAVs, taking into account both distance and safety. Experimental results show that the proposed algorithm can obtain up to 80% pareto optimal solutions (see definitions 1 and 2 in Section 3) when compared with NSGA-II under simulated urban environments.

Battery saving is an important concern when using UAVs. An offline path planning algorithm is proposed in [26] to ensure that UAVs have permanent connectivity and can always reach the base station to recharge their batteries. Different approaches for heterogeneous UAVs and multi-base stations were analysed to obtain safe paths using simulations. In [27], an energy-efficient algorithm to optimise fixed-wing UAVs trajectories is proposed. The authors derive a theoretical model of the propulsion energy consumption of fixed-wing UAVs to define the efficiency of communications. They conclude that an optimised simple circular trajectory maximises the energy efficiency and apply their findings to more general itineraries. In our study, quad-rotor UAVs move inside circular rings using unpredictable zigzag trajectories and can change the flying ring during their mission. As we are proposing a surveillance system, area coverage is taken into account in the optimisation process and our results are obtained using a simulator. In [28], the authors present a novel framework for stochastic UAV-assisted surveillance which considers battery constraints. The system uses energy-efficient random walks for flying patterns and probabilistic inspections. A centralised algorithm based on iterative geometric programming approximation was used to solve the problem and the experimentation was conducted using simulations. We propose random walks for sampling the problem's solution landscape to confirm that it is a multi-objective problem. Our UAVs' trajectories always depend on surveillance rings, random bounce angles, and the possible collaborations between drones.

Surveillance systems using UAVs is one of the most discussed applications of these "eyes in the sky". Coordination between UAVs using chaotic mobility combined with pheromone trails is proposed in [29]. In this approach, vehicles explore the surveillance area using detection cells arranged into concentric square rings, stochastically avoiding high concentrations of pheromones. An Evolutionary Algorithm (EA) optimises pheromone amounts and assigns rings to maximise early intruder detection and protect the base in the centre of the analysed scenarios. The advantages of using surveillance rings are discussed in [29], where that approach is also compared with other strategies. In our present study, we are using actual circular rings and UAVs dynamically change rings during the mission by collaborating each other using different strategies. Our proposed trajectories are not using pheromones and they were evaluated using a simulator that includes the full UAV dynamics. Finally, a multi-objective optimisation and game approach were implemented for our current study.

Path planning and coordination of multiple UAVs to provide convoy protection to ground vehicles is proposed in [30]. Different scenarios with stationary and moving convoys were analysed using UAVs, modelled as Dubins vehicles flying at constant altitude. The authors proposed a coordination strategy and optimal paths, also calculating the minimal number of UAVs required. In [31], an optimal navigation algorithm allowing UAVs to determine their movement locally with a minor use of a central station is presented. UAVs perform surveillance task for a group of moving targets while avoiding obstacles. Simulations results using case studies having up to 600 targets are presented to confirm the system's performance. Heterogeneous multiswarm approaches comprise the use of patrolling UAVs and UGVs (Unmanned Ground Vehicles) to serve as mobile refuelling stations [32]. Emerging UAV-UGV inter-swarm collaborations are analysed in [15], and in [33], UAVs, UGVs and UMVs (Unmanned Maritime Vehicles) are combined to improve detections and area coverage on the land and sea.

Finally, we analyse some research works using Evolutionary Game Theory (EGT) [22] approaches to model problems as well as to discover evolving strategies for optimisation [34]. In EGT, players are populations of individuals who follow mixed strategies when playing bimatrix games. In contrast to classical game theory, the surviving strategy after a group of interactions achieves higher benefits than the case of players making rational choices. In [35], packet forwarding strategies are optimised in a mobile wireless ad hoc network (MANET). A Prisoner's Dilemma (PD) model (see Section 3.3) and an EA are proposed to enforce cooperation between nodes. In [16], competition or cooperation are both analysed as possible strategies for drones mapping a disaster area. The problem is modelled as allocation tasks to robots in a swarm under limited communications and partial information, and is solved by a competitive algorithm and a cooperative one. The latter was reported to allocate more tasks in all the analysed scenarios. A modified binary log-linear learning (BLLL) algorithm is proposed in [36] to solve the covering problem using multiple UAVs. The cooperative search problem is modelled as a potential game and a novel action selection strategy for UAVs is proposed. Experiments simulating different mission environments were designed to evaluate the effectiveness and feasibility of the proposed learning algorithm.

Our research work has parts in common with these articles, e.g., cooperation using EGT, results obtained via simulations and bio-inspired multi-objective optimisation of UAVs' trajectories. However, we propose a surveillance area arranged in concentric circular rings, having each one its own flying constraints, plus three approaches using multi-objective and single-objective evolutionary algorithms to calculate the configurations of the UAVs. Some analysed articles involving UAVs focus on visiting predefined targets frequently. Our surveillance proposal in turn, uses unpredictable trajectories to explore the whole area, balancing coverage and battery consumption, which is something critical for every system using UAVs. To the best of our knowledge, no previous work has proposed the evaluation and use of these five evolving strategies for UAV cooperation in a surveillance system using concentric circular rings and trajectories that cannot be easily predicted by trespassers.

3. Surveillance System Enhanced by Games of Drones (SuSy-EnGaD)

The proposed surveillance architecture is based on M security rings where UAVs patrol the area of interest protecting a central base as illustrated in Figure 1. The UAVs are equipped with a video camera directed downwards to scan the area and detect possible intruders. The altitude of the UAVs is fixed, so that the area scanned is about 10×10 m (assumed to be square instead of a 16:9 rectangle for simplicity). Since we are not analysing intruder detection rates in our study, we have calculated the area explored using the current position of each UAV, instead of using a specific camera model. This simplifies the simulation model increasing the overall efficiency without losing accuracy. The innermost ring, R_0, is a no-fly zone from where the UAVs depart and eventually return to recharge their batteries. Initially, all the UAVs are assigned to R_1 and they can move to another ring $R_i \leq R_M, i > 0$ depending on their battery level and a series of autonomous decisions (strategy). Since each ring has its own speed restrictions (V_R), such as $V_{R_1} < V_{R_2} < V_{R_M}$, UAVs in inner rings consume less battery than the others in outer rings (and explore a smaller area). When a UAV reaches the external border it bounces back to the surveillance area with an angle θ_b, randomly calculated such as $-\frac{\pi}{6} \leq \theta_b \leq \frac{\pi}{6}$. On the contrary, when a UAV is on the border of a ring, the bouncing angle is θ_r, randomly calculated in the range $-\frac{\pi}{3} \leq \theta_r \leq \frac{\pi}{3}$. The use of a smaller range for θ_b (compared with θ_r) is to prevent UAVs to patrol the surface out of the surveillance area (secant trajectories) while the use of random values adds unpredictability to the UAVs' trajectories, which is desired in surveillance missions.

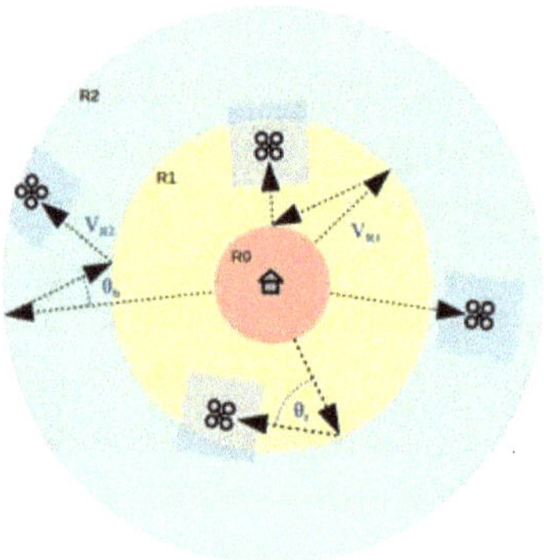

Figure 1. Schema of the Surveillance System Enhanced by Games of Drones (SuSy-EnGaD) where rings, velocities and angles for UAVs are depicted.

The surveillance system as an optimisation problem presents two metrics to be maximised in our study: the area coverage and the vehicles' flying time. By staying in inner rings, UAVs save battery and extend their flying time, but the area coverage is small. On the contrary, if there are too many UAVs in outer rings flying at higher speeds, their batteries will be drained too soon, reducing the system's efficiency.

To dynamically control how UAVs spread by the surveillance area, we propose three approaches for UAVs to decide if they wish to cooperate with their partners by changing rings, or defects and stay in the current ones (Figure 2). Indeed, we call cooperation when a UAV moves to a different flying ring after interacting with another UAV, and defection when it decides to ignore the proposal and stay in its current ring. There exist also a coordination between UAVs when they are in a collision trajectory which is mandatory to keep a safe distance between them (see Section 5). Interactions occur when two UAVs are in their respective communication range (limited to 20 m in our study) where a new ring, in case of cooperation, is calculated according to the relative battery state and the UAV position in the map. Consequently, depending on the decision rules, if both UAVs are in the same ring, the UAV with lower battery charge will move to the next inner ring (if any) and the other will move to the next outer ring (if it cooperates). Otherwise, they will change rings independently of the battery state, which enforces the use of strategies to improve the global metrics of the system by making local decisions (mini-games) that sometimes might be adverse for one or both players (altruism). Despite that, if one UAV (or both) decides not to cooperate, it (they) will stay in the same ring, regardless their battery charge. For efficiency reasons, there is a minimum time between games, i.e., 20 s, and a minimum amount of battery is also required, i.e., 20%. When one UAV has begun a game with another, any other game proposal is also discarded during the next 20 s.

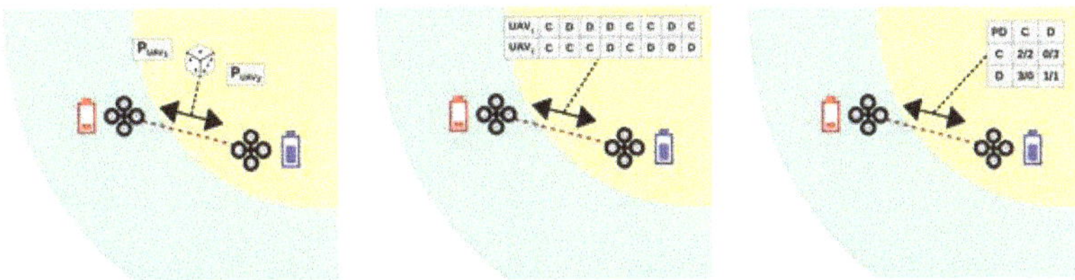

(**a**) Probability (**b**) Strategy (**c**) Game

Figure 2. The three proposed approaches for UAVs to decide whether to cooperate and change ring or to defect and stay in the current ring.

The first two approaches to optimising SuSy-EnGaD consist in modelling and solving a multi-objective optimisation problem (MOP). In the *Probability* and *Strategy* approaches, instead of a single solution, a set of pareto optimal solutions are obtained for the two metrics to be maximised: flying time and area coverage. Flying time corresponds to the time at which one of the UAVs in the swarm reaches 10% of battery (giving it enough time for a safe return to base for recharging). Area coverage was measured as the percentage of the area scanned by the onboard cameras (an area of 10 × 10 m).

The main goal of solving a MOP is to find a set of feasible solutions that maximises (in our case, although it can be also defined in terms of minimisation) the objective function vector $\vec{f} = (f_1, \dots, f_D)$, where D is the number of objectives (two in our study), according to the notion of Pareto-optimality, defined as follows:

Definition 1. *Given the vectors $\vec{x_1}$ and $\vec{x_2}$, we define the solution dominance in a maximisation problem as*

$$\vec{x_1} \prec \vec{x_2} \; if \; \forall i \in \{1, \dots, D\} f_i(\vec{x_1}) \geq f_i(\vec{x_2}) \tag{1}$$

Definition 2. *The set of solutions not dominated by any other in the solution space S is called the Pareto optimal set (P), defined as*

$$P = \{\vec{x} \in S | \nexists \vec{x\prime} \in S : \vec{x\prime} \prec \vec{x}\} \tag{2}$$

The third approach, *Game*, consists in maximising the average score (outcome) achieved by the UAV swarm as a method for obtaining a solution to the surveillance problem. In this case, it is clearly a single-objective optimisation problem. In the following sections, we describe each approach in detail.

3.1. Probability Approach

In the *Probability* approach each UAV i has a real parameter (p_i) in the range $0.0 \leq p_i \leq 1.0$ defining its cooperation probability. The problem representation is the vector $\vec{x_p}$ composed of N probability values (Equation (3)), where N is the number of UAVs. Different cooperation probability values modify the way UAVs interact, allowing strategies that save battery as well as others that cover a bigger area.

$$\vec{x_p} = \{p_1, p_2, p_3, \dots, p_N\} \tag{3}$$

3.2. Strategy Approach

The *Strategy* approach proposes for each UAV a strategy defined as a list of bits where "1" means cooperation and "0", defection. The problem representation is the vector $\vec{x_s}$ containing the list of B strategy bits (b_i) for each one of the N UAVs (Equation (4)). In our study, we have set $B = 6$ as a balance between problem complexity (similar to the *Probability* approach) and strategy accuracy. This is a cyclic approach, different from *Probability*, where after six games the strategy of one UAV starts over.

$$\vec{x_s} = \{b_{1_1}, \ldots, b_{B_1}, \ldots, b_{1_N}, \ldots, b_{B_N}\} \tag{4}$$

3.3. Game Approach

We propose a third approach based on two-person games. In this case, when two UAVs have to decide a possible cooperation, they will do so by trying to maximise their outcome according to the reward matrices shown in Figure 3, where the Nash equilibria [37] are marked with asterisks.

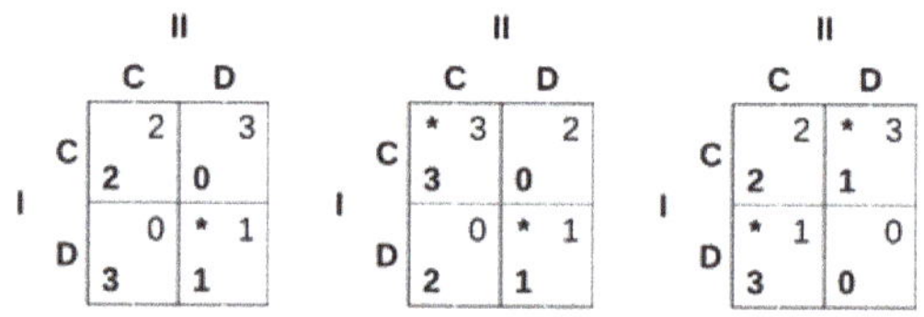

(**a**) PD (**b**) AG (**c**) CG

Figure 3. Three two-person games: prisoner's dilemma (PD), assurance game (AG), and game of chicken (CG) as proposed in [38]. I and II are the respective players and, C and D stand for cooperation and defection, respectively. Nash equilibria are marked with asterisks.

In the prisoners' dilemma (PD) [38], two prisoners are given the choice of testifying against each other or keeping silent. The best possible outcome is defecting when the other player cooperates (DC). Then comes mutual cooperation (CC) and finally mutual defection (DD) as represented in Figure 3a. The assurance game (AG) (Figure 3b) represents the situation in which a person would be willing to cooperate as long as he/she is assured that their partner would cooperate as well. The better outcome is obtained when there is mutual cooperation (CC), although mutual defection (DD) is also one of the game's equilibria as one person will defect if it thinks that the other will also defect [38]. And finally, the game of chicken (CG), also known as the game of dare, in which two cars are driven one towards the other. The driver who turns away is "chicken" and loses while the other wins. Of course, if no one turns, both drivers die and lose (DD) [38]. There are also two equilibria in CG, unilateral cooperation (CD) and unilateral defection (DC) as shown in Figure 3c.

In the *Game* approach, the UAVs are engaged in a finite two-person game with limited interaction [39]. As the number of iterations is given by how frequently the UAVs meet each other during the simulation, we propose to calculate the best strategy for each player using a bio-inspired meta-heuristic: a genetic algorithm. The *Game* approach uses the same problem representation as in the *Probability* approach (Equation (3)), although, in this case, the objective is maximising the average score of UAVs by modifying their cooperation probability. The evaluation function for this single-objective problem is shown in Equation (5). There, *outcome* refers to the average score achieved by a given UAV after a number of iterations (games) with the other UAVs, obtained from the simulation.

$$F(\vec{x}) = \frac{1}{N} \sum_{i}^{N} outcome(UAV_i) \tag{5}$$

4. Optimisation Algorithms

Multi-objective optimisation problems are usually solved by using one of the following techniques: (i) converting the multi-objective problem into a single-objective one by using a weighted-sum function and (ii) getting a set of Pareto optimal solutions and selecting the most appropriated according to an expert's criterium. The former, despite its simplicity, implies a difficult weight selection that restricts the working solutions during the optimisation process. On the other hand, the latter involves the use of a more complex algorithm, although it increases the versatility and does not require the normalisation of the objectives into the same order of magnitude.

We propose two evolutionary bio-inspired techniques to optimise the parameters of our three approaches. These are efficient methods for solving combinatorial optimisation problems which simulate processes present in evolution such as natural selection, gene recombination after reproduction, gene mutation, and the dominance of the fittest individuals over the weaker ones. Our proposed optimisation algorithms are the Non-dominated Sorting Genetic Algorithm (NSGA-II) for the *Probability* and *Strategy* approaches (multi-objective) and a generational Genetic Algorithm (GA) for the *Game* approach (single-objective).

These optimisation algorithms are to be run in a computer cluster to obtain the optimal solutions for each case study in an initial configuration stage (offline). The solutions achieved, consisting of the optimal configuration of the UAV swarm for each case study, are to be used as the parameters for each UAV to rule the possible collaborations during the surveillance missions (online).

4.1. Non-Dominated Sorting Genetic Algorithm (NSGA-II)

NSGA-II [40] is a multi-objective optimisation algorithm that implements a non-dominated sorting approach to rank solutions based on their Pareto dominance relation. Following the pseudocode in Algorithm 1, after initialising the population composed by $N_i = 28$ individuals, the main loop is executed while the termination condition holds (3000 evaluations in our case). The operators used were chosen according to our problem characteristic and representation. The selection operator is Binary Tournament, Single Point Crossover ($P_c = 0.9$), Integer Polynomial Mutation for the *Probability* approach and Bit Flip Mutation for the *Strategy* approach (for both $P_m = \frac{1}{L}$, where L is the length of the configuration vector), and finally, the replacement operator uses the Ranking and Crowding distance selection for maintaining solution diversity [40].

Algorithm 1. Pseudocode of NSGA-II.

```
procedure NSGA-II(N_i, P_c, P_m)
    t ← 0
    Q(0) ← ∅                                        ▷ Q=auxiliary population
    P(0) ← Initialisation(N_i)                      ▷ P=population
    while not Termination_Condition() do
        Q(t) ← Selection(P(t))
        Q(t) ← Crossover(Q(t), P_c)
        Q(t) ← Mutation(Q(t), P_m)
        Evaluation(Q(t))
        R ← Ranking_and_Crowding(Q(t), P(t))
        P(t+1) ← Select_Best_Individuals(R)
        t ← t+1
    end while
end procedure
```

4.2. Genetic Algorithm (GA)

The proposed GA [41,42] is a single-objective optimisation algorithm featuring a population of individuals which evolve to maximise (in our case) their fitness value. The pseudocode of GA is presented in Algorithm 2. Initially, 28 individuals (μ) for the population P are generated and while the termination condition holds the main loop is executed. This is a generational GA where the working population Q has the same number of individuals as P ($\lambda = \mu$). We used Binary Tournament as selection operator, Single Point Crossover as recombination operator, Integer Polynomial Mutation as mutation operator, and an elitist replacement. The parameters of GA are the same as in NSGA-II, i.e., 3000 evaluations, $Pc = 0.9$ and $Pm = \frac{1}{L}$.

Algorithm 2. Pseudocode of GA.

```
procedure GA(N_i, P_c, P_m)
    t ← 0
    Q(0) ← ∅                                          ▷ Q=auxiliary population
    P(0) ← Initialisation(N_i)                        ▷ P=population
    while not Termination_Condition() do
        Q(t) ← Selection(P(t))
        Q(t) ← Crossover(Q(t), P_c)
        Q(t) ← Mutation(Q(t), P_m)
        Evaluation(Q(t))
        P(t+1) ← Replacement(Q(t), P(t))
        t ← t+1
    end while
end procedure
```

4.3. Genetic Operators

In this section, we describe the operators used in both algorithms. We have used the implementation provided by the jMetalPy package [43] for the operators and both optimisation algorithms.

4.3.1. Selection

We have used Binary Tournament [44] as selection operator. It is described in Algorithm 3 where two random samples are taken from the population Q. If it is the multi-objective case (NSGA-II), each sample comprises a pareto from the population while in the case of single-objective optimisation (GA), each sample is an individual representing one configuration of the surveillance system. Then, the samples are compared in terms of dominance (pareto fronts) or fitness (system configurations) and the best of them is included in Q', which will become the working population for the current generation. This process is repeated until the λ required individuals are obtained.

4.3.2. Crossover

Single Point Crossover [45] was used as crossover operator in both algorithms. As shown in Algorithm 4, two individuals $\vec{x}$ and $\vec{y}$ are taken from the population Q to be recombined subject to the crossover probability $P_c = 0.9$. After selecting the crossing point p, the components of $\vec{x}$ and $\vec{y}$ are swapped from the position p to the end of each solution vector. The new resulting vectors $\vec{x}'$ and $\vec{y}'$ are now added to the new working population Q'. By doing so, the optimisation algorithm explores different areas of the solution space in parallel, searching for new promising configurations for the swarm.

Algorithm 3. Pseudocode of the Binary Tournament Selection Operator.

```
function SELECTION(Q)
    Q' ← ∅
    for i ← 1, λ do                          ▷ Q' will have λ individuals
        s1 ← random_sample(Q)                ▷ randomly takes two individuals from Q
        s2 ← random_sample(Q)                ▷ they are pareto fronts in NSGA-II
        if better(s1, s2) then               ▷ compare in terms of fitness/dominance
            Q' ← Q' ∪ {s1}
        else
            Q' ← Q' ∪ {s2}
        end if
    end for
    return Q'
end function
```

Algorithm 4. Pseudocode of the Single Point Crossover Operator.

```
function CROSSOVER(Q, Pc)
    Q' ← ∅
    for all {x⃗, y⃗} ∈ Q do                   ▷ all the individuals in Q, taken in pairs
        x⃗' = x⃗
        y⃗' = y⃗
        if rnd() < Pc then                   ▷ crossover probability
            p ← randInt(1, L)                ▷ crossing point p, L is the length of the solution vector
            for i ← p, L) do                 ▷ swaps vector's components from p to L
                x⃗'[i] = y⃗[i]
                y⃗'[i] = x⃗[i]
            end for
        end if
        Q' ← Q' ∪ {x⃗', y⃗'}                  ▷ adds the new vectors to the result
    end for
    return Q'
end function
```

4.3.3. Mutation

Two mutation operators were used. Bit Flip Mutation for the binary representation (*Strategy* approach) and Integer Polynomial Mutation for the *Probability* and *Game* approaches. They are meant to make small variations to the configuration vectors and explore the neighbourhood of the good solutions already found. The selected value for $P_m = \frac{1}{L}$ stochastically selects one position of each solution vector (individual of the working population) to be changed. The Bit Flip Mutation [46] consists in randomly selecting bits in the solution vector to be flipped as shown in Algorithm 5. After all the individuals were considered for a possible mutation the resulting population Q' is returned.

The Integer Polynomial Mutation [47] was selected to work with the numeric values in the vector of probabilities. To simplify the search process we have worked with integer values between 0 and 100 to represent probabilities between 0.0 and 1.0 with an accuracy of two decimal places. The Algorithm 6 shows the pseudocode of this operator. It can be seen that for each position in the solution vector subject to be mutated, four values are calculated depending on the current position's value, the maximum allowed value (100), and a new random value (ρ) as well. A Δ_q value is calculated depending on ρ, to modify the original value, subject to the right range of values.

Algorithm 5. Pseudocode of the Bit Flip Mutation Operator.

```
function BITFLIPMUTATION(Q, P_m)
    Q' ← ∅
    for all {x⃗} ∈ Q do                                    ▷ all the individuals in Q
        x⃗' ← x⃗
        for i ← 1, L do          ▷ i goes from 1 to L (the length of the solution vector)
            if rnd() < P_m then                                ▷ mutation probability
                x⃗'[i] ← not(x⃗[i])                                         ▷ bit flip
            end if
        end for
        Q' ← Q' ∪ {x⃗'}
    end for
    return Q'
end function
```

Algorithm 6. Pseudocode of the Integer Polynomial Mutation Operator.

```
function INTEGERPOLYNOMIALMUTATION(Q, P_m)
    Q' ← ∅
    for all {x⃗} ∈ Q do                                    ▷ all the individuals in Q
        x⃗' ← x⃗
        for i ← 1, L do          ▷ i goes from 1 to L (the length of the solution vector)
            if rnd() < P_m then                                ▷ mutation probability
                δ1 ← x⃗'[i]/100; δ2 ← (100 − x⃗'[i])/100; π = 1/(σ+1); ρ ← rnd()   ▷ σ = 0.2 by default
                if ρ ≤ 0.5 then                                        ▷ equiprobable
                    Δq = (2 × ρ + (1 − 2 × ρ) × (1 − δ1)^(σ+1))^(π−1)
                else
                    Δq = 1 − (2 × (1 − ρ) + (2 × (ρ − 0.5) × (1 − δ2)^(σ+1))^π
                end if
                y ← Bounds(x⃗'[i] + Delta_q × 100)_{0,100}     ▷ keeps it in range [0, 100]
                x⃗'[i] ← round(y)                              ▷ nearest integer value
            end if
        end for
        Q' ← Q' ∪ {x⃗'}
    end for
    return Q'
end function
```

5. Case Studies

To evaluate SuSy-EnGaD using the three approaches previously discussed, four case studies are proposed featuring different numbers of UAVs and map dimensions. The characteristics of the case studies are detailed in Table 1, as well as the nomenclature used, i.e., UAVS.MAPSIZE. The innermost ring corresponds to the area from where the UAVs depart, and after this initial stage, it becomes a no-fly zone. The evaluation of the UAVs configuration, according to each approach, was performed using the ARGoS simulator [48]. ARGoS is a multi-physics robot simulator that can simulate large-scale swarms of robots of any kind, efficiently. In our study, we simulate UAVs using the eye-bot drones [49] provided by ARGoS, including their communication and battery consumption models. This allows us to efficiently test multiple configurations keeping in mind reliability and safety since a wrong configuration does not end in a catastrophic collision while many scenarios can be evaluated in parallel. Depending on the number of robots, the departing point and the initial heading angle was predefined to avoid excessive use of the collision avoidance algorithm at the beginning of the simulation. Figure 4 shows the initial configuration for 4, 8, and 12 UAVs, all departing from the innermost ring (in red).

Table 1. Characteristics of the Four Case Studies.

Case Study	UAVs	Map Size (m)	Rings	Radii
4.100	4	100 × 100	3	5/28/50
8.100	8	100 × 100	3	5/28/50
8.200	8	200 × 200	3	5/53/100
12.200	12	200 × 200	3	5/53/100

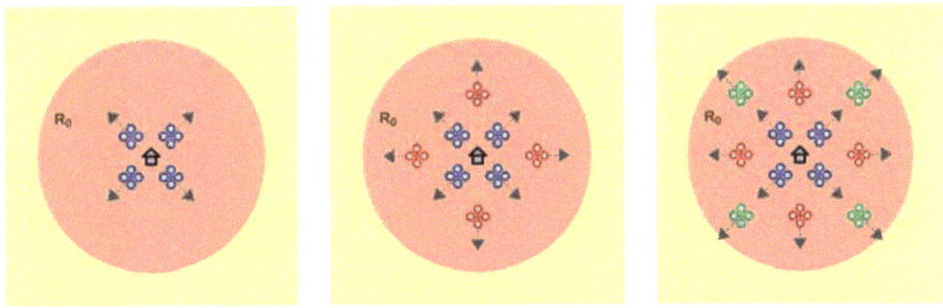

(**a**) 4 UAVs (**b**) 8 UAVs (**c**) 12 UAVs

Figure 4. Departure points and initial directions for case studies comprising 4, 8 and 12 UAVs.

The designed collision avoidance algorithm relies on repelling forces between UAVs as described in Algorithm 7. Given $u \in UAVs$, the distances between u and the rest of vehicles in $UAVs$ are calculated. Those UAVs closer than a minimum distance δ_{min} (a fixed parameter, e.g., 9 m) modify the vector $\vec{r}$, which will contain the resultant repelling force. As a result of this coordination, $\vec{r}$ is to be used to modify the original trajectory of uav.

Algorithm 7. Collision Avoidance Algorithm.

```
function COLLISIONAVOIDANCE(uav, UAVs)
    r⃗ ← (0,0)
    for all u, ∈ UAVs do                                    ▷ all the UAVs
        if u ≠ uav then
            δ ← √((uav_x − u_x)² + (uav_y − u_y)²)
            if δ < δ_min then                               ▷ δ_min = 9
                r⃗ ← r⃗ − (u_x, u_y)
            end if
        end if
    end for
    return r⃗
end function
```

6. Simulation Results

First, we analyse the proposed metrics, flying time and area coverage, to confirm that the use of a multi-objective optimisation is appropriated. Second, we have optimised the four case studies using NSGA-II for the *Probability* and *Strategy* approaches. Third, we have used the proposed GA to maximise the average score obtained by the UAVs in the four case studies using the *Game* approach (three two-person games). Finally, an analysis of the UAVs' trajectories is given.

We have used the jMetalPy package [43] to implement both optimisation algorithms. Our experiments were performed doing parallel runs in the HPC facilities of the University of Luxembourg [50]. Since the use of the ARGoS simulator implies long evaluation times to preserve the realism (10 simulation ticks per second), we have evaluated each algorithm' generation (28 individuals) in parallel on computing nodes equipped with Intel Xeon Gold

6132 2.6 GHz and 128 GB of RAM. The total optimisation time (600 runs in total) was equivalent to 1750.8 h (about 73 days).

6.1. Metric Analysis

In this experiment, we analyse the existence of correlations between the metrics used in our study. We have performed 30 random walks of 20 steps each, for 4, 8 and 12 UAVs, (we have considered that 1800 steps in total are enough for getting a significant sample of the problem's solutions) and calculated the Pearson correlation coefficients [51] shown in Table 2. Note that the random walks are performed on the system configuration, i.e., a vector of probabilities as defined in Equation (3), to sample the solution landscape of the problem. The UAVs are always following the SuSy-EnGaD trajectories based on surveillance rings.

It can be seen that 4 UAVs present some negative correlation between flying time and area coverage, so that one increases when the other decreases and vice versa. On the contrary, the metrics of 8 and 12 UAVs are not correlated (Pearson coefficient approximately equal to 0). Consequently, we can confirm that the use of a multi-objective optimisation algorithm (NSGA-II) is appropriate to obtain not one but many non-dominated solutions to our surveillance problem. After that, an expert can decide which solution suits better its needs and whether higher coverage levels could be prioritised against flying times or the other way round.

Table 2. Correlation Matrix for 4, 8 and 12 UAVs.

Metric	4 UAVs		8 UAVs		12 UAVs	
	F. Time	Coverage	F. Time	Coverage	F. Time	Coverage
F. Time	1.000	−0.599	1.000	−0.051	1.000	−0.006
Coverage	−0.599	1.000	−0.051	1.000	−0.006	1.000

6.2. Multi-Objective Optimisation

The next experiment consisted of the optimisation of the four case studies using a multi-objective algorithm (NSGA-II). We have performed 30 runs for the *Probability* and *Strategy* approaches on each case study (240 runs of NSGA-II in total) to get the results shown in Figure 5. We have plotted the empirical attainment functions (EAF) [52] describing the probabilistic distribution of the outcomes obtained by NSGA-II. Additionally, the hypervolume of the union of all sets using the reference point 0,0 is reported to compare the results of both approaches on each case study.

We can see that both strategies obtained similar results and that differences between hypervolumes are always under 4%. One optimisation run for 8.100 using the *Probability* approach seems to have suffered from stagnation, which can be clearly seen through the worst-case depicted in Figure 5b. That is one of the reasons why we performed 30 runs of stochastic algorithms. Now, one result from the set of results in each Pareto front could be selected and the associated configuration (whether it is composed by a set of probabilities or a list of bits) is to be used to set up the surveillance system prioritising coverage, flying times or a compromise solution. A different approach based on games is used in the next set of experiments described in the following section.

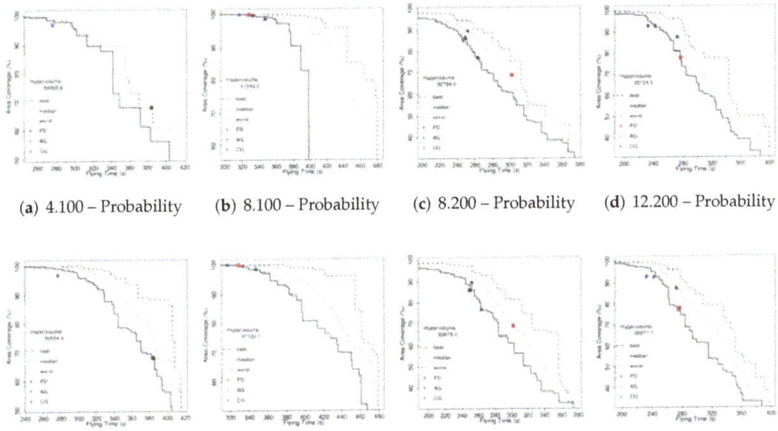

(**a**) 4.100 – Probability (**b**) 8.100 – Probability (**c**) 8.200 – Probability (**d**) 12.200 – Probability

(**e**) 4.100 – Strategy (**f**) 8.100 – Strategy (**g**) 8.200 – Strategy (**h**) 12.200 – Strategy

Figure 5. Plots of the empirical attainment functions obtained after optimising our case studies using NSGA-II (*Probability* and *Strategy* approaches). Additionally, the best values from *Game* (PD, AG and CG) obtained by GA are drawn. The respective hypervolume values are also reported.

6.3. Game Results

A different optimisation approach consists in using games to model each UAV interaction. As the objective is the maximisation of the average score of the swarm, we have used a single-objective GA. Table 3 shows the result of this optimisation process where 360 runs of GA were performed (30 per game and case study) to perform a reliable statistical analysis since we are working with a stochastic optimisation algorithm. It can be seen that UAVs playing the assurance game (AG) have maximised the average score of the swarm (fitness value). The game of chicken (CG) achieved the best maximum and minimum results for 4.100 although the median is still the same as in AG. All the results are statistically significant as shown by the reported Friedman ranks and Wilcoxon *p*-values.

Table 3. Results of the optimisation of the case studies using GA in the Game approach. Statistical tests are also included. Best values are in bold.

Case Study	Game	Fitness Min.	Fitness Median	Fitness Max.	Friedman Rank	Wilcoxon *p*-Value
4.100	PD	23.0	23.0	24.0	1.00	2.1×10^{-7}
	AG	25.3	**26.7**	26.7	2.40	3.5×10^{-7}
	CG	**26.7**	**26.7**	**30.7**	2.60	—
8.100	PD	22.1	24.0	25.3	1.00	1.8×10^{-6}
	AG	**28.6**	**34.0**	**34.0**	3.00	—
	CG	24.6	24.7	28.0	2.00	1.7×10^{-6}
8.200	PD	12.1	13.9	16.1	1.18	2.0×10^{-6}
	AG	**16.0**	**18.3**	**18.3**	2.90	—
	CG	13.1	15.4	16.6	1.92	2.3×10^{-6}
12.200	PD	12.8	14.2	16.4	1.10	1.8×10^{-6}
	AG	**16.0**	**19.3**	**20.2**	2.93	—
	CG	13.8	15.8	18.5	1.97	3.0×10^{-6}

We have plotted the corresponding metric points (flying time vs. area coverage) over the Pareto fronts graphics in Figure 5. It is interesting to see that AG favours high area coverage values in all case studies, while PD's results are obtaining better flying times in three of the case studies. UAVs playing CG are usually in a middle point between the other two games. Some games present different metric values for the same maximum scores (different trajectories, same game results) as depicted in Figure 5.

As a piece of complementary information about the optimisation process performed by the GA, we show in Figure 6 the boxplots with the distribution of the results achieved for each game and case study. We can see that the scores for AG are higher in most of the cases, except by 4.100 where CG presents some outliers revealing a higher average score (fitness) for the swarm. Looking at the scores in Figure 3, we see that the highest values obtained from cooperation (CC) are provided by AG. Consequently, we can assume that this was the strategy followed in the majority of cases by the UAVs.

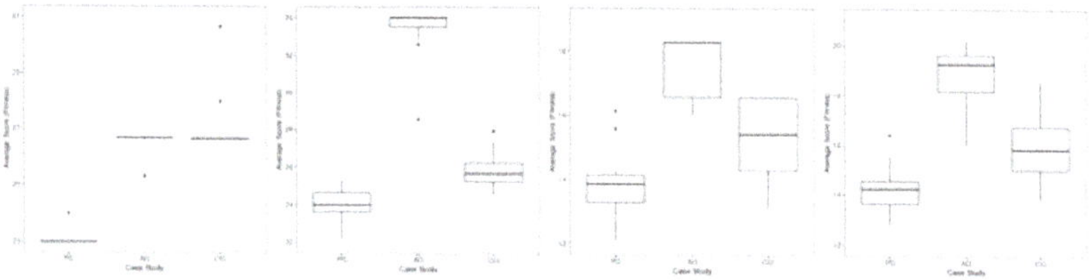

(**a**) 4.100 (**b**) 8.100 (**c**) 8.200 (**d**) 12.200

Figure 6. Boxplots showing the distribution of the results from the 30 independent runs of GA for each case study and game.

The extreme solutions for all the approaches are shown in Table 4. Note that we have used the default consumption model provided by ARGoS for all our experiments. Hence, maximum flying times ought to be longer than 480 s if other models of drones are used. In the next section, an analysis of the UAVs' trajectories is presented to better understand their behaviour under different configurations.

Table 4. Maximum values obtained after the optimisation of the three approaches. Probability and Strategy values are taken from the Pareto front (NSGA-II) whereas Game values (PD, AG and CG) are the best result obtained by GA. Best values are in bold.

Case Study	Metric	Approach				
		Probability	Strategy	Game		
				PD	AG	CG
4.100	F. Time (s)	**415.6**	**415.6**	384.4	276.1	384.4
	Coverage (%)	**100.0**	**100.0**	68.1	96.9	68.1
8.100	F. Time (s)	**480.1**	479.1	328.7	333.4	347.2
	Coverage (%)	**100.0**	**100.0**	99.9	**100.0**	98.5
8.200	F. Time (s)	373.9	**376.4**	301.9	251.9	263.4
	Coverage (%)	97.8	**98.5**	68.9	89.6	86.3
12.200	F. Time (s)	**399.3**	397.6	276.7	242.2	272.3
	Coverage (%)	99.6	**99.9**	77.8	92.9	87.5

6.4. *Analysis of Games' Trajectories*

The objective of this last study is to better understand the implications of each configuration and approach and how a cooperation/defection decision modifies the swarm behaviour. We have already presented the numerical results but now in Figures 7 and 8 the UAVs trajectories for the extreme solutions are depicted.

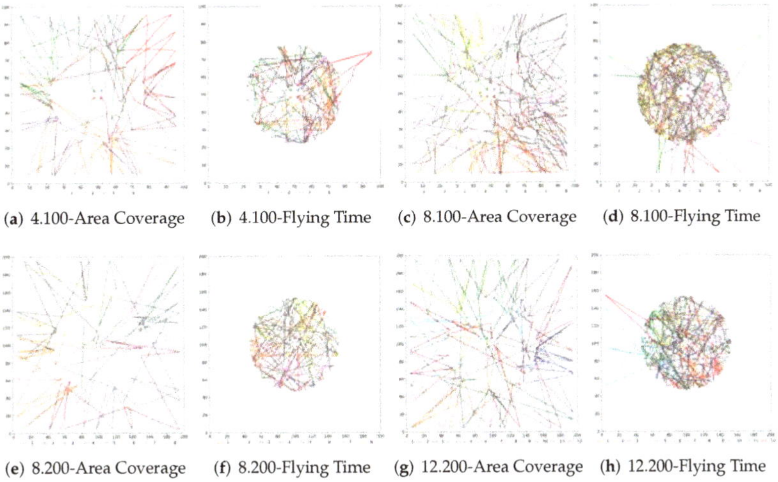

(**a**) 4.100-Area Coverage (**b**) 4.100-Flying Time (**c**) 8.100-Area Coverage (**d**) 8.100-Flying Time

(**e**) 8.200-Area Coverage (**f**) 8.200-Flying Time (**g**) 12.200-Area Coverage (**h**) 12.200-Flying Time

Figure 7. UAVs trajectories for maximum area coverage and maximum flying time using the *Probability* approach.

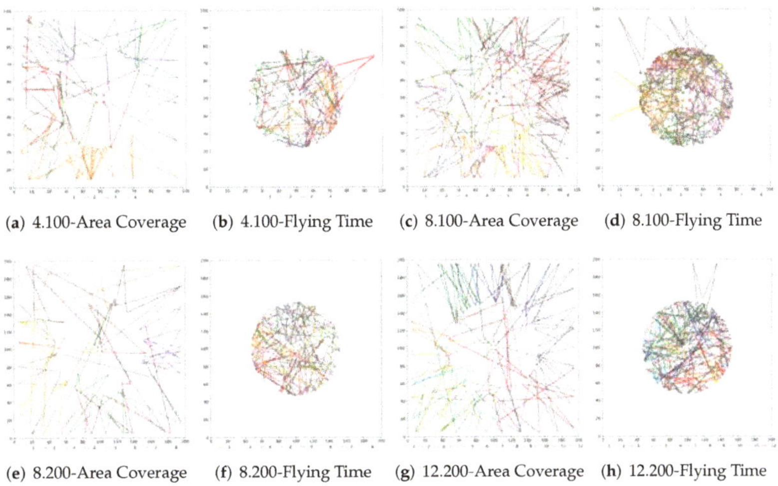

(**a**) 4.100-Area Coverage (**b**) 4.100-Flying Time (**c**) 8.100-Area Coverage (**d**) 8.100-Flying Time

(**e**) 8.200-Area Coverage (**f**) 8.200-Flying Time (**g**) 12.200-Area Coverage (**h**) 12.200-Flying Time

Figure 8. UAVs trajectories for maximum area coverage and maximum flying time using the *Strategy* approach.

We can see that there exist many similarities in the trajectories when using *Probability* and *Strategy*. In the configurations maximising area coverage, the UAVs depart from the innermost ring and visit the other two rings regularly, where the surveillance mission takes place. On the contrary, when the objective of the swarm is maximising the flying time, most of the UAVs stay in the middle ring after departing from the centre. It is interesting to remark that the maximum flying time was sometimes achieved when some UAVs have a short excursion to the outer ring. That was unexpected as we believed that flying at a lower speed all the time would have been better for saving energy. As we can see, the less congested cases (Figures 7f and 8f) are indeed according to our preliminary assumptions, since the rings' areas are bigger. We believe that multiple iterations in a limited space, plus the effects of the collision avoidance algorithm, have made NSGA-II to find better solutions (in terms of battery consumption) by temporarily sending some UAVs to the outer ring.

7. Conclusions

In this research work, we have presented a new Surveillance System Enhanced by Games of Drones (SuSy-EnGaD), based on cooperative UAVs that explore the surveillance area modelled as concentric rings. We have proposed three different approaches to obtain optimal strategies focusing on maximum flying time and area coverage. The first two approaches, *Probability* and *Strategy* are addressed as multi-objective optimisation problems using the well-known NSGA-II. The third approach is related to the evolutionary game theory where prisoner's dilemma, assurance game and game of chicken are proposed to rule possible cooperation between UAVs. This approach was optimised using a genetic algorithm maximising the average score of the swarm and the results were compared with the multi-objective approaches. We have tested SuSy-EnGaD on four different case studies featuring two map sizes and swarms of four, eight and twelve UAVs. Finally, the UAVs trajectories were analysed to better understand the best configurations in terms of flying time and area coverage.

Our results show that flying times up to 480 s can be achieved with a low area coverage, while on the other hand, a full area coverage is possible when the flying time is roughly under 250 s (depending on the case study). Vehicles' trajectories have indicated a high concentration of UAVs in the inner surveillance ring to save battery (they fly at a reduced speed there), and a more balanced distribution of UAVs throughout the surveillance area to achieve high coverage rates. The single-objective *Game* approach obtained optimal solutions by maximising the swarm's average score. Although the extreme values were obtained by using the multi-objective approaches, the *Game* approach turned out to be a different strategy to obtain particular solutions depending on the game selected.

In our simulations performed using a multi-physics robot simulator (ARGoS), we have observed UAVs' trajectories visiting the entire surveillance scenario when area coverage was prioritised. On the contrary, almost all UAVs stayed in the inner ring when the objective was to keep a maximum flying time. Under this condition, we have observed some UAVs momentarily leaving the inner ring, which we believe is due to the collision avoidance algorithm as several UAVs were flying in a reduced area. These trajectories confirm the expected UAV behaviour and are in accordance with the numeric results obtained.

As a matter of future work we plan to test our proposal using a different number of security rings and UAVs. Previously, we have experimented with square rings and pheromones, in this current article we have improved the surveillance scenario using actual circular rings, and we wish to take advantage of this new architecture in future research works. We would like to try other different games to study the various solutions that can be achieved and also defining our own payoff matrices to obtain different single solution points from the pareto front. Despite the accuracy provided by ARGoS, we are working on increasing the realism of our approach testing the trajectories using actual drones where weather conditions, vision algorithms, and communication restrictions are to be considered. Another interesting future work would consist in using a communication layer featuring different packet loss rates to analyse how it would affect the swarm behaviour, especially the collision avoidance algorithm.

Author Contributions: Conceptualisation, D.H.S.; methodology, D.H.S.; software, D.H.S.; validation, D.H.S. and G.D.; formal analysis, D.H.S.; writing—original draft preparation, D.H.S.; writing—review and editing, D.H.S., M.R.B. and G.D.; visualisation, D.H.S.; supervision, G.D. and P.B.; project administration, G.D. and P.B.; funding acquisition, G.D. and P.B. All authors have read and agreed to the published version of the manuscript.

Funding: This research was partially funded by the Office of Naval Research, grant number N62909-18-1-2176.

Institutional Review Board Statement: Not applicable.

Informed Consent Statement: Not applicable.

Data Availability Statement: The data presented in this study are openly available in https://gitlab.uni.lu/hunted/susy-engad-surveillance-system-enhanced-by-games-of-drones (accessed on 5 January 2021).

Acknowledgments: This work relates to Department of Navy award N62909-18-1-2176 issued by the Office of Naval Research. The United States Government has a royalty-free license throughout the world in all copyrightable material contained herein. This work is partially funded by the joint research programme UL/SnT-ILNAS on Digital Trust for Smart-ICT. The experiments presented in this paper were carried out using the HPC facilities of the University of Luxembourg [50]—see https://hpc.uni.lu (accessed on 5 January 2021).

Conflicts of Interest: The authors declare no conflict of interest.

References

1. Skorobogatov, G.; Barrado, C.; Salamí, E. Multiple UAV Systems: A Survey. *Unmanned Syst.* **2020**, *08*, 149–169. [CrossRef]
2. Lin, Z.; Liu, H.H.T.; Wotton, M. Kalman Filter-Based Large-Scale Wildfire Monitoring with a System of UAVs. *IEEE Trans. Ind. Electron.* **2019**, *66*, 606–615. [CrossRef]
3. Lottes, P.; Khanna, R.; Pfeifer, J.; Siegwart, R.; Stachniss, C. UAV-based crop and weed classification for smart farming. In Proceedings of the 2017 IEEE International Conference on Robotics and Automation (ICRA), Singapore, 29 May–3 June 2017; pp. 3024–3031. [CrossRef]
4. Bozcan, I.; Kayacan, E. UAV-AdNet: Unsupervised anomaly detection using deep neural networks for aerial surveillance. In Proceedings of the 2020 IEEE/RSJ International Conference on Intelligent Robots and Systems (IROS), Las Vegas, NV, USA, 24 October–24 January 2020; pp. 1158–1164. [CrossRef]
5. He, X.; Bourne, J.R.; Steiner, J.A.; Mortensen, C.; Hoffman, K.C.; Dudley, C.J.; Rogers, B.; Cropek, D.M.; Leang, K.K. Autonomous Chemical-Sensing Aerial Robot for Urban/Suburban Environmental Monitoring. *IEEE Syst. J.* **2019**, *13*, 3524–3535. [CrossRef]
6. Yakushiji, K.; Fujita, H.; Murata, M.; Hiroi, N.; Hamabe, Y.; Yakushiji, F. Short-Range Transportation Using Unmanned Aerial Vehicles (UAVs) during Disasters in Japan. *Drones* **2020**, *4*, 68. [CrossRef]
7. Kerle, N.; Nex, F.; Gerke, M.; Duarte, D.; Vetrivel, A. UAV-Based Structural Damage Mapping: A Review. *ISPRS Int. J. Geo-Inf.* **2020**, *9*, 14. [CrossRef]
8. Keller, J.; Thakur, D.; Likhachev, M.; Gallier, J.; Kumar, V. Coordinated Path Planning for Fixed-Wing UAS Conducting Persistent Surveillance Missions. *IEEE Trans. Autom. Sci. Eng.* **2017**, *14*, 17–24. [CrossRef]
9. Altan, A. Performance of metaheuristic optimization algorithms based on swarm intelligence in attitude and altitude control of unmanned aerial vehicle for path following. In Proceedings of the 2020 4th International Symposium on Multidisciplinary Studies and Innovative Technologies (ISMSIT), Istanbul, Turkey, 22–24 October 2020; pp. 1–6. [CrossRef]
10. Yang, F.; Ji, X.; Yang, C.; Li, J.; Li, B. Cooperative search of UAV swarm based on improved ant colony algorithm in uncertain environment. In Proceedings of the 2017 IEEE International Conference on Unmanned Systems (ICUS), Beijing, China, 27–29 October 2017; pp. 231–236. [CrossRef]
11. Wang, T.; Qin, R.; Chen, Y.; Snoussi, H.; Choi, C. A reinforcement learning approach for UAV target searching and tracking. *Multimed. Tools Appl.* **2018**, *78*, 4347–4364. [CrossRef]
12. Weibull, J.W. *Evolutionary Game Theory*; MIT Press: Cambridge, MA, USA, 1997.
13. Zhou, Y.; Rao, B.; Wang, W. UAV Swarm Intelligence: Recent Advances and Future Trends. *IEEE Access* **2020**, *8*, 183856–183878. [CrossRef]
14. Sims, K. Evolving 3D Morphology and Behavior by Competition. *Artif. Life* **1994**, *1*, 353–372. [CrossRef]
15. Stolfi, D.H.; Brust, M.R.; Danoy, G.; Bouvry, P. Emerging Inter-Swarm Collaboration for Surveillance Using Pheromones and Evolutionary Techniques. *Sensors* **2020**, *20*, 2566. [CrossRef] [PubMed]
16. Jesus Roldan, J.; Del Cerro, J.; Barrientos, A. Should we compete or should we cooperate? Applying game theory to task allocation in drone swarms. In Proceedings of the 2018 IEEE/RSJ International Conference on Intelligent Robots and Systems (IROS), Madrid, Spain, 1–5 October 2018; pp. 5366–5371. [CrossRef]
17. Bäck, T. *Evolutionary Algorithms in Theory and Practice: Evolution Strategies, Evolutionary Programming, Genetic Algorithms*; Oxford University Press: Oxford, UK, 1996.
18. Li, P.; Duan, H. A potential game approach to multiple UAV cooperative search and surveillance. *Aerosp. Sci. Technol.* **2017**, *68*, 403–415. [CrossRef]
19. Ruan, L.; Wang, J.; Chen, J.; Xu, Y.; Yang, Y.; Jiang, H.; Zhang, Y.; Xu, Y. Energy-efficient multi-UAV coverage deployment in UAV networks: A game-theoretic framework. *China Commun.* **2018**, *15*, 194–209. [CrossRef]
20. Ma, Y.; Wang, G.; Hu, X.; Luo, H.; Lei, X. Cooperative Occupancy Decision Making of Multi-UAV in Beyond-Visual-Range Air Combat: A Game Theory Approach. *IEEE Access* **2020**, *8*, 11624–11634. [CrossRef]
21. Shivshankar, S.; Jamalipour, A. An Evolutionary Game Theory-Based Approach to Cooperation in VANETs under Different Network Conditions. *IEEE Trans. Veh. Technol.* **2015**, *64*, 2015–2022. [CrossRef]

22. Peters, H. *Game Theory*; Springer Texts in Business and Economics; Springer: Berlin/Heidelberg, Germany, 2015; Volume 228, p. 36. [CrossRef]
23. Yin, C.; Xiao, Z.; Cao, X.; Xi, X.; Yang, P.; Wu, D. Offline and Online Search: UAV Multiobjective Path Planning Under Dynamic Urban Environment. *IEEE Internet Things J.* **2018**, *5*, 546–558. [CrossRef]
24. Hayat, S.; Yanmaz, E.; Brown, T.X.; Bettstetter, C. Multi-objective UAV path planning for search and rescue. In Proceedings of the 2017 IEEE International Conference on Robotics and Automation (ICRA), Singapore, 29 May–3 June 2017; pp. 5569–5574. [CrossRef]
25. Ren, Q.; Yao, Y.; Yang, G.; Zhou, X. Multi-objective path planning for UAV in the urban environment based on CDNSGA-II. In Proceedings of the 2019 IEEE International Conference on Service-Oriented System Engineering (SOSE), San Francisco, CA, USA, 4–9 April 2019; pp. 350–3505. [CrossRef]
26. Scherer, J.; Rinner, B. Persistent multi-UAV surveillance with energy and communication constraints. In Proceedings of the 2016 IEEE International Conference on Automation Science and Engineering (CASE), Fort Worth, TX, USA, 21–25 August 2016; pp. 1225–1230. [CrossRef]
27. Zeng, Y.; Zhang, R. Energy-Efficient UAV Communication with Trajectory Optimization. *IEEE Trans. Wirel. Commun.* **2017**, *16*, 3747–3760. [CrossRef]
28. Hosseinalipour, S.; Rahmati, A.; Eun, D.Y.; Dai, H. Energy-Aware Stochastic UAV-Assisted Surveillance. *IEEE Trans. Wirel. Commun.* **2021**, *20*, 2820–2837. [CrossRef]
29. Stolfi, D.H.; Brust, M.R.; Danoy, G.; Bouvry, P. CONSOLE: Intruder detection using a UAV swarm and security rings. *Swarm Intell.* **2021**, *15*, 205–235. [CrossRef]
30. Ding, X.C.; Rahmani, A.R.; Egerstedt, M. Multi-UAV Convoy Protection: An Optimal Approach to Path Planning and Coordination. *IEEE Trans. Robot.* **2010**, *26*, 256–268. [CrossRef]
31. Huang, H.; Savkin, A.V. Navigating UAVs for Optimal Monitoring of Groups of Moving Pedestrians or Vehicles. *IEEE Trans. Veh. Technol.* **2021**, *70*, 3891–3896. [CrossRef]
32. Jayavelu, S.; Kandath, H.; Sundaram, S. Dynamic area coverage for multi-UAV using distributed UGVs: A two-stage density estimation approach. In Proceedings of the 2018 Second IEEE International Conference on Robotic Computing (IRC), Laguna Hills, CA, USA, 31 January–2 February 2018; pp. 165–166. [CrossRef]
33. Stolfi, D.H.; Brust, M.R.; Danoy, G.; Bouvry, P. UAV-UGV-UMV Multi-Swarms for Cooperative Surveillance. *Front. Robot. AI* **2021**, *8*, 616950. [CrossRef]
34. Samuelson, L. *Evolutionary Games and Equilibrium Selection*; MIT Press: Cambridge, MA, USA, 1997.
35. Seredynski, M.; Bouvry, P. Evolutionary game theoretical analysis of reputation-based packet forwarding in civilian mobile Ad Hoc networks. In Proceedings of the 2009 IEEE International Symposium on Parallel & Distributed Processing, Rome, Italy, 23–29 May 2009; pp. 1–8. [CrossRef]
36. Ni, J.; Tang, G.; Mo, Z.; Cao, W.; Yang, S.X. An Improved Potential Game Theory Based Method for Multi-UAV Cooperative Search. *IEEE Access* **2020**, *8*, 47787–47796. [CrossRef]
37. Nash, J. Non-cooperative games. *Ann. Math.* **1951**, *54*, 286–295. [CrossRef]
38. Kollock, P. Social Dilemmas: The Anatomy of Cooperation. *Annu. Rev. Sociol.* **1998**, *24*, 183–214. [CrossRef]
39. Seredynski, F. Competitive Coevolutionary Multi-Agent Systems: The Application to Mapping and Scheduling Problems. *J. Parallel Distrib. Comput.* **1997**, *47*, 39–57. [CrossRef]
40. Deb, K.; Agrawal, S.; Pratap, A.; Meyarivan, T. A fast elitist non-dominated sorting genetic algorithm for multi-objective optimization: NSGA-II. In *Parallel Problem Solving from Nature PPSN VI*; Schoenauer, M., Deb, K., Rudolph, G., Yao, X., Lutton, E., Merelo, J.J., Schwefel, H.P., Eds.; Springer: Berlin/Heidelberg, Germany, 2000; pp. 849–858.
41. Goldberg, D.E. *Genetic Algorithms in Search, Optimization and Machine Learning*, 1st ed.; Addison-Wesley Longman Publishing Co., Inc.: Boston, MA, USA, 1989.
42. Holland, J.H. *Adaptation in Natural and Artificial Systems*; The MIT Press: Cambridge, MA, USA, 1992; p. 228. [CrossRef]
43. Benítez-Hidalgo, A.; Nebro, A.J.; García-Nieto, J.; Oregi, I.; Ser, J.D. jMetalPy: A Python framework for multi-objective optimization with metaheuristics. *Swarm Evol. Comput.* **2019**, *51*, 100598. [CrossRef]
44. Goldberg, D.E.; Deb, K. A Comparative Analysis of Selection Schemes Used in Genetic Algorithms. *Found. Genet. Algorithms* **1991**, *1*, 69–93. [CrossRef]
45. De Jong, K.A. An Analysis of the Behavior of a Class of Genetic Adaptive Systems. Ph.D. Thesis, University of Michigan, Ann Arbor, MI, USA, 1975.
46. Gen, M.; Cheng, R. *Genetic Algorithms and Engineering Optimization*; John Wiley & Sons: Hoboken, NJ, USA, 1999; Volume 7.
47. Deb, K. *Multi-Objective Optimization Using Evolutionary Algorithms*; John Wiley & Sons, Inc.: Hoboken, NJ, USA, 2001.
48. Pinciroli, C.; Trianni, V.; O'Grady, R.; Pini, G.; Brutschy, A.; Brambilla, M.; Mathews, N.; Ferrante, E.; Di Caro, G.; Ducatelle, F.; et al. ARGoS: A modular, parallel, multi-engine simulator for multi-robot systems. *Swarm Intell.* **2012**, *6*, 271–295. [CrossRef]
49. Dorigo, M.; Floreano, D.; Gambardella, L.M.; Mondada, F.; Nolfi, S.; Baaboura, T.; Birattari, M.; Bonani, M.; Brambilla, M.; Brutschy, A.; et al. Swarmanoid: A Novel Concept for the Study of Heterogeneous Robotic Swarms. *IEEE Robot. Autom. Mag.* **2013**, *20*, 60–71. [CrossRef]

50. Varrette, S.; Bouvry, P.; Cartiaux, H.; Georgatos, F. Management of an academic HPC cluster: The UL experience. In Proceedings of the 2014 International Conference on High Performance Computing & Simulation (HPCS), Bologna, Italy, 21–25 July 2014; IEEE: Bologna, Italy, 2014; pp. 959–967. [CrossRef]
51. Sheskin, D.J. *Handbook of Parametric and Nonparametric Statistical Procedures*; CRC Press: New York, NY, USA, 2003.
52. López-Ibáñez, M.; Paquete, L.; Stützle, T. Exploratory analysis of stochastic local search algorithms in biobjective optimization. In *Experimental Methods for the Analysis of Optimization Algorithms*; Springer: Berlin/Heidelberg, Germany, 2010; pp. 209–222. [CrossRef]

drones

MDPI

Article

Designing a User-Centered Interaction Interface for Human–Swarm Teaming

Mohammad Divband Soorati [1,*], Jediah Clark [1], Javad Ghofrani [2], Danesh Tarapore [1] and Sarvapali D. Ramchurn [1]

[1] School of Electronics and Computer Science, University of Southampton, Southampton SO17 1BJ, UK; j.r.clark@soton.ac.uk (J.C.); d.s.tarapore@soton.ac.uk (D.T.); sdr1@soton.ac.uk (S.D.R.)
[2] Institute of Computer Engineering, University of Lübeck, 23562 Lübeck, Germany; ghofrani@iti.uni-luebeck.de
* Correspondence: m.divband-soorati@soton.ac.uk; Tel.: +44-23-8059-5000

Abstract: A key challenge in human–swarm interaction is to design a usable interface that allows the human operators to monitor and control a scalable swarm. In our study, we restrict the interactions to only one-to-one communications in local neighborhoods between UAV-UAV and operator-UAV. This type of proximal interactions will decrease the cognitive complexity of the human–swarm interaction to $O(1)$. In this paper, a user study with 100 participants provides evidence that visualizing a swarm as a heat map is more effective in addressing usability and acceptance in human–swarm interaction. We designed an interactive interface based on the users' preference and proposed a controlling mechanism that allows a human operator to control a large swarm of UAVs. We evaluated the proposed interaction interface with a complementary user study. Our testbed and results establish a benchmark to study human–swarm interaction where a scalable swarm can be managed by a single operator.

Keywords: human–swarm interaction; situation awareness; user-centered interface design

Citation: Divband Soorati, M.; Clark, J.; Ghofrani, J.; Tarapore, D.; Ramchurn, S. Designing a User-Centered Interaction Interface for Human–Swarm Teaming. *Drones* **2021**, *5*, 131. https://doi.org/10.3390/drones5040131

Academic Editor: Diego González-Aguilera

Received: 30 September 2021
Accepted: 2 November 2021
Published: 5 November 2021

1. Introduction

Unmanned aerial vehicles (UAVs) are now commonly used to rapidly gather situational awareness in emergency responses or security applications. However, most applications are limited to high operator (n) to vehicle ratios (m), i.e., $n \geq m$. This severely limits the use of inexpensive, low-power, UAVs at scale. Hence, a number of works have focused on using a swarm of autonomous UAVs [1–3], where the UAVs have high degrees of autonomy and can coordinate with each other using simple rules and low-power communication. These limitations on computation and communication (e.g., to avoid using up their battery or to avoid detection by threats) introduce huge challenges when it comes to monitoring and managing the collective actions of the swarm [4]. Specifically, as the size of the swarm grows, so does the complexity that human operators face in terms of directing individual UAVs, orienting to the, possibly delayed, information relayed by the swarm, and monitoring the health of the swarm (e.g., UAVs may be taken down by threats such as fires or enemy attack).

Against this background, in this paper, we aim to the improve operator to vehicle ratios (to $n < m$) for large scale deployments of UAV swarms. We develop the notion of human–swarm teaming with proximal interactions, whereby, operators can direct the swarm via coarse instructions. The operator can have one-to-one interactions (in the worst case) with members of the swarm. Similarly, individual UAVs can only communicate with their neighbours (UAVs or human) within a limited range. By 'coarse instructions' we mean that they cannot individually instruct each UAV where to go but set a target for the swarm to achieve. This control approach is in a similar vein to the indirect parameter setting that exists in the literature [5,6]. Indeed, each UAV maintains a high degree of autonomy and can decide where to go and what to do based on its local inputs.

Our approach builds upon interactive user interfaces to allow operators to visualize the state of the swarm (state estimation), direct the swarm (control), and monitor the information it generates (monitoring). Our interfaces combine direct manipulation with autonomous UAV decision-making. We do not force the swarm to maintain a communication channel with the human UAVs at all times [7,8]. The operator, instead, uses the messages that the UAVs broadcast (to their nearest neighbor) to estimate the swarm's state. Lindner et al. [9] evaluated the complexity of supervising a swarm of UAVs with a manned aerial vehicle by displaying the swarm as an avatar to the plane pilots. Despite the effectiveness in reducing the cognitive complexity, the resolution of the swarm's coverage display is greatly sacrificed. The pilots in the study reported that the monitoring of the swarm was complex and that more swarm autonomy and control options are required. Patel et al. [10] use an interface with augmented reality to specify the overall goal for the swarm with a mixture of robot and environment-oriented modalities. We use a more informative visualization compared to [9] and there is flexible autonomy that allows the operator to adjust the autonomy of the swarm by modifying the messages that are exchanged with the UAVs [11]. The operator can choose to not intervene and grant full autonomy to the swarm, which makes our study different from [10]. There are also other studies that focus on the autonomy level of the swarm [12,13], different human–swarm interaction methods [10,14,15] and visualization techniques for the human–swarm systems [8,16] but this paper advances the state of the art in the following ways:

1. In a user study with 100 participants, we evaluated the effect of different visualization techniques on the usability of human–swarm interaction interface and reported the result;
2. The preferred visualization method is then used to build an interaction interface. This method reduces the number of visualizations that an operator has to use to control and monitor the swarm;
3. We propose a human-in-the-loop collective decision-making method that governs the human–swarm decisions. Our model is task-generic for human–swarm teaming (i.e., the operator is treated as an agent as well) with proximal interactions that allows for state estimation and control;
4. Through simulation, we demonstrate the effectiveness of the swarm in tracking an unfolding event (a fire spreading) through minimal interactions by a single operator.

When taken together, our methods and results provide a new benchmark for the development of human–swarm teaming models and interaction mechanisms that embed the notion of flexible autonomy.

2. User Evaluation of Human–Swarm Visualization Methods

To explore the basis in which display methods should be designed, this section presents findings on how users' acceptance and usability vary between individual UAV and heat map swarm coverage visualization methods. This user-interaction study provides the basis in which models, maps, and simulations are constructed in the following sections.

2.1. Design and Procedure

One hundred participants (62 female, 37 male, 1 non-binary) took part in an online visualization evaluation. Ethical approval was given by the University of Southampton ethics committee (ERGO number: 66360.A1). To be included in the user-study, participants were required to hold a BSc degree or an equivalent qualification. Participants watched a video of a swarm display showing individual UAVs, followed by a heat map to represent coverage.

Videos were created by recording a rudimentary UAV simulation displaying only swarm coverage with no human-interaction involved. Each video consisted of 35 s, where 50 drones were represented using black-point dots or a density heat-map indicating the density of UAVs in 2D space. After viewing each video, participants were asked questions from the System Usability Scale [17] and the System Acceptance Scale [18] to measure how easily each could be interpreted. After viewing both videos, participants were asked to provide their preferred display method given a variety of contextual factors related to swarm size, communication-constrained environments, coverage, time-criticality, error detection and transparency (see Table 1).

2.2. Results

Responses to the post-study questionnaire were analyzed using chi-square goodness of fit tests [19] to identify any preferences for heat map or individual drone displays for a variety of swarm/environmental contexts. The alpha level was corrected using the Bonferroni method [20], according to the number of analyses conducted. The chi-square goodness of fit analyses showed that heat maps were preferred when a larger swarm is being displayed, to display motion and coverage, and when time was a critical factor to task success (see Table 1). On the other hand, individual drone displays would be preferred for detecting errors in the swarm. There was no significant difference when communication would be constrained, when time was not a critical factor of task success, and for being more transparent with the operator. Table 1 shows the frequency of participants that reported either individual or heat map displays as their preferred display method, given each contextual factor.

Table 1. Table to show the frequency of display preferences among users for each contextual factor and respective Chi-square analysis results.

Contextual Factor	Individual	Heat Map	χ^2	p
Larger swarm size	21	79	33.64	0.001 *
Communication constrained	43	57	1.96	0.162
Displaying motion and coverage	31	69	14.44	0.001 *
Time critical	28	72	19.36	0.001 *
Time non-critical	43	57	1.96	0.162
Detecting errors	74	26	23.04	0.001 *
Transparency	44	56	1.44	0.23

Note. Bonferroni Corrected $\alpha = 0.0071$. * = $p < 0.0071$.

Two repeated measured ANOVAs [21] were conducted to identify whether the coverage display method had an effect on the System Usability Scores and System Acceptance Scores reported by participants. The analyses found a significant effect of display-type on usability scores (F(1,99) = 22.53, $p < 0.001$, $\eta_p{}^2 = 0.185$) and acceptance scores (F(1,99) = 29.89, $p < 0.001$, $\eta_p{}^2 - 0.232$) indicating that the heat map display method had a greater level of usability and acceptance among participants when compared with displaying individual UAVs. These results show that, overall, heat map displays are perceived as being more usable, acceptable, and more effective in displaying information when time is less available and when large numbers of UAVs are being represented.

2.3. Conclusions

This user-study provides evidence that heat map methods are more effective in addressing usability and acceptance in human–swarm interaction (see Figure 1). Further, in situations of larger swarm sizes, time-criticality, and displaying motion and coverage, a heat map has shown to have a higher preference among participants. Conversely, for detecting errors within the swarm, individual drone displays may be more appropriate. Due to 79 of the participants stating that a larger swarm sizes would be better displayed via a heat map, this display method shows greater promise for scalability. However, under conditions where detecting errors with individual drones is crucial to task performance, an individual drone display may be more appropriate, perhaps added as a diagnostic tool in addition to the heat map display, rather than being the core display method. Given these results, a heat map interaction method forms the basis for the development of human–swarm teaming interactions in the following sections.

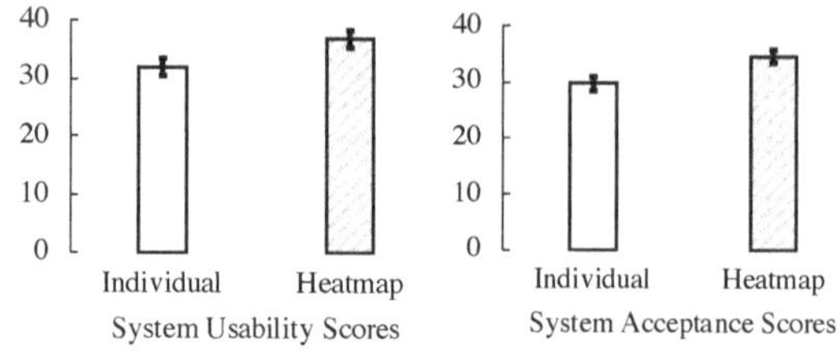

(a) (b)

Figure 1. Bar graphs to show mean usability (**a**) and acceptance (**b**) scores from subjective measures. Error bars = 95% CIs.

3. Human–Swarm Teaming Model

In this section, we describe the UAV model and the interactive user interface. We go on and describe our human–swarm collective decision-making process. We will use the term UAV and agent interchangeably in the rest of this paper.

3.1. UAV Model

We consider that agents can be heterogeneous in terms of their capabilities (e.g., influence, dynamics, sensor). This means that they may gather information differently and at different speeds. The speed of the UAVs impacts on the delays in gathering information. In terms of completing a mission, the sensing capabilities may determine the number of dimensions that an operator has to consider and the size of messages that are exchanged. Here, we model UAVs that only return information about their own health; for instance, whether they are flying, the sensors are on or not, and whether they have detected a target or not. The approach can be extended to cover the detection of different kinds of targets, and we will leave this for future work. For now, we will only consider UAVs that can return an integer value denoting the state of their position, for example, the temperature that may be indicative of a fire, or the number of people in their area. We use a point model and we do not consider external environmental factors such as gravity, wind, etc.

3.2. Operator and Swarm Interactions

UAVs move around the mission zone, over the human operator, and over the disaster zone (see Figure 2). We can add an arbitrary number of operators and disaster zones, but in this paper, our scenario contains one operator, one disaster area, and 15 UAVs. The operator can communicate with the UAVs that are in their local neighborhood and send/receive information. Similar to other UAVs, the operator can also move around and explore the environment, but in this study we assume that the operator is fixed and the UAVs move at most 5 pixels at each simulation step. In our scenario of disaster management, we set the position of the human operator and the disaster as important points that need to be noted and stored by the UAVs. The motivation for marking both operators and disasters is to facilitate further use cases in disaster management, such as forming a safe path between the operator and the disaster area (not discussed in this paper).

Our interface design contains two heat maps, a belief and a confidence map, to comply to the findings of our initial user study (see Section 2). The operator uses these two maps to observe and control the swarms.

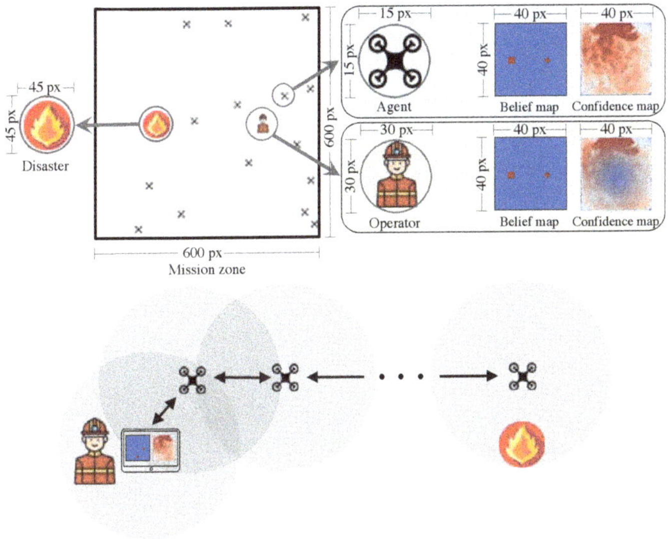

Figure 2. A map of a mission zone with UAVs, a disaster, and a human operator. All agents (UAVs and the operator) have belief and confidence maps with a lower resolution (40×40) compared to the actual images (600×600). The operator uses the interface to communicate with the UAVs in their local neighborhood.

3.2.1. The Belief Map

UAVs continuously explore and observe the mission zone for potential disasters or the areas with the human operators. A matrix—a belief map—is used as a map representation of the mission zone. UAVs use the belief map to store their observation of the area. Figure 3a–c show the formation of the operator's belief map over time, where the location of the human operator and the disaster zone are highlighted. The location of the interest areas in the belief map emerge very early in the process. Over time, the UAVs obtain a stronger belief on the area that they explore. This is shown by intensity of the belief that increases over time and turns into red.

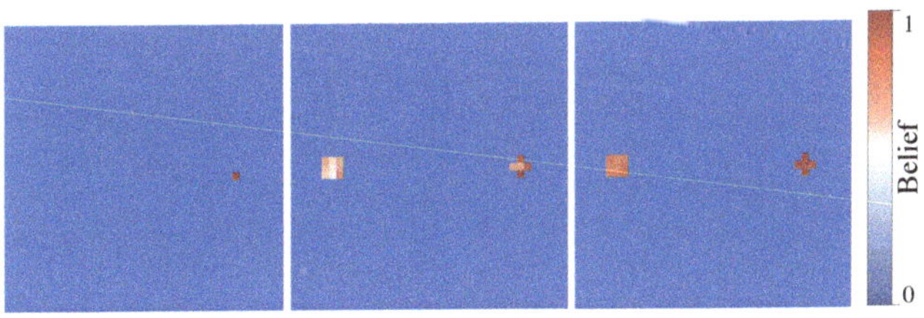

(**a**) $t = 10$ (**b**) $t = 500$ (**c**) $t = 1000$

Figure 3. The operator's belief map at several simulation steps.

3.2.2. The Confidence Map

Another heat map—a confidence map—not only represents the distribution of the swarm as seen by individual agents (see Figure 4) but also determines the agents' certainty level of the information in the corresponding cell of the belief map. We later show how the same map is used for controlling the swarm as well.

(**a**) $t = 10$ (**b**) $t = 500$ (**c**) $t = 1000$

Figure 4. Confidence map at t = 10, 500, 1000 simulation steps.

Using the belief and confidence maps, the agents store their representation of the environment and keep their confidence level about its correctness. As agents move and explore the mission zone, the cells in the belief maps are updated. The corresponding confidence values of the recently visited cells in the confidence map also increase.

3.3. *Human–Swarm Collective Decision-Making Model*

Each drone has a limited perception and communication range that makes it hard for them to individually solve a complex task. The swarm can establish a distributed sensing on a large area of the environment and when working together, the swarm will be able to construct an opinion about a certain property of the environment. In our earlier work, we have shown how a swarm can collectively decide which half of an environment is brighter and stay adaptive to changes [22]. This class of problems in swarm robotics is called the *best-of-n problem*, that is, the capability of the swarm to find the best option among a finite set of alternatives [23]. An unsolved challenge is to provide a collective decision-making strategy that is capable of solving the best-of-n problem with spatial constraints [24]. In our context, the swarm is exploring the environment and the swarm's motion is limited (e.g., not permitted in no-fly zone). The second challenge is that in the literature, the collective perception of the features in an environment is limited to few options [25]. To the best of our knowledge, none of the existing collective decision-making strategies including Direct Modulation of Voter-based Decisions [26], Direct Modulation of Majority-based Decisions [27], Direct Comparison [25] that are widely common in the swarm robotics applications can be directly applied to our scenario, as UAVs need to continuously measure the environment and disseminate whenever another UAV or an operator is in range. We are building upon our previous work on adaptive decision-making in robot swarms and add a human in the loop to find a choice among a large set of options in a dynamic environment [22].

The operator, similar to the UAVs, has a confidence that is updated based on the neighboring UAVs. The operator can decide to manipulate their own map, regardless of the current state, and propagate it back to the swarm. The modified confidence map, once received by the UAVs, enforces an attraction/repulsion mechanism that controls the swarm. The operator can focus on the overall goal and implement strategies abstracted from the detailed interactions and behaviors of each UAV, making it an ideal multi-robot organizational scheme. The cognitive complexity of such a form of interaction between the human and the swarm, where the human has to operate a maximum C number of UAVs (the number of agents that can fit into a communication range of an operator), regardless of the swarm size, is $O(1)$ [28].

The operator receives the maps from the UAVs in their neighborhood and may decide to temporarily explore a certain region in the mission zone as external information may indicate a disaster. The UAVs can be, temporarily or permanently, repelled from a prohibited region (e.g., high voltage cables). The operator can manipulate their own confidence map and introduce a major decrease/increase in the certainty of an interest zone, and constantly share it with the swarm. This forces the swarm to react to the new forces and collectively explore the areas with low confidence or avoid the areas with high confidence.

Agents continuously disseminate their belief and confidence maps as they meet. Equations (1) and (2) determine the update for each cell in the maps of agent i after communication with agent j.

$$c_{i,t}(m,n) = \begin{cases} c_{i,t-1}(m,n) & c_{i,t-1}(m,n) \geq \theta_{high} \\ c_{j,t-1}(m,n) & c_{j,t-1}(m,n) \geq \theta_{high} \\ c_{i,t-1}(m,n) & c_{i,t-1}(m,n) \leq \theta_{low} \\ c_{j,t-1}(m,n) & c_{j,t-1}(m,n) \leq \theta_{low} \\ \frac{\varphi\rho_i c_{i,t-1}(m,n)+(1-\varphi)\rho_j c_{j,t-1}(m,n)}{\rho_i+\rho_j} & o.w. \end{cases} \quad (1)$$

$$b_{i,t}(m,n) = \frac{\varphi\rho_i b_{i,t-1}(m,n) + (1-\varphi)\rho_j b_{j,t-1}(m,n)}{\rho_i + \rho_j} \quad (2)$$

where $b_{x,t}(m,n)$ and $c_{x,t}(m,n)$ are the cells in the belief and confidence maps of agent x at the time t. θs represent the higher and lower bounds for the higher priority information stored in the confidence map based on human operator's influence. In each simulation step, the confidence map ages with the factor $\alpha < 1$ to account for information quality decay. For a while, the operator's influence is locked and will circulate among the swarm until the quality decay brings the values back to the modifiable range of $\{\theta_{low}, \theta_{high}\}$. The idea behind this locking mechanism is to overrule the swarm's decision and reserve a larger impact for the operator's commands. Each agent carries a reliability factor (ρ) that indicates to which extent its passing information is reliable. This can depend on the hardware capacity, battery level, authority to influence other agents, etc. ρ can be extracted from the incoming messages but it may pose a potential security threat if additional security measures are not put in place. Alternatively, the receiving agent can keep a list of ρ values and internally decide on whether an incoming message is from a reliable UAV or not (decreases scalability). φ determines how much each agent is open for influence from other agents. In our experiments, for simplicity, all agents have $\varphi = 0.5$ and $\rho = 1$.

3.4. Swarm's Path Planning

The operator can create an artificial potential field where high confidence is repulsive and low confidence is attractive. Using potential field-based algorithms to repel UAVs from each other and from obstacles while being attracted to the target zones was studied before [29,30]. As opposed to these studies, we use one interface for state estimation, visualization, and indirect parameter setting with proximal interactions.

The swarm continuously follows the downhill gradient of the potential calculated by applying Sobel filter on the confidence map [31]. UAVs move towards the least confidence area and, therefore, try to maximize their confidence about the environment.

The confidence of the boundaries is set to 1 to repel the UAVs. This also prevents them from getting locked in clusters around the corners or near the boundaries and keeps the UAVs inside the mission zone. The two forces applied to the UAVs are the local ($\overrightarrow{f_{l,i}}(t) = \nabla_{local}(c_{i,t})$) and global gradients ($\overrightarrow{f_{g,i}}(t) = \nabla_{global}(c_{i,t})$) of their own confidence map. The total force is given by

$$\overrightarrow{f_i}(t) = \omega\overrightarrow{f_{l,i}}(t) + (1-\omega)\overrightarrow{f_{g,i}}(t) : \ \overrightarrow{f} \in \mathbb{R}^2, \quad (3)$$

where $\overrightarrow{f_i}$ determines the motion of the UAV i and ω specifies the weighting factor between local and global forces.

By defining the dynamics of each UAV as given in Equation (3), the mobility of the swarm can be controlled by adjusting the weight coefficient ω. Higher ω pushes the UAVs to explore their local neighborhood and stay in a small exploration range, while lower ω prioritizes the global certainty and the entire swarm moves towards the common targets. When the UAVs prefer to stay in their local neighborhood, their belief map will only be precise about the area around them and the UAVs across the swarm will have diverse belief and confidence maps that might mislead the operator. On the contrary, lower ω may lead to a high congestion around the common targets. In this case, UAVs perform a flocking behavior that decreases the swarm utilization as all UAVs will have common targets and will not distributively explore the mission zone. Assigning different ω values to the swarm may prove useful for developing a diversity in the swarm behavior for exploration and exploitation. In this paper, $\omega = 0.99$ and fixed for all experiments. The UAVs move in a 2D space and all forces are applied on a plane with constant altitude and parallel to the xy-plane.

4. Simulation Platform

There are several multi-UAV/swarm simulation environments based on commercial (e.g., MATLAB [32], X-Plane flight simulator [33]) and non-commercial platforms (e.g., ARGoS [34]). We designed our own simulation environment based on an open-source video game library, Arcade [35], to be able to freely modify the simulator and also use the high usability characteristic of a game environment. The parameters are optimized to run on low computational power machines. For our experiments, we used a local machine with a 1.4 GHz Intel Core i5 processor and 16 GB 2133 MHz of memory. The images taken from the mission zone have the resolution of 600×600 pixels. We use the term pixels in images or cells in matrices to refer to the same concept of information stored in the high- or low-resolution images taken from the mission zone. The UAVs decrease the resolution of the images to 40×40 pixels to speed up the computation and decrease the load on the communication channels.

Belief and Confidence Maps Setup

The belief and confidence maps are initialized to zero. As the UAVs move, they observe a radius of 2 pixels from their current position ($r = 2$ px in belief and confidence maps and $r = 30$ px in high resolution) and communicate within a radius of 4 pixels ($r = 60$ px in high resolution). The confidence of radius 2 px (a disk with a diameter of 5 pixels) is set to one. The belief map gets updated based on the observed images. If there is a disaster or an operator within the vision range (r), the corresponding cells of the map will be set to one. Changes in the environment may make the stored information obsolete. Therefore, we associate time to the confidence by multiplying the confidence matrix with the decay factor ($\alpha = 0.0001$) at each simulation step. The parameters can be tuned based on the experiments and the domain of application. For instance, a lower decay factor is suitable for dynamic environments as UAVs quickly lose their confidence and frequently need to explore the areas again. Here, we chose our parameters based on empirical observations for proving the concept. Cells with higher confidence clearly have greater impacts on the neighbors. As the UAVs have limited sensor range, the communication enables them to exchange information and broaden their perception to the boundaries of the complete zone.

Another advantage of using confidence maps as a basis for motion pattern is collision avoidance, as the attraction/repulsion mechanism indirectly results into a motion with low collision rates, assuming that there is no communication failure. Recently visited cells have high confidence values and as neighboring UAVs approach each other and share their confidence maps, they repel from the cells that their neighbor recently visited and collisions will be avoided. This also leads to spatial division of the mission zone between the neighboring UAVs (indirect coordination) as the UAVs prefer not to use the same paths that their neighbors recently took.

5. Empirical Evaluation

We designed four experiments to examine our interface and the swarm's behavior with different levels of autonomy. First, the swarm's performance in exploring the mission zone with full autonomy—no human intervention—is examined (Experiment I). We evaluate how the swarm reacts to dynamic environments with evolving disasters in Experiment II.

In Experiment III, the level of autonomy decreases and we measure the swarm's response to human interactions. Finally, we perform a separate interactive user study to evaluate our interface in Experiment IV.

We ran 15 simulation runs with 1000 simulation steps and for each experiment the mission zone consisted of a disaster area, a human operator, and 15 UAVs that are positioned randomly at start of each trial. The swarm is examined with three different locations of the disaster area to make sure that the results are not biased to a certain setup.

5.1. *Experiment I: Autonomous Exploration*

In the first experiment, the operator monitors the swarm while the UAVs explore the area, gain confidence over the entire mission zone, build a belief map, and inform the operator about the situation. We measure the accuracy of the resulting belief map by comparing it to the actual situation.

Confidence maps evolve as the UAVs explore the environment and receive information from their neighbors. This evolution also depends on the decay factor (α). In this case, the confidence map does not age fast and the certainty accumulates until the entire area is marked with high confidence.

Figure 5 shows the area coverage accumulated during one simulation run (see Figure 5a) and an averaged coverage for all 15 simulation runs (see Figure 5b). The distribution of the swarm is almost uniform, except near the boundaries (white), where there are denser footprints due to boundary avoidance. Figure 5b also shows a circular area at the right side of the zone (dark blue) with sparser coverage. This area is the communication range of the operator with high confidence values. The reason for having a high confidence in this region is that the confidence maps that the operator receives always have high values around the operator, as this is where the UAVs visited most recently. When an UAV enters this region, it receives an update from the operator and by averaging its confidence with the one from the operator, the UAV notices an area with high confidence values and refuses to explore it. By averaging over all simulation runs we see that this region is less explored.

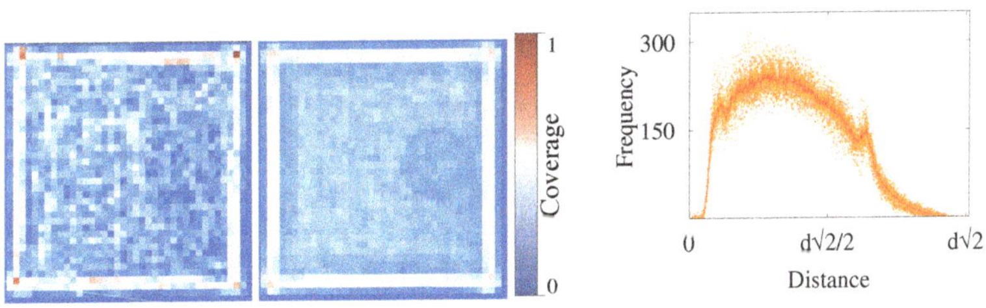

(**a**) One run (**b**) All runs (**c**) Frequency of distances

Figure 5. The area coverage with the swarm at the end of a randomly selected run (**a**) and the average over all 15 simulation runs (**b**) in Experiment I. The frequency of distances between the UAVs in autonomous exploration experiment is also shown (**c**) where *d* is width/length of the mission zone.

Figure 5c shows the frequency of distances between the UAVs. The low frequency of distances around zero shows that the collisions are rare. The two minor peaks before and after the global maximum are caused by the clusters of UAVs around the boundaries (white box in Figure 5b).

Figure 6 shows the accuracy of the mapping over time. The error in the belief map is the absolute difference between the UAVs' belief maps and the actual disaster scenario in the mission zone. Using a belief map of size 40×40, the theoretical maximum level of error is 1600. Nonetheless, the highest error we see during our experiments is 21 and for visibility we use that as our maximum error in the figures. Similarly, with the human intervention, the confidence can theoretically reach a very high or a very low value, but in our experiments we bound the range to $[0, 2000]$. Errors in the belief map start at around 14 pixels, which is the number of cells in the belief map that cover the disaster zone and the operator (normalized to $0.\bar{6}$; maximum is 21). The swarm explores the area and gradually constructs an accurate map and the error in the belief map decreases. The confidence level increases over time and with all the cells in a confidence map of size 40×40 set to one, the confidence can rise up to 1600 (normalized to 0.8; maximum is 2000). Figure 6 shows the precision of the belief map that the operator receives (see Figure 6c) and the average belief error among the UAVs (see Figure 6b). We can also observe the growth in the confidence level (see Figure 6d–f. We separately plot the progress in the maps of the swarm and the operator to compare the quality of the swarm's perception and the information that is passed on to the operator (monitoring quality).

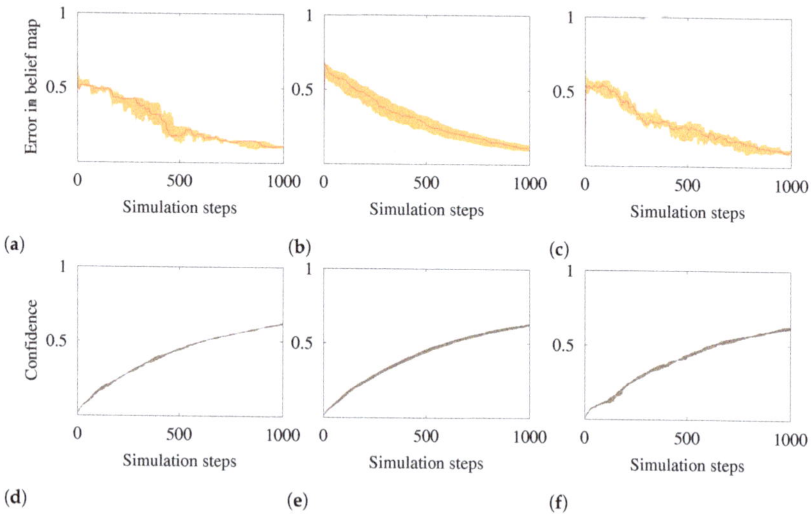

Figure 6. The precision of mapping is shown in the error level of the belief map. This figure shows the belief error and the confidence level for the autonomous exploration experiment. The top row shows the errors in the belief map and the bottom row depicts the confidence levels over time. (**a**,**d**) are from one simulation run. (**b**,**e**) show the median values over all simulation runs. (**c**,**f**) are the belief error and the confidence level of the operator.

5.2. *Experiment II: Response to Evolving Disaster*

In this experiment, we examine the adaptivity of the mapping in a dynamic environment with an evolving disaster. As opposed to Experiment I, here we assume that the disaster zone is not fixed and it can move to another point in the area. The disaster zone moved between $150 < t < 350$ simulation steps. We observe how this change is reflected in the performance of the swarm and the operator's understanding of the environment. Note that the swarm still has full autonomy and the operator only monitors the swarm. Figure 7a shows the direction of the disaster displacement and Figure 7b shows the belief map when the motion is stopped. The operator's belief map gradually adapts to the new position of the disaster (see Figure 7b,c). Figure 8a,b show the effect of a dynamic environment on the error in the belief of the swarm and the operator over time. The error increases at first as the environment starts to change, but after a while the swarm successfully adapts itself.

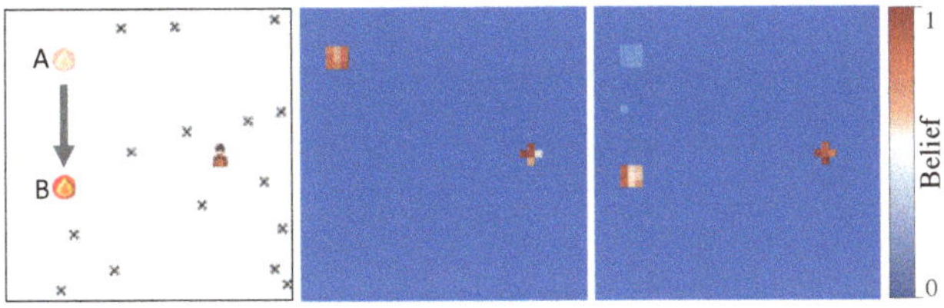

(a) moving disaster (b) before change (c) after change

Figure 7. In Experiment II, the disaster area moves from point A to B (**a**) and the belief of the swarm gradually adapts to the change (**b**,**c**).

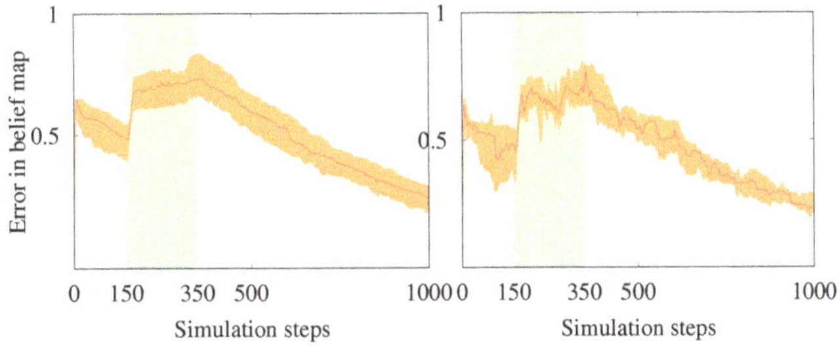

(a) Swarm's belief error (b) Operator's belief error

Figure 8. Median errors in the belief maps of the swarm (**a**) and the operator (**b**) in Experiment II. The disaster zone moves between 150 and 350 simulation steps (green area) as shown in Figure 7.

This behavior can also be seen in Figure 7b,c by the former positions of the disaster slowly fading away and the new positions appearing in the belief maps. Shortly after, the UAVs successfully map the environment and notify the operator. The results from the evolving disaster experiment show that the swarm is resilient to changes in the environment and the operator's belief map stays updated.

5.3. Experiment III: Human Interaction

After investigating the swarm's autonomous behavior, we now focus on the human–swarm communication link.

The decision of whether to influence the map or not can entirely affect the autonomy of the swarm (i.e., flexible autonomy). Leaving the confidence map untouched allows a fully autonomous swarm approach while manipulating the entire map would create an almost fully manual control over the swarm's behavior.

In this experiment, we observe the behavior of the swarm with the operator's control and evaluate the task completion. The operator can lead the swarm to explore an area of interest (attraction) or prevent them from flying over a no-fly zone (repulsion) by manipulating its own confidence map and disseminating it to the neighboring UAVs. For example, the operator can send a new confidence map with attraction in the middle of the mission zone to its neighboring UAVs. As the UAVs continuously update their confidence map based on their neighbors, some of them may receive the operator's confidence map and move according to the potential field.

Figure 9 shows the influence of the operator's command on the confidence of an UAV and how the UAV is controlled. Figure 9a illustrates the confidence map that the operator introduces to the swarm to attract them to the middle of the arena. After receiving the command, the UAV in this example computes the mean of its previous confidence map (see Figure 9b) and the command to update its map (see Figure 9c). As a result of the attraction force, the UAVs move towards the area in the middle and this influence diminishes. The UAVs' presence in the middle of the arena can be inferred as the values in the middle of the confidence map increase (see Figure 9d). The operator's communication range determines the controlling impact of the operator on the swarm.

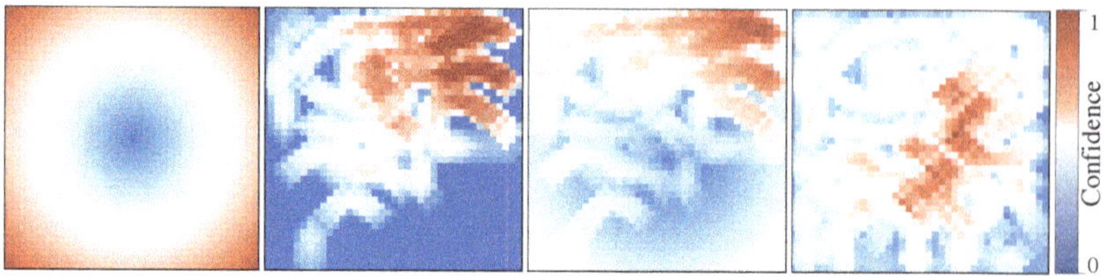

(**a**) command (**b**) $t = t_0 - 1$ (**c**) $t = t_0 + 1$ (**d**) $t = t_0 + 50$

Figure 9. An example of the operator's command at t_0 (**a**), an UAV's confidence map before (**b**) and after receiving the command (**c**), and after swarm's reaction to the influence by moving to the center (**d**) in Experiment III.

Using wider communication ranges, the operator influences more UAVs and therefore the effect is more significant. Figure 10 shows the effect of a human interaction on the confidence level of the swarm. At $t = 300$ the operator disseminates a command once and it takes 100 to 200 simulation steps to significantly influence the entire swarm. In case of attraction, UAVs quickly move to the cells with low confidence and immediately compensate the drop in the certainty level (see the valley in Figure 10a), while in repulsion the UAVs cannot visit the cell to recover from the added value to the confidence as they are repelled from that region. This explains the recovery of the confidence levels in Figure 10a and the bias that stays in the swarm for the rest of the experiment in Figure 10b.

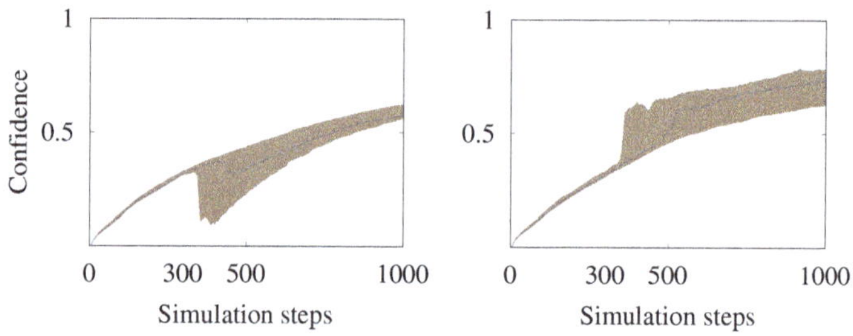

(**a**) Attraction (**b**) Repulsion

Figure 10. Influence of the operator's command at $t = 300$ on the swarm's confidence.

So far, we have covered the temporary attraction and repulsion. We also assume a disaster situation where UAVs must avoid a certain area at all times. An example can be the frequent flying zone of an airport or a no-fly area with high voltage cables. For that, the operator can continuously apply a repulsion force from the prohibited zone. Figure 11a shows the confidence bias that the operator introduces to the swarm. The cells highlighted with red mark the barrier or the prohibited zone. The blue area in Figure 11a can be either empty, which allows the swarm to keep the knowledge about the confidence and only manipulate the values in the interest zone or it can be set to zero, which may gradually decrease the swarm's confidence level on all the other areas. Figure 11b shows the average distribution of the UAVs in all simulation runs as a result of the permanent repulsion at the center. From the low coverage at the center of Figure 11b, it is clear that the operator successfully controlled the swarm to avoid the area in the middle of the mission zone. As the communication range is limited, there is a delay until all UAVs become aware of the new force introduced by the operator. In the meantime, even though the operator command is already issued, some UAVs are unaware and cross the prohibited zone. The ring in the distribution around the repellent zone is caused by the area with the equilibrium between different repulsion forces.

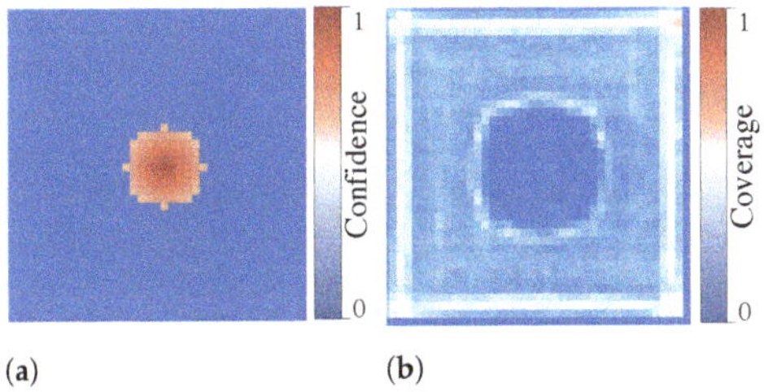

Figure 11. Operator's confidence map for introducing a permanent repulsion (**a**) and the swarm's distribution (**b**).

We increased the task complexity and simulated the influence of five permanent no-fly zones of length L to 1.8 L (imposed restrictions by the operator) in the middle of the arena (see Figure 12a,b). We assume that the messages cannot travel through the obstacles. Figure 12c shows the error in the operator's belief map with and without barriers. As the size of the barrier grows, the precision of the mapping decreases. For 1.8 L, the barrier is too spacious for the swarm to adequately travel around it and keep the operator updated about the situation, and the precision of the belief map decreases (i.e., the belief error does not significantly decline over time). Our results show that the swarm is able to follow a wide range of control commands and continue to offer a precise mapping of the environment.

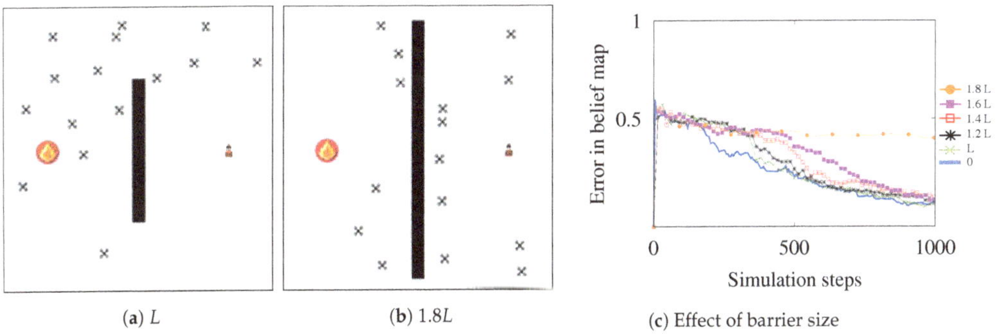

Figure 12. Error in the operator's belief map for different sizes of the repulsion zones.

5.4. Experiment IV: Human–Swarm Interaction User Study

In this final experiment, we performed an additional technology-probe user study. We compared the situation awareness of a fully autonomous swarm, a human–swarm system, where only the target is specified by the operator (Task I). Despite the preference of the users, discussed in Section 2, we conducted another experiment where the operator controls individual drones (Task II). We posted a call for participation in a small online community. Fifteen computer scientists with BSc and MSc degrees in Computer Science or Engineering were recruited. The users were given a remote access to the machine with the simulation environment. We briefed the users about the goal of the experiment in a 10- to 15-minute video call before the experiment. We made another call for 10–15 minutes to collect their feedback after they performed the experiment. Users could observe the belief and confidence maps. They were asked to predict the significance of the areas that needed to be further explored. The number of clicks on the confidence map (number of commands) was left as a choice to the users. In Task II, the users had access to the list of UAVs. They could select the UAVs and assign a target to it by clicking on a cell of the confidence map. The Black line-point in Figure 13 shows the median belief error using this approach.

For Task I, we asked the users to operate the swarm using our interface. They needed to only click on a cell in the confidence map that they wanted to be explored. They did not need to select single UAVs themselves. The blue points in Figure 13 show the median belief error using our interface. In both experiments, the disaster moves in a random direction and the human operator waits for a while to let the swarm explore the area (neglect benevolence [36]). This explains the overlapping performance between all experiments (autonomous exploration, operator controlling a single UAV, and operator controlling the swarm) at the early stage of the experiment. Once the new location of the disaster emerges, we observe that the operator frequently commands the swarm to narrow the exploration area to the points with higher belief values. As shown in Figure 13, the accuracy with human intervention is better than the results from the fully autonomous exploration. We also observe that the performance with our interface (Task I) and the individual UAV control (Task II) is almost identical. However, the cognitive workload of having to select individual UAVs first and then selecting the target area is much higher than only picking the target areas. For larger swarm sizes, controlling individual UAVs is not feasible. On the contrary, our interface is scalable. The operator can select areas of interest, regardless of the swarm size and the swarm will self-organize to assign individual target points.

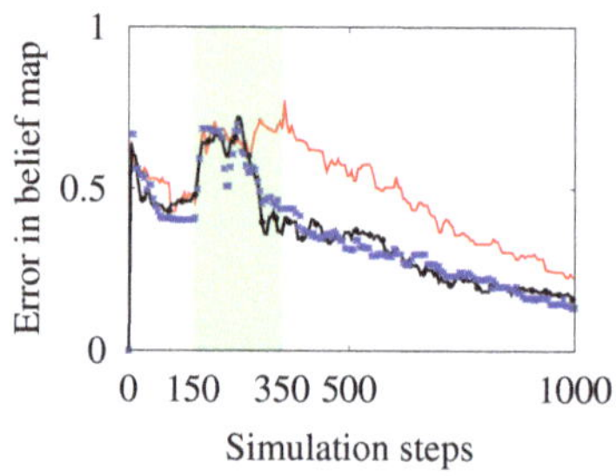

Figure 13. Median errors in the operator's belief map with autonomous exploration (Experiment II, red line), human control of individual UAVs (Experiment IV: Task II, black line-point), and human–swarm teaming with our interface (Experiment IV: Task I, blue points).

6. Discussion and Future Work

Establishing a reliable interaction between human operators and a swarm comes with plenty of issues. In this paper, we focused on the issue of situation awareness with a swarm and an operator when all interactions are local without imposing a high cognitive complexity. Our results show that the same map can be used for both visualization and control of the swarm. The operator has to interact with the swarm via this map and there is no need for one-to-one interaction with each member of the swarm, which can be overwhelming for a human operator. We made a few assumptions to prove the concept. We have not performed a parameter sensitivity analysis and our parameters were chosen experimentally. For instance, our swarm size was fixed when studies show that the swarm density has an effect on performance [37]. Another point of discussion is the scalability of our approach. Our method relies on local interactions and allows for scalable human–swarm interaction in mapping an unexplored environment. The only limit for scalability is that as the exploration area grows, the mapping resolution decreases. A potential solution for this can be registering local images and constructing a global image at the human operator's side.

There are several directions that this research can be extended: (1) We plan to investigate the effect of multiple dynamic disaster areas and human operators with collaborative/competitive management strategies. Besides the competing operator commands, there may also be intruder attacks by other UAVs or operators that need to be detected and eliminated from the decision-making process; (2) The effect of noise (communication, positioning, and sensing) on the performance of the swarm and detecting malfunctioning/lost UAVs need to be investigated as well; (3) Speed versus accuracy can be studied with several messaging qualities. (4) One of our next steps will be to co-create an interaction interface for human–swarm teams with experts from the industry and perform a more comprehensive user study; (5) Variations that may be induced by the operator bias in the human-in-the-loop experimentation; (6) In some use-cases it might prove useful to direct a subset of the swarm by a more complex communication protocol that allows access to certain members of the swarm without affecting the entire swarm; (7) We also plan to address some of these challenges and implement our approach on physical UAVs and evaluate the performance of the swarm in real world applications.

7. Conclusions

A user study with 100 participants showed that users prefer to have a global overview of the swarm's coverage instead of monitoring individual agents, especially for large swarm sizes and when the time is critical. We showed that a human–swarm interaction based on a global view of the swarm with a human-in-the-loop decision-making can assist the swarm to map a dynamic environment. Using our approach, human operators do not require a fully-connected communication network with all UAVs to be able to monitor the swarm and control them. In our experiments, an operator controlled a group of simulated UAVs and successfully explored the environment even when it was dynamic. Our approach is easily reproducible and can serve as a basis for monitoring and controlling the swarm in various application domains such as agricultural missions, space exploration, etc.

Author Contributions: Conceptualization, M.D.S., D.T. and S.D.R.; methodology, M.D.S., D.T. and S.D.R.; software, M.D.S. and J.G.; validation, M.D.S. and J.C.; formal analysis, M.D.S. and J.C.; investigation, M.D.S. and J.C.; resources, M.D.S.; data curation, M.D.S. and J.C.; writing—original draft preparation, M.D.S., J.C., J.G., D.T., and S.D.R.; writing—review and editing, M.D.S. and J.C.; visualization, M.D.S. and J.C.; supervision, D.T. and S.D.R.; project administration, S.D.R.; funding acquisition, M.D.S. and S.D.R.; All authors have read and agreed to the published version of the manuscript.

Funding: We acknowledge funding from the UKRI Trustworthy Autonomous Systems Hub (EP/V00784X/1).

Institutional Review Board Statement: The study was conducted according to the Data Protection Act 1998, and approved by the Ethics Committee of University of Southampton (Ethics and Research Governance Online number 66360.A1 on 23 August 2021).

Informed Consent Statement: Informed consent was obtained from all subjects involved in the study.

Data Availability Statement: Not applicable.

Acknowledgments: Not applicable.

Conflicts of Interest: The authors declare no conflict of interest.

References

1. Innocente, M.S.; Grasso, P. Self-organising swarms of firefighting drones: Harnessing the power of collective intelligence in decentralised multi-robot systems. *J. Comput. Sci.* **2019**, *34*, 80–101. [CrossRef]
2. Gkotsis, I.; Kousouraki, A.C.; Eftychidis, G.; Kolios, P.; Terzi, M. Swarm of UAVs as an emergency response technology. Risk Analysis Based on Data and Crisis Response Beyond Knowledge. In Proceedings of the 7th International Conference on Risk Analysis and Crisis Response (RACR 2019), Athens, Greece, 15–19 October 2019; p. 353.
3. Busnel, Y.; Caillouet, C.; Coudert, D. Self-organized Disaster Management System by Distributed Deployment of Connected UAVs. In Proceedings of the ICT-DM 2019-6th International Conference on Information and Communication Technologies for Disaster Management, Paris, France, 18–20 December 2019; pp. 1–8.
4. Chung, S.J.; Paranjape, A.A.; Dames, P.; Shen, S.; Kumar, V. A survey on aerial swarm robotics. *IEEE Trans. Robot.* **2018**, *34*, 837–855. [CrossRef]
5. Hexmoor, H.; McLaughlan, B.; Baker, M. Swarm Control in Unmanned Aerial Vehicles. In Proceedings of the IC-AI, Vegas, NV, USA, 27–30 June 2005; pp. 911–917.

6. Liu, R.; Jia, F.; Luo, W.; Chandarana, M.; Nam, C.; Lewis, M.; Sycara, K. Trust-Aware Behavior Reflection for Robot Swarm Self-Healing. In Proceedings of the 18th International Conference on Autonomous Agents and MultiAgent Systems, Montreal, QC, Canada, 13–17 May 2019; pp. 122–130.
7. Matsuka, K.; Feldman, A.O.; Lupu, E.S.; Chung, S.J.; Hadaegh, F.Y. Decentralized Formation Pose Estimation for Spacecraft Swarms. *Adv. Space Res.* **2020**, *67*, 3527–3545. [CrossRef]
8. Cain, M.S.; Wendell, D.M. Human perception and prediction of robot swarm motion. Micro-and Nanotechnology Sensors, Systems, and Applications XI. *Int. Soc. Opt. Photonics* **2019**, *10982*, 1098226.
9. Lindner, S.; Schulte, A. Evaluation of Swarm Supervision Complexity. In *International Conference on Intelligent Human Systems Integration*; Springer: Berlin/Heidelberg, Germany, 2021; pp. 50–55.
10. Patel, J.; Xu, Y.; Pinciroli, C. Mixed-granularity human-swarm interaction. In Proceedings of the 2019 International Conference on Robotics and Automation (ICRA), IEEE, Montreal, QC, Canada, 20–24 May 2019; pp. 1059–1065.
11. Ramchurn, S.D.; Wu, F.; Jiang, W.; Fischer, J.E.; Reece, S.; Roberts, S.; Rodden, T.; Greenhalgh, C.; Jennings, N.R. Human–agent collaboration for disaster response. *Auton. Agents Multi-Agent Syst.* **2016**, *30*, 82–111. [CrossRef]
12. Nam, C.; Walker, P.; Li, H.; Lewis, M.; Sycara, K. Models of trust in human control of swarms with varied levels of autonomy. *IEEE Trans. Hum.-Mach. Syst.* **2019**, *50*, 194–204. [CrossRef]
13. Ashcraft, C.C. *Moderating Influence as a Design Principle for Human-Swarm Interaction*; Brigham Young University: Provo, UT, USA, 2019.
14. Oliveira, T.L.; Batista, M.R.; Romero, R.A. Analysis of human-swarm interaction through potential field manipulation. In Proceedings of the 2017 Latin American Robotics Symposium (LARS) and 2017 Brazilian Symposium on Robotics (SBR), IEEE, Curitiba, Brazil, 8–10 November 2017; pp. 1–6.
15. Levillain, F.; St-Onge, D.; Zibetti, E.; Beltrame, G. More than the sum of its parts: Assessing the coherence and expressivity of a robotic swarm. In Proceedings of the 2018 27th IEEE International Symposium on Robot and Human Interactive Communication (RO-MAN), IEEE, Nanjing, China, 27–31 August 2018; pp. 583–588.
16. Capelli, B.; Secchi, C.; Sabattini, L. Communication through motion: Legibility of multi-robot systems. In Proceedings of the 2019 International Symposium on Multi-Robot and Multi-Agent Systems (MRS), IEEE, New Brunswick, NJ, USA, 22–23 August 2019; pp. 126–132.
17. Brooke, J. SUS - A Quick and Dirty Usability Scale. *Usability Eval. Ind.* **1996**, *189*, 4–7.
18. Van Der Laan, J.D.; Heino, A.; De Waard, D. A simple procedure for the assessment of acceptance of advanced transport telematics. *Transp. Res. Part C: Emerg. Technol.* **1997**, *5*, 1–10. [CrossRef]
19. Lancaster, H.O.; Seneta, E. Chi-square distribution. *Encycl. Biostat.* **2005**, *2*. [CrossRef]
20. Shaffer, J.P. Multiple hypothesis testing. *Annu. Rev. Psychol.* **1995**, *46*, 561–584. [CrossRef]
21. Girden, E.R. *ANOVA: Repeated Measures*; Number 84; Sage: Newcastle upon Tyne, UK, 1992.
22. Soorati, M.D.; Krome, M.; Mora-Mendoza, M.; Ghofrani, J.; Hamann, H. Plasticity in Collective Decision-Making for Robots: Creating Global Reference Frames, Detecting Dynamic Environments, and Preventing Lock-ins. In Proceedings of the 2019 IEEE/RSJ International Conference on Intelligent Robots and Systems (IROS), IEEE, Macau, China, 3–8 November 2019; pp. 4100–4105.
23. Valentini, G.; Ferrante, E.; Dorigo, M. The best-of-n problem in robot swarms: Formalization, state of the art, and novel perspectives. *Front. Robot. AI* **2017**, *4*, 9. [CrossRef]
24. Dorigo, M.; Theraulaz, G.; Trianni, V. Reflections on the future of swarm robotics. *Sci. Robot.* **2020**, *5*, eabe4385. [CrossRef] [PubMed]
25. Valentini, G.; Brambilla, D.; Hamann, H.; Dorigo, M. Collective perception of environmental features in a robot swarm. In *International Conference on Swarm Intelligence*; Springer: Berlin/Heidelberg, Germany, 2016; pp. 65–76.
26. Valentini, G.; Hamann, H.; Dorigo, M. Self-organized collective decision making: The weighted voter model. In Proceedings of the AAMAS, Paris, France, 5–9 May 2014; pp. 45–52.
27. Valentini, G.; Hamann, H.; Dorigo, M. Efficient decision-making in a self-organizing robot swarm: On the speed versus accuracy trade-off. In Proceedings of the 2015 International Conference on Autonomous Agents and Multiagent Systems, Istanbul, Turkey, 4–8 May 2015; pp. 1305–1314.
28. Kolling, A.; Walker, P.; Chakraborty, N.; Sycara, K.; Lewis, M. Human interaction with robot swarms: A survey. *IEEE Trans. Hum.-Mach. Syst.* **2015**, *46*, 9–26.
29. Kolling, A.; Sycara, K.; Nunnally, S.; Lewis, M. Human swarm interaction: An experimental study of two types of interaction with foraging swarms. *J. Hum.-Robot Interact.* **2013**, *2*, 104–129. [CrossRef]
30. Brown, D.S.; Kerman, S.C.; Goodrich, M.A. Human-swarm interactions based on managing attractors. In Proceedings of the 2014 ACM/IEEE International Conference on Human-Robot Interaction, Bielefeld, Germany, 3–6 March 2014; pp. 90–97.
31. Choset, H.; Lynch, K.M.; Hutchinson, S.; Kantor, G.; Burgard, W.; Kavraki, L.E.; Thrun, S. *Principles of Robot Motion: Theory, Algorithms, and Implementations*; Chapter 7; The MIT Press: Cambridge, MA, USA, 2005.
32. Soria, E.; Schiano, F.; Floreano, D. SwarmLab: A Matlab Drone Swarm Simulator. *arXiv* **2020**, arXiv:2005.02769.
33. Garcia, R.; Barnes, L. Multi-UAV Simulator Utilizing X-Plane. *J. Intell. Robot. Syst.* **2010**, *57*, 393–406. [CrossRef]
34. Pinciroli, C.; Trianni, V.; O'Grady, R.; Pini, G.; Brutschy, A.; Brambilla, M.; Mathews, N.; Ferrante, E.; Di Caro, G.; Ducatelle, F.; et al. ARGoS: A modular, parallel, multi-engine simulator for multi-robot systems. *Swarm Intell.* **2012**, *6*, 271–295. [CrossRef]
35. The Python Arcade Library. Project Website, 2019. Available online: http://arcade.academy/ (accessed on 1 November 2021).

36. Walker, P.; Nunnally, S.; Lewis, M.; Kolling, A.; Chakraborty, N.; Sycara, K. Neglect benevolence in human control of swarms in the presence of latency. In Proceedings of the IEEE International Conference on Systems, Man, and Cybernetics (SMC), Seoul, Korea, 14–17 October 2012; pp. 3009–3014.
37. Hamann, H.; Reina, A. Scalability in computing and robotics. *IEEE Trans. Comput.* **2021**. [CrossRef]

drones

MDPI

Article

A Single-Copter UWB-Ranging-Based Localization System Extendable to a Swarm of Drones

Christoph Steup *,†, Jonathan Beckhaus † and Sanaz Mostaghim

Faculty of Computer Science, Otto von Guericke University Magdeburg, 39106 Magdeburg, Germany; jonathan.beckhaus@ovgu.de (J.B.); sanaz.mostaghim@ovgu.de (S.M.)
* Correspondence: steup@ovgu.de; Tel.: +49-391-67-51021
† These authors contributed equally to this work.

Abstract: This paper presents a single-copter localization system as a first step towards a scalable multihop drone swarm localization system. The drone was equipped with ultrawideband (UWB) transceiver modules, which can be used for communication, as well as distance measurement. The location of the drone was detected based on fixed anchor points using a single type of UWB transceiver. Our aim is to create a swarm localization system that enables drones to switch their role between an active swarm member and an anchor node to enhance the localization of the whole swarm. To this end, this paper presents our current baseline localization system and its performance regarding single-drone localization with fixed anchors and its integration into our current modular quadcopters, which was designed to be easily extendable to a swarm localization system. The distance between each drone and the anchors was measured periodically, and a specially tailored gradient descent algorithm was used to solve the resulting nonlinear optimization problem. Additional copter and wireless-specific adaptations were performed to enhance the robustness. The system was tested with a Vicon system as a position reference and showed a high precision of 0.2 m with an update rate of <10 Hz. Additionally, the system was integrated into the FINken copters of the SwarmLab and evaluated in multiple outdoor scenarios. These scenarios showed the generic usability of the approach, even though no accurate precision measurement was possible.

Keywords: localization; communication; drone swarm

Citation: Steup, C.; Beckhaus, J.; Mostaghim, S. A Single-Copter UWB-Ranging-Based Localization System Extendable to a Swarm of Drones. *Drones* **2021**, *5*, 85. https://doi.org/10.3390/drones5030085

Academic Editor: Vishal Sharma

Received: 2 July 2021
Accepted: 18 August 2021
Published: 30 August 2021

1. Introduction

Drones are gaining increasing importance in industrial and research applications. Applications range from inspection tasks for bridges [1] and wind turbines [2] to natural disaster management [3]. For these tasks, reliability, autonomy, and stability are the most relevant properties in the current research [4–6]. Currently, there is a trend towards the combination of multiple drones into swarms to enhance the performance, stability, and mission times [7–9]. Independent of the number of drones, the quality of position information is highly relevant, because it enables reliable and robust movement of the swarm. However, such reliable position information is hard to achieve. Global Positioning System (GPS) localization is not always available or precise. Indoor scenarios especially have no access to GPS, but even in outdoor environments, GPS may be unreliable or unavailable. Consequently, a general-purpose drone swarm needs a localization mechanism that is compatible with indoor and outdoor scenarios, without specific deployment.

The transition from singular drones to swarms of drones creates new challenges, especially regarding the localization, because the precision of the provided position information needs to be higher to prevent crashes among swarm members. Additionally, the localization needs to be reliable and uncertainty-aware. Reliable localization allows the swarm to execute its mission, while uncertainty awareness allows the swarm behavior to react to a degradation in the localization precision. The last important factor is scalability because most localization systems require communication among the swarm members, which is performed through a shared medium. Consequently, using the available resources of the communication system efficiently is of utmost importance for any localization system to be scalable to any swarm size.

In a swarm of drones, more options are available for the localization of individual drones. Drones without a connection to static anchors or GPS may use multihop localization to acquire the localization information of neighbors to enhance their own localization.

Our vision is to create a swarm localization system that enables drones to switch their role between active swarm member and anchor node to enhance the localization of the whole swarm. To this end, this paper present our current baseline localization system, which was designed to be easily extendable to a swarm localization system. Additionally, its performance regarding single-drone localization with fixed anchors and its integration into our current modular quadcopters is presented.

The following section (Section 2) discusses some relevant work in the area of robot and quadcopter localization, followed by the description of the hardware used (Section 3) and our algorithms in Section 4 and the evaluation experiments and results in Section 5. The paper closes with the conclusion and future work in Section 6.

2. Related Work

The area of localization is filled with a plethora of approaches, because the application field is also very vast. The most prominent solution for outdoor applications is GPS. Baseline GPS is a receiver-only system, which uses satellites around the globe to provide Time-of-Flight (TOF) distances between the receiver and sending satellite. The existing known orbital data of the satellites allow the reconstruction of the position using these TOF distances and an accurate time stamp. The typical precision of GPS positions is in the order of 5 m (https://www.gps.gov/systems/gps/performance/accuracy/, accessed on 25 August 2021). However, GPS is prone to distortions if the line-of-sight is blocked. Assisted GPS (AGPS) is one approach to enhance the quality of the localization information. AGPS uses additional satellites, which are deployed regionally and compensate for the atmospheric distortions of the GPS signals. Most commercially available GPS receivers already support AGPS, which provides these receivers with a maximum accuracy of 30 cm.

Further localization enhancements of GPS for high-precision applications are available through Differential GPS (D-GPS), which uses an additional local receiver to compensate for local signal distortions. Even though D-GPS at best can reach millimeter precision, the hardware is costly and requires a long time to be deployed (more than 20 min of initial convergence time [10]). An example is the DJI D-RTK2, which costs EUR $\approx$ 3000 (https://store.dji.com/de/product/d-rtk-2-high-precision-gnss-mobile-station, accessed on 25 August 2021). An overview of D-GPS systems and their performance can be found in [11].

Another approach more tailored to indoor scenarios is camera-based localization systems. In this area, multiple approaches exist. The most famous one is the Vicon system, which uses infrared reflective passive markers and multiple infrared cameras to track an object with high precision (typically less than 1 mm with an update rate of 100 Hz [12]). The performance of the Vicon system comes at a very high cost of typically more than EUR 50,000. Other approaches are based on single cameras and use no markers, such as that in [13]. These provide far less precision (typically 10 cm, but also with a high update rate of 50 Hz or more). These systems are less expensive because fewer cameras and no markers are needed. A good overview of vision-based localization systems is given in [14].

The third type of system uses radiofrequency (RF) or sonar waves to detect the distance between two objects. In this area, sonar waves have been used for a long time. However, currently, the trend is moving towards extensions of existing wireless communication standards. As the information, ToF, Time-Difference of Arrival (TDoA), Received Signal Strength (RSSI), or Phase Difference (PD) is used. Among these approaches, RSSI methods provide the worst precision of $\approx$1.2 m for Bluetooth [15] and 1 m for WiFi [16], with the benefit that they can be applied to all existing wireless hardware without any modification. ToF,

TDoA, and PD provide better distance estimations, but typically require specialized hardware, such as the DWM1000 (https://www.decawave.com/product/dwm1000-module, accessed on 25 August 2021) modules. These modules are typically used with two-way-ranging, which sends 3 to 4 packets between two nodes, which cancel out internal timing errors and increase the precision at the cost of more communication per measurement [17]. A system using this approach was presented in [18]. The proposed system showed an average precision of 0.4 m with a 20 Hz update rate using 27 packets per measurement. This disqualifies the approach for large amounts of swarm members or anchor nodes.

The last type of approach is swarm-based, which exploits the existence of multiple objects of similar type to be localized to share information and increase precision. One such approach is called Simultaneous Localization and Optimization [19], which exploits movement command information in each object together with distance measurements between the objects to enhance the positioning precision and resolve ambiguities, as well as enhance the execution of the movement goals. The most complete approach in this area is OmniSwarm [20]. This approach provides up to 0.02 m precision for a three-drone system and 0.14 m precision for a swarm of two drones. The system uses visual odometry together with Ultra-Wideband (UWB) communication and distance estimation to determine the position of the swarm members. Even though the OmniSwarm approach provides superior localization quality, it does not scale well, due to the communication necessary to exchange information. Additionally, the system uses information from many subcomponents of a drone, which make its integration into an existing drone setup very tedious and difficult.

3. Hardware Design and Integration

This section gives an overview of the used hardware components and their software integration.

The localization system is integrated into our swarm of quadcopters FINken [21], which are self developed and modular consisting of a basic copter frame with four motors in X-setup. The central unit of control is a LisaMX Autopilot (https://wiki.paparazziuav.org/wiki/Lisa/MX, accessed on 25 August 2021) running Paparazzi as flight software. The autopilot also provides all necessary sensors. During our tests, the drone was equipped with a TeraRanger, which is an optical height sensor with a range of 14 m and an update rate of 1 kHz (Figure 1b). The drone is additionally equipped with a GPS module (Figure 1a), a ranging module capable of UWB communication and a micro-SD card reader used for logging.

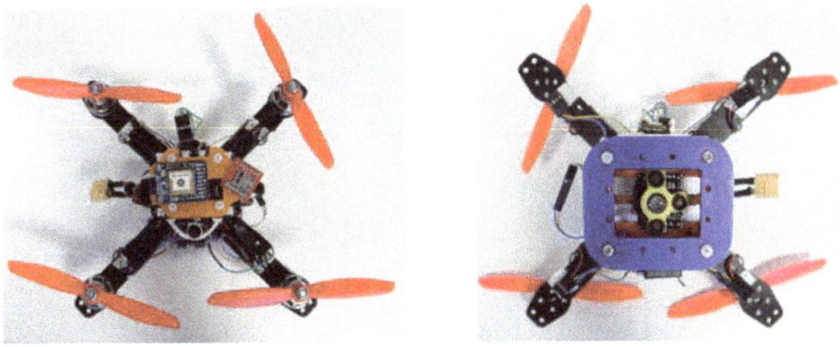

(**a**) Top view showing GPS receiver and SD card mount.

(**b**) Bottom view showing the TeraRanger distance sensor.

Figure 1. Images of the custom-built drones of the FINken swarm.

The communication and localization of the drones are powered by a DWM1000 UWB-compliant wireless transceiver. The DWM1000 module supports a wide frequency range from 3.5 GHz to 6.5 GHz and directly supports ToF measurement. The modules are used to provide drones with communication capabilities between each other and to the ground station. Additionally, they provide the distance measurements for the localization system. To integrate these DWM1000 modules into our drones, a PCB was designed; see Figure 2. Besides the DWM1000 module, it contains a STM32 microcontroller, a micro-USB port and several status LEDs. In terms of connectivity, two Universal Serial Asynchronous Receiver Transmitter (USART) connections and one Inter-Integrated-Circuit (I^2C)-Bus, along with several input and output pins, are accessible. The modules are used as components of the copters as well as anchor nodes. This enables an easy integration of dynamic anchors created by stationary copters in later stages of the localization system. The modules do not use any specific directional antenna to enhance communications to the anchors because in the later stages of the development of the swarm localization system, the anchors will be moving.

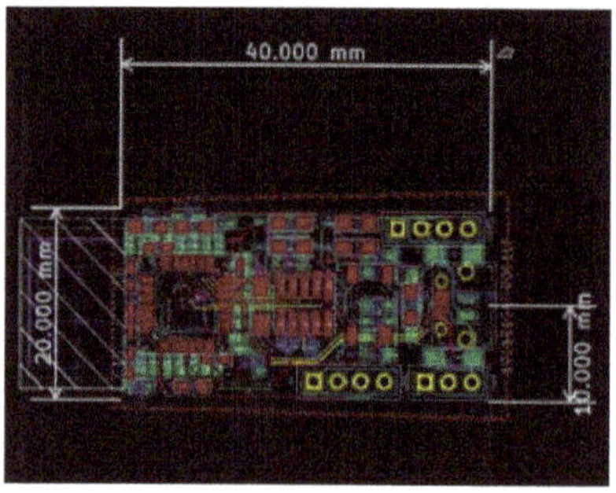

(a) The designed PCB of the custom interface module used for integration into the copters.

(b) Picture of the final manually assembled interface board.

Figure 2. Illustrations of the custom DWM1000 PCBs used for copters and anchors.

In addition to the distance measurements, the modules are used to transmit telemetry data of the copters to the ground station through the UWB link. The ground station is equipped with an anchor node to receive all telemetry data. To communicate with the drones during operation, the drones receive and send Paparazzi messages to the microcontroller on the DWM1000-board via USART. The microcontroller then handles the sending and receiving of data using the DWM1000 module. Furthermore, double-sided two-way ranging is implemented on the microcontroller. The modules mounted on the copters will periodically initiate two-way ranging to other nodes, while modules acting as anchor nodes will only respond to ranging requests. The microcontroller on the module then calculates a position according to our proposed localization Algorithm 1. Once the calculation is performed, a National Marine Electronics Association (NMEA)-GPS message based on that position is generated and sent to a dedicated GPS port on the autopilot.

By emulating a real GPS module, the module is easily exchangeable against a real GPS. Furthermore, existing hardware and software structures, especially the Inertial Navigation System (INS) filters of Paparazzi, can be reused. To support other researchers using Paparazzi as autopilot software, we provide the hardware description of our interface board and the software of the interface board as Open Hardware/Source through our GitHub Repository (https://github.com/ovgu-FINken/DWM1000_Copter_Integration, accessed on 25 August 2021).

Algorithm 1: Continuous Localization.

Data: $p^{(0)}, v_{max}, \Delta T$
Input: $d^{(t-1)}, \Delta t^{(t)},$
Output: $p^{(t)}$
begin
  $p^{(t)} \longleftarrow p^{(0)};$
  $m \longleftarrow 0;$
  while *True* **do**
    $p^{(t-1)} \longleftarrow p^{(t)};$
    $w \longleftarrow b^{\left(1-\frac{\Delta t^{(t)}}{\Delta T}\right)};$
    $w \longleftarrow w\frac{\#w}{\sum w};$
    $p^{(t)} \longleftarrow \texttt{localization\_step}\left(p^{(t-1)}, d^{(t)}, w^{(t)}\right);$
    $v \longleftarrow \frac{\|p^{(t)}-p^{(t-1)}\|}{\Delta T};$
    if $v > a^m \cdot v_{max}$ **then**
      $m \longleftarrow 0;$
      $p^{(t)} \longleftarrow p^{(t-1)};$
    else
      $m \longleftarrow m+1;$

Our localization modules differ in the generic design relating to other approaches. We combined the ranging capability of the modules with the communication capability to use the module for localization and for drone to drone and drone to ground station communication. Additionally, we did not develop a new software interface to the autopilot of the copter, but reused the existing NMEA-capable GPS interface of Paparazzi, which enables an integration of the modules to any Paparazzi-enabled copter. Finally, we did not use any specific antenna setup to not jeopardize the reception quality as soon as anchors are moving.

4. UWB-Ranging-Based Localization Algorithm

In the following sections, we describe the problem of the localization based on distances acquired through the UWB-Ranging modules (Section 4.1). Afterwards, we present the algorithm to compute a single localization step in Section 4.2, followed by the algorithm using the single step localization for continuous localization in Section 4.3.

4.1. Problem Statement

Given the observed distance values $d^{(t)} = \left(\hat{d}_0^{(t)} + e_0^{(t)}, \ldots, \hat{d}_m^{(t)} + e_m^{(t)}\right)$ between the node to be localized and each anchor node A_i at time step t, compute the position of the node to be localized $p^{(t)}$ at time step t. The observed distances are composed of the correct distance $\hat{d}_i^{(t)} = \|\hat{p}^t - A_i\|$ and an unknown time-varying error $e_i^{(t)}$. Considering we have enough anchor nodes available, the resulting system of equations is over-determined. Therefore, the problem is an optimization problem as follows:

Problem 1. *Considering the observed distances $d^{(t)}$, we want to compute a position $p^{(t)}$, which minimizes the difference between the observed distances $d^{(t)}$ and the distance given by the estimated position and each anchor node $\|p^{(t)} - A_i\|$.*

$$\min_{p^{(t)}} \left(\sum_{i=0}^{m} \left(\|p^{(t|)} - A_i\| - d_i^{(t)} \right)^2 \right)$$

This problem is a nonlinear least squares optimization problem.

4.2. Single Step Localization Algorithm

To solve the optimization Problem 1 stated in Section 4.1, various approaches are possible. We decided to use a gradient-based optimization, due to its speed and the low-resource consumption. This approach is based on the work of Mantilla-Gaviria et al. [22] and Murphy and Hereman [23].

Algorithm 2 uses static ($A, \eta, z_{min}, \Delta p_{min}$ and n) and dynamic ($p^{(t-1)}$, $w^{(t)}$ and $d^{(t)}$) inputs. A represents a $m \times 3$ matrix of the position of the m anchor nodes. η controls the speed of convergence of the gradient descent and is typically called step size. z_{min} is a parameter defining the minimum height of the trajectory of the node to be tracked. This parameter is necessary to decide between multiple possible solutions in the case of planar setups of anchor nodes; see Section 5.2. Δp_{min} and n are parameters controlling the iteration of the algorithm. Δp_{min} defines the minimum movement of the position that the algorithm needs to execute on $p^{(t)}$. If the movement is smaller, the estimated position is assumed to have converged for time step t and the iteration is stopped. The iteration is stopped anyway if n number of iterations are executed. $p^{(t-1)} = \left(x^{(t-1)}, y^{(t-1)}, z^{(t-1)}\right)^T$ is the estimated position of the last time step $t-1$, and $d^{(t)}$ is the observed distance of the current time step, which is weighted by the age weight vector $w^{(t)} \in (0,1]^m$. The output of the algorithm is the estimated position $p^{(t)} = \left(x^{(t)}, y^{(t)}, z^{(t)}\right)^T$ for time step t.

Algorithm 2: Single Localization Step.

```
Data: A, η, z_min, Δp_min, n
Input: p^(t−1), d^(t), w^(t)
Output: p^(t)
begin
    p^(t) ⟵ p^(t−1);
    i ⟵ 0;
    repeat
        for i in 0, ..., m do
            G_i ⟵ p^(t) − A_i;
            R_i ⟵ ||G_i||;
            G_i ⟵ G_i / R_i;
        Δp ⟵ G^−1((d^(t) − R) w^(t));
        p^(t) ⟵ p^(t) + η · Δp;
        if z^(t) < z_min then
            z^(t) ⟵ 2z_min − z^(t);
        j ⟵ j + 1;
    until j ≥ n or ||Δp|| < Δp_min;
```

The algorithm starts by generating a gradient matrix through the computation of the vector distance between the current estimation $p^{(t)}$ and the position of each anchor node A_i. By computing the norm of each line G_i within G, we obtain the estimated distance vector $R = \left(\|p^t - A_0\|, \ldots, \|p^t - A_m\|\right)^T$. We now normalize G_i with R_i. Afterwards, the next movement Δp of the estimated position $p^{(t)}$ is computed by pseudo-inverting G and multiplying with the weighted difference of observed distances and estimated distances $\left(d^{(t)} - R\right)w^{(t)}$. If the weighting of nodes is not wanted, the weight vector $w^{(t)}$ can be replaced with a vector of ones. The use of the pseudo-inverse guarantees a correct weighting of the individual anchors to minimize the overall error. The resulting movement Δp is scaled by the step size η and added to the current estimated position $p^{(t)}$. If the resulting z-coordinate is below the minimum z-coordinate defined by z_{min}, we invert the coordinate by computing $z' = z_{min} + z_{min} - z = 2z_{min} - z$. This is repeated until either the maximum number of iterations n is reached or the current movement of the estimated position Δp is smaller than the defined minimum movement Δp_{min}.

4.3. Continuous Localization

Localizing a node is a continuous process. Therefore, the Localization Step described in Algorithm 2 needs to be executed repeatedly. Algorithm 1 describes the process. This algorithm takes the current observed distances $d^{(t)}$ and the age of each distance value $\Delta t^{(t)}$ as input because we cannot expect to receive distances from all anchors at every time step. Additionally, v_{max} defines the maximum expected speed of the node to be tracked, a defines the base of the exponential growth of v_{max} and ΔT indicates the desired update frequency of the estimated position. b controls the decay of the influence of an anchor based on the age of its last received distance measurement. Higher values of b increase the decay, whereas the minimal value of 1 disables it.

The algorithm starts with initializing the position of the tracked node to $p^{(0)} = (0, 0, 0)^T$. This initialization is arbitrary and can be adapted as necessary based on environmental circumstances. However, for our relative localization, it worked well. For each update of distance observations $d^{(t)}$, the algorithm executes `localization_step`, supplying the observed distances and the current position estimate. The weight vector regarding the age of the measurements w is computed as exponentially decreasing with growing age. To prevent different sizes of update steps due to delays in communication, the weight vector is normalized to have the sum $\sum w$ equal to its length $\#w$. Afterwards, the speed of the node is estimated based on the results of the localization of the current $p^{(t)}$ and the previous time step $p^{(t-1)}$. If the estimated speed $v = \|p^{(t)} - p^{(t-1)}\| / \Delta T$ is smaller than the maximum speed v_{max} multiplied by the exponential growth factor a^m, the position is updated. Otherwise, the old position is kept because the newly estimated position is considered an outlier. The exponential growth factor a^m increases every time a position is not updated to prevent the localization from becoming stuck in local optima far away from the real trajectory due to a sequence of bad measurements.

Compared to existing systems using only ranging information, our approach is specially tailored towards drone localization. First, we assume the localization to output 3D-coordinates because drones typically move in all 3 dimensions. Additionally, we assume anchors to often be located in a plane, which poses additional challenges regarding the z-coordinate estimation of the algorithm. To overcome this, we use the additional z_{min} parameter and the associated logic to select the correct solution. The v_{max} and a parameters handle the specific movement speed of drones by filtering out unreasonable values based on the possible movement speed of the tracked object. This parameter may even be delivered by the control software of the drone to further enhance the filter characteristics. To prevent the localization from becoming stuck in local optimum far away from the real trajectory, the parameter a enables exponential growth of the allowed v_{max} enabling resynchronization of the localization and the real trajectory.

5. Experimental Evaluation

To evaluate the performance of the proposed localization system, we conducted multiple experiments. We started with an experiment to evaluate the performance of the GPS receiver used in the copters to evaluate their usage as position reference, see Section 5.1. Afterward, we evaluated the system in an indoor environment equipped with a Vicon indoor localization system; see Section 5.2. Finally, we tested the system in-flight in an outdoor scenario; see Section 5.3.

5.1. GPS Experiments

Our first experiment was designed to test the quality of our used GPS receivers as a position reference. Additionally, we wanted to detect the level of quality the localization system needs to be beneficial to the copters in an outdoor scenario. According to our vision, our swarm of copters flies between indoor and outdoor scenarios, which places a large quantity of the mission trajectory close to buildings. Therefore, for the GPS quality assessment, we decided to put the receiver close to the outer wall of our lab. The receiver was connected to a Raspberry Pi to collect data and write the data to an SD card. We let the receiver collect data for 7 days. The resulting distribution of positions is visible in Figure 3. Additionally, we show the distribution of the individual coordinates and the number of satellites in Figure 4a,b.

The results of the experiments show a deviation of typically ±40 m in the GPS position, even through we received data from at least 5 satellites in 99.9% of the data points. Another disadvantage is the maximum deviation of the positions, which can be as large as ±150 m. Especially noteworthy is the biased distribution of data points. The deviation is far larger for the latitude than for the longitude. This is to be expected as the receiver was placed to the north of a building, which probably reflected the incoming satellite signals and disturbed the distance measurement. This is also an explanation for the offset of the center of the measured positions against the location of the receiver.

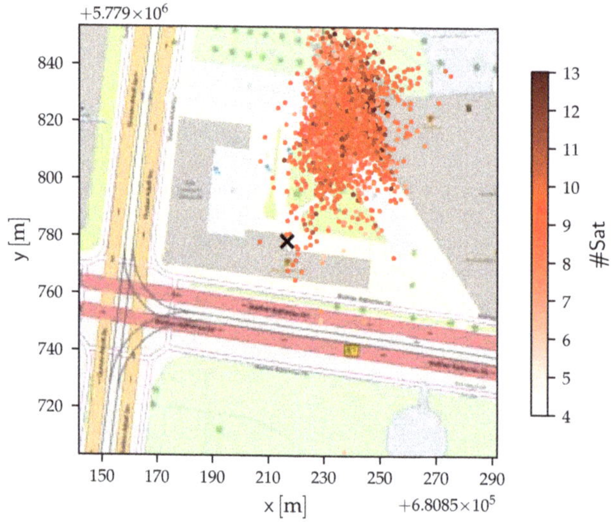

Figure 3. Visualization of every 100th coordinate of the static GPS receiver experiment. All coordinates are transformed into Universal Transverse Mercator (UTM) coordinates (Zone 32U) for better assessment of the distance to the center. The color indicates the number of satellites perceived by the receiver. The black **X** marks the position of the receiver.

In consequence, even a medium precision (±0.5 m) localization system used within a swarm will provide major benefits regarding swarm stability if the swarm is flying in an urban scenario. This is caused by the usage of distances between swarm members in most swarm algorithms. These distances are computed from positions by periodically calculating the difference between positions, which mathematically resembles a numerical differentiation. Stochastic variations in the position data will be amplified by the numerical differentiation, leading to large movements within the swarm, which may jeopardize swarm stability.

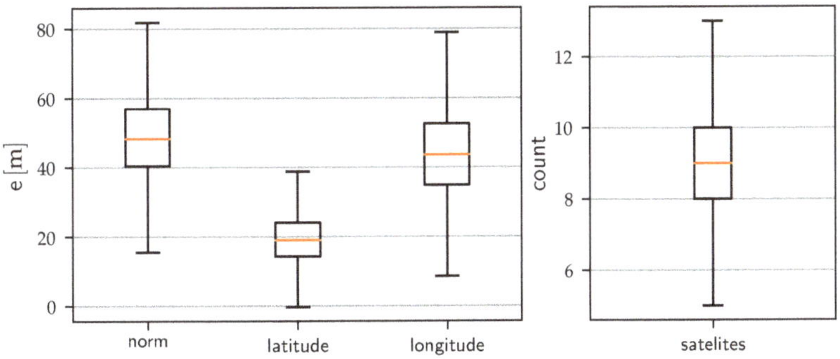

(**a**) Coordinate Precision. (**b**) Satellites.

Figure 4. Box–Whisker plots showing the distribution of the longitude and latitude coordinates after transformation to UTM, as well as the distribution of satellite reception.

5.2. *Indoor Evaluation with Precise Reference*

For the precise evaluation of the localization system, we used an existing Vicon system. The Vicon camera-based tracking system delivers 100 Hz updates with a precision of 1 mm. The observable area is a cube of $3\,\mathrm{m} \times 3\,\mathrm{m} \times 3\,\mathrm{m}$. We set up 8 static anchor nodes labeled 128–135 in the area, as shown in Figure 5. Due to the flat ground in the test area, all anchors were placed at 0.2 m height. This setup is typical for manually placed anchors in flat environments. This presents a problem for the localization system, as two solutions exists, which are equally likely: one above the anchors and one below the anchors. Consequently, the z_{min} parameter of Algorithm 2 was set to 0.2 m to mitigate this problem for the following experiments. We executed two random trajectories by moving a single node in the area manually. The resulting trajectories are visualized in Figure 5. The node was attached to a stick of 1 m length to avoid blocking the line of sight between the node and the anchor nodes with our bodies.

Our localization system was configured with the parameters described in Table 1. The step size parameter η was deduced as iterative through preliminary experiments. The origin of the local coordinate system was chosen as the initial position $p^{(0)}$. The minimum localization height z_{min} was chosen based on the average height of the anchors. The maximum speed v_{max} is defined by our copters, even though a much smaller value is defined as the maximum speed in software. The a parameter was empirically deduced. ΔT was chosen based on the capabilities of our hardware modules. We derived some relevant parameter configurations to test different components of the algorithm. The default configuration contains the typical standard parameters without the use of age information. The age configuration adds the capability of weighting the individual anchors based on the age of the information. The best parameter set is the combination of all positive parameter changes in comparison to the default. The heavy configuration tests if additional computational resources may enhance the performance of the algorithm. The ground configuration checks the performance of the algorithm in case no assumption on anchor positions is made. To this end, the minimum z-height is defined as 0 m in the local coordinate system of the localization. The slow configuration tests the influence of the speed limitation through v_{max}.

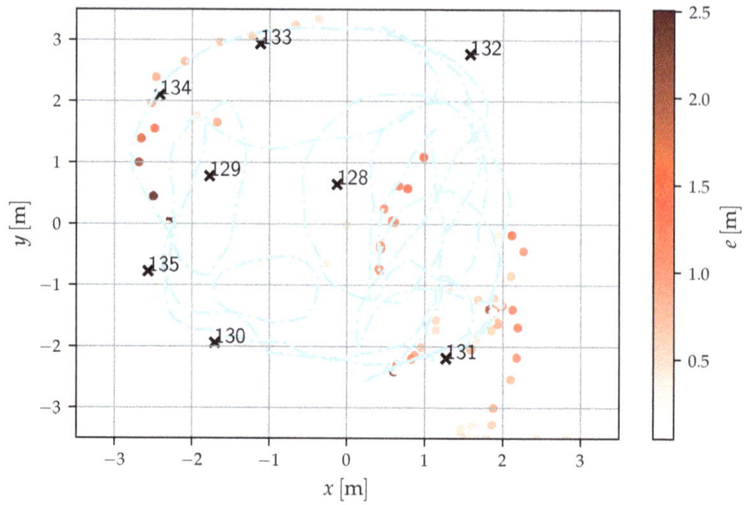

(**a**) Scenario 1

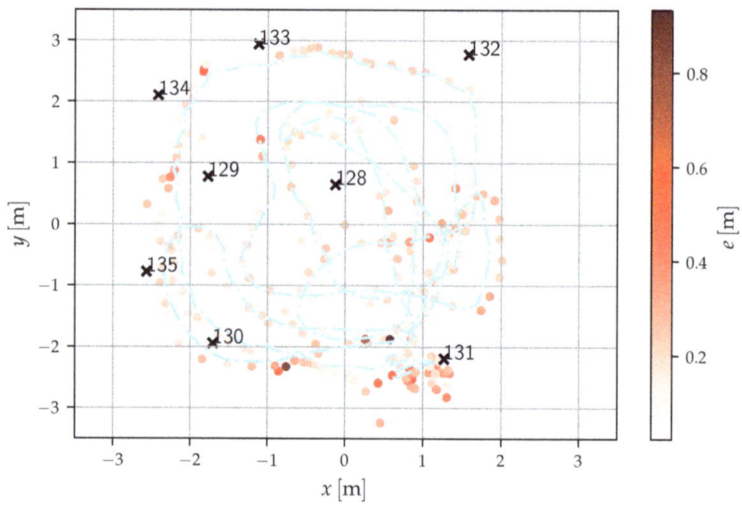

(**b**) Scenario 2

Figure 5. Visualizations of the difference between the Vicon reference positions (light blue dashed trajectory) and the estimated position by the localization system (red dots) in the xy-plane. The colors of the dots indicate the differences between the reference position and the estimated position.

Table 1. Localization algorithm parameter configurations. The table shows the evaluated configurations of the algorithm parameters for both algorithms. The differences of the individual configuration against the default configuration is highlighted in bold.

Config	Algorithm 2 Parameters					Algorithm 1 Parameters			
	$p^{(0)}$ [m]	η	z_{min} [m]	Δp [m]	n	v_{max} [m/s]	a	b	ΔT [s]
default	$(0,0,0)^T$	0.1	0.2	0.1	100	20	1.01	1	0.1
age	$(0,0,0)^T$	0.1	0.2	0.1	100	20	1.01	**2**	0.1
ground	$(0,0,0)^T$	0.1	**0.0**	0.1	100	20	1.01	1	0.1
slow	$(0,0,0)^T$	0.1	0.2	0.1	100	**5**	**1.1**	1	0.1
heavy	$(0,0,0)^T$	0.1	0.2	**0.01**	**1000**	20	1.01	1	0.1
best	$(0,0,0)^T$	0.1	**0.0**	0.1	100	**5**	**1.1**	**2**	0.1

First, we look at the difference in distance measurements between the Vicon system and our UWB-Ranging nodes. Figure 6a,b show the distance error e for each trajectory for each anchor node. Interestingly, the distribution of errors is close to a zero mean with generally less than 0.1 m deviation. However, the spread of the distribution is quite high. The 90% quantile reaches ±0.3 m for both experiments. Anchor 134 is an exception with a negative 90% quantile of over −0.5 m. In general, the distance estimation can be assumed to provide an accuracy of ±0.3 m in most cases.

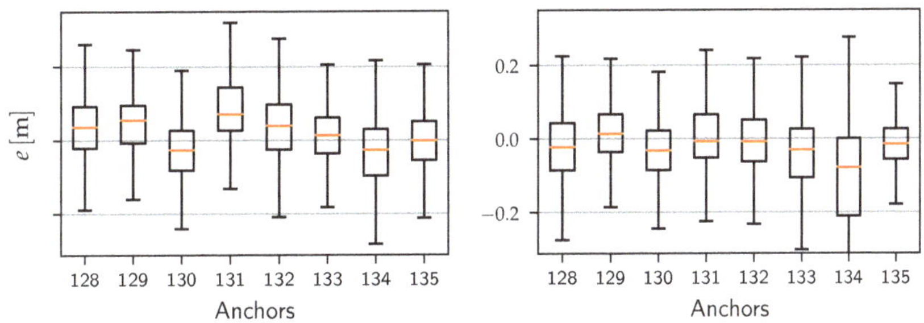

(**a**) Scenario 1. (**b**) Scenario 2.

Figure 6. Box–Whisker of distance errors for each anchor in both indoor evaluation scenarios.

The resulting trajectories of applying Algorithms 1 and 2 to the acquired distance values extracted from the UWB modules during the execution of trajectory 1 and 2 are shown in Figure 7. As shown in Figure 7, the error e between the true positions and the estimated positions from the localization system is typically in the range of ±0.25 m. However, some outliers exist with large errors beyond ±2 m. Looking at the detailed composition of the errors in Figure 8a,b, we observe that the worst outliers are not visible because they make up less than 5% of the data. However, for the first trajectory, an increased error is visible for the z-coordinate, which also increases the overall error. This is probably created through the location of the anchor nodes in a plane parallel to the xy-plane.

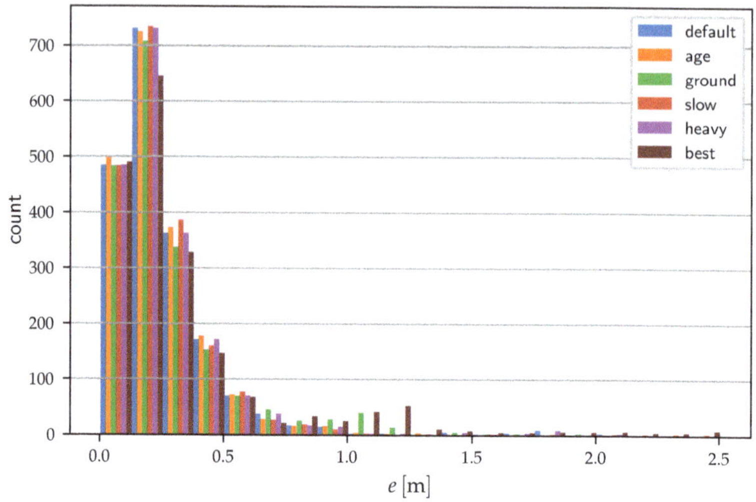

(**a**) Scenario 1.

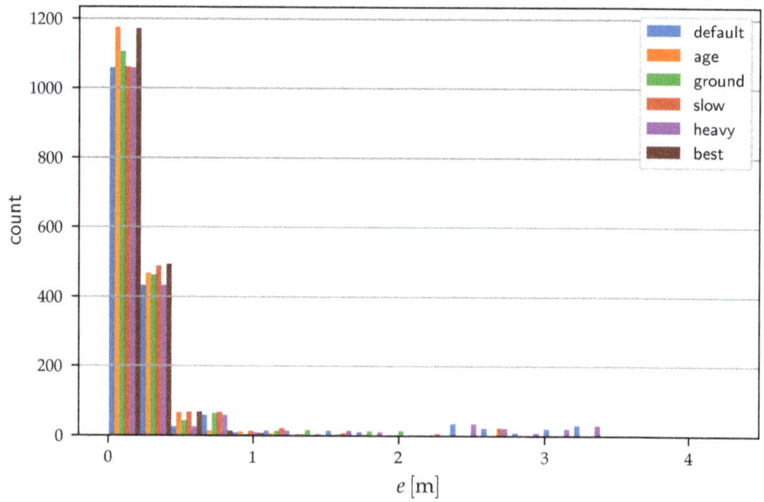

(**b**) Scenario 2.

Figure 7. Histogram of localization errors for both indoor evaluation scenarios. Each parameter configuration of the localization algorithm is shown in a separate bar.

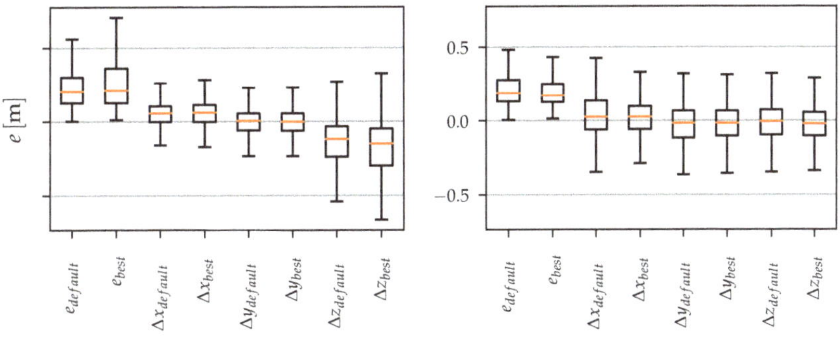

(**a**) Scenario 1. (**b**) Scenario 2.

Figure 8. Box–Whisker plot of the localization error of both indoor scenarios. e shows the norm of the error vector, whereas x, y and z show the error for each component of the error vector. The values are shown for the *default* and the *best* parameter configuration of the localization algorithm.

Figure 8a shows that the error in the z-coordinate is different from the error in the other components. Based on this situation, we suspected a correlation between the localization error and z-coordinate (height). To verify this, we conducted a Pearson correlation analysis between the observed position error Δe and the z-coordinate of the Vicon reference position. The results are visible in Table 2.

For most algorithm configurations, a correlation coefficient $r_1 \approx -0.4$ for scenario 1 and 2 can be observed. To verify the statistical soundness of the result, we conducted a two-way permutation of 100,000 permutations of the input data for each algorithm configuration for each scenario, which resulted in a p-value of 0.0 for all combinations. Consequently, there is a weak negative linear correlation between the position error and the z-coordinate of the object to be localized. It appears that the typical ground reflects the wireless signals and generates additional errors through multipath effects. Interestingly, the age configuration using the age-awareness extension of the algorithm reduced the correlation for scenario 2 to ≈0.37. The ground detection with a $z_{min} = 0.0$ m also shows a deduction in the correlation for the second scenario. The slow configuration also reduced the correlation for both scenarios to the minimally observed correlations of −0.274 and −0.25. However, this comes at the cost of reducing the maximum movement speed of the copters. Interestingly, there is a large deviation in correlation coefficients between scenario 1 and 2 for the best configuration. This is caused by the larger variation of the error in z-coordinate for the first scenario with this configuration. In the first scenario, the z_{min} parameter has a large influence and filters out some wrong z-coordinates if it is set to 0.2 m. In the second scenario, it does not have any significant impact at all. The best configuration omits the parameter because this makes the resulting algorithm setup more general without any assumptions on anchor placement. Therefore, we decided to use $z_{min} = 0.0$ for best, even though it does not provide the best possible localization precision for both scenarios.

Table 2. Pearson correlation coefficients between the error e of localization and z-coordinate of the tracked object.

Parameters / Configuration	r_1	p_1	r_2	p_2
default	−0.426	0.0	−0.439	0.0
age	−0.459	0.0	−0.370	0.0
ground	−0.447	0.0	−0.327	0.0
slow	−0.274	0.0	−0.250	0.0
heavy	−0.426	0.0	−0.439	0.0
best	−0.188	0.0	−0.398	0.0

Figure 9 shows the time difference between received distance measurements of the different anchors. As shown, the time difference is very similar for all nodes. The median is at ≈0.23 s, which indicates that either the receiver module was overloaded with the reception and forwarding of the measurements or wireless packages were lost between the nodes. As the spread of the distribution of time differences is rather high, we assume that we have approx. 50% package loss in our scenario. Consequently, the parameters of the modules need to be optimized to enhance the reliability of the communication.

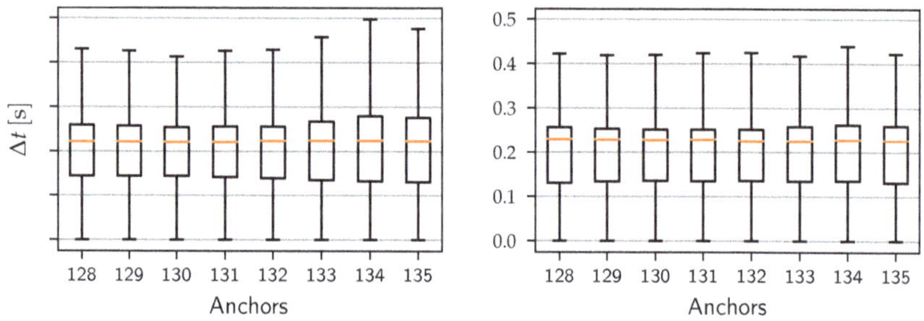

(**a**) Scenario 1. (**b**) Scenario 2.

Figure 9. Box–Whisker plot of the age between received measurements from the different anchors for both indoor scenarios.

The results of our indoor evaluation show that our systems provides similar precision, even though we observed a large loss of packets mid-flight. This, of course, reduces the update rate of the system to less than 10 Hz, which may easily be compensated by the internal INS filters of the autopilot software. Additionally, we achieved an average precision of 0.2 m, which is better than the work of Kempke et al. [18], but with much less communication overhead. This is important because our later extension to a swarm-based localization will increase the necessary communication further. We also achieved slightly less precision than OmniSwarm [20] for two drones. However, this approach used way more sensors and was integrated deeper into the drone system. Our approach can easily be used as an extension module to existing copter builds.

5.3. Outdoor Evaluation

In addition to the indoor experiments with a high-precision reference, we conducted two outdoor experiments. These experiments have no high-precision reference. Therefore, we analyzed the behavior based on the flight patterns.

The first experiment used the anchor setup shown in Figure 10. The anchors are all placed in a xy-plane because the experiment was conducted on flat ground and no variation in height was possible. The receiving node was attached to a FINken quadcopter, which was manually controlled. The node transmitted all distance measurements to a laptop for logging purposes. The localization algorithm was executed afterwards with configuration best. We executed two trajectories with the same quadcopter. In the first trajectory, the quadcopter hovered over each anchor node with slow movement speed between the anchors. The second trajectory executed multiple counter-clockwise circles over the area.

As shown in Figure 10a, the generic flight behavior can be reconstructed by the localization system. The visible instability is not necessarily caused by the localization system, but may also be caused by the manual control of the copter. The second trajectory, visible in Figure 10b, also shows the expected behavior of counter-clockwise circles in the area.

Similar to the results of the indoor experiments, Figure 11a,b show the higher probability of localization errors close to the ground. In the case of trajectory 1, the z-coordinate of the copter increased to ≈2.5 m, even though the flight had not started yet. The final z-coordinate after landing (after ≈250 s and ≈200 s, respectively) also shows an offset of ≈1 m. The offset of the z-coordinate cannot be evaluated mid-flight because no reliable reference is available for these flights.

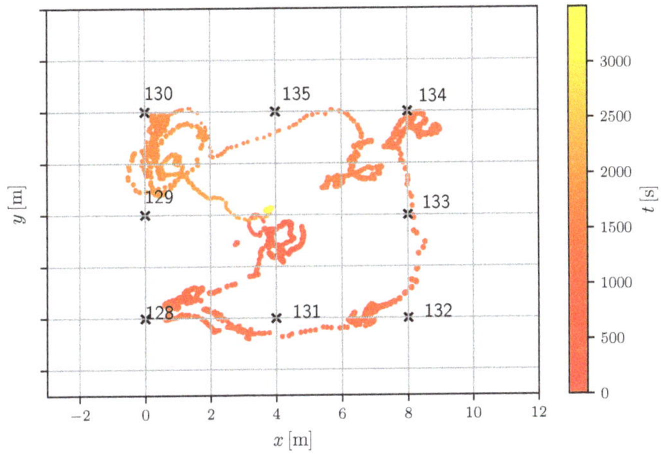

(**a**) Scenario 1

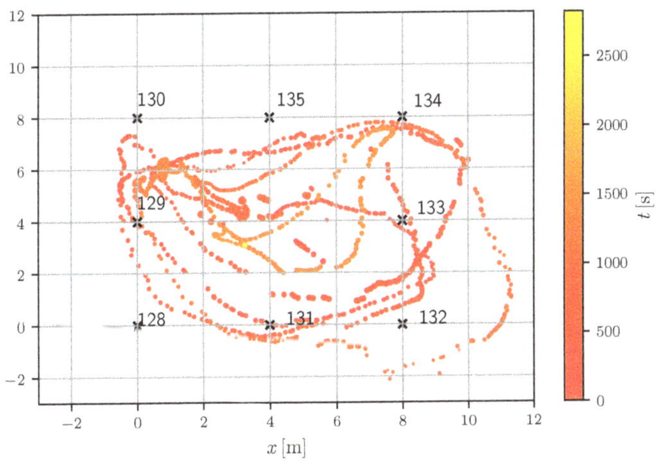

(**b**) Scenario 2

Figure 10. Visualizations of the reconstructed trajectories from the first outdoor evaluation for both scenarios. The flight time at each position is shown as a color from red to yellow according to the attached color bar.

As shown in Figure 12, the distribution of the time difference between the reception of distance measurements in both trajectories show similar median values to the indoor experiments. However, the variance between the individual anchor nodes is higher. This may be caused by shadowing effects of the components of the copter, especially the battery, which prevent line-of-sight communication to certain anchors.

The second outdoor experiment used the fully integrated system. The receiver node was attached to a copter. The node transformed the received distance measurements to positions in the local coordinate system using configuration ground without any speed limitation. The resulting relative coordinate was then transformed to GPS coordinates based on hard-coded reference points. The GPS information was packed into the NMEA format and directly transmitted to the copter via UART. The copter used the provided data, similar to real GPS data. In this experiment, we set up anchors along a walkway in a park close to the faculty. This allowed us to easily obtain the GPS coordinates of the anchors with relatively high precision from open street maps. The resulting anchor position overlaid on top of an OpenStreetMap is shown in Figure 13. The copter was controlled manually and again flew over all anchor nodes, while always hovering above each one.

Unfortunately, the used hard-coded reference points deviated from the anchor positions. Therefore, the references need to be modified after the experiment. To this end, a nonlinear gradient descent on an affine transformation consisting of scaling, rotation and translation was executed. The loss function of the gradient descent was formed by the sum of squared distances between each trajectory position and the position of the associated anchor. The position–anchor association was performed manually and is shown in Figure 14. The resulting trajectory after executing optimized affine transformation on the original trajectory is shown in Figure 13.

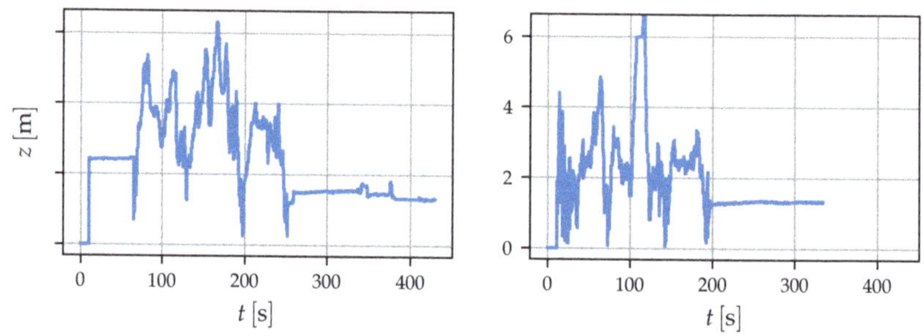

(a) Scenario 1. (b) Scenario 2.

Figure 11. Line plot of the reconstructed z-coordinates of the first outdoor evaluation experiment for both scenarios.

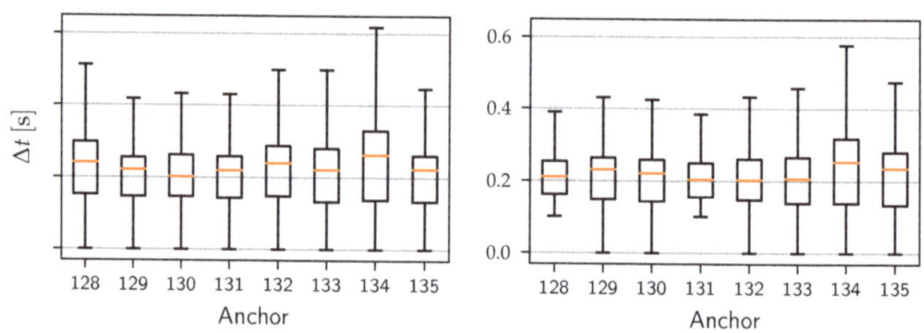

(a) Scenario 1 (b) Scenario 2

Figure 12. Box–Whisker plot of the distribution of time difference between reception of distance measurements from the different anchors for both scenario of the first outdoor evaluation.

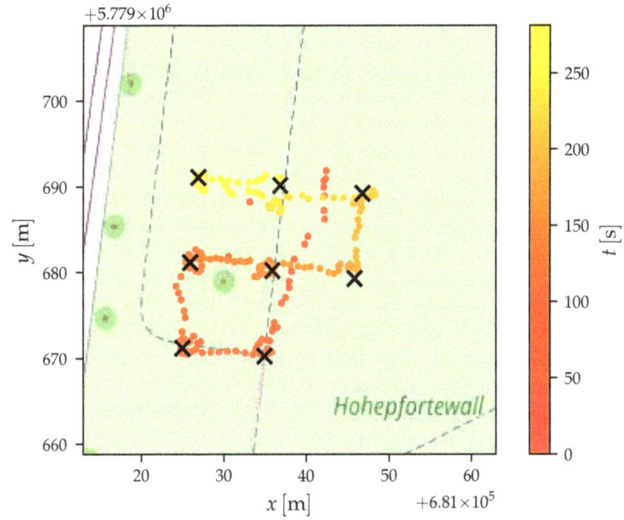

Figure 13. Visualization of the anchors, node trajectory and localization error for the second outdoor experiment. The anchors are marked with **X**. The node trajectory is colored to indicate time. The flight time is visible in seconds in the color bar.

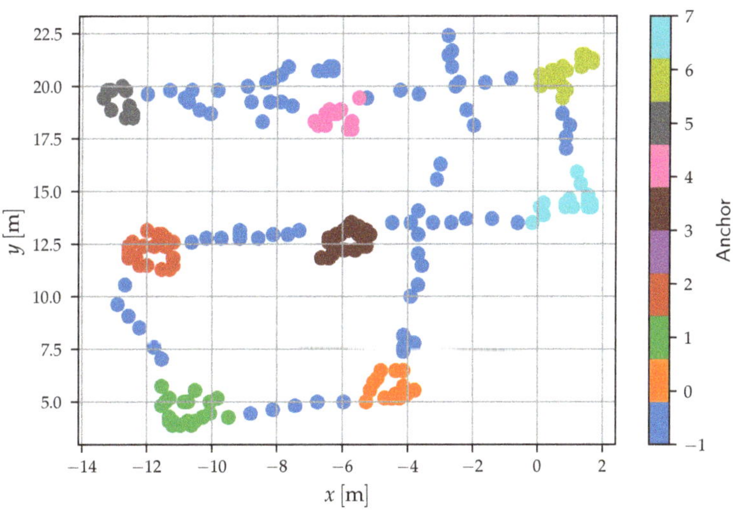

Figure 14. Visualization of the anchor assignment for evaluation of the outdoor performance. Each position is assigned to the anchor with the respective color visible in the color bar. The anchor with number −1 is a virtual anchor, capturing all positions not assigned to any anchor.

Figure 15 shows the resulting errors of the associated anchors and the positions. As shown, the general positions of the copter fit the expected trajectory. However, Figures 13 and 15 still show larger deviations from the correct positions. We expect the rotation of the copter together with the antenna characteristics of the node to have a high influence in this scenario.

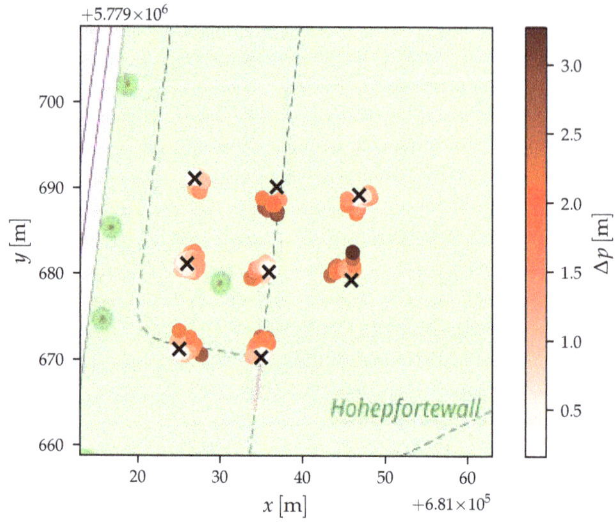

Figure 15. Visualization of the anchors and localization error for the second outdoor experiment. The anchors are marked with **X**. The estimated position error of each position is indicated using colors visible in the color bar.

Additionally, the embedded software did not yet support all parameters of the localization algorithm, which is the reason for the usage of a suboptimal configuration.

6. Conclusions and Future Work

In this paper, we showed that GPS may be unreliable, and additional local relative localization systems are necessary to support the robust behavior of drones. To this end, we developed a localization system based on UWB-Ranging to fixed anchor nodes. We showed that without additional hardware besides the DWM1000 module, the distance measurements between the quadcopter and the anchors are typically in the range of ± 0.2 m. We developed two algorithms, which together form a continuous localization mechanism usable in autonomous quadcopters. The algorithms were tested with different parameters in indoor and outdoor scenarios. In the indoor scenario, the achievable precision has a median value of ≈ 0.2 m. The outdoor scenarios showed the system capability regarding the tracking, even though no precise error estimation was possible. An interesting finding of the approach was that the error in the localization has a weak correlation to the height of the object being tracked. This seems to be created by reflection and line-of-sight obstruction created by the ground.

In summary, the described approach showed reasonable performance on par or better than state-of-the-art approaches, but with minimal integration to the drone system. This enables easy integration to other copters as long as they are using the same autopilot software. Additionally, the approach is very efficient regarding communications, which is beneficial for the later extension to a swarm-localization system.

For future work, we want to integrate information from the INS of the copters. This enables the additional usage of movement commands and local sensors on the copter to further increase the precision of the localization. While the used NMEA protocol is rather simple, it is limited in update rate and precision. However, an implementation of the UBLOX protocol is planned. This will provide a higher precision and update rate. In the next step, the embedded localization software will be enhanced to support the *best* parameter configuration to enhance localization accuracy. The third modification that we aim to evaluate is the usage of external antenna for the DWM100 modules to mitigate the line-of-sight issues in close-to-ground scenarios.

Our final goal is the extension of the current system to multihop localization to enable swarm localization of a swarm of drones. To this end, the localization node of each swarm member needs to be able to freely switch between the tracked node and anchor node based on its possible contribution to the localization quality of other localization nodes. The benefit of this approach is that even a copter with depleted energy, which cannot continue flying, may serve the swarm as an additional anchor for localization. To this end, a decision-making algorithm needs to be developed, which can manage the different goals of such an integrated swarm localization system, such as minimizing movement times, maximizing localization precision, ensuring safety of each drone and minimizing energy cost.

Author Contributions: Conceptualization, C.S.; methodology, C.S. and J.B.; software, C.S. and J.B.; validation, C.S. and J.B.; formal analysis, C.S.; resources, S.M.; data curation, J.B.; writing—original draft preparation, C.S. and J.B.; writing—review and editing, C.S. and S.M.; visualization, C.S.; supervision, S.M.; project administration, S.M. All authors have read and agreed to the published version of the manuscript.

Funding: This research received no external funding.

Institutional Review Board Statement: Not applicable.

Informed Consent Statement: Not applicable.

Data Availability Statement: The raw data generated in the experimental evaluation are available in the GitHub repository `ovgu-FINken/DWM1000_Copter_Integration` in the branch `drones21` starting from commit `ab2a713f28c8440dbf8cdc8bbb45144af2c7ad26` in folder `data`.

Conflicts of Interest: The authors declare no conflicts of interest.

Abbreviations

The following abbreviations are used in this manuscript.

UWB	Ultrawideband
GPS	Global Positioning System
AGPS	Assisted GPS
D-GPS	Differential GPS
ToF	Time-of-Flight
TDoA	Time-Difference of Arrival
RSSI	Received Signal Strength
SD-Card	Secure Digital Card
UAV	Unmanned Aerial Vehicle
USART	Universal Serial Asynchronous Receiver Transmitter
I^2C	Inter-Integrated Circuit
NMEA	National Marine Electronics Association
UTM	Universal Transverse Mercator

References

1. Seo, J.; Duque, L.; Wacker, J. Drone-enabled bridge inspection methodology and application. *Autom. Constr.* **2018**, *94*, 112–126. [CrossRef]
2. Shihavuddin, A.; Chen, X.; Fedorov, V.; Nymark Christensen, A.; Andre Brogaard Riis, N.; Branner, K.; Bjorholm Dahl, A.; Reinhold Paulsen, R. Wind turbine surface damage detection by deep learning aided drone inspection analysis. *Energies* **2019**, *12*, 676. [CrossRef]
3. Erdelj, M.; Król, M.; Natalizio, E. Wireless Sensor Networks and Multi-UAV systems for natural disaster management. *Comput. Netw.* **2017**, *124*, 72–86. [CrossRef]
4. Dorigo, M.; Floreano, D.; Gambardella, L.M.; Mondada, F.; Nolfi, S.; Baaboura, T.; Birattari, M.; Bonani, M.; Brambilla, M.; Brutschy, A.; et al. Swarmanoid: A novel concept for the study of heterogeneous robotic swarms. *IEEE Robot. Autom. Mag.* **2013**, *20*, 60–71. [CrossRef]

5. Zelazo, D.; Franchi, A.; Bülthoff, H.H.; Robuffo Giordano, P. Decentralized rigidity maintenance control with range measurements for multi-robot systems. *Int. J. Robot. Res.* **2015**, *34*, 105–128. [CrossRef]
6. Stirling, T.; Roberts, J.; Zufferey, J.C.; Floreano, D. Indoor navigation with a swarm of flying robots. In Proceedings of the IEEE International Conference on Robotics and Automation, Saint Paul, MN, USA, 14–18 May 2012; pp. 4641–4647.
7. Jatmiko, W.; Jovan, F.; Dhiemas, R.; Sakti, A.M.; Ivan, F.M.; Fukuda, T.; Sekiyama, K. Robots implementation for odor source localization using PSO algorithm. *WSEAS Trans. Circuits Syst.* **2011**, *10*, 115–125.
8. Marchant, W.; Tosunoglu, S. Rethinking Wildfire Suppression With Swarm Robotics. In Proceedings of the 29th Florida Conference on Recent Advances in Robotics, Miami, FL, USA, 12–13 May 2016; pp. 186–191.
9. Bayat, B.; Crasta, N.; Crespi, A. Environmental Monitoring using Autonomous Vehicles: A Survey of Recent Searching Techniques. *Curr. Opin. Biotechnol.* **2017**, *45*, 76–84. [CrossRef] [PubMed]
10. Landau, H.; Chen, X.; Klose, S.; Leandro, R.; Vollath, U. *Trimble's Rtk and Dgps Solutions in Comparison with Precise Point Positioning. Observing Our Changing Earth*; Sideris, M.G., Ed.; Springer: Berlin/Heidelberg, Germany, 2009; pp. 709–718. [CrossRef]
11. Shao, M.; Sui, X. Study on Differential GPS Positioning Methods. In Proceedings of the 2015 International Conference on Computer Science and Mechanical Automation (CSMA), Hangzhou, China, 23–25 October 2015; pp. 223–225. [CrossRef]
12. Merriaux, P.; Dupuis, Y.; Boutteau, R.; Vasseur, P.; Savatier, X. A Study of Vicon System Positioning Performance. *Sensors* **2017**, *17*, 1591. [CrossRef] [PubMed]
13. Hoyer, L.; Steup, C.; Mostaghim, S. A Robot Localization Framework Using CNNs for Object Detection and Pose Estimation. In Proceedings of the 2018 IEEE Symposium Series on Computational Intelligence (SSCI), Bengaluru, India, 18–21 November 2018; pp. 1388–1395.
14. Mautz, R.; Tilch, S. Survey of optical indoor positioning systems. In Proceedings of the 2011 International Conference on Indoor Positioning and Indoor Navigation, Guimaraes, Portugal, 21–23 September 2011; pp. 1–7. [CrossRef]
15. Wang, Y.; Yang, X.; Zhao, Y.; Liu, Y.; Cuthbert, L. Bluetooth positioning using RSSI and triangulation methods. In Proceedings of the 2013 IEEE 10th Consumer Communications and Networking Conference (CCNC), Las Vegas, NV, USA, 11–14 January 2013; pp. 837–842. [CrossRef]
16. Yang, C.; Shao, H. WiFi-based indoor positioning. *IEEE Commun. Mag.* **2015**, *53*, 150–157. [CrossRef]
17. Sidorenko, J.; Schatz, V.; Scherer-Negenborn, N.; Arens, M.; Hugentobler, U. Decawave UWB clock drift correction and power self-calibration. *Sensors* **2019**, *19*, 2942. [CrossRef] [PubMed]
18. Kempke, B.; Pannuto, P.; Dutta, P. PolyPoint: Guiding Indoor Quadrotors with Ultra-Wideband Localization. In Proceedings of the 2nd International Workshop on Hot Topics in Wireless, Paris, France, 11 September 2015; ACM Press: Paris, France, 2015; pp. 16–20. [CrossRef]
19. Mai, S. Simultaneous Localisation and Optimisation Towards Swarm Intelligence in Robots. Master's Thesis, Otto-von-Guericke University, Magdeburg, Germany, 2018.
20. Xu, H.; Zhang, Y.; Zhou, B.; Wang, L.; Yao, X.; Meng, G.; Shen, S. Omni-swarm: A Decentralized Omnidirectional Visual-Inertial-UWB State Estimation System for Aerial Swarm. *arXiv* **2021**, arXiv:2103.04131.
21. Steup, C.; Mai, S.; Mostaghim, S. *Evaluation Platform for Micro Aerial Indoor Swarm Robotics*; Technical Report FIN-003-2016; Otto-von-Guericke-Universität: Magdeburg, Germany, 2016.
22. Mantilla-Gaviria, I.A.; Leonardi, M.; Balbastre-Tejedor, J.V.; de los Reyes, E. On the application of singular value decomposition and Tikhonov regularization to ill-posed problems in hyperbolic passive location. *Math. Comput. Model.* **2013**, *57*, 1999–2008. [CrossRef]
23. Murphy, W.; Hereman, W. *Determination of a Position in Three Dimensions Using Trilateration and Approximate Distances*; Technical Report 7; Department of Mathematical and Computer Sciences, Colorado School of Mines: Golden, CO, USA, 1995.

drones

Article

Inter-UAV Routing Scheme Testbeds

Georgios Amponis [1], Thomas Lagkas [1,*], Panagiotis Sarigiannidis [2], Vasileios Vitsas [3] and Panagiotis Fouliras [4]

[1] Department of Computer Science, International Hellenic University, 65404 Kavala, Greece; gamponis@ihu.gr
[2] Department of Electrical and Computer Engineering, University of Western Macedonia, 50100 Kozani, Greece; psarigiannidis@uowm.gr
[3] Department of Information and Electronic Engineering, International Hellenic University, 57400 Thessaloniki, Greece; vitsas@ihu.gr
[4] Department of Applied Informatics, University of Macedonia, 54636 Thessaloniki, Greece; pfoul@uom.edu.gr
* Correspondence: tlagkas@cs.ihu.gr

Abstract: With the development of more advanced and efficient control algorithms and communication architectures, UAVs and networks thereof (swarms) now find applications in nearly all possible environments and scenarios. There exist numerous schemes which accommodate routing for such networks, many of which are specifically designed for distinct use-cases. Validation and evaluation of routing schemes is implemented for the most part using simulation software. This approach is however incapable of considering real-life noise, radio propagation models, channel bit error rate and signal-to-noise ratio. Most importantly, existing frameworks or simulation software cannot sense physical-layer related information regarding power consumption which an increasing number of routing protocols utilize as a metric. The work presented in this paper contributes to the analysis of already existing routing scheme evaluation frameworks and testbeds and proposes an efficient, universal and standardized hardware testbed. Additionally, three interface modes aimed at evaluation under different scenarios are provided.

Keywords: FANETs; ad hoc networking; drone swarms; efficient routing

Citation: Amponis, G.; Lagkas, T.; Sarigiannidis, P.; Vitsas, V.; Fouliras, P. Inter-UAV Routing Scheme Testbeds. *Drones* **2021**, *5*, 2. https://dx.doi.org/10.3390/drones5010002

Received: 30 November 2020
Accepted: 23 December 2020
Published: 28 December 2020

1. Introduction

The birth of Mobile Ad hoc Networks (MANETs) has enabled the decentralization of a massive amount of applications and services. Ad hoc networks can now function in a great variety of environments and conditions without necessarily interfacing with the rest of the internet. All cooperating nodes in such networks are considered peers, with a master–slave relationship between them being nonexistent due to the networks fundamental architecture: each node is both an end-device and a router. Vehicular Ad hoc Networks (VANETs) were the first type of MANETs to be deployed, followed by Flying Ad hoc Networks (FANETs) which are the main focus of this paper. Another type of ad hoc networks which present great interest are cooperative VANET-FANET networks, in which UAVs function essentially as relays between vehicles. According to [1,2], FANETs differ from VANETs and other types of (ad hoc) networks in matters of:

- Node mobility (in terms of Velocity/Acceleration/Jerk)
- Topology (FANETs may function in a 3D grid in addition to the more traditional single-layered swarm topology [3]; additionally, a UAV-comprised network shall alter its topology with a significantly higher frequency, due to its nature)
- Availability of energy (arguably the most significant constrain of autonomy in an ad hoc network)

FANETs find application in both military and civilian fields, such as the ones described in [4–15]:

- Real time monitoring and surveillance
- Search-and-rescue operations

- Agriculture
- Relaying of communication

As envisioned in [16], while the challenges of swarm formation are numerous (routing, path planning, QoS provision, UAV coordination, etc.), multi-UAV deployments are the future of aerial communications. In surveillance and monitoring applications, routing of information is a means to an end; such is the case in search and rescue-oriented FANET deployments as well as the ones centered around agriculture. High route selection efficiency is mandatory for an acceptable quality of service (QoS) in all mentioned use-cases. As a general rule, routed data shall incorporate desired payload and control messages which forces the scheme developer to take into account the desired application class and packet overhead alike. When considering the relaying of communication between distant parties, routing (and high efficiency thereof) becomes not just a means to an end but rather a goal in itself. Such a use-case, centered around communication relaying as well as VANET-to-FANET cooperation is mentioned in [17]; the researchers have even proposed an application-specific cross-layer routing protocol aimed at enabling close cooperation between VANETs and FANETs, as mentioned previously. The importance of routing efficiency becomes even more highlighted when considering military or emergency/first response operations [4].

There exist a vast number of routing protocols, with many being developed for a very specific use-case and others functioning as a foundation for other schemes. Evaluation of existing/under development protocols is done traditionally using classical simulation software (more often than not, developed for non-ad hoc networks). To counteract the limitations arising from a such approach and obtain usable results, developers resort to the addition of mobile properties to the individual nodes comprising their networks. Thus, in most cases, developed routing schemes are evaluated using conventional networking simulators, which can in many cases, efficiently approximate a mobile ad-hoc behavior. Different mobility models are included as libraries for discreet event network simulators. Support for the following mobility models is inherent in the NS-3 network simulator:

- Constant Position Mobility Model (NS-3 specific mobility model)
- Constant Velocity Mobility Model (NS-3 specific mobility model)
- Constant Acceleration Mobility Model (NS-3 specific mobility model)
- Gauss Markov Mobility Model [18,19]
- Hierarchical Mobility Model (NS-3 specific mobility model; a combination of specified "parent" and "child" mobility models)
- 2D Random Direction Mobility Model [19]
- 2D Random Walk Mobility Model [20]
- Random Waypoint Mobility Model [21]
- Steady State Random Waypoint Mobility Model (NS-3 specific mobility model; uses stationary instead of random distribution for initial speed, pause and position)
- Waypoint Mobility Model (NS-3 specific mobility model)

For the majority of the existing routing protocols, a discreet event network simulator such as the NS series may prove sufficient. However, non-classically layered schemes cannot be as effectively simulated due to inherent limitations of said software. To address this issue, there was developed the FINS framework capable of considering flexibility in the layer stack and thus enable nonadjacent interlayer data exchange during simulation [22–24]. Opnet Modeler has also been used by researchers to evaluate routing schemes of both classical and cross-layer architectures in numerous occasions.

Another approach to evaluating routing schemes' efficiency, is directly using them in a testbed-network, usually comprised of Raspberry Pi modules. Great examples of research incorporating hardware-based testbeds are [25–28]. Despite the existence of related work in the field of ad-hoc routing and testbed implementation, no published papers present freely available testbed frameworks. The present paper aims to fill this void by:

- Engaging in FANET-specific routing protocol and mobility model analysis

- Proposing a universal routing scheme testbed
- Making said universal testbed completely open-source for other researchers to utilize and improve

Table 1 compares the present work to already existing surveys and testbed-oriented papers. The present paper is focused not only on surveying protocols but also on proposing a testbed. Thus, only papers of a similar nature were considered for the following comparison. Our approach proves to be the most complete in covering ad hoc routing and providing an evaluation platform open for all interested researchers to use.

Table 1. Related work and papers.

Related Work	FANET-Specific	Routing Protocols	Mobility Models	Testbed	Open-Source
Miya et al. [25]	X	✓	X	✓	X
Kaysina et al. [26]	✓	✓	X	✓	X
Hamdaoui et al. [29]	X	✓	X	✓	X
Brown et al. [30]	✓	X	X	✓	X
Kim et al. [27]	✓	X	X	✓	X
Rosati et al. [31]	✓	✓	X	✓	X
Our approach	✓	✓	✓	✓	✓

The layout of the rest of this paper is as follows; Section 2 analyzes and explains matters of inter-UAV routing, while considering some of the most important and "influential" routing protocols. Section 3 is divided into two subsections; one surveying existing routing scheme testbeds (Section 3.1) and one proposing the aforementioned prototype testbed (Section 3.2). Section 4 is dedicated to the analysis of the proposed universal routing testbed and its modes of function: standalone, supplementary and stationary mode. Each of the aforementioned modes are explained in full detail in Sections 4.1–4.3 respectively.

2. Ad Hoc Routing

This section is dedicated to the analysis of the fundamental principles governing ad hoc networking, with an emphasis on FANETs. Additionally, some of the most important ad hoc routing protocols are explained and examined, in the context of evaluating them on a physical level, along with mobility models commonly used either to emulate high-mobility FANETs or as a route computation metric. The way in which mobility models affect route selection and are thus considered a route computation metric is link quality estimation (highly dependent on node mobility) accompanied by node position-awareness, which directly affects next hop selection in geographic or movement-predictive routing protocols (e.g., Greedy Perimeter Stateless Routing (GPSR) [32], Greedy Load Share Routing (GLSR) [33], Mobility Prediction based Geographic Routing (MPGR) [34], Directional Optimized Link State Routing (D-OLSR) [35], Geographic Contention-based Forwarding and Cooperative Communications (CoopGeo) [36] and their derived schemes).

2.1. Testbed-Friendly Ad Hoc Routing Protocols

Amongst the most commonly used ad hoc routing protocols in the scope of emulation of FANET deployments are:

- OLSR (usually implemented via OLSRd)
- Ad-hoc On-demand Distance Vector Routing (AODV)
- Hybrid Wireless Mesh Protocol (HWMP)
- BATMAN

2.1.1. OLSR

OLSR is one of the most popular proactive routing protocols, mostly due to the open-source nature of OLSRd enabling researchers to utilize it. Due to OLSR's popularity and configurability, it has become the foundation of a spectrum of routing protocols: Predictive OLSR (POLSR) [37], Directional OLSR (DOLSR) [35], Energy-aware Mobility Prediction OLSR (EMP-OLSR) [38], OLSR Fuzzy Cost (OLSR-FC) [39], Cross-layer OLSR (C-OLSR) [40], QoS Aware Link Defined OLSR (LD-OLSR) [41] and many more.

OLSRd can function either as classical OLSR or additionally consider the Expected Transmission Count (ETX) metric, essentially implementing OLSR-ETX to obtain link quality (*LQ*, network-wide metric) as well as neighbor link quality (*NLQ*, node-specific metric) information [29]. The ETX metric is defined as follows:

$$ETX = \frac{1}{1 - e_{pt}} \quad (1)$$

$$e_{pt} = 1 - LQ \times NLQ \quad (2)$$

$$ETX = \frac{1}{LQ \times NLQ} \quad (3)$$

where e_{pt} is the packet error probability.

When used in an actual scenario, OLSRd is responsible for detecting neighboring network nodes and performing a link quality evaluation using received signal strength indication (RSSI) information. It therefore does not handle packet transmissions, yet populates and updates nodes' routing tables as the network topology is updated by broadcasting HELLO messages and selecting MPRs. In addition to HELLO messages, another type of messages used by OLSRd is Topology Control (TC) messages, responsible for informing the network for each node's accessibility by an MPR.

Route selection in classical OLSR is implemented by considering the path consisting of the smallest number of hops. Since OLSRd can be configured to consider the ETX metric, in most cases route selection also considers the possibility of packet loss and end-to-end throughput. The ETX metric can be utilized to enable QoS awareness using OLSR [41]. The only drawback stemming from the usage of the ETX metric is the protocol's inability to sense the difference between links' data rates (it assumes that all sensed links have an identical throughput). This is counteracted by introducing the Expected Transmission Time (ETT) metric (natively implemented by OLSR-ETT) [42].

$$ETT = ETX \times \frac{S}{B} \quad (4)$$

where S = packet size in bytes and B = link capacity (measured using packet-pair technique).

2.1.2. AODV

AODV [43] is a reactive topology-based protocol as it calculates packet routes on-demand. It is based on the Destination-Sequenced Distance Vector routing (DSDV) [44]. AODV cannot optimize routes by default. As with OLSR, there exist -ETX variants/extensions of AODV. Researchers in [45] measured and compared the performance of AODV and AODV-ETX; they concluded that compared to AODV, AODV-ETX's average throughput is increased by 15.84%, packet delivery ratio is increased by 5.80%, average overhead is reduced by 4.31%, at the expense, however, of a 10.19% increased end-to-end delay and a by 3.09% increased route discovery delay. AODV is one of the most popular ad hoc routing protocols. Authors of [46] have compared AODV to Dynamic Source Routing (DSR) in an outdoors environment. They concluded that AODV performs measurably better in terms of packet loss and end-to-end delay. AODV is implemented in Linux kernel v. 3.8 as AODV-UU [47] created by the Uppsala University, hence the suffix.

2.1.3. HWMP

Using AODV as a foundation, a Hybrid Wireless Mesh Protocol (HWMP) has been developed. HWMP is a hybrid routing protocol, capable of using both proactive and reactive path selection algorithms. HWMP offers a higher packet delivery ratio, at the expense of a higher packet overhead. Similarly to AODV, HWMP uses path request (PREQ) messages to obtain optimal packet paths. Authors of [48] engage in performance comparison of AODV, DSDV, DSR, OLSR, AOMDV and HWMP alike: their measurements and simulations prove that under their test conditions, HWMP offers the best performance, followed by OLSR. HWMP's better performance is comparison to AODV stems from the aforementioned fact: usage of both proactive and reactive route path discovery methods. HWMP can route on layer 2 of the Open Systems Interconnection (OSI) stack using Media Access Control (MAC) instead of Internet Protocol (IP) addresses.

Both AODV and HWMP exist as installation for Linux-based platforms such as Raspberry pi. Due to AODV's library age, between the two, HWMP is generally preferred as a routing scheme.

2.1.4. BATMAN

The Better Approach To Mobile Ad hoc Networking (BATMAN) protocol is a decentralized routing protocol whose main selling point is the equal distribution of the optimal route information. This approach eliminates the need to broadcast possible topology changes to the entire network, effectively eliminating overhead. Nodes do not maintain full source-destination route information and instead only know their next hop. Route optimization is implemented in a hop-by-hop approach. Packets are routed to their destination individually, via constantly recomputed routes (that applies to packets comprising a single message as well). The BATMAN-adv variant of this protocol enables layer 2 routing (similarly to HWMP), in contrast to the classical layer-3 routing scheme normally implemented by BATMAN [49]. BATMAN-adv is currently implemented as a kernel driver and improves CPU performance by minimizing the required machine cycles [50].

Periodically, Originator Messages (OGM) are broadcasted to update neighboring node information; they contain the following information [51]:

- Originator (i.e., source) address
- Destination address
- Link flag
- Time To Live (TTL)

The information of the destination address is updated every time the OGM packet is rebroadcasted. Information regarding reachable nodes is stored in an originator table, carried by each node. BATMAN uses the Transmission Quality metric (TQ) to compare link quality between possible next hops and perform route selection [51]. Authors in [26] used BATMAN in a physical Raspberry Pi-based network. They propose a testbed framework for this protocol and additionally tested video content transmission in a BATMAN-enabled ad hoc network.

Table 2 contains and summarizes the main characteristics of the testable routing protocols which were analyzed in this subsection: OLSR Section 2.1.1, AODV Section 2.1.2, HWMP Section 2.1.3 and BATMAN Section 2.1.4.

Table 2. Routing scheme summary table.

Routing Protocol	Proactive	Reactive	Open-Source	Kernel Implementation
OLSR	✓	X	✓	OLSRd
AODV	X	✓	✓	AODV-UU
HWMP	✓	✓	✓	HWMP
BATMAN	✓	X	✓	BATMAN-adv

2.2. Mobility Models

Of the existing mobility models those supported by the NS3 simulator are mentioned in Section 1. Depending on the requirements set by the simulated UAV network use-case, as well as the behaviour defined by the routing protocol, different, and in some occasions more unique mobility models are used to describe the FANETs mobility; e.g., despite its general effectiveness, the Gauss-Markov Mobility Model does not describe circular swarm mobility (common in monitoring use-cases) as effectively as the Semi Random Circular Mobility Model, which is otherwise of minimal emulative value. Generally, the most effective (and thus popular) ones in terms of accurate FANET mobility emulation are the following:

- Gauss-Markov Mobility Model [18]
- Paparazzi Mobility Model [19]
- Semi Random Circular Mobility Model [52]
- Random Waypoint Mobility Model [18]

2.2.1. Gauss-Markov

Gauss-Markov is a time-dependent mobility model capable of accurately describing a FANET's movement, with individual node's paths being independent from one another. Random initial positions are defined for each UAV and the velocity vector is recomputed at each time instance depending solely on previous UAV speed and position [53].

2.2.2. Paparazzi

Paparazzi is a path-planned stochastic mobility model. In order for the model to approximate all possible UAV maneuvers, it supports the following fundamental movements: stay-at, way-point, eight, scan and oval [54].

2.2.3. Semi Random Circular

Semi Random Circular is a path-planned mobility model centered around UAV-enabled surveillance and target monitoring of an already known area [52]. This model makes the examined FANET circulate an area of interest, rendering it realistic for the emulation of e.g., search-and-rescue missions.

2.2.4. Random Waypoint

Random Waypoint is a randomized mobility model. At every time instance, each node has a distinct and random velocity, destination and a randomized stop time. The UAVs' destination and stop time both get re-updated upon each randomization iteration, depending on which event precedes the other: reaching of node destination or timeout of waypoint validity.

When simulating a FANET deployment, it is of utmost importance to select a fitting mobility model. As previously stated, the overall mobility of a node and its network strongly affects networking metrics and route computation-related parameters. Consideration of mobility models as a routing metric is therefore a valid reason to avoid or resort to the usage of a mobility model. Location or prediction-dependent schemes especially, may present an altered/deteriorated or generally false performance, should the UAVs' locations be altered in an unexpected/rapid manner (e.g., random definition of velocity vectors instead of a collectively coordinated mobility plan and movement path).

3. FANET Testbeds

This section is concerned with surveying existing routing scheme testbeds and the proposal of an open-source hardware testbed. Section 3.1 analyzes the work of researchers who faced the similar challenge of routing scheme performance evaluation in real life and under such environmental conditions. Virtually all existing work is either simulation-focused or uses Raspberry Pis on which the tested protocol is installed. The benefits of implementing a hardware-based evaluation instead of simulating a routing scheme are numerous:

- Measurement of computational capacity required for routing and load balancing
- Measurement of environmental effects on communication and link quality
- Measurement of transmission-related node energy expenditure
- Measurement of mobility-stemming effects on overhead and computational demand
- Measurement of security methods' and packet encryption/decryption (CPU-intensive tasks) effects on availability of processing power

On a few occasions, researchers resorted to the usage of actual mobile UAVs, which they deployed as a FANET. Respectively, Section 3.2 proposes a multifunctional solution capable of both extending the capabilities of existing testbeds and functioning as a standalone on-board flight controller. The hardware design is explained on a medium abstraction layer and design files are provided.

3.1. Existing Routing Scheme Testbeds

Authors of [25] performed an end-to-end packet delay measurement and evaluation of OLSR using a network comprised of stationary Raspberry Pi Zeros. Their approach proves the significant benefits of the real-life testing approach, as it enables the measurement of CPU load during routing and packet encryption. Furthermore, usage of power efficient Raspberry Pi Zeros gives insight to how a similar network would perform, should it become mobile. Additionally, usage of this platform as a routing testbed constitutes an indirect evaluation of the Linux kernel efficiency and possible bottlenecking. Random "dummy" data is generated using the *ipref* traffic generator and then embedded in a UDP frame. The source node is configured to transmit the randomly generated data to another Raspberry Pi Zero functioning as a relay between the source and destination. Due to the network being comprised of only three nodes (source, proxy, destination), no actual routing table is being made use of or updated. Therefore, usage of either a reactive/hybrid/proactive protocol has minimal effect on network metrics. Thus, this testbed implementation is more of a kernel bottlenecking evaluation rather than an actual routing protocol one. Nevertheless, repetition of this experiment using the same setup, only with a greater number of nodes would potentially wield rather important results. Similarly to this work, the researchers in [55] used five Raspberry Pis to evaluate OLSR in a content-centric network (CCN). The difference between this testbed and the one described in [25] is the usage of the networking-oriented Open-WRT operating system. Nodes maintained a constant line of sight (LOS) and proved OLSR's potential as an enabler of a CCN overlay in TCP/IP networks.

Researchers in [26] implemented a Raspberry Pi-based testbed for the BATMAN protocol. The hardware testbed was accompanied by NS-3 simulations in order for the authors to validate their results. In this case, actual UAVs were used as sources, relays and end devices alike for each tested scenario. The FANET topologies utilized were the following:

- Infrastructure topology (Figure 1)
- Mesh topology (Figure 2)
- UAV-as-a-relay topology (Figure 3)

The infrastructure topology (UAVs as end devices) (Figure 1) dictates that an intermediate ground station functions as a relay, forwarding packets to their respective destinations. Computational load of nodes is thus minimized, to the expense of the network's mobility (limited due to it being a direct function of the ground station's maneuverability). UAVs can only reach a strictly defined distance before link quality between themselves and the ground station falls below a usable threshold. The mesh topology (Figure 2) dictates that at least a UAV is constantly connected to the ground station and functions as a gateway. The difference between this topology and the previously described one is that individual UAVs are networked in a such manner that enables inter-UAV packet relaying. This topology enables greater area coverage and a higher node mobility. The UAV-as-a-relay topology (Figure 3) is technically not a FANET topology, rather a direct UAV-enabled relaying use-case which can be extended to a FANET. It is however noteworthy due to the researchers evaluation of a such relaying method's efficiency and possible bandwidth-saving improvements.

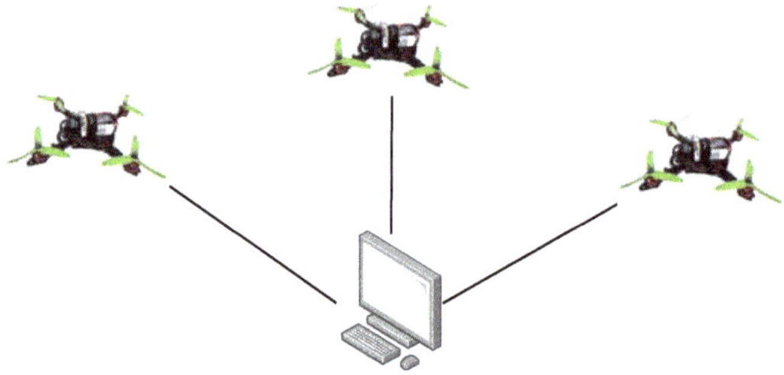

Figure 1. UAVs as end devices.

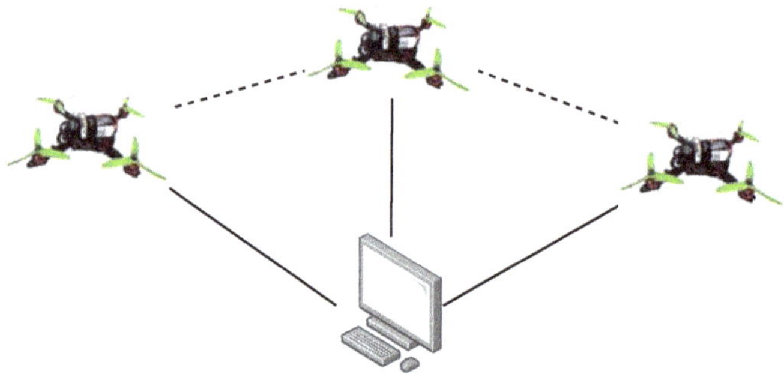

Figure 2. UAVs in a mesh topology.

Figure 3. UAV as a mobile relay.

The authors of [29] used Alix.3C2 modules equipped with a Voyage (Debian-based) Linux distribution. OLSRd was installed onto the modules and the researchers evaluated both OLSR-HOPS and OLSR-ETX, with they enriched using the ETT metric as mentioned in Section 2.1.1 for higher route computation efficiency.

Researchers in [30] implemented a fixed-UAV ad hoc networking testbed and tested it using the DSR protocol. The proposed testbed is comprised of the UAVs themselves, fixed ground stations and mobile ground stations as well. The utilized hardware is comprised of the Soekris net4511 computer module accompanied by the Orinoco 802.11b Gold radio module which has a small (PCMCIA) form factor. Fidelity Comtech RF amplifiers were used to amplify the radio module's signal which normally has an approximately 150 m range. The computer modules were equipped with the Click routing software.

3.2. Proposal of an Embedded Routing Scheme Testbed

There exist a variety of routing testbeds, both hardware and software in nature. However, there has been little to no progress in terms of standardization and compartmentalization of said systems. Furthermore, all hardware-based solutions seem to fall under the two following categories:

- Meant for evaluation using stationary nodes
- Specifically designed for a type of UAV the developers saw fit

This subsection is thus dedicated to the proposal of a universal and standardized hardware testbed meant for both stationary and aerial usage. Said testbed shall incorporate:

- A computing platform on which the desired routing protocol shall be installed
- A dedicated 802.11 WiFi antenna
- A gyroscope/accelerometer which shall enable the platform to function as a standalone flight controller
- A companion microcontroller implementing the flight control logic without burdening the main computing platform with unnecessary tasks

Considering the amount of existing work and availability of routing protocol installations for Linux-based operating systems, the Raspberry Pi Zero seems to be the best computing platform for aerial usage; it combines a small and lightweight form factor with impressive computing capabilities. As for the dedicated WiFi antenna and the companion microcontroller, the ESP32 proves more than sufficient in both aspects. The embedded WiFi antenna has a mediocre range, which can be significantly extended (up to approximately 1 km, assuming a direct LOS) using the Epsressif-patented 802.11 LR (long range) mode. This mode in combination with a MAC-layer peer-to-peer transmission handling protocol (e.g., ESP-NOW) can enable low latency and low overhead communication in an impressive range using a small amount of power in comparison to other commercially available 802.11 antenna modules; in this case, packet routing can be implemented using MAC addresses. The networking sacrifice needed to be made is throughput. Entering the LR mode is a simple function call and does not affect program functionality:

```
esp_wifi_set_protocol(ifx, WIFI_PROTOCOL_LR);
```

Table 3 showcases the specifications of both the computer and complementary microcontroller of choice. The gyroscope/accelerometer module of choice is the MPU6050 due to the amount and quality of existing libraries enabling it to interface with ESP32. More technical details and design choices are thoroughly explained below.

Table 3. Raspberry Pi Zero W & ESP32 specifications.

	Raspberry Pi Zero W	**ESP32**
CPU	ARM1176JZF-S (1 GHz, single-core)	Xtensa LX6 (240 MHz, dual-core)
Memory	RAM 512 MB (LPDDR2)	520 KB (SRAM) 4 MB (flash)

The proposed testbed architecture makes use of affordable networking and computing equipment to create a mobile ad hoc network and evaluate routing efficiency of a given scheme. In addition to route computation and transmission handling, this implementation can be used as a standalone flight controller. The embedded gyroscope/accelerometer enables each module to not only function as a self-contained flight controller but also consider orientation and angular velocity of each axis as a routing metric; this enables support for location-predictive routing protocols. Since the module is centered around and uses the Raspberry Pi as a computing unit, all the respectively supported routing protocols may be used with this design. Each module is a functional node in itself and does not require additional components.

The hardware has been designed so that each one of said testbed modules can be directly attached to an existing UAV. The main board complies to the PC/104 standard, which is the hardware specification many UAV autopilots aim to conform to as well, as mentioned in [56]. Conformity to this electromechanical standard ensures flawless mechanical interfacing and even safe physical stacking of said modules onto the same UAV. PC/104 is commonly used in specialized high-reliability embedded applications, aiming for a small and rugged form-factor. The PC/104 standard dictates the following:

- Usage of a (traditionally 104-pin) connector
- Strictly defined PCB dimensions (90 × 96 mm) with 13 mm optional extra space for additional connectors
- Four mounting holes in predefined positions

4. Findings and Discussion

The Findings and Discussion section is dedicated to the explanation of the three main modes in which the proposed module shall operate. The first mode is a result of direct usage of the proposed module as both a flight controller and a routing evaluation medium. As for the second mode, it likely constitutes the most directly usable mode of operation: the module only implements routing (and the evaluation thereof), while an external flight controller, native to the UAV, handles all aircraft control, PID tuning and external sensor interfacing. Similarly, the last mode of operation is the one in which no aircraft is utilized for scheme evaluation, rather a stationary set of networked modules. The last mode is suitable for both VANET deployments as well as stationary (e.g., sensor network) networks. Figure 4 showcases a 3D render of the design.

Figure 4. Rendering of proposed testbed module.

The design files for the proposed testbed module have been made publicly available in GitHub: [57]. The design software used is KiCAD. Along with the design files are the production file, which can be used for fast prototyping and testing. The work of engineers in [58] was utilized as a PC/104 template.

4.1. *Standalone Aerial Mode*

In standalone aerial mode, mesh modules are directly attached to each UAV comprising a FANET. The main concept behind the design is the compartmentalization of task handling so as to avoid unnecessary computational burden of each component. This design idea is directly translated to hardware design implementation. The Raspberry Pi Zero is attached directly to the main board with an exposed header configured to match its pinout. As seen in Figure 5, one of the PC/104 holes and two extra nonplated 3.2 mm ⌀mounting holes offer extra mechanical stability to the interconnection of the main components. As the Raspberry Pi is placed onto the main board, it is internally powered by the main board's regulator. The module has full support for an amplitude of commercially available GNSS modules. Node location can therefore be used as a route selection metric, in addition to node velocity and acceleration as previously mentioned. Thus, in addition to predictive routing schemes, evaluation/implementation of location aware or location-based routing is made possible.

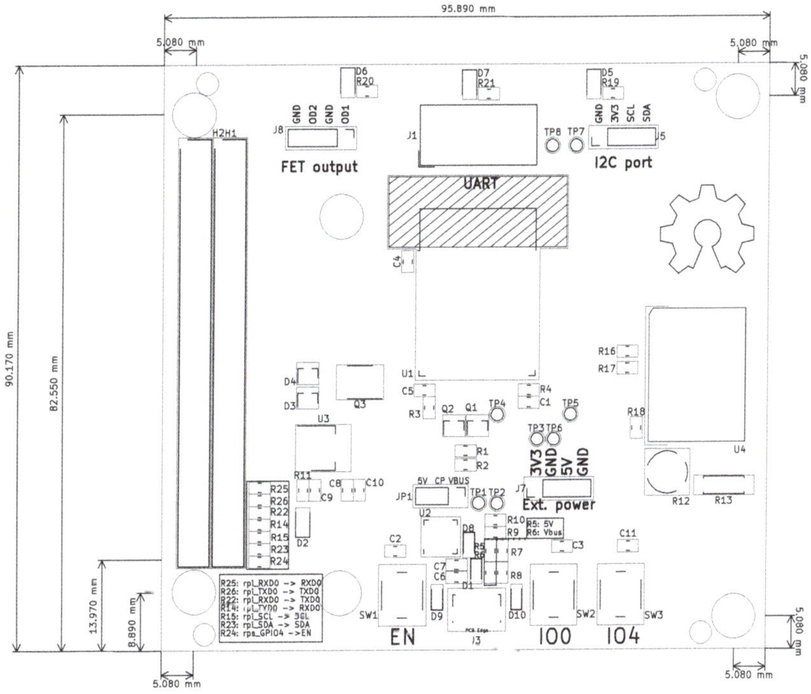

Figure 5. Layout of the module's main board.

Figure 6 illustrates the interfacing between the Raspberry Pi and the ESP32 as well as tasks each module handles.

The Raspberry Pi Zero shall only implement route-related computations. After a route has been established, the packet shall be serially transmitted to the ESP32, which shall in turn implement transmission handling. In turn, upon packet reception, the ESP32 shall serially transmit the data to the Raspberry Pi. Packet transmission/reception handling shall constitute a distinct nonblocking RTOS task, so that the ESP32 can simultaneously implement the flight control logic. It becomes obvious that the ESP32 now has two functions, which are defined as distinct RTOS tasks handled by the two cores of the Xtensa LX6 microprocessor. All packet transmission and reception is done via the embedded WiFi antenna which ensures a low power consumption as well as the possibility of interfacing with conventional 802.11 networks. The Raspberry Pi is thus successfully only occupied with route handling.

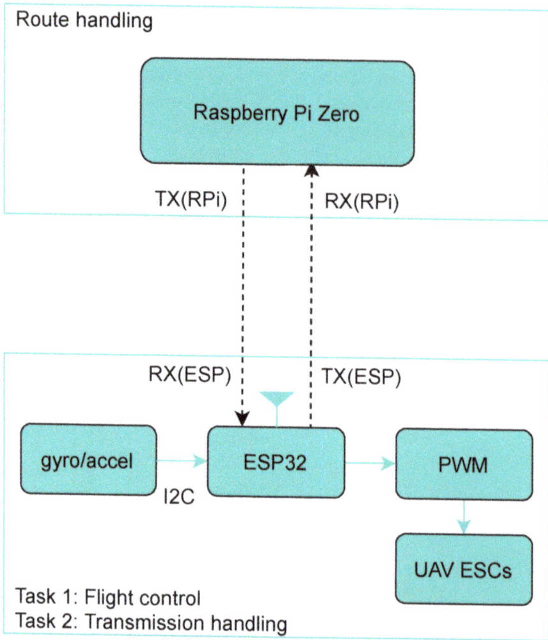

Figure 6. Standalone aerial node components.

Table 4 showcases the serial interface between said modules in standalone aerial mode. The mentioned gyroscope/accelerometer processed data shall only be transmitted to the Raspberry Pi if the evaluated routing protocol requires them as a metric. Similarly, RSSI (measured between) shall be shared if link quality or power-awareness is desired and considered by the respective protocol. GNSS data is again, optional and shall be transmitted in the case geolocation-based routing is implemented.

Table 4. Raspberry Pi Zero W—ESP serial interfacing.

Raspberry Pi Zero W		**ESP32**	
Serial Tx	**Serial Rx**	**Serial Tx**	**Serial Rx**
Packet to be transmitted	Received packet	Received packet	Packet to be transmitted
	Gyroscope/accelerometer (processed data)	Gyroscope/accelerometer (processed data)	
	RSSI	RSSI	
	GNSS data	GNSS data	

4.2. *Supplementary Aerial Mode*

In supplementary aerial mode, the module does not function as a fight controller. It may be attached to an already existing flight controller/autopilot and solely implement routing to other networked nodes on-board. The authors of [56] survey a number of UAV autopilots, a substantial number of whom offer serial interface connectivity. Flight controllers typically transmit telemetry data to a ground station, which can be made available to a serially connected interface on demand. Telemetry data can be forwarded to the Raspberry Pi to be used in route calculations. In this case, the ESP does not need to handle two distinct RTOS tasks and can fully commit to the task of transmission handling. The interfacing node components can be described as seen in Figure 7.

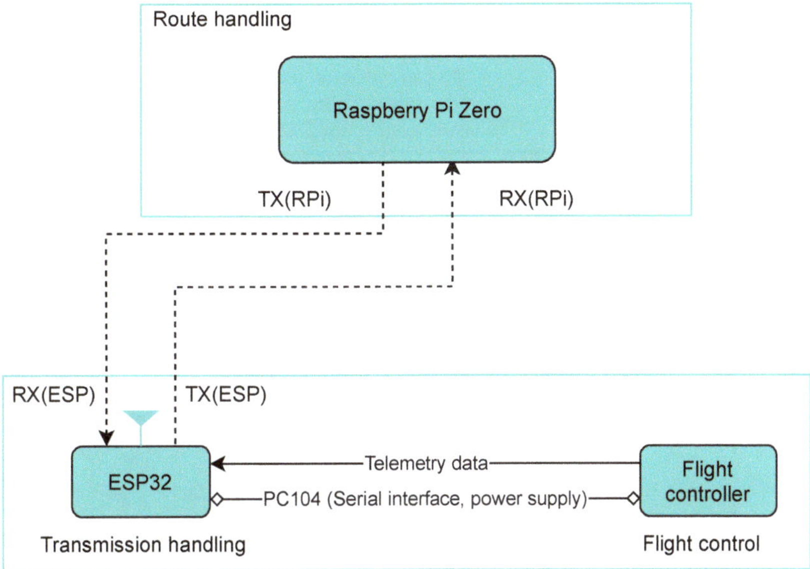

Figure 7. Supplementary aerial mode components.

As virtually all flight controller inherently implement interfacing with GNSS modules, and contain an on-board gyroscope/accelerometer module, those components need not be placed/connected to the main board.

4.3. Stationary/VANET Mode

In stationary/VANET mode, the data from the MPU6050 gyroscope/accelerometer module may be utilized to form packets of a given length, defined by the protocol's structure. In this mode, the modules shall function independently from an external flight controller. With no connection to a flight controller or an embedded flight control RTOS task, the module can still implement routing and be used as a means of evaluation thereof. This may prove to be a useful implementation for non MANET-specific ad hoc routing schemes. The advantages stemming from the usage of a compartmentalized device and a standard form factor are still there, though exploited to a lesser degree. Usage of this mode is also a possibility for the evaluation of routing schemes in nonaerial ad hoc networks.

5. Conclusions and Future Work

This paper has discussed industrial, military and agricultural use-cases for FANETs, surveyed directly applicable mobile ad hoc routing protocol to be used in such deployments, as well as existing testbed/simulation frameworks which can be used to evaluate the efficiency of routing protocols. Additionally, matters of network topology and mobility modeling are addressed in the context of scheme simulation. This work is concluded with the presentation of:

- A standardized hardware prototype capable of supporting all routing protocols currently implemented in the Linux kernel
- Component interface modes as a function of desired evaluation framework

Figure 8 constitutes the result of the proposed testbed module's evaluation using an ESP-Now link and the embedded PCB antenna of the ESP. Even when using the PCB antenna, power consumption was kept below 2.5 Watts, while the link did not fail to maintain maximum bandwidth at 85 m with a direct and constant LOS. The module's evaluation consisted of:

- Link establishment between two modules
- Formation of ESP-Now packets using dummy data harvested from the MPU6050 gyroscope/accelerometer
- Communication between a constantly increasing distance (5 m–85 m)
- Measurement of the transmitting module's power consumption at 5-m intervals

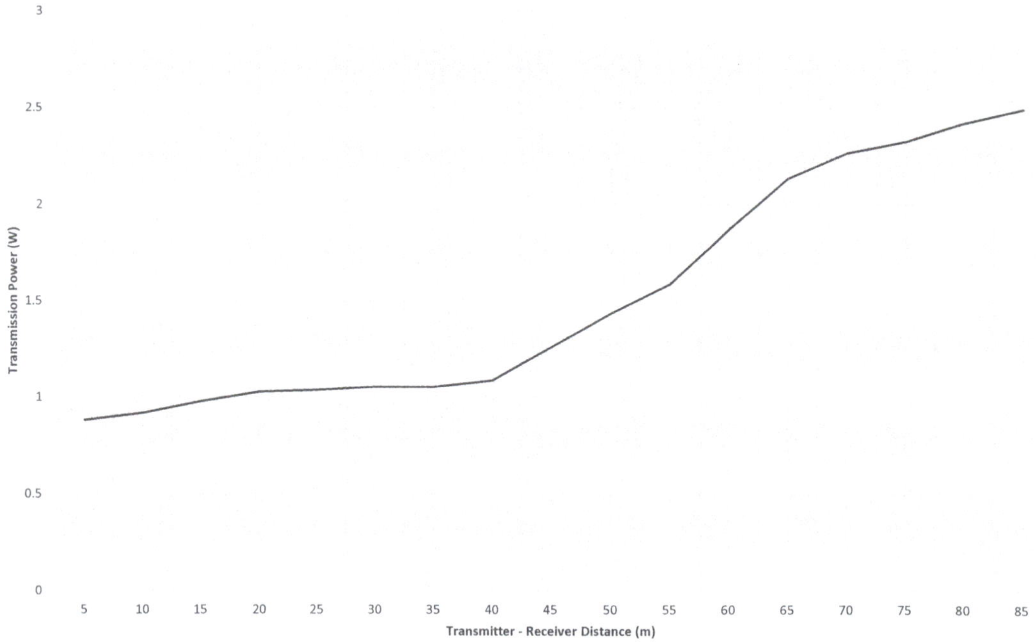

Figure 8. Communication power of an ESP-Now direct link between two modules.

Future work shall be constituted of test results originating from the usage of the proposed testbed with OLSRd and BATMAN. A comparison between said routing schemes in terms of: (a) scalability and (b) mobility shall be part of said future work. NS-3 simulations shall be used to validate the performance of the proposed testbed under the mobility models mentioned in Section 2.2. Due to the current lack of functional open-source flight controllers meant to be used with the ESP-series, such firmware shall be developed and made openly available, as is the case with the proposed module's hardware.

Author Contributions: Conceptualization, T.L., P.S., V.V., P.F. and G.A.; methodology, G.A.; software, G.A.; validation, T.L., G.A. and P.S.; formal analysis, G.A., T.L., P.S.; investigation, G.A.; resources, T.L., P.S., V.V.; data curation, T.L., P.F., V.V.; writing—original draft preparation, G.A.; writing—review and editing, T.L., G.A.; visualization, G.A.; supervision, T.L. and P.S.; project administration, T.L. and P.S. All authors have read and agreed to the published version of the manuscript.

Funding: This research was cofunded by the European Union and Greek national funds through the Operational Program Competitiveness, Entrepreneurship, and Innovation, grant number T1EDK-04759.

Informed Consent Statement: Not applicable.

Data Availability Statement: No new data were created or analyzed in this study. Data sharing is not applicable to this article.

Conflicts of Interest: The authors declare no conflict of interest.

References

1. Oubbati, O.S.; Atiquzzaman, M.; Lorenz, P.; Tareque, M.H.; Hossain, M.S. Routing in flying Ad Hoc networks: Survey, constraints, and future challenge perspectives. *IEEE Access* **2019**, *7*, 81057–81105. [CrossRef]
2. Khan, M.F.; Yau, K.L.A.; Noor, R.M.; Imran, M.A. Routing schemes in FANETs: A survey. *Sensors* **2020**, *20*, 38. [CrossRef]
3. Nawaz, H.; Ali, H.M.; Laghari, A.A. UAV Communication Networks Issues: A Review. *Arch. Comput. Methods Eng.* **2020**, 1–21. [CrossRef]
4. Di Felice, M.; Trotta, A.; Bedogni, L.; Chowdhury, K.R.; Bononi, L. Self-organizing aerial mesh networks for emergency communication. In Proceedings of the IEEE International Symposium on Personal, Indoor and Mobile Radio Communications, PIMRC, Washington, DC, USA, 2–5 September 2014; pp. 1631–1636. [CrossRef]
5. Maza, I.; Caballero, F.; Capitán, J.; Martínez-De-Dios, J.R.; Ollero, A. Experimental results in multi-UAV coordination for disaster management and civil security applications. *J. Intell. Robot. Syst. Theory Appl.* **2011**, *61*, 563–585. [CrossRef]
6. Nikitas, A.; Michalakopoulou, K.; Njoya, E.T.; Karampatzakis, D. Artificial intelligence, transport and the smart city: Definitions and dimensions of a new mobility era. *Sustainability* **2020**, *12*, 2789. [CrossRef]
7. Cameron, S.; Hailes, S.; Julier, S.; McClean, S.; Parr, G.; Trigoni, N.; Ahmed, M.; McPhillips, G.; de Nardi, R.; Nie, J.; et al. SUAAVE: Combining aerial robots and wireless networking. In Proceedings of the 25th Bristol International UAV Systems Conference, Department of Aerospace Engineering, University of Bristol, 1-9 Old Park Hill, Bristol, Avon, UK, 12–14 April 2010; pp. 1–14.
8. Ollero, A.; Kondak, K.; Previnaire, E.; Maza, I.; Caballero, F.; Bernard, M.; Martinez, J.R.; Marrón, P.; Herrmann, K.; Van Hoesel, L.; et al. Integration of aerial robots and wireless sensor and actuator networks. The AWARE project. In Proceedings of the IEEE International Conference on Robotics and Automation, Anchorage, AK, USA, 3–7 May 2010; pp. 1104–1105. [CrossRef]
9. Lagkas, T.; Argyriou, V.; Bibi, S.; Sarigiannidis, P. UAV IoT framework views and challenges: Towards protecting drones as "things". *Sensors* **2018**, *18*, 4015. [CrossRef] [PubMed]
10. Radoglou-Grammatikis, P.; Sarigiannidis, P.; Lagkas, T.; Moscholios, I. A compilation of UAV applications for precision agriculture. *Comput. Netw.* **2020**, *172*, 107148. [CrossRef]
11. Kakamoukas, G.; Sariciannidis, P.; Livanos, G.; Zervakis, M.; Ramnalis, D.; Polychronos, V.; Karamitsou, T.; Folinas, A.; Tsitsiokas, N. A Multi-collective, IoT-enabled, Adaptive Smart Farming Architecture. In Proceedings of the IST 2019—IEEE International Conference on Imaging Systems and Techniques, Abu Dhabi, UAE, 9–10 December 2019. [CrossRef]
12. Triantafyllou, A.; Sarigiannidis, P.; Bibi, S. Precision agriculture: A remote sensing monitoring system architecture. *Information* **2019**, *10*, 348. [CrossRef]
13. Tsouros, D.C.; Bibi, S.; Sarigiannidis, P.G. A review on UAV-based applications for precision agriculture. *Information* **2019**, *10*, 349. [CrossRef]
14. Egarguin, N.J.; Jackson, D.R.; Onofrei, D.; Leclerc, J.; Becker, A. Adaptive beamforming using scattering from a drone swarm. In Proceedings of the 2020 IEEE Texas Symposium on Wireless and Microwave Circuits and Systems: Making Waves in Texas, WMCS 2020, Waco, TX, USA, 26–28 May 2020. [CrossRef]
15. Mann, S.; Pierce, C.; Hernandez, J.; Li, Q., Zheng, B.C.; Xin Xiang, Y. Drone Swarms for Sensing-of-Sensing. *Proc. IEEE Sens.* **2019**, *2019*, 23–26. [CrossRef]
16. Sahingoz, O.K. Networking models in flying Ad-hoc networks (FANETs): Concepts and challenges. *J. Intell. Robot. Syst. Theory Appl.* **2014**, *74*, 513–527. [CrossRef]
17. Guo, Y.; Li, X.; Yousefi'zadeh, H.; Jafarkhani, H. UAV-aided cross-layer routing for MANETs. In Proceedings of the IEEE Wireless Communications and Networking Conference, WCNC, Paris, France, 1–4 April 2012; pp. 2928–2933. [CrossRef]
18. Kuiper, E.; Nadjm-Tehrani, S. Mobility models for UAV group reconnaissance applications. In Proceedings of the 2006 International Conference on Wireless and Mobile Communications (ICWMC'06), Bucharest, Romania, 29–31 July 2006; pp. 2–8. [CrossRef]
19. Camp, T.; Boleng, J.; Davies, V. A survey of mobility models for ad hoc network research. *Wirel. Commun. Mob. Comput.* **2002**, *2*, 483–502. [CrossRef]
20. Guillen-Perez, A.; Cano, M.D. Flying ad hoc networks: A new domain for network communications. *Sensors* **2018**, *18*, 3571. [CrossRef] [PubMed]
21. Broch, J.; Maltz, D.A.; Johnson, D.B.; Hu, Y.C.; Jetcheva, J. A performance comparison of multi-hop wireless ad hoc network routing protocols, In Proceedings of the 4th Annual ACM/IEEE International Conference on Mobile Computing and Networking New York, NY, USA, 25 October 2001; pp. 85–97. [CrossRef]
22. Reed, J.M.; Abdallah, A.S.; Thompson, M.S.; MacKenzie, A.B.; DaSilva, L.A. The FINS Framework: Design and Implementation of the Flexible Internetwork Stack (FINS) Framework. *IEEE Trans. Mob. Comput.* **2016**, *15*, 489–502. [CrossRef]
23. Thompson, M.; Abdallah, A.; Reed, J.; MacKenzie, A.; Dasilva, L. The FINS framework: An open source userspace networking subsystem for linux. *IEEE Netw.* **2014**, *28*, 32–37. [CrossRef]
24. Abdallah, A.S.; Horvath, M.D.; Thompson, M.S.; MacKenzie, A.B.; Dasilva, L.A. Facilitating experimental networking research with the FINS framework. In Proceedings of the Annual International Conference on Mobile Computing and Networking, MOBICOM, New York, NY, USA, 19 September 2011; pp. 103–104. [CrossRef]

25. Miya, T.; Ohshima, K.; Kitaguchi, Y.; Yamaoka, K. Experimental Analysis of Communication Relaying Delay in Low-Energy Ad-hoc Networks *arXiv* **2020**, arXiv:2010.15572.
26. Kaysina, I.A.; Vasiliev, D.S.; Abilov, A.; Meitis, D.S.; Kaysin, A.E. Performance evaluation testbed for emerging relaying and coding algorithms in Flying Ad Hoc Networks. In Proceedings of the Moscow Workshop on Electronic and Networking Technologies, MWENT, Moscow, Russia, 14–16 March 2018; pp. 1–5. [CrossRef]
27. Kim, G.H.; Nam, J.C.; Mahmud, I.; Cho, Y.Z. Multi-drone control and network self-recovery for flying Ad Hoc Networks. In Proceedings of the International Conference on Ubiquitous and Future Networks, ICUFN, Vienna, Austria, 5–8 July 2016; pp. 148–150. [CrossRef]
28. Shan, F.; Luo, J.; Xiong, R.; Wu, W.; Li, J. Looking before Crossing: An Optimal Algorithm to Minimize UAV Energy by Speed Scheduling with a Practical Flight Energy Model. In Proceedings of the IEEE INFOCOM, Toronto, ON, Canada, 6–9 July 2020; pp. 1758–1767. [CrossRef]
29. Hamdaoui, B.; Sinky, H. Optimized link state routing for quality-of-service provisioning: Implementation, measurement, and performance evaluation. *Wirel. Commun. Mob. Comput.* **2015**, *14*, 421–430. [CrossRef]
30. Brown, T.; Doshi, S.; Jadhav, S.; Henkel, D.; Thekkekunnel, R. A full scale wireless ad hoc network test bed. *Access* **2005**, *7*, 51–60.
31. Rosati, S.; Kruzelecki, K.; Heitz, G.; Floreano, D.; Rimoldi, B. Dynamic Routing for Flying Ad Hoc Networks. *IEEE Trans. Veh. Technol.* **2016**, *65*, 1690–1700. [CrossRef]
32. Karp, B.; Kung, H.T. Greedy Perimeter Stateless Routing for Wireless Networks(GPSR). In Proceedings of the Annual International Conference on Mobile Computing and Networking, MOBICOM, Boston, MA, USA, 1 August 2000. [CrossRef]
33. Medina, D.; Hoffmann, F.; Rossetto, F.; Rokitansky, C.H. North atlantic inflight internet connectivity via airborne mesh networking. In Proceedings of the IEEE Vehicular Technology Conference, San Francisco, CA, USA, 5–8 September 2011. [CrossRef]
34. Lin, L.; Sun, Q.; Wang, S.; Yang, F. A geographic mobility prediction routing protocol for Ad Hoc UAV Network. In Proceedings of the 2012 IEEE Globecom Workshops, GC Wkshps 2012, Anaheim, CA, USA, 3–7 December 2012; pp. 1597–1602. [CrossRef]
35. Alshabtat, A.I.; Dong, L. Low latency routing algorithm for unmanned aerial vehicles ad-hoc networks. *World Acad. Sci. Eng. Technol.* **2011**, *80*, 705–711.
36. Aguilar, T.; Ghedira, M.C.; Syue, S.J.; Gauthier, V.; Afifi, H.; Wang, C.L. A Cross-Layer Design Based on Geographic Information for Cooperative Wireless Networks. In Proceedings of the 2010 IEEE 71st Vehicular Technology Conference, Taipei, Taiwan, 16–19 May 2010; pp. 1–5. [CrossRef]
37. Rosati, S.; Kruzelecki, K.; Traynard, L.; Rimoldi, B. Speed-aware routing for UAV ad-hoc networks. In Proceedings of the 2013 IEEE Globecom Workshops, GC Wkshps 2013, Atlanta, GA, USA, 9–13 December 2013; pp. 1367–1373. [CrossRef]
38. Liu, J.; Ji, X.; Huang, C.; Tan, X. A novel OLSR protocol with mobility prediction. In Proceedings of the Third International Conference on Cyberspace Technology (CCT 2015), Beijing, China, 17–18 October 2015. [CrossRef]
39. José, D.A.M.; Bulcão-Neto, R.F.; Patto, V.S.; Sene Júnior, I.G. OLSR fuzzy cost (OLSR-FC): An extension to olsr protocol based on fuzzy logic and applied to avoid selfish nodes. *Rev. Inform. Teórica Apl.* **2019**, *26*, 60–77. [CrossRef]
40. Fan, X.; Lin, J.; Zhang, C.; Zhang, S. A cross-layer anti-jamming routing protocol for FANETS. In Proceedings of the 2018 IEEE 4th International Conference on Computer and Communications, ICCC 2018, Chengdu, China, 7–10 December 2018; pp. 301–305. [CrossRef]
41. Jain, R.; Kashyap, I. An QoS Aware Link Defined OLSR (LD-OLSR) Routing Protocol for MANETS. *Wirel. Pers. Commun.* **2019**, *108*, 1745–1758. [CrossRef]
42. Lavén, A.; Hjärtquist, P. *Multimetric OLSR and ETT Case Study Final Report Multimetric OLSR and ETT 1 Introduction*; Karlstad University: Karlstad, Sweden, 2014.
43. Perkins, C.E.; Royer, E.M. Ad-hoc on-demand distance vector routing. In Proceedings of the WMCSA'99: 2nd IEEE Workshop on Mobile Computing Systems and Applications, New Orleans, LA, USA, 25–26 February 1999; pp. 90–100. [CrossRef]
44. Perkins, C.E.; Watson, T.; Bhagwat, P. Highly Dynamic Destination-Sequenced Distance-Vector Routing (DSDV) for Mobile Computers. *ACM SIGCOMM Comput. Commun. Rev.* **1994**, *24*, 234–244. [CrossRef]
45. Purnomo, A.; Widyawan; Najib, W.; Hartono, R.; Hartatik. Performance Comparison of Modified AODV-ETX with AODV and AODV-ETX Routing Protocol in an MANET. *IOP Conf. Ser. Mater. Sci. Eng.* **2019**, *578*, 012082. [CrossRef]
46. Hou, S.; Wu, M.; Liao, W.; Wang, D. Performance comparison of AODV and DSR in MANET test-bed based on Internet of Things. In Proceedings of the 2015 IEEE 82nd Vehicular Technology Conference, VTC Fall 2015, Boston, MA, USA, 6–9 September 2015; pp. 3–7. [CrossRef]
47. Jung, S.; Kim, B.; Kim, K.; Roh, B.; Ham, J. Implementation of AODV-UU on Linux 4.15 Kernel. In Proceedings of the 2019 IEEE 16th International Conference on Mobile Ad Hoc and Smart Systems Workshops, MASSW 2019, Monterey, CA, USA, 4–7 November 2019; pp. 160–161. [CrossRef]
48. Nayyar, A. Flying Adhoc Network (FANETs): Simulation Based Performance Comparison of Routing Protocols: AODV, DSDV, DSR, OLSR, AOMDV and HWMP. In Proceedings of the 2018 International Conference on Advances in Big Data, Computing and Data Communication Systems, icABCD 2018, Durban, South Africa, 6–7 August 2018; pp. 1–9. [CrossRef]
49. Chissungo, E.; Blake, E.; Le, H. Performance Comparison of B.A.T.M.A.Nd and B.A.T.M.A.N-adv. In Proceedings of the Southern Africa Telecommunication Networks and Applications Conference, East London, South Africa, 3–7 September 2011; pp. 6–7.
50. Lagkas, T.; Imeri, A.; Eleftherakis, G. *A Pragmatic Evaluation of Distance Vector Proactive Routing in MANETs via Open Space Real-World Experiments*; Springer: Cham, Switzerland, 2015.

51. Chissungo, E.; Blake, E.; Le, H. Investigation into Batmand-0.3.2 protocol performance in an indoor Mesh Potato testbed. In Proceedings of the 26th IEEE International Conference on Advanced Information Networking and Applications Workshops, WAINA 2012, Fukuoka, Japan, 26–29 March 2012; pp. 526–532. [CrossRef]
52. Wang, W.; Guan, X.; Wang, B.; Wang, Y. A novel mobility model based on semi-random circular movement in mobile ad hoc networks. *Inf. Sci.* **2010**, *180*, 399–413. [CrossRef]
53. Meghanathan, N. Impact of the Gauss-Markov mobility model on network connectivity, lifetime and hop count of routes for mobile Ad hoc networks. *J. Netw.* **2010**, *5*, 509–516. [CrossRef]
54. Bouachir, O.; Abrassart, A.; Garcia, F.; Larrieu, N. A mobility model for UAV ad hoc network. In Proceedings of the 2014 International Conference on Unmanned Aircraft Systems, ICUAS 2014, Orlando, FL, USA, 27–30 May 2014; pp. 383–388. [CrossRef]
55. Barolli, A.; Oda, T.; Barolli, L.; Takizawa, M. Experimental results of a raspberry Pi and OLSR based wireless content centric network testbed considering OpenWRT OS. In Proceedings of the International Conference on Advanced Information Networking and Applications, AINA, Crans-Montana, Switzerland, 23–25 March 2016; pp. 95–100. [CrossRef]
56. Chao, H.; Cao, Y.; Chen, Y. Autopilots for Small Fixed-Wing Unmanned Air Vehicles: A Survey. In Proceedings of the 2007 International Conference on Mechatronics and Automation, Harbin, China, 5–8 August 2007; Technical Report.
57. Amponis, G. Mesh Module GitHub Repository. 2020. Available online: https://github.com/g-ampo/mesh_module (accessed on 27 December 2020).
58. LibreCube. LibreCube PC/104 Template. 2019. Available online: https://gitlab.com/librecube/support/templates/librecube-board (accessed on 27 December 2020).

drones

Technical Note

Geo-Referenced Mapping through an Anti-Collision Radar Aboard an Unmanned Aerial System

Lapo Miccinesi, Luca Bigazzi, Tommaso Consumi, Massimiliano Pieraccini *, Alessandra Beni, Enrico Boni and Michele Basso

Department of Information Engineering, University of Florence, Via Santa Marta 3, 50139 Firenze, Italy; lapo.miccinesi@unifi.it (L.M.); luca.bigazzi@unifi.it (L.B.); tommaso.consumi@unifi.it (T.C.); alessandra.beni@unifi.it (A.B.); enrico.boni@unifi.it (E.B.); michele.basso@unifi.it (M.B.)
* Correspondence: massimiliano.pieraccini@unifi.it

Abstract: Unmanned aerial systems (UASs) have enormous potential in many fields of application, especially when used in combination with autonomous guidance. An open challenge for safe autonomous flight is to rely on a mapping system for local positioning and obstacle avoidance. In this article, the authors propose a radar-based mapping system both for obstacle detection and for path planning. The radar equipment used is a single-chip device originally developed for automotive applications that has good resolution in azimuth, but poor resolution in elevation. This limitation can be critical for UAS application, and it must be considered for obstacle-avoidance maneuvers and for autonomous path-planning selection. However, the radar-mapping system proposed in this paper was successfully tested in the following different scenarios: a single metallic target in grass, a vegetated scenario, and in the close proximity of a ruined building.

Keywords: anti-collision radar; autonomous flight unmanned aerial system; obstacle detection; obstacle avoidance; radar mapping; unmanned aerial system (UAS); unmanned aerial vehicle (UAV)

Citation: Miccinesi, L.; Bigazzi, L.; Consumi, T.; Pieraccini, M.; Beni, A.; Boni, E.; Basso, M. Geo-Referenced Mapping through an Anti-Collision Radar Aboard an Unmanned Aerial System. *Drones* **2022**, *6*, 72. https://doi.org/10.3390/drones6030072

Academic Editors: Diego González-Aguilera and Pablo Rodríguez-Gonzálvez

Received: 31 January 2022
Accepted: 7 March 2022
Published: 9 March 2022

1. Introduction

Recent technological advances made in unmanned aerial systems (UASs), especially in combination with autonomous flights, could have a major impact on civilian tasks [1]. For instance, they could not only enable the provision of social services, such as the delivery of goods, drugs and sanitary equipment, but could also perform environmental monitoring in remote regions [2]. Since these applications would involve flights performed in complex outdoor scenarios, capabilities for autonomous obstacle detection and avoidance are of paramount importance. Researchers have recently devoted great effort to the challenging task of autonomous UAS flights [3].

An open challenge for autonomous flights is to develop a reliable mapping system of the surrounding environment. A mapping procedure of the surrounding area would not only enable the detection and avoidance of obstacles, but also the elaboration of effective decommitment strategies.

Currently, vision systems are often used with this aim [4–6]. However, their maximum detectable distance (usually lower than tens of meters) is a serious drawback [7], which does not make them the optimal solution for mapping outdoor environments. Indeed, for a UAS flying at a relatively high speed, this distance may not be sufficient for implementing a suitable decommitment strategy. Moreover, optical sensors are dramatically affected by light exposure and weather conditions.

Another way to detect possible obstacles and overcome limitations due to weather conditions is to use radar equipment. In fact, radars allow for the extension of the maximum distance of detection and are only slightly affected by environmental conditions (e.g., light, fog and rain). Given the advantages provided by radar sensors, many research groups have already worked in this direction.

In [8] Sacco G. et al. proposed a MISO (multiple-input single-output) system based on an FMCW radar, which worked at 24 GHz and was specifically optimized for drone detection. The transmitting and receiving serial arrays of patch antennas have been suitably designed to operate up to 150 m. Experimental tests performed on the ground in a controlled scenario evidenced the correct estimation of the target position in the range–azimuth plane.

Recently, a new class of radars has been developed for obstacle avoidance in automotive applications. These radars are single-chip devices that work at high frequency (W-band [9]). Today, these sensors are largely employed in the automotive field, but also for indoor mapping through terrestrial rovers.

S. Dogru et al. [10] studied the mapping performance of a radar in indoor environments to support mobile robots employed in search and rescue operations in low visibility areas. They used an FMCW radar working in the frequency range 76–81 GHz, with two transmitting and three receiving antennas to construct two-dimensional maps of the surrounding area. The radar was mounted on a robot, which moved across the investigated area. The quality of radar-based maps was compared to that of light detection and ranging (LIDAR) maps by using a quantitative map-quality metric. Their results evidence that even though LIDAR still outperforms radar, when mapping low visibility environments, i.e., with fog or smoke, radar provides better results in terms of mapping.

In [11], S. Lee et al. illustrated the mapping results of an indoor environment obtained with a dual-mode radar. The sensor was a multiple-input multiple-output (MIMO) system working at a central frequency of 62 GHz. It was capable of alternatively transmitting two waveforms with different bandwidths, optimized for long-range and short-range detection, respectively. The radar was mounted on a robot that moved in the area to be mapped and sent information about its own position. By combining the radar data with the information for the robot's position, they successfully constructed a map of the surrounding environment.

The small size and light weight of these millimeter-wave radars make them promising candidates for operating aboard a drone. Some research groups have already investigated this possibility [12–14]. These works also present strategies to overcome and mitigate the problem of poor elevation resolution, a common characteristic of these radars.

Authors of [12] implemented an active drone detection system. They mounted a millimeter-wave radar on a drone with the aim of detecting, tracking, and pursuing other target drones. Although the tracking radar performed only 2D measurements, the 3D data are recovered by complex maneuvering the pursuer drone. Despite the promising results, the proposed technique has limited applications as it assumes the target drone as the only other airborne object and requires the drone to perform specific complex maneuvers.

A millimeter-wave MIMO radar capable of three-dimensional sensing for applications in UAS formation flights and obstacle avoidance is presented in [13]. Specifically, commercially derived millimeter-wave radar technology was integrated with a custom MIMO antenna array that was optimized for specific flight dynamics. The results of the experimental test were promising, as the system confirmed the three-dimensional detection capabilities of the investigated target.

Authors of [14] present a system for mapping the environment surrounding a UAS flight that is based on the fusion of a millimeter-wave radar and a monocular camera. Specifically, monocular camera data are used to provide a reference for mapping, and to identify targets in the surrounding environment. Then, by using an extended Kalman filter, the radar data are fused with that of the camera, thus enabling the local mapping of targets. A possible drawback of this system is that the reference used for mapping is provided by the monocular camera itself. Therefore, in certain scenarios, the system may lose this reference, thus, leading to possible errors, which may compromise the mapping process.

Therefore, the aim of this article is to present some preliminary tests of a radar aboard a UAS that is able to both operate as obstacle-avoidance equipment, and is able to provide reliable mapping of its surrounding area that could be used to elaborate effective decommitment strategies. This is a challenging goal that other radar systems presented in scientific literature are not designed to accomplish.

The single-chip radar, mounted aboard the drone, provides the direction of arrival (DOA) of the scattered signal [15], and the detected targets can be correctly disposed on a local map by using the pose (position and attitude) provided by the UAS flight controller. The system was developed in the robot operating system (ROS) environment, so as to enable integration with other sensors [16]. The complete system was experimentally tested in a controlled scenario, with a single target, and in the following complex scenarios: in a wood and in close proximity to a ruined building. This article also proposes a path-planning strategy for taking into account the poor elevation resolution of this kind of radar.

2. Materials and Methods

The mapping system was developed in the ROS environment in order to be easily integrated with other sensors or to be implemented on different UAS platforms. A block scheme of ROS architecture is shown in Figure 1. The key task is the *3D mapping* node, which converts the position of detected targets from the radar frame to the fixed frame and processes the voxel map. This node communicates with the physical sensors through the *Radar* node and *Telemetry* node. The *Radar* node handles the radar and publishes the information about the detected targets on a proper topic (for example position, speed, signal amplitude). The drone interfaces with the *3D mapping* node through the *Telemetry* node. This node publishes real time kinematic (RTK) position and attitude (quaternion) in standard ROS-telemetry messages. It is noteworthy that the whole system is designed to work in real time, so it is able to provide timely alerts or the re-planning of paths.

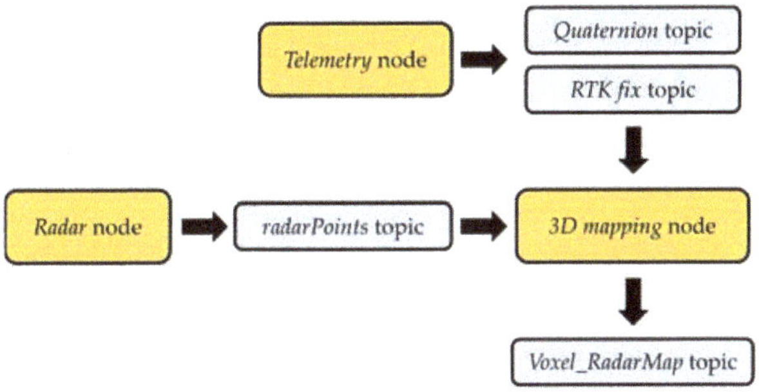

Figure 1. Block scheme of ROS architecture.

The functional scheme is shown in Figure 2. This scheme is independent from the specific drone, as it allows for the system to be easily changed by maintaining the same ROS messages. Therefore, the mapping system could be implemented on different UASs with minor changes.

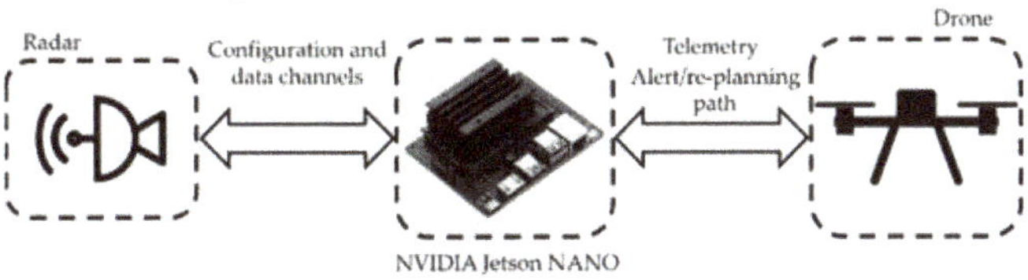

Figure 2. Functional scheme of the proposed architecture.

The ROS workspace was implemented on a NVIDIA Jetson Nano board. This board was connected to the drone to obtain the telemetry data through the *Telemetry* node, and to provide alerts or path re-planning to the drone pilot (not yet implemented). The radar was connected to the computer board, and it was controlled by the *Radar* node. The *Radar* node was used for configuring and receiving data from the radar. The NVIDIA Jetson Nano was also used for the *3D mapping* node. Indeed, the 3D mapping algorithm could be very expensive in computational terms.

2.1. 3D Mapping Node

The 3D map was created in the *3D mapping* ROS node. Within this node, both the Telemetry messages, coming from the drone, and the position of detected objects, from the radar, were received. Each object's position was moved from the radar frame to the fixed frame. Subsequently, it was mapped in a voxel map using the OctoMap library [5,17]. Indeed, the position of the detected targets was referred to the radar frame as shown in Figure 3.

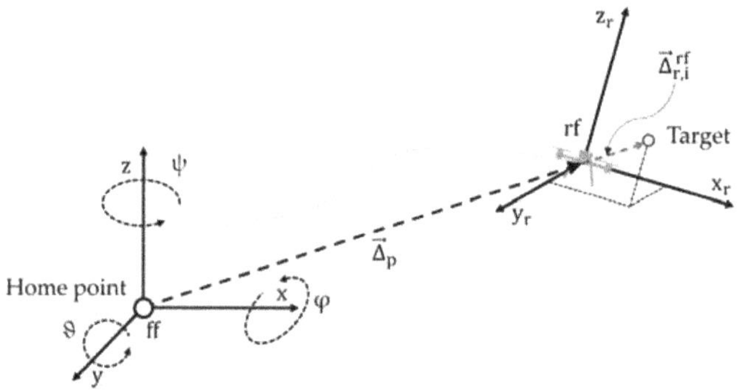

Figure 3. The image shows the two reference frames, where ff is the fixed frame and rf represents the radar frame.

The fixed reference frame has its origin corresponding to the take-off point (or home point). The orientation of the fixed frame is also coherent with the attitude quaternion provided by the *Telemetry* node during the initialization phase.

The radar reference frame is fixed with respect to the drone as shown in Figure 3. This means that the radar frame origin and orientation depend on the traveled trajectory of the drone during the mission.

First, to correctly map the radar target in the fixed frame, it is necessary to obtain the local path travelled by the drone during the mission in cartesian coordinates. Indeed, the Global Navigation Satellite System (GNSS) provides the position in terms of latitude and longitude. Equation (1) defines the difference between the latitudinal and longitudinal coordinates of the take-off point and the current ones.

$$\begin{cases} \Delta_\alpha = \alpha_c - \alpha_h \\ \Delta_\beta = \beta_c - \beta_h \end{cases} \tag{1}$$

where α_c and α_h are the latitude of the current position and the home point, while β_c and β_h represent the longitude of the current position and the home point, respectively.

To convert the trajectory from latitude-longitude coordinates into cartesian coordinates, it is possible to use the relations (2).

$$\vec{\Delta}_p = \begin{cases} \Delta_x = r \times \sin \Delta_\beta \times \cos \alpha_c \\ \Delta_y = r \times \sin \Delta_\alpha \\ \Delta_z = z_c - z_h \end{cases} \tag{2}$$

where r is the average radius of the Earth, and $(\Delta_x, \Delta_y, \Delta_z)$ are the coordinates in the fixed frame of Figure 3. The relative height of the drone is given only by the difference between the initial and the current height, which are both provided by the GNSS. In fact, the GNSS system provides height in respect to sea level in meters.

The radar frame and fixed frame also have different orientations. For this reason, we can consider the complete rotation matrix, $R_{xyz}(\vartheta, \varphi, \psi)^{-1}$, to orient the detected targets as a function of the fixed frame:

$$R_{xyz}(\vartheta, \varphi, \psi)^{-1} = R_{xyz}(\vartheta, \varphi, \psi)^{T} = \begin{bmatrix} c_\vartheta c_\psi & -c_\vartheta s_\psi & s_\vartheta \\ c_\varphi s_\psi + s_\vartheta c_\psi s_\varphi & c_\varphi c_\psi - s_\varphi s_\vartheta s_\psi & -s_\varphi c_\vartheta \\ s_\varphi s_\psi - c_\varphi s_\vartheta c_\psi & s_\varphi c_\psi + c_\varphi s_\vartheta s_\psi & c_\varphi c_\vartheta \end{bmatrix}^{T} \tag{3}$$

where ϑ, φ and ψ are the pitch angle, the roll, and the heading angle, respectively, and the symbols c and s are the cosine and sine of the respective angle. This rotation matrix was evaluated by considering the axis of the selected drone according to the right-hand rule.

General speaking, the radar is not able to provide both the azimuth and the elevation of the target. In this specific case we decided to use the azimuth resolution and to always consider zero as the elevation of the object ($z_{r,i} = 0$ m). This is equivalent to assuming that each target is on a horizontal plane at the same height of the drone. This hypothesis is not as strong as it seems, because usually the target at $z_{r,i} = 0$ m is the most reflective.

Under this hypothesis the rotation in (3) can be reduced to a matrix that considers only the rotation along the heading angle:

$$R_z(\psi)^{-1} = R_z(\psi)^{T} = \begin{bmatrix} c_\psi & -s_\psi & 0 \\ s_\psi & c_\psi & 0 \\ 0 & 0 & 1 \end{bmatrix}^{T} \tag{4}$$

where c_ψ and s_ψ are the cosine and sine of the yaw angle.

Finally, to obtain the coordinates of the targets referred to the fixed frame, it is necessary to consider the rotation (4) and the offset $\vec{\Delta}_p$. Equation (5) shows the complete equation of the desired target position in fixed frame, $\vec{\Delta}^{ff}_{r,i}$:

$$\vec{\Delta}^{ff}_{r,i} = R_z(\psi)^{-1} \times \vec{\Delta}^{rf}_{r,i} + \vec{\Delta}_p \tag{5}$$

where $\vec{\Delta}^{rf}_{r,i}$ is the object coordinates in radar frame:

$$\vec{\Delta}^{rf}_{r,i} = \begin{bmatrix} x_{r,i} \\ y_{r,i} \\ z_{r,l} \end{bmatrix} \tag{6}$$

The target coordinates obtained in (5) could be located on the same map using OctoMap library. This library allows the creation of a voxel map, where each voxel has the coordinates of the detected target defined in the fixed frame. Since the OctoMap library provides functions to search for occupied points, it is possible to use the radar map as an anti-collision and obstacle-avoidance system.

2.2. The Radar Sensor

The radar used for this article is an AWR1843BOOST by Texas Instruments [18] (Figure 4). A radar detects the distance of the target by sending and receiving an electromagnetic signal through at least a couple of antennas. Using a MIMO array, it is also able to retrieve the direction of arrival.

Figure 4. AWR1843BOOST by Texas Instruments, US.

The sensor used is comprised of 3 TX and 4 RX antennas, which correspond to 12 virtual antennas, disposed as shown in Figure 5. In Figure 5, the z axis represents the altitude, while the x axis is left to right, and λ is the wavelength of the electromagnetic signal. This arrangement of virtual antennas achieves a good azimuth resolution and a poor elevation resolution [19]. Indeed, resolution is related to the inverse of the z-distance between the antennas.

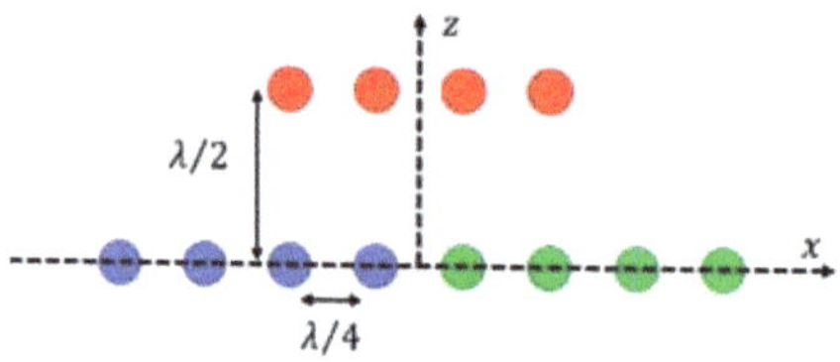

Figure 5. Virtual antennas position of AWR1843BOOST [18].

For the current application, the elevation resolution is used only as angular cut-off. In other words, the value of elevation measured by the radar is not used for mapping and it was fixed equal to zero for each target, but it is used as a spatial filter for rejecting the target outside a selected angular area. We set the angular field as ± 45 deg in azimuth and [0, 20] deg in elevation. Therefore, all targets outside this interval are not used for mapping.

The radar provides a frequency-modulated signal from 77 GHz to 81 GHz (the whole bandwidth was not used for the experiments) from each TX antenna and registers the echo from all RX as Figure 6 shows. The frequency sweeps (chirps) are collected in a structure called frame.

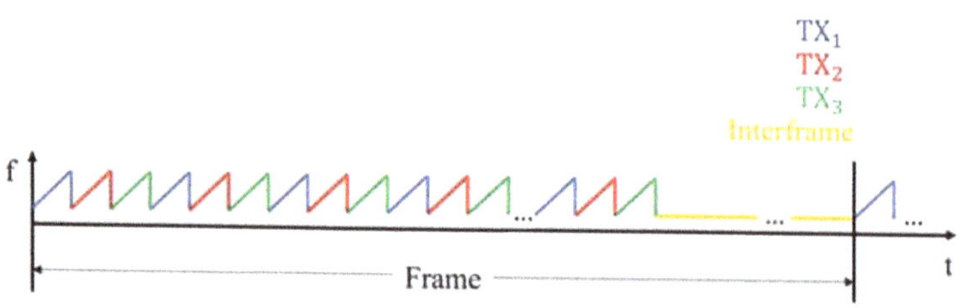

Figure 6. Example of a signal provided by AWR1843BOOST [18].

The radar is able to retrieve the position of detected targets using an internal computational unit. Indeed, all the operations in Figure 7 were carried out by the hardware and the library onboard the radar. For each single chirp, the range of the fast Fourier transform (FFT) is calculated for retrieving the distance of possible targets. All the range

FFTs are collected in a matrix. Another FFT was calculated along the doppler direction of the range-FFT matrix. This FFT, called doppler FFT, is used for detecting the target speed. Here, the targets are confused with clutter and noise. To discriminate the targets from false alarms, a two constant false alarm rate (CFAR) processes is applied. The CFAR is an algorithm that, using a threshold, compares the amplitude of each single pixel with the average amplitude of the neighbors [20,21] and selects the pixels over the threshold. For range direction, we used a cell-average smallest of (CASO)-CFAR algorithm [21]. Indeed, the CASO-CFAR is particularly suitable for detecting objects surrounded by clutter. For the doppler direction, a cell-average (CA)-CFAR algorithm [20] was used. The CA-CFAR in doppler direction allows for the selection of targets with well-defined speed. Finally, the position of targets is estimated by considering the antenna pattern. This estimation is performed on the target that exceeds the threshold of both CFAR algorithms to reduce computational load.

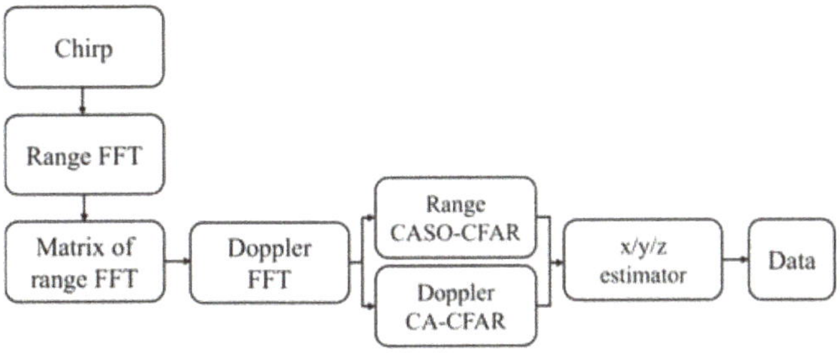

Figure 7. Detection chain.

For each detected target, the radar provides x/y/z position, speed, signal-to-noise ratio of CFAR and target index.

The radar was connected to the NVIDIA Jetson Nano through a USB cable with the following two separate communication channels: a configuration port and a data port. The *Radar* node publishes a topic for each detected point with all the information reported above and the time stamp of the frame. It is important to note that the topic is published only if at least one target is detected. The radar topic is subscribed by the mapping node, as described in Section 2.1.

2.3. The UAS and Telemetry Node

As explained in Section 2.1, the radar map is produced by considering the GNSS position of the drone. For a reliable map, the authors used a real-time kinematic positioning (RTK) system. Indeed, this system is able to provide the position with an accuracy of about 50 mm, which is enough for radar mapping (the typical range resolution is about 500 mm).

The GNSS-RTK was equipped on a DJI Matrice300RTK and provides the position at 5 Hz. The drone also provides the filtered quaternion, at 100 Hz, for retrieving the attitude.

This drone can communicate with an external computer using an UART port and the C++/ROS library provided by DJI. Using this library, it is possible to access navigation data (telemetry, battery, and navigation status, etc.) and to implement some basic piloting commands, e.g., it is possible to set new waypoints.

The NVIDIA Jetson Nano, with the DJI library, was installed on the drone as shown in Figure 8. The radar was located below the drone, and it was locked to maintain the same orientation as the drone.

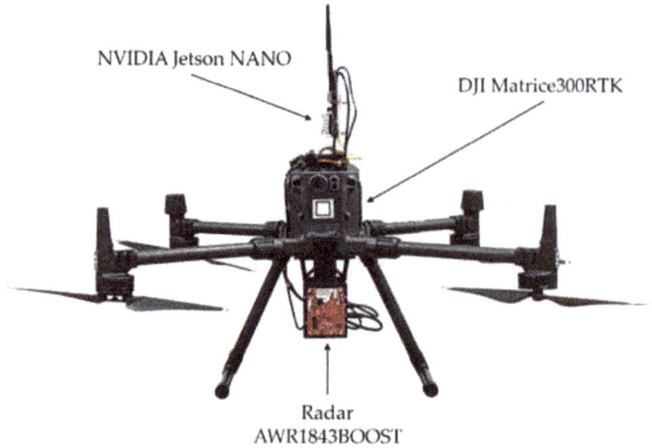

Figure 8. DJI Matrice300RTK provided by the radar AWR1843BOOST and by NVIDIA Jetson Nano.

The *Telemetry* node is a modified version of the one provided by DJI. Indeed, we have generalized the DJI telemetry messages by converting them into standard telemetry ROS messages.

2.4. Path-Planning Strategy

Using this mapping system, it is also possible to design an avoidance strategy based on radar data. Furthermore, as the map is georeferenced, it could be used also for planning the return path by considering the obstacles detected in the outbound flight. Figure 9 shows the proposed obstacle-avoidance strategy, based on radar map. There are two possible maneuvers, as follows: (1) go around sideways, (2) fly over (see Figure 9). Fly below is a forbidden maneuver. A word of caution needs to be given regarding the second maneuver (fly over). The radar has poor elevation resolution, so for prudential reasons we assume that any detected target is positioned at the same altitude of the drone. When the drone increases the altitude, it continues to detect the obstacle as long as it is inside the vertical view-angle of the radar. Therefore, a single target could be represented by a sort of vertical column, but this is not a problem in path planning.

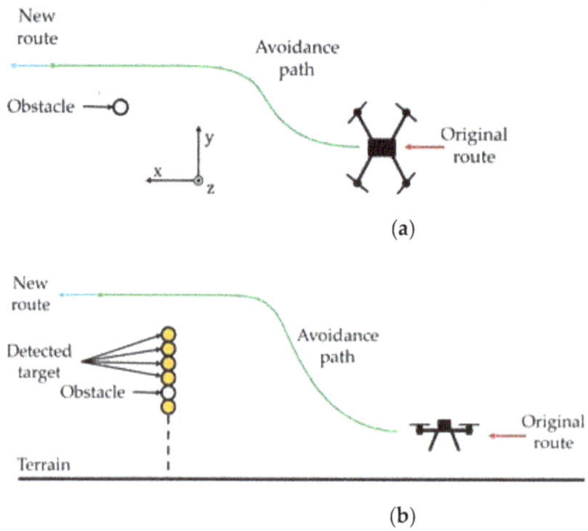

Figure 9. Proposed avoidance strategies: (**a**) go around sideways, (**b**) fly over.

3. Results

The equipment was tested in a controlled scenario with only one electromagnetic target (a naval corner reflector) and in the following two realistic scenarios: a small wooded area and an almost ruined building.

The radar parameters were the same for all the tests. The maximum range was fixed at 120 m, with 0.5 m of range resolution. The azimuth angle was within ±45°, and the elevation within [0, 20] deg. This value was selected to filter out possible artifacts due to radar side lobes or other sources of noise. The maximum speed that the radar was able to detect was ±10.4 m/s, with speed resolution of 0.31 m/s. The frame periodicity was 10 Hz, but the radar shares its topic only if at least one target is detected.

3.1. Controlled Scenario

Figure 10 shows the setup used during the test in a controlled scenario. The target was at 68 m in front of the drone. It is important to note that the target was located on a small hill and difference of the altitude was about 3 m. The drone path is shown in Figure 11. The drone flew towards the target and made a lateral movement, first going towards the left and after towards the right. After this lateral movement, the drone flew back to the home point. As shown in Figure 11, altitude was also changed during the flight.

Figure 10. Picture of the controlled scenario.

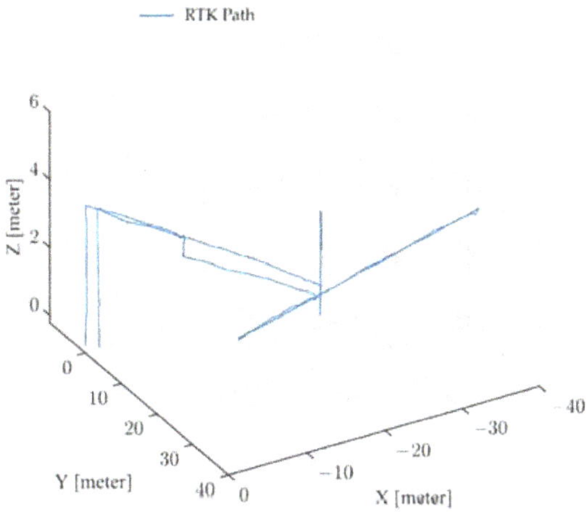

Figure 11. Three-dimensional path of drone in fixed frame.

Figure 12 shows the detected target in the radar frame during the whole flight. We can notice a series of targets that come close to the drone starting form $Y_{radar} = 70$ m to 30 m (spotlight with orange). The group of targets at 30 m corresponds to the lateral movement. These signals in orange area correspond to the naval corner reflector. Hence, it is evident that a single target in a radar frame could be seen as a "wall" in front of the drone.

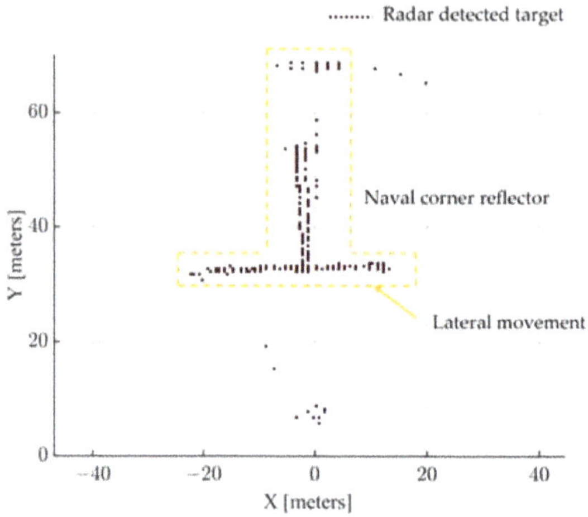

Figure 12. Detected target in radar frame.

The map in Figure 13 was evaluated using (6) and rotation matrix (5). Most of the targets in the fixed frame as shown in Figure 12 are grouped in a cluster that corresponds to the corner reflector. The other sparse targets visible on the map are probably related to the ground.

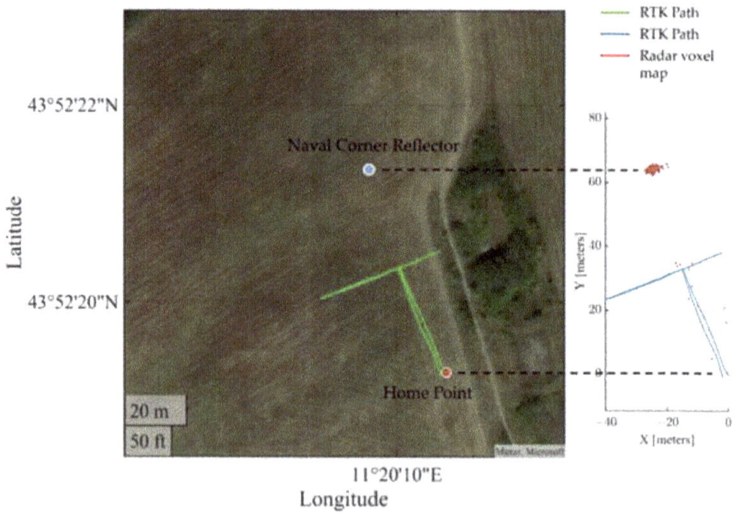

Figure 13. Radar map in fixed frame seen from above.

The vertical profile of the map is shown in Figure 14. As described in Section 2.3, the objects were mapped using only the drone's altitude. It is interesting to note the sort of spurious targets along the path. These are a consequence of the fact that the detected targets are always positioned at the flight altitude, and also when the radar detects ground targets at the lowest edge of the view cone. This is a prudential measure, but it has the unavoidable drawback of producing these false obstacles along the path. This is not a critical problem for the following two reasons: (1) these spurious targets appear only at very low altitudes, where the drone should not normally fly; (2) path planning could avoid these targets without affecting the flight. A possible problem could occur close to the landing zone, which could be interdicted by a great number of these false obstacles. Nevertheless, this radar system is not intended for operating in landing operations (as the radar is pointed to the front and not pointed down) and could be disactivated during the landing maneuver (when other sensors are operating).

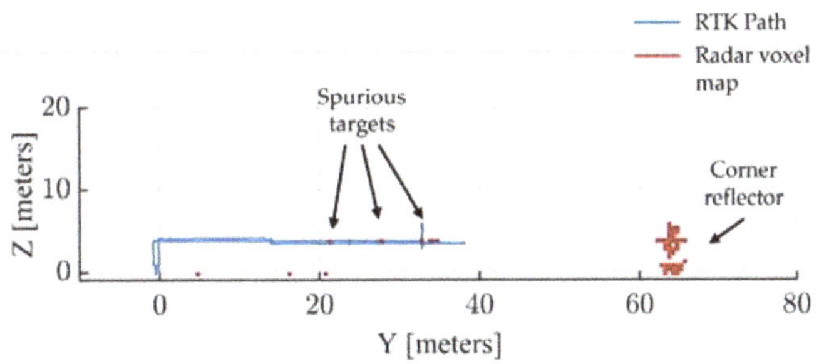

Figure 14. Radar map in fixed frame in Y–Z plane.

As mentioned above, the targets are (conventionally) mapped at the same elevation as the flying UAS. This is a rough assumption, but effective. Indeed, because of the poor angular resolution of the radar, the elevation detected by the radar could even worsen the radar mapping, as shown in Figure 15a. Figure 15a was obtained taking into account both the pitch angle (that UAS provides) and vertical positioning of the detected targets provided by the MIMO. Figure 15b was obtained considering only the pitch angle (that UAS provides), but not the vertical positioning of the detected targets provided by the MIMO. In both cases, the result is a sort of a large swipe of the target (a single naval corner reflector).

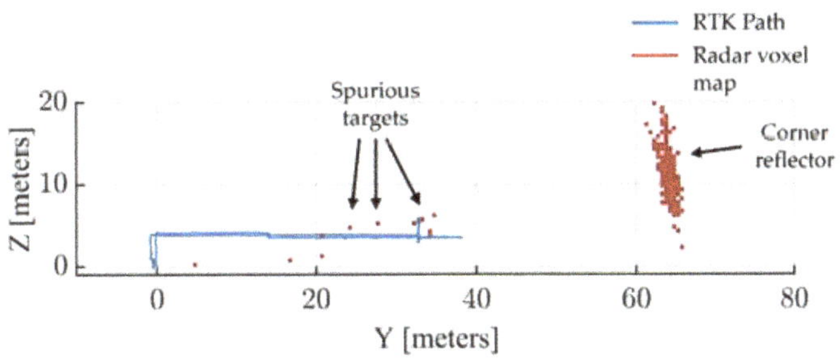

(**a**)

Figure 15. *Cont.*

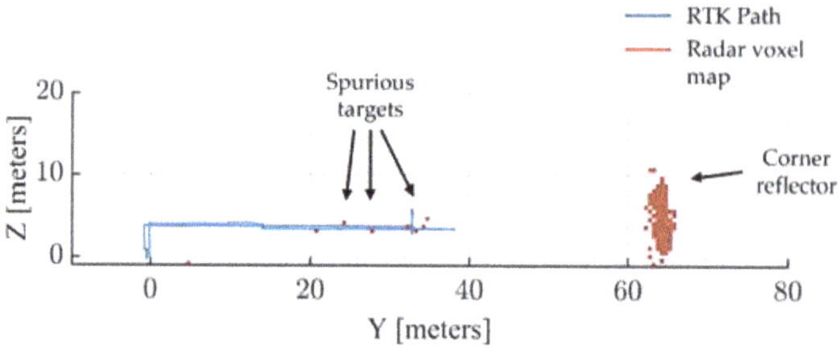

(b)

Figure 15. Radar map in fixed frame in Y–Z plane obtained: (**a**) using (4) as rotation matrix and considering the elevation detected by the radar, and (**b**) using (4) as rotation matrix and $z_{r,i} = 0$.

3.2. Wooded Area

The wooded scenario is shown in Figure 16a. The drone was located about 60 m away from the wood. The path during the test is shown in Figure 16b. We moved towards the wooded area by increasing elevation progressively. When the drone was close to the wood, we moved laterally in order to scan a portion of the wood.

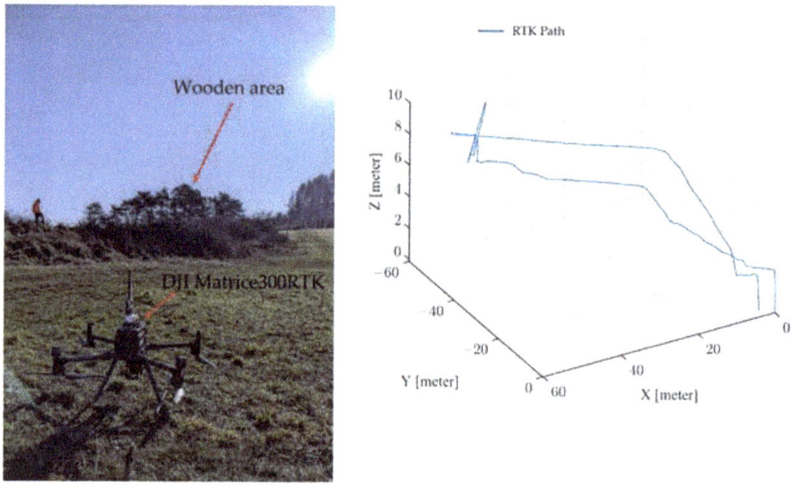

(a) (b)

Figure 16. Picture of the wooded scenario (**a**) and the flight path (**b**).

The mapped area is shown in Figure 17. By comparing the satellite view with the radar map, we can recognize the shape of the wood marked with A. The other sparse targets inside the dotted area correspond to some bushes.

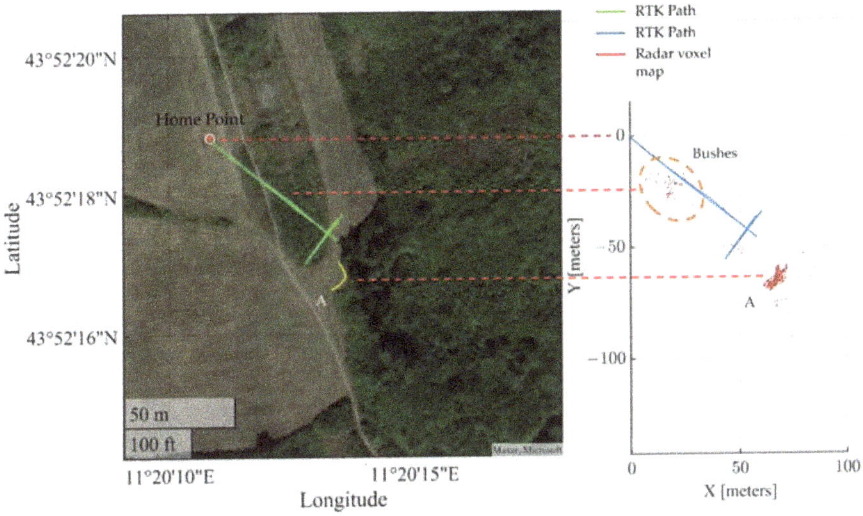

Figure 17. Radar map of the wooded scenario seen from above.

The vertical profile of the map is reported in Figure 18. The bushes were located under the flight altitude.

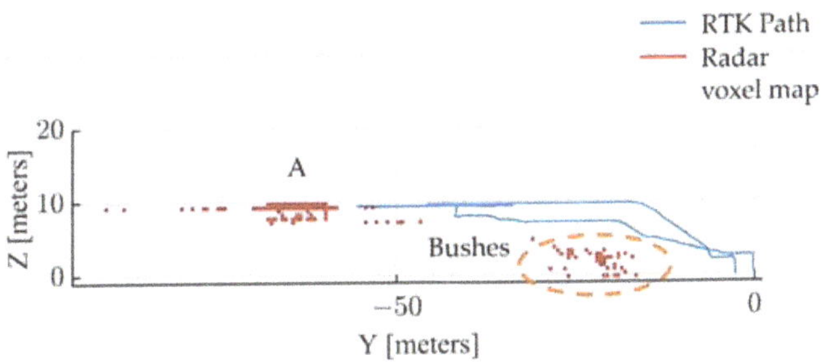

Figure 18. Radar map of wooded scenario in Y–Z plane.

3.3. Ruined Building

The final test was performed on a ruined building. The ruined complex consists of a main building and two separate courtyard buildings (on the left in Figure 19). We performed two flights in this area. During the first flight the drone flew close to the main building and scanned the front face from left to right. When the drone was at the extremity of its lateral path, we rotated the drone in overing mode. During the second flight, the drone flew closer to the front facade.

Figure 19. Picture of the scenario with a ruined building.

The results of these flights are shown in Figures 20 and 21. The front face of the main building is well clear in both cases. In Figure 20 we mapped a portion of the courtyard buildings.

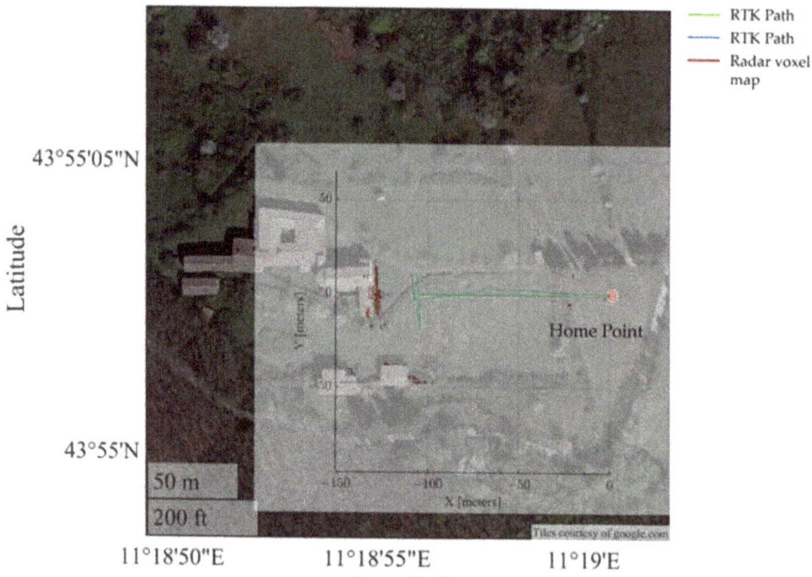

Figure 20. Radar map and trajectory superimposed on an aerial picture (from Google Maps).

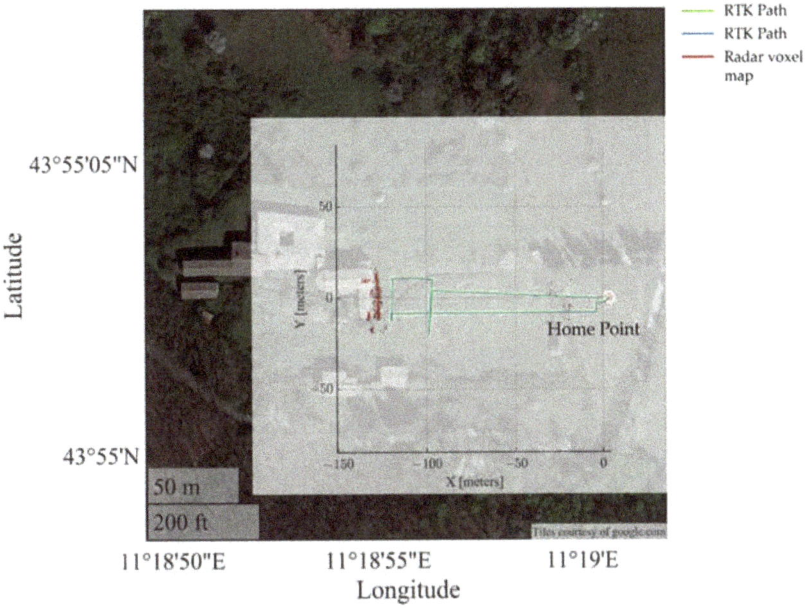

Figure 21. Radar map and trajectory superimposed on an aerial picture (from Google Maps).

4. Discussion

This is the first attempt to perform entirely radar-based mapping for autonomous UAS flights. Unfortunately, the radar is not able to provide obstacle elevation. This may introduce artifacts, i.e., objects mapped at wrong elevations (Figure 15). To overcome this limitation, the authors propose setting the elevation of detected objects equal to the drone's flight altitude. Thus, the object can be avoided by flying around it or by increasing the drone's altitude (Figure 9).

In scientific literature, the radar was used for terrestrial mapping [10,11], or for UAS obstacle detection and avoidance [12,13], but no one used the same radar for both tasks. This is the main achievement of this work. Other authors [14] proposed data fusion between a monocular camera and radar for obstacle avoidance. The latter is an interesting approach that could be a possible development of the present work. Indeed, the idea we are working on is a system that uses radar for long/medium range detection operations and a visual system based on stereoscopic cameras able to provide a 3D map for short range operations. The two systems should operate in cooperation on the same global map. Figure 22 shows a sketch of the complete system, which integrates data from the radar, vision system, global navigation satellite system (GNSS), and inertial measurement unit (IMU).

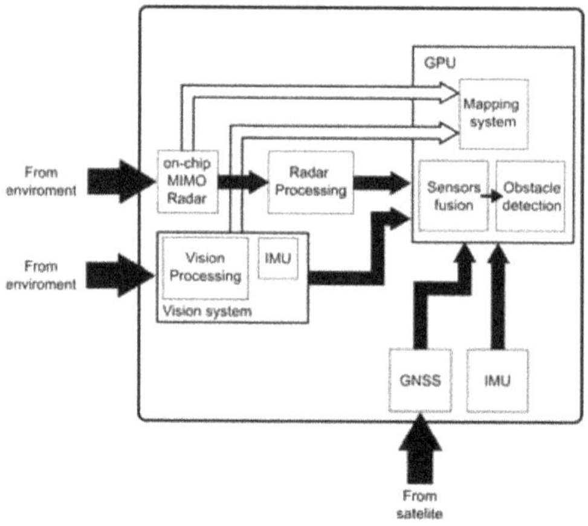

Figure 22. Block scheme of the complete obstacle-avoidance and mapping system that integrates radar and visual data.

5. Conclusions

In this article the authors presented a mapping system using an anti-collision radar. The proposed method allows the creation of 3D maps of the environment in front of the drone. The coordinates of the objects on the flight plane (left/right position and the frontal distance) were provided by the radar, while the elevation was estimated using the vertical position of the drone.

The mapping system was tested in the following three different scenarios: a single metallic target in a grass, a wooded area, and a ruined building. For each of those scenarios, the map was correctly retrieved. Indeed, the maps are always in good accordance with the satellite view, even if spurious targets appear to be at the same elevation of the flight trajectory, and even if no objects were at that elevation. These false targets are related to ground reflections and disappear by increasing the flight altitude.

By considering the results of this paper, the authors elaborated a possible path-planning method for obstacle avoidance. In particular, they proposed flying sideways or above detected targets. In this second case, the drone must increase flight altitude until the obstacle disappears below the drone. The success of this strategy is particularly visible during the test in the wooded area, when the radar detected some bushes below the drone.

The main achievement of these preliminary tests was to demonstrate the working principle of this radar technique. Nevertheless, effective performances (in terms of range, false alarm probability and undetected targets, etc.) must be assessed in specifically designed experimental tests.

Author Contributions: Conceptualization, L.M., L.B., M.P., E.B. and M.B.; methodology, L.B. and L.M.; software, L.B. and T.C.; validation, L.B., L.M. and M.P.; formal analysis, L.M. and L.B.; investigation, L.B., L.M, T.C., M.B., A.B.; resources, L.M., L.B. and T.C.; data curation, L.M.; writing—original draft preparation, L.M. and L.B.; writing—review and editing, E.B., M.P.; visualization, L.M., L.B.; supervision, M.P., M.B.; project administration, M.P.; funding acquisition, M.P. All authors have read and agreed to the published version of the manuscript.

Funding: This research was co-funded by Horizon 2020, European Community (EU), AURORA (Safe Urban Air Mobility for European Citizens) project, ID number 101007134.

Institutional Review Board Statement: Not applicable.

Informed Consent Statement: Not applicable.

Data Availability Statement: Data available on request.

Conflicts of Interest: The authors declare no conflict of interest.

References

1. Floreano, D.; Wood, R.J. Science, technology and the future of small autonomous drones. *Nature* **2015**, *521*, 460–466. [CrossRef] [PubMed]
2. Bollard, B.; Doshi, A.; Gilbert, N.; Poirot, C.; Gillman, L. Drone Technology for Monitoring Protected Areas in Remote and Fragile Environments. *Drones* **2022**, *6*, 42. [CrossRef]
3. Baytas, M.A.; Çay, D.; Zhang, Y.; Obaid, M.; Yantaç, A.E.; Fjeld, M. The Design of Social Drones: A Review of Studies on Au-tonomous Flyers in Inhabited Environments. In Proceedings of the 2019 CHI Conference on Human Factors in Computing Systems; Association for Computing Machinery: New York, NY, USA, 2019; pp. 1–2.
4. Lu, L.; Redondo, C.; Campoy, P. Optimal Frontier-Based Autonomous Exploration in Unconstructed Environment Using RGB-D Sensor. *Sensors* **2020**, *20*, 6507. [CrossRef] [PubMed]
5. Rho, E.; Jo, S. OctoMap-Based Semi-Autonomous Quadcopter Navigation with Biosignal Classification. In Proceedings of the 2018 6th International Conference on Brain-Computer Interface (BCI), Gangwon, Korea, 15–17 January 2018; pp. 1–4.
6. Vujasinović, S.; Becker, S.; Breuer, T.; Bullinger, S.; Scherer-Negenborn, N.; Arens, M. Integration of the 3D Environment for UAV Onboard Visual Object Tracking. *Appl. Sci.* **2020**, *10*, 7622. [CrossRef]
7. Carfagni, M.; Furferi, R.; Governi, L.; Santarelli, C.; Servi, M.; Uccheddu, F.; Volpe, Y. Metrological and Critical Characterization of the Intel D415 Stereo Depth Camera. *Sensors* **2019**, *19*, 489. [CrossRef]
8. Sacco, G.; Pittella, E.; Pisa, S.; Piuzzi, E. A MISO Radar System for Drone Localization. In Proceedings of the 2018 5th IEEE International Workshop on Metrology for AeroSpace (MetroAeroSpace), Rome, Italy, 20–22 June 2018; pp. 549–553.
9. W-Band—An Overview | ScienceDirect Topics. Available online: https://www.sciencedirect.com/topics/engineering/w-band (accessed on 20 January 2022).
10. Dogru, S.; Marques, L. Using Radar for Grid Based Indoor Mapping. In Proceedings of the 2019 IEEE International Conference on Autonomous Robot Systems and Competitions (ICARSC), Porto, Portugal, 24–26 April 2019; pp. 1–6. [CrossRef]
11. Lee, S.; Kwon, S.-Y.; Kim, B.-J.; Lim, H.-S.; Lee, J.-E. Dual-Mode Radar Sensor for Indoor Environment Mapping. *Sensors* **2021**, *21*, 2469. [CrossRef] [PubMed]
12. Dogru, S.; Marques, L. Pursuing Drones with Drones Using Millimeter Wave Radar. *IEEE Robot. Autom. Lett.* **2020**, *5*, 4156–4163. [CrossRef]
13. Tierney, B.B.; Rodenbeck, C.T. 3D-Sensing MIMO Radar for UAV Formation Flight and Obstacle Avoidance. In Proceedings of the 2019 IEEE Radio and Wireless Symposium (RWS), Orlando, FL, USA, 20–23 January 2019; pp. 1–3.
14. Yu, H.; Zhang, F.; Huang, P.; Wang, C.; Yuanhao, L. Autonomous Obstacle Avoidance for Based on Fusion of Radar and Monocular Camera. In Proceedings of the 2020 IEEE/RSJ International Conference on Intelligent Robots and Systems (IROS), Las Vegas, NV, USA, 24 October–24 January 2020; pp. 5954–5961.
15. Wang, Z.; Li, W.; Peng, H.; Liu, C.; Liao, X. The Development of Compact Patch Antenna Array with High Angular Resolu-tion for 77GHz FMCW Radar Applications. In Proceedings of the 2021 IEEE Electrical Design of Advanced Packaging and Systems (EDAPS), Urbana, IL, USA, 13–15 December 2021; pp. 1–3.
16. Bigazzi, L.; Basso, M.; Boni, E.; Innocenti, G.; Pieraccini, M. A Multilevel Architecture for Autonomous UAVs. *Drones* **2021**, *5*, 55. [CrossRef]
17. Hornung, A.; Wurm, K.M.; Bennewitz, M.; Stachniss, C.; Burgard, W. OctoMap: An efficient probabilistic 3D mapping framework based on octrees. *Auton. Robot.* **2013**, *34*, 189–206. [CrossRef]
18. AWR1843BOOST Evaluation Board | TI.Com. Available online: https://www.ti.com/tool/AWR1843BOOST (accessed on 20 January 2022).
19. Pieraccini, M.; Luzi, G.; Atzeni, C. Terrain mapping by ground-based interferometric radar. *IEEE Trans. Geosci. Remote Sens.* **2001**, *39*, 2176–2181. [CrossRef]
20. Hyun, E.; Oh, W.; Lee, J.-H. Two-Step Moving Target Detection Algorithm for Automotive 77 GHz FMCW Radar. In Proceedings of the 2010 IEEE 72nd Vehicular Technology Conference-Fall, Ottawa, ON, Canada, 6–9 September 2010; pp. 1–5.
21. Rohling, H. Radar CFAR Thresholding in Clutter and Multiple Target Situations. *IEEE Trans. Aerosp. Electron. Syst.* **1983**, *AES-19*, 608–621. [CrossRef]

drones

MDPI

Article

Long Distance Ground Target Tracking with Aerial Image-to-Position Conversion and Improved Track Association

Seokwon Yeom

School of AI, Daegu University, Gyeongsan 38453, Korea; yeom@daegu.ac.kr

Abstract: A small drone is capable of capturing distant objects at a low cost. In this paper, long distance (up to 1 km) ground target tracking with a small drone is addressed for oblique aerial images, and two novel approaches are developed. First, the coordinates of the image are converted to real-world based on the angular field of view, tilt angle, and altitude of the camera. Through the image-to-position conversion, the threshold of the actual object size and the center position of the detected object in real-world coordinates are obtained. Second, the track-to-track association is improved by adopting the nearest neighbor association rule to select the fittest track among multiple tracks in a dense track environment. Moving object detection consists of frame-to-frame subtraction and thresholding, morphological operation, and false alarm removal based on object size and shape properties. Tracks are initialized by differencing between the two nearest points in consecutive frames. The measurement statistically nearest to the state prediction updates the target's state. With the improved track-to-track association, the fittest track is selected in the track validation region, and the direction of the displacement vector and velocity vectors of the two tracks are tested with an angular threshold. In the experiment, a drone hovered at an altitude of 400 m capturing video for about 10 s. The camera was tilted 30° downward from the horizontal. Total track life (TTL) and mean track life (MTL) were obtained for 86 targets within approximately 1 km of the drone. The interacting multiple mode (IMM)-CV and IMM-CA schemes were adopted with varying angular thresholds. The average TTL and MTL were obtained as 84.9–91.0% and 65.6–78.2%, respectively. The number of missing targets was 3–5; the average TTL and MTL were 89.2–94.3% and 69.7–81.0% excluding the missing targets.

Keywords: small drone; long-distance surveillance; ground target tracking; track-to-track association; image-to-position conversion

Citation: Yeom, S. Long Distance Ground Target Tracking with Aerial Image-to-Position Conversion and Improved Track Association. *Drones* **2022**, *6*, 55. https://doi.org/10.3390/drones6030055

Academic Editors: Diego González-Aguilera and Pablo Rodríguez-Gonzálvez

Received: 7 February 2022
Accepted: 22 February 2022
Published: 23 February 2022

1. Introduction

Small unmanned aerial vehicles (UAVs), or drones, are useful for security and surveillance [1,2]. One important task is to track moving vehicles with aerial video. A small drone captures video from a distance at a low cost [3]. No highly trained personnel are required to generate the video.

Ground targets can be tracked with a small drone with visual, nonvisual, or combined methods. Various deep learning methods with camera motion models were studied in [4]. Tracking performance can be degraded by small objects, large numbers of targets, and camera motion [5]. Deep learning-based object detection was combined with multi-object tracking and 3D localization in [6]. In [7], YOLO and Kalman filter were used to detect and track high-resolution objects. The object detector and tracker with deep learning may require heavy computation with massive training data [8,9].

The background subtraction and the adaptive mean-shift and optical flow tracking were developed for the video sequences captured by a drone in [10]. The mean-shift tracker based on particle filtering was utilized to track a small and fast-moving object in [11]. The SIFT feature-based tracker was developed for fast processing in [12]. The kernelized correlation filter-based target tracking was studied in [13]. In [14], object tracking was performed by handing over the camera from one drone to another. Ground targets were tracked with road geometry recovery in [15]. Usually, trackers based on video sequences transmit high-resolution video streams to the ground, or a high computational burden is imposed on the drone. Aerial video was processed with high-precision GPS data from vehicles in [16]. Bayesian fusion of vision and radio frequency sensors were studied for ground target tracking in [17]. In [18], computer vision-based airborne target tracking with GPS signals was studied. Object tracking from drones with nonvisible cameras can be found in literature. A boat was captured and tracked with the Kalman filter by a fixed-wing drone in [19]. A small vessel was tracked by adopting a colored-noise measurement model in [20]. In the nonvisual approach, high-cost sensors add more payload to the drone, or infrastructure is required on the ground or in a vehicle.

In this paper, moving vehicle tracking with a small drone at long distances (up to 1 km) is addressed. In the previous work [21–24], the drone's camera was pointed directly at the ground, or the altitude of the drone was very low. When the camera is tilted, the field of view (FOV) can be extended, but constant scaling from pixel coordinates to actual position can no longer be applied. Therefore, an image-to-position conversion is developed to change the integer coordinates of a pixel to its actual positions, assuming that the angular field of view (AFOV), tilt angle, and camera (drone) altitude are known.

Moving object detection consists of frame subtraction and thresholding, morphological operation, and false alarm removal; falsely detected objects are removed using the object's actual size and two shape properties: squareness and rectangularity [23,24]. The minimum size of the extracted object is set constant, but the converted value in pixels changes depending on the distance from the drone.

Target tracking consists of three stages: track initialization, track maintenance, and track termination. Tracks are initialized with the difference between the two nearest measurements in consecutive frames following speed gating. Tracks are maintained by state estimation, measurement-to-track association (abbreviated as measurement association), and track-to-track association (abbreviated as track association). The nearest neighbor (NN) measurement association updates the state of a target by combining the measurements that are statistically closest to the prediction. The interacting multiple model (IMM) filters with a constant velocity (CV) or constant acceleration (CA) motion models are adopted to handle the various maneuvers of the target [25,26]. Track association fuses multiple tracks into a single track [27]. In this paper, the track association scheme is improved to fuse multiple tracks by sequentially searching for the nearest track in a dense track environment. Figure 1 shows a schematic block diagram of object detection and target tracking by a small drone at a long distance. First there is the image-to-position conversion, and finally there is the reverse process, position-to-image conversion.

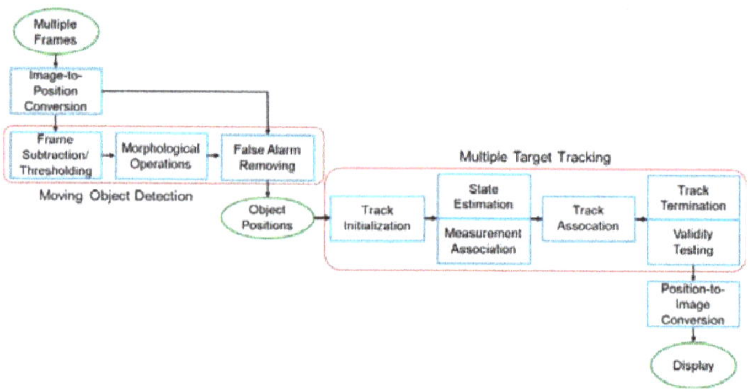

Figure 1. Block diagram of moving object detection and multiple target tracking.

In the experiment, the drone hovered from a fixed position at a height of 400 m. The title angle was 60° producing video for 10 s. Figure 2 shows three sample scenes extracted from a sample frame. The extracted area is 100 × 60 pixels and shows 2–3 vehicles with different resolutions. Overall, the frame has low resolution, and the targets are sometimes occluded by trees, structures, and other cars. Road lanes, traffic signs, and shadows can be included in backgrounds. A total of 86 targets within approximately 1 km of the drone were investigated with total track life (TTL) and mean track life (MTL). The average TTL and MTL were obtained as 84.9–91.0% and 65.6–78.2%, respectively, by various angular thresholds for directional track association. The number of missing targets was 3–5; the average TTL and MTL were 89.2–94.3% and 69.7–81.0%, respectively, if the missing targets are excluded.

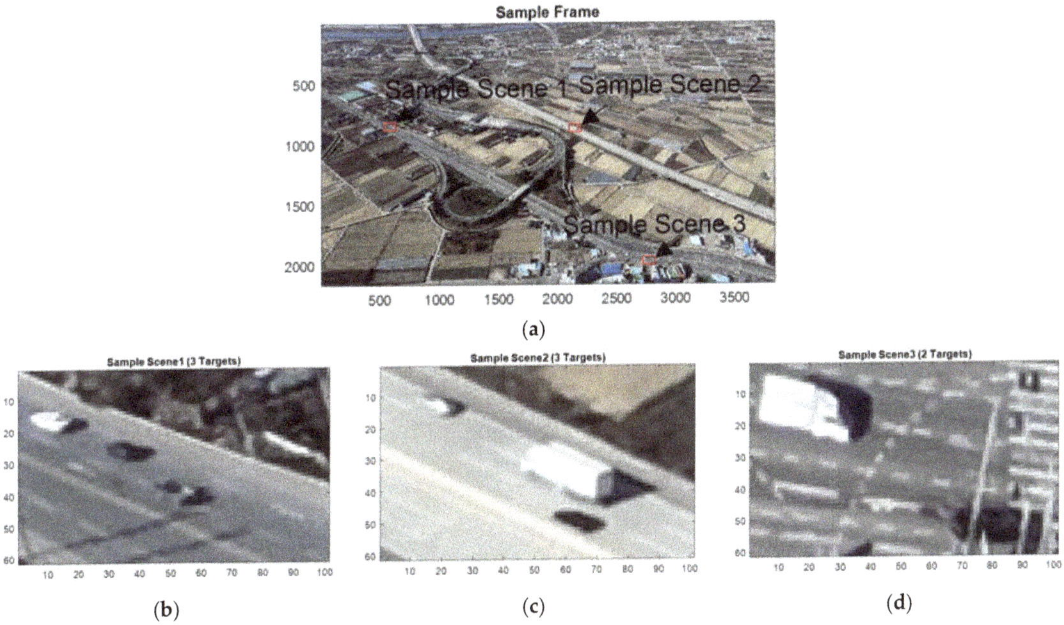

Figure 2. Three sample scenes showing targets: (**a**) sample frame; (**b**) sample scene 1; (**c**) sample scene 2; (**d**) sample scene 3.

The rest of the paper is organized as follows: the image-to-position conversion is described in Section 2. Section 3 explains multiple target tracking. Section 4 details the improved track association. Section 5 presents the experimental results. Discussion and conclusions follow in Sections 6 and 7, respectively.

2. Image-Position Conversion

Imaging is the projection of a three-dimensional real-world onto a two-dimensional plan. Thus, when the camera is not pointed directly at the ground, the discrepancy between the relative pixel positions in the image and the real world becomes irregular. Since the coordinates of the image are not indexed in proportion to the actual distance, constant scaling generates more discrepancy between the coordinates in the image and the coordinates in the real world. Therefore, integer coordinates in x and y directions of the image are converted to real-world positions based on the AFOV, tilt angle, and altitude of the camera. As shown in Figure 3, the drone camera is positioned at (0,0,h) with a tilt angle θ_T. The image size is $W \times H$ pixels. The actual position vector x_{ij} of the (i, j) pixel is approximated as

$$x_{ij} = (x_i, y_j) \approx \left(d_{\frac{H}{2}} \cdot \tan\left[\left(i - \frac{W}{2} + 1\right)\frac{a_x}{W}\right],\ h \cdot \tan\left[\theta_T + \left(\frac{H}{2} - j\right)\frac{a_y}{H}\right] \right), \quad i = 0, \ldots, W-1,\ j = 0, \ldots, H-1, \tag{1}$$

where a_x and a_y are the view angles of the camera in x and y directions, respectively; $d_{H/2}$ is the distance from the camera to $(x_{W/2-1}, y_{H/2}, 0)$, which is $\sqrt{y_{H/2}^2 + h^2}$. It is noted that land is assumed to be flat. Thus, the position in z direction is zero. The actual pixel size is calculated as

$$\Delta(i,j) = |(x_i - x_{i+1}) \cdot (y_i - y_{j+1})|. \quad (2)$$

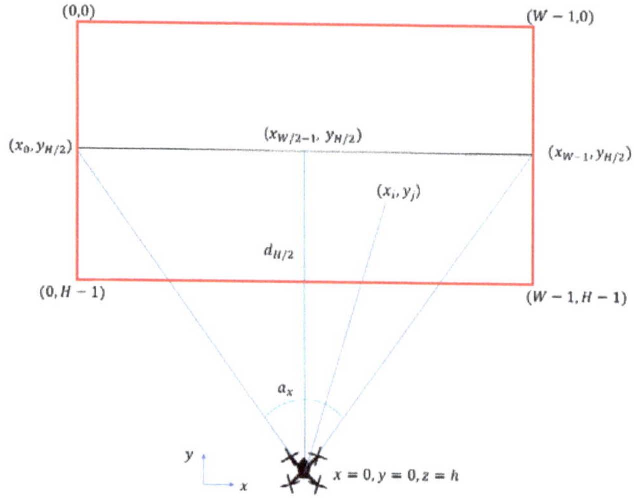

(a)

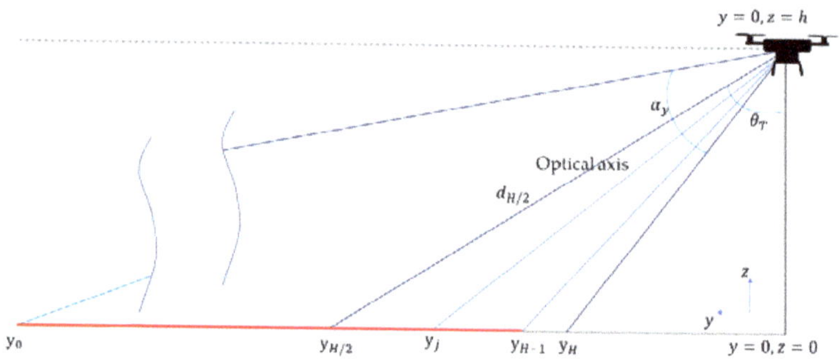

(b)

Figure 3. Coordinate conversion from image to real-world: (**a**) x direction; (**b**) y direction.

In the experiments, the altitude h is set to 400 m; W and H are 3840 and 2160 pixels; a_x and a_y are set to AFOV of the camera, 70° and 40°, respectively; θ_T is set to 60°; thus $y_{H/2}$ is calculated at 692.8 m by Equation (1), and $d_{H/2}$ is 800 m accordingly.

Figure 4 provides the visualization of the coordinate conversion and actual pixel size according to Equations (1) and (2), respectively. In Figure 4a, every 50th pixel is shown for better visualization. The maximum, median, and minimum pixel sizes in Figure 4b are 1.623 m^2, 0.1506 m^2, and 0.0561 m^2, respectively. This approximate conversion considers x and y directions separately. Thus, a simple reverse process is possible from position to image, and it will be shown that the detection result is significantly improved in the experiments although the coordinate conversion can be accompanied by various inevitable errors [28,29].

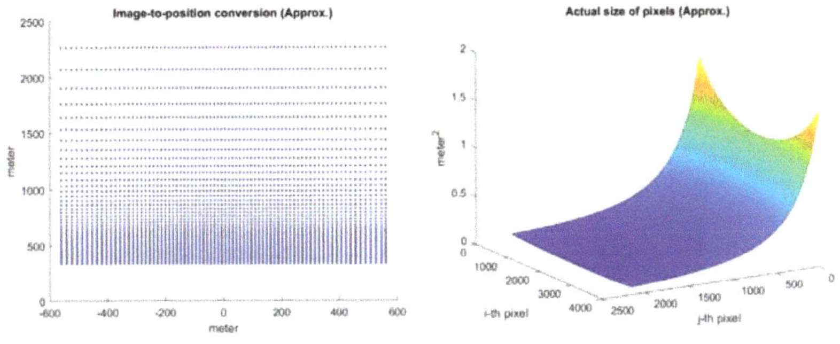

(a) (b)

Figure 4. Approximated conversion from image to real world, (**a**) visualization of the coordinate conversion; (**b**) actual pixel size.

3. Multiple Target Tracking

A block diagram of multiple target tracking is shown in Figure 5.

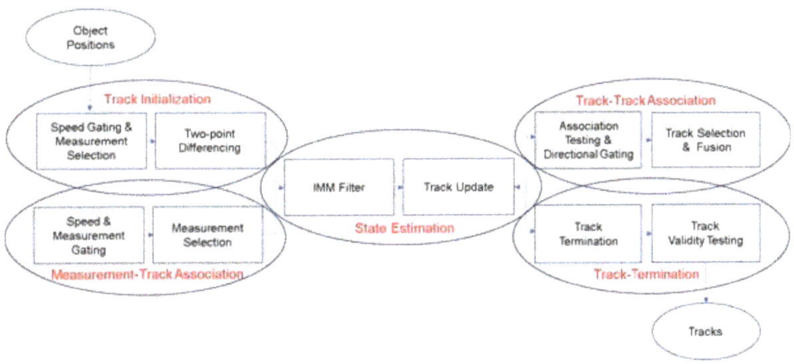

Figure 5. Block diagram of multiple target tracking.

Tracks are initialized with two-point differencing between the nearest neighbor measurements following maximum speed gating. For the measurement association, the speed gating process is first performed, followed by the measurement gating based on a chi-square hypothesis test. Then, the NN measurement selection follows. The NN rule is very effective in computing and was successfully applied to multiple ground target tracking by a drone [21–24]. The IMM filter is adopted to estimate the kinematic state of the target. It can efficiently handle various maneuvers on multiple targets. The IMM with a combined CV and CA scheme was contrived to track a single target in [24]. The IMM-CV scheme was applied to track various maneuvering 120 aerial targets for the aerial early warning system in [30]. The motion models are analyzed in detail in [31].

In a multisensor environment, a track fusion method was developed assuming the target is undergoing a common process noise [27]. A practical approach for track association was developed for the Kalman filter in [22]. This approach has been extended with the IMM filter and the directional track association was proposed to consider the moving direction of the target in [24]; directional gating tests the maximum deviation in the directions of the tracks and the direction of the displacement vector between the tracks. In this paper, the NN track selection scheme is proposed and will be described in the next section.

There are three criteria for track termination in this paper. One is associated but not selected during the track association. The other cases are the maximum frame number without measurements and the minimum target speed. The criterion of the minimum target speed is very effective when high clutter occurs on nonmoving false targets [23]. After track termination, its validity is tested with the track life length. If the track life length is shorter than the minimum track life length, the track is removed as a false track. More detailed processes of target tracking are described in [22–24].

4. Improved Track Association

In this paper, the track association procedure is developed to select the fittest track in a dense track environment. For track s, the fittest track is selected as follows:

$$\hat{c}(s) = \underset{t=1,\ldots,N_T(k),\ s\neq t}{argmin} \left[\hat{x}^s(k|k) - \hat{x}^t(k|k)\right]^T \left[T^{st}(k)\right]^{-1} \left[\hat{x}^s(k|k) - \hat{x}^t(k|k)\right], \quad s = 1,\ldots,N_T(k), \tag{3}$$

$$T^{st}(k) = P^s(k|k) + P^t(k|k) - P^{st}(k|k) - P^{ts}(k|k) \tag{4}$$

$$P^{st}(k|k) = \left[I - b^s(k)W^s(k)H\right]\left[FP^{st}(k-1|k-1)F^T + Q\right]\left[I - b^t(k)W^t(k)H\right], \tag{5}$$

where $\hat{x}^s(k|k)$ and $\hat{x}^t(k|k)$ are the state vector of tracks s and t, respectively, at frame k, $N_T(k)$ is the number of tracks at frame k, $P^s(k|k)$ and $P^t(k|k)$ are the covariance matrix of tracks s and t, respectively, at frame k, $b^s(k)$ and $b^t(k)$ are binary numbers that become one when track s or t is associated with a measurement, otherwise they are zero [27]. F, H, and Q are the transition matrix, measurement matrix and covariance of the process noise, respectively. It is noted that $T^{st}(k)$ is meaningless if its determinant is not positive. The fused covariance in Equation (5) is a linear recursion and its initial condition is set at $P^{st}(0|0) = [0]_{N_x\times N_x}$, N_x is the dimension of the state vector, which is 4 and 6 for the CV and CA motion models, respectively, and $W^t(k|k)$ is obtained as the combined filter gain of track t as [24]:

$$W^t(k|k) = \sum_{j=1}^{M} W_j^t(k|k)\mu_j^t(k). \tag{6}$$

where M is the number of modes of the IMM filter; $W_j^t(k|k)$ is the filter gain of the j-th mode matched filter at frame k; and $\mu_j^t(k)$ is the mode probability of the j-th mode- matched filter at frame k. The following chi-square hypothesis test should be satisfied between tracks s and t since multiple tracks of the same target have error dependencies on each other [27]:

$$\left[\hat{x}^s(k|k) - \hat{x}^t(k|k)\right]^T \left[T^{st}(k)\right]^{-1} \left[\hat{x}^s(k|k) - \hat{x}^t(k|k)\right] \leq \gamma_g, \tag{7}$$

where γ_g is a gate threshold for the track validation region. The directional gating process tests the maximum deviation in the direction of the displacement vector and the directions of the track velocity as [24]:

$$max\left(\cos^{-1}\frac{\left|\left\langle \hat{d}^{st}(k|k),\ \hat{v}^s(k|k)\right\rangle\right|}{\|\ \hat{d}^{st}(k|k)\ \|\|\ \hat{v}^s(k|k)\ \|},\ \cos^{-1}\frac{\left|\left\langle \hat{d}^{st}(k|k),\ \hat{v}^t(k|k)\right\rangle\right|}{\|\ \hat{d}^{st}(k|k)\ \|\|\ \hat{v}^t(k|k)\ \|}\right) \leq \theta_g, \tag{8}$$

$$\hat{d}^{st}(k|k) = \begin{bmatrix} \hat{x}^t(k|k) - \hat{x}^s(k|k) \\ \hat{y}^t(k|k) - \hat{y}^s(k|k) \end{bmatrix},\ \hat{v}^s(k|k) = \begin{bmatrix} \hat{v}^{sx}(k|k) \\ \hat{v}^{sy}(k|k) \end{bmatrix},\ \hat{v}^t(k|k) = \begin{bmatrix} \hat{v}^{tx}(k|k) \\ \hat{v}^{ty}(k|k) \end{bmatrix}, \tag{9}$$

where $\langle\cdot\rangle$ denotes the inner product operation, θ_g is an angular threshold; and $\hat{x}^t(k|k)$ and $\hat{y}^t(k|k)$ are the position components of $\hat{x}^t(k|k)$ in x and y directions, respectively; $\hat{v}^{tx}(k|k)$ and $\hat{v}^{ty}(k|k)$ are the velocity components of $\hat{x}^t(k|k)$ in x and y directions, respectively.

After the fittest track is selected, the current state of track s is replaced with a fused estimate and covariance if $|P^s(k|k)| \leq \left|P^{\hat{c}(s)}(k|k)\right|$ as

$$\begin{aligned}\hat{x}^s(k|k) = \hat{x}^s(k|k) & \\ &+ [P^s(k|k) \\ &- P^{st}(k|k)]\left[P^s(k|k) + P^t(k|k) - P^{st}(k|k)\right. \\ &\left.-P^{ts}(k|k)\right]^{-1}\left[\hat{x}^t(k|k) - \hat{x}^s(k|k)\right],\end{aligned} \tag{10}$$

$$\begin{aligned} P^s(k|k) = P^s(k|k) & \\ & - [P^s(k|k) \\ & - P^{st}(k|k)] [P^s(k|k) + P^t(k|k) - P^{st}(k|k) \\ & - P^{ts}(k|k)]^{-1} [P^s(k|k) - P^{ts}(k|k)]. \end{aligned} \tag{11}$$

The track selection process proposed in this paper is as follows: after track s becomes a fused track, track $\hat{c}(s)$ becomes a potentially terminated track. That is, fusion only occurs if the determinant of the covariance matrix of track s is less than the determinant of the selected track. It is noted that a more accurate track has less error (covariance). In the previous directional track association, track $\hat{c}(s)$ was instantly terminated, but in the procedure proposed in this paper, it is still eligible to be associated with other tracks that have not yet been fused. The detailed procedure of the track association is illustrated in Figure 6.

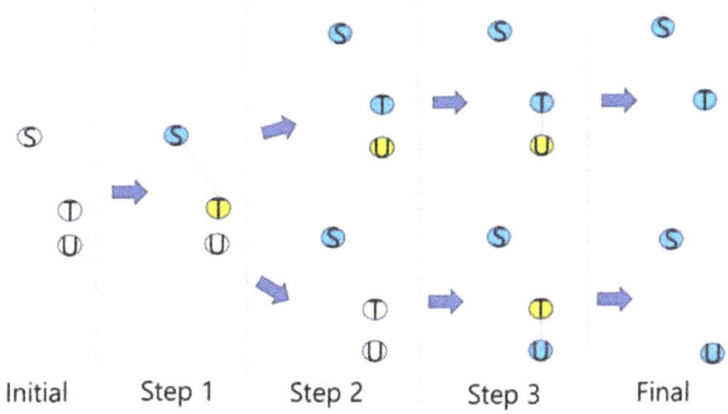

Figure 6. Illustration of track association at a frame: fused tracks in blue and potentially terminated tracks in yellow.

In Figure 6, there are initially three tracks, s, t, and u, at a certain frame. In step 1, Track s searches for the fittest track for itself. Once track t is satisfied with Equations (3), (7), (8), and $|P^s(k|k)| \leq |P^t(k|k)|$, tracks s and t are fused, track s becomes the fused track, and track t becomes the potentially terminated track. Otherwise, no change occurs, and we go to the next step. In Step 2, track t searches for the fittest track, except for any already fused track, here track s. If track t becomes the fused track after fusion with track u, then track u becomes the potentially terminated track and is terminated at the final stage because no tracks remain to be considered for track u. Otherwise, in Step 3, track u searches and can be fused with track t. Finally, all potentially terminated tracks are terminated. In the above row, tracks s and t and tracks t and u are fused, and track u is terminated. In the bottom row, tracks s and t and tracks u and t are fused, and track t is terminated. In the next frame, the remaining tracks s and t or tracks s and u can be fused if they originate from a single target. This track association procedure allows multiple fusions of one track, and fusion occurs at most once for a track in a frame. It will be shown that it can reduce the number of tracks significantly in the experiments.

5. Results

In this section, experimental results will be detailed through video description, parameter setting, and moving vehicle tracking along with the proposed strategy.

5.1. Video Description and Moving Object Detection

A video was captured by a Mavic Air 2 hovering from a fixed position at a frame rate of 30 fps. The frame size is 3840 × 2160 pixels. The scenes in the video include a highway interchange, a toll gate, road bridges, buildings, and trees. The drone was at an altitude of

400 m and the tilt angle was set to 60°. Every second, frames were processed for efficient processing, thus the actual frame rate was 15 fps. A total of 152 frames were considered for about 10 s. A total of 86 moving vehicles within a range of approximately 1 km of the drone appear in the entire video. The number of frames subtracted was 151, and the life length of the target over the entire period was 150 frames due to the two-point differencing initialization. However, the life lengths of Targets 5, 8, 29–32, 54, 62, 83, and 84 in the video are 92, 124, 42, 68,104, 100, 146, 146, 136, and 100 frames, respectively because they started late or stopped early. Some targets are occasionally occluded by a bridge, a toll gate, tress, and other vehicles. Some of them happen to be invisible because of shadows. The minimum target speed was set to 1 m/s. Thus, very slow targets were not considered as targets of interest. Figure 7a shows Targets 1–53, and Figure 7b shows Targets 54–86 in the first frame. Figure 7c shows the 1 km range with an outer circle and the view angles with blue and red lines and arcs according to distance. It is noted that Figure 7c was obtained manually to show the approximate coverage on a commercially available aerial map [32].

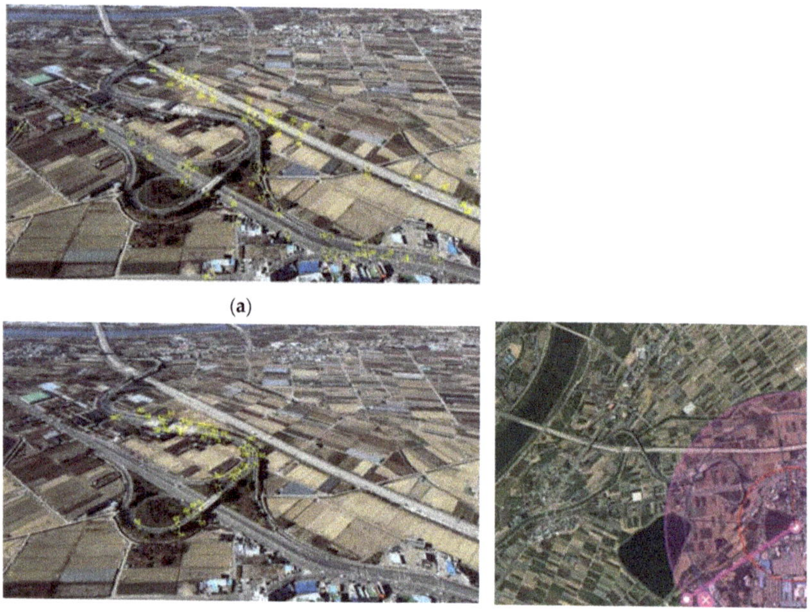

(a) (b) (c)

Figure 7. (**a**) Targets 1–53 at the first frame; (**b**) Targets 54–86 at the first frame; (**c**) 1 km range of the drone with approximated view angles.

For the object detection, the thresholding after the frame subtraction was set to 30. The structure element for the morphological operation (closing) was set at $[1]_{2\times 2}$. The minimum size of a basic rectangle for false alarm removal was set to 6 m^2, and the minimum squareness and rectangularity were set to 0.2 and 0.3, respectively. Figure 8a is a thresholded binary image after frame subtraction between Figure 7a and the next frame.

Figure 8b is the result of the morphological operation of Figure 8a. Figure 8c is basic rectangles of Figure 8b after removing false alarms. Figure 8d shows the center of the basic rectangles, indicated by the blue dot. The number of detections in Figure 8d is 127 including false alarms.

Figure 9a shows the detection results for all frames with the image-to-position conversion. Most blue dots of Figure 9a are along roads that coincided with the trajectories of the vehicles. Figure 9b–d show the detection results for all frames that have not undergone the image-position conversion process; the pixel sizes were chosen to be constant as the maximum, median, and minimum sizes in Figure 9b–d, respectively. The maximum, median, and minimum sizes of the pixels in Figure 4b are 1.623 m^2, 0.1506 m^2, and 0.0561 m^2, respectively. Thus, the pixel sizes of a basic rectangle of 6 m^2 are 4 pixels, 40 pixels, and 107 pixels in Figure 9b–d, respectively. If the threshold is too low, as shown in Figure 9b, more false alarms are detected; if the threshold is too high, as shown in Figure 9d, more missing detections are obtained. With a median threshold as shown in Figure 9c, some long-distance vehicles fail to be detected.

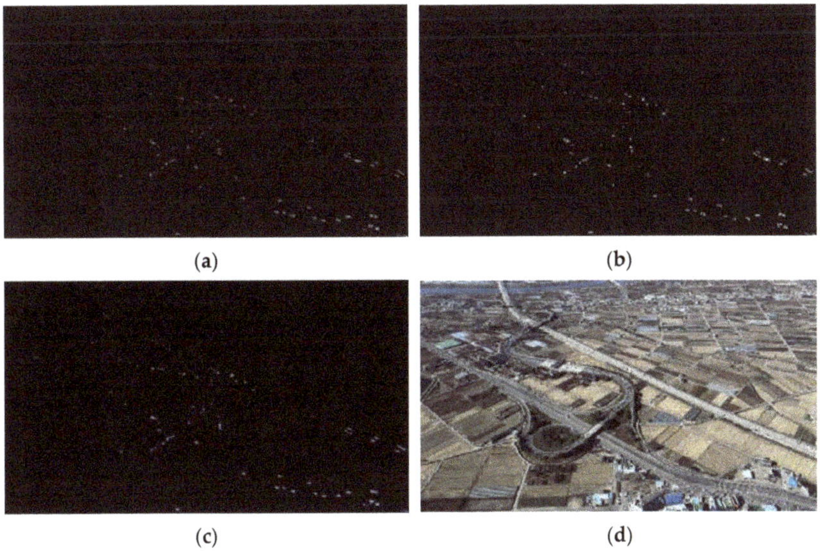

Figure 8. Object detection: (**a**) frame subtraction and thresholding; (**b**) morphological operation (closing) of (**a**); (**c**) basic rectangles after false alarm removal of (**b**); (**d**) 127 centers of basic rectangles of (**c**), indicated as blue dots.

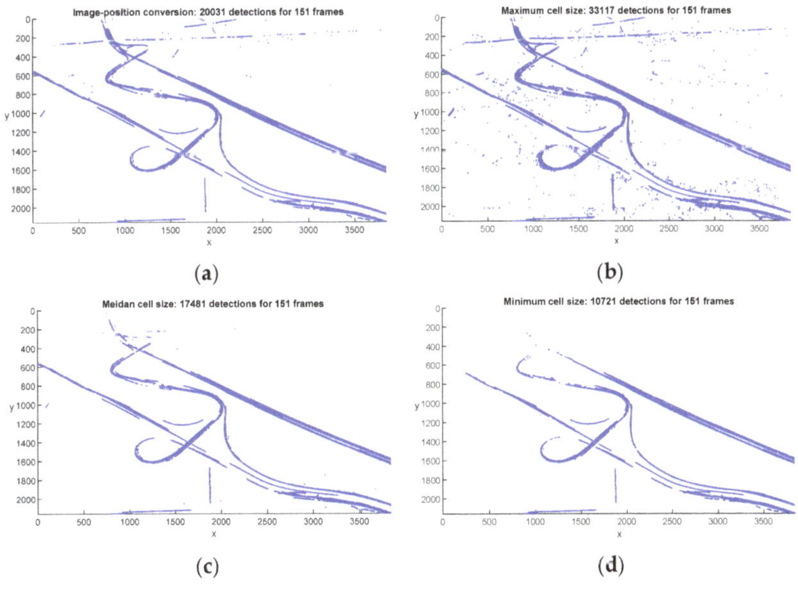

Figure 9. Object detection results of 151 frames with: (**a**) image-to-position conversion; (**b**) maximum pixel size; (**c**) median pixel size; (**d**) minimum pixel size.

5.2. Multiple Target Tracking

The positions of Figure 9a become inputs to the target-tracking stage. The sampling time is 1/15 s since every second frame is processed. The IMM-CV and IMM-CA are adopted with the image-to-position conversion and the proposed directional track association procedure. Table 1 shows the parameters for target tracking; it is noted that the angular threshold θ_g of 90° is equivalent to the track association without the directional gating.

Table 1. Parameters for target tracking.

Parameters		IMM-CV	IMM-CA
Sampling time		1/15 s	
Max. target speed for initialization, V_{max}		60 m/s	
Process noise variance	$\sigma_{1x} = \sigma_{1y}$	1 m/s^2	0.01 m/s^2
	$\sigma_{2x} = \sigma_{2y}$	10 m/s^2	0.1 m/s^2
Mode transition probabiltiy p_{ij}		$\begin{bmatrix} 0.8 & 0.2 \\ 0.3 & 0.7 \end{bmatrix}$	
Measurement noise variance, $r_x = r_y$		1.5 m	
Measurent association	Gate threshold, γ_f	8	
	Max. target speed, S_{max}	80 m/s	
Track association	Gate threshold, γ_g	100	
	Anguar threhold, θ_g	90°, 30°, 20°	
Track termination	Max. searching number	20 frames (1.33 s)	
	Min. target speed	1 m/s	
Min. track life length for track validity		20 frames (1.33 s)	

For IMM-CV and IMM-CA without track association, a total of 340 and 314 valid tracks are generated, as shown in Figure 10a,b. The number of tracks of IMM-CV is reduced to 173, 185, and 192 by the directional track association when θ_g is 90°, 30°, and 20°, respectively, as shown in Figure 11a–c. For IMM-CA, 196, 208, and 209 tracks are generated for the directional track associations when θ_g is 90°, 30°, and 20°, respectively, as shown in Figure 12a–c.

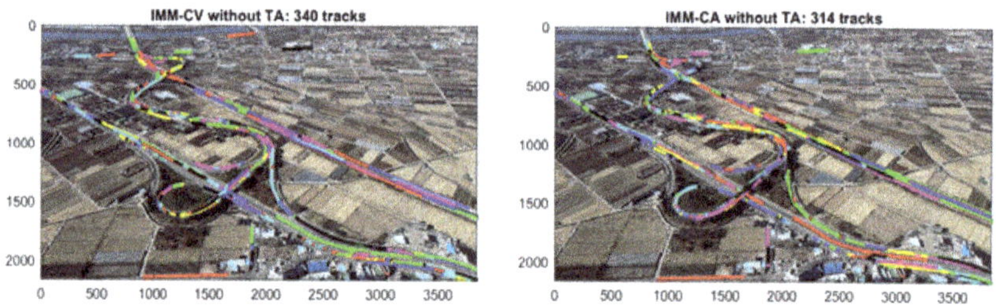

(a) (b)

Figure 10. All tracks without track association: (**a**) IMM-CV; (**b**) IMM-CA.

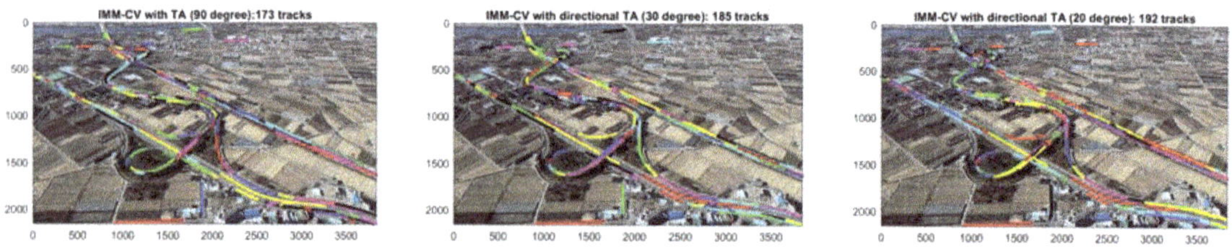

(a) (b) (c)

Figure 11. All tracks with track association, IMM-CV: (**a**) $\theta_g = 90°$; (**b**) $\theta_g = 30°$; (**c**) $\theta_g = 20°$.

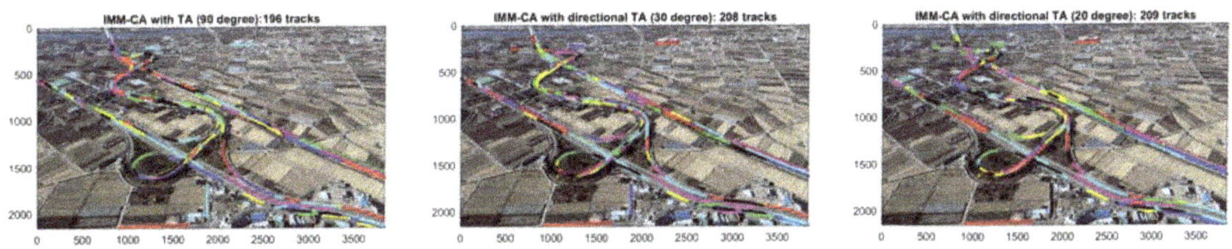

(**a**) (**b**) (**c**)

Figure 12. All tracks with track association, IMM-CA: (**a**) $\theta_g = 90°$; (**b**) $\theta_g = 30°$; (**c**) $\theta_g = 20°$.

Two metrics, TTL and MTL are employed to evaluate the tracking performance. The TTL and MTL are defined, respectively, as [30]:

$$\text{TTL} = \frac{\text{Sum of lengths of tracks which have the same target ID}}{\text{Target life length} - 1}, \tag{12}$$

$$\text{MTL} = \frac{\text{TTL}}{\text{Number of tracks associated in TTL}}. \tag{13}$$

A track's target ID is defined as the target with the most measurements on the track. The MTL becomes less than the TTL in case of track breakage or overlap. The TTL and MTL are the same if only one track is generated for one target. The TTL and MTL are 0 if no track is generated for a target when the target is missing. Figures 13 and 14 show the TTL and MTL of Figures 11 and 12, respectively. Three targets (11, 17, 27) are missing in all cases. In addition, no tracks are set on Target 55 in Figure 13a and on Targets 55 and 84 in Figure 14a–c.

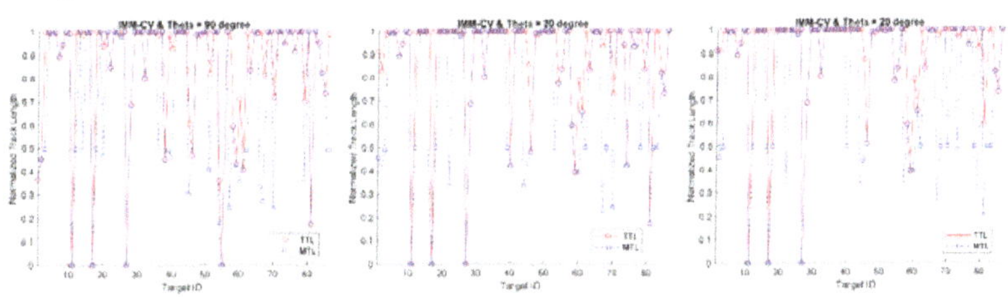

(**a**) (**b**) (**c**)

Figure 13. TTL and MLT of IMM-CV: (**a**) $\theta_g = 90°$; (**b**) $\theta_g = 30°$; (**c**) $\theta_g = 20°$.

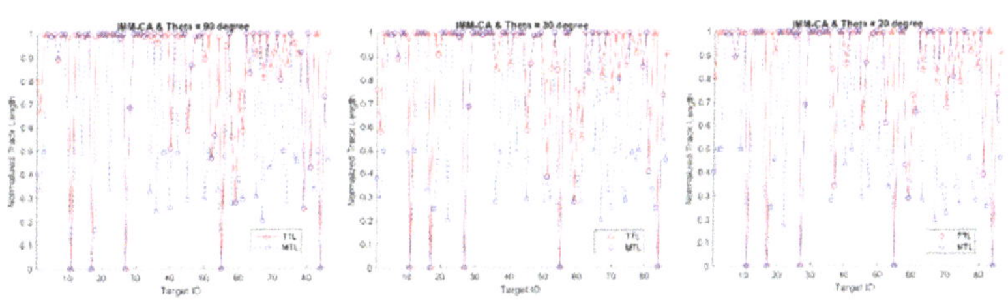

(**a**) (**b**) (**c**)

Figure 14. TTL and MLT of IMM-CA: (**a**) $\theta_g = 90°$; (**b**) $\theta_g = 30°$; (**c**) $\theta_g = 20°$.

Tables 2 and 3 show the overall tracking performance of IMM-CV and IMM-CA, respectively. They show the number of tracks, the number of tracks associated with the targets of interest, average TTL and MTL, average TTL and MTL excluding missing targets, and number of missing targets. The average TTL and MTL are, respectively, 84.9–91.0% and 65.6–78.2%. Excluding missing targets, the average TTL and MTL accounted for 89.2–94.3% and 69.7–81.0, respectively.

Table 2. Tracking performance of IMM-CV.

	IMM-CV		
	θ_g=90°	θ_g=30°	θ_g=20°
Number of tracks	173	185	192
Number of associated tracks	106	108	111
Avg. TTL	0.851	0.885	0.910
Avg. MTL	0.747	0.770	0.782
Avg. TTL w/o missing targets	0.892	0.9176	0.943
Avg. MTL w/o missing targets	0.783	0.798	0.810
Number of missing targets	4	3	3

Table 3. Tracking performance of IMM-CA.

	IMM-CA		
	θ_g=90°	θ_g=30°	θ_g=20°
Number of tracks	196	208	209
Number of associated tracks	129	133	133
Avg. TTL	0.849	0.858	0.861
Avg. MTL	0.656	0.660	0.668
Avg. TTL w/o missing targets	0.901	0.911	0.914
Avg. MTL w/o missing targets	0.697	0.700	0.709
Number of missing targets	5	5	5

Eight supplementary multimedia files (MP4 format) for Figures 10–12 is available online. The first is the IMM-CV without the track association for Figure 10a (Supplementary Material Video S1), and the second is the IMM-CA without the track association for Figure 10b (Supplementary Material Video S2). The third is the IMM-CV with $\theta_g = 90°$ for Figure 11a (Supplementary Material Video S3). The fourth is the IMM-CV with $\theta_g = 30°$ for Figure 11b (Supplementary Material Video S4). The fifth is the IMM-CV with $\theta_g = 20°$ for Figure 11c (Supplementary Material Video S5). The sixth is the IMM-CA with $\theta_g = 90°$ for Figure 12a (Supplementary Material Video S6). The seventh is the IMM-CA with $\theta_g = 30°$ for Figure 12b (Supplementary Material Video S7). The eighth is the IMM-CA with $\theta_g = 20°$ for Figure 12c (Supplementary Material Video S8). The black squares and numbers in the MP4 files are position estimates and track numbers, respectively, in the order they were initialized. For better visualization, odd numbers are shown in white, and even numbers in yellow. The blue dots are the detection positions including false alarms.

6. Discussion

The image-to-position conversion is an approximation showing a significant improvement in object detection. The reverse process is also possible, and the track is displayed in the frame after the reverse process.

The stability of the drone (camera) is important especially for oblique images where the position of the target can be concentrated or easily occluded; it can prevent false detections that result in false tracks and false track associations.

The proposed track selection can reduce the number of tracks from 340 to 173–192 for IMM-CV and from 314 to 196–209 for IMM-CA. Smaller angular threshold yields higher TTL and MTL, producing more tracks. The highest TTL and MTL excluding missing targets are, respectively, 94.3% and 81.0% for IMM-CA with $\theta_g = 20°$. Some targets are still detected and tracked outside the range of interest as shown in the videos.

The average number of missing targets is 3.33 and 5 for IMM-CV and IMM-CA, respectively. Targets 11 and 27 move too slowly. Target 17 is occluded by trees and shadows. In the experiments, the surveillance area was around 0.53 km^2. It was more than twice as large as the area when the camera was pointed directly at the ground.

7. Conclusions

In this paper, two strategies were developed for multitarget tracking by a small drone. One is the image-to-position conversion based on the AFOV, tilt angle, and altitude of the camera. The other is the improved track association for densely distributed track environments. Both the IMM-CV and IMM-CA schemes achieve robust results in TTL and MTL.

The overall process is computationally efficient as it does not require high-resolution video streaming or storage and training on large-scale data. This system is suitable for security and surveillance for civil and military applications such as threat detection, vehicle counting and chasing, and traffic control. This method can be also applied to tracking other objects such as people or animals over long distances. Target tracking using moving drones from various perspectives remains a subject of future study.

Supplementary Materials: The following are available online at https://zenodo.org/record/5932718, Video S1: IMM-CV, Video S2: IMM-CA, Video S3: IMM-CV-90, Video S4: IMM-CV-30, Video S5: IMM-CV-20, Video S6: IMM-CA-90, Video S7: IMM-CA-30, Video S8: IMM-CA-20.

Funding: This research was supported by Daegu University Research Grant 2019.

Institutional Review Board Statement: Not applicable.

Informed Consent Statement: Not applicable.

Data Availability Statement: Not applicable.

Acknowledgments: Not applicable.

Conflicts of Interest: The author declared no conflict of interest.

References

1. Alzahrani, B.; Oubbati, O.S.; Barnawi, A.; Atiquzzaman, M.; Alghazzawi, D. UAV assistance paradigm: State-of-the-art in applications and challenges. *J. Netw. Comput. Appl.* **2020**, *166*, 102706. [CrossRef]
2. Zaheer, Z.; Usmani, A.; Khan, E.; Qadeer, M.A. Aerial surveillance system using UAV. In Proceedings of the 2016 Thirteenth International Conference on Wireless and Optical Communications Networks (WOCN), Hyderabad, India, 21–23 July 2016; pp. 1–7.
3. Theys, B.; Schutter, J.D. Forward flight tests of a quadcopter unmanned aerial vehicle with various spherical body diameters. *Int. J. Micro Air Veh.* **2020**, *12*, 1–8. [CrossRef]
4. Li, S.; Yeung, D.-Y. Visual object tracking for unmanned aerial vehicles: A benchmark and new motion models. In Proceedings of the Thirty-Frist AAAI Conference on Artificial Intelligence (AAAI-17), San Francisco, CA, USA, 4–9 February 2017; pp. 4140–4146.
5. Du, D.; Qi, Y.; Yu, H.; Yang, Y.; Duan, K.; Li, G.; Zhang, W.; Huang, Q.; Tian, Q. The Unmanned Aerial Vehicle Benchmark: Object Detection and Tracking. *Lect. Notes Comput. Sci.* **2018**, 375–391. [CrossRef]
6. Zhang, H.; Lei, Z.; Wang, G.; Hwang, J. Eye in the Sky: Drone-Based Object Tracking and 3D Localization. In Proceedings of the 27th ACM International Conference on Multimedia, Nice, France, 21–25 October 2019; pp. 899–907. [CrossRef]
7. Lo, L.-Y.; Yiu, C.H.; Tang, Y.; Yang, A.-S.; Li, B.; Wen, C.-Y. Dynamic Object Tracking on Autonomous UAV System for Surveillance Applications. *Sensors* **2021**, *21*, 7888. [CrossRef]
8. Zhang, S.; Zhuo, L.; Zhang, H.; Li, J. Object Tracking in Unmanned Aerial Vehicle Videos via Multifeature Discrimination and Instance-Aware Attention Network. *Remote. Sens.* **2020**, *12*, 2646. [CrossRef]
9. Kouris, A.; Kyrkou, C.; Bouganis, C.-S. Informed Region Selection for Efficient UAV-Based Object Detectors: Altitude-Aware Vehicle Detection with Cycar Dataset. In Proceedings of the IEEE/RSJ International Conference on Intelligent Robots and Systems (IROS), Macau, China, 4–9 November 2019; pp. 51–58.
10. Kamate, S.; Yilmazer, N. Application of Object Detection and Tracking Techniques for Unmanned Aerial Vehicles. *Proc. Comput. Sci.* **2015**, *61*, 436–441. [CrossRef]
11. Fang, P.; Lu, J.; Tian, Y.; Miao, Z. An Improved Object Tracking Method in UAV Videos. *Procedia Eng.* **2011**, *15*, 634–638. [CrossRef]
12. Jianfang, L.; Hao, Z.; Jingli, G. A novel fast target tracking method for UAV aerial image. *Open Phys.* **2017**, *15*, 420–426. [CrossRef]
13. Yang, J.; Tang, W.; Ding, Z. Long-Term Target Tracking of UAVs Based on Kernelized Correlation Filter. *Mathematics* **2021**, *9*, 3006. [CrossRef]
14. Mueller, M.; Sharma, G.; Smith, N.; Ghanem, B. Persistent Aerial Tracking system for UAVs. In Proceedings of the 2016 IEEE/RSJ International Conference on Intelligent Robots and Systems (IROS), Daejeon, Korea, 9–14 October 2016; pp. 1562–1569. [CrossRef]

15. Li, Y.; Doucette, E.A.; Curtis, J.W.; Gans, N. Ground target tracking and trajectory prediction by UAV using a single camera and 3D road geometry recovery. In Proceedings of the 2017 American Control Conference (ACC), Seattle, WA, USA, 24–26 May 2017; pp. 1238–1243. [CrossRef]
16. Guido, G.; Gallelli, V.; Rogano, D.; Vitale, A. Evaluating the accuracy of vehicle tracking data obtained from Unmanned Aerial Vehicles. *Int. J. Transp. Sci. Technol.* **2016**, *5*, 136–151. [CrossRef]
17. Rajasekaran, R.K.; Ahmed, N.; Frew, E. Bayesian Fusion of Unlabeled Vision and RF Data for Aerial Tracking of Ground Targets. In Proceedings of the IEEE/RSJ International Conference on Intelligent Robots and Systems (IROS), Las Vegas, NV, USA, 25–29 October 2020; pp. 1629–1636.
18. Upadhyay, J.; Rawat, A.; Deb, D. Multiple Drone Navigation and Formation Using Selective Target Tracking-Based Computer Vision. *Electronics* **2021**, *10*, 2125. [CrossRef]
19. Leira, F.S.; Helgensen, H.H.; Johansen, T.A.; Fossen, T.I. Object detection, recognition, and tracking from UAVs using a thermal camera. *J. Field Robot.* **2021**, *38*, 242–267. [CrossRef]
20. Helgesen, H.H.; Leira, F.S.; Johansen, T.A. Colored-Noise Tracking of Floating Objects using UAVs with Thermal Cameras. In Proceedings of the 2019 International Conference on Unmanned Aircraft Systems (ICUAS), Atlanta, GA, USA, 11–14 June 2019; pp. 651–660. [CrossRef]
21. Yeom, S.; Cho, I.-J. Detection and Tracking of Moving Pedestrians with a Small Unmanned Aerial Vehicle. *Appl. Sci.* **2019**, *9*, 3359. [CrossRef]
22. Yeom, S.; Nam, D.-H. Moving Vehicle Tracking with a Moving Drone Based on Track Association. *Appl. Sci.* **2021**, *11*, 4046. [CrossRef]
23. Yeom, S. Moving People Tracking and False Track Removing with Infrared Thermal Imaging by a Multirotor. *Drones* **2021**, *5*, 65. [CrossRef]
24. Yeom, S. Long Distance Moving Vehicle Tracking with a Multirotor Based on IMM-Directional Track Association. *Appl. Sci.* **2021**, *11*, 11234. [CrossRef]
25. Blom, H.A.P.; Bar-shalom, Y. The interacting multiple model algorithm for systems with Markovian switching coefficients. *IEEE Trans. Autom. Control* **1988**, *33*, 780–783. [CrossRef]
26. Houles, A.; Bar-Shalom, Y. Multisensor Tracking of a Maneuvering Target in Clutter. *IEEE Trans. Aerospace Electron. Syst.* **1989**, *25*, 176–189. [CrossRef]
27. Bar-Shalom, Y.; Li, X.R. *Multitarget-Multisensor Tracking: Principles and Techniques*; YBS Publishing: Storrs, CT, USA, 1995.
28. Babinec, A.; Jiří, A. On accuracy of position estimation from aerial imagery captured by low-flying UAVs. *Int. J. Transp. Sci. Technol.* **2016**, *5*, 152–166. [CrossRef]
29. Cai, Y.; Ding, Y.; Zhang, H.; Xiu, J.; Liu, Z. Geo-Location Algorithm for Building Targets in Oblique Remote Sensing Images Based on Deep Learning and Height Estimation. *Remote Sens.* **2020**, *12*, 2427. [CrossRef]
30. Yeom, S.-W.; Kirubarajan, T.; Bar-Shalom, Y. Track segment association, fine-step IMM and initialization with doppler for improved track performance. *IEEE Trans. Aerosp. Electron. Syst.* **2004**, *40*, 293–309. [CrossRef]
31. Li, X.R.; Jilkov, V.P. Survey of maneuvering target tracking, part I: Dynamic models. *IEEE Trans. Aerosp. Electron. Syst.* **2003**, *39*, 1333–1364.
32. Available online: https://map.naver.com/v5/search/%EA%B2%BD%EC%82%B0ic?c=14336187.1023075,4283701.1549134,15,0,0,1,dh (accessed on 31 January 2022).

drones

MDPI

Article

Acceleration-Aware Path Planning with Waypoints

Rudolf Ortner, Indrajit Kurmi and Oliver Bimber *

Institute of Computer Graphics, Johannes Kepler University Linz, 4040 Linz, Austria; rudolf.ortner@jku.at (R.O.); indrajit.kurmi@jku.at (I.K.)
* Correspondence: oliver.bimber@jku.at

Abstract: In this article we demonstrate that acceleration and deceleration of direction-turning drones at waypoints have a significant influence to path planning which is important to be considered for time-critical applications, such as drone-supported search and rescue. We present a new path planning approach that takes acceleration and deceleration into account. It follows a local gradient ascend strategy which locally minimizes turns while maximizing search probability accumulation. Our approach outperforms classic coverage-based path planning algorithms, such as spiral- and grid-search, as well as potential field methods that consider search probability distributions. We apply this method in the context of autonomous search and rescue drones and in combination with a novel synthetic aperture imaging technique, called Airborne Optical Sectioning (AOS), which removes occlusion of vegetation and forest in real-time.

Keywords: search and rescue; autonomous drones; path planning; acceleration; deceleration

Citation: Ortner, R.; Kurmi, I.; Bimber, O. Acceleration-Aware Path Planning with Waypoints. *Drones* **2021**, *5*, 143. https://doi.org/10.3390/drones5040143

Academic Editors: Diego González-Aguilera and Pablo Rodríguez-Gonzálvez

Received: 29 October 2021
Accepted: 26 November 2021
Published: 27 November 2021

1. Introduction

Autonomous UAVs are becoming more and more adept at handling complex tasks and are thus used in various civil and commercial applications [1]. However, autonomous and adaptive path planning still poses serious challenges due to numerous constraints, such as limited energy, speed, and payload. An additional constraint in search and rescue scenarios can be considered of locating the target as fast as possible. This type of problem is well studied in literature and is termed as a minimum time search problem (MTS) [2–14]. The most prominent objective in these approaches is to optimize the expected time of target detection [3–6]; however, other alternative approaches involve optimizing the probability of target detection [7–9,15], minimizing its counterpart, i.e., probability of non-detection [10,11] or maximizing the information gain [12,13,16]. Various sub optimal and heuristics-based algorithms such as gradient-based approaches [7,10–12,15], cross-entropy optimization [2,5], Bayesian optimization algorithms [4], ant colony optimization [6], or genetic algorithms [3] have been proposed to address the NP-hard complex problem [13]. These approaches can also be differentiated based on the considered UAV dynamics models, where they either do not consider velocity at all [2,4–9,15], or only consider simple linear velocity models [3,10,11] but not acceleration or deceleration.

In addition to efficient path planning, effective imaging is also essential for wilderness search and rescue operations. Large depth of field of conventional cameras (resultant of having a narrow aperture) often project sharply the entire occlusion volumes (such as forests) into the images captured. Objects of interest (people in search and rescue scenario) at a particular distance often remain occluded by the occluders (such as forests). Airborne optical sectioning (AOS) is a wide synthetic-aperture aerial imaging technique that applies camera drones for the real-time removal of occlusion caused by vegetation, such as forests [15,17–25]. It has been demonstrated as a capable and effective tool in various applications (such as archaeology [17], wildlife observation [21], and search and rescue [15,24]). AOS' efficiency concerning the occlusion density, occluder sizes, number of integrated samples, and size of the synthetic aperture has been explained by employing a randomly distributed statistical model [19,25]. By computationally integrating individual images captured over a large scan area (possibly hundreds to thousands of square meters) with narrow aperture camera optics, AOS generates integral images of an extremely shallow depth of field below an occluding volume (cf. Figure 1b). These images enable optical slicing through dense occlusion (caused by leaves, branches, and bushes) and reveal focused targets in each slice (such as artefacts, objects, wildlife, or persons) which would remain occluded for regular cameras (cf. Figure 1c). A fully autonomous and classification-driven UAV (cf. Figure 1a) has been developed and deployed for carrying out wilderness search and rescue operations [15]. The system presented in [15] comprises of two essential and independent modules (AOS based imaging and classification module and an adaptive path planning module). Thermal images are acquired in a 1D sampling pattern and integrated for achieving an occlusion free view of the ground/target. A pre-trained deep learning network achieves an average precision of 86% for detecting person in integral images. In [24] we have already demonstrated how classification of partially occluded persons in forests using aerial thermal images is significantly more effective when AOS is used to integrate single images before classification rather than combined classification results of single images. However, certain environmental conditions (warm background temperature, precipitation, fog, etc.) can affect the performance of the thermal imaging system and thus limiting the efficiency of AOS.

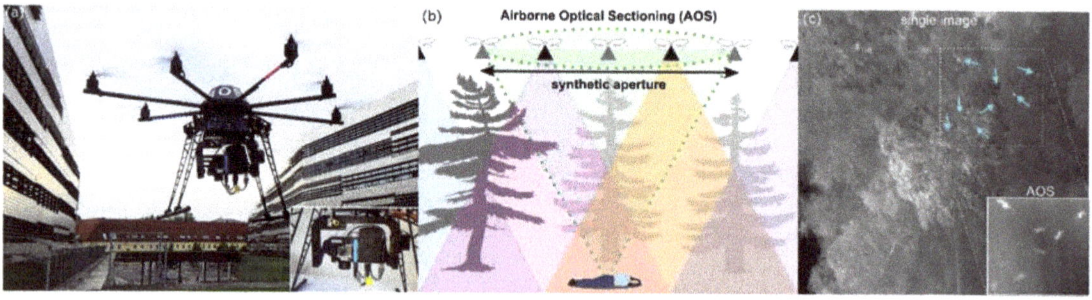

Figure 1. Airborne Optical Sectioning (AOS). (**a**) An autonomous drone [15] was developed and deployed for search and rescue with its payload (thermal camera, Raspberry Pi, Intel Neural Compute Stick, LTE hat) shown in the inset. (**b**) Wide synthetic aperture imaging principle with AOS for search and rescue purposes. Single images captured through a large scan are computationally integrated (registered to the ground surface and averaged) to remove occlusion [15,17–25]. (**c**) Occlusion removal result with AOS: thermal signature of occluded people in single images is quite similar to that of trees. With AOS, we achieve an unoccluded view (integral image) of the people in real-time. A pre-trained deep learning classifier automatically detects them with >90% average precision [15,24].

In our previous work [15], a potential-field [14] based adaptive path planning was applied that was driven by confidences from a deep-learning person classifier collected during the flight. The probability map used by the path-planning algorithm was continuously updated with classification confidence values of potential findings. However, acceleration and deceleration were also not considered, which makes path-planning highly unrealistic in practice—especially for rotor-based drone systems that navigate through waypoints. In this work, we focus on achieving a linear velocity over the scanned region while considering the UAV's acceleration and deceleration to plan its trajectories. Achieving constant velocity over the scanned region is essential for uniform sampling in AOS [15]. We also propose a gradient-based approach that maximizes the probability of target detection with the above mentioned limitations. This proposed path planning algorithm can directly replace our old potential-field based method while utilizing the other aspects of AOS (e.g., integral imaging, person classification) for target detection as it is described in [15]. Our trajectory planning, which takes a dynamic UAV model with acceleration and deceleration into account, is described in Section 2.1 whereas Section 2.2 describes our proposed algorithm to maximize the probability of target detection. Section 3 presents simulation results for various representative probability distributions, and in Section 4 we discuss limitations of our approach and the potential of future UAV models and flight controllers with more dynamic maneuver capabilities for path planning.

2. Materials and Methods

Most commercial drones (especially rotor-based drones) only support piecewise linear waypoint flights and are still not able to fly continuously at a high uniform speed. They need to stop, or slow down (due to limitations of maneuvering and tracking speed capability), before changing directions. Not considering the changes in acceleration and deceleration leads to an unrealistic path planning, as illustrated in Section 2.1. In Section 2.2, we present a new path planning approach that considers acceleration/deceleration and outperforms our previous potential field based method used in [15], as well as classic coverage-based path planning algorithms (i.e., spiral search or grid search).

2.1. Acceleration-Driven Trajectory Planning

As in [15], we want to assume the search region to be discretized into a uniform grid of 30 m × 30 m cells, and that cells are sampled horizontally, vertically, or diagonally at constant velocity to ensure full coverage of the cell within the drones' field-of-view and uniform sampling. Similar to our work in [15], we utilize calibrated cameras and GPS information to project the digital elevation model of the terrain within each cell. The drone samples multiple single images while crossing each cell and combines them to AOS integral images. Flight speed within a cell must be constant to ensure uniform sampling for AOS, since the person classifier used (YOLOv4-tiny network architecture [26]) was trained with uniformly sampled image data. A non-uniform sampling which differs from the training data would significantly reduce classification rate. Person classification is carried out for each integral image. Each cell is associated with the probability of a person being found within it (cf. Figure 2). The probability maps are initially defined by the rescue team (a uniform probability map is assumed in its absence), and is adapted during flight based on confidence scores of the person classifier. The quality of the initial probability map accounts for how much area has to be scanned until the person is potentially found. In the worst case, the whole area has to be covered. A discussion on adaptive sampling and person classification is out of the scope of this article, and is independent of the presented path-planning approach. The interested reader is referred to [15] for more details. Thus far, constant flight speeds were assumed for the entire path through search area, which does not hold in practice because of finitely fast accelerations and decelerations at the turning waypoints.

Our new trajectory planner ensures that each scanned cell is crossed exactly with the desired and constant scanning speed (see Figure 3a). This, however, introduces additional linear trajectories (for acceleration and deceleration shown in Figure 3b) at both ends (entrance and exit) of the scanned cell. Before entering it, the drone must be decelerated to the desired scan speed and after leaving it the drone can be accelerated again to the drone's maximum velocity (to progress quicker to the next scan cell, see Figure 3c). We utilize a kinematics based linear motion model to generate these piecewise linear path segments. The distance between the entrance and exit edges of the scanned cell and the auxiliary waypoints at which deceleration starts and acceleration is finished are computed as follows: $d = \left(v_2^2 - v_1^2\right)/(2a)$, where v_1, v_2 are the velocities before and after acceleration/deceleration, and a the rate of acceleration/deceleration (positive/negative).

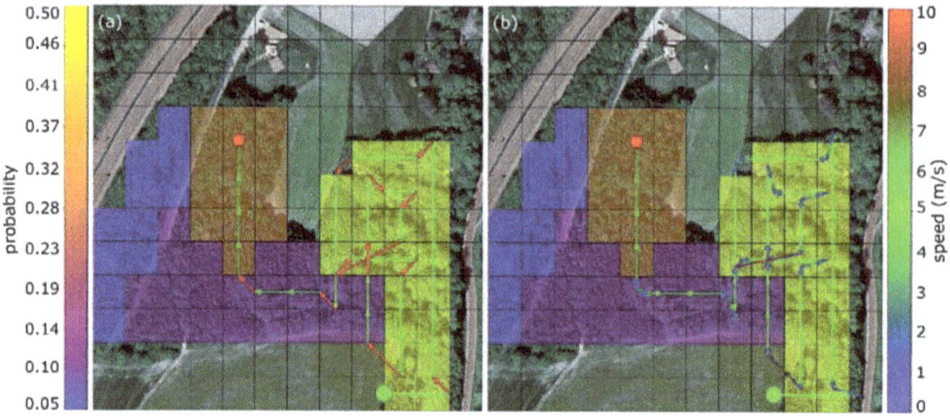

Figure 2. Probability map of a practical search and rescue scenario considered in [15]. The potential field algorithm is used for path planning, as explained in [15]. (**a**) Ignoring acceleration/deceleration leads to an unrealistically short flight path and time of 1291 m and 225 s. (**b**) Considering acceleration/deceleration increases path-length and flight-time to 1489 m and 380 s, respectively. Start point (green circle), person found (red circle), detection probabilities (colors of cells), drone speed (colors of path segments). The axillary acceleration/deceleration trajectories are the segments that gradually change colors in (**b**). Each cell is 30 m × 30 m. The search area covers 6.3 ha.

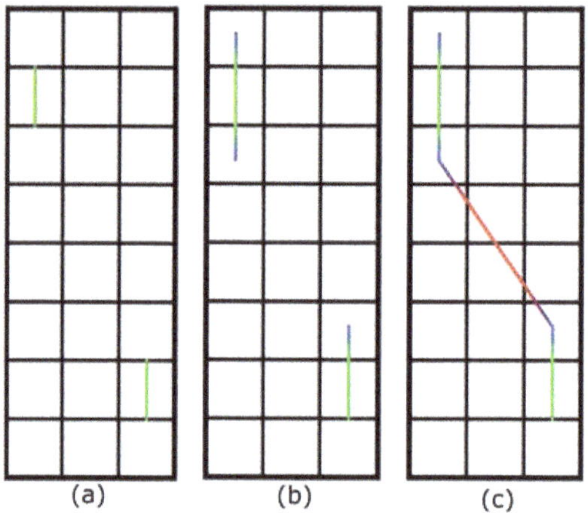

Figure 3. Trajectory planner ensuring constant velocity for scanning over the desired scan cell ((**a**), here top left cell is visited first followed by the bottom right cell). Additional acceleration/deceleration path segments generated on both sides of the scan cells (**b**). Direct line segment in-between scan cells are flown with maximum flight speed (**c**). Colors indicate velocity (blue to red = slow to fast).

Note, that if neighboring cells are scanned in the same flight direction, the drone is not accelerated or decelerated, but continues at constant scan speed. Thus, the auxiliary deceleration and acceleration trajectories are only needed to bridge distances between non-neighboring cells fast, or if flight direction must be changed. They are flown at the maximum speed that the drone supports without overshooting into the scan cell too fast.

Figure 2 illustrates, for the same probability map as used in [15], the difference in potential-field-based adaptive path-planning when acceleration/deceleration is ignored (Figure 2a), as it was the case in [15], and when it is considered (Figure 2b). With our drone prototype (a 4.5 kg MikroKopter OktoXL 6S12, measured average acceleration/deceleration of 1.4 m/s^2, a scan speed of 5 m/s, and a maximum flight speed of 10 m/s) and for a 6.3 ha search area, a flight path of 1291 m and a total of 225 s flight time was determined until the target person is found for constant flying speed (i.e., ignoring acceleration/deceleration). In practice, however, this does not hold because the acceleration/deceleration trajectories are ignored. Considering them as explained above results in a 1489 m long flight path and in a flight time of 380 s instead (an increase in required flight time by a factor of 1.7).

We conclude that for waypoint-based path-planning acceleration driven trajectory planning leads to more realistic results and in more accurate estimates of drones' limited energy and flight time. However, it should be noted that the drones´ energy consumption does not only depend on flight time and acceleration/deceleration but also on many other varying factors like wind, type of drone, gross weight, etc. Our trajectory computes an ideal path without considering these unpredictable outside forces like wind, etc. The flight controllers available on commercial rotor based drones are capable to correct any drifts that are caused by such forces.

When only minimizing flight time or flight distance while ignoring local detection probabilities but considering acceleration/deceleration, classic coverage-based path planning algorithms (i.e., spiral search or grid search) are fastest in traversing the entire search region fully. This is primarily due to a minimum number of direction changes caused by these search techniques, which also minimizes acceleration/deceleration trajectories (see also examples in Appendix A). However, classification probabilities cannot be ignored if the overall goal is to find a person as fast and as reliable as possible. Therefore, we utilize the integral (area) under the sequentially accumulated probability ($p(t)$) curve w.r.t time (APT) for evaluating the efficiency of different path planning algorithms in Section 3. The upper limit of the integral is set by the path traversal time of the fastest algorithm (t_{min}).

$$\text{APT} = \int^{t_{min}} p(t)dt. \tag{1}$$

The following section presents a new path-planning approach that considers acceleration/deceleration, and that outperforms not only our previous potential-field based method, but also spiral search and grid search for AOS-supported search and rescue applications.

2.2. Radial Gradient Accent (RGA)

Observing the need to minimize turns while utilizing an acceleration driven trajectory planner, we propose a new gradient-based method to maximize the probability of target detection while minimizing flight time by reducing turns (cf. Figure 4).

To decide for the next cell to be scanned from the drone's current position, a unique set of directions (called radials) is determined in such a way that each radial will have a unique direction and that all radials (originating all from the current cell's center) together will cross the centers of all unvisited cells. Each unvisited cell is then assigned to its corresponding radial if its center is crossed by it. How many radials exist depends on the resolution of the grid and the number of remaining unvisited cells. For each radial (i.e., the cells assigned to it), we now compute the required trajectories and resulting flight time, as explained in Section 2.1, and select the radial with the highest gradient of accumulated probability w.r.t time (APT). Note, that the APT is similar to the unnormalized cumulative distribution function and the area under this function is the inverse of the Expected Time (ET) of target detection as explained in [2]. The first unvisited cell along this radial is sampled next. Then we iteratively repeat the above process to decide for the next cells to be scanned until all cells are visited or the target is found (i.e., classification with high confidence confirmed by the rescue team, as in [15]).

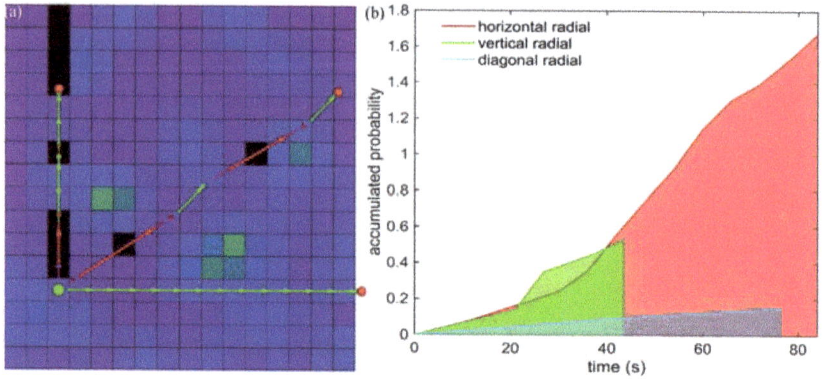

Figure 4. Radial Gradient Accent: (**a**) Three sample radials for a probability map and (**b**) their corresponding APT plots. Cell colors in (**a**) indicate probabilities (as in Figure 2). Black cells are previously visited cells. Green dot and red dots indicate current cell and end points of radials. Colors of trajectories indicate flight speed (as in Figure 2): scan speed (green line segments), max. flight speed (red line segments), acceleration/deceleration (blue line segments). Only cells with the centers located on the corresponding radial are considered for scanning. In this example, the horizontal radial is chosen next as it has the highest APT gradient (**b**).

Overall, this approach follows a local APT gradient ascend strategy with the assumption that after the last cell was scanned, the best local choice for the next scan direction is the one which contributes the highest APT gradient because it maximizes probability accumulation while reducing turns (and with that, flight time). Note that, as explained in Section 2.1, unvisited cell segments are scanned at constant scan speed, but are approached at maximum flight speed. This requires deceleration and acceleration before and after scanning.

3. Results

In this section, we evaluate our Radial Gradient Accent (RGA) method by comparing it against grid search and spiral search which, as fastest coverage-based path planning algorithms, set the speed bars for full grid coverage but do not consider probabilities; and the potential field algorithm in [15] (extended to acceleration/deceleration, as discussed in Section 2.1 and shown in Figure 2a). We apply the probability maps shown in Figure 5 as representative examples for various scenarios during search and rescue operations conducted in the field. Note, however, that the behavior of a lost person depends on many factors (including psychology, physiology, age, gender, etc.) and is not considered here while generating the representative probability maps. For evaluation, only simulations have been carried out to compare the different path planning algorithms under identical conditions. An exact comparison in the field would not be possible, since our online-path planning depends on classification scores which vary with many factors, such as local occlusion, lighting, and wind. Thus, carrying out two search flights under exactly identical conditions is not possible.

Figures 6 and 7 illustrate the performance of the potential field algorithm and our Radial Gradient Accent approach without and with considering acceleration/deceleration on the scattered-smooth probability map (Figure 5b). Results for the other probability maps can be found in Appendix A. Appendix B shows the results of our algorithm for a uniform probability map (assumed in cases where the initial probability map is unreliable or unavailable). The resulting path is more or less uniform spiral like depending on the initial starting position.

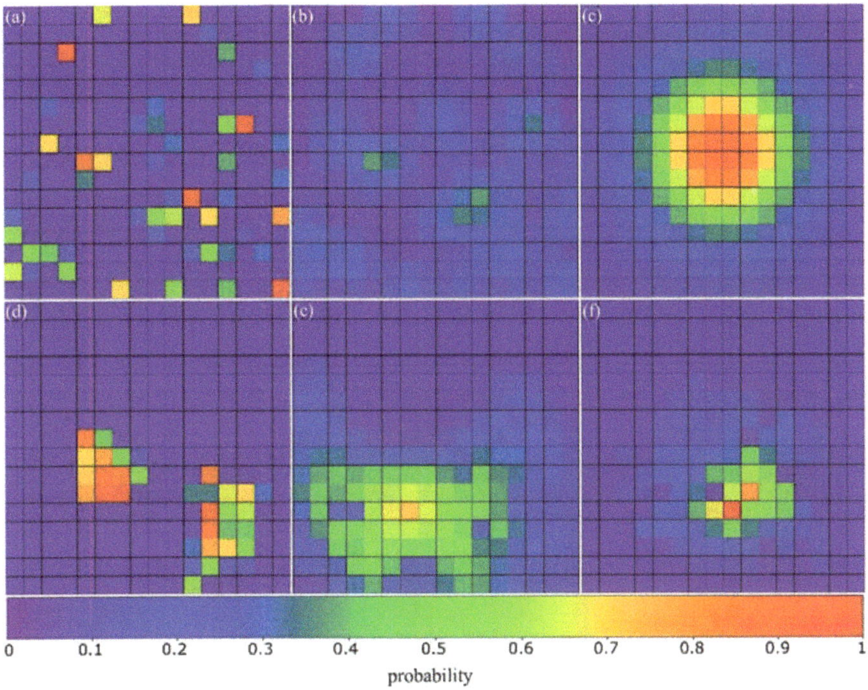

Figure 5. Representative probability maps resembling real-life search and rescue scenarios used for evaluation. (**a**) Multiple scattered locations where the target could be found (*scattered*). (**b**) *Scattered*, but with smooth probability transitions (*scattered-smooth*). (**c**) Exponentially (Gaussian) decreasing probabilities from a certain center location (*exponential*). (**d**) Multiple connected regions (*multiple-patches*). (**e**,**f**) large and small single connected region (*large-patch* and *small-patch*).

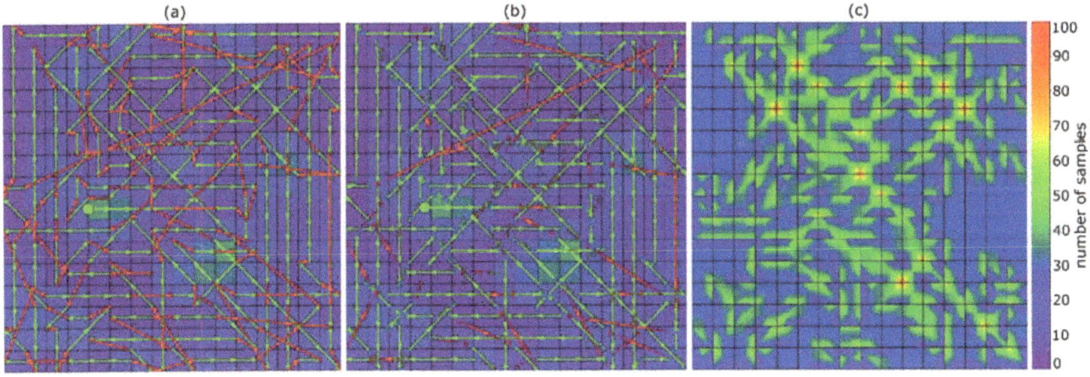

Figure 6. Potential Field. (**a**) Planned trajectory including scan legs without considering acceleration/deceleration. (**b**) Planned trajectory considering acceleration/deceleration. (**c**) Coverage map showing how many images of a certain region have been sampled. The starting point is indicated with the green dot.

Comparing the results in Figures 6 and 7 reveals visually already that RGA requires much less turns (i.e., accelerations/decelerations) and samples more uniformly than the potential field algorithm. A quantitative comparison for all probability maps shown in Figure 5 is presented in Figure 8.

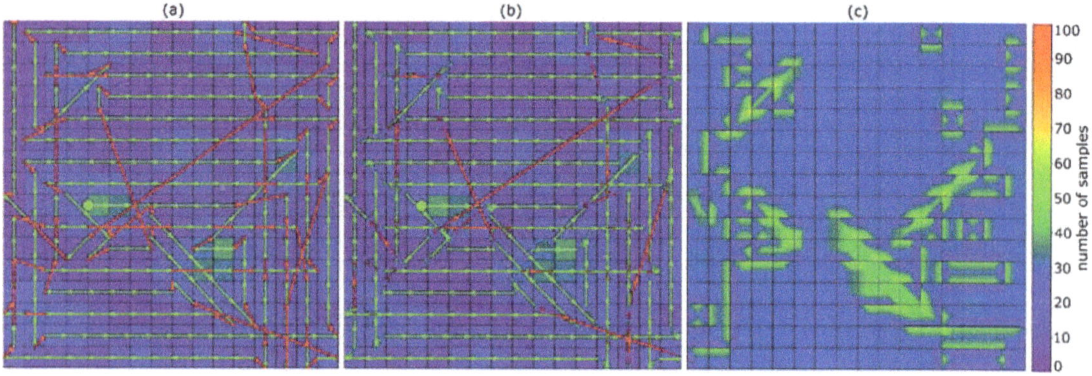

Figure 7. Radial Gradient Accent. (**a**) Planned trajectory including scan legs without considering acceleration/deceleration. (**b**) Planned trajectory considering acceleration/deceleration. (**c**) Coverage map showing how many images of a certain region have been sampled. The starting point is indicated with the green dot.

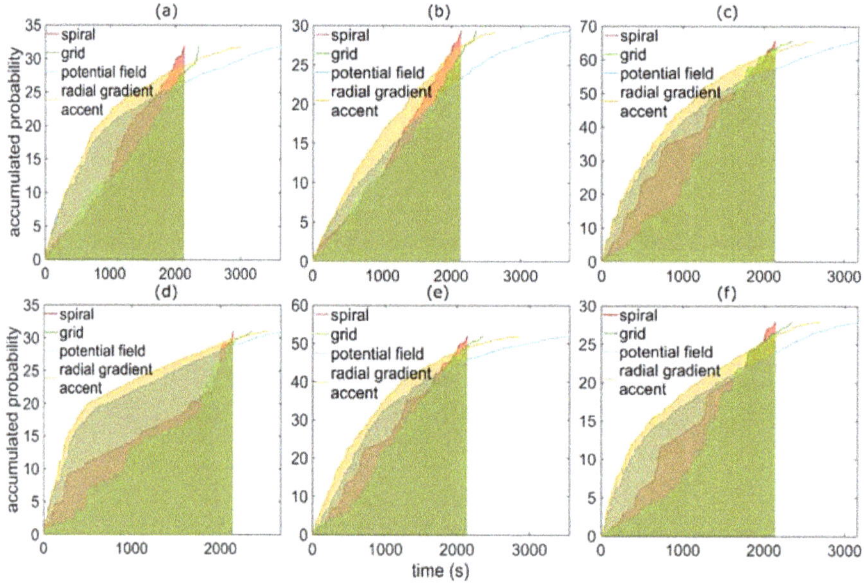

Figure 8. Accumulated probability w.r.t time (APT) plots for different probability maps. (**a**) *scattered*, (**b**) *scattered-smooth*, (**c**) *exponential*, (**d**) *multiple-patches*, (**e**) *large-patch*, (**f**) *small-patch*. The filled areas indicate the integrals up to the shortest full coverage time.

The APT plots in Figure 8 show that although spiral- and grid search always cover the full search region fastest (planned trajectories are presented in Appendix A), they do not maximize detection reliability over time. The latter corresponds to the integral of the accumulated probability w.r.t time plots up to the shortest possible full coverage time. Table 1 presents a quantitative summary.

From the results presented in Figure 8 and Table 1 it can be seen that our RGA approach always outperforms spiral- and grid-search, as well as potential field in all cases as its APT score (which is the accumulated probability w.r.t time integral) is significantly higher (28%, 50%, 11% respectively). Note, that APT combines accumulated detection probability over search time.

Table 1. Quantitative comparison (total flight time and distance, APT score) between considered path planning methods and probability maps. While spiral search covers the full search region fastest without considering probabilities, RGA outperforms all other methods in APT score (integral of the accumulated probability w.r.t time). Note that, depending on the chosen starting point, grid and spiral search lead to slightly different results, and grid search might even be faster than spiral search. However, both methods always result in a much lower APT score than RGA since probabilities are not considered.

Probability Map	Method	Time (s)	Distance (m)	APT Score
scattered	grid	2355.98	9945.46	28,523.70
	spiral	2144.08	9518.96	32,121.20
	potential field	3612.76	17,030.36	39,247.74
	RGA	3026.44	14,433.31	**42,834.65**
scattered-smooth	grid	2355.98	9945.46	26,463.08
	spiral	2144.08	9518.96	28,934.02
	potential field	3704.78	16,799.93	28,814.11
	RGA	2644.96	12,483.69	**34,513.97**
exponential	grid	2355.98	9945.46	62,702.47
	spiral	2144.08	9518.96	77,296.31
	potential field	3173.29	15,238.73	84,015.36
	RGA	2614.88	12,766.08	**91,526.18**
multiple-patches	grid	2355.98	9945.46	26,909.31
	spiral	2144.08	9518.96	32,415.52
	potential field	2678.48	12,225.33	44,628.51
	RGA	2551.18	11,624.64	**47,666.90**
large-patch	grid	2355.98	9945.46	50,121.14
	spiral	2144.08	9518.96	57,197.52
	potential field	3546.04	16,542.41	62,949.78
	RGA	2852.51	13,679.18	**70,344.42**
small-patch	grid	2355.98	9945.46	24,251.82
	spiral	2144.08	9518.96	29,707.75
	potential field	3158.86	14,456.42	34,517.50
	RGA	2691.61	12,677.97	**38,068.94**

4. Discussion and Conclusions

This article demonstrates that considering acceleration and deceleration matters for realistic path planning—especially when drones are applied, where velocity is by far not linear over the flight path. Acceleration and deceleration for waypoint sampling has a significant share of total flight time. Taking this into account is important especially for time-critical applications, such as search and rescue. Furthermore, we presented a new path planning approach, Radial Gradient Accent (RGA), which considers acceleration/deceleration. It follows a local gradient ascend strategy that locally minimizes turns while maximizing probability accumulation. In case of the 16 $\times$ 16 search grid resolutions that were chosen for our evaluations, RGA requires about 2.36 ms on a laptop equipped with Intel(R) Core(TM) i5-8250U CPU @ 1.60GHz and 8GB of RAM. Considering all radials per iteration, however, might be computationally too intensive for higher resolution grids and lower-performance on-board processors. Furthermore, many of the radials intersect only one cell at its center. Achieving a better radial coverage by considering also cells in the vicinity of the radials might lead to performance improvements. RGA is a greedy choice approach that strongly depends on the quality of local decisions. More efficient heuristics than the max. APT gradient might achieve better overall results (i.e., higher APT scores). Our algorithm could also be adapted and applied to drone swarms for conducting faster search and rescue operations. Additional constraints, such as collision and obstacle avoidance also needs to be considered. All of this has to be explored in the future. Currently, we require uniform sampling in AOS as we apply a uniformly distributed occlusion model to represent forest [19]. However, considering more complex, non-uniform occlusion model (e.g., if sparse and densely occluded regions can be measured during flight) will benefit from non-uniform sampling. This will be investigated in future.

The problem of acceleration and deceleration is caused mainly because today's flight controllers used in commercial drones follow waypoints in a piecewise linear fashion. Sharper turns at waypoints require major acceleration and deceleration—especially for rotor-based drones. The influence of acceleration and deceleration can be significantly reduced if future flight controllers (especially in combination with agile wing-based drones and fast tracking) allow polynomial instead of piece-wise linear flight paths. Figure 9 illustrates this effect for the assumption that the drone is physically able to fly continuously at constant velocity (without acceleration/deceleration). Instead of flying from waypoint to waypoint, we simulate continuous heading changes (at differential time steps of 1 s) towards the highest probability gradient (at +/−20 deg limits for smooth heading changes at realistic velocities). As shown in Figure 9 for the *scattered-smooth* probability map, a constant-velocity-based path-planning (*continuous gradient*) clearly outperforms (in APT integral and sampling uniformity) an acceleration-aware path-planning, such as RGA, that is applied for classical waypoint navigation which requires acceleration and deceleration. Thus, improvements in aerodynamics and flight-control open entirely new doors for efficient UAV-supported search and rescue missions.

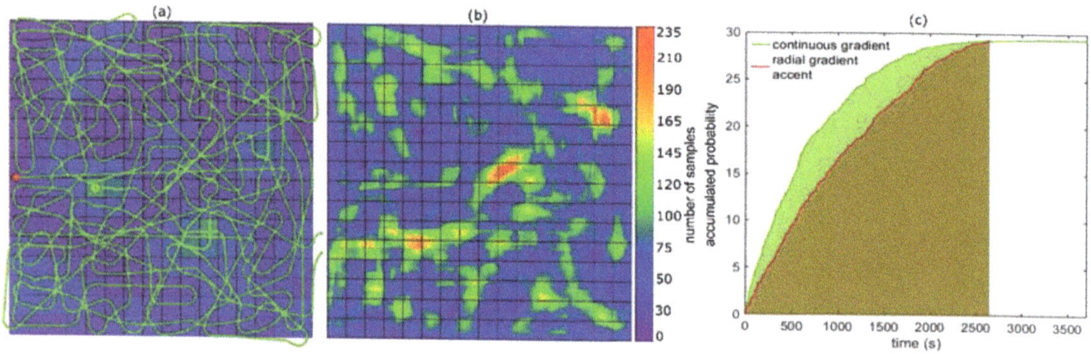

Figure 9. Constant-velocity assumption. (**a**) Planned trajectory with constant velocity and smooth heading changes. (**b**) Coverage map showing how many images of a certain region have been sampled. (**c**) Accumulated probability w.r.t time (APT) plots for acceleration-driven path planning (RGA) and constant-velocity-based path-planning (continuous gradient). The starting point is indicated with the green dot.

Author Contributions: Conceptualization, O.B. and R.O.; methodology, R.O.; software, R.O.; validation, R.O., O.B., and I.K.; formal analysis, R.O.; investigation, R.O.; resources, R.O.; data curation, R.O.; writing—original draft preparation, R.O., O.B., and I.K.; writing—review and editing, R.O., O.B., and I.K.; visualization, R.O.; supervision, O.B.; project administration, O.B.; funding acquisition, O.B. All authors have read and agreed to the published version of the manuscript.

Funding: This research was funded by the Austrian Science Fund (FWF) under grant number P 32185-NBL, and by the State of Upper Austria and the Austrian Federal Ministry of Education, Sci-ence and Research via the LIT–Linz Institute of Technology under grant number LIT-2019-8-SEE-114.

Institutional Review Board Statement: Not applicable.

Informed Consent Statement: Not applicable.

Data Availability Statement: The data and code for acceleration-aware waypoint based path planning approach along with the proposed trajectory planner is available on GitHub: https://github.com/JKU-ICG/AOS/ last accessed on 27 November 2021.

Conflicts of Interest: The authors declare no conflict of interest.

Appendix A

Planned trajectories with (a) and without (b) acceleration/deceleration and overage maps (c) for all probability maps and methods (Table 1). Examples are shown in Figures 6 and 7.

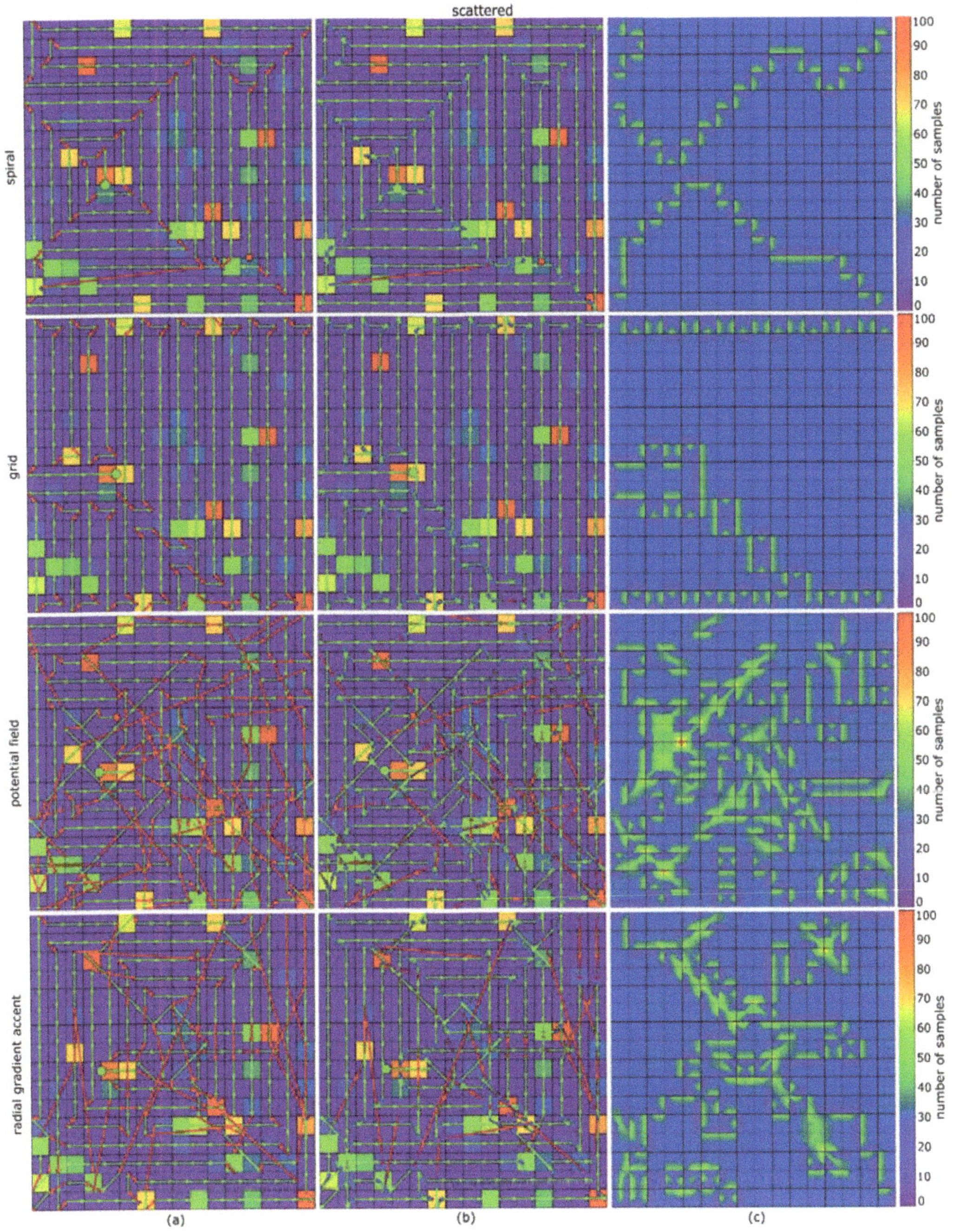

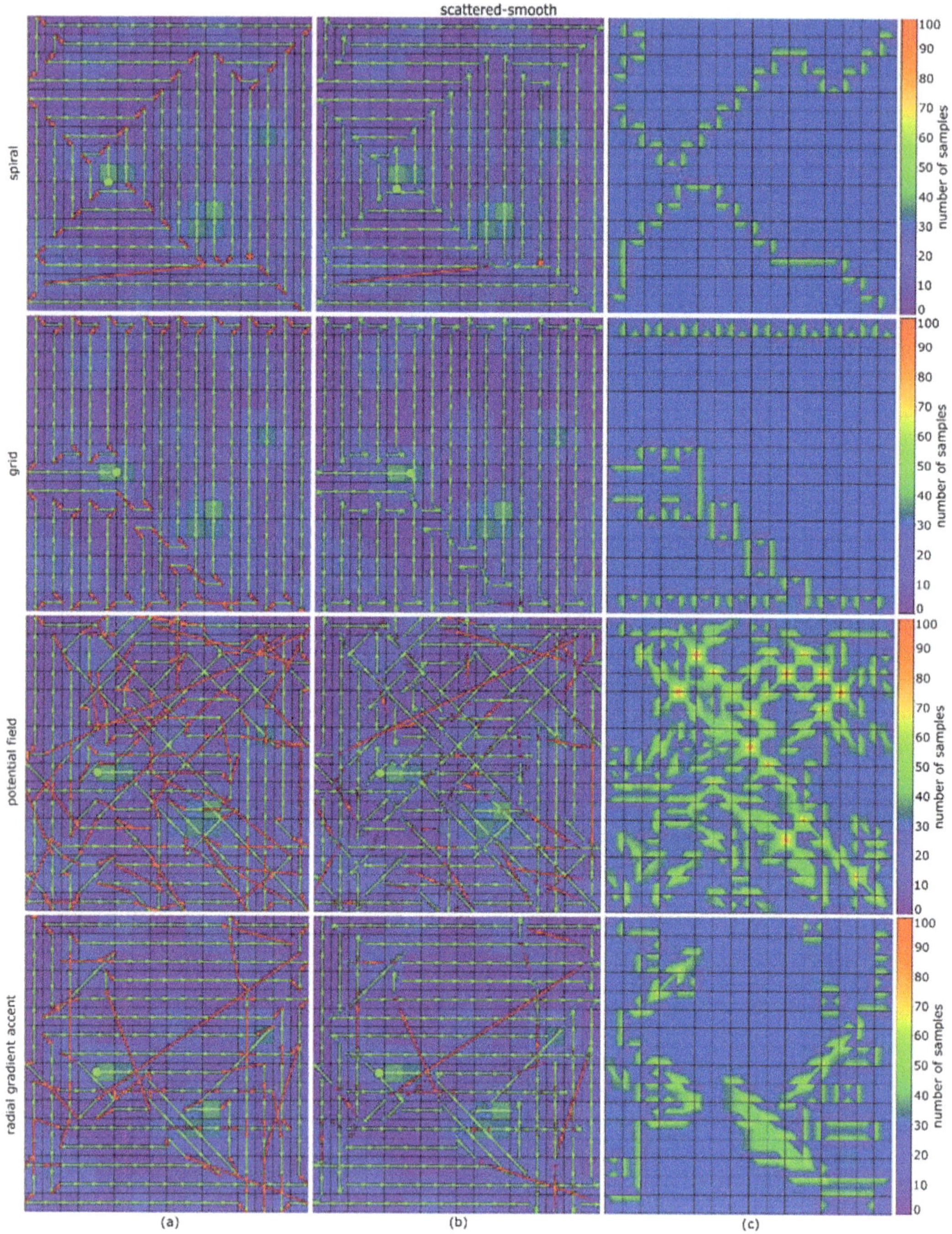
scattered-smooth
spiral
grid
potential field
radial gradient accent
number of samples
0
10
20
30
40
50
60
70
80
90
100
(a)
(b)
(c)

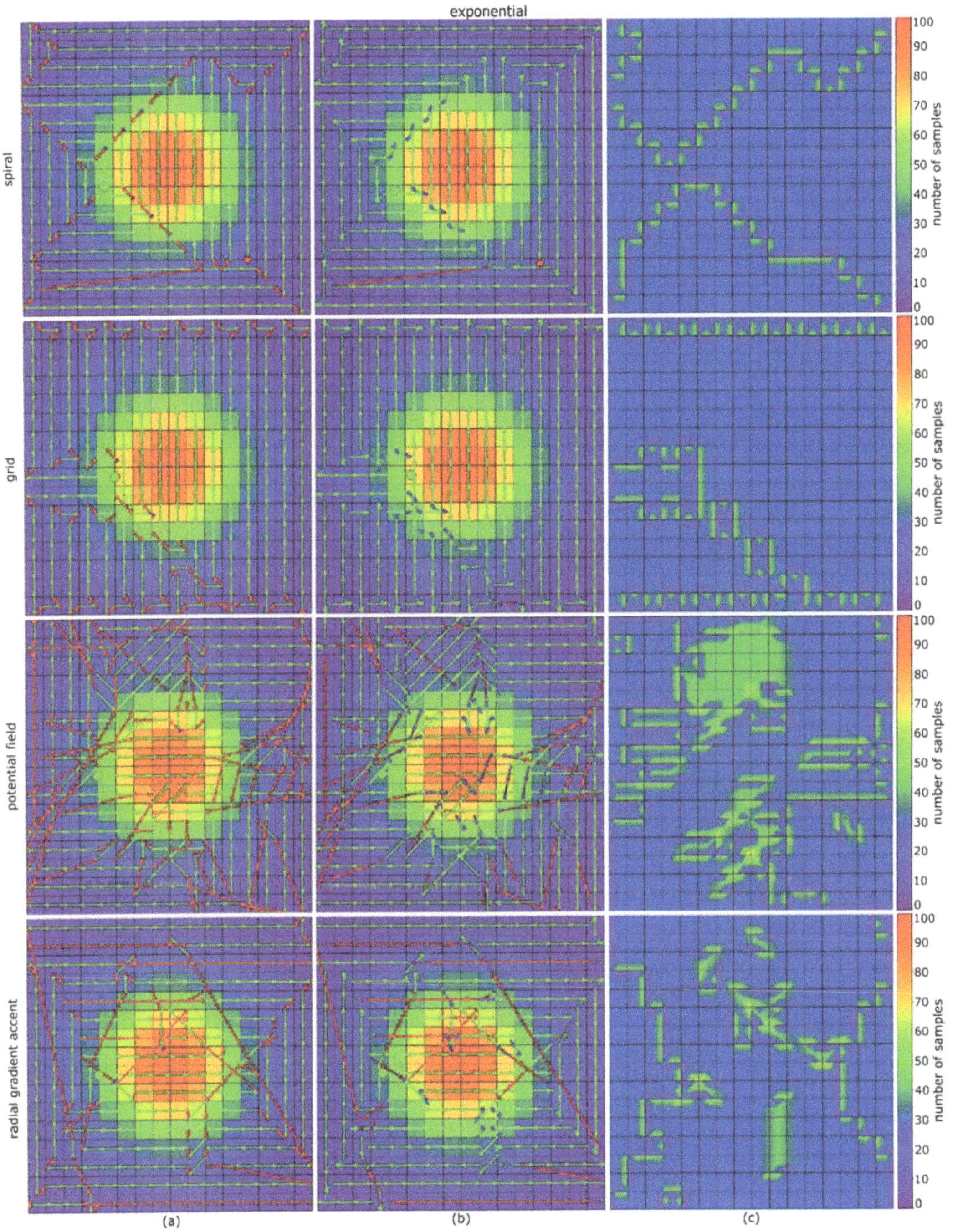
exponential
spiral
grid
potential field
radial gradient accent
number of samples
0
10
20
30
40
50
60
70
80
90
100
(a)
(b)
(c)

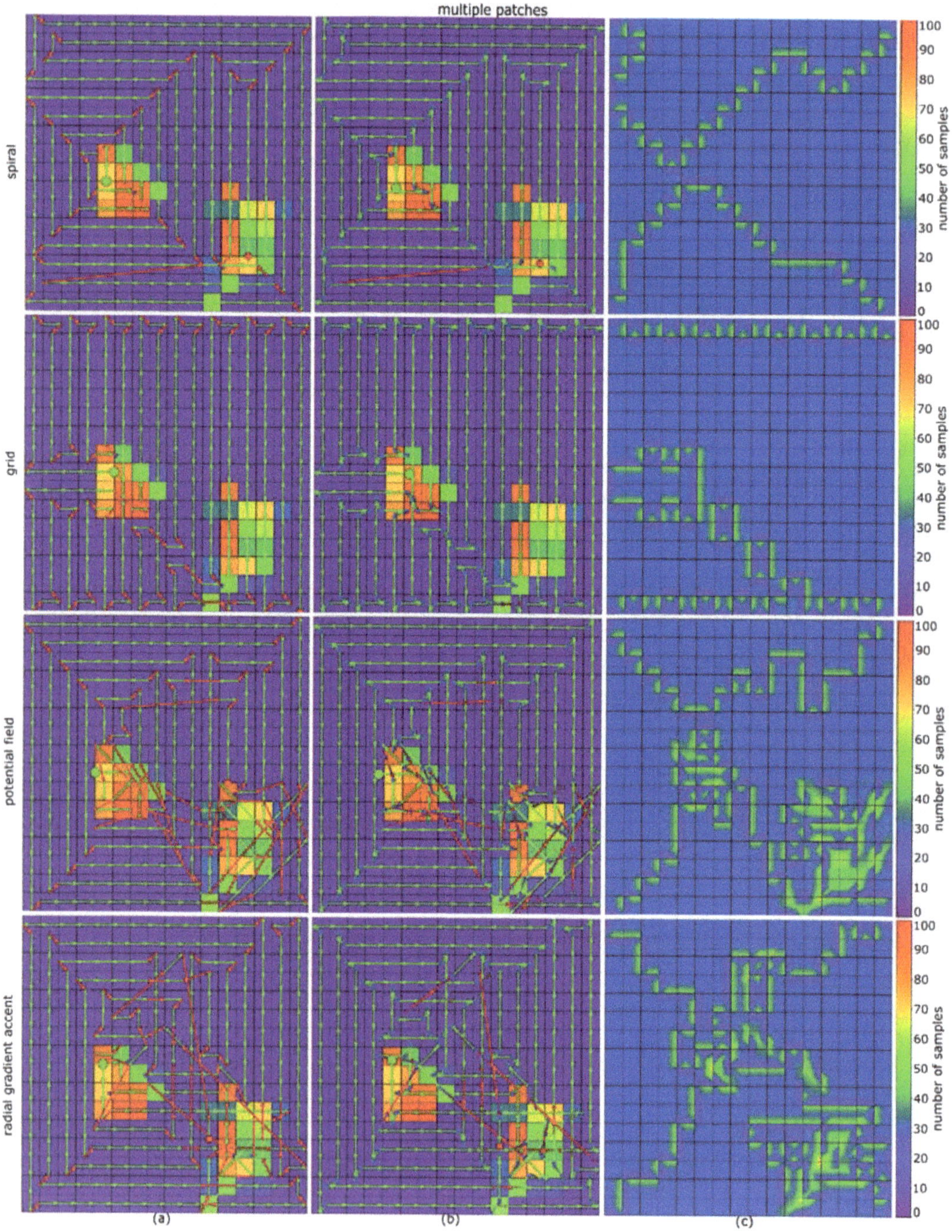
multiple patches
spiral
grid
potential field
radial gradient accent
number of samples
100
90
80
70
60
50
40
30
20
10
0
(a)
(b)
(c)

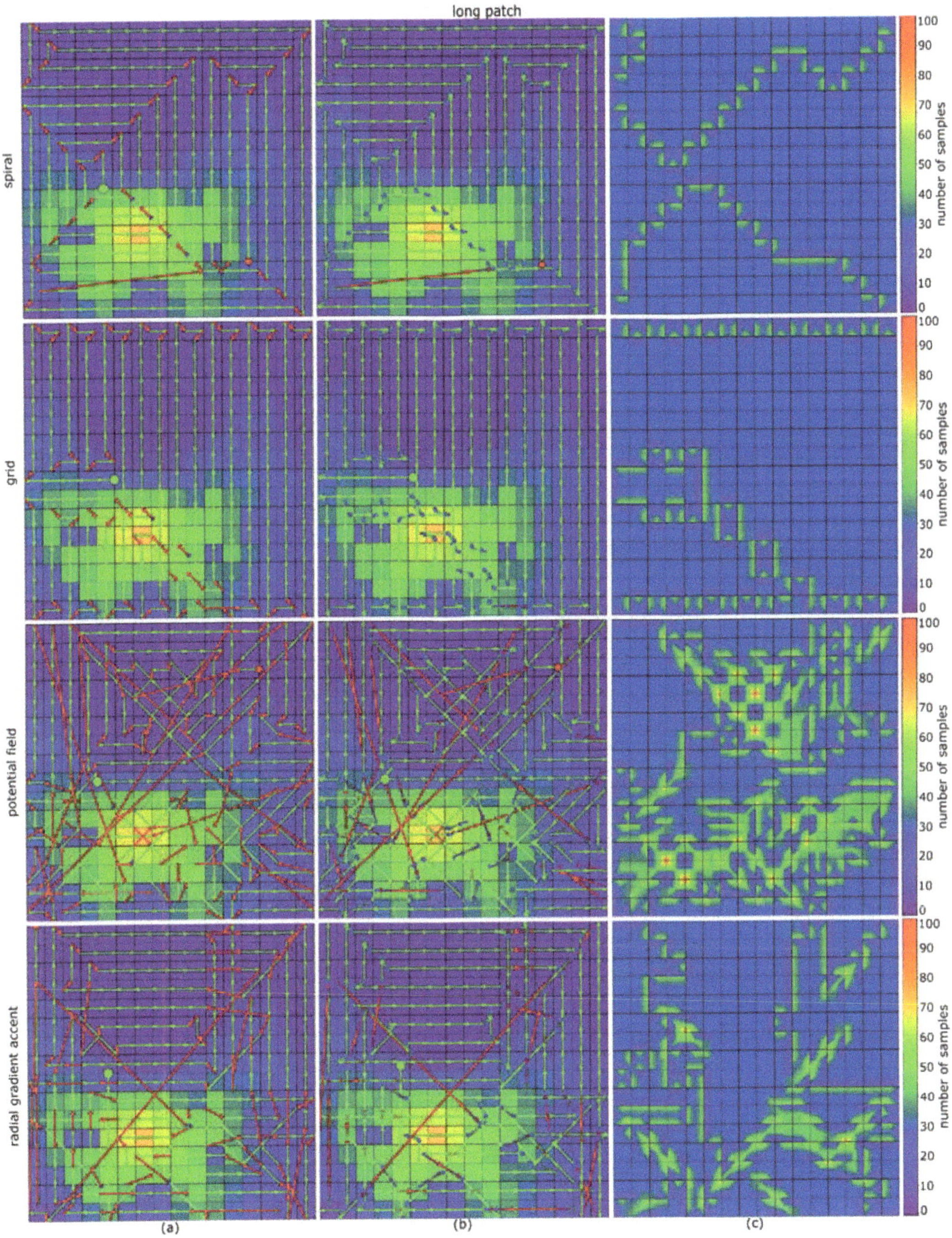
long patch
spiral
grid
potential field
radial gradient accent
number of samples
100
90
80
70
60
50
40
30
20
10
0
(a)
(b)
(c)

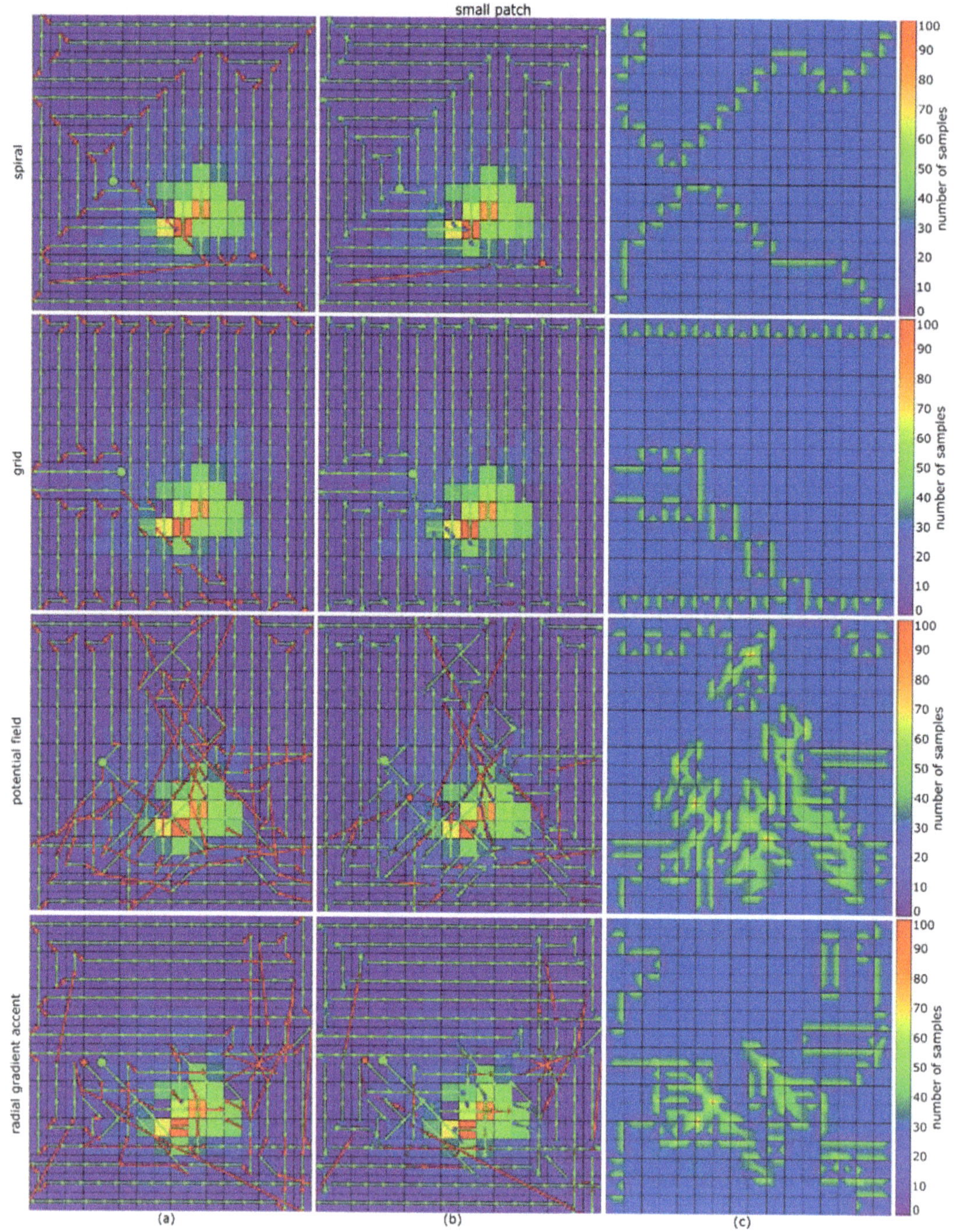
small patch
spiral
grid
potential field
radial gradient accent
number of samples
100
90
80
70
60
50
40
30
20
10
0
(a)
(b)
(c)

Appendix B

Planned trajectories with (a) and without (b) acceleration/deceleration and coverage maps (c) for a uniform probability map and methods (Table 1).

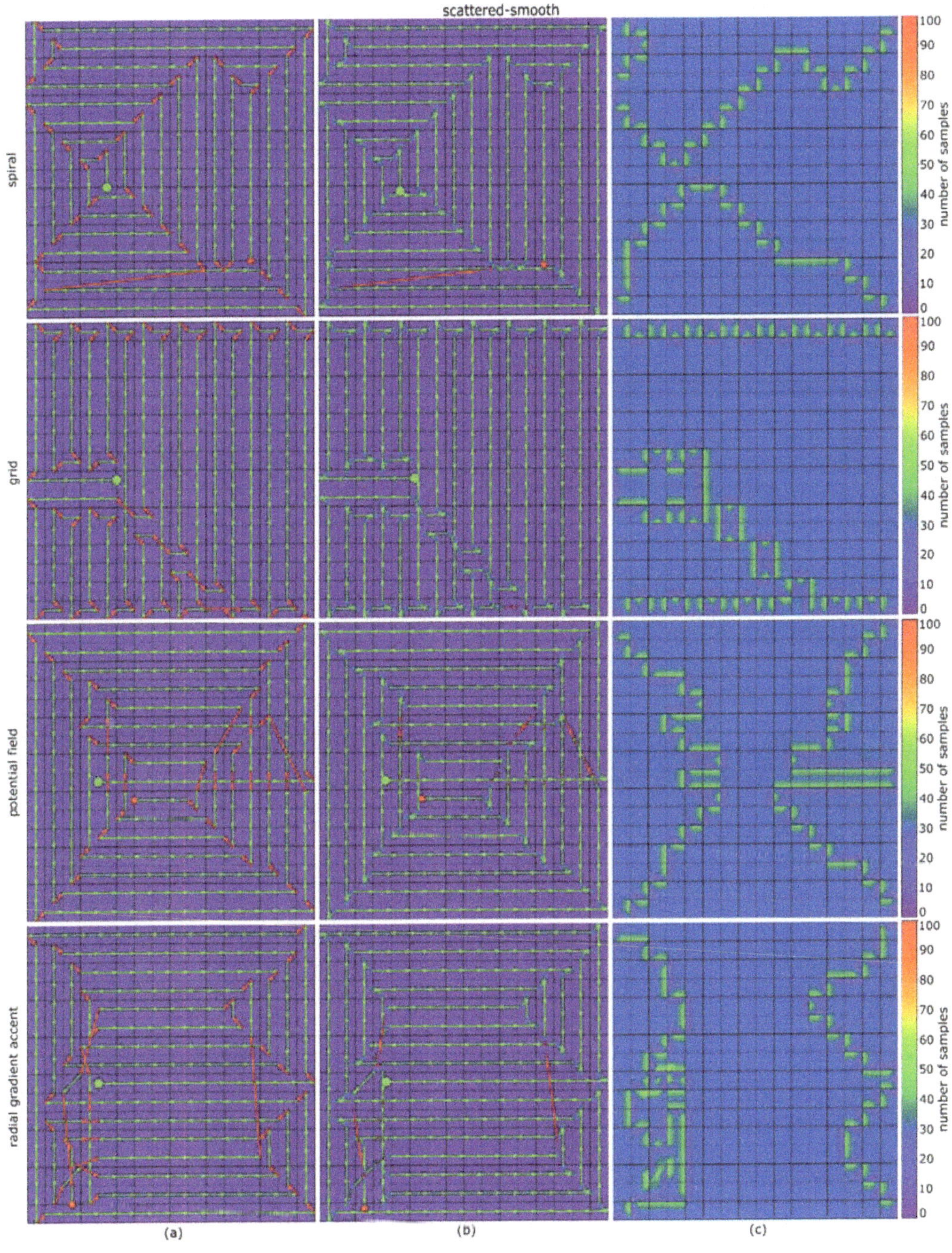

References

1. Shakhatreh, H.; Sawalmeh, A.H.; Al-Fuqaha, A.; Dou, Z.; Almaita, E.; Khalil, I.; Othman, N.S.; Khreishah, A.; Guizani, M. Unmanned Aerial Vehicles (UAVs): A Survey on Civil Applications and Key Research Challenges. *IEEE Access* **2019**, *7*, 48572–48634. [CrossRef]
2. Lanillos, P.; Besada-Portas, E.; Pajares, G.; Ruz, J.J. Minimum time search for lost targets using cross entropy optimization. In Proceedings of the 2012 IEEE/RSJ International Conference on Intelligent Robots and Systems, Vilamoura, Portugal, 7–12 October 2012; pp. 602–609. [CrossRef]
3. Pérez-Carabaza, S.; Besada-Portas, E.; Lopez-Orozco, J.A.; Pajares, G. Minimum Time Search in Real-World Scenarios Using Multiple UAVs with Onboard Orientable Cameras. *J. Sens.* **2019**, *2019*, 7673859. [CrossRef]
4. Lanillos, P.; Yañez-Zuluaga, J.; Ruz, J.J.; Besada-Portas, E. A Bayesian approach for constrained multi-agent minimum time search in uncertain dynamic domains. In Proceedings of the 2013 15th Genetic and Evolutionary Computation Conference, GECCO 2013, Amsterdam, The Netherlands, 6–10 July 2013; pp. 391–398.
5. Lanillos, P.; Besada-Portas, E.; Lopez-Orozco, J.A.; De la Cruz, J.M. Minimum Time Search in Uncertain Dynamic Domains with Complex Sensorial Platforms. *Sensors* **2014**, *14*, 14131–14179. [CrossRef] [PubMed]
6. Perez-Carabaza, S.; Bermudez-Ortega, J.; Besada-Portas, E.; Lopez-Orozco, J.A.; De La Cruz, J.M. A Multi-UAV minimum time search planner based on ACOR. In Proceedings of the 2017 Genetic and Evolutionary Computation Conference, GECCO 2017, Berlin, Germany, 15–19 July 2017; pp. 35–42.
7. Tisdale, J.; Kim, Z.W.; Hedrick, J.K. Autonomous UAV path planning and estimation: An online path planning framework for cooperative search and localization. *IEEE Robot. Autom. Mag.* **2009**, *16*, 35–42. [CrossRef]
8. Wong, E.M.; Bourgault, F.; Furukawa, T. Multi-vehicle Bayesian search for multiple lost targets. In Proceedings of the 2005 IEEE International Conference on Robotics and Automation, Barcelona, Spain, 18–22 April 2005; pp. 3169–3174.
9. Bourgault, F.; Furukawa, T.; Durrant-Whyte, H. Optimal search for a lost target in a bayesian world in Field and Service Robotics. In Proceedings of the Recent Advances in Research and Applications, Lake Yamanaka, Japan, 14–16 July 2003; pp. 209–222.
10. Gan, S.K.; Sukkarieh, S. Multi-UAV target search using explicit decentralized gradient-based negotiation. In Proceedings of the IEEE International Conference on Robotics and Automation, Shanghai, China, 9–13 May 2011; pp. 751–756.
11. Lanillos, P.; Gan, S.K.; Besada-Portas, E.; Pajares, G.; Sukkarieh, S. Multi-UAV target search using decentralized gradient-based negotiation with expected observation. *Inf. Sci.* **2014**, *282*, 92–110. [CrossRef]
12. Hu, J.; Xie, L.; Xu, J.; Xu, Z. Multi-agent cooperative target search. *Sensors* **2014**, *14*, 9408–9428. [CrossRef]
13. Trummel, K.E.; Weisinger, J.R. The complexity of the optimal searcher path problem. *Oper. Res.* **1986**, *34*, 324–327. [CrossRef]
14. Juan, V.S.; Santos, M.; Andújar, J.M. Intelligent UAV map generation and discrete path planning for search and rescue operations. *Complexity* **2018**, *2018*, 6879419. [CrossRef]
15. Schedl, D.C.; Kurmi, I.; Bimber, O. An autonomous drone for search and rescue in forests using airborne optical sectioning. *Sci. Robot.* **2021**, *6*, eabg1188. [CrossRef] [PubMed]
16. Meera, A.A.; Popović, M.; Millane, A.; Siegwart, R. Obstacle-aware Adaptive Informative Path Planning for UAV-based Target Search. In Proceedings of the 2019 International Conference on Robotics and Automation (ICRA), Montreal, QC, Canada, 20–24 May 2019; pp. 718–724. [CrossRef]
17. Kurmi, I.; Schedl, D.C.; Bimber, O. Airborne optical sectioning. *J. Imaging* **2018**, *4*, 102. [CrossRef]
18. Bimber, O.; Kurmi, I.; Schedl, D.C. Synthetic aperture imaging with drones. *IEEE Comput. Graph. Appl.* **2019**, *39*, 8–15. [CrossRef] [PubMed]
19. Kurmi, I.; Schedl, D.C.; Bimber, O. A statistical view on synthetic aperture imaging for occlusion removal. *IEEE Sensors J.* **2019**, *19*, 9374–9383. [CrossRef]
20. Kurmi, I.; Schedl, D.C.; Bimber, O. Thermal airborne optical sectioning. *Remote Sens.* **2019**, *11*, 1668. [CrossRef]
21. Schedl, D.C.; Kurmi, I.; Bimber, O. Airborne optical sectioning for nesting observation. *Sci. Rep.* **2020**, *10*, 7254. [CrossRef] [PubMed]
22. Kurmi, I.; Schedl, D.C.; Bimber, O. Fast Automatic Visibility Optimization for Thermal Synthetic Aperture Visualization. *IEEE Geosci. Remote Sens. Lett.* **2021**, *18*, 836–840. [CrossRef]
23. Kurmi, I.; Schedl, D.C.; Bimber, O. Pose Error Reduction for Focus Enhancement in Thermal Synthetic Aperture Visualization. *IEEE Geosci. Remote. Sens. Lett.* **2021**. to be published. [CrossRef]
24. Schedl, D.C.; Kurmi, I.; Bimber, O. Search and rescue with airborne optical sectioning. *Nat. Mach. Intell.* **2020**, *2*, 783–790. [CrossRef]
25. Kurmi, I.; Schedl, D.C.; Bimber, O. Combined person classification with airborne optical sectioning. *arXiv* **2021**, arXiv:2106.10077.
26. Bochkovskiy, A.; Wang, C.Y.; Liao, H.Y.M. YOLOv4: Optimal Speed and Accuracy of Object Detection. 2020. Unpublished work. Available online: https://arxiv.org/abs/2004.10934 (accessed on 15 November 2021).

drones

MDPI

Article

A Hybrid Approach for Autonomous Collision-Free UAV Navigation in 3D Partially Unknown Dynamic Environments

Taha Elmokadem and Andrey V. Savkin *

School of Electrical Engineering and Telecommunications, The University of New South Wales, Sydney 2052, Australia; t.elmokadem@unsw.edu.au
* Correspondence: a.savkin@unsw.edu.au

Abstract: In the past decades, unmanned aerial vehicles (UAVs) have emerged in a wide range of applications. Owing to the advances in UAV technologies related to sensing, computing, power, etc., it has become possible to carry out missions autonomously. A key component to achieving this goal is the development of safe navigation methods, which is the main focus of this work. A hybrid navigation approach is proposed to allow safe autonomous operations in three-dimensional (3D) partially unknown and dynamic environments. This method combines a global path planning algorithm, namely RRT-Connect, with a reactive control law based on sliding mode control to provide quick reflex-like reactions to newly detected obstacles. The performance of the suggested approach is validated using simulations.

Keywords: UAVs; aerial drones; autonomous navigation; path planning; reactive control; obstacle avoidance; sense-and-avoid; hybrid navigation method; dynamic environments

Citation: Elmokadem, T.; Savkin, A.V. A Hybrid Approach for Autonomous Collision-Free UAV Navigation in 3D Partially Unknown Dynamic Environments *Drones* **2021**, 5, 57. https://doi.org/10.3390/drones5030057

Academic Editors: Diego González-Aguilera and Pablo Rodríguez-Gonzálvez

Received: 15 June 2021
Accepted: 7 July 2021
Published: 8 July 2021

1. Introduction

Unmanned aerial vehicles (UAVs), also known as aerial drones, have emerged in many applications where it is required that some repetitive tasks are performed in a certain environment. Autonomous operation is highly desirable in these applications, which adds more requirements on the vehicle to achieve safe navigation towards areas of interest. Navigation methods can be generally classified as global path planning (deliberative), local path planning (sensor-based), and hybrid. A subset of sensor-based methods include reactive approaches [1].

Global path planning requires an overall knowledge about the environment to produce optimal and efficient paths which can be tracked by the vehicle's control system. There exist many different techniques to address global planning problems, including roadmap methods [2,3], cell decomposition [4], potential field [5,6], probabilistic roadmap (PRM) [7], rapidly exploring random tree (RRT) [8], and optimization-based techniques [9–12]. Some of these techniques become computationally challenging when dealing with unknown and dynamic environments since a complete updated map is required a priori. As a workaround to handle such environments, extensions to some of these approaches were proposed by adding an additional layer to continuously refine the initial path locally around detected obstacles. This still may be less efficient in highly complex and dynamic environments.

On the other hand, sensor-based methods generate local paths or motion commands in real-time based on a locally observed fraction of the environment interpreted directly from sensors measurements. Search-based methods, such as those used for global planning, and optimization-based methods can be used with a local map to generate paths locally where computational complexity depends on the selected map size. On the contrary, reactive methods offer better computational solutions by directly coupling sensors observations into control inputs providing quick reactions to perceived obstacles. Hence, they can be more suitable in unknown and dynamic environments. Examples of classical reactive methods used in unknown environments are dynamic window [13] and curvature velocity [14]. Other classical examples of reactive methods dealing with dynamic obstacles include collision cones [15] and velocity obstacles [16]. A class of reactive approaches adopt a boundary following paradigm to circumvent obstructing obstacles; for examples, see [17–23]. The low computational cost of such methods comes at the expense of being prone to trapping situations. Some researchers suggested a combination of a randomized behavior with the boundary following approach to escape such situations [24]. However, this may sometimes produce very unpredictable motions and even inefficient ones [25] without utilizing previous sensors observation acquired through the motion.

Hybrid strategies tend to address the afromentioned drawbacks by combining both deliberative and reactive approaches for a more efficient navigation behavior in unknown and dynamic environments. There exists a body of literature on hybrid approaches; for examples, see [25–31] and references therein. A hybrid approach was suggested in [25] for navigation in dynamic environments which combined a potential field-based local planner with the A* algorithm as a global planner based on a topological map. Similarly, the work [26] adopted the A* algorithm with binary grid maps while using a variant of the Bug algorithm as a reactive component to address navigation in partially unknown environments. The authors of [27] suggested another hybrid approach for micro aerial vehicles that uses A* for both deliberative and local planning components where the global path gets refined locally around obstacles through replanning processes. The hybrid approach presented in [28] adopted a fuzzy logic-based boundary following technique to implement the reactive layer while an optimal reciprocal collision avoidance (ORCA) algorithm was used in [29]. Sampling-based search methods were also used in some approaches such as [30], where a global planner based on the dynamic rapidly exploring random tree (DRRT) was suggested. Real-time obstacle avoidance was then dealt with by choosing a best candidate trajectory from a sampled set. In [31], the parallel elliptic limit-cycle approach was adopted to implement both global and local planning components.

Many of the existing three-dimensional (3D) hybrid navigation methods consider search-based methods to implement the local planning component. Among those that adopt reactive-based approaches, many have just considered two-dimensional methods by constraining UAV movement to a fixed altitude which does not utilize the full capabilities of UAVs. Therefore, the main contribution of this work is to propose a hybrid 3D navigation strategy for UAVs to allow efficient navigation in partially unknown/dynamic environments. The suggested strategy combines a global path planning layer with a reactive obstacle avoidance control law developed based on a general 3D kinematic model. The global path planning layer, based on RRT-Connect, can produce efficient paths based on the available knowledge about the environment. The sliding mode technique is adopted to implement a boundary following behavior in the reactive layer. This choice provides quick reactions to obstacles with a cheap computational cost compared to search-based and optimization-based local planners. To develop a proper hybrid navigation strategy, implementation of a switching mechanism is presented to handle the transition between the two control laws. Overall, the proposed method can overcome the shortcomings of relying purely on a deliberative or reactive approaches.

This paper is organized as follows. Section 2 provides a formulation of the tackled navigation problem. The suggested hybrid strategy is then presented in Section 3. The performance of this approach is confirmed through different simulation scenarios, which is shown in Section 4. Finally, concluding remarks are made in Section 5.

2. Problem Statement

A general 3D navigation problem is considered here, where a UAV is required to navigate safely in a partially known environment $\mathcal{E} \subset \mathbb{R}^3$. The environment $\mathcal{E}$ contains a set of n obstacles $\mathcal{O} = \{\mathcal{O}_1, \mathcal{O}_2, \cdots, \mathcal{O}_n\} \subset \mathcal{E}$. These obstacles can either be static and known $\mathcal{O}^k$ or unknown static/dynamic $\mathcal{O}^u$ such that $\mathcal{O} = \mathcal{O}^k \cup \mathcal{O}^u$. The main objective is to safely guide the UAV to reach some goal position $p_{goal} \in \mathcal{E} \backslash \{\mathcal{O}\}$ represented in a world coordinate frame $\mathcal{W}$ with obstacle avoidance capability. The UAV starts with an initial map of the environment containing only information about $\mathcal{O}^k$ based on previous knowledge. Unknown environments can also be considered where the UAV can start building a map as it traverses the environment. In this case, $\mathcal{O}^k = \emptyset$ initially. In addition, let $\boldsymbol{p}(t) = [x(t), y(t), z(t)]^T$ be the UAV Cartesian coordinates expressed in $\mathcal{W}$. Note that we represent vector quantities as column vectors; however, they can be written as the transpose of row vectors $[\cdot]^T$ to simplify writing. A safety requirement for the UAV is defined as keeping a safe distance $d_{safe} > 0$ from all obstacles according to the following:

$$d(t) := \min_{\boldsymbol{p}'(t) \in \mathcal{O}} ||\boldsymbol{p}(t) - \boldsymbol{p}'(t)|| \geq d_{safe} \; \forall t \tag{1}$$

where $d(t)$ is the distance to the closest obstacle, and $|| \cdot ||$ is the standard Euclidean norm of a vector in $\mathbb{R}^3$.

We consider a general 3D nonholonomic kinematic model which is applicable to different types of UAVs and autonomous underwater vehicles. A description of this model is given as follows:

$$\dot{\boldsymbol{p}}(t) = V(t)\boldsymbol{r}(t) \tag{2}$$

$$\dot{\boldsymbol{r}}(t) = \boldsymbol{u}(t) \tag{3}$$

$$\boldsymbol{r}(t) \cdot \boldsymbol{u}(t) = 0 \tag{4}$$

where $V(t) \in \mathbb{R}^+$ is the linear speed, and $\boldsymbol{u}(t) \in \mathbb{R}^3$ is a 2 degree-of-freedom (DOF) control input related to angular velocity. The motion direction at any given time instant t (i.e., orientation) is characterized by a unit vector $\boldsymbol{r}(t)$. The condition (4) indicates that the input $\boldsymbol{u}(t)$ is always perpendicular to $\boldsymbol{r}(t)$ which generates a steering-like behavior in 3D. The UAV velocities $V(t)$ and $\boldsymbol{u}(t)$ are regarded as control inputs with some upper bounds (denoted by $V_{max}, u_{max} > 0$) due to physical limitation. These constraints can be expressed as follows:

$$\begin{aligned} 0 \leq V(t) &\leq V_{max} \\ ||\boldsymbol{u}(t)|| &\leq u_{max} \end{aligned} \tag{5}$$

Notice that we consider only forward motions by not allowing negative values for $V(t)$. For constant-speed applications, $V(t)$ is kept constant at some value $\bar{V} > 0$. Additionally, the following assumptions are made.

Assumption 1. *The UAV senses a fraction of the surroundings by onboard sensors, and it can determine the distance to closest obstacle as part of its perception system. An abstract sensing model is considered where obstacles within a distance of $d_{sensing} > 0$ from the UAV can be detected, and bounding geometric primitive can be used to represent the sensed fraction of the object (ex. box, sphere, cylinder, and/or ellipsoid).*

Assumption 2. *Estimates of the UAV's position $\boldsymbol{p}(t)$ and orientation vector $\boldsymbol{r}(t)$ are available.*

The following statement summarizes the considered problem in this work.

Problem 1. *Consider a UAV whose motion can be described by the model (2)–(4). Under assumptions 1 and 2, design control laws for $V(t)$ and $\boldsymbol{u}(t)$ to ensure a collision-free navigation through an unknown or partially known environment $\mathcal{E}$ to reach a goal position $\boldsymbol{p}_{goal}$ by satisfying the safety requirement in (1) and the constraints in (5).*

Remark 1. *The model (2)–(4) can be applicable to fixed-wing UAVs, multi-rotor UAVs, and autonomous underwater vehicles.*

3. Proposed Hybrid Navigation Strategy

Generally, the design of autonomous navigation methods adopts modular structures. The suggested structure in this work for the overall system from a software perspective includes subsystems for perception, high-level navigation, and low-level control. The perception subsystem is responsible for processing onboard sensor measurements to provide meaningful information about the environment. Thus, it can generate an updated global map representation of the environment as well as providing distance and direction to the closest obstacle as required by our reactive control law.

The design of the high-level navigation subsystem is the main contribution of this work, and the proposed design combines a few main components, namely *global path planning*, *motion supervisor*, *path following control*, *reactive control*, and a *switching mechanism*. The architecture of the overall hybrid navigation strategy showing these components is presented in Figure 1.

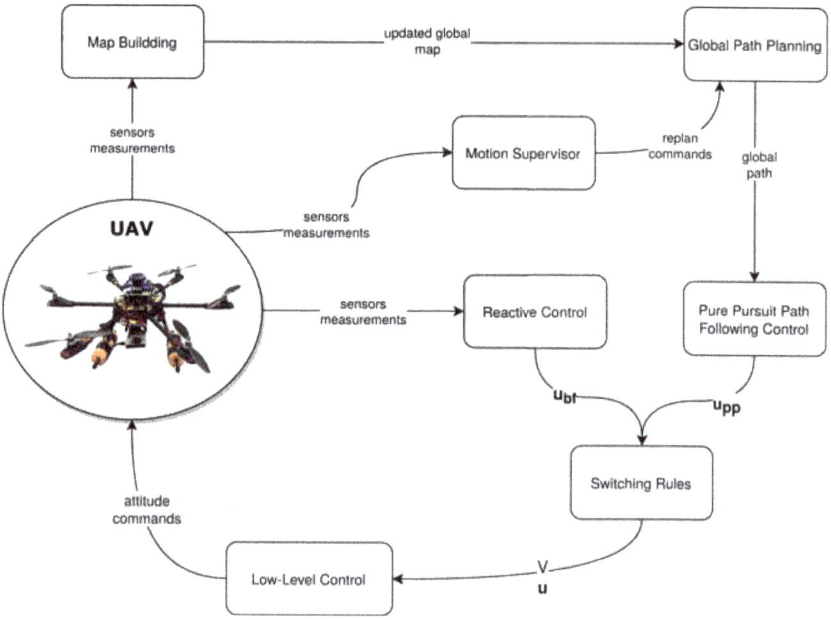

Figure 1. Architecture of the proposed navigation strategy.

The *global path planning* component is responsible for generating feasible and safe geometric paths based on the currently available global map. It is triggered to generate new paths by the *motion supervisor* whenever a new goal position is assigned or whenever a trapping situation (i.e., the UAV gets stuck) is detected as new portions of the environment are discovered. Initially, the map can be empty in unknown environments or partially filled in the case of partially known environments. As the vehicle navigates through the environment, the perception system updates the map based on sensors measurements. To execute the planned global paths, a *path following control* is adopted based on the pure pursuit guidance laws.

In unknown and dynamic environments, the vehicle may find that the planned global paths are unsafe due to the detection of new obstacles. In order to provide quick responses, a *reactive control* component is used to generate locally safe motions around unknown/dynamic obstacles by directly coupling command velocities into current sensors observations. The quick reactions to obstacles provided by reactive control is due to its low computational cost compared to replanning new paths. A mechanism is then used to switch between both path following and reactive navigation modes based on sensors measurements.

A detailed description of these components is given in the following subsections.

Note that the high-level navigation can normally be implemented on an onboard mission computer. The mission computer is normally connected to a Flight control unit (FCU) which is directly connected to all motors. A low-level control subsystem is then used to generate actuators/motors commands to execute the high-level velocity commands generated by the high-level navigation subsystem.

3.1. Global Path Planning

A global path planning algorithm requires a map representation of the environment. There exist many algorithms in the literature where the choice of appropriate method can vary according to the UAV sensing and computing capabilities. Optimality of the planned paths and the computational complexity of the overall algorithm could be key factors in determining which method to consider for a specific application. Note that it is also possible to adopt more than one path planning algorithm where each planner is used under certain conditions. A typical example for this case is the use of optimization-based algorithms to initially generate optimal paths while implementing a computationally efficient algorithm to modify the initial path when replanning is needed.

Our proposed hybrid navigation strategy is not restricted to a specific path planning algorithm. However, this work considers a sampling-based approach as a backbone for implementing the global planner which scales very well when planning in three-dimensional spaces. Namely, a variant of the rapidly exploring random tree (RRT) algorithm is used, which is called RRT-Connect [32]. This choice is due to the algorithm's popularity and low computational cost which is favorable since online replanning might be needed during the motion whenever trapping scenarios are detected. Moreover, we follow an optimistic approach in our implementation of the RRT-Connect algorithm where unknown space is considered to be unoccupied (i.e., obstacle-free). Based on this assumption, the reactive controller will be responsible for handling obstacle avoidance whenever an unknown space is found to be occupied.

RRT algorithms are randomized sampling-based path planning methods which can find collision-free paths if exist with a probability that will reach one as the runtime increases (i.e., probabilistically complete). In practice, these algorithms can provide quick solutions. The basic concept of an RRT planner can be summarized as follows. Let $\mathcal{G}$ be a search tree (a graph) initialized with an initial *configuration* (defined later). An iterative approach is used to extend the search tree through the *configuration space* until a feasible solution is found. At each iteration, a configuration is sampled either randomly or using some heuristics which can help biasing the growth of the search tree. Biasing the sampling process to select the goal configuration with some probability ($0 < pb_{goal} < 1$) was found to enhance RRT growth. The algorithm then tries to extend the search tree $\mathcal{G}$ to the sampled configuration by connecting it to the nearest one within the tree. Different methods could be used to connect configurations within the configuration space especially when trying to satisfy some constraints. However, a common approach is to use straight lines to connect two configurations especially when dealing with Euclidean spaces which is considered here. A collision checker is then used to check the feasibility of each extension based on the available environment map where each feasible extension results in growing the search tree by adding the new configuration. The sampling process gets repeated iteratively until a feasible path between the initial and goal configurations is found or until a stopping criteria is met (for example, exceeding a predefined planning time limit).

RRT-Connect follows the same idea; however, it maintains two trees originating from both initial and goal configurations. At each iteration, the planner attempts to extend one of the trees followed by an attempt to find a collision-free connection to one of the vertices in the other tree. This can provide rapid convergence to a solution in complex environments compared to the standard RRT algorithm.

Due to the sampling nature of RRT algorithms, the generated paths are non-optimal in terms of the overall length. Furthermore, these paths do not satisfy nonholonomic constraints if straight lines were considered when extending the search tree. Therefore, we follow a common practice by refining the obtained paths through two post-processing stages, namely *pruning* and *smoothing*, as was done in [33,34].

For the *pruning* stage, redundant waypoints are removed from the planned path to improve the overall path quality. Let $W = \{w_1, w_2, \cdots, w_k\}$ be a set of waypoints representing a path generated by RRT-Connect, and let W_p be the pruned path obtained after this stage. Redundant waypoints can then be removed using Algorithm 1 based on [33].

Algorithm 1 RRT Path Prunning

Input: $W = \{w_1, w_2, \cdots, w_k\}$
Output: W_p

$W_p \leftarrow \{\}$ ▷ W_p is initialized as an empty list
$i \leftarrow k$
insert w_i into W_p
repeat
    $j \leftarrow 0$
    repeat
        $j \leftarrow j + 1$
    until collisionFree(w_i, w_j) or $j = i - 1$ ▷ Stop when a collision-free segment is found
    insert w_j into W_p
    $i \leftarrow j$
until i = 1

A *smoothing* algorithm is applied next to the pruned path to ensure that the final path satisfies the vehicle's nonholonomic constraints (minimum radius of curvature). To that end, parametric Bezier curves were used to generate continuous-curvature smooth paths following the approach suggested in [34].

3.2. *Pure Pursuit Path Following Control*

A path following control design is needed to ensure that the UAV can accurately track the planned path. The proposed design adopts the pure pursuit tracking (PP) algorithm which is known for its stability and simplicity [35]. Assuming that a geometric path $\tau \subset \mathbb{R}^3$ is available, the PP algorithm steers the vehicle to follow a virtual target $\boldsymbol{p}_v = (x_v, y_v, z_v) \in \tau$ moving along the path. This target is usually selected to be at some lookahead distance L away from the closest path point $\boldsymbol{p}_c \in \tau$ to the vehicle's current position. For more stability, a modified version of the PP algorithm is considered in this work based on [36] which suggested using an adaptive lookahead distance instead of a fixed value. This PP variant provides more stability when the vehicle is further from the planned path which is needed here since the vehicle can sometimes deviate from the planned path when avoiding obstacles with the reactive control component. The following control law is used for path following:

$$V_{pp}(t) = V_{max} \tanh\Big(\gamma \|\boldsymbol{p}_v(t) - \boldsymbol{p}(t)\|\Big) \tag{6}$$

$$\boldsymbol{u}_{pp}(t) = u_{max}\, F\Big(\boldsymbol{r}(t), \frac{\boldsymbol{p}_v(t) - \boldsymbol{p}(t)}{\|\boldsymbol{p}_v(t) - \boldsymbol{p}(t)\|}\Big) \tag{7}$$

where $\gamma > 0$. The mapping function $F(\boldsymbol{w}_1, \boldsymbol{w}_2) : \mathbb{R}^3 \times \mathbb{R}^3 \rightarrow \mathbb{R}^3$ acts as a steering function which produces a vector that is perpendicular to $\boldsymbol{w}_1$ in the direction of $\boldsymbol{w}_2$, and it is defined as follows:

$$F(\boldsymbol{w}_1, \boldsymbol{w}_2) = \begin{cases} \frac{f(\boldsymbol{w}_1, \boldsymbol{w}_2)}{\|f(\boldsymbol{w}_1, \boldsymbol{w}_2)\|}, & \|f(\boldsymbol{w}_1, \boldsymbol{w}_2)\| \neq 0 \\ \boldsymbol{0}_{3\times 1}, & \|f(\boldsymbol{w}_1, \boldsymbol{w}_2)\| = 0 \end{cases} \tag{8}$$

where

$$f(\boldsymbol{w}_1, \boldsymbol{w}_2) = \boldsymbol{w}_2 - (\boldsymbol{w}_1 \cdot \boldsymbol{w}_2)\boldsymbol{w}_1$$

3.3. Reactive Control Law

Navigation in unknown and dynamic environments requires more safety measures as the planned path from the global planner can become unsafe whenever new obstacles are detected, especially if they are dynamic. Hence, a reactive control law is used to generate reflex-like reactions to detected obstacles by navigating around them until it is safe to continue following the previously planned path.

In our implementation, we adopt a reactive control law utilizing a sliding mode control technique based on [37]. The distance to closest obstacle $d(t)$ is the only information needed to implement this controller and can be obtained from onboard sensors. This reactive control is based on a boundary following paradigm, and it is given as follows:

$$V_{bf}(t) = V_{max} \tag{9}$$

$$\boldsymbol{u}_{bf}(t) = \Gamma\, u_{max}\, sgn\Big(\dot{d}(t) + \chi(d(t) - d_0)\Big)\boldsymbol{i}_n(t) \tag{10}$$

$$\boldsymbol{i}_n(t) = \boldsymbol{i}_a(t_*) \times \boldsymbol{r}(t) \tag{11}$$

where a constant forward speed is considered, $d_0 > d_{safe} > 0$ is a desired distance, $\Gamma = \pm 1$ determines the avoidance maneuver direction (i.e., clockwise or counter clockwise with respect to the axis of rotation), and $sgn(\alpha)$ is the signum function. In addition, $\chi(\beta)$ is a saturation function which is defined as:

$$\chi(\beta) = \begin{cases} \gamma\beta & if\ |\beta| \le \delta \\ \delta\gamma\, sgn(\beta) & otherwise \end{cases}, \quad \gamma, \delta > 0 \tag{12}$$

for some design parameters $\gamma, \delta > 0$. In addition, $\boldsymbol{i}_a(t_*)$ represents an avoidance plane normal associated with a certain obstacle as explained in [37]. This normal can be different for each obstacle, and it can be determined at the time instant t_* when a new obstacle is detected.

The reactive navigation law (9)–(11) ensures that the vehicle will maintain a fixed distance d_0 while navigating around the nearest obstacle under some assumptions as was mathematically proven in [20].

3.4. Switching Rules

A switching mechanism is important for hybrid navigation methods. It is responsible for deciding whether it is safe to follow the planned path or to reactively avoid a newly detected obstacle. In general, two navigation modes will be used, and a switching mechanism is adopted to switch between those two modes. The two modes are path following mode $\mathcal{M}_1$ (control law (6) and (7)) and obstacle avoidance/reactive mode $\mathcal{M}_2$ (control law (9)–(11)).

Assuming that the vehicle initially starts in mode $\mathcal{M}_1$, we consider the following switching rules:

R1: switch to the reactive mode $\mathcal{M}_2$ when the distance to the closest obstacle $\mathcal{O}_i$ drops below some threshold distance C (i.e., $d(t) = C$ and $\dot{d}(t) < 0$).

R2: switch to the path following mode $\mathcal{M}_1$ from $\mathcal{M}_2$ when $|d(t) - d_0| < \epsilon_0\ \forall i$ for some small value $\epsilon_0 > 0$ and the vehicle's heading is targeted towards the virtual target $\boldsymbol{p}_v$ on the planned path which can be determined according to the following condition:

$$\left|\cos^{-1}\left(\frac{\boldsymbol{r} \cdot (\boldsymbol{p}_v - \boldsymbol{p})}{\|\boldsymbol{p}_v - \boldsymbol{p}\|}\right)\right| \le \epsilon_1 \tag{13}$$

where $\epsilon_1 > 0$.

3.5. Motion Supervisor

The motion supervisor is responsible for detecting trapping situations whenever the vehicle gets stuck in a local minimum due to a newly discovered fraction of the environment. An example of such scenario can be seen when navigating in maze-like environments, concave obstacles, and/or long blocking walls [38]. The proposed approach to tackle this problem is by issuing a replanning command to the global path planner to acquire a new path based on the updated knowledge about the environment.

4. Simulation Results and Discussion

The developed hybrid navigation approach was tested using simulations considering two different cases, namely static and dynamic environments. Some knowledge about these environments were assumed to be known, a priori, to show the role of the global planner; however, the developed approach can work well even when no such knowledge is available. The RRT-Connect algorithm was implemented based on [32] to plan global paths.

The simulations were carried out using the MATLAB software running on a 2.6 GHz Intel Core i7 CPU. We considered a maximum linear speed of $V = 0.75$ m/s and a maximum angular velocity of $u_{max} = 1.75$ and 2.5 rad/s for the first and second cases, respectively. The design parameters were also selected as follows: $L = 0.5$, $\delta = 0.5$, $\gamma = 1$, $d_0 = 1$, and $C = 2$. An arbitrary choice was made for $\boldsymbol{i}_a(t_*)$ used in (11) which vary for each obstacle. In addition, the update rate of the control was set to be 0.01 s. A description and the results of these simulations are presented next.

4.1. Case I: Unknown Static Obstacles

In the first simulation case, we consider an environment in which the UAV performs some repetitive tasks. Hence, some knowledge about the environment is known a priori, such as wall locations. However, new static obstacles which are not known to the vehicle can be added to the environment at different times. A real-life example of this scenario is when operating in a warehouse or inside a building. The warehouse/building layout can be known in advance or from an initial mapping process while objects can be moved around all the time.

Figure 2 shows the considered environment for this simulation case. The initial map available to the vehicle includes only information about the two walls (shown in gray/black). There are also 4 unknown cylindrical shaped obstacles with different sizes. A bounding shape can usually be estimated in practice by the vehicle's perception system to represent nearby obstacles where very close obstacles can be represented by a single bounding object to satisfy the control assumptions.

It can clearly be seen from Figure 2 that the proposed hybrid navigation strategy can safely guide the vehicle starting from some initial position (green circular marker) to reach the goal location (blue star marker). This figure shows both the path generated by the global planner based on the initial knowledge about the environment as well as the actual executed path by the vehicle. The vehicle can successfully track the planned path whenever it has good clearance from obstacles. Once an obstacle is detected by the vehicle's sensors, the vehicle switches to the reactive mode $\mathcal{M}_2$ to move around the obstacle. Then, it goes back to the path following mode $\mathcal{M}_1$ whenever it is clear to do so according to the switching rule **R2** as described earlier. The distance to the closest obstacle during the motion is given in Figure 3, which clearly shows that the vehicle can satisfy the safety requirement by maintaining a proper clearance from all obstacles. In addition, the linear speed of the vehicle is shown in Figure 4. Overall, these results confirm that the proposed method can guide the vehicle safely among unknown static obstacles.

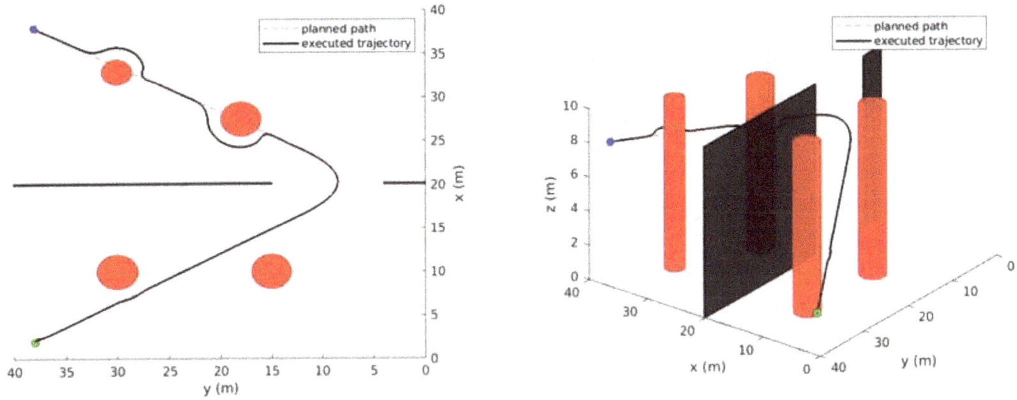

Figure 2. Simulation Case I: Initially planned path and executed motion (static environment).

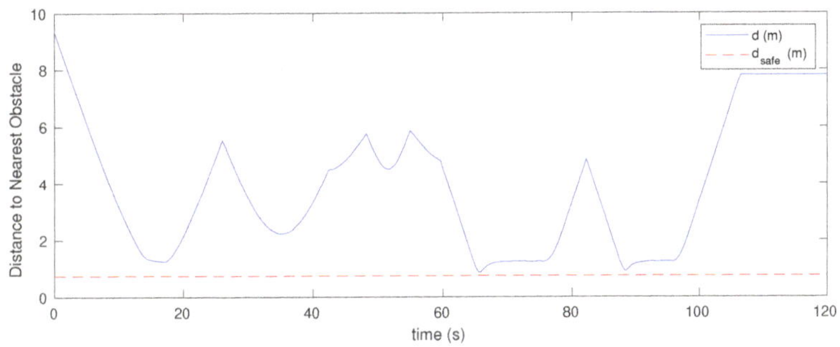

Figure 3. Simulation Case I: Distance from the UAV to obstacles.

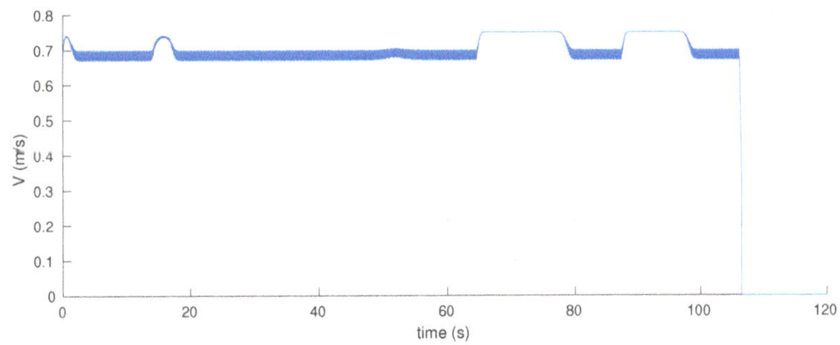

Figure 4. Simulation Case I: Linear speed of the UAV during the motion.

4.2. Case II: Unknown Dynamic Obstacles

The second simulation scenario considers the case where there are several unknown moving obstacles, which makes the environment dynamic. Similar to the previous case, some initial knowledge about the environment layout is assumed to be known, which is shown in gray/black in Figure 5. Moreover, there are multiple unknown dynamic spherical obstacles with random sizes between 0.2 and 1 m in radius and arbitrary linear speeds of 0.1, 0.2, and 0.25 m/s (less than 0.75 m/s). Note that for simplicity, collisions between different obstacles are not considered in this case.

Based on the initial map, the global planner finds a safe path using RRT-Connect, which is shown as a dashed red line in Figure 5. The vehicle then starts moving in mode $\mathcal{M}_1$ to track the planned (reference) path. However, due to the highly dynamic nature of the environment, this path becomes unsafe whenever there are obstacles approaching the vehicle, as shown in Figure 6, at different time instants during the motion. Each time a threatening obstacle is detected, the vehicle switches to navigation in the reactive mode $\mathcal{M}_2$ according to the switching rule **R1**. This can be seen in Figures 5 and 6, which show that the vehicle's actual executed path deviates from the planned path at some locations to avoid the dynamic obstacles. This is verified in Figure 7, where the distance to the closest obstacle remain above the safety threshold. In addition, the UAV's linear speed is shown in Figure 8. It is evident from these results that the proposed strategy also works well in dynamic environments. It should be mentioned that the UAV's maximum velocity must be larger than obstacles' velocities to guarantee safety. However, the reactive control law can still handle some cases where the obstacles are moving faster than the vehicle but with no safety guarantees in some aggressive scenarios.

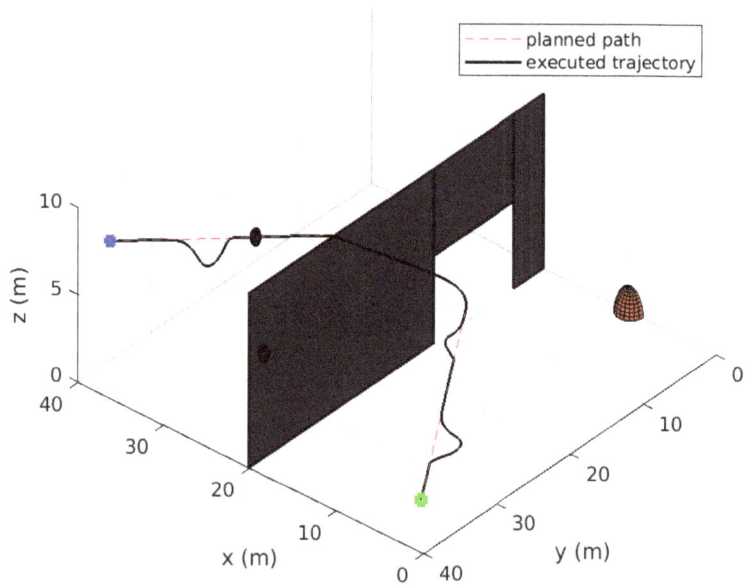

Figure 5. Simulation Case II: Initial planned path and executed motion (dynamic environment).

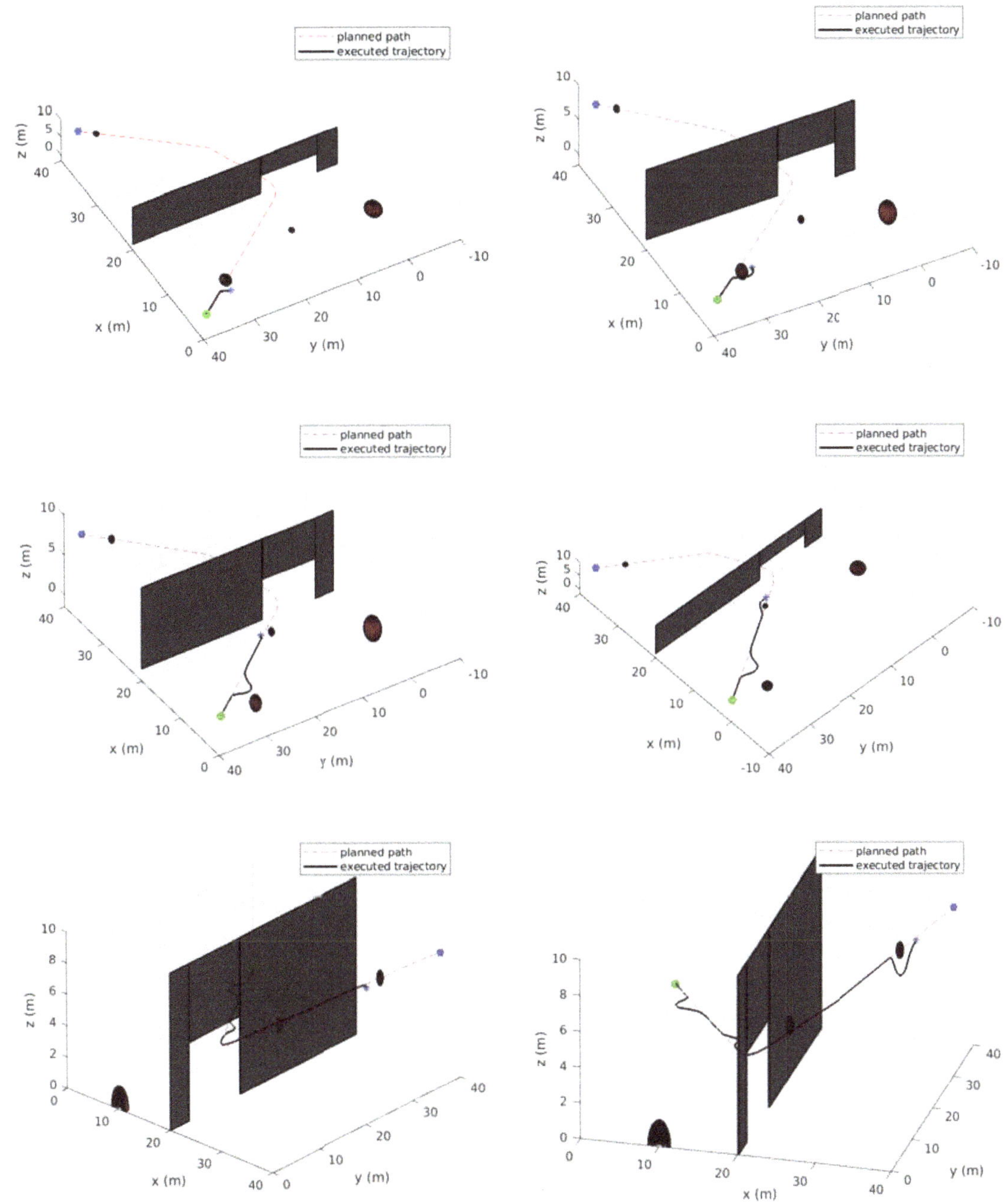

Figure 6. Simulation Case II: Different instances during motion at which switching to reactive mode was triggered.

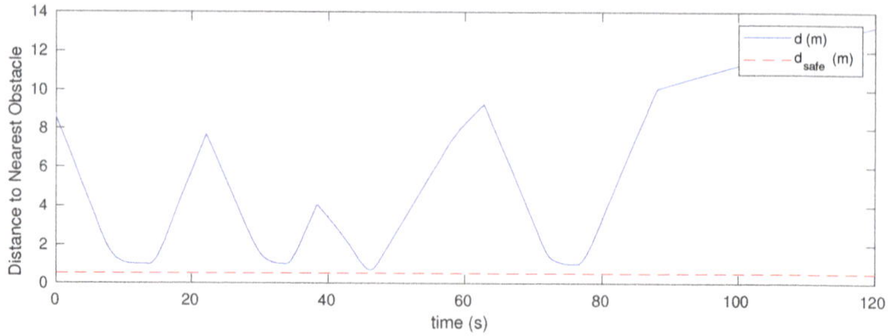

Figure 7. Simulation Case II: Distance from the UAV to obstacles.

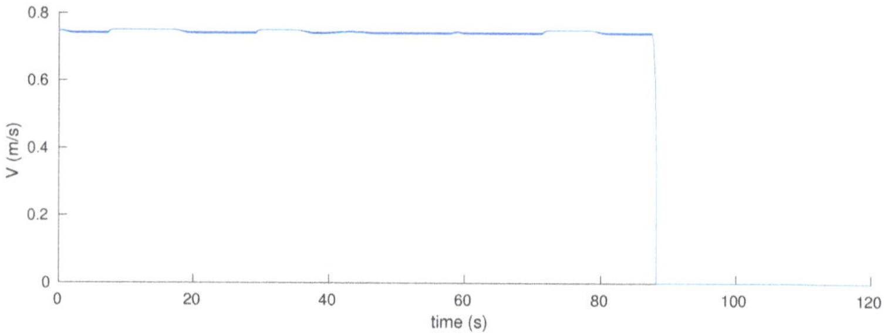

Figure 8. Simulation Case II: Linear speed of the UAV during the motion.

5. Conclusions

A hybrid three-dimensional (3D) navigation strategy for unmanned aerial vehicles (UAVs) was presented in this paper. The problem formulation considered a general 3D non-holonomic kinematic model which is applicable to different UAV types and autonomous underwater vehicles. A global planner based on the RRT-Connect algorithm was used which works well in three-dimensional spaces. A reactive control law based on sliding mode control was used to avoid unknown and dynamic obstacles. Simulation results confirm that the proposed hybrid navigation method works well in 3D environments with unknown static and dynamic obstacles.

Author Contributions: Conceptualization, T.E. and A.V.S.; methodology, T.E.; software, T.E.; validation, T.E.; formal analysis, T.E.; resources, A.V.S.; writing—original draft preparation, T.E.; writing—review and editing, A.V.S.; visualization, T.E.; supervision, A.V.S.; project administration, A.V.S.; funding acquisition, A.V.S. Both authors have read and agreed to the published version of the manuscript.

Funding: This work was supported by the Australian Research Council. This work also received funding from the Australian Government, via grant AUSMURIB000001 associated with ONR MURI grant N00014-19-1-2571.

Institutional Review Board Statement: Not applicable.

Informed Consent Statement: Not applicable.

Conflicts of Interest: The authors declare no conflict of interest.

References

1. Hoy, M.; Matveev, A.S.; Savkin, A.V. Algorithms for collision-free navigation of mobile robots in complex cluttered environments: a survey. *Robotica* **2015**, *33*, 463–497. [CrossRef]
2. Lozano-Pérez, T.; Wesley, M.A. An algorithm for planning collision-free paths among polyhedral obstacles. *Commun. ACM* **1979**, *22*, 560–570. [CrossRef]
3. Ó'Dúnlaing, C.; Yap, C.K. A "retraction" method for planning the motion of a disc. *J. Algorithms* **1985**, *6*, 104–111. [CrossRef]
4. Lozano-Perez, T. Spatial planning: A configuration space approach. *IEEE Trans. Comput.* **1983**, *C-32*, 108–120. [CrossRef]

5. Khatib, O. Real-time obstacle avoidance for manipulators and mobile robots. *Int. J. Robot. Res.* **1986**, *5*, 90–98. [CrossRef]
6. Ge, S.S.; Cui, Y.J. New potential functions for mobile robot path planning. *IEEE Trans. Robot. Autom.* **2000**, *16*, 615–620. [CrossRef]
7. Kavraki, L.E.; Svestka, P.; Latombe, J.C.; Overmars, M.H. Probabilistic roadmaps for path planning in high-dimensional configuration spaces. *IEEE Trans. Robot. Autom.* **1996**, *12*, 566–580. [CrossRef]
8. LaValle, S.M. *Rapidly-Exploring Random Trees: A New Tool for Path Planning*; The Annual Research Report; Iowa State University: Ames, IA, USA, 1998.
9. Dragan, A.D.; Ratliff, N.D.; Srinivasa, S.S. Manipulation planning with goal sets using constrained trajectory optimization. In Proceedings of the 2011 IEEE International Conference on Robotics and Automation (ICRA), Shanghai, China, 9–13 May 2011; pp. 4582–4588.
10. Zucker, M.; Ratliff, N.; Dragan, A.D.; Pivtoraiko, M.; Klingensmith, M.; Dellin, C.M.; Bagnell, J.A.; Srinivasa, S.S. Chomp: Covariant hamiltonian optimization for motion planning. *Int. J. Robot. Res.* **2013**, *32*, 1164–1193. [CrossRef]
11. Schulman, J.; Duan, Y.; Ho, J.; Lee, A.; Awwal, I.; Bradlow, H.; Pan, J.; Patil, S.; Goldberg, K.; Abbeel, P. Motion planning with sequential convex optimization and convex collision checking. *Int. J. Robot. Res.* **2014**, *33*, 1251–1270. [CrossRef]
12. Li, G.; Chou, W. Path planning for mobile robot using self-adaptive learning particle swarm optimization. *Sci. China Inf. Sci.* **2018**, *61*, 052204. [CrossRef]
13. Fox, D.; Burgard, W.; Thrun, S. The dynamic window approach to collision avoidance. *IEEE Robot. Autom. Mag.* **1997**, *4*, 23–33. [CrossRef]
14. Simmons, R. The curvature-velocity method for local obstacle avoidance. In Proceedings of the 1996 IEEE International Conference on Robotics and Automation, Minneapolis, MN, USA, 22–28 April 1996; Volume 4, pp. 3375–3382.
15. Chakravarthy, A.; Ghose, D. Obstacle avoidance in a dynamic environment: A collision cone approach. *IEEE Trans. Syst. Man Cybern. Part A Syst. Hum.* **1998**, *28*, 562–574. [CrossRef]
16. Fiorini, P.; Shiller, Z. Motion planning in dynamic environments using velocity obstacles. *Int. J. Robot. Res.* **1998**, *17*, 760–772. [CrossRef]
17. Bemporad, A.; Di Marco, M.; Tesi, A. Sonar-Based Wall-Following Control of Mobile Robots. *J. Dyn. Syst. Meas. Control* **2000**, *122*, 226–229. [CrossRef]
18. Toibero, J.M.; Roberti, F.; Carelli, R. Stable contour-following control of wheeled mobile robots. *Robotica* **2009**, *27*, 1–12. [CrossRef]
19. Teimoori, H.; Savkin, A.V. A biologically inspired method for robot navigation in a cluttered environment. *Robotica* **2010**, *28*, 637–648. [CrossRef]
20. Matveev, A.S.; Teimoori, H.; Savkin, A.V. A method for guidance and control of an autonomous vehicle in problems of border patrolling and obstacle avoidance. *Automatica* **2011**, *47*, 515–524. [CrossRef]
21. Matveev, A.S.; Wang, C.; Savkin, A.V. Real-time navigation of mobile robots in problems of border patrolling and avoiding collisions with moving and deforming obstacles. *Robot. Auton. Syst.* **2012**, *60*, 769–788. [CrossRef]
22. Savkin, A.V.; Wang, C. A simple biologically inspired algorithm for collision-free navigation of a unicycle-like robot in dynamic environments with moving obstacles. *Robotica* **2013**, *31*, 993–1001. [CrossRef]
23. Matveev, A.; Savkin, A.; Hoy, M.; Wang, C. *Safe Robot Navigation among Moving and Steady Obstacles*; Elsevier: Amsterdam, The Netherlands, 2015.
24. Savkin, A.V.; Hoy, M. Reactive and the shortest path navigation of a wheeled mobile robot in cluttered environments. *Robotica* **2013**, *31*, 323–330. [CrossRef]
25. Urdiales, C.; Pérez, E.; Sandoval, F.; Vázquez-Salceda, J. A hybrid architecture for autonomous navigation in dynamic environments. In Proceedings of the IEEE/WIC International Conference on Intelligent Agent Technology, Halifax, NS, Canada, 13–17 October 2003.
26. Zhu, Y.; Zhang, T.; Song, J.; Li, X. A new hybrid navigation algorithm for mobile robots in environments with incomplete knowledge. *Knowl. Based Syst.* **2012**, *27*, 302–313. [CrossRef]
27. Nieuwenhuisen, M.; Behnke, S. Layered mission and path planning for MAV navigation with partial environment knowledge. In *Intelligent Autonomous Systems 13*; Springer: Berlin/Heidelberg, Germany, 2016; pp. 307–319.
28. Hank, M.; Haddad, M. A hybrid approach for autonomous navigation of mobile robots in partially-known environments. *Robot. Auton. Syst.* **2016**, *86*, 113–127. [CrossRef]
29. Wzorek, M.; Berger, C.; Doherty, P. A Framework for Safe Navigation of Unmanned Aerial Vehicles in Unknown Environments. In Proceedings of the 2017 25th International Conference on Systems Engineering (ICSEng), Las Vegas, NV, USA, 22–23 August 2017; pp. 11–20.
30. D'Arcy, M.; Fazli, P.; Simon, D. Safe navigation in dynamic, unknown, continuous, and cluttered environments. In Proceedings of the 2017 IEEE International Symposium on Safety, Security and Rescue Robotics (SSRR), Shanghai, China, 11–13 October 2017; pp. 238–244.
31. Adouane, L. Reactive versus cognitive vehicle navigation based on optimal local and global PELC*. *Robot. Auton. Syst.* **2017**, *88*, 51–70. [CrossRef]
32. Kuffner, J.J.; LaValle, S.M. RRT-connect: An efficient approach to single-query path planning. In Proceedings of the 2000 ICRA, Millennium Conference, IEEE International Conference on Robotics and Automation, Symposia Proceedings (Cat. No. 00CH37065), San Francisco, CA, USA, 24–28 April 2000; Volume 2, pp. 995–1001.

33. Yang, K.; Gan, S.K.; Sukkarieh, S. An Efficient Path Planning and Control Algorithm for RUAV's in Unknown and Cluttered Environments. *J. Intell. Robot. Syst.* **2010**, *57*, 101. [CrossRef]
34. Yang, K.; Sukkarieh, S. An analytical continuous-curvature path-smoothing algorithm. *IEEE Trans. Robot.* **2010**, *26*, 561–568. [CrossRef]
35. Amidi, O.; Thorpe, C.E. *Integrated Mobile Robot Control;* Mobile Robots V. International Society for Optics and Photonics: Bellingham, WA, USA, 1991; Volume 1388, pp. 504–524.
36. Giesbrecht, J.; Mackay, D.; Collier, J.; Verret, S. *Path Tracking for Unmanned Ground Vehicle Navigation: Implementation and Adaptation of the Pure Pursuit Algorithm;* Technical Report; Defence Research and Development Suffield : Suffield, AB, Canada, 2005.
37. Elmokadem, T. A 3D Reactive Collision Free Navigation Strategy for Nonholonomic Mobile Robots. In Proceedings of the 2018 37th Chinese Control Conference (CCC), Wuhan, China, 25–27 July 2018; pp. 4661–4666.
38. Nakhaeinia, D.; Payeur, P.; Hong, T.S.; Karasfi, B. A hybrid control architecture for autonomous mobile robot navigation in unknown dynamic environment. In Proceedings of the 2015 IEEE International Conference on Automation Science and Engineering (CASE), Gothenburg, Sweden, 24–28 August 2015; pp. 1274–1281.

drones

MDPI

Article

A Multilevel Architecture for Autonomous UAVs

Luca Bigazzi, Michele Basso, Enrico Boni, Giacomo Innocenti and Massimiliano Pieraccini *

Department of Information Engineering, University of Florence, Via Santa Marta 3, 50139 Firenze, Italy; luca.bigazzi@unifi.it (L.B.); michele.basso@unifi.it (M.B.); enrico.boni@unifi.it (E.B.); giacomo.innocenti@unifi.it (G.I.)
* Correspondence: massimiliano.pieraccini@unifi.it

Abstract: In this paper, a multilevel architecture able to interface an on-board computer with a generic UAV flight controller and its radio receiver is proposed. The computer board exploits the same standard communication protocol of UAV flight controllers and can easily access additional data, such as: (i) inertial sensor measurements coming from a multi-sensor board; (ii) global navigation satellite system (GNSS) coordinates; (iii) streaming video from one or more cameras; and (iv) operator commands from the remote control. In specific operating scenarios, the proposed platform is able to act as a "cyber pilot" which replaces the role of a human UAV operator, thus simplifying the development of complex tasks such as those based on computer vision and artificial intelligence (AI) algorithms which are typically employed in autonomous flight operations.

Keywords: UAV; autonomous flight; indoor positioning; cyber pilot; sensor fusion

Citation: Bigazzi, L.; Basso, M.; Boni, E.; Innocenti, G.; Pieraccini, M. A Multilevel Architecture for Autonomous UAVs. *Drones* **2021**, *5*, 55. https://doi.org/10.3390/drones5030055

Academic Editors: George Nikolakopoulos, Diego González-Aguilera and Pablo Rodríguez-Gonzálvez

Received: 14 May 2021
Accepted: 28 June 2021
Published: 30 June 2021

1. Introduction

In the scientific community, one of the topics of major interest in the UAV field concerns autonomous navigation based on computer vision [1,2]. Many authors have proposed *motion capture systems* (mocap) [3–8], as they guarantee the highest performance. Given the high precision that these systems are able to achieve, they are often used to compare different control techniques. This approach allows for appreciating even small differences in the performance of different control algorithms.

However, the progress made in terms of research development regarding mocap systems is not reflected in the industrial sector, since most of the navigation algorithms and control techniques that exploit mocap technology (which is ground assisted) can hardly be extensively implemented on the typical environments of this field. Recently, thanks to the technological evolution and the proliferation of open source boards such as *Raspberry Pi* and *Nvidia Jetson* (running operating systems such as Linux), the research community has become increasingly interested in on-board *vision-based navigation* (VBN) techniques [9–12]. Given the possibilities currently offered by technology and the large amount of novel applications which can be envisioned in this new setting, the interest has shifted to how to solve the problem of the autonomous navigation of UAVs without ground assistance.

In order to investigate this new formulation of the problem, the classic UAV architecture shown in Figure 1 and composed of a micro-controller, sensors and a receiver capable of establishing a radio link for the drone commands is no longer sufficient. Instead, it is necessary to use a multilevel architecture, composed of both classic micro-controllers and other boards capable of processing images and complex algorithms, where each level must be able to communicate with the adjacent ones (see, e.g., [4]). As a consequence, the complexity of the drone inevitably increases.

In the multilevel paradigm, the choice of hardware is crucial, because this affects the internal communications, and this is particularly true for what concerns the flight controller device. A quite popular choice for the flight controller board in UAV robotic applications is *PixHawk*. In fact, most of the scientific papers dealing with the development of technology and algorithms for autonomous UAVs use this platform [13]. Among the most important reasons for this choice, two are prominent: (i) the PixHawk board was one of the first and most complete open source platforms for drones; and (ii) some custom versions of the firmware give native support for a *robot operating system* (ROS), a popular framework for robot development and programming [4]. Indeed, usually on UAVs based on a multilevel hardware, an additional card running ROS is used. This latter can be regarded as a "high-level" device, specifically designed for solving the original robotic problem, and that must communicate with the "lower" flight controller in order to perform the correct action. PixHawk platforms are able to receive ROS commands through a UART port, which uses the *MAVLINK* protocol [14–16]. Despite its popularity, this approach has some limitations, since the choice of flight controller is limited to PixHawk, which can only be driven via the MAVLINK protocol.

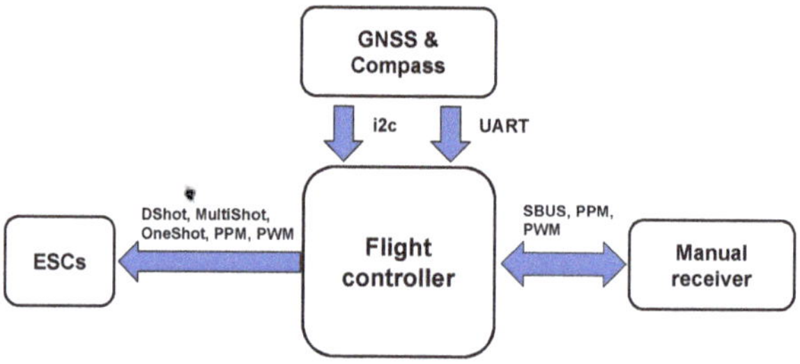

Figure 1. Standard UAV architecture: the flight controller is directly connected to the receiver, GNSS and ESCs (*electronic speed controllers*).

The approach presented in this paper, on the other hand, is conceived to increase flexibility in the choice of the hardware configuration, and thus, to avoid these limitations. Such a feature is obtained by implementing proper inter-level interfaces according to the decode–elaborate–encode paradigm, which allows, for instance, the regeneration of the signals sent to the flight controller, using any protocol used by commercially available receivers. Thanks to this feature, any flight controller can be used without the need to adapt to the hardware configuration of the software implementing the solution of the robotic problem.

This paper is organized as follows: Section 2 describes the general UAV multilevel architecture and the interfaces among the different layers, introducing the concept of a cyber pilot for autonomous navigation. Moreover, considerations of the mechanical design of the UAV with respect to the additional sensors and boards required are also presented, along with actual hardware/software implementation. Section 3 shows the experimental results of a specific autonomous mission performed by the UAV in an indoor environment using vision-based navigation. Finally, Section 4 reports concluding remarks and future developments.

2. Materials and Methods

2.1. Architecture

The proposed architecture aims to provide a flexible hardware and software platform for UAV robotic applications. The basic idea is in the subdivision of the general problem into sub-problems, which can be directly related to hardware or software implementations. Each solution is placed into a "level" comprising affine elements, and the levels are then interconnected to each other by standardized interfaces. Therefore, regarding UAV robotic applications based on VBN techniques, an architecture comprising three different abstract levels has been envisioned. The functions each single level is devoted to are listed below, and because of the hierarchical division of the problem, they are assumed to only communicate with adjacent levels.

- *Functional high-level.*
  This comprises the hardware and software necessary to implement the VBN system and to run the autonomous navigation algorithms.
- *Functional mid-level.*
  This is represented by the software necessary to manage the commands sent to the flight controller, which can be those from the ground station (human pilot remote control) or the ones generated by the autonomous navigation algorithm. This is the communication interface between the other two levels of the architecture.
- *Functional low-level.*
  It comprises the flight controller together with the hardware and software necessary for the attitude stabilization of the drone.

In Figure 2, a more detailed description of the used implementation of this architecture is reported. The low-level is mainly implemented as a hardware layer. Indeed, it accounts for a generic flight controller for drones and its equipped software, as well as the devices (i.e., sensors) necessary for attitude stabilization. Conversely, the mid-level is implemented as a software layer, which, in the presented solution, is hosted in the same Jetson Nano board used for the high-level. Indeed, this latter comprises both the hardware and software needed to implement the autonomous navigation system, that is, the previous Jetson Nano board and the related devices illustrated in Figure 2. In detail, the Jetson Nano is equipped with the following sensors:

1. A multi-sensor board, also comprising a GNSS receiver, connected through a UART3 port;
2. An Intel Realsense t-265, connected via USB-3.0 and used for the 3D localization and orientation of the drone;
3. A digital camera, connected via *Camera Serial Interface* (CSI) and used by computer vision algorithms.

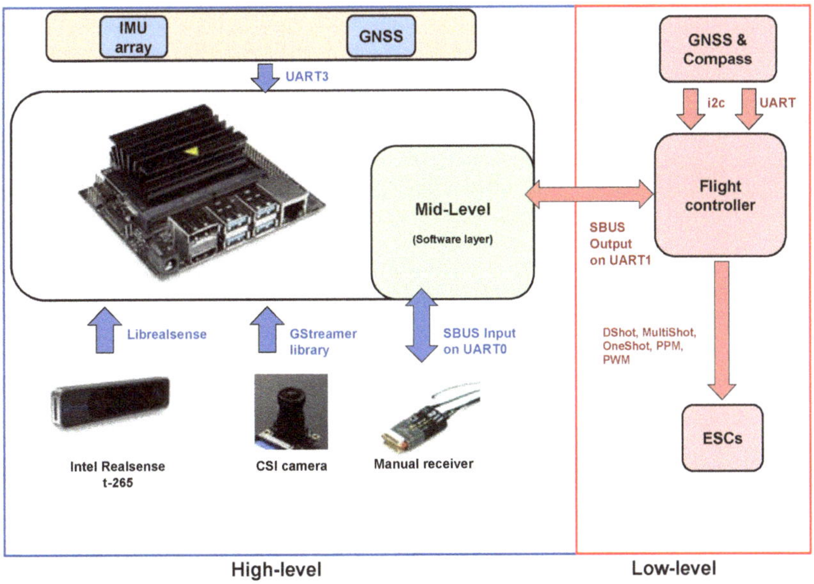

Figure 2. Proposed multilevel architecture, where the Jetson Nano computer is connected between the flight controller and the drone receiver.

The multi-sensor unit was designed to collect position and inertial data from a multitude of different sensors. The board hosts two U-Blox NEO-M9N precision GNSS receivers, which provide an absolute position at the 18 Hz rate in a dual-redundant configuration.

Roll, pitch and yaw angles and altitude values are estimated by means of *micro electro mechanical systems*-based (MEMS) 3D inertial sensors (accelerometer and gyroscope), a 3D magnetometer and an MEMS barometric pressure sensor. All the sensors are proposed in a triple-redundant configuration. Finally, a micro-controller unit (STM32L476) collects the digital samples from all the sensors and performs a sensor fusion algorithm [17,18] to provide a stable and reliable estimation of the roll, pitch and yaw angles at a rate of 200 Hz.

It is worth noting that, in this architecture implementation, the low-level is just an attitude controller equivalent to the ones used in traditional UAV configurations. As such, it accounts for a fully equipped flight controller, and indeed, it has its own sensors (e.g., a compass and a GNSS receiver) as shown in Figure 2, and it can execute the functions this kind of system is usually able to run (e.g., the *return to home* (RTH).

As it will be detailed in Section 2.2, the software layer representing the mid-level has the important role of deciding which command has to be sent to the flight controller between the one received from the receiver and the signal generated by the high-level. The actual signal passed to the flight controller depends in general on the policies of the mid-level, and possibly on the commands from the human pilot

2.2. *Mid-Level Software Layer: Cyber Pilot and Interface with the Low-Level Implementation*

As introduced in Section 2.1, the proposed architecture was conceived so that the high- and mid-level software implementations are such that any signal from the drone receiver passes through the Jetson Nano before arriving to the flight controller (see also [9,19]). This intermediate stage of processing allows for substituting the commands from the receiver with others generated by the high-level software running on the same Jetson Nano, thus replacing the human pilot with a "cyber pilot" (see also [19]). According to such a concept, the problem of driving the drone motion is "lifted" to the same sphere of the human operator, whereas the common approach, instead, "pulls down" the control strategy to the motor sphere. On the one hand, the disadvantages are that the controller is bound to act as a human, but on the other hand, the problems of elaborating the control strategy and of driving the motors are held separated, and so they can be solved and implemented as different modules.

The cyber pilot paradigm makes the system independent on the low-level implementation, as long as this latter uses standard interfaces. Indeed, as already noted, from the functional point of view, the low-level can be implemented by a traditional, fully equipped flight controller. Therefore, it is not necessary to design a custom implementation of this level, and already existing solutions can be used. The mid-level software layer manages this interface, and therefore, any flight controller can be used, provided that it can be physically interconnected to the high-level hardware layer (where the mid-level software runs), and its communication protocol is known.

In the implementation of the proposed architecture, the low-level is represented by a system based on a CC3D Revo flight controller running the LibrePilot firmware. This controller provides good performance (it has been specifically developed for racing applications) and it respects the form factor of the chosen drone frame. Moreover, it communicates using standard connectors and protocols. It is worth pointing out that the flight controller has been configured to operate in the "stabilize" mode, which only provides attitude control, leaving the altitude to be controlled manually. Therefore, the cyber pilot must be implemented to control the altitude as well.

Depending on the mid-level software layer policies, a variety of behaviors can be implemented. For instance, the autonomous flight mode can be switched back to the manual flight mode at any time with a simple strategy based on auxiliary channels in the communication with the human pilot. Indeed, the software layer implementing the mid-level can read all the channels transmitted by the ground station, and therefore, one of them can be used to carry auxiliary information specifically directed to the define the switching policy. Hence, depending on this input, the software can decide whether to replace the human commands read on the Jetson UART0 port (input) with others generated by the cyber pilot, or to transfer them to the UART1 port (output) directed to the flight controller.

The advantages of the cyber pilot approach are evident when the possible protocols for the communication between the receiver and the flight controller are taken into account. In the typical solutions, receiver and flight controller must share the communicating protocol, whereas the cyber pilot approach, introducing a signal processing stage in between the two devices, is not affected by this limitation.

On the mass market, drone receivers use three main communication protocols to send the operator commands to the flight controller:

1. *Pulse width modulation* (PWM);
2. *Pulse position modulation* (PPM);
3. *Serial bus* (SBUS),

The PWM protocol is the oldest among them and as illustrated in Figure 3, it is based on the variation of a duty cycle. The signal from a radio channel is passed through a single wire, i.e., if one needs to transmit 8 channels to the flight controller (for example, 4 channels to control the drone and 4 to transmit additional information), it is necessary to connect 8 wires between the receiver and the flight controller.

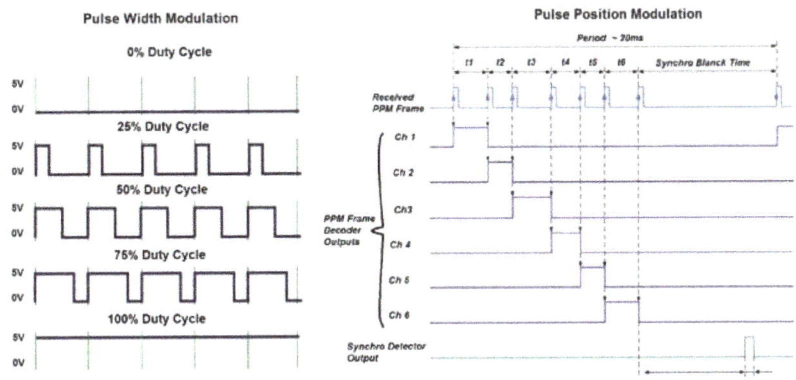

Figure 3. A pictorial representation of the two analog protocols used as a communication standard for commercial UAV receivers. On the left, it is possible to observe the signal generated through the PWM standard, while on the right, a frame generated through the PPM standard is shown [20].

The PPM protocol was introduced just to reduce the wiring needed to send the channels to the flight controller. As shown in Figure 3, this protocol modulates all the channels on a single wire. It is an analog protocol, just like PWM, and it is able to transmit multiple channels on a single wire under a bandwidth limitation, which is about 44 Hz for the transmission of eight channels.

The implementation presented in this paper exploits the SBUS protocol, since it has the following advantages over the previously mentioned ones: (i) it is the most recent protocol and represents today's state of the art for drone receivers; (ii) it is a digital protocol, so it is not affected by noise; and (iii) this standard allows for transmitting up to 16 channels with a band equal to 100 Hz. Furthermore, it is worth noting that it is also the only protocol which can be easily read and regenerated by the Jetson Nano board. Indeed, doing the same with analog protocols (PWM and PPM) would require to manage interrupts in real time, that cannot be accurately achieved on cards not running real-time operating systems—such as the Jetson Nano.

As illustrated in Figure 4, an SBUS frame is made up of 25 bytes, where Byte 0 represents the header, bytes from 1 to 22 contain the information relating to the 16 channels to be transmitted, Byte 23 contains some additional information (not always used) and the fail-safe management in case the receiver is unable to receive commands from the ground station. Finally, Byte 24 is the closing byte of the frame.

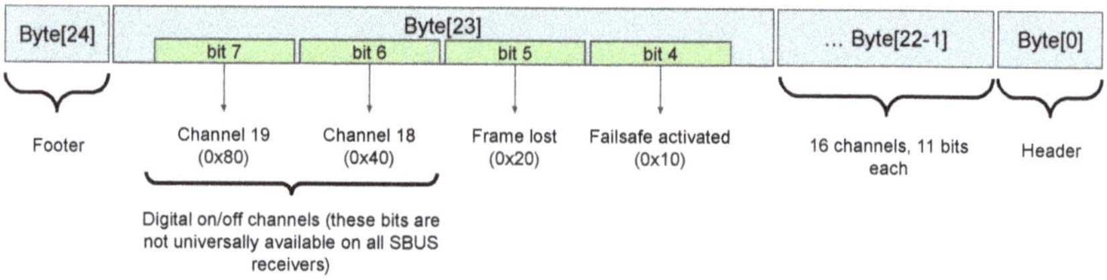

Figure 4. This figure shows the frame architecture according to the SBUS standard, which unlike the two previous standards, is a digital protocol. Being the latest protocol developed in the UAV field, it is currently the best choice for sending commands to the flight controller, as in addition to guaranteeing a greater useful band, it is not affected by noise.

Since a single channel was coded with 11 bits, the SBUS standard can associate a maximum of 2048 different values to each channel. This defines the maximum resolution obtainable with this protocol. In the mid-level of the proposed architecture, a software for decoding the SBUS frames coming from the drone receiver was implemented along with another software for (re-)coding the information to be sent to the flight controller. In the current implementation of the mid-level, the human operator commands were overwritten by those generated by the cyber pilot, when a high signal is read in channel 8 of the SBUS frame. This command was sent by the ground station, and when it was received and read by the mid-level, this latter replaces the first 44 bits contained in the byte range relating to the status of the channels. Indeed, these first 44 bits of the SBUS sub-frame contain the commands associated with the reference values for pitch, roll, yaw and thrust. This way, the cyber pilot takes the command of the drone bypassing the human pilot, who can still resume manual control at any time by sending a low signal in the same channel 8. If required, it would also be possible to program the mid-level to replace all 16 channels. This could be useful for sending additional commands to the flight controller during the autonomous flight modes managed by the cyber pilot. Figure 5 shows a principle diagram of the mid-level where its operation can be considered as a software switch, capable of assigning the drone commands to the cyber pilot or to the human one at will.

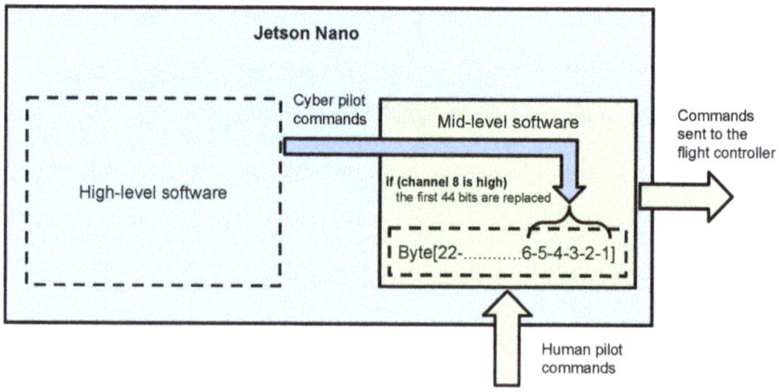

Figure 5. Diagram showing one of the mid-level operating policies. In the specific case, during the transition to the flight modes managed by the cyber pilot, the mid-level operates by replacing only the channels necessary for piloting the UAV.

2.3. *High-Level Software Layer*

In this section, we illustrate the overall structure and operating logic of the software layer pertaining to the high-level.

From the functional point of view, this layer implements all the software necessary for the cyber pilot, i.e., for the autonomous navigation system. The video stream from the cameras and the data acquired by the sensors are elaborated by sensor fusion algorithms to formulate estimates of drone position and drone speed. This information is used by the control strategy to track the reference trajectory, which can be a desired motion along a path fixed by the user or computed within a *simultaneous localization and mapping* (SLAM) problem.

The software is divided into threads, which are scheduled by the operative system of the high-level hardware, i.e., in the proposed implementation, the Jetson Nano board. For management purposes, a number of command line keywords, reported in the following list, were implemented:

- ***−−displaySampleTime*** : Shows active threads, expected and real sample rates;
- ***−−windows***: Shows the images processed by the vision algorithm inside a window;
- ***−−plot***: Makes the real-time plot of the variables of interest inside a window (it also needs the *−−windows* command);
- ***−−log***: Save all shared variables in a *.txt* file, usable for data post processing;
- ***−−rec***: Records the video stream processed by the vision algorithm and superimposes the telemetry obtained in real time;
- ***−−ssh***: If enabled, it eliminates the *OpenGL* optimization of the windows which are rendered on the remote PC.

The implemented features, which can be de/activated with appropriate commands, are meant to provide different operating modes without recompiling the software at each experimental test. For instance, if the *−−windows* and *−−rec* commands are not executed, the display thread is not initialized as it is not necessary. This way, the computational burden can be adapted to the actual situation avoiding unnecessary functions.

Figure 6 shows the overall software architecture, where it is possible to observe the division of the various jobs into different threads. Each thread can operate on shared variable blocks through the reader/writer sync method.

1. **SLAM thread.**
   Thread 1 implements SLAM algorithms. It receives the 3D position data generated with a frequency of 200 Hz by the t-265 stereoscopic camera. These data are related to the position of the UAV in the environment and are defined in fixed frames. Thread 1 also receives the video streams of the two fish-eye lenses of the stereoscopic camera at a frequency of 30 Hz;
2. **Vision thread.**
   Thread 2 receives the information computed by Thread 1 together with the additional video stream from the CSI camera. The vision thread also receives some additional data, such as the attitude, relative and absolute position given by the t-265 stereoscopic camera and by the GNSS (if available). Inside this thread, various computer vision algorithms can be implemented, such as environmental marker detection [21,22] or object recognition [23,24];
3. **Inertial and GNSS thread.**
   Thread 3 provides additional information on accelerations and position to the vision thread. This information is elaborated by sensor fusion algorithms to improve the precision in the estimation of the drone position and speed.
4. **Planner thread.**
   Thread 4 exploits the outputs of the previous threads to build the reference trajectory. It contains several routines for path generation, which are specific for different kinds of missions. For example, this thread is responsible for the generation of a specific path built from environmental markers, but it can also account for fixed trajectories in the space. The desired path is a sequence of vectors of four elements:

$$s_{hl} \equiv [x_d, y_d, z_d, \psi_d]^T \tag{1}$$

which represent set points to be tracked. In particular, x_d, y_d, z_d and ψ_d are the desired values for, respectively, the lateral position, the altitude, the longitudinal position and the yaw angle. The comparison between s_{hl} and the estimated position and attitude provide the displacements from the desired trajectory.

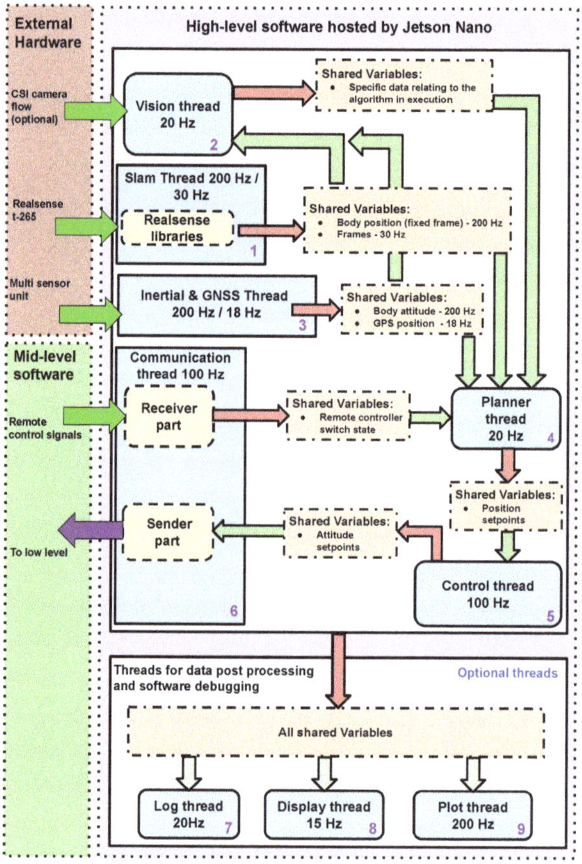

Figure 6. The figure shows the high-level software architecture, where all the threads that can be executed according to the type of mission to be performed are defined. Note that the inter-thread synchronization method implemented is based on the reader/writer model.

5. **Control thread.**
Thread 5 implements the control strategy [25–27] of the cyber pilot with respect to the planned trajectory generated by Thread 4. The result is a sequence of vectors containing the reference values, which are sent to the flight controller implementing the low-level:

$$s_{ll} \equiv [\theta_d, t_d, \varphi_d, \dot{\psi}_d]^T \, . \tag{2}$$

In each vector s_{ll}, θ_d and φ_d are the set points for roll and pitch angles, t_d is the reference value of the thrust and $\dot{\psi}_d$ is the set point for the yaw rate.
Many different control algorithms can be used here, depending on the mission specifications, whether for instance the UAV has to maintain a specific position (hovering) or track a complex trajectory. A quite simple but efficient control scheme implemented in this thread and tested in experiments consists of an array of double-nested loops of *proportional–integral–derivative* (PID) controllers that process the 3D position and

velocity errors to compute the first three components of the s_{ll} vector, i.e., roll, pitch and thrust set points, whereas the yaw rate is left as an additional degree of freedom. In the cyber pilot paradigm, the s_{ll} vector contains the commands used by the mid-level software layer to overwrite the manual commands during the autonomous flight modes. It is important to stress that the signals in the vector s_{ll} depend on the commands accepted by the flight controller implementing the low-level. Anyway, the commands generated by the cyber pilot share their nature with those received from the human pilot during manual flight. In other words, from the flight controller point of view, these two kinds of commands look exactly the same. In this sense, the cyber pilot makes the autonomous flight system independent from the flight controller, provided that its input interface can be properly replicated.

6. **Communication thread.**
   Thread 6 is devoted to sending the output of the control thread to the mid-level software layer, which, based on its policies, decides whether to use them or those coming from the human pilot. This thread also handles the inputs from the auxiliary channels of the remote controller, which regard useful information for the management of the drone and in particular for interacting with the mid-level functions.
7. **Log thread.**
8. **Display thread.**
9. **Plot Thread.**
   Threads 7, 8 and 9 are optional and can be activated if needed. They are used to log, store and plot the shared variables, and so they are useful during the experimental phase, as they allow for acquiring data for the post processing and for the debugging phases.

2.4. Frame Sizing

A relevant aspect for the development of the UAV used in this paper and shown in Figure 7 concerns its mechanical frame design and sizing.

Figure 7. The autonomous UAV prototype implementing the multilevel architecture is presented in this paper. The frame is fully carbon fiber and was designed to host the Nvidia Jetson Nano board and all the additional electronics. Given the increase in weight due to the new elements, the frame was designed to allow the installation of 7-inch propellers.

In fact, there are no small drones on the mass market capable of supporting the development of high-level algorithms. Given the high performance and robustness of racing drones, it was decided to develop the UAV frame according to the same philosophy used for the construction of racing drones [28]. Normally, this type of drone has a full carbon frame, a wheelbase of 250 mm and 5-inch propellers. These drones are combined with small motors that are able to reach high rpm. Usually, 22 mm stator motors are used which can easily exceed 40,000 rpm. The high number of revolutions is necessary to achieve the desired thrust considering the small propeller diameter. Furthermore, the use of small propellers greatly decreases the inertia of the same, making these drones very reactive.

However, the additional hardware mentioned in Section 2.1 and necessary for the realization of the cyber pilot results in a weight increase of about 300 g. In theory, it would not be a problem, given that racing drones have thrust-to-weight ratios between 6:1 and 8:1. On the other hand, their typical motor/propeller set has a relatively low energy efficiency, defined as the ratio between the maximum thrust generated and the electrical power used. A lower energy efficiency has the negative effect of decreasing the battery life, which in racing drones, is already a few minutes, therefore, this phenomenon must be avoided, trying to increase the value of efficiency as much as possible. In addition, the racing drone frames available on the market do not have enough space to be used for mounting the new parts. To solve the above issues, a carbon frame was designed that can accommodate all the additional parts, whereas the wheelbase was increased to 330 mm, providing enough space for mounting 7-inch propellers. Notice that the use of larger diameter propellers is not enough to increase the energy efficiency of a drone, but it is necessary that these are suitably matched to the motors. For this purpose, motors with a stator diameter of 25 mm (instead of the usual 22 mm) capable of reaching approximately 22,000 rpm have been chosen. Even if these motors are able to reach a lower number of revolutions than those normally used for racing applications, the larger stator diameter allows for an increase in the lever arm, which guarantees higher torque values suitable for the use of 7-inch propellers. These considerations have been made on the basis of data-sheets provided by the various motor manufacturers, which among other data, also provide the power consumed and the thrust obtained depending on the type of propeller and the type of battery used.

3. Results

The implementation illustrated thus far of the proposed architecture was tested in a challenging mission consisting of a circular trajectory of specified shape to be followed in an indoor environment. The difficulty of such a mission is due to the availability and accuracy of data from the traditional sensors (i.e., GNSS, *inertial measurement unit* (IMU), compass and barometer) in small indoor environments. Indeed, GNSS is just not available, the compass is negatively affected by the building structure, the barometer has insufficient precision in small places, and the IMU alone is, in general, not well suited for navigation because of the well-known problem of drifting. Such a situation is perfect to check whether the VBN techniques exploited in the proposed solution are able to recover from that loss of accuracy. Indeed, it is worth stressing that the cyber pilot has a three-dimensional perception of the environment, which makes it capable of following generic 3D trajectories, thanks to the t-265 stereoscopic camera.

In Figure 8, the blue line shows the trend of the UAV position on the single degrees of freedom, while the dashed black line defines the desired trajectory.

The actual configuration of the mission can be programmed within the planner thread, whose role, as mentioned in Section 2.3, is to generate the desired trajectory. In particular, the circular trajectory is defined as a relative path starting from the initial position of the drone and position on the vertical plane orthogonal to the heading. Its radius and the traveling (constant) speed are the two tuning parameters, and for these field tests, they have been set to 0.5 m and 0.1 m/s, respectively. Such a mission was performed in repeated experiments to record statistics on the drone performance. Figure 8 shows the 3D position x, y, z of the drone against the corresponding set-points. The position is directly measured by the t-265 Realsense camera in a fixed frame where the y axis represents the drone altitude while x–z lie on the horizontal normal plane.

The actual trajectory lies in the x–y plane and it follows the reference path with high accuracy. In Figure 9, the position errors are shown, confirming the small overshoots nearly visible in Figure 8. The following table reports the statistics of the errors computed on the entire set of experiments.

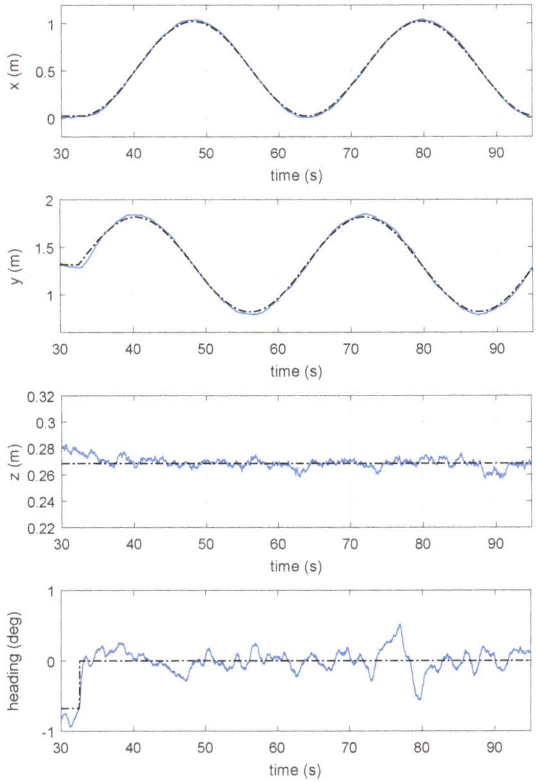

Figure 8. The black dashed lines show the desired trajectories along the 4 degrees of freedom of the UAV, while the blue lines show the trend of the position of the UAV along the three axes and its orientation.

Axis	std
horizontal (along x)	1.36 cm
altitude (along y)	2.20 cm
longitudinal (along z)	0.42 cm
yaw	0.22 deg

It should be noted that the lowest standard deviations are relative to the degrees of freedom that remained constant in the experiment (z axis and orientation angle). However, even the standard deviations obtained along the axes of the x–y plane are very small and highlight the good performance achievable by the architecture presented in this paper.

Figure 10 reports the SBUS commands sent to the flight controller and synthetically generated by the cyber pilot. These commands are the vector defined in Equation (2) which replace the manual flight commands of the human pilot. It is interesting to note the high information content they possess, thanks to the fact that the cyber pilot is able to exploit the entire bandwidth of the SBUS protocol. These commands are very different from a human operator, as the latter is unable to generate commands with a frequency higher than 2–3 Hz.

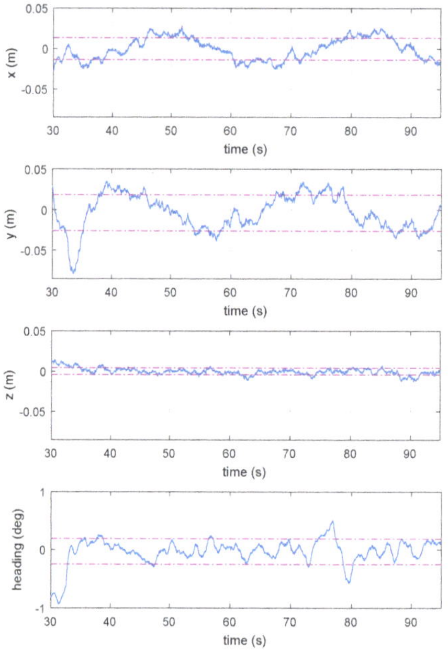

Figure 9. This figure shows the error trends as a function of time. The sinusoidal trend of the errors relating to the *x*–*y* plane—which is the plane where the drone carries out the mission with a circular trajectory—is also observed.

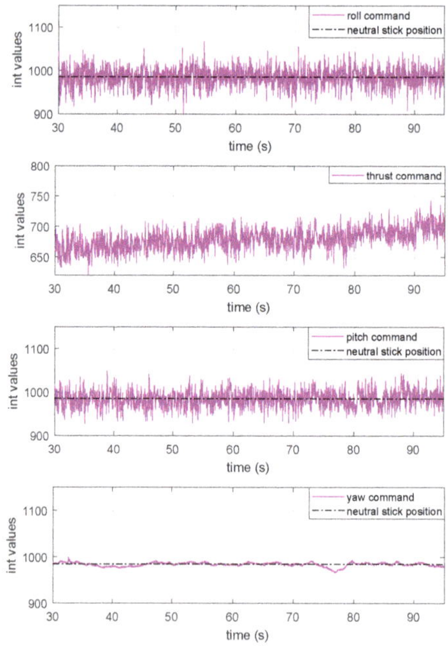

Figure 10. This figure shows the commands that the cyber pilot sends to the flight controller through the SBUS protocol managed by the mid-level. The commands are encoded in 16-bit integer variables.

The black dotted line shows the neutral position of the remote control sticks, and as can be seen in Figure 10, both the pitch, roll and yaw commands are close to the central values, while the thrust commands generated by the cyber pilot are lower than the neutral value. This is because the UAV used has a thrust-to-weight ratio higher than the usual 2:1 ratio, which is normally used in non-racing drones.

Another interesting aspect that can be observed in Figure 10 relates to the trend of the thrust commands over time, whose average tends to increase. This happens because the cyber pilot is able to compensate for the battery discharge curve, and to do so, it increases the average value of the thrust commands. This is clearly a byproduct of the feedback control strategy based on the position error.

Figure 11 shows the UAV during the execution of the preset mission. The image has been generated by superimposing several shots, taken at almost uniformly distributed shooting times. The image shows the high precision reached by the drone, and it also gives a good idea of the correct tracking of the desired velocity, since the drone is in the correct position in every shot.

Figure 11. Circular mission task: the right image shows the UAV that correctly tracks a circle with a diameter of 1 m and a constant speed of 0.1 m/s. In the left image, the positions taken by the UAV during the execution of a second lap of the trajectory have been superimposed "in transparency". As it is possible to observe, the trajectories followed by the drone are perfectly reproducible.

4. Discussion

The experimental tests show that the system presented in this paper is able to follow the desired trajectories with centimeter precision. In particular, all the experiments highlighted that the tracking accuracy is not imposed by the measurements of the onboard sensors (IMU and camera), whose error is in the order of under a millimeter, but is mainly limited by the propellers' turbulence, which act as disturbances that the controller is unable to completely reject. Figure 11 shows the repeatability of the test mission, since the UAV is able to track the same trajectory several times. Given the nature of the optical sensors used, the drone can follow 3D trajectories with controlled velocity and acceleration. In the literature, most works employ motion capture systems (mocap) [6,7] for the development of indoor autonomous navigation algorithms [8]. The achievable accuracy of these systems is very high, although they can only work in environments suitably set up by external computers and fixed cameras that frame the drone. This limits autonomous navigation to the environment monitored by the mocap system only.

In this paper, an alternative approach was conceived that allows to achieve similar performance, with the difference that all sensors and computing units are mounted onboard. This feature ensures high flexibility and the ability to operate in a wide variety of environments, without the need to use off-board sensors and computing units. The experiment shown in Section 3 was carried out with the sole purpose of demonstrating the capabilities of the developed architecture, in fact, thanks to the modular nature of the high-level software, the mission objectives can be easily modified by calling different routines within the planner thread. On the other hand, if the purpose of the research concerns the development of computer vision algorithms, it will be sufficient to write additional routines that can be called up by the vision thread.

Another aspect of fundamental importance, which will be very useful for future research, is the possibility of replicating the same communication protocol used by the radio link during manual flights. This peculiar feature, as discussed in previous sections, allows the development of cyber pilot missions in specific scenarios. In the future, this aspect can be used for the development of algorithms based on neural networks and AI, able to learn from the human pilot and to replicate similar command signals. A second development path will be that of the autonomous exploration of buildings based on the use of depth sensors, such as the Intel D4xx camera series [12]. Thanks to the addition of these devices, it will be possible to develop on-board 3D mapping algorithms of the environment, thus making the UAV capable of detecting obstacles and targets of interest in complete autonomy and without ground assistance.

Author Contributions: Conceptualization L.B., M.B., E.B., G.I. and M.P.; data curation L.B.; formal analysis L.B. and M.B.; funding acquisition M.B., E.B., G.I. and M.P.; investigation L.B., M.B. and E.B.; methodology L.B., M.B. and G.I.; project administration M.B., G.I. and M.P.; software L.B.; supervision M.B., E.B., G.I. and M.P.; writing—original draft L.B., M.B. and G.I.; writing—review and editing L.B., M.B., E.B., G.I. and M.P. All authors have read and agreed to the published version of the manuscript.

Funding: This research was co-funded by Horizon 2020, European Community (EU), AURORA (Safe Urban Air Mobility for European Citizens) project, ID number 101007134.

Institutional Review Board Statement: Not applicable.

Informed Consent Statement: Not applicable.

Data Availability Statement: The data presented in this study are available in the supplementary material of the article.

Acknowledgments: Thanks to Lapo Miccinesi for the help given in the acquisition of the photos during the laboratory tests.

Conflicts of Interest: The authors declare no conflict of interest.

References

1. Grabe, V.; Bülthoff, H.H.; Scaramuzza, D.; Giordano, P.R. Nonlinear ego-motion estimation from optical flow for online control of a quadrotor UAV. *Int. J. Robot. Res.* **2015**, *34*, 1114–1135. [CrossRef]
2. Penin, B.; Spica, R.; Giordano, P.R.; Chaumette, F. Vision-based minimum-time trajectory generation for a quadrotor UAV. In Proceedings of the 2017 IEEE/RSJ International Conference on Intelligent Robots and Systems (IROS), Vancouver, BC, Canada, 24–28 September 2017.
3. Al Habsi, S.; Shehada, M.; Abdoon, M.; Mashood, A.; Noura, H. Integration of a Vicon camera system for indoor flight of a Parrot AR Drone. In Proceedings of the 2015 10th International Symposium on Mechatronics and Its Applications (ISMA), Sharjah, United Arab Emirates, 8–10 December 2015.
4. Gargioni, G.; Peterson, M.; Persons, J.B.; Schroeder, K.; Black, J. A Full Distributed Multipurpose Autonomous Flight System Using 3D Position Tracking and ROS. In Proceedings of the 2019 International Conference on Unmanned Aircraft Systems (ICUAS), Atlanta, GA, USA, 11–14 June 2019.
5. Antonio-Toledo, M.E.; Sanchez, E.N.; Alanis, A.Y.; Flórez, J.A.; Perez-Cisneros, M.A. Real-Time Integral Backstepping with Sliding Mode Control for a Quadrotor UAV. In Proceedings of the (IFAC-PapersOnLine) 2nd IFAC Conference on Modelling, Identification and Control of Nonlinear Systems MICNON 2018, Guadalajara, Jalisco, Mexico, 20–22 June 2018.
6. Masiero, A.; Fissore, F.; Antonello, R.; Cenedese, A.; Vettore, A. A comparison of UWB and motion capture UAV indoor positioning. In Proceedings of the The International Archives of the Photogrammetry, Remote Sensing and Spatial Information Sciences, SPRS Geospatial Week 2019, Enschede, The Netherlands, 10–14 June 2019.
7. Xiao, X.; Dufek, J.; Suhail, M.; Murphy, R. Motion Planning for a UAV with a Straight or Kinked Tether. In Proceedings of the 2018 IEEE/RSJ International Conference on Intelligent Robots and Systems (IROS), Madrid, Spain, 1–5 October 2018.

8. Aguilar, W.G.; Manosalvas, J.F.; Guillén, J.A.; Collaguazo, B. Robust Motion Estimation Based on Multiple Monocular Camera for Indoor Autonomous Navigation of Micro Aerial Vehicle. In Proceedings of the International Conference on Augmented Reality, Virtual Reality, and Computer Graphics (AVR 2018), Otranto, Italy, 14 July 2018.
9. Basso, M.; Bigazzi, L.; Innocenti, G. DART Project: A High Precision UAV Prototype Exploiting On-board Visual Sensing. In Proceedings of the 15th International Conference on Autonomic and Autonomous Systems (ICAS), Athens, Greece, 2–6 June 2019.
10. Bigazzi, L.; Basso, M.; Gherardini, S.; Innocenti, G. Mitigating latency problems in vision-based autonomous UAVs. In Proceedings of the 29th Mediterranean Conference on Control and Automation (MED2021), Bari, Puglia, Italy, 22–25 June 2021.
11. Ceron, A.; Mondragon, I.; Prieto, F. Onboard visual-based navigation system for power line following with UAV. *Int. J. Adv. Robot. Syst.* **2018**, *15*, 2–12. [CrossRef]
12. Lu, L.; Redondo, C.; Campoy, P. Optimal Frontier-Based Autonomous Exploration in Unconstructed Environment Using RGB-D Sensor. *Sensors* **2020**, *20*, 6507. [CrossRef] [PubMed]
13. Ma, C.; Zhou, Y.; Li, Z. A New Simulation Environment Based on Airsim, ROS, and PX4 for Quadcopter Aircrafts. In Proceedings of the 2020 6th International Conference on Control, Automation and Robotics (ICCAR), Singapore, 20–23 April 2020.
14. Hinas, A.; Roberts, J.M.; Gonzalez, F. Vision-Based Target Finding and Inspection of a Ground Target Using a Multirotor UAV System. *Sensors* **2017**, *17*, 12.
15. Atoev, S.; Kwon, K.R.; Lee, S.H.; Moon, K.S. Data analysis of the MAVLink communication protocol. In Proceedings of the 2017 International Conference on Information Science and Communications Technologies (ICISCT), Tashkent, Uzbekistan, 2–4 November 2017.
16. Kwon, Y.M.; Yu, J.; Cho, B.M.; Eun, Y.; Park, K.J. Empirical Analysis of MAVLink Protocol Vulnerability for Attacking Unmanned Aerial Vehicles. *IEEE Access* **2018**, *6*, 43203–43212. [CrossRef]
17. Madgwick, S.O.H. An efficient orientation filter for inertial and inertial/magnetic sensor arrays. April 2010.
18. Mahony, R.; Hamel, T.; Pflimlin, J.M. Nonlinear Complementary Filters on the Special Orthogonal Group. *IEEE Trans. Autom. Control* **2008**, *53*, 1203–1218. [CrossRef]
19. Bigazzi, L.; Gherardini, S.; Innocenti, G.; Basso, M. Development of Non Expensive Technologies for Precise Maneuvering of Completely Autonomous Unmanned Aerial Vehicles. *Sensors* **2021**, *21*, 391. [CrossRef] [PubMed]
20. Gardner, W.; Brown, W.; Chen, C.-K. Spectral Correlation of Modulated Signals: Part II—Digital Modulation. *IEEE Trans. Commun.* **1987**, *35*, 595–601. [CrossRef]
21. Olson, E. AprilTag: A robust and flexible visual fiducial system. In Proceedings of the 2011 IEEE International Conference on Robotics and Automation, Shanghai, China, 9–13 May 2011.
22. Wang, J.; Olson, E. AprilTag2: Efficient and robust fiducial detection. In Proceedings of the 2016 IEEE/RSJ International Conference on Intelligent Robots and Systems (IROS), Daejeon, Korea, 9–14 October 2016.
23. Kechagias-Stamatis, O.; Aouf, N.; Nam, D. 3D Automatic Target Recognition for UAV Platforms. In Proceedings of the 2017 Sensor Signal Processing for Defence Conference (SSPD), London, UK, 6–7 December 2017.
24. Vujasinovic, S.; Becker, S.; Breuer, T.; Bullinger, S.; Scherer-Negenborn, N.; Arens, M. Integration of the 3D Environment for UAV Onboard Visual Object Tracking. *Appl. Sci.* **2020**, *10*, 7622. [CrossRef]
25. Antonelli, G.; Cataldi, E.; Giordano, P.R.; Chiaverini, S.; Franchi, A. Experimental validation of a new adaptive control scheme for quadrotors MAVs. In Proceedings of the 2013 IEEE/RSJ International Conference on Intelligent Robots and Systems, Tokyo, Japan, 3–7 November 2013.
26. Koubaa, A.; Taher Azar, A. *Unmanned Aerial Systems: Theoretical Foundation and Applications. Advances in Nonlinear Dynamics and Chaos (ANDC)*; Academic Press: Cambridge, MA, USA, 2021.
27. Sutton, A.; Fidan, B.; van der Walle, D. Hierarchical UAV Formation Control for Cooperative Surveillance. *IFAC Proc.* **2008**, *41*, 12087–12092. [CrossRef]
28. Castiblanco, J.M., García-Nieto, S.; Simarro, R.; Salcedo, J.V. Experimental study on the dynamic behaviour of drones designed for racing competitions. *Int. J. Micro Air Veh.* **2021**. [CrossRef]

drones

MDPI

Article

Efficient Reactive Obstacle Avoidance Using Spirals for Escape

Fábio Azevedo [1,2,*], Jaime S. Cardoso [1,3], André Ferreira [2], Tiago Fernandes [2,†], Miguel Moreira [2] and Luís Campos [4]

1 Electrical and Computing Engineering Department, FEUP, University of Porto, 4200-465 Porto, Portugal; jaime.cardoso@inesctec.pt
2 Beyond Vision, 3830-352 Ílhavo, Portugal; andre.ferreira@beyond-vision.pt (A.F.); tiago.fernandes@criticaltechworks.com (T.F.); miguel.moreira@beyond-vision.pt (M.M.)
3 Institute for Systems and Computer Engineering, Technology and Science, 4200-465 Porto, Portugal
4 PDMFC, 1300-609 Lisbon, Portugal; luis.campos@pdmfc.com
* Correspondence: fabio.azevedo@beyond-vision.pt
† Current address: Critical Techworks, 4000-091 Porto, Portugal.

Abstract: The usage of unmanned aerial vehicles (UAV) has increased in recent years and new application scenarios have emerged. Some of them involve tasks that require a high degree of autonomy, leading to increasingly complex systems. In order for a robot to be autonomous, it requires appropriate perception sensors that interpret the environment and enable the correct execution of the main task of mobile robotics: navigation. In the case of UAVs, flying at low altitude greatly increases the probability of encountering obstacles, so they need a fast, simple, and robust method of collision avoidance. This work covers the problem of navigation in unknown scenarios by implementing a simple, yet robust, environment-reactive approach. The implementation is done with both CPU and GPU map representations to allow wider coverage of possible applications. This method searches for obstacles that cross a cylindrical safety volume, and selects an escape point from a spiral for avoiding the obstacle. The algorithm is able to successfully navigate in complex scenarios, using both a high and low-power computer, typically found aboard UAVs, relying only on a depth camera with a limited FOV and range. Depending on the configuration, the algorithm can process point clouds at nearly 40 Hz in Jetson Nano, while checking for threats at 10 kHz. Some preliminary tests were conducted with real-world scenarios, showing both the advantages and limitations of CPU and GPU-based methodologies.

Keywords: collision avoidance; depth cameras; point cloud; UAV; CPU; GPU; spiral

Citation: Azevedo, F.; Cardoso, J.S.; Ferreira, A.; Fernandes, T.; Moreira, M.; Campos, L. Efficient Reactive Obstacle Avoidance Using Spirals for Escape. *Drones* **2021**, *5*, 51. https://doi.org/10.3390/drones5020051

Academic Editor: George Nikolakopoulos

Received: 30 April 2021
Accepted: 2 June 2021
Published: 7 June 2021

1. Introduction

In recent years, there have been research efforts in the field of aerial robotics that have led to a broader range of application scenarios [1]. This growing number of applications is closely related to the continuous change in the research focus, which tends to approach higher-level tasks (such as navigation and task planning, paying attention to visual odometry, localisation and mapping) [2].

Due to their success, the use of multirotor unmanned aerial vehicles (UAV) is migrating from an isolated robotics' research topic to a trendy tool [3] that supports several applications and enables the possibility of new ones. The general interest in this type of robot stems from their main characteristics: high manoeuvrability, hover ability, and a reasonable payload–size ratio. At the current stage of development, UAVs can be used for search and rescue missions, surveillance and inspection, mapping, among others [4], carrying task-specific sensors (for example, cameras, optical sensors, and laser scanners).

The main objective of using drones is to support or enable the execution of tasks that usually rely on human labour, while reducing risks and operational costs [3,5]. However, this usually implies a high degree of autonomy of the UAV, which significantly increases the complexity of the system and requires the use of perception sensors and algorithms to interpret them and generate appropriate control outputs. Navigation in complex and harsh environments can be challenging, which implies the need for an obstacle detection and collision avoidance module.

Obstacle detection is divided into three main steps: (1) data acquisition from the perception sensors; (2) data treatment and representation in memory; (3) and, finally, the detection of threatening obstacles. The first module is responsible for successfully communicating with the sensors and reading data from them. In the second step, the data treatment may involve some pre-filtering (such as limiting the accepted range of a point cloud), and their representation may also require some data clustering to reduce the memory requirements. Detection of the potentially risky obstacles depends on their good representation and benefits from clustering, as it reduces the computational power required to verify whether the obstacle interferes with the desired path. However, when clustering, special care has to be taken to avoid losing meaningful environmental features.

Collision avoidance is responsible for calculating an avoidance trajectory in a minimal time window. It needs to be fast enough to generate a solution without colliding with obstacles. Therefore, it works as a bridge to produce navigation control commands from the perceived obstacles in the surrounding environment.

As mentioned earlier, operating in harsh and confined environments increases the perception requirements and control complexity. UAVs must be aware of the obstacles and avoid collisions, even if this means neglecting their primary goal for self-security reasons.

Some approaches deal well with the task of path planning in known scenarios, such as PRM [6] or RRT [7]. However, when working outdoors, the assumption of known scenarios rarely holds [8], so an algorithm capable of reacting quickly enough to changes in the environment to avoid crashing is needed. Works such as that in [5] attempts to address this by using a 360-degree long-range 3D light detection and ranging (LiDAR) to perceive the environment, using Octomaps [9] to store and process large point clouds.

This work aims to contribute with a reactive approach to the collision avoidance methods, while keeping its procedure robust and straightforward. Despite the focus on the collision avoidance layer, obstacle detection is also part of the study, as the avoidance procedure clearly depends on it. The main challenge that arises as a limitation for the developed work is the use of only one depth camera with a very limited field-of-view (FOV) as a perception sensor, with a range of up to 10 m. This requires the algorithm to be extremely fast, from data acquisition to control commands generation. Adding to this, is the requirement to be able to run in a resource-constrained UAV on-board computer. To exploit the implementation possibilities in a real scenario, both a CPU and a GPU solution are developed, being tested, evaluated, and validated in a simulation environment.

The contributions of this article are then fourfold:

1. The development of an efficient and simple but robust method for reacting and avoiding collisions with static or slow-moving obstacles that are not known beforehand;
2. The implementation of the method for both CPU and GPU, evidencing their advantages and drawbacks. The evaluation is done with different hardware setups and conditions;
3. Consideration of the environment perception as part of the method, since it is one of the main bases for the avoidance procedure;
4. Usage of a spiral to search for escape point candidates. Starting from the obstacle's centre, the spiral provides a quasi-optimal solution to avoid the obstacle in terms of deviation from the original path. The first valid point is the one that deviates the least from the original path, providing a collision-free operation.

The remaining article is organised as follows: Section 2 provides some insight about literature related to the reactive collision avoidance topic; Section 3 presents some background on the frameworks used to implement the methodologies detailed in Section 4. The results are in Section 5 and their discussion in Section 6. The final conclusions and future steps envisioned are in Section 7.

2. Related Work

This section presents some of the existing works that use a reactive collision avoidance approach, mainly using point-cloud-based methods.

The addition of a third degree of freedom in trajectory planning for obstacles avoidance significantly increases the complexity of the task. To overcome this problem, some literature works [10–12] decompose the problem into a 2D analysis using 2.5D maps. However, these maps result from projecting the 3D environment onto multiple planes and are particularly useful for robots with predominant movement in 2D planes. When the robot has the freedom to navigate in 3D space, some authors [13,14] use adaptive path planners based on traditional approaches, like A* [15], D* lite [16], PRM [6], or RRT [7]. Besides their high computational cost, these types of approaches generally have the disadvantage of always computing a collision-free path to the desired final position, which can be a waste of resources in an unknown or continuously changing environment.

Given this, the use of a reactive approach is preferable to deal with the unknown, since the key idea is to navigate in complex environments using simple approaches to locally avoid the most threatening obstacles. Examples of sensors that directly generate point clouds are LiDARs, depth cameras or, more recently, event cameras (through the accumulation of events over time). A more in-depth analysis on the usage of LiDAR or camera-based approaches is done in [17].

2.1. Reactive Approaches

Merz and Kendoul [18] presented a LiDAR-based collision avoidance solution for helicopters in structure inspection scenarios. They use only a 2D LiDAR and apply the pirouette descent and waggle cruise flight modes to correctly map the 3D environment by sweeping the sensor. This approach relies heavily on the accurate estimation of the vehicle's position to properly detect obstacles. No processing time analysis performed, but the vehicle is able to perform uninterrupted flights while avoiding the obstacles, providing a real-time solution.

Hrabar, in [19] also presented a reactive approach to collision avoidance. Whenever an obstacle is detected in the planned path, the algorithm spawns an elliptical search zone, orthogonal to the vehicle-goal straight path, to find potential escape points. Those candidate escape points are considered valid only if there is a collision-free path from the vehicle to them, and from them to a predefined distance in the direction of the destination point. If no valid points are found, the ellipse expands up to 100 m, needing the pilot intervention if no valid solution is found. The safety volume is composed of spheres spaced in a way that approximate a cylinder. The escape points search typically took 17.7 ms (10 m ellipse) but can go up to 260 ms in the worst-case scenario, using a Bkd-tree search approach. Both a stereo vision and LiDAR-based approach were tested (isolated and combined), with the best results obtained when using the LiDAR-only system, even using a 15-m range, in contrast to the 35-meter range of the stereo system.

In [5], Hrabar's work is studied and extended, using a LiDAR sensor with 360-degree coverage and a range of up to 100 m (but using only 40 m). The main contributions to the previous solution were (1) limiting the search area of escape points to favour horizontal movements and prevent navigation manoeuvres with the sensor's blind spots, (2) a dynamic safety volume that increases with the vehicle's speed, (3) a local minima escape procedure if no valid escape points are found, and (4) a limited search space of 20 m around the UAV's position. The algorithm's procedure is depicted in Figure 1. The obstacle's search took, on average, 1 ms to be processed, while the escape point calculations required 0.97 ms in a desktop computer (760 obstacle points). The tests with the UAV on-board computer had only 83 obstacle points, resulting in 0.4 ms for obstacle search and 8.63 ms to calculate escape points, clearly showing a performance dependence on the number of obstacles found.

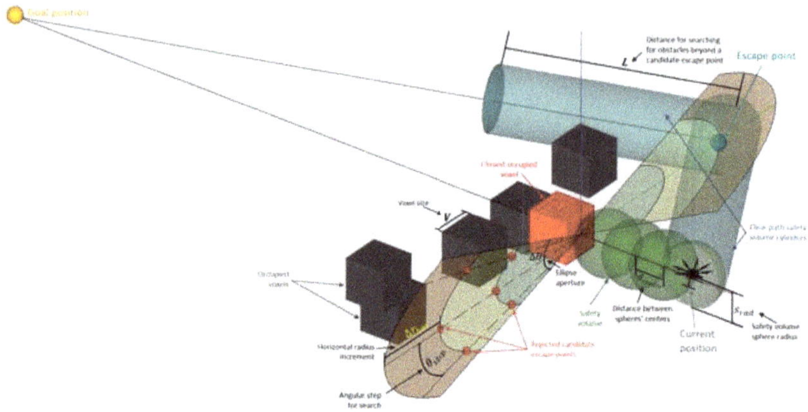

Figure 1. Algorithm representation of [5].

An alternative real-time approach proposed by Vanneste et al. [20] consists of an extrapolation of the 2D VFH+ [21] to 3D using a 2D polar histogram of the environment. The histogram is constructed based on the Octomap voxels, representing the azimuth and elevation angles between the robot position and the evaluated voxel. The generation of a new robot motion takes on average only 326 µs. The main drawback of this method is the number of tuning parameters needed, depending on the application.

Other existing methods take into account the existence of known UAVs in the surroundings for the collision avoidance trajectory. Alejo et al. [22] proposed an indoor method based on optimal reciprocal collision avoidance (ORCA), where each UAV takes responsibility in the process of avoiding others' trajectories. All the planning and trajectory generation is done by a simple central computer that knows the initial planned trajectories of each UAV (pre-inserted), their positions at every moment (using a VICON localisation system), and the parameters of the UAVs' model. Static objects are also previously known, since they are imported from a mesh file. The planning is done in the velocity space, and the vehicles have to react when the current velocities differ from the desired. Due to the existence of a centralised control and lack of perception on-board the UAVs, this method is not suitable for direct application in an outdoor scenario.

The works of [23,24] deal with collision avoidance for aircraft sharing the same space and static obstacles. In [23], they assume that UAVs know the position, speed, and heading of neighbouring aircraft, within a given sensing range. The optimal path is based on essential visibility graphs (EVG) dynamically updated when new UAVs or obstacles are detected. When in a collision course, the UAVs are forced to follow the right of way rules to comply with the existing aviation regulations. Du et al. approach [24] relies on dynamically updated potential fields with an attractive force to the original planned path, and repulsive forces to uncooperative UAVs. They also assume the timely and correct detection of the latter, as well as an accurate inference of their position and speed. Despite being valid methods for avoiding collisions with other UAVs, they both present results for detection and avoidance in the range of hundreds of meters.

The approaches in [22,23] are mainly focused on the reaction to the presence of other moving agents. There are also some works that try to deal with the short-range collision eminence with dynamic objects at low altitudes, like birds or thrown balls. These objects either have an unpredictable reaction to the eminence of collision, or are simply unable to take action for avoiding it. [3] uses a monocular camera to detect potential dynamic obstacles based on a deep learning approach. DroNet [25] also uses a deep learning solution to enable drone navigation by training the network with car driving videos. Similarly to [22], the processing is all done by an external computer that sends commands to the drone. When an obstacle crosses its path, the drone simply stops. A successful implementation to avoid dynamic obstacles as well is presented in [26], where the authors use an event camera and calculate an avoidance command depending on the direction and speed of the obstacles, by combining attractive and repulsive forces (similarly to [24]). The calculations are based on the accumulation of events over a time window and take 3.5 ms from acquisition to command generation.

2.2. Analysis

After examining some 3D reactive collision avoidance methods, it becomes apparent that it is a complex task that usually requires significant processing power. Some approaches try to deal with this issue by reducing the search space or clustering the data. However, the resources required to perform these actions are not considered in the final evaluation of the algorithm. The reconfiguration of the search space may take some time, depending on the size of the working map. The insertion of the point cloud is also a very critical step, as it can be of considerable size and should be processed as quickly as possible due to the urgency of the collision avoidance task. However, existing methods tend to neglect this step as part of the avoidance procedure.

Another very important feature that needs special care is the navigation to unknown areas, especially when the perception sensor is not able to map them while moving, e.g., moving vertically and having only a static 2D laser in the horizontal. Navigating into the blind spot of the sensor is undesirable due to the risk of collision with undetected obstacles. In the case of depth cameras, the effect of limited range and field-of-view (FOV) can be minimised by rotating the UAV to the desired point before moving linearly.

3. Background

3.1. Octomap

The Octomap framework [9] provides an efficient way to represent 3D environments using an occupancy grid based on the octree data structure. The octree provides an hierarchical structure containing multiple nodes, called voxels, that correspond to cubic structures in the 3D space. Each voxel may contain either 8 or 0 child voxels. If all 8 child voxels have the same state (free or occupied), their representation is reduced to the parent voxel only; otherwise, the 8-voxel representation is kept (Figure 2).

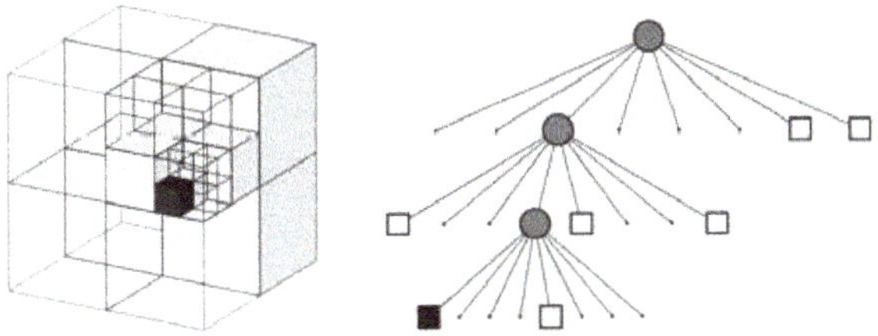

Figure 2. Octree hierarchical representation [9]. Occupied voxels in black and free in white. On the right, parent voxels represented as circles and child as squares.

The base resolution of the octree and the number of layers determine the precision and size of the voxels and allow the environment to be represented at the desired resolution (Figure 3). The Octomap framework does not contain direct information about the voxel 3D location. It is defined only for the root voxel, being the location of the child nodes obtained by descending the layers of the tree. Each leaf of the tree corresponds to a cubic structure in the 3D space. The deeper the layer, the smaller the cube and, therefore, more resolution of the map.

Due to the noise and uncertainties of the sensors, the Octomap's leaves have a probabilistic number of occupancy to represent the environment. Applying a defined threshold, the voxels can be classified as *free* or *occupied*. Unseen voxels have an *unknown* state. When using this map representation for navigation, there are two possible approaches: avoiding the occupied voxels or following the free voxels.

Figure 3. Environment voxelised representation at multiple resolutions [9].

3.2. GPU-Voxels

GPU-Voxels [27] is very similar to the Octomap framework. It outperforms Octomaps by exploring the highly-parallelised capabilities of a GPU, instead of being only CPU-based, with a task division as depicted in Figure 4.

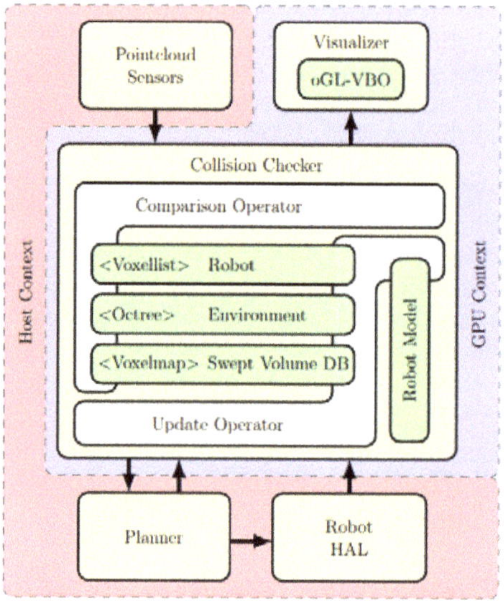

Figure 4. Structure of software components in GPU-Voxels [27]. The red area corresponds to CPU components, while the blue area is GPU's.

There are currently three implemented storage data structures: *Voxel Map*, *Octree* and *Voxel Lists*. Depending on the use case, the structure shall be chosen accordingly (Figure 5). In the insertion of new point clouds, the *Voxel Map* representation provides an additional distance map, where each location is updated with the distance to the nearest occupied voxel.

Voxel Map storage provides a very fast interface to update the voxels and get useful measurements for collision detection. However, it allocates a complete map even to unknown areas, making it less memory-efficient. In contrast, *Voxel Lists* have good memory efficiency while maintaining collision detection speed by storing data in lists of relevant data only and not dealing with empty voxels. The *Octree* representation is a more balanced structure that is especially beneficial in sparse environments due to its hierarchical nature, which has already been described in Section 3.1.

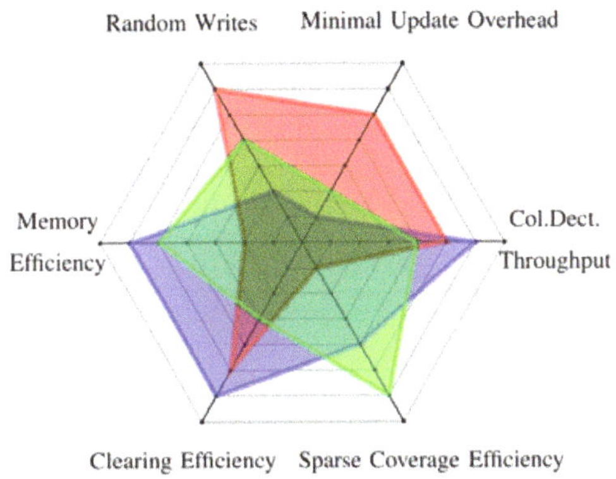

Figure 5. Properties of the implemented storage structures [27]. *Voxel Map* in red, *Octree* in green, and *Voxel Lists* in blue.

4. Materials and Methods

This section describes the work developed. This work covers the entire pipeline of a point-cloud-based collision avoidance algorithm, starting with the insertion of the perceived data (Figure 6).

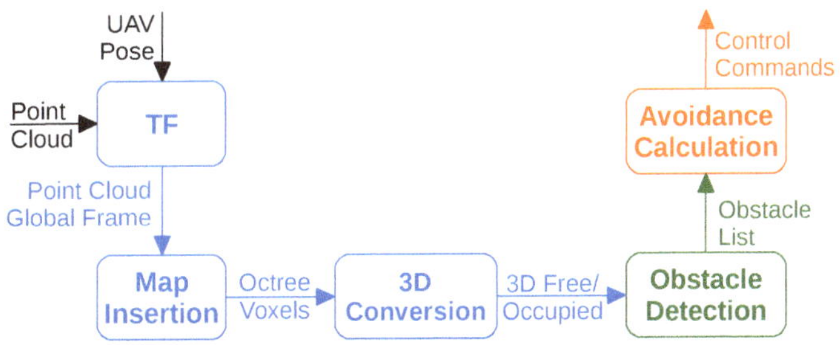

Figure 6. Collision avoidance high-level pipeline. Blue blocks are related with the point cloud insertion and representation.

Data received from sensors must be mapped into a global reference frame before being inserted into the map representation to ensure coherence of the data over time. In general, planners need the input data mapped into a 3D space. Therefore, after an octree representation of the environment, both free and occupied voxels are mapped into the 3D Euclidean space representing the input for the obstacle detection phase. From the obstacle list, the avoidance path calculation generates the control commands for the UAV. The low-level specifics of UAV control are beyond the scope of this work.

In the implementation details described here, the collision avoidance pipeline consists of three stages: the insertion of the point cloud and environment representation (in blue, Figure 6), the search and detection of obstacles (in green), and the final computation of the collision avoidance path (in orange). All phases were implemented using ROS *nodelets* [28–30] associated to the same handler to reduce both memory usage (zero-copy between nodes) and processing time, since the data does not need to be serialised and deserialised in the publish–subscribe mechanism.

The method for reactive collision avoidance was implemented to work, and being compared, in both CPU and GPU. Due to the difference in architectures, some procedures differ between implementations. Therefore, the full pipeline of the method is first presented for the CPU implementation, being then pointed out and described the differences for the GPU.

4.1. CPU Approach

4.1.1. Point Cloud Insertion

In this work, the input point cloud is directly provided by a depth camera. In the CPU-based approach, the algorithm uses Octomap [9] to represent the environment.

After the point cloud is adequately transformed into the global frame, it is inserted into the octree, updating the corresponding nodes. To speed up the processing, the map insertion is stopped at this phase, neglecting the conversion of occupied nodes in the 3D Euclidean space, as depicted in Figure 7. Instead, the obstacle detection stage directly accesses the octree keys, by inverse converting the 3D points in the evaluation only when absolutely necessary.

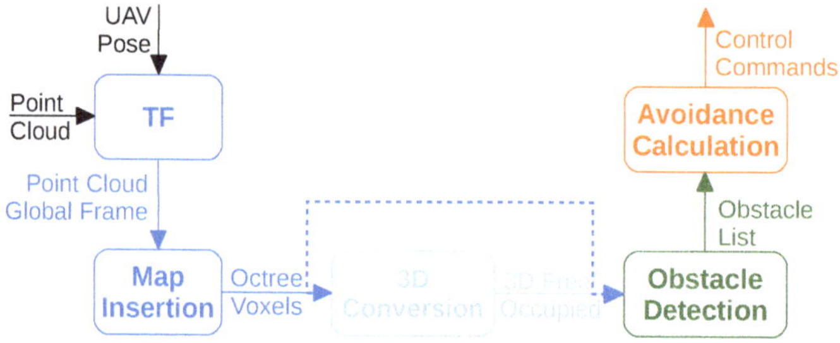

Figure 7. Collision avoidance high-level pipeline with 3D representation skip.

4.1.2. Obstacle Detection

The obstacle detection method is based on the works [5,19] by implementing a cylindrical safety volume to check threatening objects. However, the CPU and GPU-based solutions differ in the cylinder approximation strategy.

To take advantage of the very efficient *raycast* implementation of Octomap, the cylinder approximation was performed by firing n rays along the drone-waypoint direction with a length L. Figure 8 shows an orthogonal cross-section of the approximation process.

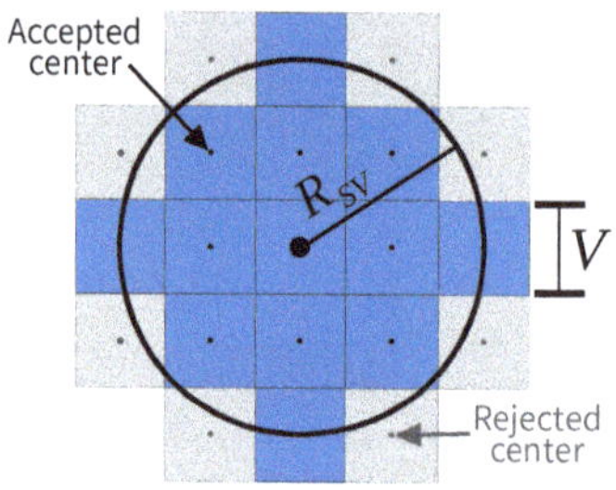

Figure 8. Orthogonal cross section of cylinder approximation.

The number of rays fired depends on the desired safety radius R_{SV} and the voxel size V. The rays are parallel and spaced by V to ensure the coverage of all internal voxels in the desired safety volume. On the external part, the cylinder volume is not fully covered, but the tolerance can be neglected since the motion of the vehicle will cover it at a later time. In a centre calculation with i horizontal steps and j vertical steps, the centres are accepted to fire a ray only if the following condition is satisfied:

$$\sqrt{(i \cdot V)^2 + (j \cdot V)^2} \leq R_{SV} \Leftrightarrow \sqrt{i^2 + j^2} \cdot V \leq R_{SV} \tag{1}$$

Being $\mathbf{r}$ the current UAV's position and $\mathbf{p}$ the goal position, the normalised direction $\mathbf{d}$ is given by $\mathbf{d} = \frac{\mathbf{p}-\mathbf{r}}{||\mathbf{p}-\mathbf{r}||}$ and the ray's starting positions (after passing the validation step of Equation (1)) are given by:

$$\mathbf{ray}_{\mathbf{center}_{i,j}} = \mathbf{r} + (i \cdot \mathbf{d}_y, -i \cdot \mathbf{d}_x, j) \cdot V, \text{ where } i, j \in \mathbb{Z} \quad (2)$$

The search with ray firing begins from the centre of the safety volume up to the radius limit and always returns the nearest obstacle that intersects the volume. The used cylinder length L is calculated by the minimum value between a predefined search range L_{search} and the distance to the waypoint plus a safety distance:

$$L = \min(L_{search}, ||\mathbf{p} - \mathbf{r}|| + R_{SV}) \quad (3)$$

4.1.3. Avoidance Calculation

This stage is only triggered when an obstacle is found and uses the closest object for reacting to the existence of obstructing obstacles.

The process of avoiding an obstacle consists of searching an escape point, similarly to [5,19], but this work uses an Archimedean spiral ($radius = a \cdot \theta$) rather than an ellipse. The spiral was chosen since it provides a continuous function that moves away from the origin with a linear angle growth. This provides an approach that tries to search for avoidance paths, giving preference to those that less affect the original desired path.

Whenever an obstacle is found, the algorithm starts the search for valid escape points using the spiral presented in Figure 9. The spiral is centred in the obstacle $\mathbf{o}$, following a horizontal direction that is orthogonal to the drone-waypoint direction $\mathbf{d} = \frac{\mathbf{p}-\mathbf{r}}{||\mathbf{p}-\mathbf{r}||}$. Therefore, the candidates' coordinates $\mathbf{e}$ will be calculated as follows:

$$\begin{aligned} radius_{hor} &= a \cdot \theta \cdot \cos(\theta) \\ \mathbf{e}_x &= \mathbf{o}_x + radius_{hor} \cdot \mathbf{d}_y \\ \mathbf{e}_y &= \mathbf{o}_y - radius_{hor} \cdot \mathbf{d}_x \\ \mathbf{e}_z &= \mathbf{o}_z + a \cdot \theta \cdot \sin(\theta) \end{aligned} \quad (4)$$

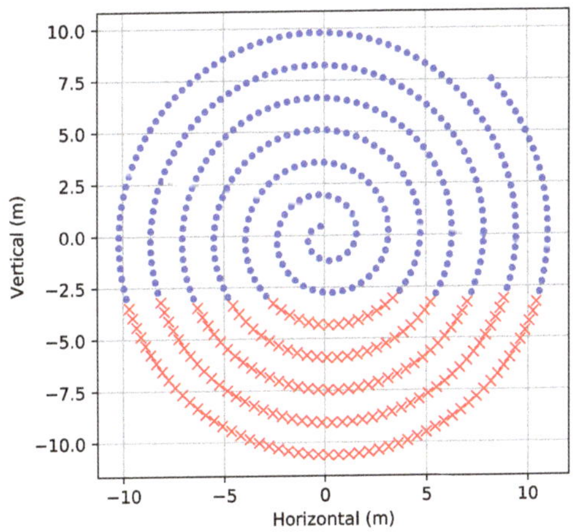

Figure 9. Archimedean spiral samples with a constant arc length. Blue points are considered valid for further calculations. Red crosses are candidate escape points rejected beforehand to avoid very low altitude flights.

If $\mathbf{e}_z - \mathbf{o}_z < -3m$, the candidate escape point is rejected, and the algorithm continues the search. This approach tries to prevent the avoidance of obstacles from below and gives preference to horizontal avoidance or altitude increasing since the probability of finding obstacles tends to increase with the decrease in altitude. In the implemented approach, the arc length used was $l_{arc} = V$ and the winding separation $w = \pi \cdot V \implies a = V/2$.

With the increasing of the spiral radius, the angle step needed to ensure a constant l_{arc} decreases. Using an iterative search approach, the following angle θ can be approximated based on the previous θ_{prev} by solving:

$$\theta = \frac{\sqrt{\left(\frac{V}{2}\cdot\theta_{prev}\right)^2 + 4\cdot\frac{V}{4}\cdot l_{arc}}}{V/2} = \frac{\sqrt{V^2\cdot\left(\frac{\theta_{prev}}{2}\right)^2 + V^2}}{V/2} = 2\cdot\sqrt{\frac{\theta_{prev}^2}{4}+1} \quad (5)$$

Considering a valid escape point **e**, the algorithm applies the procedure described in Sections 4.1.2 and 4.2.2 for detecting obstacles for CPU or GPU, according to the setup chosen. This stage tries to verify if the path is clear from the current UAV position to the escape point, and from it to the final goal point (along a predefined length, for example, 10 m). If the escape point is valid, it is sent as an intermediate goal point to the UAV; otherwise, the algorithm moves to the next candidate until it finds a solution or reaches a maximum number of iterations. The scenario depicted in Figure 1 helps in understanding this method, in which case the spiral replaces the ellipse.

4.2. GPU Approach

4.2.1. Point Cloud Insertion

Instead of Octomaps, the GPU-approach uses GPU-Voxels [27] with the *Voxel Map* storage method. Despite the large memory requirements, the *Voxel Map* storage method was chosen as it provides a faster update while keeping the collision detection throughput high (see Figure 5). It also generates a distance map that is updated on insertion, having the distance to the closest occupied voxel at a certain point. This feature is desirable to the obstacle detection stage since it keeps the searching process simple.

4.2.2. Obstacle Detection

Taking advantage of the distance map provided by the *Voxel Map* storage method, the process of searching for obstacles is the same as in [5,19], where the cylindrical shape is approximated by a set of spheres (Figure 10).

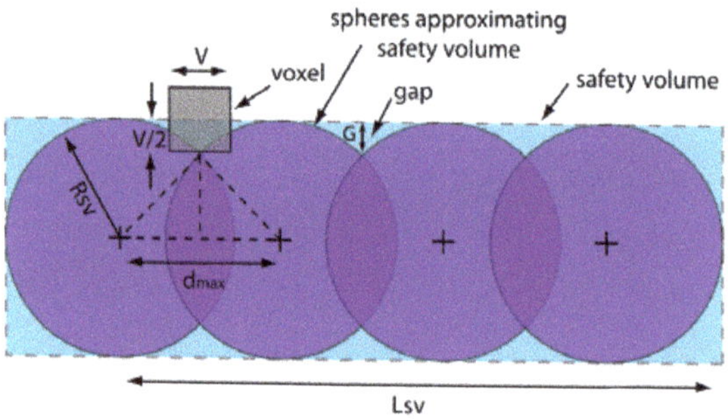

Figure 10. Sphere approximation for cylinder approximation [19].

The maximum allowed sphere distance d_{max} is defined by:

$$d_{max} = 2\sqrt{R_{SV}V - \frac{V^2}{4}} \quad (6)$$

This value is used to calculate the number of needed spheres:

$$n_{spheres} = ceil\left(\frac{L}{d_{max}}\right) + 1 \quad (7)$$

Then the sphere distance d_{sphere} is recalculated to keep a constant displacement:

$$d_{sphere} = \frac{L}{n_{spheres} - 1} \tag{8}$$

Using the normalised displacement vector $\mathbf{d} = \frac{\mathbf{p}-\mathbf{r}}{||\mathbf{p}-\mathbf{r}||}$, the spheres' centres are calculated as:

$$\mathbf{sphere}_c = \mathbf{r} + i \cdot \mathbf{d} \cdot d_{sphere}, \quad \text{where } i \in \{0, ..., n_{spheres} - 1\} \tag{9}$$

The obstacle's search is then simply getting the distance to the closest occupied voxel in the GPU-Voxel distance map and verify if it is smaller than R_{SV}. If yes, the correspondent occupied voxel position is returned. To optimise the process, the search begins in the drone's centre and expands until L, returning immediately if an obstacle is found, ignoring the verification of farther spheres.

4.2.3. Avoidance Calculation

Unlike the previous stages, the avoidance path calculation is common to the CPU and GPU approaches.

4.3. UAV Control

In this work, only a depth camera will be used to perceive the environment. Since it has a very limited FOV, when an obstacle is detected, the UAV controller stops the motion and aligns its heading to the direction of the next waypoint before resuming the motion. This is necessary to ensure a collision-free operation without navigating blindly in unknown areas.

In case no viable alternate path is found after the iteration limit, a recovery procedure has been implemented that instructs the drone to navigate to the previous waypoint and to try to navigate to the problematic area again, but now with more knowledge about the environment. To enable this approach, the UAV keeps a record of the waypoints used throughout the process. This is also of interest for return to launch manoeuvres in complex environments since a viable path (although in reverse) was already found, depending on moving obstacles.

4.4. Summary

Figure 11 presents the algorithm's state flowchart to help to understand the steps presented in this section.

The grey box contains the elements related to the UAV control. Whenever a *Final Waypoint* (desired position) is received, the drone aligns its heading and moves towards it. At each iteration, it verifies if the waypoint was reached, and stops the movement whenever it reaches the final waypoint. A waypoint is considered either a goal position received externally or an escape point from the avoidance procedure.

In parallel with the control, the depth point cloud is stored using the Octomaps or GPU-Voxels framework (blue box). Combining this information with the UAV's pose and waypoint, the algorithm keeps searching for potentially threatening obstacles (green box), triggering the avoidance path calculation (orange box) if any is found. If the algorithm is not able to get a valid avoidance path, the UAV navigates to the previous waypoint and then resumes its operation towards the *Final Waypoint*. Otherwise, an escape point is used to avoid the obstacle before going to the final position.

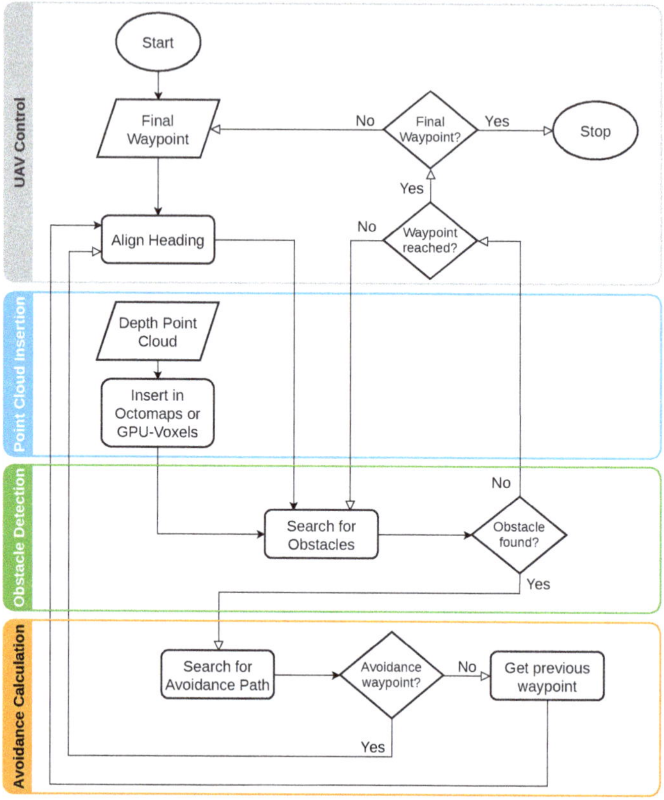

Figure 11. Flowchart of the method.

5. Results

This section presents the main results of the different components of this work. The evaluation of the method was divided into two phases: the first was an offline test using Gazebo simulator, and the second was a preliminary trial using the real hardware in a real scenario. The Gazebo simulator is a primary tool used by ROS for robotic simulations in both academic and professional environments. Since it was developed from its genesis for ROS compatibility, it is straightforward to get it ready to work alongside algorithm implementations in ROS.

5.1. Setup

5.1.1. Vehicle and Sensors

Hexa Exterior Intelligent Flying Unit (HEIFU) [31] was the UAV used to test the algorithm. It is a C3-class hexacopter equipped with a Pixhawk [32] flight controller running ArduPilot, with an RTK-enabled GNSS system. It has a Jetson Nano [33] on-board computer to receive data from the perception sensors and run the high-level control algorithms. It uses an Intel Realsense D435i [34] camera to perceive the environment, that provides both RGB images and a depth point cloud.

Jetson Nano is a low-cost small-sized computer suited for robotics applications with a Quad-core ARM A57 @ 1.43 GHz CPU, a 128-core Maxwell GPU, and 4 GB of LPDDR4 RAM. The version on HEIFU has Ubuntu 18.04 operating system running ROS Melodic Morenia. The Intel Realsense D435i camera has several working configurations. The one used has VGA resolution (640 × 480 pixels) with 75 degrees of Horizontal FOV and 62 degrees of Vertical FOV, providing data at 15 Hz.

For simulation purposes, a model of HEIFU (Figure 12) and a simulation environment were developed and integrated into Gazebo, creating an algorithm testbed similar to real-world conditions and valuable for first-stage testing. In the simulation model, the on-board computer is not simulated as the algorithms will run either in the simulation laptop or in the real on-board computer.

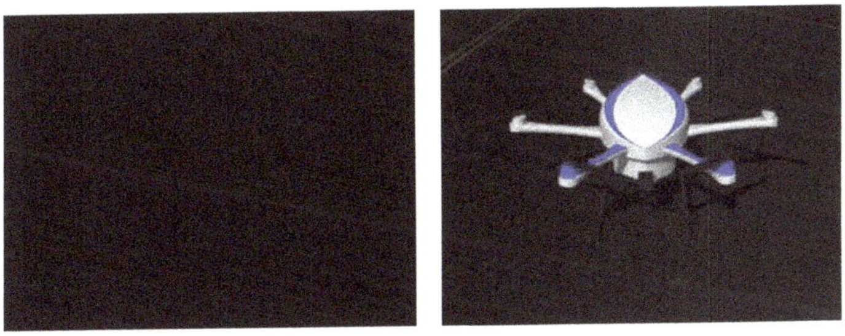

Figure 12. HEIFU prototype (**left**) and Gazebo simulation model (**right**).

5.1.2. Testing Stages

The first stage of the testing is subdivided into two parts:

1. Running on a laptop with an Intel Core i7-9750H CPU @ 2.60 GHz with 12 threads, 16 GB of RAM, and a GeForce RTX2060 GPU. The operating system is Ubuntu 18.04 running ROS Melodic Morenia;
2. Running on a Jetson Nano fully dedicated to the task. It runs only the developed algorithm, being connected to the simulation environment via LAN.

The second testing stage was a preliminary trial for a real flight with all the drone's software components running. Therefore, it only consists of an analysis of the algorithm's response to the environment perception running online in the onboard computer, but without affecting the UAV control for safety reasons.

The depth point cloud is provided at a rate of 10 Hz in the first stage, and 15 Hz in the second. In both stages, the control is set to run at 100 Hz.

5.2. First Stage Results

During the evaluation of the algorithm, a safety volume $R_{SV} = 1$ m, a search maximum length $L_{search} = 10$ m, and 500 as the maximum number of iterations for searching an escape point were used.

A set of tower cranes (Figure 13) was used to evaluate the algorithm's success, through which the UAV must be able to navigate, avoiding them when necessary.

Figure 13. Simulation scenario for evaluating the algorithm's success.

5.2.1. Laptop

Success

The success of the algorithm was tested using both approaches. Figures 14 and 15 (CPU and GPU, respectively) show the trajectories chosen to avoid collisions with the tower cranes. It can be observed that the algorithm is successful in both cases, generating similar, but not equal, avoidance paths (in yellow).

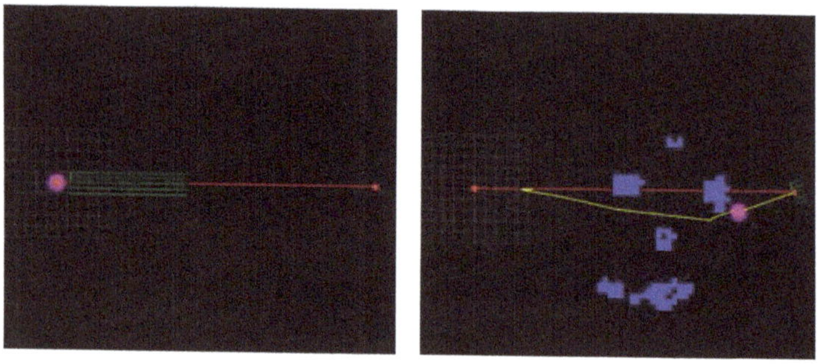

Figure 14. Success test using CPU approach in laptop (top view). UAV path in yellow. Green rays are the safety volume; Red line is the straight-line connection between the starting position and the waypoint. Pink sphere is the last escape point used. Blue cubes represent occupied voxels.

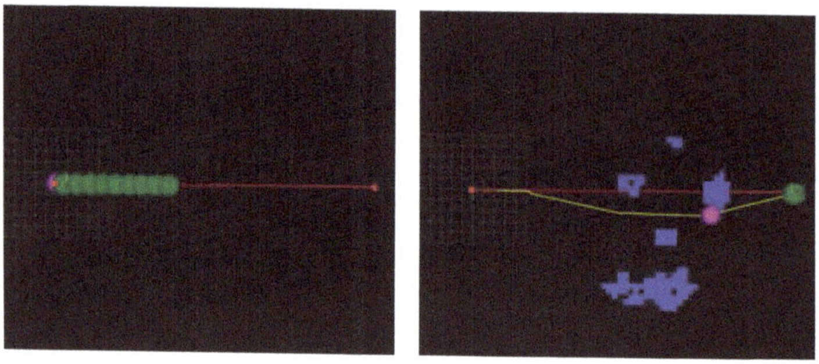

Figure 15. Success test using GPU approach in laptop (top view). UAV path in yellow. Green spheres are the safety volume; Red line is the straight-line connection between the starting position and the waypoint. Pink sphere is the last escape point used. Blue cubes represent occupied voxels.

Processing Times

The processing times presented result from two trials with the same set of waypoints. However, they are not from the exact same flight test, and some quantities may vary, like the number of environment voxels, due to the probabilistic nature of the map representation and UAV pose.

Point Cloud Insertion: Figures 16 and 17 show the times of inserting the point cloud into the map representation for the CPU and GPU approaches. The CPU approach shows a clear linear dependence of the insertion times with the increase in input points. The GPU approach also shows a linear but much smaller dependence on the number of inserted occupied points.

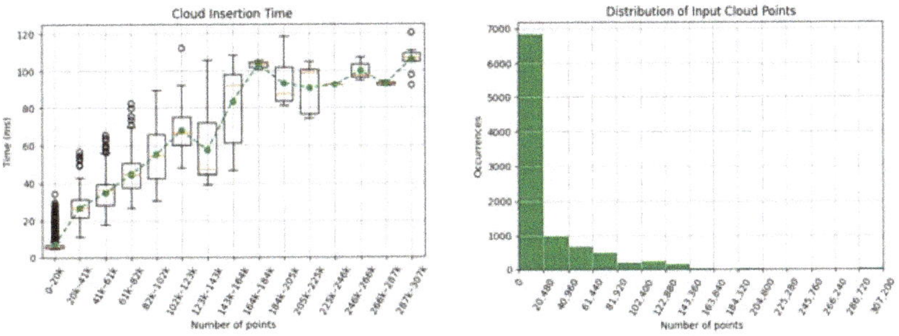

Figure 16. Point cloud insertion times using CPU approach in laptop.

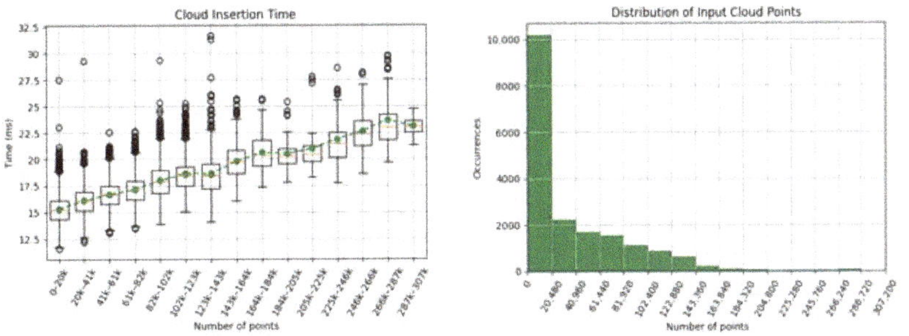

Figure 17. Point cloud insertion times using GPU approach in laptop.

Since the GPU insertion times are low and less dependent on the number of input points, on another trial, the free points were also added to the map in the directions that no data are provided, using a range of 10 m from the camera. This makes the number of input points constant (307,200 points). Figure 18 shows the new insertion times as a function of the number of occupied points. Comparing this to Figure 17, there is not much of an effect, as the linear dependence is only due to additional checking of the occupied points. This also shows that the insertion times are mainly due to CPU data treatment before insertion and CPU-GPU data streams communication overhead. The implementation of this feature increases the map quality, since it erodes outliers.

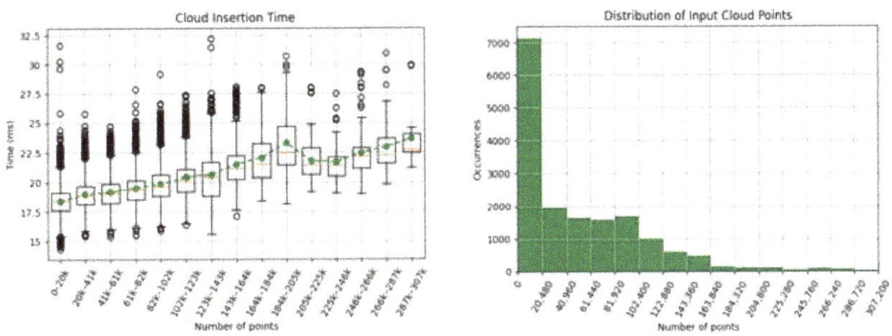

Figure 18. Point cloud and free points insertion times using GPU approach in laptop.

It is worth noting that the CPU approach tends to have faster insertion times for depth clouds below 20 thousand points, which has occurred in most of the flight.

Obstacle Detection The search for threatening obstacles relies on the cylinder approximation methods described in Sections 4.1.2 and 4.2.2.

Figures 19 and 20 depict the times used to verify the need to compute an avoidance path, given the number of occupied voxels in the environment map representation. No clear dependence on the number of occupied nodes is shown in the graphs, being the variations mainly due to the CPU load fluctuations since it was also running the simulation environment.

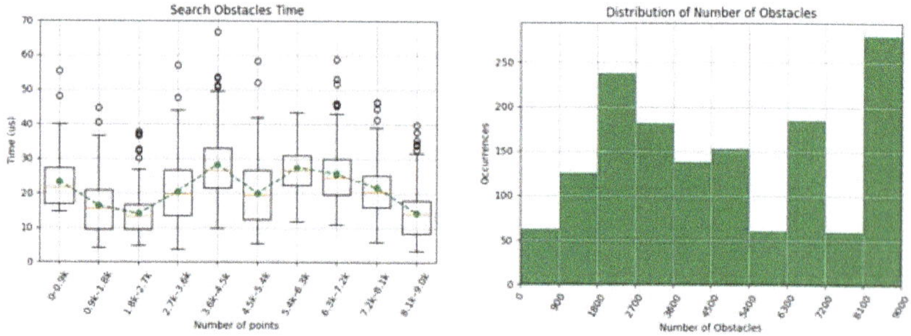

Figure 19. Obstacle search times with respect to the number of occupied voxels using CPU approach in laptop.

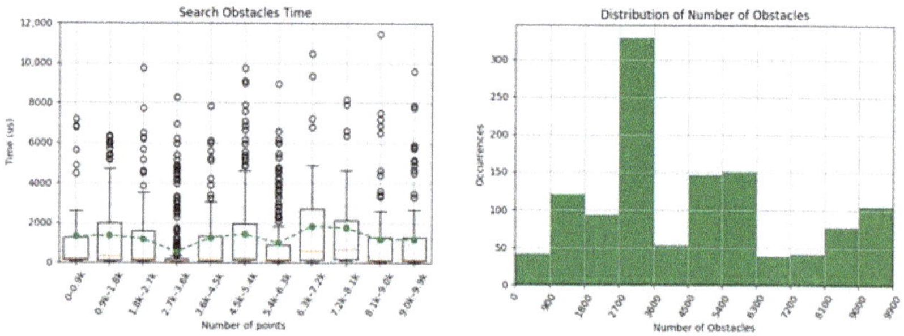

Figure 20. Obstacle search times with respect to the number of occupied voxels using GPU approach in laptop.

The same analysis was performed for the dependence on the search range L (Figures 21 and 22). The CPU approach with ray cast shows some dependence on the range, but no dependence is observed for the sphere approach.

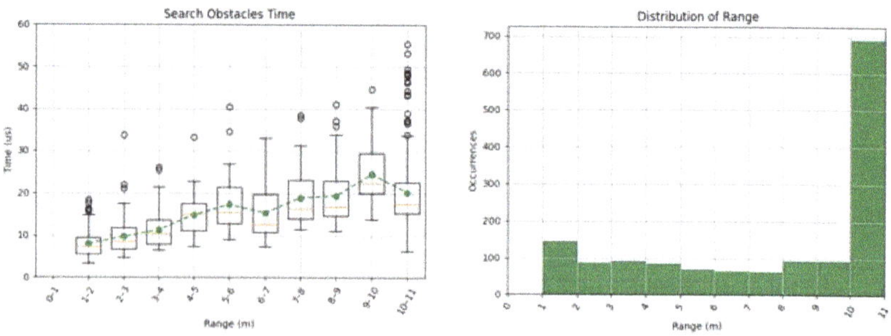

Figure 21. Obstacle search times with respect to the search range using CPU approach in laptop.

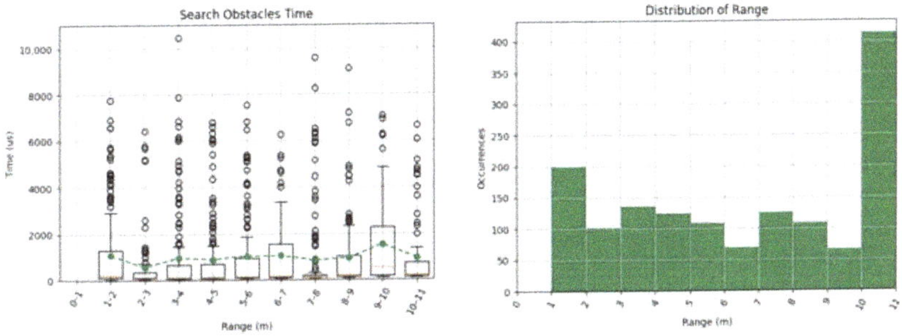

Figure 22. Obstacle search times with respect to the search range using GPU approach in laptop.

Regarding the obstacle search times, the CPU approach had by far the best results, showing that the ray firing strategy is much more efficient than the method with the sphere approximation, despite its dependence on the range.

Avoidance Calculation: Finally, the time required to calculate a valid avoidance path was analysed. In this trial, the maximum number of iterations was never reached. Due to the iterative nature of the search and the dependence on the obstacle detection procedures, a linear dependence on calculations times is expected with the number of iterations, which is confirmed in Figures 23 and 24. Additionally, the GPU approach takes more time to find a valid escape point. In Figure 24, the decreasing time for 75–100 iterations might be due to the early cycle break strategy if an obstacle is found near the drone.

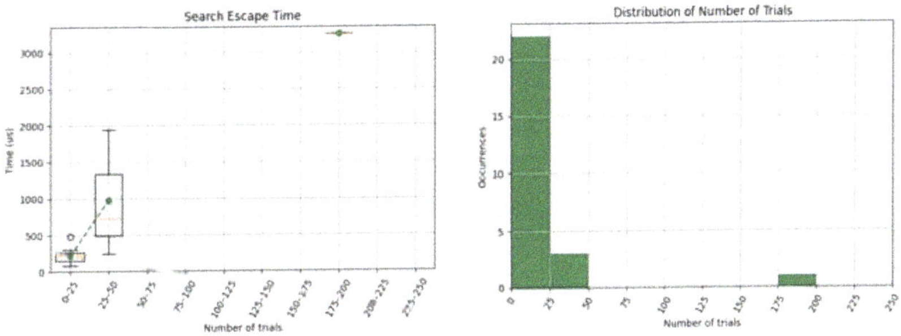

Figure 23. Avoidance path calculation times using CPU approach in laptop.

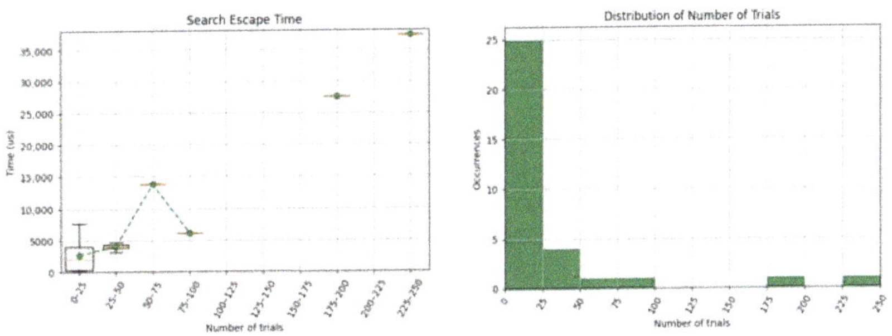

Figure 24. Avoidance path calculation times using GPU approach in laptop.

5.2.2. Jetson Nano

The evaluation steps in the Jetson Nano were the same as in the laptop; however, in this case, the laptop is responsible for running the simulation environment, while the

Jetson Nano is fully dedicated to run the developed algorithm, connected via LAN to the laptop. This configuration removes the heavy simulation from the Jetson core, but relies on the LAN speed.

Success

The first test focuses on the success of the algorithm. Figures 25 and 26 show the avoidance trajectories chosen to overpass the tower cranes. The algorithm succeeds with both implementations, generating similar safe avoidance paths.

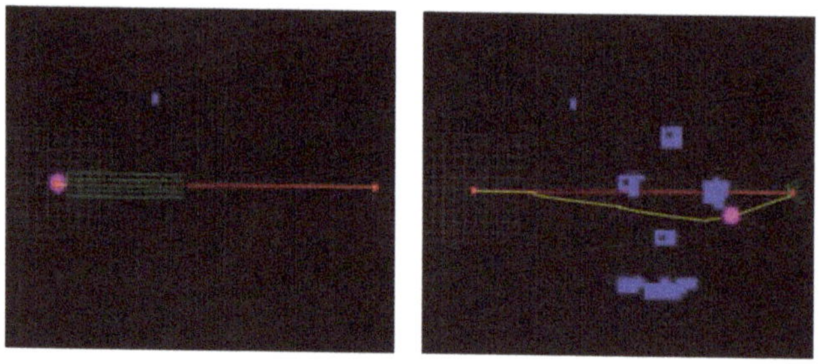

Figure 25. Success test using CPU on Jetson Nano (top view). UAV path in yellow. Green rays are the safety volume; Red line is the straight-line connection between the starting position and the waypoint. Pink sphere is the last escape point used. Blue cubes represent occupied voxels.

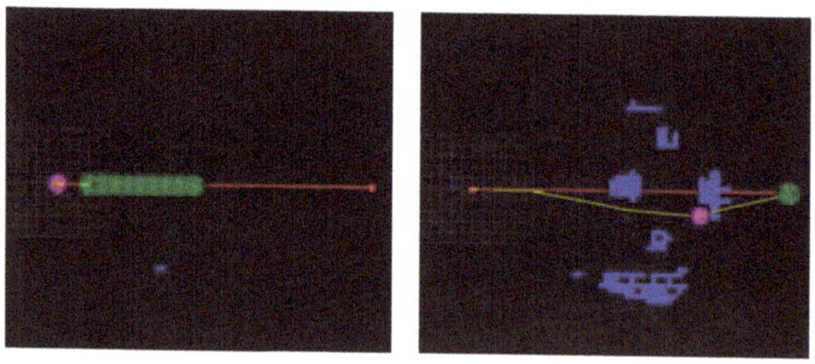

Figure 26. Success test using GPU on Jetson Nano (top view). UAV path in yellow. Green spheres are the safety volume; Red line is the straight-line connection between the starting position and the waypoint. Pink sphere is the last escape point used. Blue cubes represent occupied voxels.

Processing Times

The set of waypoints is the same used in the laptop evaluation.

Point Cloud Insertion: The times for inserting the point clouds are depicted in Figures 27 and 28 for CPU and GPU. Both approaches have a dependence on the point cloud size, being more notorious for the CPU. The results presented in Figure 28 include the insertion of free points, used to clear outliers that might exist in the map from previous insertions.

On Jetson Nano, the GPU implementation is only advantageous for depth clouds above nearly 150 thousand points, which rarely happened during the flight. This suggests that there exists a bottleneck in the CPU-GPU communication, causing the growth of the insertion times. Using a low-power CPU also affects the performance, since the pre-insertion processing tends to be slower.

In both cases, the large times (compared to the laptop's evaluation) led to the drop of some depth cloud messages, as can be confirmed by comparing the histograms of occurrences (Figures 16 and 18 for laptop).

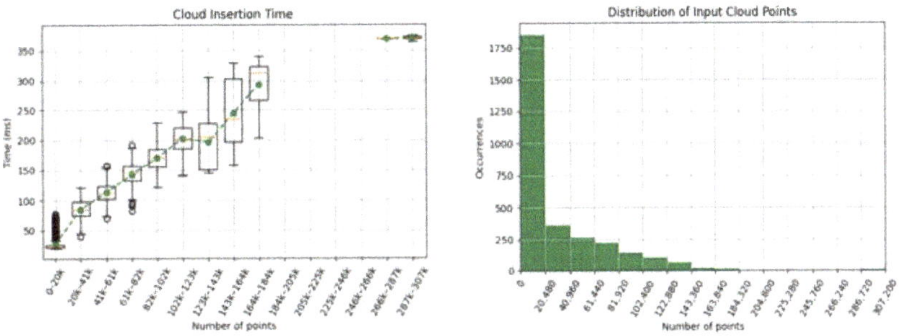

Figure 27. Point cloud insertion times using CPU approach on Jetson Nano.

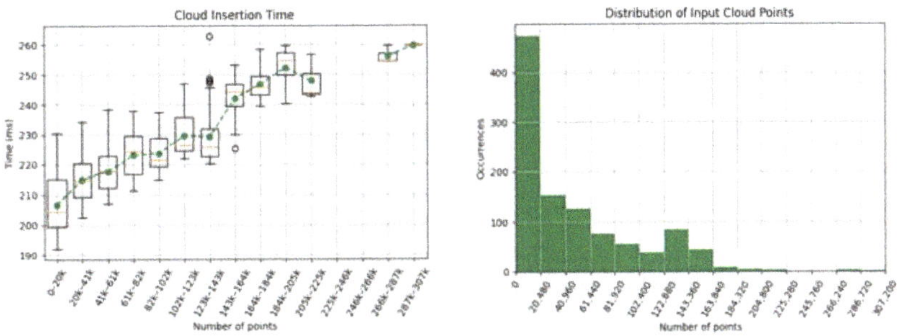

Figure 28. Point cloud insertion times using GPU approach on Jetson Nano.

Obstacle Detection: Using the respective approaches for cylinder approximation, described in Sections 4.1.2 and 4.2.2, the algorithm had the processing times depicted in Figures 29 and 30. None of the cases shows a clear time dependence on the number of obstacles, but the processing time differences (between CPU and GPU) are even more notorious here. For the ray cast method, Jetson Nano is nearly 2 to 3 times slower than the laptop. However, in the case of sphere approximation, the processing time increases 10 times.

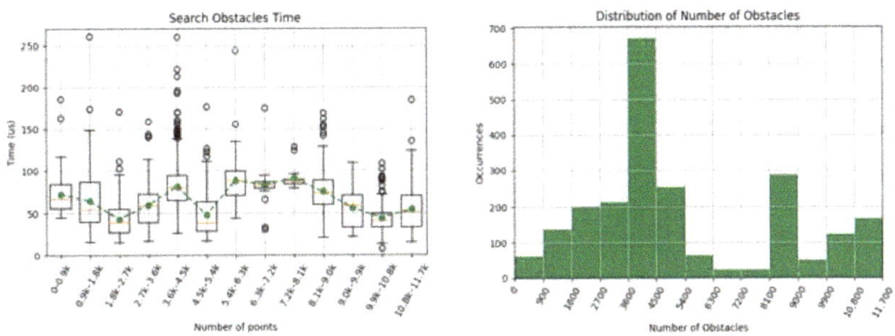

Figure 29. Obstacle search times with respect to the number of occupied voxels using CPU approach on Jetson Nano.

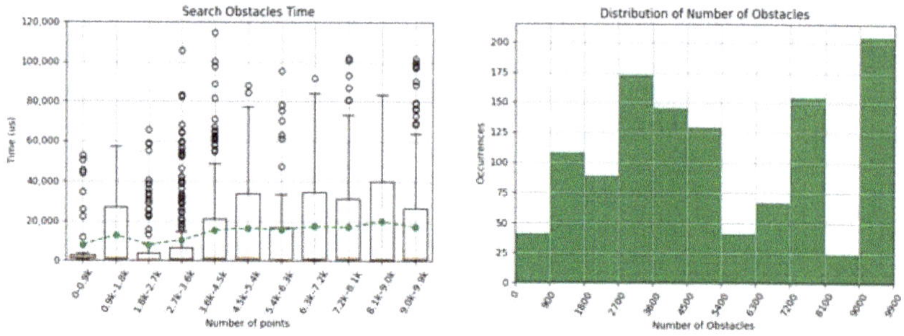

Figure 30. Obstacle search times with respect to the number of occupied voxels using GPU approach on Jetson Nano.

Regarding the dependence on the search range L (Figures 31 and 32). Similarly to the laptop evaluation, the CPU approach with ray cast shows some dependence on the range, while no dependence is observed for the sphere approach. The processing time increase, when compared to the laptop approach, is the same as for the number of obstacles analysis, 2–3 times for CPU and 10 times for GPU. This large increase in the GPU approach also suggests the CPU-GPU communication bottleneck, since the sphere method makes requests to the GPU-Voxels distance map.

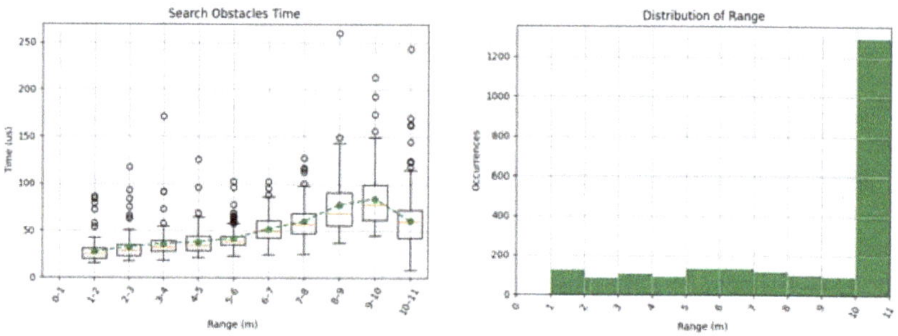

Figure 31. Obstacle search times with respect to the search range using CPU on Jetson Nano.

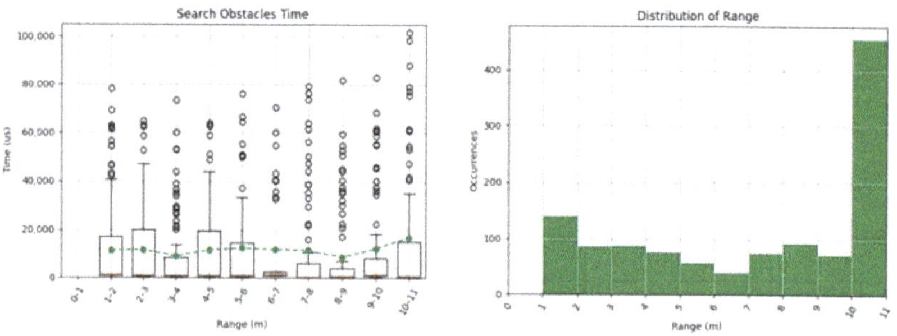

Figure 32. Obstacle search times with respect to the search range using GPU on Jetson Nano.

Avoidance Calculation: During the experimental flight, the maximum number of iterations was never reached, and the processing times are presented in Figures 33 and 34. Due to the dependency on the obstacle search strategy, the GPU method requires more time for the same number of iterations; however, the relation between CPU and GPU methods is not constant over the iterations, which again could be due to the early cycle interruption when an obstacle is near the UAV. This situation tends to occur more often with slower processing. Since the vehicle will take longer to perceive the obstacles and stop, the obstacles will be nearer when calculating the avoidance path.

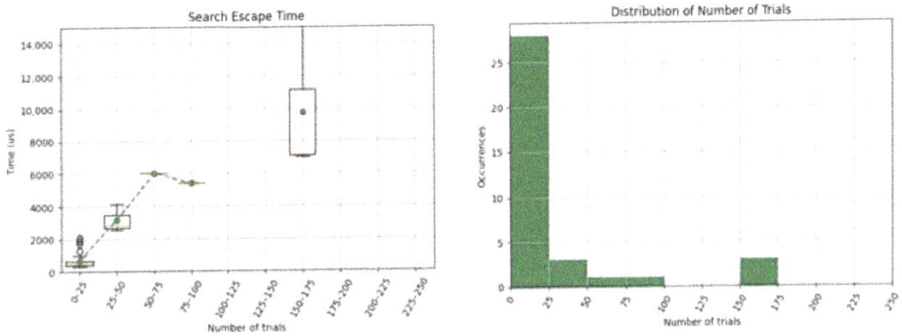

Figure 33. Avoidance path calculation times using CPU approach on Jetson Nano.

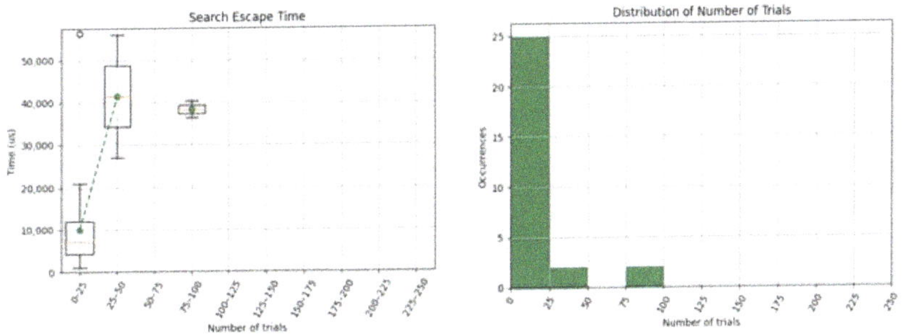

Figure 34. Avoidance path calculation times using GPU approach on Jetson Nano.

5.3. Second Stage Results

After a complete evaluation of the algorithm in the simulated scenario, the next step was to carry out a preliminary test with the real hardware in a real scenario. For safety reasons, for this first test, the algorithm had no control over the drone navigation, being the safety pilot the responsible for positioning the drone for the sensor acquisition of the environment.

Unlike the offline test, the onboard computer Jetson Nano was now running all the control algorithms, acquiring sensors, logging data to an external disk, and communicating with a ground control station (GCS).

5.3.1. Processing Times

The test setup was simple and consisted only of flying the drone facing a wall (Figure 35) to evaluate the algorithm's perception and outputs. Both CPU and GPU-based implementations were tested using the same strategy.

Figure 35. HEIFU flying in the real test scenario.

Point Cloud Insertion

Figures 36 and 37 show the times obtained for inserting the depth cloud in the map representations. The approach using CPU has shown to be entirely not viable to apply in the real scenario. The insertion times of a single medium-size point cloud can take 3 s, not being suitable for the avoidance strategy. This effect occurred due to the high computational load in the CPU, since it was managing other tasks in parallel (like acquiring the camera).

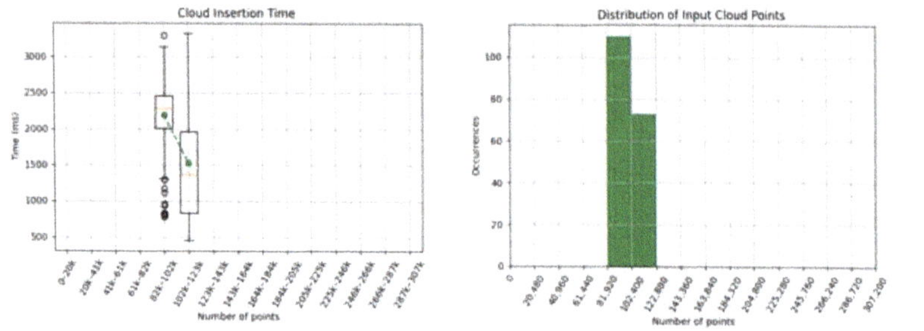

Figure 36. Point cloud insertion times using CPU approach on Jetson Nano in the real scenario.

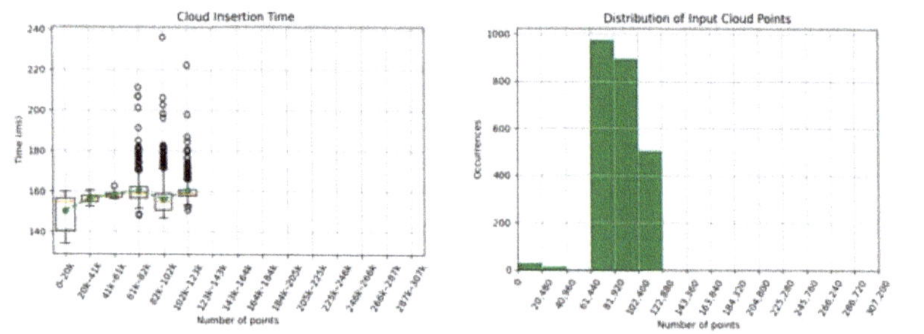

Figure 37. Point cloud insertion times using GPU approach on Jetson Nano in the real scenario.

On the other hand, the GPU insertion times were on pair (and even lower) with the results obtained in simulation (see Figure 28). The lower times are due to a difference between the simulated and real Intel Realsense cameras. The simulated camera is configured to send a dense point cloud with the pixels with no value as NaN, allowing to easily implement the strategy of cleaning the environment map by inserting free points. The real camera only sends a stream with points containing valid measures, not being possible to enable the free point insertion methodology. The implementation could be adapted by identifying the pixels that were not received through trigonometric calculations; however, they are computationally expensive and were not implemented for the first trials.

Obstacle Detection and Avoidance Calculation

These operations are mainly CPU-based and, since there was only one waypoint, they have few measurements to plot in a graph. The few values obtained followed the trend of the results presented in the Jetson Nano fully-dedicated operation (Section Processing Times). The obstacle search times using the ray cast in the CPU approach were below 100 µs, showing that the point cloud insertion increased time is due to scheduling priorities, since fast tasks keep their processing times.

In both methods, when navigating against the wall, an avoidance procedure was triggered, showing that the obstacles were detected as threats.

6. Discussion

In the results section, some implementation specificities and conditions that led to them have already been discussed. The evaluation had its focus on the processing times of each module on different platforms.

When using the laptop to process the algorithm, the CPU approach is able to correctly insert the depth cloud at a rate of 10–100 Hz, depending on the cloud size, while the GPU is less dependent on the cloud size and can process the data at nearly 50 Hz. Searching for potential obstacles in the map can be done at more than 30 kHz using the ray cast CPU method, but only at 500 Hz when using the sphere with the GPU distance map method. The times for calculating an avoidance trajectory depend strongly on the parameters used (arc length, number of steps, spiral winding) but can be done, in a general map, in less than 1 ms with Octomaps, or 5 ms with the GPU version.

In the fully-dedicated Jetson Nano results, the insertion of point clouds in Octomaps drops to a rate of 2.5–40 Hz, and in GPU-Voxels to an average of 4 Hz. The obstacle detection can be done with a rate of 10 kHz in CPU, while the GPU method only reaches 50 Hz and has some processing time peaks. Dependent on these last values, the avoidance path can be typically found in around 4 ms using ray cast to search possible paths and 50 ms using the cylinder approximation by spheres.

The test in a real scenario has shown the impossibility of using the CPU approach for point cloud insertion due to the heavy CPU load (only reaches 0.4 Hz), but the GPU-Voxels keeps the performance obtained in simulation, with an insertion rate of 6 Hz. The remaining modules had the same performance as in the simulation trials.

From the real scenario results, the straightforward approach would be to implement the GPU-Voxels approach to deal with the CPU load in operation. However, the ray cast feature of Octomaps can greatly reduce the search and avoidance calculation times, being very useful for having a quicker perception and response. Therefore, adding this feature to the GPU approach would make it a good solution for applying in Jetson Nano, not compromising the remaining UAV software modules.

Despite implemented, the strategy of moving to the previous waypoint to recover from local traps was never triggered, since the iterations' limit when searching for an escape point was never reached. Adding to this, this method needs some extra benchmark in terms of success and quality of the path generated. Being a reactive solution for preventing collisions, the path followed might not be optimal (or even near it). However, the spiral strategy tries to minimise it by deviating from the straight line only when needed, and the distance strictly needed to ensure the safety of the vehicle and structures.

7. Conclusions

In this work, a reactive approach to the collision avoidance problem was developed. Unlike other existing works in the literature, this paper also covers the problem of transforming the data from the perception sensors in the map representation, considering it part of the collision avoidance problem, since the avoidance success highly depends on an accurate map representation.

Two approaches were developed to evaluate both CPU and GPU advantages and limitations. The map representation is based on octrees, and the obstacle's search relies on a safety volume from the drone, pointing in the direction of the movement. The safety volume is cylindrical, but approximated either by a set of ray firings (CPU approach) or a set of spheres (GPU approach). If the safety volume is crossed, an avoidance procedure is triggered. The avoidance path calculation evaluates a set of candidate escape points that are generated from a spiral-shaped iterative method, centred on the closest threatening object. The escape point (and, therefore, the path) is valid if the safety volume is ensured from the drone to it and from it to the final goal point along a specified distance.

The method was implemented in ROS and validated with Gazebo simulation. It successfully made the UAV navigate in a complex environment, keeping low processing times both in a high-performance laptop and in Jetson Nano, a low-power GPU-enabled computer. The map representation in the CPU used Octomaps, and GPU-Voxels for the GPU implementation. The algorithm was able to deal with point cloud insertion rates of 10–100 Hz using Octomaps in the laptop and 2.5–40 Hz in the Jetson Nano. GPU-Voxels representation had less dependency on the point cloud size, updating the map at 50 Hz in the laptop and 4 Hz in the low-power computer. The ray cast method to approximate the cylindrical safety volume has provided a tool to search obstacles at 10 kHz in the Jetson Nano.

Due to the high CPU load caused by the other components running, the Octomaps performed poorly in a real scenario for updating the map, but GPU-Voxels kept the performance obtained in simulation. Generally, the GPU map representation provided a solution almost independent of the point cloud size, while the CPU ray cast approach was the best for searching potential obstacles.

It is worth noting that the GPU-Voxels native implementation is for Intel-based CPU architectures, while Jetson Nano has an ARM-based CPU.

7.1. Future Work

In order to improve the developed work so far, an implementation of ray cast in GPU seems to be a promising path to follow, accompanied by a deeper study of GPU-Voxels structure and implementation. An alternative to this framework is the Nvidia GVDB-Voxels [35] . Other map representations not based on octrees shall also be considered.

The implementation code can be further optimised, avoiding floating-point operations and replacing trigonometric functions with lookup tables. In pair with this, a focus on the UAV control can also help to improve the performance. If the direction change between waypoints is not significant, the drone could keep moving while rotating, having smoother behaviours.

For improving the robustness and efficiency of the algorithm, further real-scenario tests are envisioned to deal with the real-world variables and uncertainties that are often not present in the simulation environments. The testing of the algorithm in a Jetson Nvidia Xavier NX [36] is also planned to evaluate whether it is sufficient to alleviate the high CPU load suffered by Jetson Nano.

Author Contributions: Conceptualisation, F.A.; methodology, F.A.; software, F.A. and A.F.; validation, F.A., A.F., T.F. and M.M.; formal analysis, F.A.; investigation, F.A.; resources, L.C.; data curation, F.A.; writing—original draft preparation, F.A. and J.S.C.; writing—review and editing, F.A., J.S.C., A.F., T.F., M.M. and L.C.; visualisation, F.A.; supervision, J.S.C.; project administration, L.C.; funding acquisition, L.C. All authors have read and agreed to the published version of the manuscript.

Funding: This project has received funding from the ECSEL Joint Undertaking (JU) under grant agreement No 783221. This project has also received funding from the ECSEL JU under grant agreement No 783119. The JU receives support from the European Union's Horizon 2020 research and innovation programme and Austria, Belgium, Czech Republic, Finland, Germany, Greece, Italy, Latvia, Norway, Poland, Portugal, Spain, Sweden.

Institutional Review Board Statement: Not applicable.

Informed Consent Statement: Not applicable.

Data Availability Statement: The data presented in this study are available in the article.

Acknowledgments: The authors would like to thank Beyond Vision Research and Development Group and PDMFC Research and Development Team.

Conflicts of Interest: The authors declare no conflict of interest.

Abbreviations

The following abbreviations are used in this manuscript:

CPU	Central Processing Unit
EVG	Essential Visibility Graphs
FOV	Field-Of-View
GCS	Ground Control Station
GNSS	Global Navigation Satellite System
GPU	Graphics Processing Unit
LAN	Local Area Network
LiDAR	Light Detection And Ranging
ORCA	Optimal Reciprocal Collision Avoidance
P2P	Peer-to-Peer
ROS	Robot Operating System
RTK	Real Time Kinematics
UAV	Unmanned Aerial Vehicle

References

1. Liew, C.F.; DeLatte, D.; Takeishi, N.; Yairi, T. Recent Developments in Aerial Robotics: A Survey and Prototypes Overview. *arXiv* **2017**, arXiv:1711.10085.
2. Azevedo, F.; Dias, A.; Almeida, J.; Oliveira, A.; Ferreira, A.; Santos, T.; Martins, A.; Silva, E. LiDAR-Based Real-Time Detection and Modeling of Power Lines for Unmanned Aerial Vehicles. *Sensors* **2019**, *19*, 1812. [CrossRef] [PubMed]
3. Pedro, D.; Mora, A.; Carvalho, J.; Azevedo, F.; Fonseca, J. *ColANet: A UAV Collision Avoidance Dataset*; Technological Innovation for Life Improvement; Camarinha-Matos, L.M., Farhadi, N., Lopes, F., Pereira, H., Eds.; Springer International Publishing: Cham, Switzerland, 2020; pp. 53–62.
4. Vergouw, B.; Nagel, H.; Bondt, G.; Custers, B. Drone Technology: Types, Payloads, Applications, Frequency Spectrum Issues and Future Developments. In *The Future of Drone Use: Opportunities and Threats from Ethical and Legal Perspectives*; Custers, B., Ed.; T.M.C. Asser Press: The Hague, The Netherlands, 2016; pp. 21–45. [CrossRef]
5. Azevedo, F.; Oliveira, A.; Dias, A.; Almeida, J.; Moreira, M.; Santos, T.; Ferreira, A.; Martins, A.; Silva, E. Collision avoidance for safe structure inspection with multirotor UAV. In Proceedings of the 2017 European Conference on Mobile Robots (ECMR), Paris, France, 6–8 September 2017; pp. 1–7. [CrossRef]
6. Kavraki, L.E.; Svestka, P.; Latombe, J.; Overmars, M.H. Probabilistic roadmaps for path planning in high-dimensional configuration spaces. *IEEE Trans. Robot. Autom.* **1996**, *12*, 566–580. [CrossRef]
7. Lavalle, S.M. *Rapidly-Exploring Random Trees: A New Tool for Path Planning*; Technical Report TR 98-11; Computer Science Department, Iowa State University: Ames, IA, USA, 1998.
8. Huang, S.; Teo, R.S.H.; Tan, K.K. Collision avoidance of multi unmanned aerial vehicles: A review. *Annu. Rev. Control* **2019**, *48*, 147–164. [CrossRef]
9. Hornung, A.; Wurm, K.M.; Bennewitz, M.; Stachniss, C.; Burgard, W. OctoMap: An efficient probabilistic 3D mapping framework based on octrees. *Auton. Robot.* **2013**, *34*, 189–206. [CrossRef]
10. Maier, D.; Hornung, A.; Bennewitz, M. Real-time navigation in 3D environments based on depth camera data. In Proceedings of the 2012 12th IEEE-RAS International Conference on Humanoid Robots (Humanoids 2012), Osaka, Japan, 29 November–1 December 2012; pp. 692–697. [CrossRef]
11. Chestnutt, J.; Takaoka, Y.; Suga, K.; Nishiwaki, K.; Kuffner, J.; Kagami, S. Biped navigation in rough environments using on-board sensing. In Proceedings of the 2009 IEEE/RSJ International Conference on Intelligent Robots and Systems, St. Louis, MO, USA, 10–15 October 2009; pp. 3543–3548. [CrossRef]
12. Gutmann, J.S.; Fukuchi, M.; Fujita, M. 3D Perception and Environment Map Generation for Humanoid Robot Navigation. *Int. J. Robot. Res.* **2008**, *27*, 1117–1134. [CrossRef]
13. Nieuwenhuisen, M.; Behnke, S. Hierarchical Planning with 3D Local Multiresolution Obstacle Avoidance for Micro Aerial Vehicles. In Proceedings of the ISR/Robotik 2014, 41st International Symposium on Robotics, Munich, Germany, 2–3 June 2014; pp. 1–7.
14. Grzonka, S.; Grisetti, G.; Burgard, W. A Fully Autonomous Indoor Quadrotor. *IEEE Trans. Robot.* **2012**, *28*, 90–100. [CrossRef]
15. Hart, P.E.; Nilsson, N.J.; Raphael, B. A Formal Basis for the Heuristic Determination of Minimum Cost Paths. *IEEE Trans. Syst. Sci. Cybern.* **1968**, *4*, 100–107. [CrossRef]
16. Koenig, S.; Likhachev, M. Fast replanning for navigation in unknown terrain. *IEEE Trans. Robot.* **2005**, *21*, 354–363. [CrossRef]
17. Hrabar, S. An evaluation of stereo and laser-based range sensing for rotorcraft unmanned aerial vehicle obstacle avoidance. *J. Field Robot.* **2012**, *29*, 215–239. [CrossRef]
18. Merz, T.; Kendoul, F. Beyond visual range obstacle avoidance and infrastructure inspection by an autonomous helicopter. In Proceedings of the 2011 IEEE/RSJ International Conference on Intelligent Robots and Systems, San Francisco, CA, USA, 25–30 September 2011; pp. 4953–4960. [CrossRef]

19. Hrabar, S. Reactive obstacle avoidance for Rotorcraft UAVs. In Proceedings of the 2011 IEEE/RSJ International Conference on Intelligent Robots and Systems, San Francisco, CA, USA, 25–30 September 2011; pp. 4967–4974. [CrossRef]
20. Vanneste, S.; Bellekens, B.; Weyn, M. 3DVFH+: Real-Time Three-Dimensional Obstacle Avoidance Using an Octomap. In Proceedings of the CEUR Workshop Proceedings, York, UK, 21 July 2014; Volume 1319.
21. Ulrich, I.; Borenstein, J. VFH+: Reliable obstacle avoidance for fast mobile robots. In Proceedings of the 1998 IEEE International Conference on Robotics and Automation (Cat. No.98CH36146), Leuven, Belgium, 20 May 1998; Volume 2, pp. 1572–1577. [CrossRef]
22. Alejo, D.; Cobano, J.A.; Heredia, G.; Ollero, A. Optimal Reciprocal Collision Avoidance with mobile and static obstacles for multi-UAV systems. In Proceedings of the 2014 International Conference on Unmanned Aircraft Systems (ICUAS), Orlando, FL, USA, 27–30 May 2014; pp. 1259–1266. [CrossRef]
23. Blasi, L.; D'Amato, E.; Mattei, M.; Notaro, I. Path Planning and Real-Time Collision Avoidance Based on the Essential Visibility Graph. *Appl. Sci.* **2020**, *10*, 5613. [CrossRef]
24. Du, Y.; Zhang, X.; Nie, Z. A Real-Time Collision Avoidance Strategy in Dynamic Airspace Based on Dynamic Artificial Potential Field Algorithm. *IEEE Access* **2019**, *7*, 169469–169479. [CrossRef]
25. Loquercio, A.; Maqueda, A.I.; del Blanco, C.R.; Scaramuzza, D. DroNet: Learning to Fly by Driving. *IEEE Robot. Autom. Lett.* **2018**, *3*, 1088–1095. [CrossRef]
26. Falanga, D.; Kleber, K.; Scaramuzza, D. Dynamic obstacle avoidance for quadrotors with event cameras. *Sci. Robot.* **2020**, *5*, eaaz9712. [CrossRef] [PubMed]
27. Hermann, A.; Drews, F.; Bauer, J.; Klemm, S.; Roennau, A.; Dillmann, R. Unified GPU voxel collision detection for mobile manipulation planning. In Proceedings of the 2014 IEEE/RSJ International Conference on Intelligent Robots and Systems, Chicago, IL, USA, 14–18 September 2014. [CrossRef]
28. Quigley, M.; Conley, K.; Gerkey, B.P.; Faust, J.; Foote, T.; Leibs, J.; Wheeler, R.; Ng, A.Y. ROS: An open-source Robot Operating System. In Proceedings of the ICRA Workshop on Open Source Software in Robotics, Kobe, Japan, 12–13 May 2009.
29. Curran, W.; Thornton, T.; Arvey, B.; Smart, W.D. Evaluating impact in the ROS ecosystem. In Proceedings of the 2015 IEEE International Conference on Robotics and Automation (ICRA), Seattle, WA, USA, 26–30 May 2015; pp. 6213–6219. [CrossRef]
30. Damm, C. Object Detection in 3D Point Clouds. Ph.D. Thesis, Institut für Informatik der Freien Universität Berlin, Berlin, Germany, 2016.
31. HEIFU Drone. Available online: https://www.beyond-vision.pt/product/heifu-drone (accessed on 20 April 2021).
32. Pixhawk®. Available online: https://pixhawk.org (accessed on 20 April 2021).
33. Jetson Nano Developer Kit. Available online: https://developer.nvidia.com/embedded/jetson-nano-devkit (accessed on 21 April 2021).
34. Intel® Realsense™ Depth Camera D435i. Available online: https://www.intelrealsense.com/depth-camera-d435i (accessed on 20 April 2021).
35. NVIDIA® GVDB Voxels. Available online: https://developer.nvidia.com/gvdb (accessed on 29 April 2021).
36. Jetson Xavier NX Developer Kit. Available online: https://developer.nvidia.com/embedded/jetson-xavier-nx-devkit (accessed on 29 April 2021).

drones

MDPI

Article

Simulation and Characterization of Wind Impacts on sUAS Flight Performance for Crash Scene Reconstruction

Tianxing Chu [1,2], Michael J. Starek [1,2,*], Jacob Berryhill [1], Cesar Quiroga [3] and Mohammad Pashaei [1,2]

1 Conrad Blucher Institute for Surveying and Science, Texas A&M University-Corpus Christi, Corpus Christi, TX 78412, USA; tianxing.chu@tamucc.edu (T.C.); jacob.berryhill@tamucc.edu (J.B.); mpashaei@islander.tamucc.edu (M.P.)
2 Department of Computing Sciences, Texas A&M University-Corpus Christi, Corpus Christi, TX 78412, USA
3 Texas A&M Transportation Institute, San Antonio, TX 78213, USA; c-quiroga@tti.tamu.edu
* Correspondence: michael.starek@tamucc.edu; Tel.: +1-361-825-3978

Abstract: Small unmanned aircraft systems (sUASs) have emerged as promising platforms for the purpose of crash scene reconstruction through structure-from-motion (SfM) photogrammetry. However, auto crashes tend to occur under adverse weather conditions that usually pose increased risks of sUAS operation in the sky. Wind is a typical environmental factor that can cause adverse weather, and sUAS responses to various wind conditions have been understudied in the past. To bridge this gap, commercial and open source sUAS flight simulation software is employed in this study to analyze the impacts of wind speed, direction, and turbulence on the ability of sUAS to track the pre-planned path and endurance of the flight mission. This simulation uses typical flight capabilities of quadcopter sUAS platforms that have been increasingly used for traffic incident management. Incremental increases in wind speed, direction, and turbulence are conducted. Average 3D error, standard deviation, battery use, and flight time are used as statistical metrics to characterize the wind impacts on flight stability and endurance. Both statistical and visual analytics are performed. Simulation results suggest operating the simulated quadcopter type when wind speed is less than 11 m/s under light to moderate turbulence levels for optimal flight performance in crash scene reconstruction missions, measured in terms of positional accuracy, required flight time, and battery use. Major lessons learned for real-world quadcopter sUAS flight design in windy conditions for crash scene mapping are also documented.

Keywords: small unmanned aircraft systems; photogrammetry; structure-from-motion; wind impact; turbulence; crash scene reconstruction; traffic incident management

Citation: Chu, T.; Starek, M.J.; Berryhill, J.; Quiroga, C.; Pashaei, M. Simulation and Characterization of Wind Impacts on sUAS Flight Performance for Crash Scene Reconstruction. *Drones* **2021**, *5*, 67. https://doi.org/10.3390/drones5030067

Academic Editor: Diego González-Aguilera

Received: 14 June 2021
Accepted: 17 July 2021
Published: 23 July 2021

1. Introduction

A motor vehicle crash can cause considerable economic loss, serious bodily injuries and loss of human life. Crash scene investigation and reconstruction are considered crucial being part of the major concerns in traffic incident management (TIM) [1]. Traditional coordinate and triangulation methods have long been adopted by investigators at a crash scene. They use mechanical measurement tools such as tape measures and roller wheels to acquire baseline measurements and delineate crash scene diagrams [2]. While relatively low cost, these methods have limited efficiency to document measurements and pose safety risks to investigators due to possible exposure to traffic. In order to automate accurate documentation of distance and angle measurements, total stations have started to play a key role at crash scenes since the early 1990s [3]. The ability to collect digital data off the roadway eases investigators' exposure risk to traffic and reduces entire surveying time. Close-range photogrammetry, which emerged around the same time in accident investigation, is able to recover accurate two-dimensional (2D) and three-dimensional (3D) measurements and diagrams by taking overlapping photographs from different viewpoints around crash scenes [4]. Over the past two decades, the potential of terrestrial laser scanning (TLS) has been validated in various crash scene scenarios [5–7]. Enormous scene details can be scanned and captured in a relatively short period of time. However, the costs of TLS equipment are usually high, and multiple scan locations may be needed to minimize scan occlusions in scenes where terrain and crash are complex.

With rapid advances in microelectronics, radio communication, miniaturized imaging lenses and positioning modules, small unmanned aircraft systems (sUASs) have pioneered a series of TIM applications, such as traffic monitoring, flow analysis, crash detection and response, and situational awareness [8–10]. Advantages of using sUAS platforms for TIM include: (1) allowing for customization of onboard sensing systems and observing parameters, (2) offering adequate flexibility in data collection above the scene to be investigated, (3) reducing the exposure of investigators to the dangers of traffic in the roadway, and (4) providing detailed 2D and/or 3D measurement documentation and imagery for post-crash scene investigation conducted in the office. Nowadays, the potential of multirotor sUASs being low-cost and robust crash scene recovery platforms has been manifested via structure-from-motion (SfM) photogrammetry. SfM converts overlapped image sequences taken by a consumer-grade digital camera into 2D orthorectified image products and reconstructed 3D scenes (dense 3D point cloud data and textured 3D meshes). SfM photogrammetry with an octocopter platform was reported to save up to 90% of data collection time compared with traditional coordinate method [11]. Measurements obtained from the point cloud were in accordance with sketches drawn by the investigators, and centimeter-level differences were found in the entire scene. Above the accident scene, orthophotos can be generated using a sequence of individual photos converted from the 4 K-resolution video taken by the quadcopter camera [12]. The results demonstrated a horizontal accuracy of 5–8 cm in scene documentation compared with a real time kinematic (RTK) global navigation satellite system (GNSS) survey.

While growing attention has been paid to SfM photogrammetric surveys with sUAS platforms in crash scene investigation and recovery, it is important to realize that nearly 21% of the crashes are weather-related every year in the United States [13] and performing flight missions under hazardous weather conditions remains a difficult task due to safety and data quality concerns. For example, wind is a frequent natural phenomenon, but high winds tend to increase the risks of freight truck crashes on the roadway [14,15]. In such a scenario, before an sUAS is dispatched to conduct the crash scene reconstruction mission, it is essential to ensure that the wind speed does not exceed the aircraft's operation limit specified by the vendor. Some sUASs are less susceptible to the wind disturbance, but their battery life and flight time is reduced as wind speeds increase [16,17]. High wind speed can also negatively affect the SfM photogrammetry and derived mapping products due to disturbed waypoint targeting and image orientation. Wind direction and turbulence are also important variables to consider as they affect the flight path geometry, energy consumption and overall flight safety [18–21].

Some studies have documented preliminary findings on the variations of sUAS flight stability due to wind forces. Wang conducted an in-house simulation to assess the wind impacts on sUAS flight stability at low altitudes [22]. The differences in flight speed and attitude were summarized when various types of wind were examined. Siqueira mathematically created a wind model and evaluated its effects on sUAS trajectory tracking by looking into 2D/3D error and control activities [23]. The results suggested rapid trajectory tracking degradation as a response to the increased magnitude of wind dynamics.

Initial research efforts have been made in recent years to use open-source mission planning tools such as Mission Planner to conduct sUAS simulation runs in windy conditions [24]. However, comprehensively characterizing sUAS responses to various wind conditions has been understudied but is considered crucial before planning flight operations for crash scene reconstruction. While there are multiple types of suboptimal weather conditions that raise safety and data quality concerns for flying sUASs, the main aim of this study is to set wind as an exclusive factor to parameterize its impacts on flight performance of a representative quadcopter sUAS platform type via realistic flight simulations using a standard gridded flight design for SfM image acquisition. Simulation results are applied to document and generalize lessons learned for platform-independent quadcopter sUAS flight design under windy conditions for crash scene mapping. A series of quadcopter simulations with incremental increases in wind speed, direction, and turbulence are performed to model suboptimal weather conditions. Positional error, battery use, and flight time are used as statistical metrics to characterize the wind impacts on flight performance.

2. Materials and Methods

2.1. Test Environment Setup

As the most popular open-source autopilot software suite adopted by a variety of autonomous vehicles, ArduPilot was employed as the underlying simulation framework in this study [25–27]. ArduPilot enables modeling a wide range of unmanned vehicle characteristics with regard to mission planning, remote control, communication and navigation. To realistically characterize wind impacts on sUAS behavior in a simulated environment, the software-in-the-loop (SITL) simulator was run on a host computer without risking an actual aircraft platform. ArduPilot on SITL can compile source code based on a sophisticated sUAS flight dynamics model and perform code execution in software environment for development and testing purposes. In such a simulation framework, ArduPilot supports the MAVLink protocol for real-time telemetry between the simulated sUAS and the ground control station (GCS) such as Mission Planner. Mission Planner as an open-source GCS includes major flight planning functions similar to that in commercial software packages such as Pix4Dcapture and Map Pilot Pro. In this work, Mission Planner along with ArduPilot on SITL was used for flight simulation, which enabled: (1) editing various parameters regarding changes in environment type, sensor failure and vehicle platform, (2) simulating an sUAS as a virtual flight control unit (FCU) to conduct pre-defined simulation runs on a computer without any special hardware, and (3) storing and downloading the log files of a mission for post-flight analysis.

In the United States, quadcopter sUAS platforms have been widely chosen and increasingly used for TIM and crash scene reconstruction by law enforcement agencies [9,28–32]. Compared with other frame types, such as hexacopters and octocopters, quadcopters can be designed and developed relatively cheap and small in size for carrying positioning and non-metric camera payloads to perform SfM photogrammetric tasks. This study, therefore, selected a quadcopter frame type to run realistic simulations in ArduPilot SITL to characterize wind impacts on flight performance for crash scene reconstruction. This quadcopter frame type has been employed by some commercial platforms including 3DR Solo and Parrot Bebop 2 [33,34]. It is possible to edit a list of behavior controlling parameters through the MAVLink protocol. A complete set of such parameters is available in [35].

A rectangular area at the Texas A&M Flight Test Station Airport in Bryan, Texas, USA, was identified to define the boundary of the simulated crash scene. The rectangular area was 105 × 70 m and centered over an airfield intersection at 30°38′16.50″ N, 96°28′54.70″ W. In this work, a DJI Mavic 2 Pro camera model was selected in Mission Planner due to its current popularity in use by law enforcement and transportation agencies for crash scene mapping [28–30]. In this study, two flight paths were planned in simulation as follows:

I. A single flight height of 80 m above ground level (AGL) was determined, targeting a ground sample distance (GSD) of 2.0 cm/px to enable capturing crash scene details. The flight mission intended to achieve 80% frontal and 80% side overlap to facilitate creation of adequate SfM photogrammetric mapping products [36,37], resulting in 48 photo locations taken along four flight lines in the East-West course (i.e., flight direction) with 12 photo locations along each (Figure 1). The flight plan included the sUAS stopping at each waypoint location to capture an image before moving to the next waypoint. The spacing between adjacent waypoints was 12.5 m and the spacing between flight lines was 18.8 m.
II. Simulations ran at dual altitudes of 80 m and 10 m AGL. In this flight path, the simulated quadcopter first completed an entire set of actions at 80 m AGL as defined

in flight path I, then it descended and continued the mission at 10 m AGL (Figure 2). At this lower altitude, the flight courses were kept the same, and 80% frontal and 80% side overlap settings remained. At 10 m AGL, an additional 48 photos were acquired along four flight lines that were scaled down. At 10 m AGL, the spacing between waypoints was 2.5 m and the spacing between flight lines was 3.7 m.

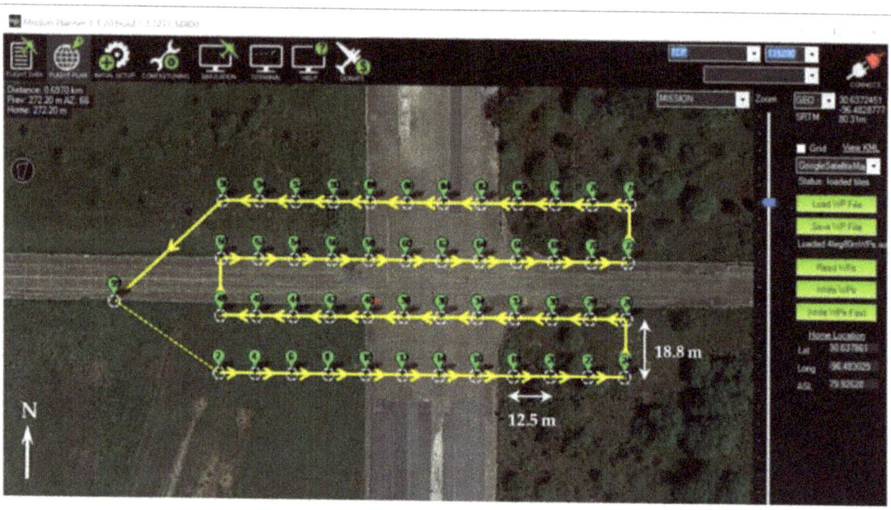

Figure 1. Overview of flight lines and waypoints of the simulated crash scene within Mission Planner.

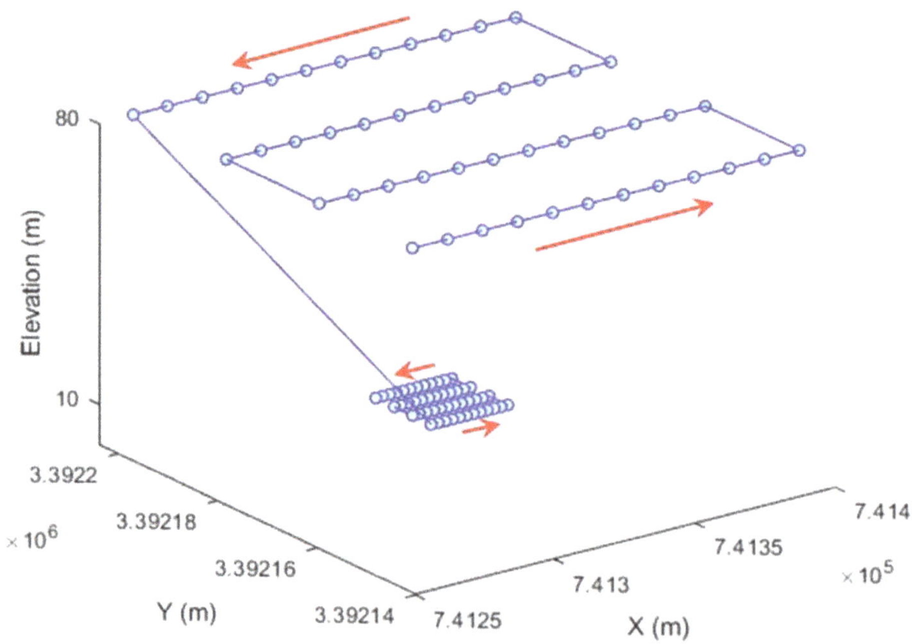

Figure 2. Flight path for dual altitudes of 80 m and 10 m AGL. The simulated flight started the mission at 80 m AGL and then descended to continue the mission at 10 m AGL. The red arrows indicate flight courses. 80% frontal and 80% side overlap settings remained at both 80 m and 10 m AGL. There were 48 images taken at each AGL level.

It is worth noting that in an SfM photogrammetric survey mission, the flight height relates closely to the length of a mission and waypoint locations given specific overlap and camera model settings. However, the flight height is independent to simulated wind effects to be defined in Section 2.2. In other words, the choice of flight height is generic in this study and another flight altitude will not vary wind impacts compared with that of 80 m or 10 m AGL.

Mission Planner placed the simulated aircraft at the home position near the mission scene. The SITL ran a simulated FCU within the virtual aircraft and the planned actions were then uploaded into the aircraft via the simulated GCS link the same way it would be done in a real flight with an actual aircraft. The interface then allowed setting environmental factors for each simulation run such as wind speed, direction, and turbulence. In this article, these three wind parameters were the primary variables to study the sUAS responses.

2.2. *Creation of Simulation Runs*

2.2.1. Wind Speed

For single-altitude flight path at 80 m AGL (Figure 1), five values were evaluated with regard to wind speed, which were 0 m/s, 3.5 m/s, 7.0 m/s, 10.5 m/s, and 14 m/s. The value of 14 m/s was chosen as the upper bound because after this level, the respective quadcopter frame selected in Mission Planner's SITL would not be able to maintain its position and started to drift off of the simulated crash scene. For dual-altitude flight path (Figure 2), wind speeds of 10.5 m/s and 3.5 m/s were employed at 80 m and 10 m AGL, respectively.

2.2.2. Wind Direction

Wind direction is defined as the direction from which the wind is coming with respect to North in a clockwise fashion. For example, 0° represents North wind blowing from North to South and 90° represents East wind blowing from East to West. As described in Figures 1 and 2, the flight lines were oriented in the East-West direction, therefore, without loss of generality, simulating wind directions in the first quadrant (i.e., from 0° to 90°) was sufficient to depict distinct aircraft-wind angular relationships. This angular relationship is defined as α angle, which refers to the included angle from wind direction to flight course moving in a clockwise motion. In this work, the wind directions of 0°, 22.5°, 45°, 67.5°, and 90° were chosen for single-altitude flight path at 80 m AGL (Figure 1), the wind directions of 0°, 45°, and 90° were chosen for dual-altitude flight path at 80 m and 10 m AGL (Figure 2). The α angle and its relation to the wind direction and flight course are summarized in Table 1. In the case of wind direction of 90°, wind blew in or against the direction of aircraft travel, yielding pure tailwind or headwind, respectively. A 0° wind direction generated pure crosswind scenarios where the wind blew perpendicular to the flight path. All the other wind directions were able to decompose the force into both crosswind and headwind/tailwind components, and the corresponding α angles are illustrated in Figure 3.

Table 1. α angle and its relation to the wind direction and flight course.

α Angle	Wind Direction (°)	Flight Course
0° (Headwind)	90°	E
180° (Tailwind)	90°	W
90° (Crosswind)	0°	E
270° (Crosswind)	0°	W
67.5°	22.5°	E
247.5°	22.5°	W
45°	45°	E
225°	45°	W
22.5°	67.5°	E
202.5°	67.5°	W

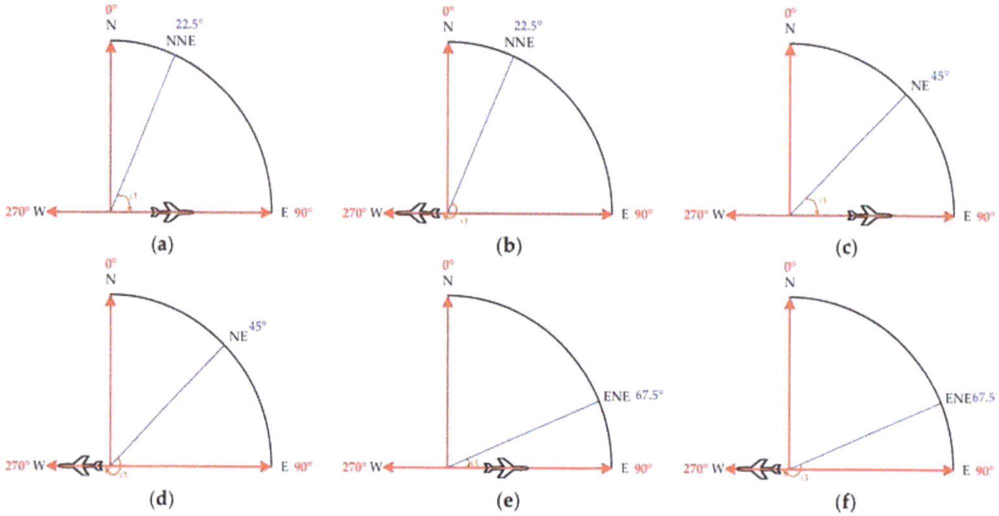

Figure 3. Examples of α angles when an aircraft does not fly in headwind, tailwind or crosswind. (**a**) α = 67.5° when wind direction is 22.5° and flight course is East. (**b**) α = 247.5° when wind direction is 22.5° and flight course is West. (**c**) α = 45° when wind direction is 45° and flight course is East. (**d**) α = 225° when wind direction is 45° and flight course is West. (**e**) α = 22.5° when wind direction is 67.5° and flight course is East. (**f**) α = 202.5° when wind direction is 67.5° and flight course is West.

2.2.3. Turbulence

Turbulence takes place by adding 3D random vectors and magnitudes to the existing wind conditions. According to the National Weather Service, turbulence for operating aircraft is classified as [38]:

- Light (less than 7.2 m/s, and less than 5.7 m/s vertically),
- Moderate (7.2–12.3 m/s, and 5.7–11.3 m/s vertically),
- Severe (greater than 12.3 m/s, and 11.3–14.9 m/s vertically), and
- Extreme (greater than 12.3 m/s, and greater than 14.9 m/s vertically).

Within ArduPilot on SITL, turbulence is modeled as a combination of high pass and low pass filters in both horizontal and vertical directions satisfying the following conditions [39]

$$t_h^n = 0.98\, t_h^{n-1} + 10 \cdot (1 - 0.98) t_i r_g \quad (1)$$

$$t_v^n = 0.98\, t_v^{n-1} + 10 \cdot (1 - 0.98) t_i r_g \quad (2)$$

where t_h and t_v represent horizontal and vertical turbulences in the unit of m/s, respectively, n and $n-1$ denote current and previous time instances, r_g is a random number following a $\mathcal{N}(0,\ 1)$ Gaussian distribution, and t_i is a turbulence index value set to 0, 5, 10, or 20 in this work. As the turbulence index increased, the turbulence magnitude increased. More specifically, a turbulence index of five (i.e., $t_i = 5$ in Equations (1) and (2)) involved randomized horizontal and vertical air speed fluctuations as great as 5 m/s. Likewise, turbulence indices of 10 and 20 (i.e., $t_i = 10$ and 20 in Equations (1) and (2)) involved randomized horizontal and vertical air fluctuations as great as 10 m/s and 20 m/s, respectively. For single-altitude flight path at 80 m AGL as shown in Figure 1, light (i.e., turbulence indices of 0 and 5), moderate (i.e., turbulence index of 10), and extreme (i.e., turbulence index of 20) turbulence conditions were simulated. It was not necessary to simulate severe turbulence because the only difference between severe and extreme turbulences was the vertical turbulence component. For dual-altitude flight path at 80 m and 10 m AGL as shown in Figure 2, light (i.e., turbulence index of 0) and moderate (i.e., turbulence index of 10) turbulence conditions were simulated.

It is worth noting that turbulence does not equate to wind gust. Gust denotes a local maximum above a certain threshold above the mean wind speed within a certain amount of time such as one or two minutes [40].

2.2.4. Overview of Simulation Runs

The incrementally increased wind speed, direction, and turbulence values resulted in 58 simulation runs at single altitude of 80 m AGL to evaluate flight stability and endurance (Table 2). A wind speed of 14 m/s was only used under no turbulence conditions because adding turbulence at this level would have resulted in failed flight missions for the quadcopter frame selected in Mission Planner's SITL. Simulation runs for a turbulence index of five only included wind speeds of 10.5 m/s because the results for turbulence indices of 0 and 5 were similar.

Table 2. A total of 58 simulation runs with incrementally increased wind speed, direction and turbulence values at single altitude of 80 m AGL.

Wind Direction (°)	Wind Speed (m/s)	Turbulence Index	Flight Height AGL (m)
N/A	0.0	0	80
0	3.5	0	80
0	7.0	0	80
0	10.5	0	80
0	14.0	0	80
22.5	3.5	0	80
22.5	7.0	0	80
22.5	10.5	0	80
22.5	14.0	0	80
45	3.5	0	80
45	7.0	0	80
45	10.5	0	80
45	14.0	0	80
67.5	3.5	0	80
67.5	7.0	0	80
67.5	10.5	0	80
67.5	14.0	0	80
90	3.5	0	80
90	7.0	0	80
90	10.5	0	80
90	14.0	0	80
0	10.5	5	80
22.5	10.5	5	80
45	10.5	5	80
67.5	10.5	5	80
90	10.5	5	80
0	0.0	10	80
0	3.5	10	80
0	7.0	10	80
0	10.5	10	80
22.5	3.5	10	80
22.5	7.0	10	80
22.5	10.5	10	80
45	3.5	10	80
45	7.0	10	80
45	10.5	10	80
67.5	3.5	10	80
67.5	7.0	10	80

Table 2. *Cont.*

Wind Direction (°)	Wind Speed (m/s)	Turbulence Index	Flight Height AGL (m)
67.5	10.5	10	80
90	3.5	10	80
90	7.0	10	80
90	10.5	10	80
0	0.0	20	80
0	3.5	20	80
0	7.0	20	80
0	10.5	20	80
22.5	3.5	20	80
22.5	7.0	20	80
22.5	10.5	20	80
45	3.5	20	80
45	7.0	20	80
45	10.5	20	80
67.5	3.5	20	80
67.5	7.0	20	80
67.5	10.5	20	80
90	3.5	20	80
90	7.0	20	80
90	10.5	20	80

In addition, a total of six simulation runs were generated for dual altitudes of 80 m and 10 m AGL (Table 3). At 10 m AGL, the simulation included a wind speed of 3.5 m/s and a turbulence index of 0. At 80 m AGL, the simulation included a wind speed of 10.5 m/s and two possible turbulence index values: 0 and 10.

Table 3. A total of six simulation runs for dual altitudes of 80 m and 10 m AGL.

Wind Direction (°)	Wind Speed (m/s)	Turbulence Index	Flight Height AGL (m)
0	10.5	0	80
	3.5	0	10
45	10.5	0	80
	3.5	0	10
90	10.5	0	80
	3.5	0	10
0	10.5	10	80
	3.5	0	10
45	10.5	10	80
	3.5	0	10
90	10.5	10	80
	3.5	0	10

After setting up wind parameters in each individual run, the simulated aircraft was then armed and flown in autopilot mode. Once the simulated aircraft landed and disarmed its motors within Mission Planner, the flight log was downloaded from the flight controller via the simulated radio link. Each flight log was stored in a folder for dissemination.

2.3. *Flight Log Dissemination and Parsing*

When a simulation run completed the mission, the flight log downloaded contained the status of the FCU recorded at 5 Hz throughout the flight. This led to a log containing 150,000 to 200,000 lines of information in each simulation run. For this study, the position and attitude of the craft at the time of image acquisition and the battery power consumed at the end of the mission were of particular interest. To meet the needs, a Python script that read the flight log and output a text file containing the desired information was written. The script looked for flight log messages beginning with the "CAM" tag (i.e., camera shutter information) and wrote them to a text file in a designated format. Then the script found the last flight log message containing a "BAT" tag (i.e., gathered battery data) and wrote it to a separate text file. The "CAM" messages contained the attitude and positional information of the airframe when the camera was triggered during the mission (Figure 4a). The position coordinates were recorded in the World Geodetic System 1984 (WGS-84) ellipsoidal model, and they were converted to the Universal Transverse Mercator (UTM) projected coordinate system to facilitate statistical analysis (Figure 4b) [41]. All altitudes were left in their original AGL format. The "BAT" message contained information on the total power consumed during the mission.

```
CAMFMT,TimeUS,GPSTime,GPSWeek,Lat,Lng,RelAlt,Roll,Pitch,Yaw
CAM,115588746,250292200,2099,30.6376503,-96.4826566,79.99,-10.37,12.62,124.12
CAM,122248581,250299000,2099,30.6376472,-96.4825271,80,-12.58,5.71,93.72
CAM,128888424,250305600,2099,30.6376445,-96.4823975,80,-12.42,5.97,95.34
CAM,135488283,250312200,2099,30.6376417,-96.4822672,79.99,-12.35,6.13,95.81
CAM,142088975,250318800,2099,30.6376388,-96.4821372,80,-12.31,6.2,95.78
CAM,148688834,250325400,2099,30.6376359,-96.482007,79.99,-12.37,6.29,95.92
CAM,155308685,250332000,2099,30.637633,-96.4818768,79.99,-12.34,6.31,96.08
CAM,161928536,250338600,2099,30.6376302,-96.4817467,79.99,-12.27,6.3,96.25
CAM,168548387,250345200,2099,30.6376273,-96.4816167,80,-12.19,6.28,95.96
CAM,175148246,250351800,2099,30.6376243,-96.4814865,80,-12.19,6.55,96.84
CAM,181748938,250358400,2099,30.6376214,-96.4813565,80,-12.18,6.42,96.46
CAM,188348797,250365000,2099,30.6376186,-96.4812263,79.99,-12.14,6.55,96.77
CAM,197088633,250373800,2099,30.6377863,-96.4812189,80,0.74,-7.89,355.46
CAM,203988372,250380600,2099,30.6377906,-96.4813484,79.99,12.86,5.42,269.01
CAM,210628215,250387400,2099,30.6377917,-96.4814785,80,13.98,3.29,276.91
CAM,217228907,250394000,2099,30.6377945,-96.4816088,80,14.14,3.05,278.33
CAM,223848758,250400600,2099,30.6377975,-96.4817388,80,14.17,3.27,277.97
CAM,230448617,250407200,2099,30.6378002,-96.4818689,79.99,14.27,3.06,278.69
CAM,237048476,250413800,2099,30.6378031,-96.4819991,79.99,14.33,3.07,279
CAM,243668327,250420400,2099,30.637806,-96.4821292,80,14.36,3.06,279.18
CAM,250288178,250427000,2099,30.6378088,-96.4822593,79.99,14.47,3.02,279.43
CAM,256908862,250433600,2099,30.6378118,-96.4823895,79.99,14.41,3.07,279.52
CAM,263548705,250440200,2099,30.6378146,-96.4825198,79.99,14.46,3.04,279.97
CAM,270208540,250446800,2099,30.6378174,-96.4826498,79.99,14.5,2.97,279.93
CAM,278788440,250455400,2099,30.637987,-96.4826473,80.01,-1.09,-8.35,12.63
CAM,285788972,250462400,2099,30.6379872,-96.4825186,80,-11.33,6.34,102.29
CAM,292168919,250468800,2099,30.6379841,-96.4823867,80,-11.42,7.02,99.83
CAM,298808762,250475400,2099,30.6379813,-96.4822563,79.99,-11.32,7.29,100.81
```

(a)

```
CAMFMT,TimeUS,GPSTime,GPSWeek,Lat,Lng,RelAlt,Roll,Pitch,Yaw
CAM,115588746,250292200,2099,3392147.63932,741262.33462,79.99000,-10.37,12.62,124.12
CAM,122248581,250299000,2099,3392147.57378,741274.75748,80.00000,-12.58,5.71,93.72
CAM,128888424,250305600,2099,3392147.55282,741287.18892,80.00000,-12.42,5.97,95.34
CAM,135488283,250312200,2099,3392147.52230,741299.68773,79.99000,-12.35,6.13,95.81
CAM,142088975,250318800,2099,3392147.48005,741312.15803,80.00000,-12.31,6.2,95.78
CAM,148688834,250325400,2099,3392147.43825,741324.64750,79.99000,-12.37,6.29,95.92
CAM,155308685,250332000,2099,3392147.39647,741337.13697,79.99000,-12.34,6.31,96.08
CAM,161928536,250338600,2099,3392147.36557,741349.61661,79.99000,-12.27,6.3,96.25
CAM,168548387,250345200,2099,3392147.32338,741362.08692,80.00000,-12.19,6.28,95.96
CAM,175148246,250351800,2099,3392147.27056,741374.57664,80.00000,-12.19,6.55,96.84
CAM,181748938,250358400,2099,3392147.22840,741387.04695,80.00000,-12.18,6.42,96.46
CAM,188348797,250365000,2099,3392147.19777,741399.53618,79.99000,-12.14,6.55,96.77
CAM,197088633,250373800,2099,3392165.80670,741399.82881,80.00000,0.74,-7.89,355.46
CAM,203988372,250380600,2099,3392166.00514,741387.40298,79.99000,12.86,5.42,269.01
CAM,210628215,250387400,2099,3392165.84751,741374.92758,80.00000,13.98,3.29,276.91
CAM,217228907,250394000,2099,3392165.87795,741362.42879,80.00000,14.14,3.05,278.33
CAM,223848758,250400600,2099,3392165.93122,741349.95826,80.00000,14.17,3.27,277.97
CAM,230448617,250407200,2099,3392165.95103,741337.47889,79.99000,14.27,3.06,278.69
CAM,237048476,250413800,2099,3392165.99281,741324.98944,79.99000,14.33,3.07,279
CAM,243668327,250420400,2099,3392166.03483,741312.50957,80.00000,14.36,3.06,279.18
CAM,250288178,250427000,2099,3392166.06577,741300.02996,79.99000,14.47,3.02,279.43
CAM,256908862,250433600,2099,3392166.11868,741287.54026,79.99000,14.41,3.07,279.52
CAM,263548705,250440200,2099,3392166.14922,741275.04148,79.99000,14.46,3.04,279.97
CAM,270208540,250446800,2099,3392166.18042,741262.57146,79.99000,14.5,2.97,279.93
CAM,278788440,250455400,2099,3392184.98946,741262.38984,80.01000,-1.09,-8.35,12.63
CAM,285788972,250462400,2099,3392185.28808,741274.72775,80.00000,-11.33,6.34,102.29
CAM,292168919,250468800,2099,3392185.22772,741287.38065,80.00000,-11.42,7.02,99.83
CAM,298808762,250475400,2099,3392185.19741,741299.88900,79.99000,-11.32,7.29,100.81
```

(b)

Figure 4. A section of flight log downloaded in a simulation run. (**a**) Original flight log; (**b**) Flight log after coordinate conversion to the UTM system.

2.4. Data Aggregation and Statistical Metrics

The "CAM" and "BAT" files were imported into a master file in Microsoft Excel format to determine the wind impacts. The metrics used included average 3D error, standard deviation, flight time, and battery use.

The average 3D error and the standard deviation of 3D error provided metrics for the differences between intended waypoints and the corresponding camera trigger locations. The average 3D error in each individual simulation run is defined as

$$\bar{d} = \frac{\sum_{n-1}^{m} \|p_n - q_n\|}{m} \tag{3}$$

where $\bar{d}$ is the average 3D error between intended waypoints and actual camera trigger locations, m is the number of waypoints where the camera was supposed to be triggered ($m = 48$ and 96 in the single-altitude and dual-altitude cases, respectively), p and q are the intended waypoint location vector and corresponding actual camera trigger location vector, respectively, and $p - q$ is the 3D Euclidean distance.

The standard deviation in each individual simulation run can be expressed as

$$\sigma = \sqrt{\frac{\sum_{n=1}^{m} \left(d_m - \bar{d}\right)^2}{m-1}} \tag{4}$$

where σ is standard deviation of 3D errors between intended waypoints and actual camera trigger locations, m is the number of waypoints where the camera is supposed to be triggered, and d_m is the 3D error between an intended waypoint and its associated camera trigger location.

Flight time provided a metric for the total time needed to complete a mission (i.e., from the first to the last image taken). Battery use provided a metric for the cumulative use of battery power.

3. Results

3.1. Simulation Runs for Single Altitude at 80 m AGL

3.1.1. Average 3D Error and Standard Deviation

As shown in Table 4, the average 3D error between intended waypoints and actual camera trigger locations increased as the wind speed increased. The magnitude of the impacts varied significantly as a function of the turbulence level. Specifically, as the turbulence level increased, the impact on the average 3D error became more noticeable. This observation is not surprising because the sUAS tried to compensate for the wind and stabilize its platform to where the waypoints were intended.

Table 4. Average 3D errors (m) between intended waypoints and actual camera trigger locations for 58 simulation runs at single altitude of 80 m AGL.

Wind Speed (m/s)	**Wind Direction (°)**				
	0	**22.5**	**45**	**67.5**	**90**
			Turbulence index = 0		
0.0			0.19		
3.5	0.19	0.19	0.20	0.19	0.19
7.0	0.19	0.19	0.20	0.19	0.20
10.5	0.19	0.20	0.20	0.21	0.24
14.0	0.24	0.34	0.66	0.71	0.66
			Turbulence index = 5		
10.5	0.26	0.26	0.26	0.28	0.30
			Turbulence index = 10		
0.0			0.26		
3.5	0.27	0.28	0.27	0.27	0.25
7.0	0.36	0.36	0.33	0.37	0.39
10.5	0.55	0.56	0.45	0.53	0.53
			Turbulence index = 20		
0.0			0.66		
3.5	0.67	0.62	0.56	0.60	0.54
7.0	0.78	0.9	0.71	0.80	0.83
10.5	1.25	1.28	1.36	1.21	1.09

For low turbulence levels (i.e., turbulence indices of 0 and 5), the average 3D error did not increase significantly if the wind speed was up to 10.5 m/s. If the speed increased to 14 m/s, the 3D error was at least 26% higher compared to the average 3D error for the speed up to 10.5 m/s. For a turbulence index of 10, the average 3D error began to vary significantly at lower speeds. For instance, compared to the reference zero-speed with no-turbulence wind scenario, a 42% higher average 3D error was observed for a wind speed of 3.5 m/s. 89% and 189% higher values were observed for wind speeds of 7.0 m/s and 10.5 m/s, respectively. For a turbulence index of 20, the average 3D error began to vary at even lower speeds. For instance, compared to the reference zero-speed with no-turbulence wind scenario, the average 3D errors were found to be 253%, 311%, and 558% higher for wind speeds of 3.5 m/s, 7.0 m/s, and 10.5 m/s, respectively.

Based on the simulation runs, maximum average 3D error reached up to 1.36 m when wind speed was set to 10.5 m/s with a turbulence index of 20. This was the worst (i.e., most extreme) wind condition evaluated in the study that still allowed the simulated quadcopter to maintain travel along the planned mission route. Greater wind disturbance could be programmed in Mission Planner's SITL. However, that would result in failed flight missions for the selected quadcopter model and divergent 3D errors.

In addition to displaying wind direction as shown in Table 4, Table 5 further investigates average 3D errors with respect to aircraft-wind angular relationship (i.e., α value) for all 58 simulation runs at single altitude of 80 m AGL and divides each run up into two separate segments that correspond to the East and West flight lines. For instance, the columns of "α = 180° (Tailwind)" and "α = 0° (Headwind)" together correspond to that of wind direction of 90° in Table 1. The columns of "α = 90° (Crosswind)" and "α = 270° (Crosswind)" together correspond to that of wind direction of 0° in Table 1. It is worth reiterating that the crosswind refers to the wind movement occurring perpendicular to the flight path. The columns of "α = 202.5°" and "α = 22.5°" together reflect the scenario where wind direction equated to 67.5° (Figure 3e,f).

Table 5. Average 3D errors (m) between intended waypoints and actual camera trigger locations with respect to aircraft-wind angular relationship (i.e., α value) for 58 simulation runs at single altitude of 80 m AGL.

Wind Speed (m/s)	Aircraft-Wind Angular Relationship (i.e., α Value)									
	α = 0° (Headwind)	α = 180° (Tailwind)	α = 90° (Crosswind)	α = 270° (Crosswind)	α = 67.5°	α = 247.5°	α = 45°	α = 225°	α = 22.5°	α = 202.5°
	Turbulence index = 0									
0.0					0.19					
3.5	0.20	0.19	0.19	0.19	0.19	0.19	0.19	0.20	0.19	0.19
7.0	0.18	0.21	0.18	0.19	0.18	0.20	0.18	0.21	0.18	0.21
10.5	0.19	0.29	0.18	0.20	0.17	0.22	0.17	0.23	0.17	0.24
14.0	0.96	0.37	0.19	0.29	0.36	0.32	0.93	0.39	0.96	0.46
	Turbulence index = 5									
10.5	0.30	0.30	0.23	0.30	0.25	0.27	0.22	0.30	0.30	0.26
	Turbulence index = 10									
0.0					0.26					
3.5	0.24	0.25	0.27	0.28	0.28	0.29	0.23	0.31	0.28	0.26
7.0	0.37	0.41	0.34	0.30	0.37	0.35	0.28	0.37	0.37	0.37
10.5	0.53	0.52	0.54	0.55	0.59	0.53	0.43	0.48	0.51	0.55
	Turbulence index = 20									
0.0					0.66					
3.5	0.54	0.54	0.72	0.62	0.53	0.72	0.55	0.57	0.60	0.61
7.0	0.84	0.81	0.70	0.86	1.06	0.73	0.74	0.67	0.82	0.78
10.5	1.14	1.03	1.35	1.16	1.14	1.42	1.20	1.52	1.10	1.31

For a turbulence index of 0, if the wind speed was up to 7.0 m/s, the variation in average 3D error due to differences in the α value was up to $(0.21-0.18)/0.18 = 17\%$. Compared to a reference zero-speed with no-turbulence wind scenario, the average 3D error varied from −5% to 11%. The impact was more noticeable as the wind speed increased. If the wind speed was 10.5 m/s, the variation in average 3D error due to differences in the α value was up to 64%. At 14 m/s, the variation in average 3D error due to differences in the α value was up to 405%.

For a turbulence index of 10, if the wind speed was up to 7.0 m/s, the variation in average 3D error due to differences in the α value was up to 69%. If the wind speed was 10.5 m/s, the variation in average 3D error due to differences in the α value was up to 85%.

For a turbulence index of 20, if the wind speed was up to 7.0 m/s, the variation in average 3D error due to differences in the α value was up to 205%. If the wind speed was 10.5 m/s, the variation in average 3D error due to differences in the α value was up to 258%.

Figure 5 creates wind rose charts that give a view of how wind conditions impact average 3D error in all 58 simulation runs. Results display average 3D error without turbulence (Figure 5a) and that impacted by turbulence (Figure 5b). The angles written outside the circles depict the α angles. Each wind speed/turbulence scenario assigns a distinct color for representation. The radius of the wind rose reflects the magnitude of the average 3D error for a particular wind condition. Adjacent concentric circles have an interval of 0.5 m in radius.

Figure 5a suggests that when no turbulence was involved, least average 3D error was generated in crosswind scenarios, i.e., the aircraft and wind directions were perpendicular (i.e., α = 90° or 270°). Average 3D error tended to enlarge in all other scenarios where a headwind or tailwind component contributed during the flight mission. Observations in Figure 5a also reveal that pure headwind scenario (i.e., α = 0°) created worst flight stabilities and positional errors than any other scenarios when wind speed was up to 14 m/s (i.e., average 3D error was 0.96 m). Figure 5b demonstrates that when turbulence existed (i.e., turbulence index of 5, 10 or 20), average 3D error remarkably increased in all possible α values, indicating that average 3D error was less sensitive to the differences of aircraft-wind angular relationship than turbulence.

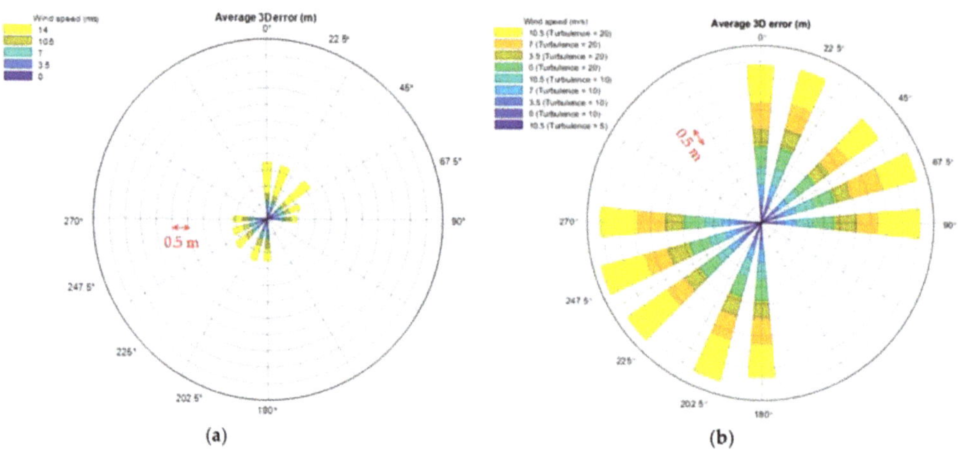

Figure 5. Wind rose plots of average 3D errors between intended waypoints and actual camera trigger locations at single altitude of 80 m AGL. (**a**) Wind rose plot of average 3D errors for no-turbulence runs; (**b**) Wind rose plot of average 3D errors for all simulation runs with turbulence.

Table 6 shows the standard deviation of 3D distance (i.e., 3D error) between intended waypoints and actual camera trigger locations for all 58 simulation runs at single altitude of 80 m AGL, and Table 7 summarizes standard deviation with respect to aircraft-wind angular relationship. Figure 6 shows wind rose plots of standard deviations for all 58 simulation runs. Compared with wind impacts on average 3D error, similar conclusions can be drawn to the impacts on standard deviation.

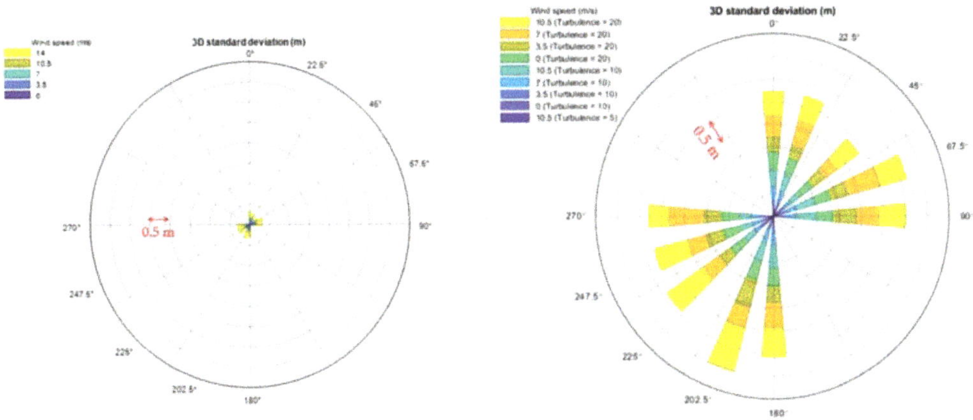

(a) (b)

Figure 6. Wind rose plots of standard deviation of 3D errors between intended waypoints and actual camera trigger locations at single altitude of 80 m AGL. (**a**) Wind rose plot of standard deviations for no-turbulence runs; (**b**) Wind rose plot of standard deviations for all simulation runs with turbulence.

Table 6. Standard deviations (m) of 3D errors between intended waypoints and actual camera trigger locations for 58 simulation runs at single altitude of 80 m AGL.

Wind Speed (m/s)	Wind Direction (°)				
	0	22.5	45	67.5	90
	Turbulence index = 0				
0.0	0.04				
3.5	0.03	0.03	0.04	0.04	0.03
7.0	0.03	0.04	0.04	0.04	0.04
10.5	0.04	0.05	0.05	0.05	0.07
14.0	0.18	0.18	0.31	0.29	0.34
	Turbulence index = 5				
10.5	0.11	0.09	0.15	0.11	0.16
	Turbulence index = 10				
0.0	0.10				
3.5	0.11	0.11	0.10	0.11	0.1
7.0	0.18	0.19	0.14	0.24	0.26
10.5	0.39	0.37	0.27	0.42	0.43
	Turbulence index = 20				
0.0	0.39				
3.5	0.48	0.42	0.32	0.49	0.34
7.0	0.65	0.62	0.41	0.50	0.50
10.5	0.55	0.71	0.82	0.8	0.59

Table 7. Standard deviations (m) of 3D errors between intended waypoints and actual camera trigger locations with respect to aircraft-wind angular relationship (i.e., α value) for 58 simulation runs at single altitude of 80 m AGL.

Wind Speed (m/s)	Aircraft-Wind Angular Relationship (i.e., α Value)									
	α = 0° (Headwind)	α = 180° (Tailwind)	α = 90° (Crosswind)	α = 270° (Crosswind)	α = 67.5°	α = 247.5°	α = 45°	α = 225°	α = 22.5°	α = 202.5°
Turbulence index = 0										
0.0	0.04									
3.5	0.03	0.04	0.04	0.03	0.04	0.03	0.04	0.03	0.04	0.04
7.0	0.03	0.04	0.04	0.02	0.04	0.02	0.04	0.03	0.04	0.04
10.5	0.04	0.06	0.05	0.03	0.06	0.03	0.05	0.03	0.04	0.04
14.0	0.19	0.12	0.14	0.20	0.16	0.20	0.07	0.20	0.10	0.19
Turbulence index = 5										
10.5	0.17	0.16	0.09	0.11	0.10	0.09	0.10	0.18	0.14	0.08
Turbulence index = 10										
0.0	0.10									
3.5	0.10	0.11	0.09	0.12	0.11	0.11	0.09	0.10	0.12	0.10
7.0	0.28	0.25	0.20	0.16	0.22	0.17	0.13	0.14	0.22	0.26
10.5	0.36	0.51	0.48	0.28	0.43	0.31	0.27	0.27	0.34	0.50
Turbulence index = 20										
0.0	0.39									
3.5	0.33	0.36	0.55	0.40	0.42	0.40	0.35	0.29	0.29	0.64
7.0	0.46	0.55	0.49	0.78	0.75	0.41	0.34	0.47	0.56	0.44
10.5	0.54	0.65	0.62	0.46	0.61	0.79	0.60	0.98	0.55	1.00

3.1.2. Flight Time

Table 8 shows total flight times (i.e., from the first image to the last image) for each simulation run at single altitude of 80 m AGL. In general, the total time to complete a mission was less sensitive to changes in wind speed, direction, and turbulence conditions than the average 3D error and standard deviation. If the wind speed was up to 7.0 m/s and the turbulence index was up to 10, the flight time increased approximately 2.5% compared to the reference zero-speed with no-turbulence wind simulation run. At this wind speed, even if the turbulence index was 20, the total flight time increased less than 25% compared to the reference zero-speed with no-turbulence simulation run. If the wind speed was 10.5 m/s and the turbulence index was up to 10, the total flight time increased approximately 12% compared to the reference zero-speed with no-turbulence wind simulation run. However, if the turbulence level was 20, the total flight time increased 109% compared to the reference zero-speed with no-turbulence run.

Table 9 shows flight times disaggregated by aircraft-wind angular relationship for all 58 simulation runs at single altitude of 80 m AGL. The time required to transit from the last waypoint of one flight line to the first waypoint of the next line was not included (i.e., 18.8 m as shown in Figure 1). If the wind speed was up to 7.0 m/s and the turbulence index was up to 10, the variation in flight time due to differences in aircraft-wind angular relationship (i.e., α value) was no greater than 6%. Unsurprisingly, headwind conditions produced a higher total flight time than tailwind conditions. The effect due to differences in aircraft-wind angular relationship was more noticeable for higher wind speeds and turbulence levels. If the wind speed was 10.5 m/s and the turbulence index was 10, the total flight time was 18% greater under headwind condition than under tailwind condition. However, if the wind speed was 14 m/s and the turbulence index was 20, the total flight time was 182% higher under headwind condition than under tailwind condition. In Table 9, it is worth nothing that "N/A" was marked as flight times for zero-wind-speed simulation runs where the aircraft-wind angular relationships were not formed. The directionality of wind impacts on flight times is also quantified in the wind rose plots in Figure 7.

Table 8. Total flight times (s) for 58 simulation runs at single altitude of 80 m AGL.

Wind Speed (m/s)	Wind Direction (°)				
	0	22.5	45	67.5	90
Turbulence index = 0					
0.0	318				
3.5	318	318	318	318	318
7.0	318	318	318	318	318
10.5	319	318	318	318	319
14.0	358	359	416	448	456
Turbulence index = 5					
10.5	320	323	325	325	325
Turbulence index = 10					
0.0	320				
3.5	320	320	320	323	321
7.0	322	323	323	326	324
10.5	339	341	344	357	356
Turbulence index = 20					
0.0	351				
3.5	360	346	357	353	362
7.0	377	374	390	391	397
10.5	635	635	665	635	657

Table 9. Flight times (s) with respect to aircraft-wind angular relationship (i.e., α value) for 58 simulation runs at single altitude of 80 m AGL.

Wind Speed (m/s)	Aircraft-Wind Angular Relationship (i.e., α Value)									
	α = 0° (Headwind)	α = 180° (Tailwind)	α = 90° (Crosswind)	α = 270° (Crosswind)	α = 67.5°	α = 247.5°	α = 45°	α = 225°	α = 22.5°	α = 202.5°
Turbulence index = 0										
0.0	N/A									
3.5	146	146	146	146	146	146	146	146	146	146
7.0	146	146	146	146	146	146	146	146	146	146
10.5	146	147	146	146	146	146	146	146	146	146
14.0	284	145	146	146	152	147	218	147	262	146
Turbulence index = 5										
10.5	153	146	146	146	147	146	151	145	151	148
Turbulence index = 10										
0.0	N/A									
3.5	148	146	146	147	148	146	148	146	151	146
7.0	153	145	148	147	150	146	151	145	152	147
10.5	178	151	150	158	156	152	170	146	177	150
Turbulence index = 20										
0.0	N/A									
3.5	172	160	159	164	159	159	158	164	168	157
7.0	202	163	168	172	177	166	197	163	196	166
10.5	442	177	313	256	340	219	397	194	403	174

Note: This table excluded transit times between flight lines. This table did not include flight time information for all zero-speed simulation runs due to unavailability of the α angle.

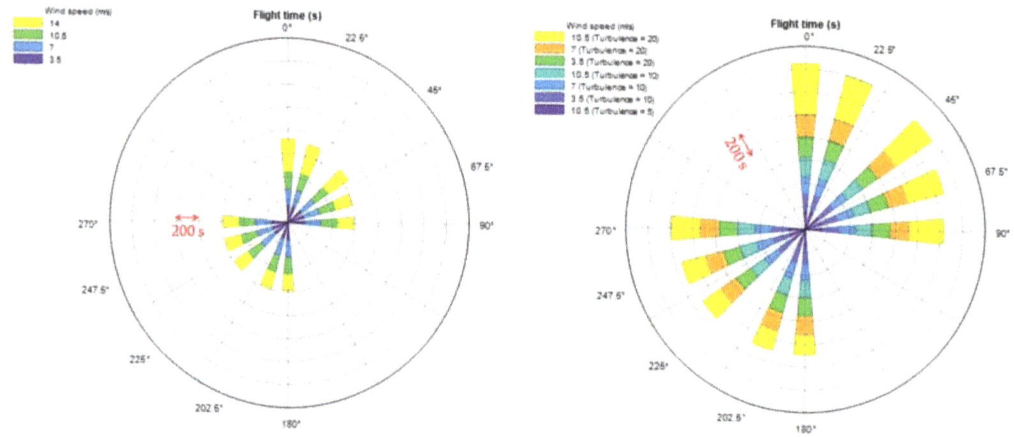

(a) (b)

Figure 7. Wind rose plots of total flight time at single altitude of 80 m AGL. (**a**) Wind rose plot of flight times for no-turbulence runs; (**b**) Wind rose plot of flight times for all simulation runs with turbulence.

3.1.3. Battery Use

Similar to analysis performed with the flight time metric, battery use (Table 10) was also less sensitive to changes in wind speed, direction, and turbulence conditions than the average 3D error and standard deviation. If the wind speed was up to 7.0 m/s and the turbulence index was up to 10, the flight time increased less than 14% compared to the reference zero-speed with no-turbulence simulation run. At this wind speed, even if the turbulence index was 20, the total battery use increased less than 34% compared to the reference zero-speed with no-turbulence run. If the wind speed was up to 10.5 m/s and the turbulence index was up to 10, the battery use increased less than 35% compared to the reference zero-speed with no-turbulence run. However, if the turbulence level was 20, the battery use increased 125%.

3.2. Simulation Runs for Dual Altitudes at 80 m and 10 m AGL

As displayed in Table 3, the simulation included a wind speed of 10.5 m/s with two possible turbulence index values of 0 and 10 at 80 m AGL, and the simulation also included a wind speed of 3.5 m/s with a turbulence index of 0 at 10 m AGL. These were evaluated at three wind directions resulting in a total of six simulation runs taking into account both the 80 m and 10 m AGL flight paths per a simulation as shown in Figure 2. Table 11 summarizes evaluation results for the six simulation runs. Each value in the table indicates the evaluation result that took both 80 m and 10 m AGL into account. When no turbulence was involved at both 80 m and 10 m AGL, each evaluation metric was found to vary slightly between scenarios with different wind directions. The average 3D error and standard deviation statistics were consistent with that reported in Tables 4 and 6 when looking into the cells with a turbulence index of 0 and wind speed of 10.5 m/s. This emphasizes that the choice of flight height will not affect simulated wind impacts on flight performance as stated in Section 2.1. However, the flight endurance and battery use in Table 11 increased compared with that reported in Tables 8 and 10 when looking into the cells with a turbulence index of 0 and wind speed of 10.5 m/s. This is because Table 11 involved dual altitudes in the mission. Compared with the scenario where the turbulence index was 0 at both 10 m and 80 m AGL, when the turbulence index was increased up to 10 at 80 m AGL, the average 3D error and standard deviation values were found to increase significantly while the total flight time and battery use were observed to rise by up to 7.6% and 6.1%, respectively (Table 11).

Table 10. Battery use (mAh) for 58 simulation runs at single altitude of 80 m AGL.

Wind Speed (m/s)	Wind Direction (°)				
	0	22.5	45	67.5	90
Turbulence index = 0					
0.0	2601				
3.5	2661	2660	2660	2659	2659
7.0	2917	2910	2915	2921	2928
10.5	3214	3212	3217	3220	3224
14.0	3874	3883	4377	4707	4774
Turbulence index = 5					
10.5	3233	3177	3188	3206	3274
Turbulence index = 10					
0.0	2572				
3.5	2644	2639	2643	2657	2658
7.0	2953	2955	2943	2970	2952
10.5	3385	3323	3339	3511	3492
Turbulence index = 20					
0.0	2726				
3.5	2904	2816	2865	2941	2893
7.0	3248	3219	3411	3413	3483
10.5	5537	5641	5849	5746	5822

Table 11. Evaluation results for six simulation runs at dual altitudes of 80 m and 10 m AGL. Each value indicates the evaluation result that took both 80 m and 10 m AGL into account.

Evaluation Metric	Wind Direction (°)		
	0	45	90
Turbulence index = 0 at both 10 m and 80 m AGL			
Average 3D error (m)	0.20	0.21	0.23
Standard deviation (m)	0.05	0.05	0.06
Total flight time (s)	473	471	473
Battery use (mAh)	4418	4418	4425
Turbulence index = 0 at 10 m, turbulence index = 10 at 80 m AGL			
Average 3D error (m)	0.38	0.34	0.37
Standard deviation (m)	0.33	0.23	0.35
Total flight time (s)	494	497	509
Battery use (mAh)	4590	4539	4693

Note: The total flight time values in the table excluded transit times from 80 m descending to 10 m AGL.

Table 12 shows evaluation results with respect to aircraft-wind angular relationship (i.e., α value) for the six simulation runs at dual altitudes of 80 m and 10 m AGL. Each value indicates the evaluation result that took both 80 m and 10 m AGL into account. Similar to observations in Section 3.1.1, when turbulence existed at 80 m AGL (i.e., turbulence index of 10), flight stability remarkably decreased in all possible α values compared to the no-turbulence scenarios indicating that average 3D error and standard deviation were more sensitive to turbulence than the differences of aircraft-wind angular relationship.

Table 12. Evaluation results with respect to aircraft-wind angular relationship (i.e., α value) for six simulation runs at dual altitudes of 80 m and 10 m AGL. Each value indicates the evaluation result that took both 80 m and 10 m AGL into account.

Evaluation Metric	Aircraft-Wind Angular Relationship (i.e., α Value)					
	α = 0° (Headwind)	α = 180° (Tailwind)	α = 90° (Crosswind)	α = 270° (Crosswind)	α = 45°	α = 225°
Turbulence index = 0 at both 10 m and 80 m AGL						
Average 3D error (m)	0.21	0.25	0.19	0.21	0.19	0.23
Standard deviation (m)	0.05	0.07	0.06	0.04	0.05	0.04
Flight time (s)	212	213	212	212	212	212
Turbulence index = 0 at 10 m AGL and turbulence index = 10 at 80 m AGL						
Average 3D error (m)	0.38	0.36	0.37	0.38	0.32	0.35
Standard deviation (m)	0.30	0.39	0.38	0.26	0.22	0.23
Flight time (s)	244	218	216	224	236	212

Note: The total flight time values in the table excluded transit times from 80 m descending to 10 m AGL. The flight time values in the table also excluded transit times between flight lines.

Different from Table 12, each value in Table 13 indicates the disaggregated evaluation result that considered partial flight operation conducted at each altitude (i.e., either 80 m or 10 m AGL) separately. Expectedly, the worst flight performance was achieved when the turbulence index was up to 10 at 80 m AGL. However, when no turbulence was involved, nearly identical flight performances were obtained between the 80 m and 10 m AGL simulation scenarios in terms of the average 3D error and standard deviation. This is because the flight height in the study is independent to simulated wind effects and, therefore, the choice of flight height will not affect wind impacts as explained earlier in Section 2.1. On the other hand, the flight time at 10 m AGL was decreased remarkably compared to that at 80 m AGL for any given α angle. This is because the flight lines were scaled down at 10 m AGL as shown in Figure 2.

Table 13. Evaluation results with respect to aircraft-wind angular relationship (i.e., α value) for six simulation runs at dual altitudes of 80 m and 10 m AGL. This table shows disaggregated results by considering partial flight operation conducted at each altitude (i.e., either 80 m or 10 m AGL) separately.

Evaluation Metric	Aircraft-Wind Angular Relationship (i.e., α Value)					
	α = 0° (Headwind)	α = 180° (Tailwind)	α = 90° (Crosswind)	α = 270° (Crosswind)	α = 45°	α = 225°
Turbulence index = 0 at 80 m AGL						
Average 3D error (m)	0.19	0.29	0.18	0.20	0.17	0.23
Standard deviation (m)	0.04	0.06	0.05	0.03	0.05	0.03
Flight time (s)	146	147	146	146	146	146
Turbulence index = 10 at 80 m AGL						
Average 3D error (m)	0.53	0.52	0.54	0.55	0.43	0.48
Standard deviation (m)	0.36	0.51	0.48	0.28	0.27	0.27
Flight time (s)	178	151	150	158	170	146
Turbulence index = 0 at 10 m AGL						
Average 3D error (m)	0.22	0.20	0.19	0.21	0.22	0.22
Standard deviation (m)	0.05	0.04	0.07	0.05	0.05	0.04
Flight time (s)	66	66	67	67	66	66

Note: The total flight time values in the table excluded transit times from 80 m descending to 10 m AGL. The flight time values in the table also excluded transit times between flight lines.

4. Lessons Learned and Conclusion

A timely investigation and reconstruction at the motor vehicle crash scene plays a pivotal role in identifying the cause and severity of the accident, assessing roadway safety risks, clarifying insurance liabilities, and facilitating legal proceedings. SfM photogrammetry with sUAS in crash scene reconstruction has gained increasing attention in recent years because of its reliability and flexibility in offering quality geospatial surveying products. One of the main deficiencies in past literature has been the failure of demonstrating its effectiveness in adverse weather conditions, under which the risk of motor vehicle crashes is exacerbated. In this article, wind was chosen as the primary factor that drove adverse weather conditions, and the sUAS was presented in the form of a simulated quadcopter in the ArduPilot SITL environment. The main objectives were to: (1) characterize the impacts of wind speed, direction, and turbulence on the positional accuracy, required flight time, and battery use of the simulated quadcopter type via realistic flight simulations, and (2) generalize lessons learned for platform-independent quadcopter flight design under wind conditions for crash scene mapping. Real-world quadcopters were not tested and compared due to flight safety and compliance concerns. The simulation also provided a method to assess wind impact on flight design in a systematic and controlled fashion, which is generally not feasible in real-world operating conditions. A total of 58 simulation runs with incremental increases in wind speed, direction, and turbulence were created and analyzed at a single altitude of 80 m AGL. Six more simulation runs were created and analyzed at dual altitudes of 80 m and 10 m AGL.

The wind disturbance settings were deliberately chosen to ensure that the simulated quadcopter was able to maintain the planned route and avoid divergent 3D errors between intended waypoints and actual camera trigger locations. The results indicated that as wind conditions departed from the ideal zero-speed with no-turbulence scenario, there was an adverse impact on sUAS stability performance, measured in terms of positional accuracy, required flight time, and battery use. Several lessons learned related to real-world quadcopter sUAS flight design under wind conditions for crash scene reconstruction are discussed and summarized below.

The average 3D error and corresponding standard deviation of 3D error between planned waypoints and the actual camera trigger locations increased as the wind speed increased. The research findings suggested operating the simulated quadcopter when the wind is not greater than 11 m/s. It is important to note that this ArduPilot quadcopter module has been run on real-world systems such as 3DR Solo and Parrot Bebop 2, and this conclusion accords well with the accepted wind tolerance that the respective user guides recommend for these platforms [33,34,42]. This recommendation on wind tolerance is also consistent with that of other popular quadcopter sUAS platforms in current operation of similar size such as DJI Phantom 4 Pro/Pro+, DJI Inspire 2 and Skydio 2 [43–45]. However, it should be realized that wind tolerance is platform dependent and 11 m/s as the documented value in this work may not be generalized for all types of quadcopter sUAS platforms. For the average 3D error and corresponding standard deviation of 3D error between planned waypoints and the actual camera trigger locations, the magnitude of the impact varied significantly as a function of the turbulence level. The simulation runs also demonstrated that the flight performance remained relatively stable under light to moderate turbulence levels (i.e., turbulence level $\leq$5).

Flying perpendicular to the wind direction is a well-known practice for fixed-wing sUAS survey missions. It helps maintain stable ground speed during imagery data collection [46,47]. Results obtained in this study suggest that this rule of thumb for fixed-wing aircrafts applies well to quadcopter sUAS platforms. Under high wind and low turbulence scenarios (e.g., a wind speed of 14 m/s and turbulence level of 0), statistical results proved that flying in crosswind (i.e., $\alpha = 90°$ or 270°) and against headwind (i.e., $\alpha = 0°$) were the most and least favorable flight patterns, respectively. While the average 3D error metric was not sensitive to the increase of wind speed in crosswind scenarios (i.e., $\alpha = 90°$ or 270°), the average 3D error sharply grew four times if wind speed rose from 0 to 14 m/s for the headwind scenario (i.e., $\alpha = 0°$). Similar results were observed in the flight time metric. The sUAS spent nearly twice as much time on headwind flight segments (i.e., $\alpha = 0°$) than those segments where flight lines were perpendicular to the wind direction (i.e., $\alpha = 90°$ or $270°$). As shown in the simulation runs, the total flight time metric was less sensitive to changes in wind speed, direction, and turbulence conditions than the average 3D error and standard deviation metrics. The ideal flight paths for quadcopter sUAS platforms are, therefore, supposed to stay nearly perpendicular to horizontal wind. However, in real-world scenarios, this goal may not be achievable in every single flight attempt due to road geometry at the crash location, overhead safety concerns, airspace restrictions and so forth. A general recommendation to the crash investigation team and sUAS remote pilot within high wind and low turbulence environments is to apply this conclusion when pertinent on-site conditions permit.

Under strong wind conditions (e.g., wind speed $\geq$ 11 m/s and turbulence level >5 for the simulated quadcopter platform in ArduPilot SITL), flight stability degraded dramatically, resulting in disrupted frontal and side overlap settings due to large waypoint targeting errors. This overlap disruption is expected to get amplified for flight missions at a low altitude (e.g., below 10 m AGL) due to smaller camera field of view (FOV). This can result in a decreased number of detected and matched features in a pair of overlapping images in SfM processing routine, potentially affecting the overall quality of 2D orthomosaic image and 3D point cloud products. Even if overlaps are not disrupted in some cases, remote pilots are supposed to take battery usage into careful consideration prior to conducting high-wind flight missions. Nowadays, many commercially available multirotor sUAS platforms need a 2S to 6S lithium-ion polymer (LiPo) battery with the energy capacity less than 6000 mAh to achieve best balance between performance, flight time, and weight [43–45,48,49]. Assuming this battery specification applies to the examined quadcopter platform in ArduPilot SITL, Table 10 implies that when strong wind and turbulence exist (i.e., wind speed of 10.5 m/s and turbulence level of 20), the fully charged LiPo battery can potentially run out of power before completion of the mapping mission over the simulated crash scene area of 105 × 70 m. The rapid degradation in flight efficiency as observed from the simulation runs and platform demonstrates the need for adequate battery backup under high wind conditions, even when the scene is within a limited geographic extent. However, while this statement is generalized, it needs to be pointed out that battery consumption is platform dependent. Some lighter weight quadcopters may potentially complete a mission with one LiPo battery should they withstand as much wind as the simulated quadcopter type.

At a crash scene, the quadcopter to be deployed on the ground usually stay in close proximity to the on-site investigators and nearby vehicles. If the wind speed is marginally below the aforementioned maximum wind tolerance (e.g., 11 m/s for the simulated quadcopter in ArduPilot SITL), the remote pilot is supposed to pay extra attention during quadcopter take-off and landing phases. This is because a quadcopter sUAS is likely more susceptible to the air movement when it is flat on the ground and sitting still, causing an increased risk of flipping over in windy conditions. When the platform comes to hover in the air or traverse pre-defined waypoints to collect images for a photogrammetric survey, it should largely be able to withstand the maximum wind tolerance in a relatively steady state. In such phase, the gimbal stabilizer, which is nowadays attached to many commercially available quadcopter sUAS platforms, also helps maintain favorable image quality by compensating for angular motions to the onboard camera.

It is worth emphasizing that as a popular and reliable open-source autopilot software suite, ArduPilot has been used on a wide spectrum of autonomous systems with full-featured autonomous capabilities. ArduPilot on SITL features a high-fidelity simulation environment running a realistic flight dynamics model, and therefore, the major simulation findings summarized in this study are expected to be applicable to real-world flight scenarios. In addition, the choice of 80 m and 10 m AGL in the article was generic. Changing flight altitude did not vary wind and turbulence effect and hence, did not affect the simulated results other than flight endurance and battery consumption. The wind and turbulence parameters were defined and adjusted in Mission Planner interface, so the same wind impact results apply to other pre-defined flight altitude for the simulated quadcopter without loss of generality. However, in a real-world flight mission for crash scene reconstruction, flight height is usually determined by considering factors such as desired GSD and camera model settings, flight time, geographic extent of the mission, and waypoint geometry given specific overlap settings.

In the United States, public and commercial sUAS operations within the national airspace system (NAS) require obtaining either a Certificate of Waiver or Authorization (COA), with the option of including a Section 333 exemption, or Part 107 license from the Federal Aviation Administration (FAA). For public safety and transportation agencies that administer auto crash investigation, having a COA may warrant sUAS operations across vast airspace regions. More importantly, the COA may grant these agencies privilege of flying in suboptimal flight conditions, such as during nighttime and foggy hours to carry out an emergency survey immediately after a crash takes place. On the other hand, getting a Part 107 license usually requires less time and effort, but it is generally considered more restrictive in various dimensions compared with the COA. Thanks to the FAA's recent efforts, waivers to Part 107 rules can be applied to gain approvals of certain sUAS operations outside the limitations defined in the rules [50]. For example, a waiver can request easing the restrictions of conducting routine sUAS operations above standard Part 107 flight height limits (e.g., 121.9 m (400 ft) AGL at the time of this writing). This type of special permission allows pilots of public safety and transportation agencies to operate sUASs with adequate flexibility at a crash scene while ensuring safety and compliance. Recently, the FAA provided rule amendments to Part 107 to enable routine operations of sUAS over people, from moving vehicles, and at night under certain conditions. Nowadays, sUASs have become an integral part of our daily life, and technologies to advance sUAS safety evolve rapidly. This leaves open the possibilities of exploring impacts of multiple suboptimal weather conditions on sUAS flight performance for crash scene reconstruction in the near future. The intended future work also includes: 1) investigating wind impacts on fixed-wing sUAS platforms, and 2) conducting real-world flight missions with different commercially available quadcopters under varying wind conditions and assessing the flight performance as well as SfM photogrammetric products.

Author Contributions: Methodology, M.J.S.; data collection, J.B.; formal analysis, T.C., J.B., M.J.S. and M.P.; investigation, T.C., J.B. and M.J.S.; writing—original draft preparation, T.C., J.B. and M.J.S.; writing—review and editing, T.C., M.J.S., J.B. and C.Q.; supervision, M.J.S.; project administration C.Q. All authors have read and agreed to the published version of the manuscript.

Funding: This research was funded by the Texas Department of Transportation (TxDOT) in cooperation with the Federal Highway Administration (FHWA), Research Project 0-7063. The contents of this paper reflect the views of the authors, who are responsible for the facts and the accuracy of the data presented herein. The contents do not necessarily reflect the official view or policies of FHWA or TxDOT.

Data Availability Statement: Not applicable.

Acknowledgments: The authors thank ArduPilot development staff for their continued dedication to providing a powerful and versatile open-source autopilot platform for a wide array of autonomous systems. The authors also gratefully acknowledge and thank TxDOT. Finally, the authors thank Melanie Gingras for her contributions to project management.

Conflicts of Interest: The authors declare no conflict of interest. The funders had no role in the design of the study; in the collection, analyses, or interpretation of data; or in the writing of the manuscript.

References

1. Li, X.; Khattak, A.J.; Wali, B. Role of multiagency response and on-scene times in large-scale traffic incidents. *Transp. Res. Rec.* **2017**, *2616*, 39–48. [CrossRef]
2. Agent, K.R.; Pigman, J.G. *Traffic Accident Investigation (Report No. KTC-93-10)*; Kentucky Transportation Center-University of Kentucky: Lexington, KY, USA, 1993.
3. Jacobson, L.N.; Legg, B.; O'Brien, A. *Incident Management using Total Stations;* Report No. WA-RD 284.1; Washington State Department of Transportation: Olympia, WA, USA, 1992.
4. Duignan, P.; Griffiths, M.; Lie, A. Photogrammetric Methods in Crash Investigation. In Proceedings of the 15th International Technical Conference on Enhanced Safety of Vehicles, Melbourne, Australia, 13–26 May 1996.
5. Forman, P.; Parry, I. Rapid Data Collection at Major Incident Scenes using Three Dimensional Laser Scanning Techniques. In Proceedings of the IEEE 35th Annual 2001 International Carnahan Conference on Security Technology, London, UK, 16–19 October 2001.
6. Pagounis, V.; Tsakiri, M.; Palaskas, S.; Biza, B.; Zaloumi, E. 3D Laser Scanning for Road Safety and Accident Reconstruction. In Proceedings of the XXIIIth international FIG Congress, Munich, Germany, 8–13 October 2006.
7. James, W.; McKinzie, S.; Benson, W.; Heise, C. *Crash Investigation and Reconstruction Technologies and Best Practices;* Report No.FHWA-HOP-16-009; U.S. Department of Transportation: Washington, DC, USA, 2015.
8. Stevens, C.R.; Blackstock, T. *Demonstration of Unmanned Aircraft Systems Use for Traffic Incident Management (UAS-TIM);* Report No. PRC 17-69 F; Texas A&M Transportation Institute: College Station, TX, USA, 2017.
9. John Hopkins University. *Operational Evaluation of Unmanned Aircraft Systems for Crash Scene Reconstruction;* Report No. AOS-17-0078; National Institute of Justice: Washington, DC, USA, 2017.
10. Khan, M.A.; Ectors, W.; Bellemans, T.; Janssens, D.; Wets, G. Unmanned aerial vehicle-based traffic analysis: A case study for shockwave identification and flow parameters estimation at signalized intersections. *Remote Sens.* **2018**, *10*, 458. [CrossRef]
11. Kamnik, R.; Perc, M.N.; Topolšek, D. Using the scanners and drone for comparison of point cloud accuracy at traffic accident analysis. *Accid. Anal. Prev.* **2020**, *135*, 105391. [CrossRef] [PubMed]
12. Pérez, J.A.; Gonçalves, G.R.; Rangel, J.M.G.; Ortega, P.F. Accuracy and effectiveness of orthophotos obtained from low cost UASs video imagery for traffic accident scenes documentation. *Adv. Eng. Softw.* **2019**, *132*, 47–54. [CrossRef]
13. How Do Weather Events Impact Roads? Available online: https://ops.fhwa.dot.gov/weather/q1_roadimpact.htm (accessed on 30 April 2021).
14. Young, R.K.; Liesman, J. Intelligent transportation systems for operation of roadway segments in high-wind conditions. *Transp. Res. Rec.* **2007**, *2000*, 1–7. [CrossRef]
15. Naik, B.; Tung, L.W.; Zhao, S.; Khattak, A.J. Weather impacts on single-vehicle truck crash injury severity. *J. Saf. Res.* **2016**, *58*, 57–65. [CrossRef] [PubMed]
16. Slocum, R.K.; Wright, W.; Parrish, C.; Costa, B.; Sharr, M.; Battista, T.A. *Guidelines for Bathymetric Mapping and Orthoimage Generation using sUAS and SfM: An Approach for Conducting Nearshore Coastal Mapping;* NOAA Technical Memorandum: Silver Spring, MD, USA, 2019.
17. Joyce, K.E.; Duce, S.; Leahy, S.M.; Leon, J.; Maier, S.W. Principles and practice of acquiring drone-based image data in marine environments. *Mar. Freshw. Res.* **2019**, *70*, 952–963. [CrossRef]
18. Ding, L.; Wang, Z. A robust control for an aerial robot quadrotor under wind gusts. *J. Robot.* **2018**, *2018*, 5607362. [CrossRef]
19. Ware, J.; Roy, N. An Analysis of Wind Field Estimation and Exploitation for Quadrotor Flight in the Urban Canopy Layer. In Proceedings of the 2016 IEEE International Conference on Robotics and Automation (ICRA), Stockholm, Sweden, 16–21 May 2016.
20. Stepanyan, V.; Krishnakumar, K. Estimation, Navigation and Control of Multi-Rotor Drones in an Urban Wind Field. In Proceedings of the AIAA Information Systems-AIAA Infotech @ Aerospace, Grapevine, TX, USA, 9–13 January 2017.
21. Wang, J.Y.; Luo, B.; Zeng, M.; Meng, Q.H. A wind estimation method with an unmanned rotorcraft for environmental monitoring tasks. *Sensors* **2018**, *18*, 4504. [CrossRef] [PubMed]
22. Wang, B.H.; Wang, D.B.; Ali, Z.A.; Ting, B.T.; Wang, H. An overview of various kinds of wind effects on unmanned aerial vehicle. *Meas. Control* **2019**, *52*, 731–739. [CrossRef]
23. Siqueira, J. Modeling of Wind Phenomena and Analysis of Their Effects on UAV Trajectory Tracking Performance. Master's Thesis, West Virginia University, Morgantown, WV, USA, 2017.
24. Biradar, A.S. Wind Estimation and Effects of Wind on Waypoint Navigation of UAVs. Master's Thesis, Arizona State University, Tempe, AZ, USA, 2014.
25. Yoon, S.; Shin, D.; Choi, Y.; Park, K. Development of a flexible and expandable UTM simulator based on open sources and platforms. *Aerospace* **2021**, *8*, 133. [CrossRef]
26. Koubaa, A.; Allouch, A.; Alajlan, M.; Javed, Y.; Belghith, A.; Khalgui, M. Micro air vehicle link (mavlink) in a nutshell: A survey. *IEEE Access* **2019**, *7*, 87658–87680. [CrossRef]
27. Baidya, S.; Shaikh, Z.; Levorato, M. FlyNetSim: An Open Source Synchronized UAV Network Simulator based on ns-3 and Ardupilot. In Proceedings of the 21st ACM International Conference on Modeling, Analysis and Simulation of Wireless and Mobile Systems (MSWiM'18), Montréal, QC, Canada, 28 October–2 November 2018.
28. Eyerman, J.; Mooring, B.; Catlow, M.; Datta, S.; Akella, S. *Low-Light Collision Scene Reconstruction Using Unmanned Aerial Systems;* North Carolina Department of Transportation: Raleigh, NC, USA, 2018.

29. Transforming Accident Investigation with Drones. Available online: https://medium.com/aerial-acuity/transforming-accident-investigation-with-drones-edec7162d8ce (accessed on 1 June 2021).
30. Police Use of Unmanned Aircraft Systems (sUAS). Available online: http://records.tukwilawa.gov/weblink/1/edoc/285955/TIC%202017-02-28%20Item%202D%20-%20Discussion%20-%20Police%20Use%20of%20Unmanned%20Aircraft%20Systems%20 (accessed on 1 June 2021).
31. Montgomery, K. Documenting Vehicle Crashes with Drone Photogrammetry. Available online: https://www.evidencemagazine.com/index.php?option=com_content&task=view&id=2437&Itemid=49 (accessed on 7 July 2021).
32. Dukowitz, Z. Drones in Accident Reconstruction: How Drones Are Helping Make Traffic Crash Site Assessments Faster, Safer, and More Accurate. Available online: https://uavcoach.com/drones-accident-reconstruction/ (accessed on 7 July 2021).
33. *Solo User Manual V9*; 3D Robotics Inc: Berkeley, CA, USA, 2015.
34. Parrot Bebop Autopilot. Available online: https://ardupilot.org/copter/docs/parrot-bebop-autopilot.html (accessed on 1 June 2021).
35. ArduPilot Complete Parameter List. Available online: https://ardupilot.org/copter/docs/parameters.html (accessed on 30 April 2021).
36. Javadnejad, F.; Slocum, R.K.; Gillins, D.T.; Olsen, M.J.; Parrish, C.E. Dense point cloud quality factor as proxy for accuracy assessment of image-based 3D reconstruction. *J. Surv. Eng.* **2021**, *147*, 04020021. [CrossRef]
37. Mölg, N.; Bolch, T. Structure-from-motion using historical aerial images to analyse changes in glacier surface elevation. *Remote Sens.* **2017**, *9*, 1021. [CrossRef]
38. Turbulence—The Definitive Edition. Available online: https://www.weather.gov/media/zhu/ZHU_Training_Page/turbulence_stuff/turbulence2/turbulence.pdf (accessed on 30 April 2021).
39. ArduPilot SIM_Aircraft.cpp Source Code. Available online: https://github.com/ArduPilot/ardupilot/blob/master/libraries/SITL/SIM_Aircraft.cpp (accessed on 30 April 2021).
40. Seregina, L.S.; Haas, R.; Born, K.; Pinto, J.G. Development of a wind gust model to estimate gust speeds and their return periods. *Tellus A: Dyn. Meteorol. Oceanogr.* **2014**, *66*, 22905. [CrossRef]
41. NOAA Vertical Datum Transformation. Available online: https://vdatum.noaa.gov/ (accessed on 30 April 2021).
42. *Parrot Bebop 2 Drone User Guide*; Parrot SA: Paris, France, 2016.
43. *DJI Phantom 4 Pro/Pro+ Series User Manual*; SZ DJI Technology Co., Ltd.: Shenzhen, China, 2020.
44. *DJI Inspire 2 User Manual*; SZ DJI Technology Co., Ltd: Shenzhen, China, 2016.
45. *Skydio 2 User Guide*; Skydio Inc: Redwood City, CA, USA, 2019.
46. Chu, T.; Starek, M.J.; Brewer, M.J.; Murray, S.C.; Pruter, L.S. Assessing lodging severity over an experimental maize (*Zea mays* L.) field using UAS images. *Remote Sens.* **2017**, *9*, 923. [CrossRef]
47. Chu, T.; Starek, M.J.; Brewer, M.J.; Murray, S.C.; Pruter, L.S. Characterizing canopy height with UAS structure-from-motion photogrammetry—results analysis of a maize field trial with respect to multiple factors. *Remote Sens. Lett.* **2018**, *9*, 753–762. [CrossRef]
48. *DJI Mavic 2 Pro/Zoom User Manual*; SZ DJI Technology Co., Ltd.: Shenzhen, China, 2018.
49. *Parrot ANAFI Drone User Guide v2.6.2*; Parrot SA: Paris, France, 2019.
50. Operations of Small Unmanned Aircraft over People. Available online: https://www.faa.gov/news/media/attachments/OOP_Final%20Rule.pdf (accessed on 12 July 2021).

MDPI
St. Alban-Anlage 66
4052 Basel
Switzerland
Tel. +41 61 683 77 34
Fax +41 61 302 89 18
www.mdpi.com

Drones Editorial Office
E-mail: drones@mdpi.com
www.mdpi.com/journal/drones